ALLE · ZEIT · WACH 1842

Maurice Roseau

Vibrations in Mechanical Systems

Analytical Methods and Applications

With 112 Figures

Springer-Verlag
Berlin Heidelberg New York
London Paris Tokyo

Maurice Roseau
Université Pierre et Marie Curie (Paris VI)
Mécanique Théorique, Tour 66
4, place Jussieu
F-75230 Paris Cedex 05
France

Translator:

H. L. S. Orde
Bressenden, Biddenden
Ashford, Kent TN27 8DU
England

Title of the French original edition:
Vibrations des systèmes mécaniques. Méthodes analytiques et applications.
© Masson, Editeur, Paris, 1984

Mathematics Subject Classification (1980): 70

ISBN 3-540-17950-X Springer-Verlag Berlin Heidelberg New York
ISBN 0-387-17950-X Springer-Verlag New York Berlin Heidelberg

Typesetting: Thomson Press (India) Ltd., New Delhi
Printing: Druckhaus Beltz, Hemsbach/Bergstr.
Bookbinding: J. Schäffer GmbH & Co. KG, Grünstadt
2141/3140-543210

Preface

The familiar concept described by the word "vibrations" suggests the rapid alternating motion of a system about and in the neighbourhood of its equilibrium position, under the action of random or deliberate disturbing forces. It falls within the province of mechanics, the science which deals with the laws of equilibrium, and of motion, and their applications to the theory of machines, to calculate these vibrations and predict their effects.

While it is certainly true that the physical systems which can be the seat of vibrations are many and varied, it appears that they can be studied by methods which are largely indifferent to the nature of the underlying phenomena. It is to the development of such methods that we devote this book which deals with free or induced vibrations in discrete or continuous mechanical structures. The mathematical analysis of ordinary or partial differential equations describing the way in which the values of mechanical variables change over the course of time allows us to develop various theories, linearised or non-linearised, and very often of an asymptotic nature, which take account of conditions governing the stability of the motion, the effects of resonance, and the mechanism of wave interactions or vibratory modes in non-linear systems.

Illustrated by numerous examples chosen for their intrinsic interest, and graduated in its presentation of parts involving difficult or delicate considerations, this work, containing several chapters which have been taught to graduate students at the Pierre and Marie Curie University in Paris, includes unpublished results and throws a new light on several theories.

A glance at the table of contents will convince the reader of the variety of subjects covered. They were selected primarily with an eye to forming a coherent whole, but no doubt the choice also reflects some personal preferences which would be hard to justify, but which we hope may give some grounds for believing that the reader will derive as much pleasure from reading the book as its author had in writing it.

Paris, October 1983 *Maurice Roseau*

Contents

Chapter VII. Modal Analysis and Vibrations of Structures

Chapter VIII. Synchronisation Theory

Chapter XII. Non-Linear Waves and Solitons

Chapter I. Forced Vibrations in Systems Having One Degree or Two Degrees of Freedom

The study of linear vibrations of systems with one degree or two degrees of freedom allows certain essential ideas, and in particular the concept of a response curve, to be introduced by means of a few simple calculations. Similarly it allows us to appreciate the influence of damping on the system under conditions in the neighbourhood of resonance. This is of relevance to many widely-used mechanisms, such as shock-absorbers, two-stage suspensions for vehicles or machinery where the characteristics for optimum performance can be determined by using an analytic approach.

Elastic Suspension with a Single Degree of Freedom

A mass m is in contact with a horizontal floor, $z = 0$, through a spring of natural length l_0, of stiffness k. The length of the spring, in its equilibrium position, under compression is l_1 and

(1.1) $$-mg - k(l_1 - l_0) = 0.$$

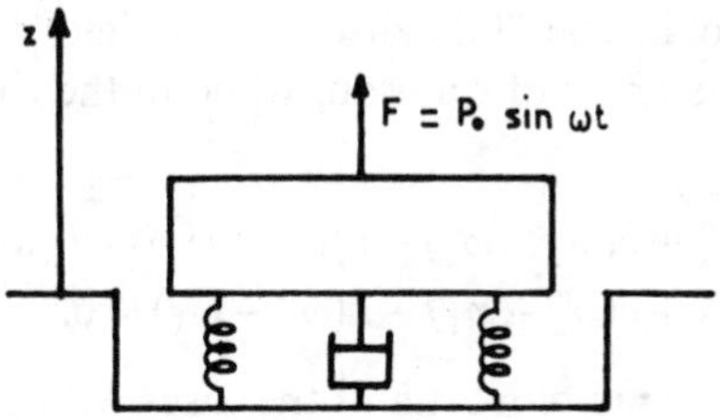

Figure 1.1

We now suppose the mass m to be acted upon by a vertical alternating force of magnitude $P_0 \sin \omega t$; denoting by z the height above ground (the length of the spring) the equation of motion is:

(1.2) $$mz'' + cz' = -mg - k(z - l_0) + P_0 \sin \omega t,$$

where cz' is the viscous damping term. Writing $x = z - l_1$, we obtain, taking (1.1) into account:

(1.3) $$mx'' + cx' + kx = P_0 \sin \omega t.$$

Applying the laws of mechanics to the system comprising the mass, spring and damping device, assuming the masses of the two latter to be negligible, we can write:

$$mx'' = -R + P_0 \sin \omega t, \tag{1.4}$$

where R measures the vertical force exerted by the system on the foundation, or by (1.3):

$$R = cx' + kx. \tag{1.5}$$

Torsional Oscillations

If torsional stresses, whose torque about the axis is $T_0 \sin \omega t$, are exerted on a disc, then in the case where the shaft is embedded, we have

$$I\varphi'' + c\varphi' + k\varphi = T_0 \sin \omega t,$$

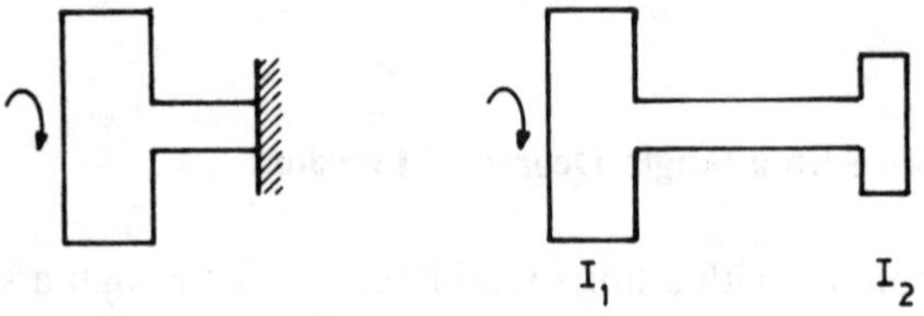

Figure 1.2

where I is the moment of inertia of the disc, k, c are rigidity and damping constants of the shaft, and φ is the angle of rotation, while in the case of a free system with two discs we have:

$$\begin{aligned} I_1\varphi_1'' + c(\varphi_1' - \varphi_2') + k(\varphi_1 - \varphi_2) &= T_0 \sin \omega t \\ I_2\varphi_2'' + c(\varphi_2' - \varphi_1') + k(\varphi_2 - \varphi_1) &= 0, \end{aligned} \tag{1.6}$$

where I_1, I_2 are the moments of inertia of the discs with respect to their common axis, and φ_1, φ_2 their angle of rotation.

The relative motion is described by the co-ordinate $\psi = \varphi_1 - \varphi_2$, which, by (1.6) is the solution of:

$$\frac{I_1 I_2}{I_1 + I_2}\psi'' + c\psi' + k\psi = \frac{I_2 T_0}{I_1 + I_2}\sin \omega t.$$

In the absence of damping ($c = 0$) the natural frequency of the oscillations ($T_0 = 0$) is

$$\omega = \sqrt{k \cdot \frac{I_1 + I_2}{I_1 I_2}}.$$

Natural Oscillations

We consider the model described by (1.3), on the supposition that $P_0 = 0$. In the absence of damping ($c = 0$), the frequency of the natural oscillations is

$$\omega_n = \sqrt{\frac{k}{m}}. \tag{1.7}$$

If $c \neq 0$, the solutions of (1.3) are given by:

$$x = ae^{s_1 t} + be^{s_2 t}, \quad s_{1,2} = -\frac{c}{2m} \pm \sqrt{\left(\frac{c}{2m}\right)^2 - \frac{k}{m}}, \quad c > 0.$$

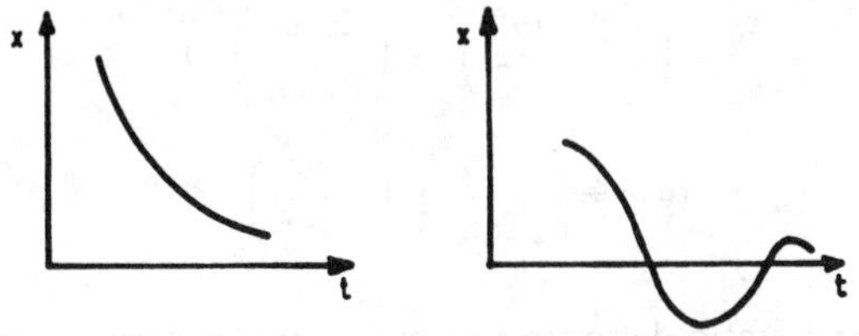

Figure 1.3

In the strongly-damped case where $\left(\frac{c}{2m}\right)^2 > \frac{k}{m}$, there is no oscillation; if the damping is weak and $\left(\frac{c}{2m}\right)^2 < \frac{k}{m}$, we write $s = -\frac{c}{2m} \pm iq$,

$$q = \sqrt{\frac{k}{m} - \left(\frac{c}{2m}\right)^2} \quad \text{and} \quad x = \exp\left(-\frac{ct}{2m}\right)\cdot(a_1 \cos qt + a_2 \sin qt). \tag{1.8}$$

It will be seen that the amplitudes of the oscillations have relative maxima at intervals of $T = \frac{\pi}{q}$, which decrease like the terms of a geometric progression of ratio $\exp\left(-\frac{\pi c}{2qm}\right)$. The frequency of vibration

$$\omega = \omega_n \sqrt{1 - \left(\frac{c}{c_*}\right)^2}, \quad c_* = 2\sqrt{mk} \tag{1.9}$$

diminishes when the damping factor increases.

Forced Vibrations

We consider once more the model described by (1.3), with a forced excitation due to the load $P_0 \sin \omega t$; ignoring the transient case just discussed, we can write the periodic solution of (1.3) in the form:

$$x = A \sin \omega t + B \cos \omega t = x_0 \sin(\omega t - \varphi) \tag{1.10}$$

and calculate A, B or x_0, φ from

$$(k - m\omega^2)A - c\omega B = P_0$$
$$c\omega A + (k - m\omega^2) B = 0,$$

whence

$$A = \frac{P_0(k - m\omega^2)}{(k - m\omega^2)^2 + c^2\omega^2}, \quad B = -\frac{P_0 c\omega}{(k - m\omega^2)^2 + c^2\omega^2}$$

and

$$\frac{x_0}{x_{st}} = \left(\left(1 - \frac{\omega^2}{\omega_n^2}\right)^2 + \left(\frac{2c}{c_*}\frac{\omega}{\omega_n}\right)^2\right)^{-1/2} \tag{1.11}$$

$$\operatorname{tg} \varphi = \frac{2c}{c_*} \cdot \frac{\omega}{\omega_n} \cdot \left(1 - \frac{\omega^2}{\omega_n^2}\right)^{-1}, \tag{1.12}$$

where $x_{st} = \frac{P_0}{k}$ is the static deformation which the system would undergo under the effect of a stationary load P_0.

We deduce from (1.11) that at resonance $\omega = \omega_n$, we have $\frac{x_0}{x_{st}} = \frac{c_*}{2c} = \frac{1}{2\varepsilon}$, with $\varepsilon = \frac{c}{c_*}$. We shall say that $\frac{1}{2\varepsilon}$ is the 'overtension' of the system, which thus appears as the ratio, at resonance, of the amplitude of the forced vibration to the static deformation, when the corresponding ratio of the excitative forces is equal to 1.

The formula (1.11) which expresses for a given c the amplitude $\frac{x_0}{x_{st}}$ of the forced vibration as a function of $\frac{\omega}{\omega_n}$ defines the response curve; the maximum amplitude is obtained when $\frac{\omega}{\omega_n} = \eta$ is such that $(1 - \eta^2)^2 + \left(\frac{2c}{c_*}\eta\right)^2$ is a minimum, i.e. when $\eta^2 = 1 - 2\left(\frac{c}{c_*}\right)^2$, if $\frac{c}{c_*} < 2^{-1/2}$ and when $\eta = 0$ otherwise. The preceding discussion has thus led us to consider in turn:

1. the frequency of the natural or free oscillations:

$$\omega_n = \sqrt{\frac{k}{m}}$$

2. that of the damped oscillations

$$q = \omega_n \sqrt{1 - \left(\frac{c}{c_*}\right)^2}$$

3. that for which the amplitude of the forced vibration is a maximum:

$$\tilde{\omega} = \omega_n \sqrt{1 - 2\left(\frac{c}{c_*}\right)^2}$$

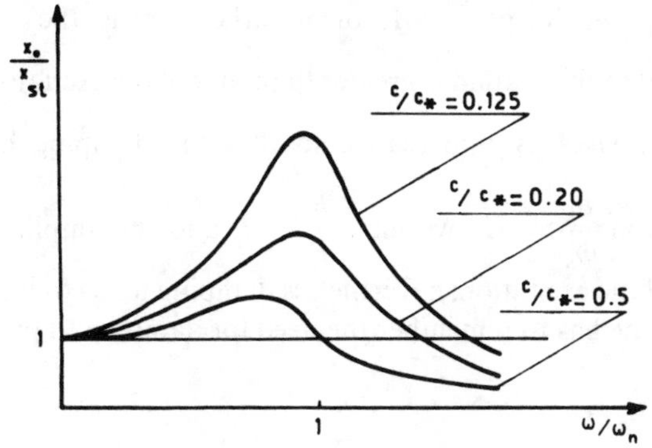

Figure 1.4

Vibration Transmission Factor

We deduce from (1.5) and (1.10) that the amplitude of the periodic force R exerted on the foundation is

$$R_0 = x_0\sqrt{k^2 + (c\omega)^2},$$

so that by (1.11) the transmittivity coefficient is:

(1.13)
$$\frac{R_0}{P_0} = \left(\frac{1 + \left(\dfrac{2c\omega}{c_*\omega_n}\right)^2}{\left(1 - \dfrac{\omega^2}{\omega_n^2}\right)^2 + \left(\dfrac{2c}{c_*}\dfrac{\omega}{\omega_n}\right)^2}\right)^{1/2}$$

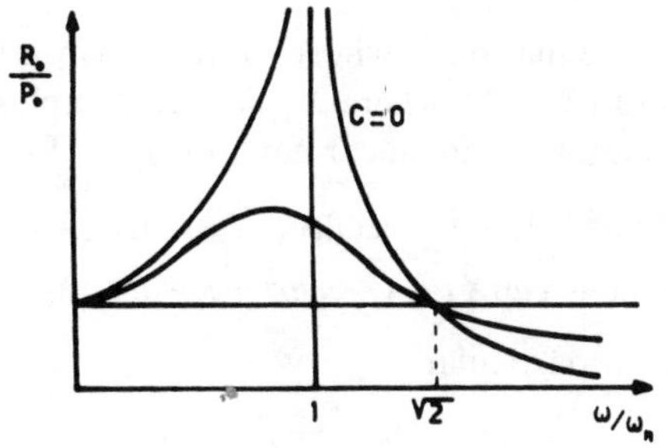

Figure 1.5

All the representations of (1.13), when c varies, pass through the points $\frac{\omega}{\omega_n} = 0$, $\frac{R_0}{P_0} = 1$ and $\frac{\omega}{\omega_n} = \sqrt{2}$, $\frac{R_0}{P_0} = 1$.

For $\frac{\omega}{\omega_n} < \sqrt{2}$, we have $\frac{R_0}{P_0} > 1$, or in other words the amplitude of the thrust transmitted to the ground is greater than that of the excitative force; but for a fixed $\frac{\omega}{\omega_n}$, $\frac{R_0}{P_0}$ decreases as c increases, so that the damping has a favourable effect. In the region $\frac{\omega}{\omega_n} > \sqrt{2}$, we have $\frac{R_0}{P_0} < 1$, and the amplitude of the thrust is less than that of the excitation; nevertheless damping tends to increase transmittivity. In any case one has to remember the need for damping to prevent breakdown at resonance.

Elastic Suspension with Two Degrees of Freedom. Vibration Absorber

We consider once more a body of mass M, in vertical translational motion, which is in contact with the ground through an elastic device of stiffness K, and subject to a gravity force of intensity g. Suspended from this main body by a spring of stiffness k, in parallel with a viscous damping device, is a body of considerably smaller mass m. The positions of these bodies are defined with respect to the upward vertical axis z by the co-ordinates of their centres of inertia z_1 for M, z_2 for m. Assuming a force $P_0 \sin \omega t$, exerted on the mass M, the equations are

$$\begin{aligned} Mz_1'' &= -Mg - c(z_1' - z_2') - K(z_1 - L) - k(z_1 - z_2 - l) + P_0 \sin \omega t \\ mz_2'' &= -mg + c(z_1' - z_2') + k(z_1 - z_2 - l), \end{aligned}$$

where l, L are the lengths of the elastic components of the system in the rest state and $c > 0$ is the damping coefficient. Writing $x_1 = z_1 - \xi_1$, $x_2 = z_2 - \xi_2$ where ξ_1, ξ_2 are the values at equilibrium of the co-ordinates z_1, z_2 in the absence of excitation, we obtain finally.

$$(1.14) \qquad \begin{aligned} Mx_1'' + c(x_1' - x_2') + Kx_1 + k(x_1 - x_2) &= P_0 \sin \omega t \\ mx_2'' + c(x_2' - x_1') + k(x_2 - x_1) &= 0, \end{aligned}$$

a set of differential equations whose forced vibration mode solution: $x_1 = \operatorname{Im}(X_1 e^{i\omega t})$, $x_2 = \operatorname{Im}(X_2 e^{i\omega t})$, where X_1, X_2 are complex amplitudes, is easily found by solving the simultaneous linear equations

$$\begin{aligned} (-M\omega^2 + K + k + ic\omega)X_1 - (k + ic\omega)X_2 &= P_0 \\ -(k + ic\omega)X_1 + (-m\omega^2 + k + ic\omega)X_2 &= 0 \end{aligned}$$

from which we derive, in particular

$$(1.15) \qquad X_1 = P_0 \frac{(k - m\omega^2) + i\omega c}{[(M\omega^2 - K)(m\omega^2 - k) - m\omega^2 k] + i\omega c[K - (M + m)\omega^2]}.$$

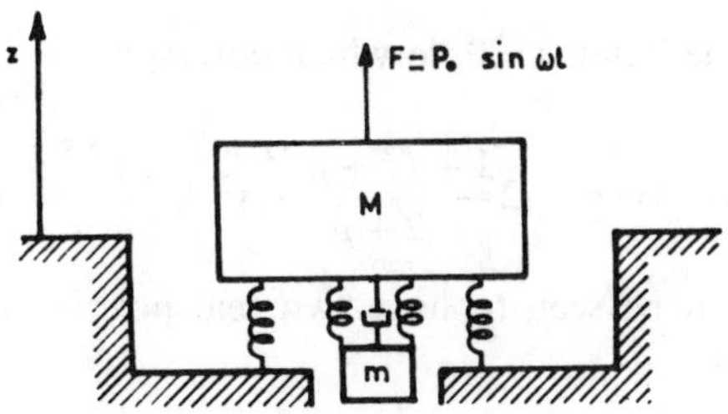

Figure 1.6

Response Curve of an Elastic System with Two Degrees of Freedom

From (1.15) we can easily deduce the real amplitude $|X_1|$, which, by a slight abuse of notation, we shall from now on denote by x_1. With

$$\mu = \frac{m}{M}$$

(1.16)
$$\omega_a = \sqrt{\frac{k}{m}}, \quad \Omega_n = \sqrt{\frac{K}{M}},$$

the natural frequencies of the absorber and the main system,

$$f = \frac{\omega_a}{\Omega_n}, \quad x_{st} = \frac{P_0}{K}, \quad c_* = 2m\Omega_n$$

we represent $\frac{x_1}{x_{st}}$ in terms of the variable $s = \frac{\omega}{\Omega_n}$ by

$$(1.17) \quad \frac{x_1}{x_{st}} = \left(\frac{\left(2\frac{c}{c_*}s\right)^2 + (s^2 - f^2)^2}{\left(2\frac{c}{c_*}s\right)^2 (s^2 - 1 + \mu s^2)^2 + [\mu f^2 s^2 - (s^2 - 1)(s^2 - f^2)]^2} \right)^{1/2}$$

which shows that the response curves depend on three parameters $\mu, f, \frac{c}{c_*}$.

Noting that (1.17) can be written in the form

$$\frac{x_1}{x_{st}} = \left(\frac{A\left(\frac{c}{c_*}\right)^2 + B}{C\left(\frac{c}{c_*}\right)^2 + D} \right)^{1/2}$$

where A, B, C, D are independent of $\frac{c}{c_*}$, we look for values of s such that $\frac{A}{C} = \frac{B}{D}$.

We find, on the one hand, $s = 0$, to which corresponds $\dfrac{x_1}{x_{st}} = 1$, and then:

$$\psi(s^2) = s^4 - 2\frac{1 + f^2 + \mu f^2}{2 + \mu}s^2 + \frac{2f^2}{2 + \mu} = 0 \tag{1.18}$$

an equation which can be seen to have two real positive roots s_1, s_2 to which correspond the values:

$$\frac{x_1}{x_{st}} = \left(\frac{A}{C}\right)^{1/2} = \frac{1}{|1 - (1 + \mu)s_{1,2}^2|} \tag{1.19}$$

and the points P and Q:

$$P = \left(s_1, \frac{1}{|1 - (1 + \mu)s_1^2|}\right), \quad Q = \left(s_2, \frac{1}{|1 - (1 + \mu)s_2^2|}\right)$$

through which pass all the response curves obtained when c is made to vary. Moreover it is easily checked by (1.18) that

$$s_1^2 < \frac{1}{1 + \mu} < s_2^2. \tag{1.20}$$

In the absence of damping, $c = 0$, it can be seen from (1.17) that x_1 becomes infinite (resonance occurs) for the values s such that

$$\phi(s^2) = s^4 - (1 + (1 + \mu)f^2)s^2 + f^2 = 0$$

This is an equation which always has two real positive roots s_{1c}, s_{2c}, which can be ordered in relation to s_1, s_2 by calculating, with the help of (1.18)

$$\phi(s_1^2) = -\frac{\mu}{2 + \mu}[(1 + (1 + \mu)f^2)s_1^2 - f^2]$$

$$\phi(s_2^2) = -\frac{\mu}{2 + \mu}[(1 + (1 + \mu)f^2)s_2^2 - f^2].$$

From (1.20) and $\phi\left(\dfrac{1}{1 + \mu}\right) = -\dfrac{\mu}{(1 + \mu)^2} < 0$ we deduce:

$$s_{1c} < (1 + \mu)^{-1/2} < s_2 < s_{2c}.$$

Noting from (1.18) that

$$\psi\left(\frac{f^2}{1 + (1 + \mu)f^2}\right) = \frac{f^4}{(1 + (1 + \mu)f^2)^2} > 0,$$

and bearing in mind that

$$\frac{f^2}{1 + (1 + \mu)f^2} < \frac{1}{1 + \mu} < s_2^2,$$

we conclude that $\dfrac{f^2}{1 + (1 + \mu)f^2} < s_1^2$ or $\phi(s_1^2) < 0$ and thus finally that:

(1.21) $$s_{1c} < s_1 < (1+\mu)^{-1/2} < s_2 < s_{2c}.$$

Writing the equation (1.18) in the form

(1.22) $$(s^2(1+\mu)-1)^2 + 2\frac{1-(1+\mu)^2 f^2}{2+\mu}(s^2(1+\mu)-1) - \frac{\mu}{2+\mu} = 0$$

we see that the ordinates $\xi_1, \xi_2\left(=\dfrac{x_1}{x_{st}}\right)$ of the points P, Q, defined by (1.19) satisfy the relation

(1.23) $$\xi_1\xi_2 = 1 + \frac{2}{\mu},$$

independent of f. With the object of obtaining a response which varies little with the frequency of excitation, at least over a certain band of frequencies, it may seem appropriate to arrange matters so that the points P and Q have the same ordinate. It will be seen from (1.19) and (1.22) that this result is obtained if

(1.24) $$f = \frac{1}{1+\mu}$$

and that the value common to these co-ordinates is $\sqrt{1+\dfrac{2}{\mu}}$.

A careful study of the equation (1.22), with

$$\xi_1 = \frac{1}{1-(1+\mu)s_1^2}, \quad \xi_2 = \frac{1}{(1+\mu)s_2^2 - 1},$$

leads to the following conclusions

$$f < \frac{1}{1+\mu} \qquad 1 < \xi_1 < \xi_2$$

$$\frac{1}{1+\mu} < f < \frac{\sqrt{2}}{1+\mu} \qquad 1 < \xi_2 < \xi_1$$

$$\frac{\sqrt{2}}{1+\mu} < f \qquad \xi_2 < 1 < \xi_1.$$

We give two examples:

1. $f = 1$, we have $s_1^2 = 1 - \sqrt{\dfrac{\mu}{2+\mu}}$, $s_2^2 = 1 + \sqrt{\dfrac{\mu}{2+\mu}}$ and

$$\xi_1 = \frac{1}{-\mu + (1+\mu)\sqrt{\dfrac{\mu}{2+\mu}}}, \quad \xi_2 = \frac{1}{\mu + (1+\mu)\sqrt{\dfrac{\mu}{2+\mu}}}.$$

2. $f = 0$, $(k = 0)$, $s_1 = 0$, $s_2^2 = \dfrac{2}{2+\mu}$ et $\xi_1 = 1$, $\xi_2 = 1 + \dfrac{2}{\mu}$.

Returning to the representation (1.17), having chosen $f = \frac{1}{1+\mu}$ so that $\xi_1 = \xi_2 = \sqrt{1 + \frac{2}{\mu}}$, one is tempted to adjust the damping $\frac{c}{c_*}$ so that the tangent to the response curve at P or at Q becomes parallel to the axis of the s; we thus obtain: for zero slope at P:

$$\left(\frac{c}{c_*}\right)^2 = \mu \frac{3 - \left(\frac{\mu}{\mu+2}\right)^{1/2}}{8(1+\mu)^3}$$

for zero slope at Q:

$$\left(\frac{c}{c_*}\right)^2 = \mu \frac{3 + \left(\frac{\mu}{\mu+2}\right)^{1/2}}{8(1+\mu)^3}.$$

A satisfactory intermediate value might be $\left(\frac{c}{c_*}\right)^2 = \frac{3\mu}{8(1+\mu)^3}$. Lastly, even if f differs from $\frac{1}{1+\mu}$, it may be advisable to arrange, by a suitable choice of $\frac{c}{c_*}$, to have a horizontal tangent at P or at Q.

For the cases already quoted, we find:

$f = 1$, zero slope at P

$$\left(\frac{c}{c_*}\right)^2 = \frac{\mu(\mu+3)\left(1 + \left(\frac{\mu}{\mu+2}\right)^{1/2}\right)}{8(1+\mu)}$$

$f = 0$, zero slope at Q

$$\left(\frac{c}{c_*}\right)^2 = \frac{1}{2(2+\mu)(1+\mu)}.$$

In general a dynamic absorber designed on the basis indicated above is quite effective. A value commonly chosen for μ is $\mu \sim 1/10$; However it should be noted that the small amplitudes of the oscillations of the principal mass are obtained at the expense of large amplitudes and correspondingly heavy loading of the absorber spring, a factor which complicates matters in the practical realisation of the design.

To clarify this point, we note that the work done by the excitation force during a cycle is:

$$\int P_0 \sin \omega t \cdot d(\tilde{x}_1 \sin(\omega t - \varphi)) = \pi P_0 \tilde{x}_1 \sin \varphi, \tag{1.25}$$

where φ denotes the phase-lag of the height of the principal mass with respect to the excitation force, $\tilde{x}_1$ is the amplitude of the motion of M, and where we revert to the original notation of x_1 and x_2 to represent the motions: $x_1 = \tilde{x}_1 \sin(\omega t - \varphi)$.

Now the work dissipated by the absorber during a cycle can easily be calculated from the equations (1.14); after multiplying by x_1', x_2', we obtain by addition and integration with respect to time over a complete period

$$c\int (x_1' - x_2')^2 \,dt = \int P_0 \sin \omega t \cdot x_1' \,dt. \tag{1.26}$$

This equation expresses the fact that the work done by the excitation force is entirely dissipated by the absorber. Starting from the representation $x_1 - x_2 = \zeta \sin(\omega t + \alpha)$, we deduce from (1.25) and (1.26): $\zeta^2 = \dfrac{P_0 \tilde{x}_1 \sin \varphi}{c\omega}$.

In general we have $\varphi \sim 90°$, so that $\zeta^2 \sim \dfrac{P_0 \tilde{x}_1}{c\omega}$ and in dimensionless variables

$$\left|\frac{\zeta}{x_{st}}\right|^2 = \left(\frac{2c}{c_*}\mu s\right)^{-1} \cdot \left|\frac{\tilde{x}_1}{x_{st}}\right|. \tag{1.27}$$

To construct the response curves (1.17) corresponding to the given values of μ and f, it is convenient to begin by drawing those corresponding to $c = 0$ and $c = \infty$.

For $c = 0$ the response curve cuts the axis $\dfrac{x_1}{x_{st}} = 1$ in four points, the first with abscissa $s = 0$, the third with abscissa $s = f\sqrt{1+\mu}$. It is easily verified that for $f < \dfrac{\sqrt{2}}{1+\mu}$ we have $s_1 < f\sqrt{1+\mu} < s_2$, whereas for $f > \dfrac{\sqrt{2}}{1+\mu}$, $s_1 < s_2 < f\sqrt{1+\mu}$.

Vehicle Suspension

1. A vehicle suspension mechanism can be represented schematically by a system comprising two solids of mass M, m linked to each other by a spring mechanism of stiffness k, and a shock absorber with viscous damping coefficient c. The solid of mass m representing the wheel is in direct contact with the ground through a spring of stiffness K representing the tyres.

We denote by z_1, z_2 the heights above the horizontal reference plane of the reference points, fixed in relation to the suspended part and to the wheel respectively, on the assumption that the vehicle is moving at a horizontal speed v over undulating ground whose surface can be represented by the equation: $z = \eta \sin \dfrac{2\pi y}{\lambda}$, y being the horizontal co-ordinate.

The system has a vertical translational motion with respect to axes moving with the vehicle at a horizontal speed of v, which can be described with the help of the co-ordinates z_1, z_2; the excitation being caused by the vertical movement of the plane on which the system reposes, namely the plane represented by $z = \eta \sin \omega t$, $\omega = \dfrac{2\pi v}{\lambda}$. Denoting the natural lengths of the springs by l, L, we obtain easily the equations of motion

$$\begin{aligned} Mz_1'' &= -k(z_1 - z_2 - l) - c(z_1' - z_2') - Mg \\ mz_2'' &= k(z_1 - z_2 - l) + c(z_1' - z_2') - mg - K(z_2 - L - \eta \sin \omega t) \end{aligned}$$

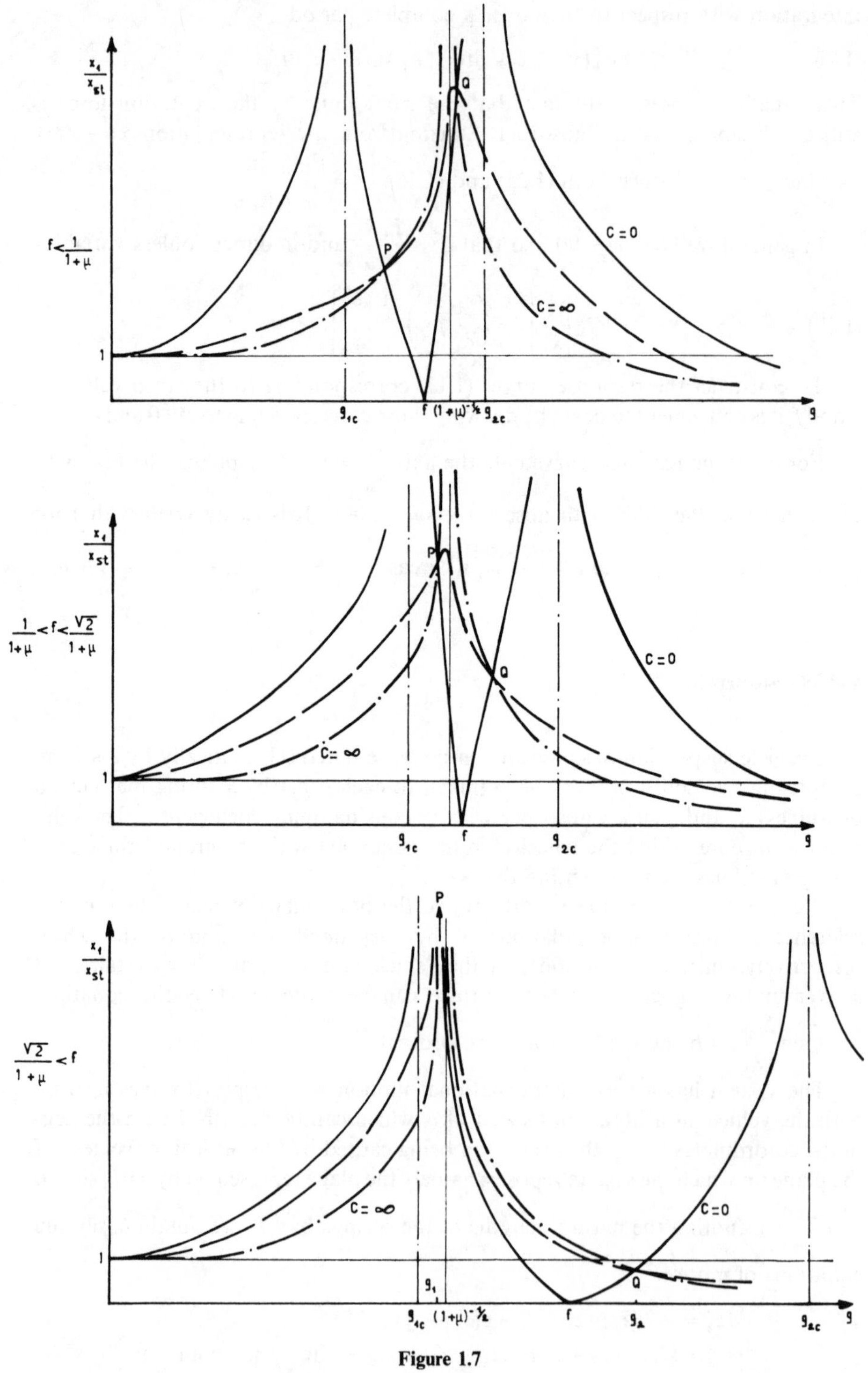
x1/xst
f<1/(1+μ)
P
Q
C=0
C=∞
1
g1c
f (1+μ)^-½ g2c
g
x1/xst
1/(1+μ)<f<√2/(1+μ)
P
Q
C=0
C=∞
1
g1c
f
g2c
g
P
x1/xst
√2/(1+μ)<f
C=∞
C=0
1
Q
g1
g1c (1+μ)^-½
f
g2
g2c
g

Figure 1.7

or, expressed in terms of the new variables $x_1 = z_1 - \xi_1$, $x_2 = z_2 - \xi_2$, where ξ_1, ξ_2 are the values of z_1, z_2 corresponding to the equilibrium position in the absence of excitation:

$$\begin{aligned} &Mx_1'' + c(x_1' - x_2') + k(x_1 - x_2) = 0 \\ &mx_2'' - c(x_1' - x_2') - k(x_1 - x_2) + Kx_2 = K\eta \sin \omega t. \end{aligned} \tag{1.28}$$

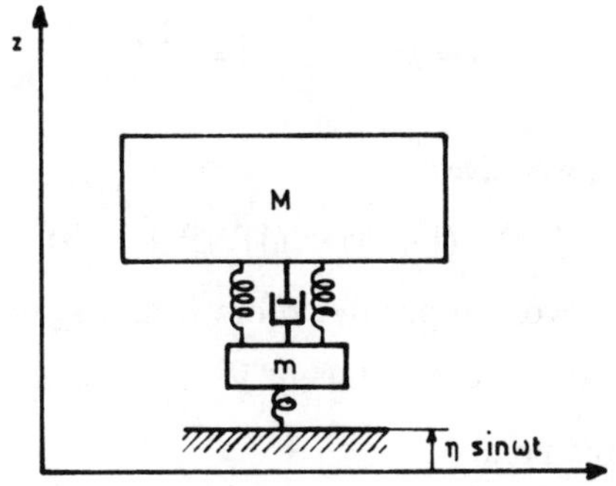

Figure 1.8

We now look for the periodic solution of (1.28) of the form $x_1 = \operatorname{Im} X_1 e^{i\omega t}$, $x_2 = \operatorname{Im} X_2 e^{i\omega t}$, where X_1, X_2 are complex amplitudes, and obtain

$$\begin{aligned} (k + ic\omega - \omega^2 M)X_1 - (k + ic\omega)X_2 &= 0 \\ -(k + ic\omega)X_1 + (k + K - m\omega^2 + ic\omega)X_2 &= K\eta \end{aligned}$$

from which X_1 can be found; with

$$\begin{aligned} &\omega_a = \sqrt{\frac{K}{m}} \quad \text{tyre frequency} \\ &\Omega_n = \sqrt{\frac{k}{M}} \quad \text{suspension frequency} \\ &c_* = 2M\Omega_n, \quad f = \frac{\Omega_n}{\omega_a}, \quad M = m\mu, \quad \mu > 1 \end{aligned} \tag{1.29}$$

we can represent, in terms of the variable $s = \dfrac{\omega}{\Omega_n}$:

(1.30)

$$\left|\frac{X_1}{\eta}\right| = \left(\frac{1 + \left(\dfrac{2c}{c_*}s\right)^2}{[s^2(s^2f^2 - 1) + (1 - (1+\mu)s^2f^2)]^2 + \left(\dfrac{2c}{c_*}s\right)^2 [1 - (1+\mu)s^2f^2]^2} \right)^{1/2}$$

It will be seen that the family of response curves obtained by varying c, but keeping μ and f fixed, has several points in common which can be determined from the equation:

$$s^2(s^2f^2-1)+1-(1+\mu)s^2f^2=\pm(1-(1+\mu)s^2f^2).$$

Taking the + sign, we find:

$$s=0, \qquad \left|\frac{X_1}{\eta}\right|=1$$

$$s=f^{-1}, \qquad \left|\frac{X_1}{\eta}\right|=\mu^{-1}$$

while, with the − sign, we have:

(1.31) $$f^2s^4-(1+2(1+\mu)f^2)s^2+2=0$$

an equation which has two real positive roots s_1 and s_2 such that:

(1.32) $$0<s_1<f^{-1}(1+\mu)^{-1/2}<f^{-1}<s_2$$

to which correspond the points

$$P=\left\{s_1,\frac{1}{1-(1+\mu)f^2s_1^2}\right\}, \quad Q=\left\{s_2,\frac{1}{(1+\mu)f^2s_2^2-1}\right\}$$

which belong to all the response curves obtained by making c vary.

It can be verified from (1.32) that $(1+\mu)f^2s_2^2-1>\mu>1$, whence

$$\left|\frac{X_1}{\eta}\right|_{s_2}<\mu^{-1}=\left|\frac{X_1}{\eta}\right|_{s=f^{-1}}<1,$$

and thus $0<1-(1+\mu)f^2s_1^2<1$, or $\left|\frac{X_1}{\eta}\right|_{s_1}>1$.

The values of s for which there is resonance in the absence of damping are the solutions of:

(1.33) $$\Phi(s^2)=f^2s^4-(1+(1+\mu)f^2)s^2+1=0.$$

Calculating $\Phi(s_1^2)$, $\Phi(s_2^2)$ with the help of (1.31) we find

$$\Phi(s_1^2)=(1+\mu)f^2s_1^2-1<0$$

$$\Phi(s_2^2)=(1+\mu)f^2s_2^2-1>0 \qquad \text{by (1.32)}$$

from which can be deduced, on the one hand, that (1.33) has two positive roots s_{1c}, s_{2c} and on the other hand, that $s_{1c}<s_1<s_{2c}<s_2$. After verifying that $\Phi(f^{-2})<0$ we obtain the more precise result

$$s_{1c}<s_1<f^{-1}(1+\mu)^{-1/2}<f^{-1}<s_{2c}<s_2$$

and also $s_1<\sqrt{2}<s_{2c}$.

This information enables the shape of the response curves to be determined remembering that any two of them, for distinct values of c, have no other points in common apart from those already found.

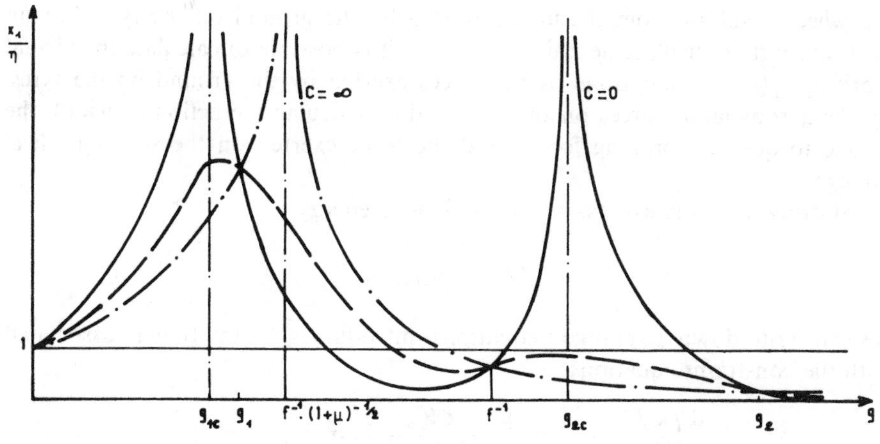

Figure 1.9

2. It is important to study the motion of a vehicle, at the design stage, in order to find out what effect certain parameters have on its road-holding properties.

Simulation techniques are commonly used to calculate the motion numerically with quite complex models, working in an open loop mode, i.e. the rotation of the steering wheel, the engine torque, and the pressure on the brakes are all prescribed as functions of the time.

The position of the body of the vehicle with respect to the system of reference is defined by 6 parameters; and similarly the position of each wheel in relation to the body is defined with the help of 6 parameters. The engine is identified in the model with a longitudinally or transversely mounted flywheel; and the rack and pinion steering mechanism is represented by a point moving transversely across the body. The angle of rotation of the flywheel, and the displacement of the point representing the steering are specified by two additional parameters, so that this description requires 32 parameters q_i, $1 \leqslant i \leqslant 32$. However, once the steering parameter has been fixed, the position of each wheel relative to the body of the vehicle depends only on two parameters, one for the height and one for the angle of rotation. In other words there are 16 'linkage' relations between the parameters q_i:

$$\mathscr{L}_k(q) = 0, \quad 1 \leqslant k \leqslant 16;$$

and an additional 'transmission' relation, between the rotational velocities of the flywheel and the road wheels:

$$\sum_{j=1}^{32} r_j q'_j = 0.$$

The deformation of each tyre depends on the forces exerted upon it by the ground; and these forces can be measured on specialised test-equipment on certain assumptions. These are that a knowledge of the 6-road-wheel-position parameters, and their derivatives with respect to time, suffices to describe the state of deformation of the tyre and that it is thus possible to represent, in terms of these 12 parameters, the components of the reduced system of forces at the centre of the wheel, resulting from the forces exerted by the ground on the tyre. For an arbitrary virtual displacement δq_i, $1 \leqslant i \leqslant 32$, it is possible to calculate the virtual work $\sum_{i=1}^{32} Q_i(q, q', t)\delta q_i$ done by the forces exerted by the ground on the tyres, by the aerodynamic forces on the vehicle (in particular the action of wind), the engine torque, the braking forces, and the force exerted on the steering wheel linkage.

Starting from the expression for the kinetic energy

$$2T = \sum_{i,j}^{32} a_{ij} q_i' q_j',$$

we can write down Lagrange's equations, introducing 17 multipliers associated with the constraint equations

$$\frac{\mathrm{d}}{\mathrm{d}t}\left(\frac{\partial T}{\partial q_i'}\right) - \frac{\partial T}{\partial q_i} = \sum_{p=1}^{16} \lambda_p \frac{\partial \mathscr{L}_p}{\partial q_i} + \lambda_{17} r_i + Q_i(q, q', t).$$

The full set of equations is then solved numerically, starting from prescribed initial conditions and data which consists basically of the angular position of the steering wheel or the force exerted on the steering linkage, the torque exerted by the engine, and the braking pressure.

Numerical analysis of the system for various given situations, such as braking on a bend in the road, changing lanes, etc. enables one to obtain a simulation based on a model with fifteen degrees of freedom which can be very useful in assessing the effects of such and such a structural modification or design change.

Whirling Motion of a Rotor-Stator System with Clearance Bearings

The diagram in Figure 1.10 represents part of a turbo-engine consisting of a rotor at the end of a shaft of radius r rotating at an angular velocity ω and supported in bearings (not shown) rigidly attached to the structure. An additional bearing of radius $r_1 = r + \delta$, $\delta > 0$ at the centre of a grid suspended from the stator and placed in front of the rotor, can come into contact with the shaft, if the latter's

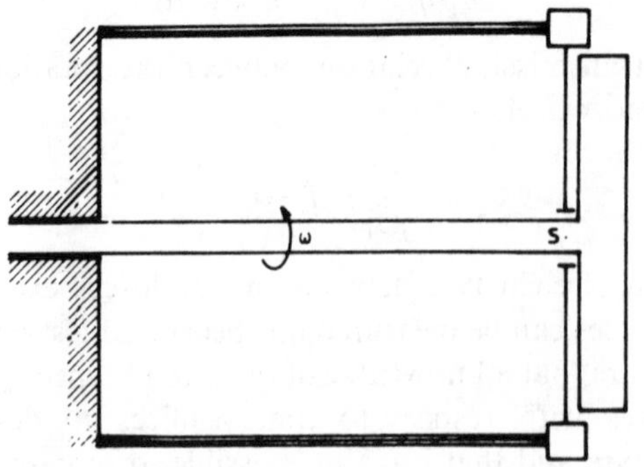

Figure 1.10

deflection, due to a defect in the balancing of the rotor, exceeds δ at this point. In that case the interaction between the elastic structure of the stator, and the rotating shaft which bears the rotor, generates vibrations which can be studied under suitable assumptions, following [62], by using a simplified model with two degrees of freedom.

We assume that, from the dynamic standpoint, all the masses can be concentrated in a plane Π perpendicular to the axis of rotation at a point S in the centre of the bearing with clearance. Because of the bending of the shaft the geometric centres S_1 and C of the grid-bearing and rotor respectively (or more precisely their projections on Π) will be distinct from S. We shall make the hypothesis that the centre of inertia of the grid/stator system of mass m_1 coincides with S_1, and that G, the centre of inertia of the rotor of mass m is distinct from C, the want of balance being measured by $CG = \varepsilon$.

We shall ignore gyroscopic effects and consider the case of synchronous motion in which the points C, S, G remain aligned with S_1.

Denoting by S_1y the axis carrying these points, oriented so that $\overline{CG} = \varepsilon > 0$ there are two cases to be considered for the contact between shaft and bearing, Case I with $\overline{CS_1} = \delta$ and Case II with $\overline{S_1C} = \delta$.

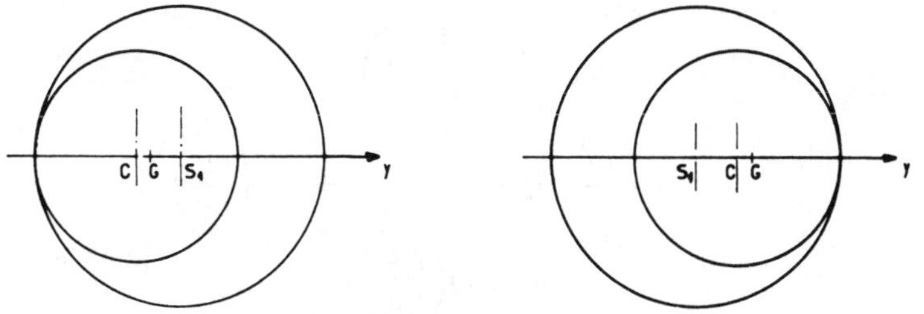

Figure 1.11

We introduce generally the co-ordinate $y{:}\overline{SC}$, $y_1 = \overline{SS_1}$, and refer to the case of motion without contact as Case III.

If R represents the radial component along Sy of the reaction exerted by the bearing on the shaft in the two cases of contact, then maintenance of contact requires $R > 0$ in Case I and $R < 0$ in Case II.

Denoting by k, k_1 the radial stiffnesses of the elastic system comprising on the one hand the shaft, and on the other hand the stator and grid, we obtain, by applying the theorem on the motion of the centre of inertia:

Case I:

$$y_1 - y = \delta$$

(1.34)
$$R - ky = -m\omega^2(y + \varepsilon)$$

$$-R - k_1 y_1 = -m_1\omega^2 y_1, \quad R > 0.$$

Case II:

$$y - y_1 = \delta$$

(1.35)
$$R - ky = -m\omega^2(y + \varepsilon)$$

$$-R - k_1 y_1 = -m_1\omega^2 y_1, \quad R < 0.$$

Case III:

(1.36)
$$y_1 = 0$$
$$-ky = -m\omega^2(y+\varepsilon)$$
$$|y| < \delta.$$

It is appropriate to introduce the natural frequencies of the rotor $\omega_r = \left(\dfrac{k}{m}\right)^{1/2}$, of the stator $\omega_s = \left(\dfrac{k_1}{m_1}\right)^{1/2}$, and of the system rotor/stator without bearing clearance $\omega_n = \left(\dfrac{k+k_1}{m+m_1}\right)^{1/2}$, the latter being intermediate in value between the two former.

With $\mu = \dfrac{m_1}{m}$, we have $\omega_n = \left(\dfrac{\omega_r^2 + \mu\omega_s^2}{1+\mu}\right)^{1/2}$ and we can distinguish the two possible cases $\omega_r < \omega_n < \omega_s$ or $\omega_s < \omega_n < \omega_r$; the first, which corresponds to a stator of high rigidity is the more interesting physically, and is the one which we shall consider in the analysis below.

The solution of the systems (1.34) to (1.36) leads to:

Case I:
$$\frac{y_1}{\varepsilon} = \frac{\left(\dfrac{\omega}{\omega_n}\right)^2(\delta-\varepsilon) - \left(\dfrac{\omega_r}{\omega_n}\right)^2\delta}{\varepsilon(1+\mu)\left(\left(\dfrac{\omega}{\omega_n}\right)^2 - 1\right)}$$

(1.37)
$$\frac{y}{\varepsilon} = \frac{-\left(\dfrac{\omega}{\omega_n}\right)^2(\mu\delta+\varepsilon) + \delta\left(1+\mu-\left(\dfrac{\omega_r}{\omega_n}\right)^2\right)}{\varepsilon(1+\mu)\left(\left(\dfrac{\omega}{\omega_n}\right)^2 - 1\right)}$$

$$\frac{R}{k_1\varepsilon} = \left(\frac{\omega^2}{\omega_s^2} - 1\right)y_1, \quad R > 0.$$

Case II: y, y_1, R are again obtained from the formulae (1.37) by changing δ into $-\delta$; the inequality becomes $R < 0$.

Case III:
$$y_1 = 0$$

(1.38)
$$\frac{y}{\varepsilon} = \frac{\left(\dfrac{\omega}{\omega_n}\right)^2}{\left(\dfrac{\omega_r}{\omega_n}\right)^2 - \left(\dfrac{\omega}{\omega_n}\right)^2}$$

$$\left|\frac{y}{\varepsilon}\right| < \frac{\delta}{\varepsilon}.$$

The response curves which represent $\dfrac{y}{\varepsilon}, \dfrac{y_1}{\varepsilon}$ as a function of $\dfrac{\omega}{\omega_n}$ depend on three parameters $\mu, \dfrac{\delta}{\varepsilon}, \dfrac{\omega_r}{\omega_n}$; the behaviour of the system shows some interesting features if the clearance is large enough, particularly if $\frac{\delta}{\varepsilon} > \left(1 - \left(\frac{\omega_r}{\omega_n}\right)^2\right)^{-1}$, the case to which we shall from now on restrict ourselves. Incidentally, this hypothesis, taken in conjunction with that of a rigid stator, is equivalent to

$$\left(\frac{\omega_r}{\omega_n}\right)^2 \cdot \frac{\delta}{\delta - \varepsilon} < 1 < \left(\frac{\omega_s}{\omega_n}\right)^2, \quad \delta > \varepsilon.$$

Taking the inequalities in (1.37) and (1.38) into account, it will be seen that the various possible modes of vibration can be established only under the following conditions:

Case I: $\quad \left(\frac{\omega_r}{\omega_n}\right)^2 \cdot \frac{\delta}{\delta - \varepsilon} < \left(\frac{\omega}{\omega_n}\right)^2 < 1 \quad \text{or} \quad \left(\frac{\omega_s}{\omega_n}\right)^2 < \left(\frac{\omega}{\omega_n}\right)^2$

Case II: $\quad \left(\frac{\omega_r}{\omega_n}\right)^2 \cdot \frac{\delta}{\delta + \varepsilon} < \left(\frac{\omega}{\omega_n}\right)^2 < 1 \quad \text{or} \quad \left(\frac{\omega_s}{\omega_n}\right)^2 < \left(\frac{\omega}{\omega_n}\right)^2$

Case III: $\quad \left(\frac{\omega}{\omega_n}\right)^2 < \frac{\delta}{\delta + \varepsilon}\left(\frac{\omega_r}{\omega_n}\right)^2 \quad \text{or} \quad \frac{\delta}{\delta - \varepsilon}\left(\frac{\omega_r}{\omega_n}\right)^2 < \left(\frac{\omega}{\omega_n}\right)^2.$

The form of the curve in Figure 1.12 shows a hysteresis phenomenon and a jump in amplitude in the neighbourhood of the resonance $\omega = \omega_n$, which can be

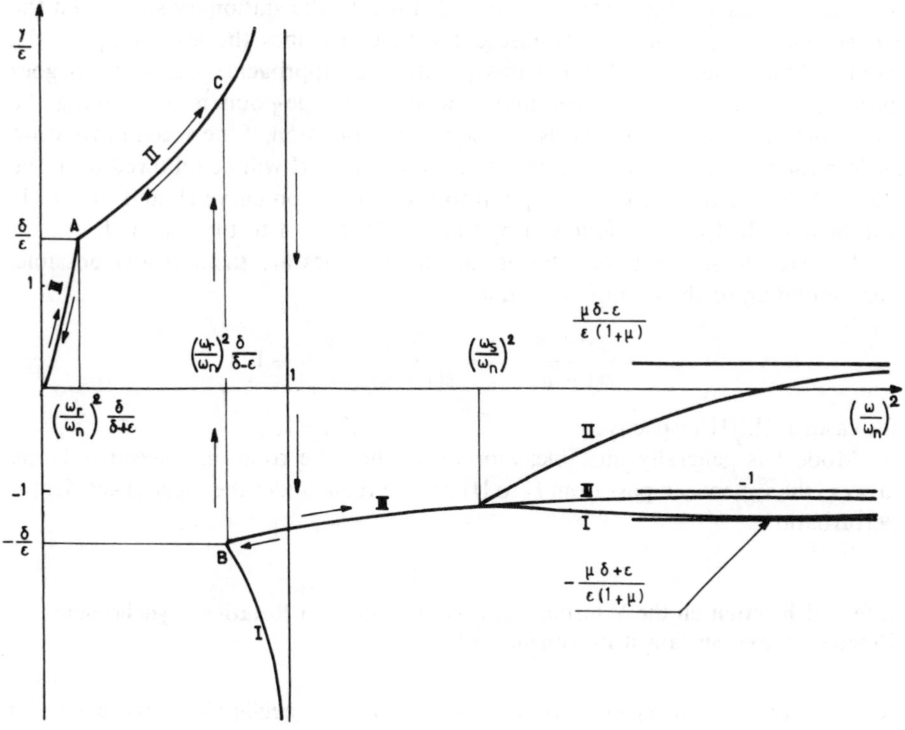

Figure 1.12

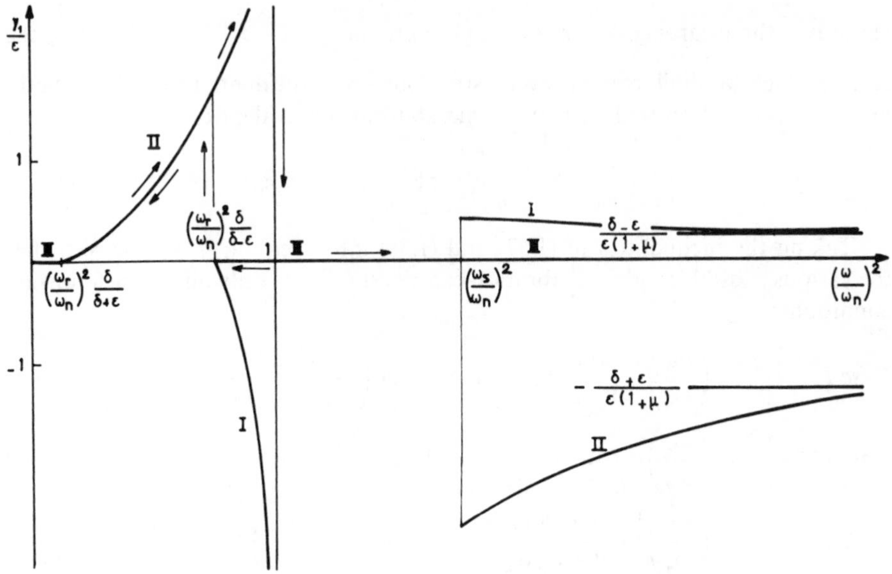

Figure 1.13

described as follows: if the speed ω is increased enough to validate the assumption that at each instant the system has settled down to the stationary state, then the image point of y/ε on the response curve first describes the arc III up to the point A, then the arc II beyond this point, as ω approaches ω_n; when ω goes past ω_n, y/ε undergoes a discontinuity, with the image-point moving along the branch III, from $\omega = \omega_n$ onwards. In the reverse direction, if the speed of rotation is decreased, starting from a value $\omega > \omega_n$, the curve III will be followed until the point B, and then there will be a jump to the point C on curve II, and the arc II will be described until A, followed by the arc III from A to the origin O.

For very high speeds of rotation, all these modes are theoretically possible, corresponding to the asymptotic values

$$\frac{y}{\varepsilon} \sim \frac{\mu\delta + \varepsilon}{\varepsilon(1+\mu)}, \quad \frac{y}{\varepsilon} \sim \frac{\mu\delta - \varepsilon}{\varepsilon(1+\mu)}, \quad \frac{y}{\varepsilon} \sim -1$$

for cases I, II, III respectively.

Mode I is generally unstable; moreover when the rotational speed is large, $\omega > \omega_s$, the system can pass from II to III or vice versa under the effect of accidental perturbations.

Effect of Friction on the Whirling Motion of a Shaft in Rotation; Synchronous Precession, Self-sustained Precession [32]

A shaft, of circular cross-section of radius r and negligible mass carries a centrally-mounted disc D. We investigate the motion by reference to orthonormal axes $Oxyz$, with Ox along the axis of revolution of the undeformed shaft and Oz vertical.

The shaft is supported by two bearings such that we may assume that during the motion, the centres of their cross-sections lying in the planes $x = \pm l$ have their y, z co-ordinates remain equal to zero.

Secondary bearings, provided with a certain amount of clearance, connected to an elastic structure and lying in the planes $x = \pm l_1$, $l_1 < l$, may come into contact with the shaft and may, in particular, be affected by the effects of the latter's flexion induced by any want of balance in the disc.

Disregarding for the moment the secondary bearings, the motion of C, the geometric centre of the disc, whose complex co-ordinates $\zeta = y + iz$, in the plane $x = 0$, is given by the equation:

$$m\zeta'' + c\zeta' + k\zeta = m_0 a\omega^2 e^{i(\omega t + \delta)}, \tag{1.39}$$

where ω is the speed of rotation of the shaft ($\omega > 0$), k the coefficient of flexural rigidity, c the damping factor, m the total mass in rotation and a the distance from the axis of rotation of the mass m_0 representing the want of balance.

The motion of synchronous precession corresponds to the periodic solution of (1.39) of frequency ω, i.e.:

$$\zeta = \frac{m_0 a\omega^2}{k - m\omega^2 + ic\omega} e^{i(\omega t + \delta)}.$$

We now consider the case where contact with the secondary bearings takes place, on the assumption of motion which has yOz as a plane of symmetry. Denoting by ζ_1, ξ the complex co-ordinates in the planes $x = \pm l_1$, of the geometric centre of the circular cross-section of the bearing of radius $r_1 > r$ and of the centre of the shaft respectively, we can write down the condition for contact as

$$\xi - \zeta_1 = (r_1 - r)e^{i\alpha}, \tag{1.40}$$

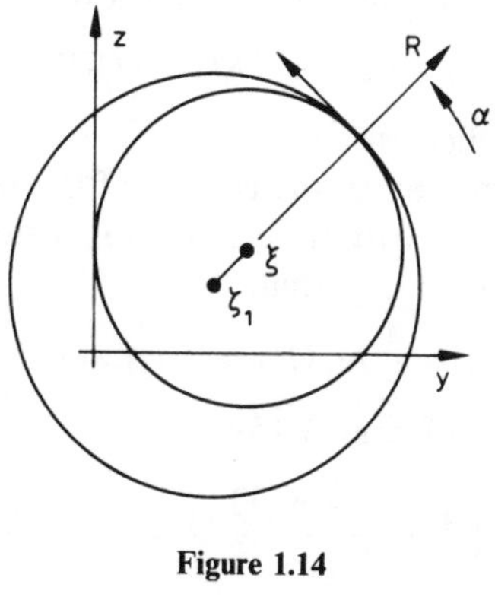

Figure 1.14

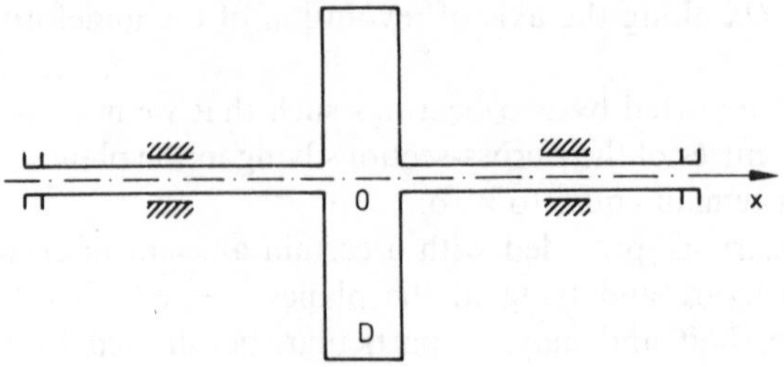

Figure 1.15

where α is the polar angle of the outward normal to the shaft at the point of contact with the bearing. With R denoting the normal component of the reaction exerted by the shaft on the bearing, and f the coefficient of friction, the equations of motion are:

$$m\zeta'' + c\zeta' + k(1+i\eta)\zeta + \sigma R(1+if)e^{i\alpha} = m_0 a\omega^2 e^{i(\omega t+\delta)} \tag{1.41}$$

$$m\zeta_1'' + c_1\zeta_1' + k_1(1+i\eta_1)\zeta_1 - R(1+if)e^{i\alpha} = 0 \tag{1.42}$$

where m_1, k_1, c_1 are the mass, coefficient of rigidity and damping factor of the structure connected to the secondary bearings respectively and η, η_1 are loss factors. The equation (1.41) simply expresses the fact that the deflection ζ of the shaft at $x=0$ depends linearly on the forces exerted on it, namely the inertial forces of the disc at $x=0$ on the one hand, and the reactions of the secondary bearings at $x=\pm l_1$ on the other. The coefficient σ thus makes its appearance, a coefficient which we shall be able to take as positive in what follows.

To these have to be added the inequalities:

$$R > 0 \tag{1.43}$$

$$v_r = \omega r + (r_1 - r)\frac{d\alpha}{dt} > 0 \tag{1.44}$$

and we can now express ξ, the deflection of the shaft at $x=\pm l_1$, in terms of the deflection ζ at $x=0$ and the forces $-R(1+if)e$ applied to the shaft at $x=\pm l_1$. In linearised form this response can be written in the form:

$$\xi = b\zeta - \varkappa^{-1}R(1+if)e^{i\alpha} \tag{1.45}$$

where $b, \varkappa$ are positive coefficients.

If we assume that the motion of the centres of the secondary bearings can be represented by $\zeta_1 = B_1 e^{i\Omega t}$ we see from (1.42), after ignoring in a first approximation the damping and loss factors c_1, η_1, that:

$$\zeta_1(k_1 - m_1\Omega^2) = R(1+if)e^{i\alpha} \tag{1.46}$$

whence we deduce, using (1.40) and (1.45):

$$b\zeta - \left(1 + \frac{k_1 - m_1\Omega^2}{\varkappa}\right)\zeta_1 = (r_1 - r)e^{i\alpha} \tag{1.47}$$

and also:

(1.48) $$\frac{\zeta_1}{|\zeta_1|}=\varepsilon\frac{1+if}{\sqrt{1+f^2}}e^{i\alpha},\quad \varepsilon=\pm 1, \text{ such that}$$

(1.49) $$\varepsilon(k_1-m_1\Omega^2)>0 \qquad \text{by (1.43).}$$

Lastly, having regard to (1.47), we arrive at:

(1.50) $$\frac{\zeta_1}{|\zeta_1|}=\varepsilon\frac{1+if}{(r_1-r)\sqrt{1+f^2}}\left(b\zeta-\left(1+\frac{k_1-m_1\Omega^2}{\varkappa}\right)\zeta_1\right)$$

from which we obtain $|\zeta_1|$ as a function of $|\zeta|$:

(1.51) $$|\zeta_1|=\frac{\pm\Delta-\varepsilon(r_1-r)}{\left(1+\dfrac{k_1-m_1\Omega^2}{\varkappa}\right)\sqrt{1+f^2}}$$

with

(1.52) $$\Delta=\sqrt{(1+f^2)b^2|\zeta|^2-f^2(r_1-r)^2}$$

and the condition for annular contact:

(1.53) $$|\zeta|>f\cdot\frac{r_1-r}{b\sqrt{1+f^2}}$$

We can therefore calculate ζ_1 in terms of ζ by (1.50), (1.51):

(1.54) $$\zeta_1=\frac{\zeta}{1+\dfrac{k_1-m_1\Omega^2}{\varkappa}}(\psi_1(\zeta)-i\psi_2(\zeta))$$

(1.55) $$\psi_1(\zeta)=b-\frac{r_1-r}{b|\zeta|^2(1+f^2)}[(r_1-r)f^2\pm\varepsilon\Delta]$$

$$\psi_2(\zeta)=\frac{f(r_1-r)}{b|\zeta|^2(1+f^2)}[(r_1-r)\mp\varepsilon\Delta]$$

We choose ε, using (1.49), and then the sign of Δ ensures that the right-hand side of (1.51) is positive. We can now, reverting to (1.41), (1.46), (1.54) and introducing the operator defined by: $\phi(\zeta_1)=m_1\frac{\zeta_1''}{\zeta_1}+c_1\frac{\zeta_1'}{\zeta_1}+k_1(1+i\eta_1)$, write the equations of motion in the form:

(1.56) $$m\zeta''+c\zeta'+k(1+i\eta)\zeta+\frac{\sigma\zeta\phi(\zeta_1)}{1+\varkappa^{-1}\phi(\zeta_1)}\cdot(\psi_1(\zeta)-i\psi_2(\zeta))=m_0a\omega^2e^{i(\omega t+\delta)}$$

$$\zeta_1 = \frac{\zeta}{1 + \varkappa^{-1}\phi(\zeta_1)} (\psi_1(\zeta) - i\psi_2(\zeta)) \tag{1.57}$$

subject to the provision that the condition (1.53) be satisfied.

Synchronous Motion

Solutions of the form $\zeta = Be^{i(\omega t + \delta + \gamma)}$, $\zeta_1 = B_1 e^{i(\omega t + \delta_1 + \gamma_1)}$ with B, B_1 positive constants, can be constructed starting from (1.56), (1.57). Noting that ψ_1, ψ_2 only depend on B, we obtain by (1.56):

$$Be^{i\gamma}\left\{k(1 + i\gamma) + ic\omega - m\omega^2 + \sigma \cdot \frac{k_1(1 + i\eta_1) + ic_1\omega - m_1\omega^2}{1 + \varkappa^{-1}(k_1(1 + i\eta_1) + ic_1\omega - m_1\omega^2)} \cdot (\psi_1(B) - i\psi_2(B))\right\} = ma_0\omega^2 \tag{1.58}$$

We choose ε, using (1.49) (with $\omega = \Omega$), and then the sign of Δ on the basis of certain assumptions regarding B. We solve (1.58) in terms of B, γ and then obtain B_1, γ_1 using (1.57). Depending on the particular values of the speed of rotation ω and the various parameters of the system, the problem may have no solution, one solution or several solutions.

Self-maintained Precession

Suppose there is perfect balance, so that $a = 0$; the equations of the problem can be written as:

$$m\zeta'' + c\zeta' + k(1 + i\eta)\zeta + \sigma R(1 + if)e^{i\alpha} = 0 \tag{1.59}$$

$$m_1\zeta_1'' + c_1\zeta_1' + k_1(1 + i\eta_1)\zeta_1 - R(1 + if)e^{i\alpha} = 0 \tag{1.60}$$

$$\xi - \zeta_1 = (r_1 - r)e^{i\alpha} \tag{1.61}$$

$$\xi = b\zeta - \varkappa^{-1}R(1 + if)e^{i\alpha} \tag{1.62}$$

$$R > 0 \tag{1.63}$$

$$\omega r + (r_1 - r)\frac{d\alpha}{dt} > 0 \tag{1.64}$$

Seeking a solution of the form:

$$\zeta = Ce^{i\Omega t}, \tag{1.65}$$

$$\zeta_1 = C_1 e^{i(\Omega t + \beta)} \tag{1.66}$$

we obtain, from (1.59), (1.60):

$$C = -\frac{\sigma R(1 + if)e^{i(\alpha - \Omega t)}}{k - m\Omega^2 + i(c\Omega + k\eta)} \tag{1.67}$$

(1.68) $$C_1 = \frac{R(1+if)e^{i(\alpha-\beta-\Omega t)}}{k_1 - m_1\Omega^2 + i(c_1\Omega + k_1\eta_1)}$$

(1.69) $$\alpha = \Omega t + \text{constant}$$

Eliminating ξ from (1.61), (1.62) we arrive at:

$$b\zeta - \zeta_1 - \varkappa^{-1}R(1+if)e^{i\alpha} = (r_1 - r)e^{i\alpha}$$

whence by (1.65)...(1.69):

(1.70) $$-R(1+if)[\sigma b(k - m\Omega^2 + i(c\Omega + k\eta))^{-1} + (k_1 - m_1\Omega^2 + i(c_1\Omega + k_1\eta_1))^{-1} + \varkappa^{-1}] = r_1 - r$$

which defines the real quantities R and Ω.

Writing

(1.71) $$\delta = (k - m\Omega^2)^2 + (c\Omega + k\eta)^2,$$
$$\delta_1 = (k_1 - m_1\Omega^2)^2 + (c_1\Omega + k_1\eta_1)^2$$

we obtain, on equating to zero the imaginary part of (1.70):

(1.72)

$$\sigma b\delta_1[f(k - m\Omega^2) - c\Omega - k\eta] + \delta[f(k_1 - m_1\Omega^2) - c_1\Omega - k_1\eta_1] + f\varkappa^{-1}\delta\delta_1 = 0$$

which enables us to obtain the admissible values of Ω, independent of ω. The inequality $R > 0$ imposes furthermore

$$\sigma b\delta_1[k - m\Omega^2 + f(c\Omega + k\eta)] + \delta[k_1 - m_1\Omega^2 + f(c_1\Omega + k_1\eta_1)] + \varkappa^{-1}\delta\delta_1 < 0$$

which, in view of (1.72), can be written as:

(1.73) $$\sigma b\delta_1(c\Omega + k\eta) + \delta(c_1\Omega + k_1\eta_1) < 0$$

Let us examine the simplified case where $\eta = \eta_1 = c_1 = 0$; the roots of the equation (1.72) can be interpreted as the abscissae Ω of the points common to

(1.74) $$u = f\frac{(k - m\Omega^2)^2 + c^2\Omega^2}{k_1 - m_1\Omega^2} + \varkappa^{-1}f[(k - m\Omega^2)^2 + c^2\Omega^2] + fb(k - m\Omega^2)$$

(1.75) $$u = \sigma bc\Omega$$

and the condition (1.73) requires that $\Omega < 0$, while (1.64) means that the speed of rotation of the shaft must exceed a certain threshold. Taken together these conditions can be written:

$$\omega > -(r_1 - r)\Omega > 0.$$

Inspection of the graphs of (1.74), (1.75) makes it clear that there always exists at least one solution with negative Ω which shows that self-sustained motions with retrograde precession can occur, a possibility which is confirmed by experiment.

Chapter II. Vibrations in Lattices

We shall examine the nature of the vibrations which can arise in a lattice structure, that is to say a system of point masses which in their state of rest have a periodic spatial distribution, under a variety of different hypotheses. These will include uni-dimensional and three-dimensional lattices, and interactions which may be closely or remotely coupled, and linear or non-linear. In the case of non-linear interactions between the elements of an uni-dimensional lattice, analysis of the process of propagation leads, on the basis of an approximation of long-wave type, to a standard partial differential equation, known as the Korteweg-de Vries equation, which will form the subject of detailed study in Chapter 12.

A Simple Mechanical Model

We consider a system of particles of mass m along a straight line, linked by identical springs of natural length a, and stiffness k. In the rest state the abscissae of the positions of the particles are $x_n = na$, n being an integer of any sign, but if the chain is disturbed, the particle of rank n undergoes a displacement u_n and the equations of motion are:

(2.1) $$m\frac{d^2u_n}{dt^2} = k(u_{n+1} + u_{n-1} - 2u_n).$$

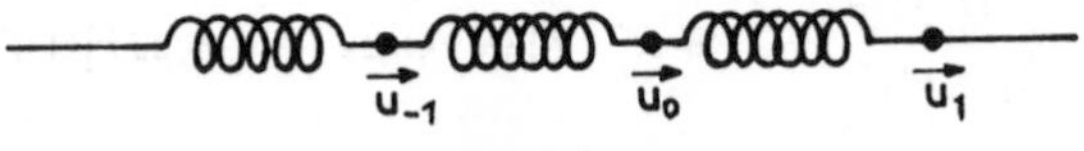

Figure 2.1

We can, supposing the chain to be infinite, introduce a closure condition $u_{n+N} = u_n$, $\forall n$, where N is a given positive integer, which means that we propose to study motions which are periodic with respect to the space co-ordinate, of wavelength Na.

Introducing the new co-ordinates

(2.2) $$v_p = \frac{1}{\sqrt{N}} \sum_{\nu=0}^{N-1} e^{2\pi i p\nu/N} \cdot u_\nu,$$

multiplying (2.1) by $e^{2\pi ipn/N}$ and summing with respect to n from 0 to $N-1$, we arrive at

(2.3)
$$\frac{d^2 v_p}{dt^2} + \omega_p^2 v_p = 0$$

where

(2.4)
$$\omega_p = 2\left(\frac{k}{m}\right)^{1/2} \sin\left(\frac{\pi p}{N}\right).$$

(2.2) can easily be inverted and written in the form:

(2.5)
$$\frac{1}{\sqrt{N}} \sum_{p=0}^{N-1} e^{-2\pi ipn/N} \cdot v_p = \frac{1}{N} \sum_{\nu,p} e^{2\pi ip(\nu-n)/N} \cdot u_\nu = u_n.$$

Hence, writing $v_p = A_p e^{i\omega_p t}$, where A_p is the complex amplitude, for the solution of (2.3), we have

$$u_n = \frac{1}{\sqrt{N}} \sum_{p=0}^{N-1} A_p \exp i(\omega_p t - 2\pi pn/N)$$

or, with $A_p = |A_p| e^{i\alpha_p}$, and $\varkappa_p = \dfrac{2\pi p}{Na}$ the wave-number, we have after taking the real part:

(2.6)
$$u_n = \frac{1}{\sqrt{N}} \sum_{p=0}^{N-1} |A_p| \cos[(\omega_p t - \varkappa_p x_n) + \alpha_p].$$

The dispersion relation, or relation between ω and the wave-number is described by (2.4); we can restrict ourselves to values $\dfrac{p}{N} \in \left[-\dfrac{1}{2}, +\dfrac{1}{2}\right]$ and we note that the highest harmonic of the system has a frequency $\omega_{max} = 2\left(\dfrac{k}{m}\right)^{1/2}$.

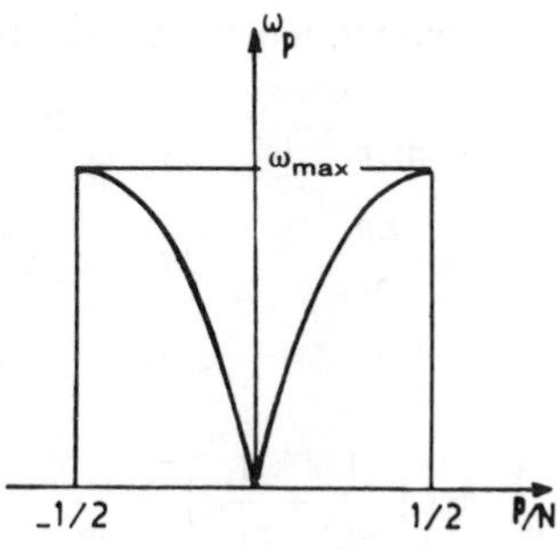

Figure 2.2

The Alternating Lattice Model

A more complex model is obtained by considering an infinite chain of particles of masses m_1 and m_2 arranged alternately and connected to each other by identical springs of stiffness k.

The equations of motion are:

$$(2.7)\qquad \begin{aligned} m_1 \frac{d^2 u_{2n}}{dt^2} &= k(u_{2n+1} + u_{2n-1} - 2u_{2n}) \\ m_2 \frac{d^2 u_{2n+1}}{dt^2} &= k(u_{2n+2} + u_{2n} - 2u_{2n+1}). \end{aligned}$$

Assuming a closure condition $u_{n+2N} = u_n, \forall n$ and introducing the variables

$$(2.8)\qquad \begin{aligned} v_p &= \frac{1}{\sqrt{N}} \sum_{\nu=0}^{N-1} e^{2i\pi p\nu/N} \cdot u_{2\nu} \\ w_p &= \frac{1}{\sqrt{N}} \sum_{\nu=0}^{N-1} e^{i\pi p(2\nu+1)/N} \cdot u_{2\nu+1} \end{aligned}$$

or, by inversion

$$(2.9)\qquad \begin{aligned} u_{2n} &= \frac{1}{\sqrt{N}} \sum_{p=0}^{N-1} v_p e^{-2i\pi np/N} \\ u_{2n+1} &= \frac{1}{\sqrt{N}} \sum_{p=0}^{N-1} w_p e^{-i\pi(2n+1)p/N} \end{aligned}$$

we deduce from (2.7) after multiplying by $e^{2i\pi pn/N}$, $e^{i\pi p(2n+1)/N}$ respectively and adding the corresponding equations for $n = 0$ to $N-1$:

$$(2.10)\qquad \begin{aligned} m_1 \frac{d^2 v_p}{dt^2} &= -2kv_p + 2k\cos\frac{\pi p}{N} \cdot w_p \\ m_2 \frac{d^2 w_p}{dt^2} &= 2k\cos\frac{\pi p}{N} \cdot v_p - 2kw_p \end{aligned}$$

an uncoupled system, whose eigenfrequencies are the roots of the equation

$$\det \begin{vmatrix} 2k - m_1\omega^2 & -2k\cos\frac{\pi p}{N} \\ -2k\cos\frac{\pi p}{N} & 2k - m_2\omega^2 \end{vmatrix} = 0$$

that is to say:

$$(2.11)\qquad \omega^2 = k\left[\frac{1}{m_1} + \frac{1}{m_2} \pm \left(\left(\frac{1}{m_1} + \frac{1}{m_2}\right)^2 - \frac{4}{m_1 m_2}\sin^2\frac{\pi p}{N}\right)^{1/2}\right]$$

the amplitudes A_p, B_p of v_p, w_p respectively, being connected by the equation

(2.12) $$(2k - m_1\omega^2)A_p = 2k\cos\frac{\pi p}{N}\cdot B_p.$$

We shall confine ourselves to the examination of two particular cases.

1. v_p, w_p have the value zero except when $p=0$. By (2.11) we have $\omega = 0$, $\omega^2 = 2k\left(\frac{1}{m_1} + \frac{1}{m_2}\right) = \omega^2_{\max}$

$$\begin{matrix} v_0 = A_0 e^{i\omega t} \\ w_0 = B_0 e^{i\omega t} \end{matrix} \quad \frac{A_0}{B_0} = \frac{2k}{2k - m_1\omega^2} = \frac{2k - m_2\omega^2}{2k}.$$

The solution $\omega = 0$ corresponds to a translational movement of the chain; for $\omega = \omega_{\max}$, we have $\frac{A_0}{B_0} = -\frac{m_2}{m_1}$ and

$$u_{2n} = A_0 e^{i\omega_{\max}t}, \quad u_{2n+1} = B_0 e^{i\omega_{\max}t}$$

to within a factor of $N^{-1/2}$, so that the masses m_1, m_2 vibrate in unison.

2. v_p, w_p have the value zero except when $p = \frac{N}{2}$, N even. We have, by (2.11):

$$\omega^2 = \frac{2k}{m_2} = \omega_L^2, \quad \text{or} \quad \omega^2 = \frac{2k}{m_1} = \omega_H^2.$$

With $\omega = \omega_L$, we have $A_0 = 0$, B_0 has any non-zero value, the masses m_1 are fixed, while the masses m_2 have a vibratory motion described by

$$u_{2n+1} = B_0 e^{i\omega_L t}.$$

The situation is reversed if $\omega = \omega_H$.

The dispersion relation (2.11) between the frequency ω_p and the wave number $\varkappa_p = \frac{p}{Na}$ may be represented in the $\omega, \varkappa$ plane, taking care to distinguish the two branches

$$\omega^2 = k\left(\frac{1}{m_1} + \frac{1}{m_2}\right)\left(1 \pm \left(1 - \frac{4m_1m_2}{(m_1+m_2)^2}\sin^2 p\frac{\pi}{N}\right)^{1/2}\right).$$

With the + sign, supposing $m_1 > m_2$, it will be seen that

$$\omega_L \leqslant \omega \leqslant \omega_{\max}$$

where

$$\omega_{\max} = \sqrt{2k\left(\frac{1}{m_1} + \frac{1}{m_2}\right)}, \quad \omega_L = \sqrt{\frac{2k}{m_2}}$$

and taking the $-$ sign, $0 \leqslant \omega \leqslant \omega_H$, $\omega_H = \sqrt{\frac{2k}{m_1}} < \omega_L$.

The upper branch corresponds to the optical frequency band, the lower branch to the acoustic band, and it will be noticed that there is a forbidden band of

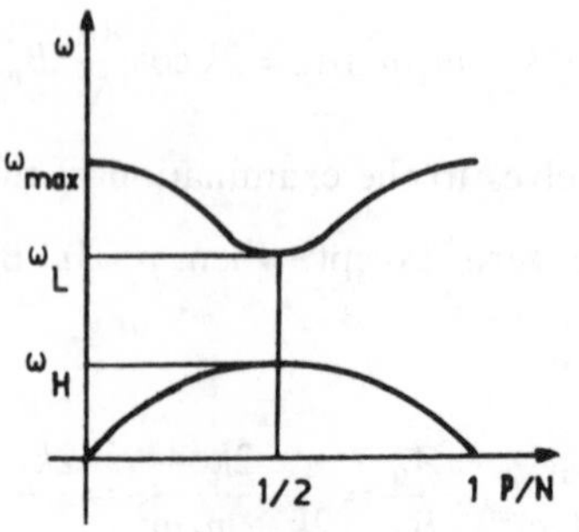

Figure 2.3

frequencies $[\omega_H, \omega_L]$. This analysis could be generalised to a model comprising s different kinds of particles with a periodic spatial distribution, i.e. such that the spacing between two consecutive particles of the same kind should be equal to $s \cdot a$; in that case one would obtain for the dispersion curve, $s-1$ optical branches, one acoustic branch corresponding to the lowest frequencies, and $s-1$ forbidden bands. Numerous fruitful developments on this subject will be found in reference [7].

Vibrations in a One-Dimensional Lattice with Interactive Forces Derived from a Potential

1. Let us again consider the system of particles of mass m, situated at $x_n = na$ in their equilibrium positions; we set $\xi_n = na + u_n$ where u_n is the displacement and we shall suppose the forces of interaction to be derived from a potential $V(\ldots, \xi_{-1}, \xi_0, \xi_1, \ldots)$ so that the force exerted on the nth particle is $F_n = -\dfrac{\partial V}{\partial \xi_n}$, on the understanding that at equilibrium $\xi = x$ we have $\left.\dfrac{\partial V}{\partial \xi_n}\right|_x = 0$.

Now using the Taylor expansion up to the terms of second order, which suffices for a linearised theory, we can write this as:

$$V = \frac{1}{2} \sum_{l,l'} \left.\frac{\partial^2 V}{\partial \xi_l \partial \xi_{l'}}\right|_x \cdot u_l u_{l'}.$$

Let us assume that the interactions between two particles of the system depend only upon their positions; this leads to postulating a V of the form

$$V = \frac{1}{2} \sum_{l,l',l \neq l'} V_{ll'}(\xi_l, \xi_{l'})$$

where $V_{ll'}$ depends only on the variables $\xi_l, \xi_{l'}$, and satisfies the symmetry condition

$$V_{ll'}(\xi_l, \xi_{l'}) = V_{l'l}(\xi_{l'}, \xi_l).$$

Under these conditions

$$l \neq l': \frac{\partial^2 V}{\partial \xi_l \partial \xi_{l'}}\bigg|_x = \frac{\partial^2 V_{ll'}}{\partial \xi_l \partial \xi_{l'}}\bigg|_x = W_{ll'} \quad \text{and} \quad W_{ll'} = W_{l'l}$$

$$\frac{\partial^2 V}{\partial \xi_l^2} = \sum_{l', l' \neq l} \frac{\partial^2 V_{ll'}}{\partial \xi_l^2} = v_l.$$

In addition we shall assume the property of invariance $V_{l+q,l'+q}((l+q)a+u, (l'+q)a+u') = V_{ll'}(la+u, l'a+u')$, $\forall$ integer q whence $W_{l+q,l'+q} = W_{ll'}$, $W_{ll'} = W_{l-l',0} = W_{0,l'-l} = W_{l'-l,0}$ which suggests introducing the notation $W_{q,0} = W_{0,q} = W_q$; finally it seems natural to suppose that the forces internal to the system should not be altered by a rigid-body displacement; starting from the representation

(2.13) $$V = \frac{1}{2}\sum_l v_l u_l^2 + \frac{1}{2}\sum_{l \neq l'} W_{ll'} u_l u_{l'}$$

we are led to write that $V(u_l + \delta)$ is independent of real δ, which finally can be expressed by

$$v_l + \sum_{q, q \neq 0} W_q = 0.$$

Thus $v_l = v$ does not depend on the suffix l and

(2.14) $$v + \sum_{q, q \neq 0} W_q = 0 \quad \text{or} \quad v + 2\sum_{q > 0} W_q = 0.$$

Taking account of (2.13), the equation of motion can be written as

$$m\frac{\mathrm{d}^2 u_n}{\mathrm{d}t^2} = -v u_n - \sum_{n', n' \neq n} W_{n'-n} u_{n'}$$

or, by (2.14)

(2.15) $$m\frac{\mathrm{d}^2 u_n}{\mathrm{d}t^2} = \sum_{q, q > 0} k_q (u_{n+q} + u_{n-q} - 2u_n), \quad k_q = -W_q.$$

This can be interpreted by a simple mechanical model, by imagining identical springs of stiffness k_q to be stretched between the particles of rank n and $n+q$, if one makes the assumption, which has not so far been implicit, that $k_q > 0$.

2. It is easy to make the transition from the equations (2.15) for a discrete system to those relating to a continuous medium; for the displacement $u_n(t)$ is substituted a function $u(x,t)$ of the variables x, t satisfying $u(na, t) = u_n(t)$. Expanding up to terms of the second order in a scheme where ultimately a will tend to zero, and q takes only a finite number of integer values

$$u(na \pm qa, t) = u \pm qa \cdot u_x + \tfrac{1}{2}(qa)^2 u_{xx}$$

with $na = x$, whence by (2.15)

(2.16) $$m u_{tt} = \left(\sum_{q > 0} q^2 k_q\right) a^2 u_{xx}$$

which can be written with $\rho = \frac{m}{a^3}$, the mass per unit volume, and $E = a^{-1}\sum_{q>0} q^2 k_q$:

(2.17) $$pu_{tt} = Eu_{xx}$$

which is the equation for the longitudinal vibrations of a homogeneous bar of volume density ρ, of material with modulus of elasticity E. One obtains in the obvious way, a representation of travelling waves of (2.17),

(2.18) $$u = f(\omega t - \varkappa x) \quad \text{with} \quad \omega = \sqrt{\frac{E}{\rho}} \cdot \varkappa$$

where $\varkappa$ is the wave-number, and $c = \frac{\omega}{\varkappa} = \sqrt{\frac{E}{\rho}}$ the wave velocity.

But one can also seek directly a solution of the same type for the differential equation (2.15) of the discretised model under the form $u_n = ue^{i(\varkappa x_n - \omega t)}$

whence $$u_{n \pm q} = u_n e^{\pm i\varkappa q a}$$

and the dispersion relation is:

(2.19) $$m\omega^2 - 4 \sum_{q>0} k_q \sin^2\left(\frac{\varkappa q a}{2}\right) = 0.$$

If the wave-number is not too large, $\varkappa q a \ll 1$, we can replace (2.19) by the tangent approximation

(2.20) $$\omega = \sqrt{\frac{\sum_{q>0} k_q q^2}{m}} \cdot a\varkappa = \sqrt{\frac{E}{\rho}} \cdot \varkappa$$

identical to (2.18).

It is interesting to discuss the dispersion curve (2.19) associated with (2.15), in the case $q \leqslant 2$:

$$\omega = 2\left(\frac{k_1}{m} \sin^2 \frac{\varkappa a}{2} + \frac{k_2}{m} \sin^2 \varkappa a\right)^{1/2}, \quad \varkappa \in \left(0, \frac{\pi}{a}\right).$$

If $\frac{k_1}{4k_2} > 1$, ω is an increasing function of $\varkappa$ and attains its maximum $2\sqrt{\frac{k_1}{m}}$, at $\varkappa = \frac{\pi}{a}$.

If $\frac{k_1}{4k_2} < 1$, ω is increasing in the interval $(0, \varkappa^*)$ and decreasing in $\left(\varkappa^*, \frac{\pi}{a}\right)$. The maximum is attained at $\varkappa = \varkappa^*$, where $\varkappa^*$ satisfies $\cos \varkappa^* a = -\frac{k_1}{4k_2}$ and has the value

$$\omega(\varkappa^*) = 2\left(1 + \frac{k_1}{4k_2}\right)\sqrt{\frac{k_2}{m}}.$$

It is apparent that vibrations of the same frequency but with different wave number can be produced in this case.

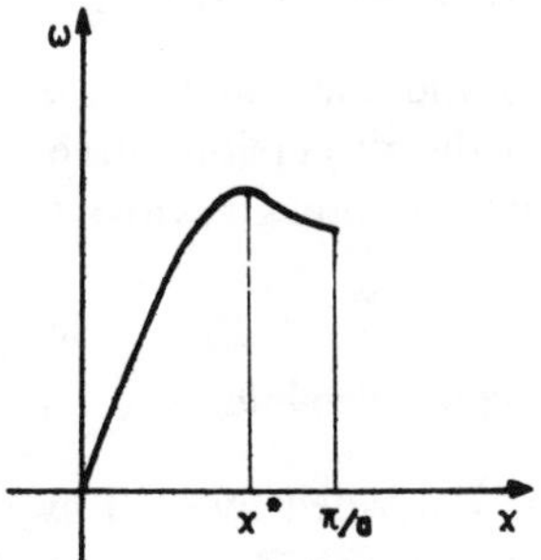

Figure 2.4

3. There are several representations of interatomic potential which depend only on the distance r separating two particles; amongst these may be mentioned

$$V = -\frac{M\rho}{r} + Be^{-r/\rho} \qquad \text{(Born-Mayer)}$$

$$V = D(e^{-2\alpha(r-a)} - 2e^{-\alpha(r-a)}) \qquad \text{(Morse),}$$

$$V = D\left(\left(\frac{a}{r}\right)^{12} - 2\left(\frac{a}{r}\right)^{6}\right) \qquad \text{(Lennard Jones).}$$

For the two latter the distance $r = a$ corresponds to $\left.\frac{dV}{dr}\right|_a = 0$.

For the iron atom, the Morse potential is acceptable with $D = 0{\cdot}4174\,\text{eV}$, $a = 2.845\,\text{Å}$, $\alpha = 1.3888\,\text{Å}^{-1}$. It is easily calculated that $k_1 = \frac{\partial^2 V}{\partial r^2} = 2D\alpha^2$, and remembering that $1\,\text{eV} = 1.6\cdot 10^{-12}\,\text{erg}$, $1\,\text{Å} = 10^{-8}\,\text{cm}$, we obtain

$$k_1 \sim 2.5\cdot 10^4\,\text{dyne/cm},\ E = \frac{k_1}{a} \sim 0.9\cdot 10^{12}\,\text{dyne/cm}^2 \approx E_{\text{exp}},$$

$$\omega_{\max} = 2\left(\frac{k_1}{m}\right)^{1/2} = \frac{2c}{a}$$

with $c = \left(\frac{k_1}{a}\right)^{1/2}\left(\frac{m}{a^3}\right)^{-1/2}$ as wave velocity; thus taking $c \sim 1000\,\text{m/s}$ as the velocity of sound propagation in the solid we have $\omega_{\max} = O(10^{13}\,\text{s}^{-1})$, and for the corresponding wave-number

$$\varkappa_{\max} = \frac{\pi}{a} = 10^8\,\text{cm}^{-1}.$$

In elastodynamics it is difficult to conceive of vibrating systems of frequency greater than $10^6\,\mathrm{s}^{-1}$. This limit, in the case considered with a velocity $c = 1000$ m/s, corresponds to a wave number $\varkappa = \frac{\omega}{c} = 10\,\mathrm{cm}^{-1}$.

These data make it understandable that, as far as macroscopic phenomena are concerned, the useful part of the dispersion curve lies near the origin, and this explains why the tangent approximation gives an adequate account of the observed effects.

Vibrations in a System of Coupled Pendulums

We can generalise the model described by (2.7) by replacing the points of mass m_1 and m_2 by pendulums, which we shall suppose to be coupled to each other by elastic links of the same stiffness k. The equations which describe the angular deviations which we denote by u_{2n}, u_{2n+1} are then:

$$\frac{\mathrm{d}^2 u_{2n}}{\mathrm{d}t^2} + \Omega_1^2 u_{2n} = \omega_1^2 (u_{2n+1} + u_{2n-1} - 2u_{2n})$$

(2.21)

$$\frac{\mathrm{d}^2 u_{2n+1}}{\mathrm{d}t^2} + \Omega_2^2 u_{2n+1} = \omega_2^2 (u_{2n+2} + u_{2n} - 2u_{2n+1})$$

with $\omega_1 = \sqrt{\frac{k}{m_1}}$, $\omega_2 = \sqrt{\frac{k}{m_2}}$, and Ω_1, Ω_2 are the frequencies of the free oscillations of each uncoupled pendulum.

The change of variables (2.8), (2.9) under the closure condition $u_{n+2N} = u_n, \forall n$, allows us to transform (2.21) into:

$$\frac{\mathrm{d}^2 v_p}{\mathrm{d}t^2} + (\Omega_1^2 + 2\omega_1^2) v_p - 2\omega_1^2 \cos p\frac{\pi}{N}\cdot w_p = 0$$

$$\frac{\mathrm{d}^2 w_p}{\mathrm{d}t^2} + (\Omega_2^2 + 2\omega_2^2) w_p - 2\omega_2^2 \cos p\frac{\pi}{N}\cdot w_p = 0$$

so that the eigenfrequencies ω of the system are the roots of:

$$\det \begin{vmatrix} \Omega_1^2 + 2\omega_1^2 - \omega^2 & -2\omega_1^2 \cos p\frac{\pi}{N} \\ -2\omega_2^2 \cos p\frac{\pi}{N} & \Omega_2^2 + 2\omega_2^2 - \omega^2 \end{vmatrix} = 0$$

i.e.

$$2\omega^2 = \Omega_1^2 + \Omega_2^2 + 2\omega_1^2 + 2\omega_2^2$$
$$\pm \left((\Omega_1^2 - \Omega_2^2 + 2\omega_1^2 - 2\omega_2^2)^2 + 16\omega_1^2\omega_2^2 \cos^2 p\frac{\pi}{N} \right)^{1/2}.$$

We obtain for the representation of ω as a function of p/N, or of the wave-number, an upper branch with ω in the band $\omega_L \leqslant \omega \leqslant \omega_{\max}$ where

$$\omega_{\max}^2 = \tfrac{1}{2}[\Omega_1^2 + \Omega_2^2 + 2\omega_1^2 + 2\omega_2^2 + ((\Omega_1^2 - \Omega_2^2 + 2\omega_1^2 - 2\omega_2^2)^2 + 16\omega_1^2\omega_2^2)^{1/2}]$$
$$\omega_L^2 = \mathrm{Sup}(\Omega_1^2 + 2\omega_1^2, \Omega_2^2 + 2\omega_2^2)$$

and a lower branch with ω belonging to $(\omega_{\min}, \omega_H)$, where

$$\omega_H^2 = \inf(\Omega_1^2 + 2\omega_1^2, \Omega_2^2 + 2\omega_2^2)$$

$$\omega_{\min}^2 = \tfrac{1}{2}[\Omega_1^2 + \Omega_2^2 + 2\omega_1^2 + 2\omega_2^2 - ((\Omega_1^2 - \Omega_2^2 + 2\omega_1^2 - 2\omega_2^2)^2 + 16\omega_1^2\omega_2^2)^{1/2}].$$

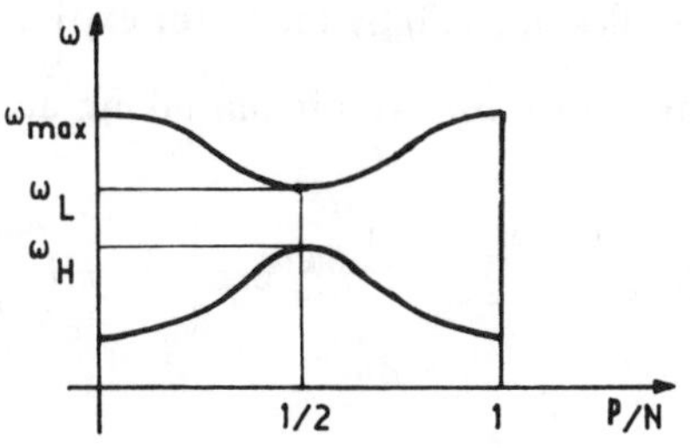

Figure 2.5

Vibrations in Three-Dimensional Lattices

We consider a system of particles of the same mass located at the nodal points of a lattice in a crystalline medium defined by the base vectors $\vec{a}, \vec{b}, \vec{c}$. Let $l\vec{a} + m\vec{b} + n\vec{c}$, where l, m, n are integers of arbitrary sign, denote the vectorial representation of these nodes. We write u_{lmn}^j for the jth component of the displacement of the point which, in the state of rest, occupies the node defined by the triplet (l, m, n) and it is assumed that the interaction force exerted on the point (l, m, n) by the point $(l + \lambda, m + \mu, n + \nu)$ is an affine function of the relative displacements, with stiffness coefficients which depend only on λ, μ, ν. Assuming the closure conditions:

$$(2.22) \qquad u_{lmn} = u_{l+L,m,n} = u_{l+L,m+M,n} = u_{l+L,m+M,n+N},$$

$\forall$ integers l, m, n, with L, M, N given positive integers, we can write the equations of motion in the form:

$$(2.23) \qquad \frac{d^2 u_{lmn}^j}{dt^2} = \sum_{\lambda=0}^{L-1} \sum_{\mu=0}^{M-1} \sum_{\nu=0}^{N-1} \sum_{k=1}^{3} \varkappa_{\lambda\mu\nu}^{jk} \cdot u_{l+\lambda,m+\mu,n+\nu}^k$$

with
$$\sum_{\lambda=0}^{L-1} \sum_{\mu=0}^{M-1} \sum_{\nu=0}^{N-1} \varkappa_{\lambda\mu\nu}^{jk} = 0.$$

The omission of this last condition would simply mean having to take account, in addition to the interactions between the particles, of the actions of external forces on each of them which were affine with respect to displacement.

To study the system (2.23) which generalises, for space of three dimensions, the models described by (2.1) and (2.15) we make the change of variable

$$(2.24) \qquad v_{l^*m^*n^*}^j = \frac{1}{\sqrt{LMN}} \sum_{l,m,n=0}^{L-1,M-1,N-1} \exp\left[2\pi i\left(\frac{ll^*}{L} + \frac{mm^*}{M} + \frac{nn^*}{N}\right)\right] \cdot u_{lmn}^j$$

which can readily be inverted

$$(2.25)\quad u^j_{lmn} = \frac{1}{\sqrt{LMN}} \sum_{l^*, m^*, n^* = 0}^{L-1, M-1, N-1} \exp\left[-2\pi i\left(\frac{ll^*}{L} + \frac{mm^*}{M} + \frac{nn^*}{N}\right)\right] \cdot v^j_{l^* m^* n^*}.$$

Multiplying the two sides of (2.23) by the factor $\exp\left[2\pi i\left(\frac{ll^*}{L} + \frac{mm^*}{M} + \frac{nn^*}{N}\right)\right]$ and summing with respect to l, m, n we obtain, taking account of (2.24) and the closure conditions (2.22):

$$(2.26)\quad \frac{d^2 v^j_{l^* m^* n^*}}{dt^2} = \sum_{k=1}^{3} v^k_{l^* m^* n^*} \sum_{\lambda, \mu, \nu = 0}^{L-1, M-1, N-1} \varkappa^{jk}_{\lambda\mu\nu} \exp\left[-2\pi i\left(\frac{l^*\lambda}{L} + \frac{m^*\mu}{M} + \frac{n^*\nu}{N}\right)\right]$$

or in matrix form:

$$\frac{d^2 v}{dt^2} = \sigma v$$

with

$$v = (v^1, v^2, v^3)^T, \quad \sigma = \begin{pmatrix} \sigma_{11} & \sigma_{12} & \sigma_{13} \\ \sigma_{21} & \sigma_{22} & \sigma_{23} \\ \sigma_{31} & \sigma_{32} & \sigma_{33} \end{pmatrix}$$

and omitting the writing of (l^*, m^*, n^*) because the variables are decoupled, but bearing in mind however that the matrix σ depends on the triplet (l^*, m^*, n^*).

The equation (2.27) may have periodic solutions of frequency ω, where ω is the real root of

$$(2.27)\quad \det \begin{vmatrix} \sigma_{11} + \omega^2 & \sigma_{12} & \sigma_{13} \\ \sigma_{21} & \sigma_{22} + \omega^2 & \sigma_{23} \\ \sigma_{31} & \sigma_{32} & \sigma_{33} + \omega^2 \end{vmatrix} = 0.$$

The matrix σ is symmetric on the assumption that $\varkappa^{jk} = \varkappa^{kj}$; in this case (2.27) considered as an equation in ω^2, has three real roots, of which only those that are positive correspond to possible modes of vibration. These roots have to be studied as functions of the triplet (l^*, m^*, n^*) on which they depend: the existence of forbidden bands of frequencies into which the lattice vibrations cannot enter follows from more detailed analysis, in [58].

Non-Linear Problems

We again consider the one-dimentional lattice of particles, equidistant at equilibrium, under the hypothesis that the forces of interaction are exerted only between two consecutive particles and are derived from a potential $\phi(r)$ which is a function of r, their distance apart. Thus the force exerted on the nth particle by the $(n+1)$th particle is $\frac{\partial \phi}{\partial r}(a + u_{n+1} - u_n)$, while that exerted on the nth particle by the $(n-1)$th particle is $-\frac{\partial \phi}{\partial r}(a + u_n - u_{n-1})$, so that the equation of motion is

(2.28) $$m\frac{d^2u_n}{dt^2}=\frac{\partial\phi}{\partial r}(a+u_{n+1}-u_n)-\frac{\partial\phi}{\partial r}(a+u_n-u_{n-1}).$$

To take account of the non-linear effects in the simplest way, we shall adopt the representation

(2.29) $$\frac{\partial\phi}{\partial r}=k(r-a)+\frac{\sigma}{2}(r-a)^2,\quad k>0$$

for the potential, so that we can now write (2.28) as:

(2.30) $$m\frac{d^2u_n}{dt^2}=\left(k+\frac{\sigma}{2}(u_{n+1}-u_{n-1})\right)(u_{n+1}+u_{n-1}-2u_n).$$

Passing from the discrete to the continuous model $u_n(t)=u(na,t)$, we represent u_{n+1} by its Taylor expansion up to terms of the fourth order:

$$u_{n+1}=u((n+1)a,t)=u(na,t)+u_x(na,t)\cdot a+\tfrac{1}{2}u_{xx}(na,t)\cdot a^2+\cdots+\tfrac{1}{24}u_{xxxx}(na,t)a^4$$

The expansion of u_{n-1} is obtained by changing a into $-a$, and (2.30) can be written in the form

(2.31) $$mu_{tt}=(k+\sigma au_x)(u_{xx}a^2+\tfrac{1}{12}u_{xxxx}a^4)$$

where in expressing $u_{n+1}-u_{n-1}$ the terms of order 3 with respect to a are omitted since after multiplication they would lead to terms of order 5 at least, in the product on the right-hand side of (2.31).

Furthermore we restrict ourselves to terms of order 4, i.e. we can write (2.31) in the form, cf. [55]:

(2.32) $$u_{tt}-c_0^2(1-\beta au_x)u_{xx}=\frac{c_0a^2}{12}u_{xxxx}$$

with $$\frac{ka^2}{m}=c_0^2,\quad \beta=-\frac{\sigma m}{k}.$$

We can seek solutions representing a progressive wave $u=v(\xi)$, $\xi=x-ct$, which implies that $v(\xi)$ has to satisfy an ordinary differential equation

(2.33) $$c^2v''-c_0^2(1-\beta av')v''=\frac{c_0^2a^2}{12}v^{IV}$$

for which one can easily construct the solution

$$v=-\frac{\alpha}{\beta}\,\mathrm{th}\left(\frac{\alpha}{a}\xi\right),\quad \frac{\alpha^2}{3}=\left(\frac{c}{c_0}\right)^2-1$$

for $c>c_0$, which has a velocity profile

$$\frac{\partial u}{\partial t}=-cv'=\frac{\alpha^2}{\beta a}\cdot\frac{c}{\mathrm{ch}^2\left[\frac{\alpha}{a}(x-ct)\right]}$$

of the solitary wave type, with an amplitude $A = \frac{3c}{\beta a}\left(\left(\frac{c}{c_0}\right)^2 - 1\right)$ which depends on the speed of the wave.

But one can also look for solutions of (2.33) that are periodic with respect to ξ. To this end we integrate (2.33),

$$\text{const} + (c^2 - c_0^2)v' + \beta\frac{ac_0^2}{2}v'^2 = \frac{c_0^2 a^2}{12}v'''$$

then after multiplying by v'' we integrate a second time, obtaining, with $v' = w$:

$$w'^2 = \frac{4\beta}{a}w^3 + \frac{12}{a^2}\left(\left(\frac{c}{c_0}\right)^2 - 1\right)w^2 + Aw + B. \tag{2.34}$$

A, B are arbitrary constants. However for v to be periodic with respect to ξ, we have to try and construct a solution ω of the equation (2.34) which will be periodic and of mean value zero.

Since the speed c is also an unknown we can, in view of the polynomial form of the right-hand side of (2.34), introduce the zeros of this polynomial, which we shall suppose real and in the order $p<q<r$, instead of the as yet unspecified quantities A, B, c.

In particular it will be convenient to normalise so that

$$p + q + r = -\frac{3}{\beta a}\left(\left(\frac{c}{c_0}\right)^2 - 1\right). \tag{2.35}$$

We shall carry out the calculations on the hypothesis that $\beta > 0$ and it will then follow from (2.34) that w will stay in the interval (p, q), which suggest the change of variables $w \to z$:

$$w = q + (p - q)z^2. \tag{2.36}$$

The equation (2.34) after being transformed by (2.36) becomes:

$$z'^2 = \frac{\beta(r-p)}{a}(1 - z^2)\left(\frac{q-p}{r-p}z^2 + \frac{r-q}{r-p}\right) \tag{2.37}$$

or, on integration:

$$\int_z^1 \frac{ds}{\sqrt{(1-s^2)(k^2s^2 + k'^2)}} = \sqrt{\frac{(r-p)\beta}{a}}\cdot\xi, \quad k^2 = \frac{q-p}{r-p}, \quad k'^2 = \frac{r-q}{r-p}.$$

The elliptic function $z = \mathrm{cn}(u, k)$ obtained by inversion of

$$u = \int_z^1 \frac{ds}{\sqrt{(1-s^2)(k^2s^2 + k'^2)}} \tag{2.38}$$

is periodic with period $4K$:

$$K = \int_0^1 \frac{ds}{\sqrt{(1-s^2)(k^2s^2 + k'^2)}} = \int_0^{\pi/2} (1 - k^2\sin^2\varphi)^{-1/2}\,d\varphi \tag{2.39}$$

and satisfies the relation:

(2.40) $$\operatorname{cn}(u,k)=\operatorname{cn}(-u,k),\quad |\operatorname{cn}(u,k)|\leqslant 1.$$

Finally one can write the solution:

(2.41) $$w=q+(p-q)\operatorname{cn}^2\left(\sqrt{\frac{(r-p)\beta}{a}}\cdot\xi\right)$$

and it now remains to be shown that the available parameters can be chosen so that w has a mean value of zero.

Let $g(k^2)$ be the average value of the periodic function $\operatorname{cn}^2(\tau,k)$, which by (2.40), satisfies $0<g<1$ (in particular $\operatorname{cn}(\tau,0)=\cos\tau$ and $g(0)=1/2$).

The condition for a zero mean is:

(2.42) $$q+(p-q)g(k^2)=0$$

which can be compared with the definition of k^2:

(2.43) $$q-p=k^2(r-p).$$

If k^2 is given, $0<k^2<1$, we can find p,q in terms of r and k^2 from (2.42), (2.43), namely:

(2.44) $$p=-\frac{(1-g)k^2}{g+(1-g)(1-k^2)}r$$
$$q=\frac{gk^2}{g+(1-g)(1-k^2)}r.$$

It remains only to express the inequality condition $p<q<r$ which is easily seen to be equivalent to $r>0$, and to recall (2.35), or:

(2.45) $$\left(\frac{c}{c_0}\right)^2=1-\frac{a\beta r}{3}\,\frac{1-(2-3g)k^2}{g+(1-g)(1-k^2)}$$

which determines c, for any given $r>0$; or at least for small enough r, this restriction depending on the sign of $1-(2-3g)k^2$.

Thus it will be seen that by integration of (2.41) a two-parameter family of solutions of (2.33) is obtained, which are periodic progressive waves $u=v(x-ct)$, of wave-length $\lambda=4K\sqrt{\dfrac{a}{(r-p)\beta}}$, or by (2.44), $\lambda=v(k^2)\sqrt{\dfrac{a}{r\beta}}$, where $v(k^2)$ is a function of k^2, the two parameters being λ, k^2.

For large wave-length motions, $r\sim 0$ and c is near c_0.

One can seek an approximate representation of this family of solutions in the form:

(2.46) $$u(x,t)=\gamma\left(\varepsilon\sin\frac{2\pi}{\lambda}\xi+\varepsilon^2\sin\frac{4\pi}{\lambda}\xi\right)$$

with $\xi = x - ct$, remembering that these solutions are odd functions of ξ, and limiting the Fourier expansion to the first two harmonics.

Carrying out the computations to the second order terms in ε, we obtain after substitution of (2.46) in (2.32):

$$\gamma = -\frac{4\pi a}{\beta\lambda}, \quad c^2 = c_0^2\left(1 - \frac{\pi^2 a^2}{3\lambda^2}\right)$$

and the representation:

$$u(x, t) = -\frac{4\pi a}{\beta\lambda}\left(\varepsilon \sin\frac{2\pi}{\lambda}\xi + \varepsilon^2 \sin\frac{4\pi}{\lambda}\xi\right),$$

$\xi = x - ct$, which depends on the two parameters λ, ε.

Equation (2.32) is a particular case of

$$u_{tt} - c_0^2(1 + \varepsilon u_x)u_{xx} = \varkappa^2 u_{xxxx}, \quad |\varepsilon| \ll 1 \tag{2.47}$$

which appears in numerous problems in hydrodynamics, in shallow water waves, and in waves in plasmas. To investigate solutions of the long wave type we shall, following [54], introduce the stretched variables:

$$\xi = \frac{\varepsilon}{c_0}(x - c_0 t), \quad \tau = \varepsilon^3 t \tag{2.48}$$

and seek solutions of (2.47) of the form $u = 2c_0 z(\xi, \tau, \varepsilon)$; on transforming (2.47) by (2.48), we obtain the equation satisfied by z:

$$z_{\xi\tau} + z_\xi \cdot z_{\xi\xi} + \frac{\varkappa^2}{2c_0^4} z_{\xi\xi\xi\xi} - \frac{\varepsilon^3}{2} z_{\tau\tau} = 0.$$

Since ε occurs only through the ε^2 term in this equation, we can look for a representation of z in the form of a power series in ε^2, or in other words, to the first order:

$$z(\xi, \tau, \varepsilon) = z(\xi, \tau) + O(\varepsilon^2).$$

To order 0, the function $z(\xi, \tau)$ must be a solution of:

$$z_{\xi\tau} + z_\xi z_{\xi\xi} + \frac{\varkappa^2}{2c_0^4} z_{\xi\xi\xi\xi} = 0$$

which, with

$$v = z_\xi, \quad \delta^2 = \frac{\varkappa^2}{2c_0^4},$$

can be written

$$v_\tau + v v_\xi + \delta^2 v_{\xi\xi\xi} = 0.$$

We thus obtain the Korteweg-de Vries equation, which can in this way easily be shown to have solutions of the solitary wave type:

$$v = v_\infty + \frac{v_0 - v_\infty}{\mathrm{ch}^2[(\xi - c\tau)/\Gamma]}$$

with
$$c = v_\infty + (v_0 - v_\infty)/3, \quad \Gamma = \delta\sqrt{\frac{12}{v_0 - v_\infty}}$$

where v_0, v_∞ are arbitrary constants.

Recent researches have enabled many of its solutions to be constructed and in particular the mechanisms of interaction between several solitary waves, or solitons to be clarified.

Chapter III. Gyroscopic Coupling and Its Applications

We shall begin our study of gyroscopic phenomena—i.e. the effects on the behaviour of a mechanical system caused by the rapid rotation of a solid of revolution forming an integral part of the system—by considering a simple example, the gyroscopic pendulum.

This model will enable us to explain the nature of the gyroscopic coupling and to show how it can be used with advantage to stabilise certain mechanical systems with perfect constraints (frictionless connections). However to get round the difficulties imposed by the existence of frictional forces, however slight, it is necessary to bring into play additional degrees of freedom, and constraints whose function is to annul the effect of the dissipative forces, or to introduce non-conservative forces designed to counteract the destabilising effect of the friction.

In more complex cases, for example the inertial platform, where the system may include several gyroscopes linked in various ways, the equations of motion may be found by Lagrange's method, and it may thus be shown that the inertial effects due to the proper rotations of the individual gyroscopes are equivalent to those which would be produced by certain torques or gyroscopic torque. The study of the gyroscopic compass, completed by an analysis of how its behaviour is affected by relative motion, and the study of the gyroscopic stabilisation of the monorail car are interesting applications among the many possible.

However to improve the performance of on-board guidance instruments on aircraft, missiles and the like, a new type of gyroscopic system has been developed in which the motor which keeps the gyroscopic element in rotation is fixed to the platform. The rotating element is held in a mounting with two degrees of freedom which allows a change of attitude with respect to a frame of reference which is fixed in relation to the motor shaft. One or more intermediate gimbals are introduced to provide the necessary connections, but in contrast with the traditional arrangement, these rotate at a high speed, and consequently play a significant role due to their inertia. The theory shows that the system functions optimally at a certain speed of rotation, which justifies the name of tuned gyroscope given to the instrument, which is designed essentially for the measurement of the instantaneous angular velocity of the platform on which it is mounted.

1. The Gyroscopic Pendulum

In addition to the gyroscope proper S, the device comprises two frameworks (gimbals) S_1 and S_2; the gimbal S_2, assumed to be mounted on ideal frictionless bearings, can turn freely about the axis $z_1'Oz_1$ of the orthonormal frame of reference $Ox_1y_1z_1$. The gimbal S_1 is mounted on the gimbal S_2 in such a way that it rotates about an axis $y'y$ fixed in relation to S_2, orthogonal to $z_1'Oz_1$ and passing through the origin O. Lastly the gyroscope S is a solid of revolution which can rotate about its axis $x'x$, fixed in S_1 orthogonal to $y'y$ and passing through O. By the aid of some suitable mechanism it is given a constant rotational velocity $\Omega\vec{x}$ with respect to S_1. (The only forces brought into play by the mechanism are forces internal to the system $S+S_1$, which consequently form a null-torsor or system of vectors equivalent to the null vector.)

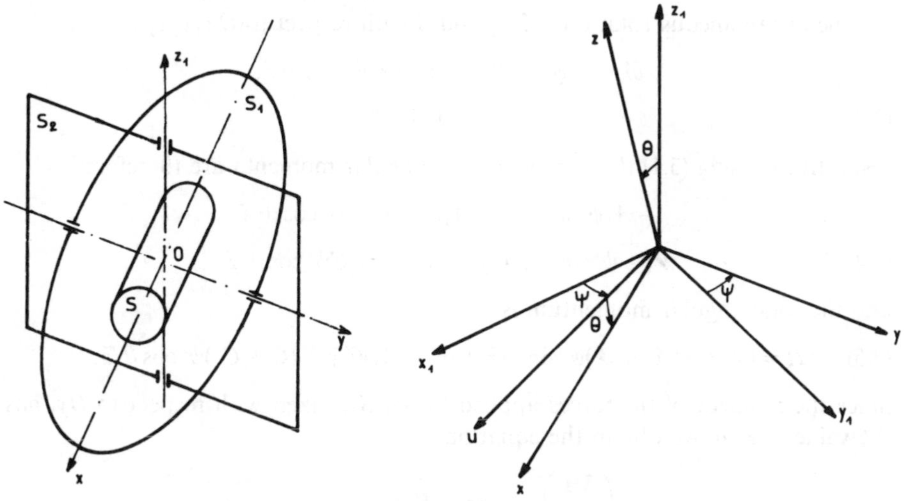

Figure 3.1

The position of the system depends on two parameters ψ, θ the first of which may be defined as the angle of the rotation which brings the orthogonal system $Ox_1y_1z_1$ into coincidence with $Ouyz_1$ (a rotation ψ about $\vec{z}_1$), and the second as the angle of the rotation which then brings $Ouyz_1$ into coincidence with $Oxyz$ (a rotation θ about $\vec{y}$). These operations moreover define the axes Ou and Oz.

S, S_1 and S_2 are all assumed to have the same centre of inertia at the point O, and it is further assumed that certain elastic forces act upon S_2 and also between S_1 and S_2, in such a way that the moment of the external forces acting on S_2 with respect to Oz_1 can be represented by $-K_2\psi$, and the moment of the forces exerted by S_2 on S_1 with respect to Oy by $-K_1\theta$, where K_1 and K_2 are positive constants (restoring couples).[1]

We write

$$\begin{pmatrix} \mathscr{I} & 0 & 0 \\ 0 & C & 0 \\ 0 & 0 & C \end{pmatrix}, \begin{pmatrix} A_1 & 0 & 0 \\ 0 & A_1 & 0 \\ 0 & 0 & C_1 \end{pmatrix} \tag{3.1}$$

[1] More generally the torques μ_1, μ_2 which are to be exerted upon the gimbals S_1, S_2 of a gyroscopic device to comply with some stability requirement will be provided by servo motors.

for the inertia tensors of S and S_1 respectively, in each case with respect to the axes $Oxyz$, and C_2 for the moment of inertia of S_2 with respect to Oz_1. The instantaneous angular velocities, with respect to the reference axes $Ox_1y_1z_1$ are

$$\theta'\vec{y}+\psi'\vec{z}_1, \quad \psi'\vec{z}_1, \quad \theta'\vec{y}+\psi'\vec{z}_1+\Omega\vec{x} \tag{3.2}$$

for S_1, S_2 and S respectively.

The equations of motion are obtained by applying the angular momentum theorem at O, first to $S+S_1$, then to $S+S_1+S_2$.

a) The instantaneous rotations of S_1 and S with respect to $Ox_1y_1z_1$ are

$$\begin{gathered}\vec{\omega}=-\psi'\sin\theta\cdot\vec{x}+\theta'\vec{y}+\psi'\cos\theta\cdot\vec{z},\\ \vec{\omega}+\Omega\vec{x}\end{gathered} \tag{3.3}$$

respectively; using (3.1) the corresponding angular momenta are therefore:

$$\begin{gathered}-A_1\psi'\sin\theta\cdot\vec{x}+A_1\theta'\vec{y}+C_1\psi'\cos\theta\cdot\vec{z}\\ \mathscr{I}(-\psi'\sin\theta+\Omega)\vec{x}+C\theta'\vec{y}+C\psi'\cos\theta\cdot\vec{z}\end{gathered} \tag{3.4}$$

and the total angular momentum is

$$\vec{H}=(\mathscr{I}\Omega-(A_1+\mathscr{I})\psi'\sin\theta)\vec{x}+(C+A_1)\theta'\vec{y}+(C+C_1)\psi'\cos\theta\cdot\vec{z}. \tag{3.5}$$

Since the moment of the forces applied to $S+S_1$, taken with respect to Oy, has the value $-K_1\theta$, we obtain the equation:

$$\vec{y}\cdot\left(\frac{\mathrm{d}\vec{H}}{\mathrm{d}t}\right)_{x_1y_1z_1}=-K_1\theta$$

or, since

$$\left(\frac{\mathrm{d}\vec{H}}{\mathrm{d}t}\right)_{x_1y_1z_1}=\left(\frac{\mathrm{d}\vec{H}}{\mathrm{d}t}\right)_{xyz}+\vec{\omega}\wedge\vec{H},$$

we have by (3.3), (3.5):

$$(C+A_1)\theta''+\mathscr{I}\Omega\psi'\cos\theta+(C+C_1-A_1-\mathscr{I})\psi'^2\sin\theta\cos\theta=-K_1\theta. \tag{3.6}$$

b) The projection on Oz_1 of the angular momentum of $S+S_1+S_2$ at O is $\vec{H}\cdot\vec{z}_1+C_2\psi'$; applying the angular momentum theorem we have:

$$\frac{\mathrm{d}}{\mathrm{d}t}(\vec{H}\cdot\vec{z}_1+C_2\psi')=-K_2\psi$$

that is

$$[(C+C_1+C_2)\psi'-(C+C_1-A_1-\mathscr{I})\sin^2\theta\cdot\psi'-\mathscr{I}\Omega\sin\theta]'=-K_2\psi. \tag{3.7}$$

The equations (3.6) and (3.7) are satisfied by $\psi=\theta=0$; to study movements in the neighbourhood of this equilibrium position, and in particular to discuss stability, one can linearise (3.6), (3.7), that is to say retain only the first-order terms, by regarding θ, ψ and their derivatives as infinitesimals of the same order, a process which leads to:

$$(3.8)\qquad \begin{aligned} I_1\theta'' + K_1\theta + \mathscr{I}\Omega\psi' &= 0 \\ I_2\psi'' + K_2\psi - \mathscr{I}\Omega\theta' &= 0 \end{aligned}$$

$$(3.9)\qquad I_1 = C + A_1, \quad I_2 = C + C_1 + C_2.$$

Discussion of the Linearised System

With ω_1, ω_2 defined by $K_1 = I_1\omega_1^2$, $K_2 = I_2\omega_2^2$, the equation which gives the frequencies of the system with gyroscopic coupling is

$$\det\begin{vmatrix} -I_1\omega^2 + K_1 & i\mathscr{I}\Omega\omega \\ -i\mathscr{I}\Omega\omega & -I_2\omega^2 + K_2 \end{vmatrix} = 0$$

or

$$(3.10)\qquad (\omega_1^2 - \omega^2)(\omega_2^2 - \omega^2) - \frac{\mathscr{I}^2}{I_1 I_2}\Omega^2\omega^2 = 0.$$

With

$$\xi = \frac{\mathscr{I}}{\sqrt{I_1 I_2}}, \quad x^2 = \frac{\xi^2\Omega^2}{\omega_1\omega_2}, \quad \frac{\omega_1}{\omega_2} + \frac{\omega_2}{\omega_1} = 2 + \beta, \quad y^2 = \frac{\omega^2}{\omega_1\omega_2}$$

(3.10) becomes

$$(3.11)\qquad x^2 + \beta = \frac{(y^2 - 1)^2}{y^2},$$

an equation which allows us to represent x as a function of y and which shows that for every given x there exist two distinct positive values of y.

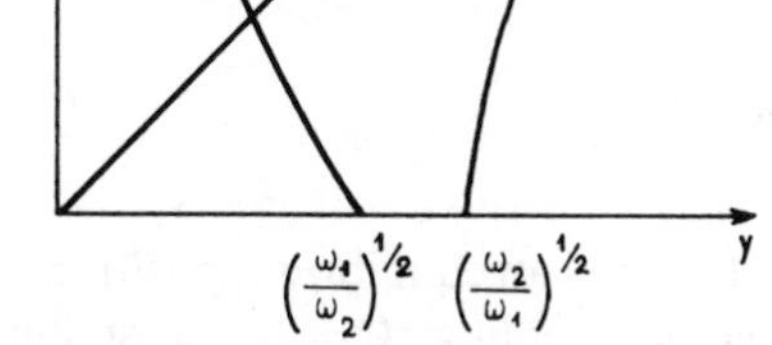

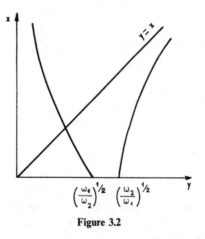

Figure 3.2

If the gyroscopic coupling is very strong (x large), one of the solutions is close to 0, and the other of the order of x so that $y \sim \dfrac{\xi\Omega}{\sqrt{\omega_1\omega_2}}$ and the corresponding vibration frequency is $\omega \sim \xi\Omega$; it no longer depends on the eigenfrequencies ω_1, ω_2 of the uncoupled system, that is to say on the stiffness of the springs which provide the restoring couple. It may be noted from (3.10) that the eigenfrequencies $\tilde{\omega}_1, \tilde{\omega}_2$ of the coupled system satisfy the equation $\tilde{\omega}_1\tilde{\omega}_2 = \omega_1\omega_2$.

Appraisal of the Linearisation Process in the Case of Strong Coupling

The linearised system (3.8) admits of a solution (θ, ψ) which can be represented by $\theta = \operatorname{Re}\lambda e^{i\omega t}$, $\psi = \operatorname{Re}\mu e^{i\omega t}$, where ω denotes one or other of the frequencies defined by (3.10). Now we have seen that when the coupling is strong, $\xi\Omega \gg \sqrt{\omega_1\omega_2}$, so that $\omega \sim \xi\Omega$, the other frequency being close to zero; it appears that $\theta', \psi', \theta'', \psi''$ might not be of an order comparable to that of θ, ψ and that this would put in doubt the validity of the linearisation process. However a closer examination of the terms neglected in (3.6), (3.7) to arrive at (3.8) shows that it is sufficient that $\frac{|C + C_1 - A_1 - \mathscr{I}|}{\mathscr{I}\Omega} |\psi'|\cdot|\theta|$ be small in comparison with 1.

Remembering that $(-I_1\omega^2 + K_1)\lambda + i\mathscr{I}\Omega\omega\mu = 0$, i.e. that for large ω:

$$\left|\frac{\mu}{\lambda}\right| \sim \frac{I_1\omega}{\mathscr{I}\Omega} \sim \xi\frac{I_1}{\mathscr{I}} = \sqrt{\frac{I_1}{I_2}} = O(1)$$

and that $|\psi'| \sim |\mu|\omega$ we can derive the estimates:

$$\frac{|C + C_1 - A_1 - \mathscr{I}|}{\mathscr{I}\Omega} |\psi'|\cdot|\theta| \sim |C + C_1 - A_1 - \mathscr{I}| \frac{|\lambda||\mu|}{\mathscr{I}\Omega}\omega$$

$$\sim |C + C_1 - A_1 - \mathscr{I}| \frac{|\lambda|^2}{\sqrt{I_1 I_2}}.$$

Very generally $\frac{|C + C_1 - A_1 - \mathscr{I}|}{\sqrt{I_1 I_2}} \ll 1$ and for amplitudes of the order of one degree $|\lambda| \sim \frac{\pi}{180}$ and $|\lambda|^2 \sim \frac{1}{2500}$.

Gyroscopic Stabilisation

The system described by (3.8) has an equilibrium position $\theta = \psi = 0$ which is stable in the absence of gyroscopic coupling ($\Omega = 0$); the stability persists when $\Omega \neq 0$ the most obvious effect of the gyroscopic forces being to enlarge the separation between the natural frequencies.

However it is interesting to discuss, for a system with two degrees of freedom a more general situation, in which damping terms are included, and where the equilibrium may be unstable when $\Omega = 0$.

Assuming conservative external forces we thus arrive at equations of the type

$$\begin{aligned} x_1'' + c_1 x_1' - \xi x_2' + a x_1 + h x_2 &= 0 \\ x_2'' + c_2 x_2' + \xi x_1' + h x_1 + b x_2 &= 0. \end{aligned} \tag{3.12}$$

On looking for a solution proportional to $e^{\omega t}$, we obtain:

$$\det \begin{vmatrix} \omega^2 + c_1\omega + a & h - \xi\omega \\ h + \xi\omega & \omega^2 + c_2\omega + b \end{vmatrix} = 0 \tag{3.13}$$

as the equation giving the permissible values of ω, and the stability of the solution $x_1 = x_2 = 0$ requires that the real parts of all solutions of (3.13) should be negative or zero.

1. No damping: $c_1 = c_2 = 0$. We write

$$\omega^4 + (a + b + \xi^2)\omega^2 + ab - h^2 = 0$$

and it can be seen that there is stability if and only if the roots ω^2 are real and negative, i.e. when

$$a + b + \xi^2 > 0. \tag{3.14}$$

$$ab - h^2 > 0. \tag{3.15}$$

$$\xi^4 + 2(a + b)\xi^2 + (a - b)^2 + 4h^2 > 0. \tag{3.16}$$

We shall suppose $ab - h^2 > 0$, and there are two cases to be considered: if $a > 0$, $b > 0$, there is stability for all values of ξ; if $a < 0$, $b < 0$ the system is unstable for $\xi = 0$, but stability appears as soon as ξ is large enough for the inequalities (3.14) and (3.16) to be satisfied. Thus in the cases where both degrees of freedom are unstable, the gyroscopic coupling allows the system to be made stable.

2. Damping present, $c_1 > 0$, $c_2 > 0$.
We apply to the expanded version of the equation (3.13):

$$\omega^4 + (c_1 + c_2)\omega^3 + (c_1 c_2 + a + b + \xi^2)\omega^2 + (c_1 b + c_2 a)\omega + ab - h^2 = 0 \tag{3.17}$$

Routh's conditions (Sect. 4), which are necessary and sufficient to ensure that all the roots of (3.17) lie in the half-plane $\operatorname{Re}\omega < 0$;

$$c_1 + c_2 > 0 \tag{3.18}$$

$$(c_1 c_2 + \xi^2)(c_1 + c_2) + a c_1 + b c_2 > 0. \tag{3.19}$$

$$(c_1 b + c_2 a)[(c_1 c_2 + \xi^2)(c_1 + c_2) + c_1 a + c_2 b] - (c_1 + c_2)^2(ab - h^2) > 0. \tag{3.20}$$

$$ab - h^2 > 0. \tag{3.21}$$

It is clear that (3.19), (3.20), (3.21) imply

$$c_1 b + c_2 a > 0 \tag{3.22}$$

while conversely (3.20), (3.21), (3.22) imply (3.19). Finally noting that (3.20) can be written in the form

$$(c_1 b + c_2 a)(c_1 c_2 + \xi^2)(c_1 + c_2) + c_1 c_2 (a - b)^2 + (c_1 + c_2)^2 h^2 > 0 \tag{3.23}$$

we conclude that (3.18), (3.21), (3.22) and (3.23) are necessary and sufficient conditions for stability.

If $a, b, ab - h^2$ are all positive these conditions are satisfied for all ξ.

If $ab < 0$, there will be instability whatever the value of ξ; but we find ourselves facing the same situation if $a < 0$ and $b < 0$ because (3.22) cannot be satisfied since the damping factors are positive. Accordingly it would be impossible to stabilise by means of a gyroscopic coupling a mechanical system with two naturally unstable degrees of freedom, which has any damping no matter how feeble, except by artificially making one of the damping factors negative.

If we suppose the conditions $a < 0$, $b < 0$, $ab - h^2 > 0$ to hold, then in order to have $c_1 + c_2 > 0$, and $c_1 b + c_2 a > 0$, that is:

$$-c_2 < c_1 < -c_2 \cdot \frac{a}{b},$$

it is clear that when $a/b > 1$, c_2 must be negative and c_1 positive, whereas these conclusions are reversed when $a/b < 1$.

Incidentally, if the structural conditions (3.18), (3.21) and (3.22) hold, it will always be possible to satisfy (3.23) by choosing ξ large enough.

To obtain negative damping, a mechanical element with an additional degree of freedom represented by a variable x_3 is introduced into the system so that the equations of motion now become

$$\text{(3.24)} \qquad \begin{aligned} x_1'' + c_1 x_1' - \xi x_2' + a x_1 + h x_2 &= x_3 \\ x_2'' + c_2 x_2' + \xi x_1' + h x_1 + b x_2 &= 0 \end{aligned}$$

and by means of a suitable feedback device the co-ordinate x_3 is controlled by x_1 in such a way that

$$\text{(3.25)} \qquad x_3 = \lambda x_1', \quad \text{where } \lambda \text{ is a positive constant.}$$

Using (3.24) and (3.25) to eliminate x_3, we get back to equations of the type (3.12) for x_1 and x_2, where the coefficients c_1 and c_2 are replaced by $\tilde{c}_1 = c_1 - \lambda$, $\tilde{c}_2 = c_2$. A suitable choice for λ is to choose it such that

$$-1 < \frac{c_1 - \lambda}{c_2} < -\frac{a}{b}, \quad \text{when} \quad \frac{a}{b} > 1;$$

so that $\lambda \sim c_1 + c_2$ and will be of the same order of magnitude as the damping factors.

Another method consists in altering the conservative character of the external forces applied to the system; for example, assuming that the co-ordinate x_3 is controlled by the condition $x_3 = -\lambda x_2$ and supposing for simplicity that $h = 0$, then the equations of motion can be written as

$$\begin{aligned} x_1'' + c_1 x_1' - \xi x_2' + a x_1 + \lambda x_2 &= 0 \\ x_2'' + c_2 x_2' + \xi x_1' + b x_2 &= 0, \end{aligned}$$

under the hypotheses $a < 0$, $b < 0$, $c_1 > 0$, $c_2 > 0$.

The equation corresponding to (3.13) is:

$$\omega^4 + (c_1 + c_2)\omega^3 + \omega^2(\xi^2 + c_1 c_2 + a + b) + \omega(c_2 a + c_1 b - \lambda\xi) + ab = 0$$

and the conditions for stability:

$$c_1 + c_2 > 0$$

$$(c_1 + c_2)(\xi^2 + c_1 c_2 + a + b) - (c_2 a + c_1 b - \lambda\xi) > 0$$

$$[(c_1 + c_2)(\xi^2 + c_1 c_2 + a + b) - (c_2 a + c_1 b - \lambda\xi)][c_2 a + c_1 b - \lambda\xi] - (c_1 + c_2)^2 ab > 0$$

$$ab > 0$$

will be satisfied if $-\lambda\xi$ is positive and $|\xi|$ is large enough.

In this case the sign of rotation of the gyroscope will depend on that of λ.

2. Lagrange's Equations and Their Application to Gyroscopic Systems

To obtain the equations of motion of systems comprising several gyroscopes, such as those to be found in certain navigational devices, it may be convenient to use Lagrange's method, suitably modified.

The system comprises platforms, frames, and the gyroscopes proper, i.e. the spinning solids of revolution $S_1, S_2, \ldots, S_p$; we use $\bar{S}$ to represent the aggregate of all the solid bodies in the system other than the gyroscopes, so that the total system is $X = \bar{S} \cup S_1 \cup S_2 \cup \cdots \cup S_p$.

If the gyroscopes were locked in their respective mountings the configuration of the system could be described by the set of n independent parameters (Lagrangian co-ordinates) $q_1, q_2, \ldots, q_n$ but since each gyroscope is free to spin about its own axis fixed in its mounting, it is necessary to introduce p additional degrees of freedom, represented by the angles $\varphi_1, \ldots, \varphi_p$, which specify the rotation of each gyroscope about its axis of revolution (in relation to the gimbal mounting containing this axis).

In normal operation there are electromagnetic forces acting between the gyroscope and the mounting (e.g. where the gyro is the armature of an electric motor and the primary coil is fixed to the mounting) to ensure that the gyro is kept spinning at a constant speed relative to its mounting. Thus $\varphi'_1 = \Omega_1, \ldots, \varphi'_p = \Omega_p$ are constants (this maintaining force is necessary to overcome the residual friction between the gyroscope and its bearings, which owing to the high speeds of rotation cannot be neglected even when the coefficients of friction are very small).

We apply to the system, in motion with respect to a Galilean frame of reference $\mathcal{R}$, the theorem on virtual work (d'Alembert's principle) which states that in any arbitrary virtual displacement the work done by the effective forces $m\vec{\gamma}$ summed over the whole system is equal to the work done by all the applied forces both internal and external.

Actually we shall choose a class of virtual displacements restricted in such a way as to simplify the calculation of the work done by these forces; we consider, in fact, the set of virtual displacements of the system which respect the indeformable nature of each solid constituent and the contact relationships between them, are compatible with the external constraints existing at time t, and finally are such that the gyroscopes remain fixed to their own gimbals.

Under these conditions the position of any arbitrary point of the system can be defined in terms of the parameters $q_1, \ldots, q_n$ and the time t

$$M = M(q_1, \ldots, q_n, t)$$

and we can define the virtual displacement by

$$\overrightarrow{\delta M} = \sum_{k=1}^{n} \frac{\partial \overrightarrow{M}}{\partial q_k} \delta q_k$$

The virtual work theorem then states that

$$\int_X \vec{\gamma} \cdot \overrightarrow{\delta M}\, dm = \int_X \vec{F} \cdot \overrightarrow{\delta M}\, dx \tag{3.26}$$

where γ denotes the acceleration of the point M with respect to $\mathcal{R}$, $\vec{F}$ is the volume density of the forces at M while dx and dm denote the volume and mass elements respectively.

Because of the assumptions made with regard to the virtual displacement, the total virtual work done by the forces which tie the various parts of $\bar{S}$ together, or which arise when $\bar{S}$ encounters obstacles external to X (in the absence of friction), and by the internal forces (those which ensure the cohesion of each solid, and those which act between the gyroscope and its mounting, including frictional and magnetic forces, which in each case amount to a null system) amounts to nothing. Consequently $\int_X \vec{F} \overrightarrow{\delta M}\, dx$ is reduced to the work done by certain explicitly known forces which can be represented by:

$$\int_X \vec{F} \cdot \overrightarrow{\delta M}\, dx = \sum_{k=1}^{n} Q_k \delta q_k, \quad \text{where} \quad Q_k(q_1, \ldots, q_n, t). \tag{3.27}$$

If one writes $\bar{S}_j$ to denote the gyroscope S_j when locked (φ_j constant) then the acceleration of a point M of S_j, with respect to $\mathcal{R}$, can be written:

$$\vec{\gamma}(M; M \in S_j) = \vec{\gamma}(M; M \in \bar{S}_j) + \vec{\gamma}_r(M) + \vec{\gamma}_c(M) \tag{3.28}$$

where $\vec{\gamma}_r(M)$ is the acceleration of $M \in S_j$ relative to a frame of reference tied to the mounting; $\vec{\gamma}_c(M) = 2\vec{\omega} \wedge \vec{v}_r$, is the Coriolis acceleration, and $\vec{v}_r$ is the velocity of $M \in S_j$ relative to the mounting, while $\vec{\omega}$ is the instantaneous angular velocity of the mounting with respect to $\mathcal{R}$. Here we have used the Coriolis theorem on rotating frames of reference. We can therefore write (3.27) as:

$$\int_{\bar{S} \cup (\cup_j \bar{S}_j)} \vec{\gamma} \cdot \overrightarrow{\delta M}\, dm = \sum_{k=1}^{n} Q_k \delta q_k - \int_{\cup_j S_j} (\vec{\gamma}_r + \vec{\gamma}_c) \overrightarrow{\delta M}\, dm. \tag{3.29}$$

The term on the left can be obtained in the classical manner from Lagrange's formula, starting from the expression

$$T(q_1, \ldots, q_n, q'_1, \ldots, q'_n, t)$$

for the kinetic energy of the total system, calculated on the hypothesis of locked gyroscopes:

$$\int_{\bar{S}U(U_j\bar{S}_j)} \vec{\gamma}\cdot\overrightarrow{\delta M}\,\mathrm{d}m = \sum_{k=1}^{n}\left(\frac{\mathrm{d}}{\mathrm{d}t}\left(\frac{\partial T}{\partial q_k'}\right)-\frac{\partial T}{\partial q_k}\right)\delta q_k.$$

To calculate the remaining terms of (3.29), let us introduce $\vec{\omega}_j$ to denote the instantaneous angular velocity relative to $\mathscr{R}$ of the mounting which holds S_j (to simplify the formulae we shall temporarily omit the suffix j):

(3.30)
$$\vec{\omega} = \sum_{k=1}^{n} \vec{P}_k q_k' + \vec{\mathscr{P}}$$

or

$$\vec{\omega}_j = \sum_{k=1}^{n} \vec{P}_{jk} q_k' + \vec{\mathscr{P}}_j, \quad \vec{P}_{jk}(q,t), \quad \vec{\mathscr{P}}_j(q,t)$$

so that the virtual displacement of an arbitrary point of the mounting or its locked gyro is represented by

$$\overrightarrow{\delta M} = \overrightarrow{\delta O} + \left(\sum_k \vec{P}_k \delta q_k\right) \wedge \overrightarrow{OM},$$

O being a reference point, for example the centre of inertia of the gyroscope. The rotation of the latter with respect to its mounting is $\vec{\Omega} = \Omega\vec{z}$ where $\vec{z}$ is the unit vector in the direction of the axis of rotation of the gyro, and consequently: $\vec{\gamma}_r = -\Omega^2\overrightarrow{HM}$, where H is the orthogonal projection of M on Oz.

It follows from this that the system of quantities of relative acceleration of S equates to zero (the material elements can be associated in pairs which are symmetrically situated with respect to Oz); their virtual work is zero since the displacement imposed does not deform the solid. Thus the contribution $\int_{U_jS_j}\vec{\gamma}_r\overrightarrow{\delta M}\,dm$ vanishes and it remains to calculate $\int_S \vec{\gamma}_c\overrightarrow{\delta M}\,dm$ with

$$\vec{\gamma}_c = 2\vec{\omega}\wedge\vec{v}_r, \quad \vec{v}_r = \vec{\Omega}\wedge\overrightarrow{OM}.$$

Observing that $\int_S \vec{\gamma}_c\cdot\overrightarrow{\delta O}\,dm = \left(2\vec{\omega}\wedge\int_S \vec{v}_r dm\right)\overrightarrow{\delta O} = 0$ as $\int_S \vec{v}_r dm = 0$ we arrive at:

$$\int_S \vec{\gamma}_c\cdot\overrightarrow{\delta M}\,dm = \int_S (2\vec{\omega}\wedge\vec{v}_r)\cdot\left(\left(\sum_k \vec{P}_k\delta q_k\right)\wedge\overrightarrow{OM}\right)dm$$

$$= -\int_S 2(\vec{\omega}\wedge(\vec{\Omega}\wedge\overrightarrow{OM})\wedge\overrightarrow{OM})\mathrm{d}m\cdot\left(\sum_k \vec{P}_k\delta q_k\right)$$

and since

$$(\vec{\omega}\wedge(\vec{\Omega}\wedge\overrightarrow{OM}))\wedge\overrightarrow{OM} = (\vec{\omega}\cdot\overrightarrow{OM})(\vec{\Omega}\wedge\overrightarrow{OM})$$

we have

$$-\int_S \vec{\gamma}_c\cdot\overrightarrow{\delta M}\,dm = 2\int_S(\vec{\omega}\cdot\overrightarrow{OM})(\vec{\Omega}\wedge\overrightarrow{OM})\mathrm{d}m\cdot\sum_k \vec{P}_k\delta q_k.$$

The integral on the right is easily calculated, using as intermediary the reference system $Oxyz$ tied to the mounting. We find, in this way, that its value is $C\,\vec{\Omega}\wedge\vec{\omega}$,

so that

$$-\int_S \vec{\gamma}_c \overrightarrow{\delta M}\, \mathrm{d}m = C(\vec{\Omega} \wedge \vec{\omega}) \cdot \sum_{k=1}^{n} \vec{P}_k \delta q_k$$

or for the set of p gyros

$$-\int_{U_j S_j} \vec{\gamma}_c \overrightarrow{\delta M}\, \mathrm{d}m = \sum_{j=1}^{p} C_j(\vec{\Omega}_j \wedge \vec{\omega}_j) \cdot \sum_{k=1}^{n} \vec{P}_{jk} \delta q_k.$$

We thus obtain finally the n equations of motion:

$$\frac{\mathrm{d}}{\mathrm{d}t}\left(\frac{\partial T}{\partial q'_k}\right) - \frac{\partial T}{\partial q_k} = Q_k + \sum_{j=1}^{p} C_j(\vec{\Omega}_j, \vec{\omega}_j, \vec{P}_{jk}) \tag{3.31}$$

or, with the help of (3.30):

$$\frac{\mathrm{d}}{\mathrm{d}t}\left(\frac{\partial T}{\partial q'_k}\right) - \frac{\partial T}{\partial q_k} = Q_k + \sum_{s=1}^{n} \sum_{j=1}^{p} C_j(\vec{\Omega}_j, \vec{P}_{js}, \vec{P}_{jk}) q'_s + \sum_{j=1}^{p} C_j(\vec{\Omega}_j, \vec{\mathscr{P}}_j, \vec{P}_{jk}),$$

$$1 \leqslant k \leqslant n. \tag{3.32}$$

We see, from (3.31) that in writing down the equations of motion, it is quite permissible to disregard the proper rotations of the gyroscopes provided that we include, among the external forces acting on each gyroscope frame, couples of moment $C_j(\vec{\Omega}_j \wedge \vec{\omega}_j)$.

As the frame of reference $\mathscr{R}$ is tied to the Earth, it may however be necessary to take into account the Earth's rotation $\vec{\varepsilon}$. It suffices to add it to $\vec{\omega}_j$ in equation (3.31), or simply to $\vec{\mathscr{P}}_j$ in equation (3.32), and where necessary to take it into account in calculating the kinetic energy of the system with stationary gyros, but this correction is usually superfluous.

Anyway the effect of the gyroscopic mechanisms manifests itself by the presence of terms in Lagrange's equation which are structurally of the type $\sum_{s=1}^{n} \Gamma_{ks} q'_s + \Gamma_k$ where the matrix $\Gamma_{ks} = \sum_{j=1}^{p} C_j(\vec{\Omega}_j, \vec{P}_{js}, \vec{P}_{jk})$, which depends only on the co-ordinates q_i and t, is skew-symmetric, i.e. $\Gamma_{ks} = -\Gamma_{sk}$. The terms Γ_k moreover vanish if the constraints are time-independent and the Earth's rotation is ignored.

A simplified approach, valid in certain circumstances, in particular in the neighbourhood of an apparent equilibrium position, consists in neglecting the kinetic energy of the system with stationary gyroscopes. The equations of motion then become:

$$Q_k + \sum_{j=1}^{k} C_j(\vec{\Omega}_j, \vec{\omega}_j, \vec{P}_{jk}) = 0 \tag{3.33}$$

which can be interpreted as follows: the set of forces acting directly on the components of the system, and the set of fictitious couples of moment $C_j(\vec{\Omega}_j \wedge \vec{\omega}_j)$ acting on the jth gyroscope frame, perform no work in any virtual displacement of the type considered.

Example: The Gyroscopic Pendulum

Reverting to the notations at the beginning of the chapter, the instantaneous rotation of the gyroscope gimbal S_1 can be written

$$\vec{\omega} = -\psi' \sin\theta \cdot \vec{x} + \theta' \vec{y} + \psi' \cos\theta \cdot \vec{z} \tag{3.34}$$

and the kinetic energy of the total system, with stationary gyro is:

$$2T = (A_1 + \mathscr{I})\psi'^2 \sin^2\theta + (C + A_1)\theta'^2 + (C + C_1)\psi'^2 \cos^2\theta + C_2\psi'^2. \tag{3.35}$$

The virtual work of the fictitious couple $\mathscr{I}(\vec{\Omega} \wedge \vec{\omega})$ is

$$(\mathscr{I}\Omega\vec{x} \wedge \vec{\omega})(\vec{y}\delta\theta + (\vec{z}\cos\theta - \vec{x}\sin\theta)\delta\psi) = \mathscr{I}\Omega\theta' \cos\theta \cdot \delta\psi - \mathscr{I}\Omega\psi' \cos\theta \cdot \delta\theta$$

and the equations of motion are:

$$\frac{\mathrm{d}}{\mathrm{d}t}\left(\frac{\partial T}{\partial \theta'}\right) - \frac{\partial T}{\partial \theta} = -K_1\theta - \mathscr{I}\Omega\psi' \cos\theta$$

$$\frac{\mathrm{d}}{\mathrm{d}t}\left(\frac{\partial T}{\partial \psi'}\right) - \frac{\partial T}{\partial \psi} = -K_2\psi + \mathscr{I}\Omega\theta' \cos\theta \tag{3.36}$$

identical to (3.6), (3.7).

3. Applications

The Gyrocompass

We suppress, in the gyroscopic pendulum device described by the equations (3.6), (3.7) or (3.36), the restoring couple associated with the co-ordinate ψ. The apparatus is assumed to be in a position such that Oz_1 represents the local upward vertical. We take account in the equations of motion of the correction due to the Earth's rotation in calculating the virtual work of the Coriolis forces, this correction proving to be necessary if the gyroscope's own rotational speed is very high.

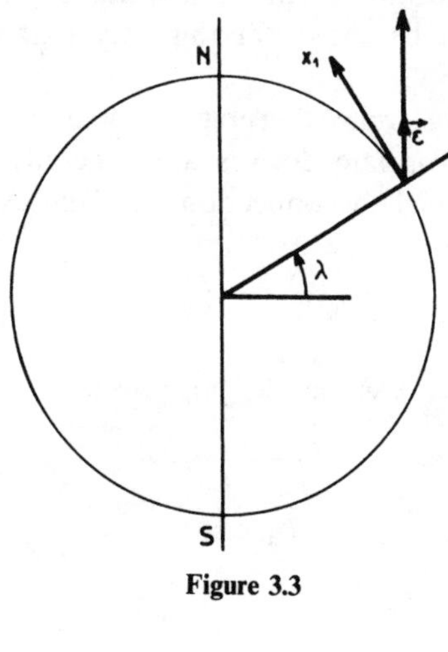

Figure 3.3

In terms of virtual work this correction amounts to $\mathscr{I}(\vec{\Omega} \wedge \vec{\varepsilon}) \times (\vec{y}\delta\theta + \vec{z}_1\delta\psi)$, $\vec{\varepsilon}$ denoting the Earth's rotation vector.

With the axis Oz_1 being, as already mentioned, directed along the upward vertical at the place under consideration, we shall define the axis Ox_1 in the meridian plane directed towards the North. At a place in the northern hemisphere, we have

$$(\vec{z}_1, \vec{\varepsilon}) = \frac{\pi}{2} - \lambda,$$

where λ is the latitude of the place concerned. The virtual work correction can be expressed as

$$\begin{aligned}&[\mathscr{I}\Omega\vec{x} \wedge \varepsilon(\sin\lambda\cdot\vec{z}_1 + \cos\lambda\cdot\vec{x}_1)]\cdot(\vec{y}\delta\theta + \vec{z}_1\delta\psi)\\ &\quad = -\mathscr{I}\Omega\varepsilon[(\sin\lambda\cos\theta + \sin\theta\cos\lambda\cos\psi)\delta\theta + \cos\theta\cos\lambda\sin\psi\cdot\delta\psi]\end{aligned}$$

and so the equations of motion are:

$$\text{(3.37)}\quad \begin{aligned}&(C + A_1)\theta'' + (C + C_1 - A_1 - \mathscr{I})\psi'^2 \sin\theta\cos\theta\\ &\quad = -K_1\theta - \mathscr{I}\Omega\psi'\cos\theta - \mathscr{I}\Omega\varepsilon(\sin\lambda\cos\theta + \sin\theta\cos\lambda\cos\psi)\end{aligned}$$

$$\{[(A_1 + \mathscr{I})\sin^2\theta + (C + C_1)\cos^2\theta + C_2]\psi'\}' = \mathscr{I}\Omega\theta'\cos\theta - \mathscr{I}\Omega\varepsilon\cos\theta\cos\lambda\sin\psi.$$

It is apparent that an equilibrium solution is possible for $\psi = 0$, $\theta = \theta_0$, with $\theta_0 = -\dfrac{\mathscr{I}\Omega\varepsilon}{K_1}(\sin\lambda\cos\theta_0 + \sin\theta_0\cos\lambda)$, or taking account of $\dfrac{\mathscr{I}\Omega\varepsilon}{K_1} \ll 1$,

$$\theta_0 = -\frac{\mathscr{I}\Omega\varepsilon\sin\lambda}{K_1 + \mathscr{I}\Omega\varepsilon\cos\lambda}.$$

The axis of the gyroscope then lies in the meridian plane and its position consequently provides an indication of geographical North.

Note that θ_0 is a very small angle; for example with $T = 2\pi\sqrt{\dfrac{\mathscr{I}}{K_1}} = 1$ s, $\Omega = 12\,500$ r.p.m., $\lambda = 45°$, we find $\theta_0 \sim (10)^{-1}$ degrees or 6 min of arc. Motion around the equilibrium position[2] can be investigated by linearisation in this neighbourhood and it can be shown in this way that the motion is stable if $\mathscr{I}\Omega\varepsilon > 0$.

The existence of damping, even very feeble damping, can jeopardise this stability; to provide a remedy appropriate devices are generally necessary. A situation of this kind will be examined in connection with the example of the monorail car.

---

[2] After linearising the equation (3.37), and ignoring the inertia terms, we obtain for the expression of the period of the oscillations

$$T_1 = 2\pi\sqrt{\frac{\mathscr{I}\Omega}{K_1\varepsilon\cos\lambda}} \sim 85\text{ min},$$

with the data of the example.

Influence of Relative Motion on the Behaviour of the Gyrocompass

We resume our analysis by supposing the instrument to be mounted on a mobile vehicle on the Earth's surface. The movement of the vehicle can be described by a knowledge of its latitude λ and longitude L (the latter measured positively from the point O in a westerly direction) as a function of time.

We shall suppose that the restoring couple associated with the co-ordinate θ is obtained with the help of an additional load of mass m fixed at a point P of the axis Oz, connected to the inner gimbal, $\overrightarrow{OP} = -l\vec{z}$, $l > 0$, which has the effect of producing a torque about the axis Oy of magnitude $-mgl\theta$. Since the motion has to be studied in relation to Galilean axes $T\xi\eta\zeta$, emanating from T the centre of the earth in directions which are invariable with respect to the fixed stars, and as $T\zeta$ can be taken along the line SN joining the Earth's poles, it is necessary firstly to introduce the effect of the Earth's rotation, and secondly to take account also, as far as the frame of reference $Ox_1y_1z_1$ is concerned, of the additional rotation induced by the motion of the point O.

Thus the instantaneous rotation of the inner gimbal with respect to $T\xi\eta\zeta$ can be expressed as a sum of two terms, the one $\vec{\omega} = \theta'\vec{y} + \psi'\vec{z}_1$ being the rotation with respect to $Ox_1y_1z_1$, the other

$$\vec{\omega}_* = \lambda'\vec{y}_1 + (\varepsilon - L')(\sin\lambda\cdot\vec{z}_1 + \cos\lambda\cdot\vec{x}_1)$$

being the rotation of $Ox_1y_1z_1$ with respect to $T\xi\eta\zeta$.

We shall use the method described earlier for obtaining an approximate equation, with however a corrective which we should explain.

We shall suppose the system to be designed so that, apart from the extra load m at P, the point O is the centre of inertia for the remainder whose total mass we denote by μ.

The kinetic energy of the system with gyroscope clamped and excluding the additional load, is the sum of the kinetic energy of the motion around O which we shall ignore, in accordance with our simplified approach, and the kinetic energy of the motion of a particle of mass μ concentrated at O, the centre of inertia, i.e.

$$\frac{\mu}{2}(R^2\lambda'^2 + R^2(\varepsilon - L')^2\cos^2\lambda),$$

where R is the Earth's radius. This expression however depends only on t and therefore its contribution under the Lagrangian operator is nil. It now remains to take into account the kinetic energy of the extra load. Certain terms of this can have a significant effect, because of the factor R.

The velocity of the mass m at P, with respect to $T\xi\eta\zeta$ is:

$$\vec{v} = R\lambda'\vec{x}_1 - R(\varepsilon - L')\cos\lambda\cdot\vec{y}_1 + l\vec{z} \wedge (\vec{\omega} + \vec{\omega}_*).$$

Since it is plain that

$$\begin{aligned}\vec{z} \wedge (\vec{\omega} + \vec{\omega}_*) &= -\theta'\vec{x} - (\psi' + (\varepsilon - L')\sin\lambda)\sin\theta\cdot\vec{y} + \lambda'\vec{z} \wedge \vec{y}_1 \\ &\quad + (\varepsilon - L')\cos\lambda\cdot(\vec{z} \wedge \vec{x}_1)\\ \vec{x} &= \cos\theta\cos\psi\cdot\vec{x}_1 + \cos\theta\sin\psi\cdot\vec{y}_1 - \sin\theta\cdot\vec{z}_1\end{aligned}$$

$$\vec{y} = -\sin\psi\cdot\vec{x}_1 + \cos\psi\cdot\vec{y}_1$$
$$\vec{z}\wedge\vec{y}_1 = -\cos\theta\cdot\vec{x}_1 + \sin\theta\cos\psi\cdot\vec{z}_1$$
$$\vec{z}\wedge\vec{x}_1 = \cos\theta\cdot\vec{y}_1 - \sin\theta\sin\psi\cdot\vec{z}_1$$

we can write

$$\begin{aligned}\vec{v} = [&R\lambda' - l\theta'\cos\theta\cos\psi + l(\psi' + (\varepsilon - L')\sin\lambda)\sin\theta\sin\psi \\ &- l\lambda'\cos\theta]\vec{x}_1 + [-R(\varepsilon - L')\cos\lambda - l\theta'\cos\theta\sin\psi \\ &- l(\psi' + (\varepsilon - L')\sin\lambda)\sin\theta\cos\psi + l(\varepsilon - L')\cos\lambda\cos\theta]\vec{y}_1 \\ &+ [\theta'\sin\theta + \lambda'\sin\theta\cos\psi - (\varepsilon - L')\cos\lambda\sin\theta\sin\psi]l\vec{z}_1\end{aligned}$$

and so

$$\begin{aligned}2T = mv^2 = 2mRl\lambda'[&-(\theta'\cos\psi + \lambda')\cos\theta \\ &+ (\psi' + (\varepsilon - L')\sin\lambda)\sin\theta\sin\psi] + 2mRl(\varepsilon - L')\cos\lambda\cdot[\theta'\cos\theta\sin\psi \\ &+ (\psi' + (\varepsilon - L')\sin\lambda)\sin\theta\cos\psi - (\varepsilon - L')\cos\lambda\cos\theta] + \cdots\end{aligned}$$

the marks of omission representing terms proportional to R^2 but depending solely on t and therefore annihilated by the Lagrangian operator, or else terms not involving R representing slow movements which can be neglected under the point of view adopted.

We now apply Lagrange's equations, simplifying the resulting formulae by an expansion to the first order in $\theta, \psi, \lambda', L'$ and their derivatives, and on the assumption however that $L' \ll \varepsilon$. This leads to:

$$\frac{\mathrm{d}}{\mathrm{d}t}\left(\frac{\partial T}{\partial\theta'}\right) - \frac{\partial T}{\partial\theta} = -mRl(\varepsilon - L')^2(\sin\lambda + \theta\cos\lambda)\cos\lambda - mRl\lambda''$$

$$\frac{\mathrm{d}}{\mathrm{d}t}\left(\frac{\partial T}{\partial\psi'}\right) - \frac{\partial T}{\partial\psi} = 0$$

and to equations of motion which can be written in explicit form, starting from the virtual work principle:

$$\left(\frac{\mathrm{d}}{\mathrm{d}t}\left(\frac{\partial T}{\partial\theta'}\right) - \frac{\partial T}{\partial\theta}\right)\delta\theta = -mgl\theta\delta\theta + \mathscr{I}(\Omega\vec{x}\wedge(\vec{\omega} + \vec{\omega}_*))\cdot(\delta\theta\vec{y} + \delta\psi\vec{z}_1)$$

or after calculation:

(3.38) $$\begin{aligned}&mRl\lambda'' + mRl(\varepsilon - L')^2(\sin\lambda + \theta\cos\lambda)\cos\lambda - mgl\theta \\ &\quad - \mathscr{I}\Omega(\psi' + (\varepsilon - L')(\sin\lambda + \theta\cos\lambda)) = 0\end{aligned}$$

(3.39) $$\theta' + \lambda' - (\varepsilon - L')\cos\lambda\cdot\psi = 0.$$

Differentiating the equation (3.39) with respect to t, and neglecting $L''\cdot\psi$, $\lambda'\cdot\psi$ in comparison with θ''... we arrive at

$$\theta'' + \lambda'' - (\varepsilon - L')\cos\lambda\cdot\psi' = 0$$

whence, after eliminating ψ' from (3.38) we obtain:

$$\theta'' + (\varepsilon - L')\frac{\cos\lambda}{\mathscr{I}\Omega}\{mgl + [\mathscr{I}\Omega - mRl(\varepsilon - L')\cos\lambda](\varepsilon - L')\cos\lambda\}\theta$$

$$+\left(1 - \frac{mRl(\varepsilon - L')\cos\lambda}{\mathscr{I}\Omega}\right)\lambda''$$

$$+\frac{(\varepsilon - L')^2}{\mathscr{I}\Omega}(\mathscr{I}\Omega - mRl(\varepsilon - L')\cos\lambda)\sin\lambda\cos\lambda = 0.$$

It is clearly advantageous to arrange matters so that the coefficient of λ'' should be zero.

(3.40) $$mRl(\varepsilon - L')\cos\lambda = \mathscr{I}\Omega.$$

In fact the equation for θ then becomes

$$\theta'' + p^2\theta = 0$$

with $p^2 = (\varepsilon - L')\dfrac{mgl\cos\lambda}{\mathscr{I}\Omega}$, which implies $\Omega > 0$ since $L' \ll \varepsilon$ and $\varepsilon, \cos\lambda$ are positive.

The period of oscillation of the system is:

$$T_1 = 2\pi\sqrt{\frac{\mathscr{I}\Omega}{(\varepsilon - L')mgl\cos\lambda}} \sim 2\pi\sqrt{\frac{\mathscr{I}\Omega}{\varepsilon mgl\cos\lambda}}$$

and condition (3.40), known as Schuler's condition, expresses the fact that

$$\frac{R}{g} = \frac{\mathscr{I}\Omega}{\varepsilon mgl\cos\lambda}, \quad \text{or in other words} \quad T_1 \sim 2\pi\sqrt{\frac{R}{g}} \sim 85\,\text{min}.$$

The equilibrium position corresponds to $\theta = 0$, $\psi \sim \dfrac{\lambda'}{\varepsilon\cos\lambda}$. For the response of the instrument to conform to the conclusions of the preceding theory, it is obviously necessary that λ', L' should be small compared to ε.

Gyroscopic Stabilisation of the Monorail Car

A car rests on a single horizontal rail $z_1'Az_1$ whose direction is represented by the unit vector $\vec{z}_1$. $\vec{y}_1$ defines the upward vertical unit vector and ψ is the angular displacement between Π the longitudinal plane of symmetry of the car, and the vertical plane through the rail; $\psi = 0$ is an equilibrium position, assuming that O, the centre of inertia of the car, lies in Π, but it is a position of unstable equilibrium because as soon as $\psi \neq 0$ the weight which has a moment about the line $z_1'Az_1$ of magnitude $Mgl\sin\psi \sim K_2\psi$ (where $l = OA$ and M is the car's mass) tends to tilt the car. The system can be stabilised by a gyroscopic mechanism similar to the one discussed at the beginning of this chapter.

About an axis $y'Oy$, lying in the plane of symmetry Π, passing through O, and perpendicular to the rail is mounted a gimbal S_1, of which the axis $x'x$ forms an integral part. This axis $x'x$ which carries the axis of revolution of the gyroscope S is designed to meet $y'y$ at right angles at the point O.

To simplify the calculations we shall suppose that O is the common centre of inertia of the car, of the gimbal mounting S_1, and of the gyroscope S.

If we denote by θ the angle between the $x'x$ axis and the vertical plane containing $\vec{y}$ and $\vec{y}_1$, and assume that elastic forces are exerted by the car on S in such a way that their moment about Oy is $K_1\theta$, then we can at once write down the equations of motion, from (3.36).

It suffices to change the signs of K_1 and K_2 in these two equations, and to obtain the kinetic energy computed with respect to a frame of reference tied to the rail, we have merely to add the term $Ml^2\psi'^2$ to the kinetic energy of the system derived from its motion around the centre of inertia calculated using (3.35). Thus after linearisation the equations of motion can be written:

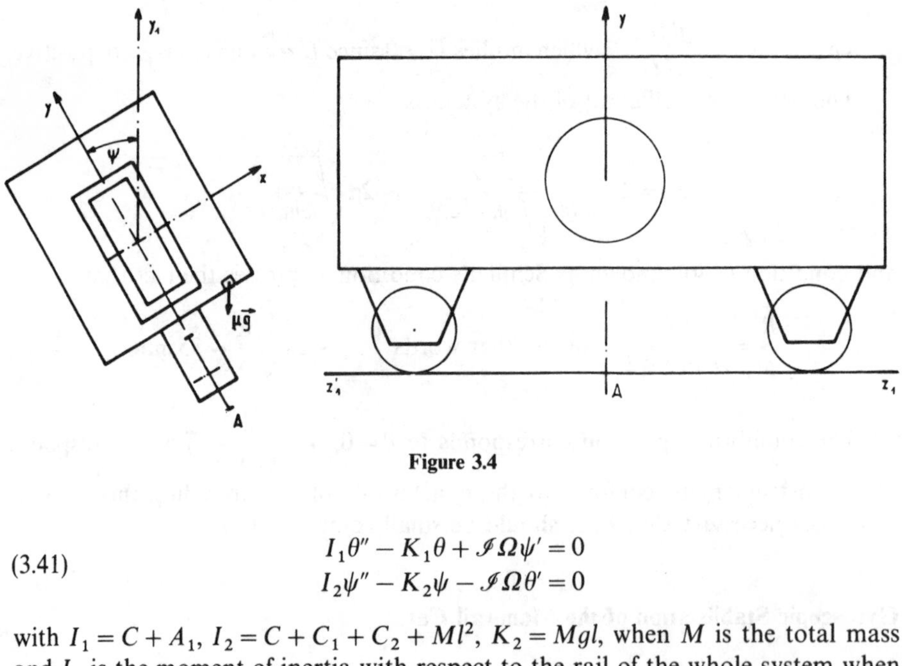

Figure 3.4

(3.41)
$$\begin{aligned} I_1\theta'' - K_1\theta + \mathscr{I}\Omega\psi' &= 0 \\ I_2\psi'' - K_2\psi - \mathscr{I}\Omega\theta' &= 0 \end{aligned}$$

with $I_1 = C + A_1$, $I_2 = C + C_1 + C_2 + Ml^2$, $K_2 = Mgl$, when M is the total mass and I_2 is the moment of inertia with respect to the rail of the whole system when $\theta = 0$.

The solution $\psi = \theta = 0$ of (3.41) is stable if $\mathscr{I}\Omega$ is large enough, more precisely if

$$\mathscr{I}|\Omega| > \sqrt{I_1K_2} + \sqrt{I_2K_1}.$$

However account has to be taken of damping though this may be very slight if there is little friction; we have seen that in such circumstances the system is unstable no matter what Ω may be. However as has been suggested earlier this difficulty can be overcome by a servo device.

We first rewrite the equations with $\sqrt{I_1}\theta$, $\sqrt{I_2}\psi$ replaced by new variables still denoted by θ, ψ but introducing the positive damping coefficient c_1, c_2:

$$\begin{aligned} &\theta'' + c_1\theta' + \xi\psi' + a\theta = 0 \\ &\psi'' + c_2\psi' - \xi\theta' + b\psi = 0 \end{aligned} \quad \text{with} \quad a = -\frac{K_1}{I_1} < 0, \quad b = -\frac{K_2}{I_2} < 0 \tag{3.42}$$

$$\xi = \frac{\mathscr{I}\Omega}{\sqrt{I_1 I_2}}.$$

We can make the equilibrium position $\theta = \psi = 0$ stable if we can succeed, by some suitable device, in altering one of the damping coefficients so that it becomes negative, either c_2 in the case where $b/a < 1$, or c_1, if $b/a > 1$. Let us imagine a counterpoise of mass μ mounted on the floor of the car, and constrained to move transversely in the plane Oyy_1; if x measures the displacement from the plane Π, the action of the force of gravity on this mass, for any virtual displacement of the system, (satisfying the constraints of the general theory) makes a contribution of $\mu g(h\psi - x)\delta\psi - \mu g\psi\delta x$, to the virtual work done by the forces, h being the distance from rail to floor.

The as yet unspecified forces exerted between the car and the mass μ are internal to the system as a whole and hence their combined total is zero; their virtual work is nil for any displacement characterised by $\delta x = 0$, $\delta\theta, \delta\psi$ arbitrary.

Thus, for such a virtual displacement, the Lagrange equations can be written in the form (3.41), provided that we remember to include the kinetic energy of the mass μ, namely

$$\tfrac{1}{2}\mu[(x^2 + h^2)\psi'^2 - 2hx'\psi' + x'^2]$$

in calculating T and take account of the additional effects of the gravity forces acting on μ on the virtual work. We obtain in this way, after linearisation

$$\begin{aligned} &I_1\theta'' - K_1\theta + \mathscr{I}\Omega\psi' = 0 \\ &(I_2 + \mu h^2)\psi'' - \mu h x'' - K_2\psi - \mathscr{I}\Omega\theta' = \mu g(h\psi - x). \end{aligned} \tag{3.43}$$

In the second of these equations we can neglect the terms $\mu h^2\psi''$ and $\mu h^2\left(\frac{x}{h}\right)''$ which are small in comparison with $I_2\psi''$ since $\mu h^2 \ll I_2$.

Hence we see, on replacing K_2 by $K_2 + \mu g h$ and changing over to the reduced variables $\sqrt{I_1}\theta \to \theta$, $\sqrt{I_2}\psi \to \psi$, $x \to \sqrt{I_2}x$, that (3.43) can be written in the form:

$$\begin{aligned} &\theta'' + c_1\theta' + \xi\psi' + a\theta = 0 \\ &\psi'' + c_2\psi' - \xi\theta' + b\psi = -\mu g x \end{aligned}$$

We shall from now on suppose that, thanks to a servo-mechanism the forces between the car and the counterbalance μ are controlled in such a way that $x = -\lambda\psi'$ where λ is a positive constant. Stability at the equilibrium position can be obtained provided that

$$\begin{aligned} &c_1 + c_2 - \lambda\mu g > 0 \\ &c_1 b + (c_2 - \lambda\mu g)a > 0 \end{aligned}$$

or
$$-c_1 < c_2 - \lambda\mu g < -c_1\frac{b}{a}$$

the inequalities being compatible since $0 < \frac{b}{a} < 1$, $c_1 > 0$.

In fact, c_1 and c_2 being small, the same can be true of $\lambda\mu$ and it is clear that the servo-mechanism can function with the help of a very small supply of energy. The structural conditions having been satisfied in the manner just mentioned, $|\xi|$ and $|\Omega|$ will have to be chosen large enough to ensure stability.

If $b/a > 1$ the device is modified as follows: an auxiliary motor attached to the car exerts a force on the mounting S_1 whose moment with respect to the axis Oy is a variable γ linked to the co-ordinate θ by the relation $\gamma = \lambda\theta'$, $\lambda > 0$.

The equations of motion are then:

$$\begin{aligned} \theta'' + c_1\theta' + \xi\psi' + a\theta &= \gamma \\ \psi'' + c_2\psi' - \xi\theta' + b\psi &= 0 \\ \gamma &= \lambda\theta' \end{aligned}$$

and stability of the equilibrium $\theta = \psi = 0$ can be obtained by a suitable choice of the parameter λ, for quite high values of $|\xi|$.

We shall discuss further on, for this problem, the case of a non-linear control described by $\gamma = (\lambda - \alpha\theta'^2)\theta'$.

4. Routh's Stability Criterion

To discuss stability of equilibrium in the context of a linearised theory, we often need to be able to express necessary and sufficient conditions that all the roots of an algebraic equation of degree n with real coefficients should have their real parts negative.

Consider the equation

(3.44) $$f(z) = a_0 z^n + a_1 z^{n-1} + \cdots + a_n = 0$$

in which the coefficients are real and we assume $a_0 > 0$. Denoting its roots, which may be real or complex and not necessarily all distinct, by $\zeta_1, \zeta_2, \ldots, \zeta_n$ we may factorise $f(z) = a_0 \prod_{j=1}^{n} (z - \zeta_j)$, from which it emerges that a necessary and sufficient condition that $\operatorname{Re}\zeta_j < 0$, $\forall j \in (1, 2, \ldots, n)$ is that the argument of $f(iy)$ be an increasing function of y and increase by $n\pi$ as y increases from $-\infty$ to $+\infty$.

Let us, always supposing y to be real, write $f(iy) = i^n (P_n(y) - iP_{n-1}(y))$, where

(3.45) $$\begin{aligned} P_n(y) &= a_0 y^n - a_2 y^{n-2} + a_4 y^{n-4} - \cdots \\ P_{n-1}(y) &= a_1 y^{n-1} - a_3 y^{n-3} + \cdots \end{aligned}$$

and let us first find some necessary conditions. Accordingly we shall make the hypothesis that $\operatorname{Re}\zeta_j < 0$, $\forall j$. Since $\sum_{j=1}^{n} \zeta_j = -\frac{a_1}{a_0}$, we must have $a_1 > 0$. On the other hand the function $R(y)=\sqrt{P_n^2(y)+P_{n-1}^2(y)}$ never vanishes for real y and we can therefore introduce the argument $\theta(y)$ defined by

$$\cos\theta=\frac{P_n}{R},\quad \sin\theta=\frac{P_{n-1}}{R}, \tag{3.46}$$

with $\theta=0$ for $y=+\infty$ or $f(iy)=i^n R(y)e^{-i\theta(y)}$; when y decreases from $+\infty$ to $-\infty$, $\theta(y)$ increases from 0 to $n\pi$ and consequently takes once and once only the $(2n-1)$ values $\frac{\pi}{2},\pi,\frac{3\pi}{2},\dots,(2n-1)\frac{\pi}{2}$ to which correspond alternately the zeros α_j of $P_n(y)$ and β_j of $P_{n-1}(y)$. The polynomial $P_{n-1}(y)$ thus has $(n-1)$ distinct real zeros which separate the n real zeros of $P_n(y)$.

Conversely suppose that $a_1>0$ and that the polynomials $P_{n-1}(y)$, $P_n(y)$ have $n-1$ and n real, distinct zeros, the former zeros β_j separating the latter zeros α_j:

$$\alpha_n<\beta_{n-1}<\alpha_{n-1}<\cdots<\alpha_2<\beta_1<\alpha_1. \tag{3.47}$$

We may write

$$\begin{aligned} P_{n-1}&=a_1(y-\beta_1)(y-\beta_2)\cdots(y-\beta_{n-1})\\ P_n&=a_0(y-\alpha_1)(y-\alpha_2)\cdots(y-\alpha_n)\end{aligned} \tag{3.48}$$

and define $\xi(y)$ by

$$\xi=\frac{P_{n-1}(y)}{P_n(y)}. \tag{3.49}$$

For $y>\alpha_1$ the sign of $\xi(y)$ is constant, and in fact positive by (3.47), (3.48) and the assumption that a_0,a_1 are positive. Furthermore $\lim_{y\to+\infty}\xi(y)=0$ and $\xi(y)\to+\infty$ if $y\to a_1$ from the right, in such a way that $\xi(y)$ assumes every positive value of the interval $]\alpha_1,+\infty[$, at least once. In every interval $]\alpha_{j+1},\alpha_j[$, $\xi(y)$ is a continuous function of y which becomes infinite at the boundaries. More precisely

$$\xi(\alpha_{j+1}+0)=+\infty,\quad \xi(\alpha_j-0)=-\infty, \tag{3.50}$$

as can be seen from (3.47), (3.48), and the positiveness of a_0,a_1. Consequently $\xi(y)$ assumes every real value at least once over this interval. Lastly it can be verified similarly that $\xi(y)$ assumes every negative value at least once over the interval $]-\infty,\alpha_n[$. Since the equation $\frac{P_{n-1}(y)}{P_n(y)}=\lambda$, with λ real, cannot have more than n real roots, it follows from the preceding argument that the function $\xi(y)$ is monotonic in each interval of continuity and that by (3.50) it decreases when y increases. Consequently when y decreases from $+\infty$ to $-\infty$, $\theta(y)$ defined by (3.46) increases by $n\pi$.

Accordingly, for the equation $f(z)=0$ to have all its roots in the half-plane $\operatorname{Re} z<0$, it is necessary and sufficient, when a_0 is positive, that a_1 should also be positive and that the roots of $P_{n-1}(y)=0$ should all be real, distinct and should separate the roots of $P_n(y)$, which must all be real.

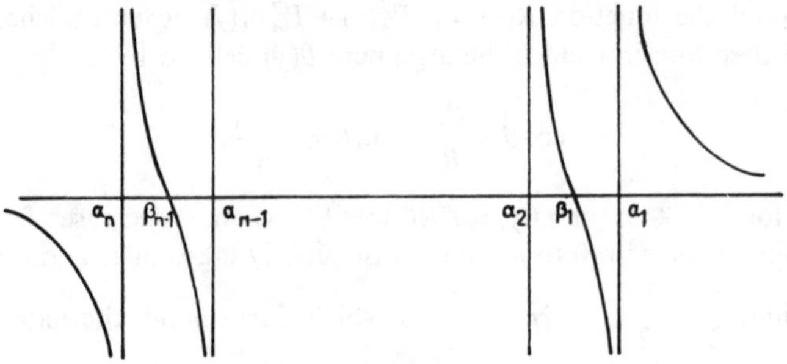

Figure 3.5

Now let $P_{n-2}(y)$ be the polynomial of degree $n-2$ defined by

(3.51) $$P_{n-2}(y) = a_0 y P_{n-1}(y) - a_1 P_n(y).$$

If the roots β_j of $P_{n-1}(y)$ are all distinct and separate the roots α_j of $P_n(y)$, it follows from

(3.52) $$\begin{aligned} P_{n-2}(\beta_{j-1}) &= -a_1 P_n(\beta_{j-1}) \\ P_{n-2}(\beta_j) &= -a_1 P_n(\beta_j) \end{aligned}$$

and from $P_n(\beta_{j-1}) \cdot P_n(\beta_j) < 0$, since $P_n(y)$ takes the value zero only once in $]\beta_j, \beta_{j-1}[$, that there must be at least one real zero of $P_{n-2}(y)$ in the interval $]\beta_j, \beta_{j-1}[$. As one obtains $n-2$ zeros in this way, and as no more can be expected, it is clear that $P_{n-2}(y)$ has $(n-2)$ distinct real zeros γ_j which separate those of $P_{n-1}(y)$:

$$\beta_{n-1} < \gamma_{n-2} < \beta_{n-2} < \cdots < \beta_2 < \gamma_1 < \beta_1.$$

Furthermore since $P_{n-2}(\beta_1) = -a_1 P_n(\beta_1)$ and $P_n(\beta_1) < 0$ by reason of $\alpha_2 < \beta_1 < \alpha_1$, it is plain that $P_{n-2}(\beta_1) > 0$, or in other words, since $\beta_1 > \gamma_1$, $P_{n-2}(y)$ is positive when $y \to +\infty$, and consequently b_0, the coefficient of y^{n-2} in $P_{n-2}(y)$ is positive.

Conversely, suppose a_0, a_1, b_0 are all positive and that the $(n-2)$ real zeros of $P_{n-2}(y)$, $\gamma_1 \cdots \gamma_{n-2}$ are distinct and separate the zeros $\beta_1 \cdots \beta_{n-1}$ of $P_{n-1}(y)$: then it can easily be shown, starting from (3.52) and $P_{n-2}(\beta_{j-1})P_{n-2}(\beta_j) < 0$ (which follows from the fact that $P_{n-2}(y)$ vanishes once at γ_{j-1} in the interval $]\beta_{j-1}, \beta_j[$) that there must be at least one zero of P_n in the same interval. Moreover, for $y = \beta_1$ we have $P_{n-2}(\beta_1) = -a_1 P_n(\beta_1)$ but because $\gamma_1 < \beta_1$, $P_{n-2}(\beta_1)$ has the sign of $P_{n-2}(y)$ for $y \to +\infty$, or in other words the sign of b_0, and is therefore positive. Thus $P_n(\beta_1) < 0$ and since it is known that $P_n(y)$ is positive for $y \to +\infty$, we deduce that there is at least one zero α_1 of $P_n(y)$ greater than β_1.

A similar argument enables us to prove that this polynomial must have at least one zero less than β_{n-1}. We can exhaust in this way the number of possible zeros of $P_n(y)$ and it can be seen that they must be separated by the zeros of $P_{n-1}(y)$.

We continue in the obvious way by defining

$$P_{n-3}(y) = a_1 y P_{n-2}(y) - b_0 P_{n-1}(y)$$

and verifying that if all the zeros of $P_{n-2}(y)$ are real and distinct and separate those of $P_{n-1}(y)$, then the zeros of $P_{n-3}(y)$ are all real and distinct and separate those of $P_{n-2}(y)$, and that furthermore the coefficient c_0 of y^{n-3} in $P_{n-3}(y)$ is positive if a_1 and b_0 are positive. Conversely the property that a_0, a_1, b_0, c_0 be positive and that the zeros of $P_{n-3}(y)$ separate those of $P_{n-2}(y)$ implies that the corresponding property is true for the pair (P_{n-2}, P_{n-1}) and then for the pair (P_{n-1}, P_n).

We see finally that we shall be led to form a sequence of polynomials $P_{n-2}, P_{n-3}, \ldots, P_0$ of degree $n-2$, $n-3, \ldots$ respectively, in which the coefficient of the term of highest degree is in each case positive. Thus at the last stage we shall have

$$P_0 = \lambda y P_1 - \mu P_2 \qquad \lambda, \mu, P_0 \text{ positive;}$$

and conversely if λ, μ, P_0 are positive, it is clear that $P_2(y)$ takes a value of opposite sign to that of P_0 when one substitutes for y the unique root of the equation $P_1(y) = 0$, or in other words it takes a negative value. But since λ, the coefficient of y^2 in $P_2(y)$, is positive, it follows that $P_2(y)$ has two zeros separated by the zero of $P_1(y)$. Thus the necessary and sufficient condition for which we have been seeking is expressed in the statement that the coefficient of the term of highest degree in each of the polynomials P_n, $P_{n-1}, \ldots, P_0$ is positive.

By way of example, we consider the case of an equation of degree 3. We have

$$\begin{aligned}
P_3(y) &= a_0 y^3 - a_2 y \\
P_2(y) &= a_1 y^2 - a_3 \\
P_1(y) &= a_0 y P_2(y) - a_1 P_3(y) = (a_1 a_2 - a_0 a_3) y = b_0 y \\
P_0 &= a_1 y P_1(y) - b_0 P_2(y) = b_0 a_3
\end{aligned}$$

whence Routh's conditions are

$$a_0 > 0, \quad a_1 > 0, \quad a_1 a_2 - a_0 a_3 > 0, \quad a_3 > 0. \tag{3.53}$$

In the case of an equation of degree 4, we have:

$$\begin{aligned}
P_4(y) &= a_0 y^4 - a_2 y^2 + a_4 \\
P_3(y) &= a_1 y^3 - a_3 y \\
P_2(y) &= a_0 y P_3(y) - a_1 P_4(y) = (a_1 a_2 - a_0 a_3) y^2 - a_1 a_4 = b_0 y^2 - a_1 a_4 \\
P_1(y) &= a_1 y P_2(y) - b_0 P_3(y) = (b_0 a_3 - a_1^2 a_4) y = c_0 y \\
P_0 &= b_0 y P_1(y) - c_0 P_2(y) = c_0 a_1 a_4
\end{aligned}$$

whence the conditions are

$$a_0 > 0, a_1 > 0, a_1 a_2 - a_0 a_3 > 0, (a_1 a_2 - a_0 a_3) a_3 - a_1^2 a_4 > 0, a_4 > 0 \tag{3.54}$$

Remark. The method described above to obtain Routh's conditions may also be applied when $f(z)$ is a polynomial with complex coefficient. We may suppose a_0 to be real and positive, which obviously involves no loss of generality, and one can still define the polynomials with real coefficients $P_n(y)$, $P_{n-1}(y)$ by $f(iy) = i^n(P_n(y) - iP_{n-1}(y))$, which for real y are polynomials of degree n and $n-1$

respectively, but which no longer enjoy the obvious parity properties which are self-evident in the case when f has real coefficients. Nevertheless the general principle still holds: we define $-P_{n-2}$ as the remainder after dividing P_n by P_{n-1}, and then iteratively $P_{n-3}, \ldots, P_0$ in the same way.

A necessary and sufficient condition for all the zeros of $f(z)$ to lie within the half-plane $\operatorname{Re} z < 0$ is that the coefficient of the term of highest degree should be positive for each of the polynomials $P_n, P_{n-1}, P_{n-2}, \ldots, P_0$.

This result can be modified if, for example, one wishes to state necessary and sufficient conditions for the zeros of $f(z)$ all to lie within the half-plane:

$$\beta - \pi < \arg z < \beta$$

with a given β. We simply express the condition that the zeros of $f(\zeta e^{(\beta - 3\pi/2)i})$ should all lie in the half-plane $\operatorname{Re}\zeta < 0$. By combining such results we could also write down necessary and sufficient conditions for the zeros of $f(z)$ all to lie within the strip $\alpha < \arg z < \beta$, where α, β are given constants satisfying $0 \leqslant \alpha < \beta < 2\pi$.

5. The Tuned Gyroscope as Part of an Inertial System for Measuring the Rate of Turn

In a gyroscopic device of traditional design, the role of the gimbals is to give freedom of orientation for the gyroscope and the motor which keeps it running by allowing them to turn freely about a point which is fixed with respect to the platform on which the instrument is mounted. The gimbals move slowly and consequently have little effect on the overall behaviour of the system, this being all the more so as their mass is usually small compared with that of the gyroscope proper.

A mechanism with very different structural design characteristics [8,9] can be achieved by fixing the motor to the platform and arranging for the gyroscope to be connected to it by a gimbal mounting with two degrees of freedom, allowing a change of attitude with respect to a frame of reference attached to the uniformly rotating shaft. This arrangement calls for at least one intermediate gimbal mounted in such a way that it turns about an axis fixed to the motor-shaft and at right angles to it. Since the motion of the intermediate gimbal has to participate in the rapid rotation of the motor-shaft, it is clear that its inertia can have an appreciable influence on the behaviour of the system. Furthermore it is possible to conceive of a device with multiple gimbals arranged in parallel, that is to say mounted on bearings whose axes are each fixed in relation to the shaft of the motor, and each orthogonal to it, but with each axis being displaced in relation to the preceding axis by the same fixed angle. Each gimbal, in addition to carrying the bearing which connects it to the motor-shaft also carries another bearing at right angles to the first around which can turn the gimbal mounting in the form of a ring or collar enveloping the mechanism. An examination of the compatibility of the kinematic constraints involved in a design of this kind in which all the axes in the equilibrium configuration lie in the same plane makes it necessary however to relax somewhat the constraints.

To build such a device with actual materials one requires, among other things, small windows to be made in the gimbal rings through which the suspension axes can pass, as shown in the diagram.

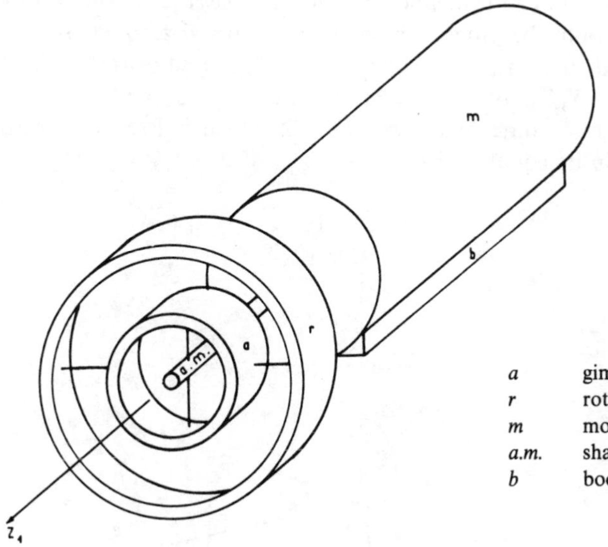

a	gimbal
r	rotor
m	motor
a.m.	shaft of motor
b	body

Figure 3.6

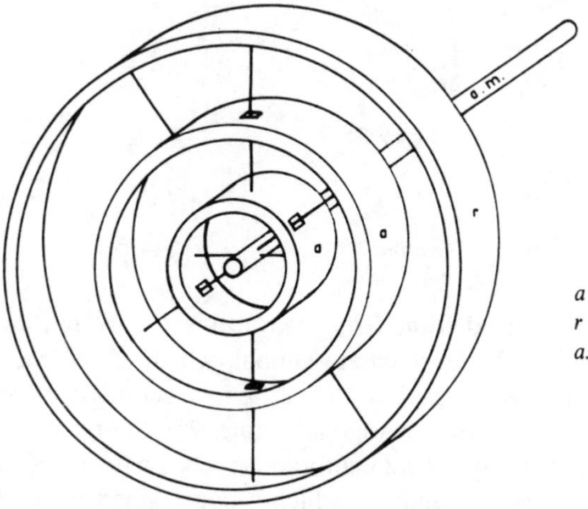

a	gimbal
r	rotor
a.m.	shaft of motor

Figure 3.7

Kinematics of the Multigimbal Suspension

a) Orientation of the Rotor

Let OZ_1 be the axis of the motor, OX_1 the inner axis of the smallest gimbal, numbered zero, and $OX_1Y_1Z_1$ the corresponding orthonormal frame of reference.

Ignoring for the moment the intermediate gimbals situated between the smallest which has just been defined and the rotor, the position of the latter with reference to the frame $OX_1Y_1Z_1$ can be defined by two variables θ_1, θ_2. The rotor effectively pivots about the outer bearing of the gimbal numbered zero whose axis denoted by OY is designed to be perpendicular to OX_1: θ_1 measures the angle of rotation which brings $OX_1Y_1Z_1$ into coincidence with OX_1YZ_2 and θ_2 the one which brings OX_1YZ_2 into coincidence with $OXYZ$ which is fixed to the rotor, with OXY contained in its equatorial plane.

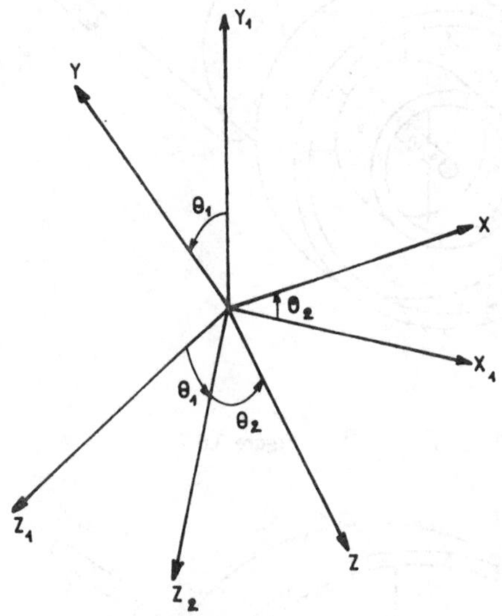

Figure 3.8

b) Co-ordinates of an Intermediate Gimbal

We write $O\xi$ for the axis derived from OX_1 by rotating it through an angle α round OZ_1 the inner axis of an intermediate gimbal, which can if necessary be identified by a suffix numbered from 1 to $N-1$; α is a geometrical parameter which stays constant. We define the orthogonal frame $O\xi\eta Z_1$ and ψ_1, the angle of rotation about $O\xi$ which brings $O\xi\eta Z_1$ into coincidence with $O\xi y\zeta$ where Oy is the outer axis of the intermediate gimbal, which is constrained by its design to lie in the equatorial plane OXY of the rotor. This arrangement can be secured by providing a circular groove lying in the equatorial plane OXY and cut in the internal face of the rotor, into which the extremities of the outer pivots of the intermediate gimbal fit and slide smoothly. In addition we define the angles of rotation ψ_2, ψ_3 by the transformation scheme:

$$O\xi\eta Z_1 \xrightarrow[\psi_1]{} O\xi y\zeta \xrightarrow[\psi_2]{} OxyZ \xrightarrow[\psi_3]{} OXYZ,$$

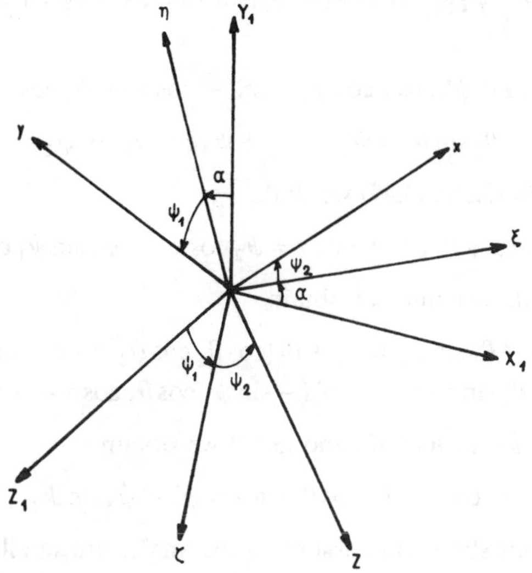

Figure 3.9

the rotation ψ_2 about Oy having the effect of bringing the axes $O\xi y$ on to Oxy in the equatorial plane of the rotor, and the rotation ψ_3 about OZ leading finally to the axes $OXYZ$ tied to the rotor.

c) *Relations Between the Parameters θ and ψ*

It is clear that the instantaneous rotation of the rotor, relative to $OX_1Y_1Z_1$ can be expressed in either of the two equivalent ways

$$\psi'_1\vec{\xi} + \psi'_2\vec{y} + \psi'_3\vec{Z} = \theta'_1\vec{X}_1 + \theta'_2\vec{Y},$$

where, as usual, the notation $\vec{\xi}, \vec{y}, \ldots$ denotes unit vectors along the axes of the same denomination. Observing that $\theta'_1\vec{X}_1 + \theta'_2\vec{Y} = \theta'_1\vec{X}_1 + \theta'_2\cos\theta_1\vec{Y}_1 + \theta'_2\sin\theta_1\vec{Z}_1$, we see that to obtain the required relations it suffices to express $\vec{\xi}, \vec{y}, \vec{Z}$, in terms of $\vec{X}_1, \vec{Y}_1, \vec{Z}_1$.

Thus we write:

$$\begin{aligned} &\vec{\xi} = \vec{X}_1\cos\alpha + \vec{Y}_1\sin\alpha, \quad \vec{\eta} = -\vec{X}_1\sin\alpha + \vec{Y}_1\cos\alpha \\ (3.55)\quad &\vec{y} = \vec{\eta}\cos\psi_1 + \vec{Z}_1\sin\psi_1 = (-\vec{X}_1\sin\alpha + \vec{Y}_1\cos\alpha)\cos\psi_1 + \vec{Z}_1\sin\psi_1 \\ &\vec{\zeta} = -\vec{\eta}\sin\psi_1 + \vec{Z}_1\cos\psi_1 = (\vec{X}_1\sin\alpha - \vec{Y}_1\cos\alpha)\sin\psi_1 + \vec{Z}_1\cos\psi_1 \end{aligned}$$

then

$$\vec{Z} = \vec{\zeta} \cos \psi_2 + \vec{\xi} \sin \psi_2 = [(\vec{X}_1 \sin \alpha - \vec{Y}_1 \cos \alpha) \sin \psi_1 + \vec{Z}_1 \cos \psi_1] \cos \psi_2 + (\vec{X}_1 \cos \alpha + \vec{Y}_1 \sin \alpha) \sin \psi_2$$

from which we deduce:

$$(3.56)\quad \theta_1' = \psi_1' \cos \alpha - \psi_2' \sin \alpha \cos \psi_1 + \psi_3'(\sin \alpha \sin \psi_1 \cos \psi_2 + \cos \alpha \sin \psi_2)$$

(3.57)

$$\theta_2' \cos \theta_1 = \psi_1' \sin \alpha + \psi_2' \cos \alpha \cos \psi_1 + \psi_3'(-\cos \alpha \sin \psi_1 \cos \psi_2 + \sin \alpha \sin \psi_2)$$

$$(3.58)\quad \theta_2' \sin \theta_1 = \psi_2' \sin \psi_1 + \psi_3' \cos \psi_1 \cos \psi_2$$

Eliminating ψ_1' from (3.56), (3.57) we find:

$$-\theta_1' \sin \alpha + \theta_2' \cos \theta_1 \cos \alpha = \psi_2' \cos \psi_1 - \psi_3' \sin \psi_1 \cos \psi_2$$

and taking (3.58) into account we obtain

$$(3.59)\quad \begin{aligned} \psi_2' &= -\theta_1' \sin \alpha \cos \psi_1 + \theta_2'(\cos \theta_1 \cos \psi_1 \cos \alpha + \sin \theta_1 \sin \psi_1), \\ \psi_3' \cos \psi_2 &= \theta_1' \sin \alpha \sin \psi_1 + \theta_2'(-\sin \psi_1 \cos \theta_1 \cos \alpha + \sin \theta_1 \cos \psi_1), \end{aligned}$$

finally, eliminating ψ_2' from (3.56) and (3.57) we obtain

$$(3.60)\quad \theta_1' \cos \alpha + \theta_2' \cos \theta_1 \sin \alpha = \psi_1' + \psi_3' \sin \psi_2$$

These calculations show that, assuming the angles are small ($\theta_1, \theta_2, \psi_1, \psi_2, \psi_3$ and their derivatives are by hypothesis small quantities) ψ_3' is of second order and consequently can be neglected in the context of the linearised theory which we shall develop.

Accordingly, on linearising (3.59), (3.60) we obtain

$$(3.61)\quad \begin{aligned} \psi_1' &= \theta_1' \cos \alpha + \theta_2' \sin \alpha \\ \psi_2' &= -\theta_1' \sin \alpha + \theta_2' \cos \alpha \end{aligned}$$

The Equations of Motion

We write $O\tilde{X}\tilde{Y}Z_1$ for an orthonormal frame of reference tied to the motor platform: denoting the assumed constant rotational speed of the motor by n, we have $(O\tilde{X}, OX_1) = nt$.

The instrument is designed to detect the rotation imposed on the vehicle on which it is installed. Now any instantaneous rotation whatsoever of the body with respect to a frame of reference $\mathscr{R}$ can always be decomposed into a rotation borne by the axis OZ_1 and another in the plane $O\tilde{X}\tilde{Y}$; the first will have no effect on the system, being in any case negligible compared with n, so that we shall confine ourselves to examining the effects of the second which, incidentally, may without inconvenience be assumed to be a rotation about the axis $O\tilde{X}$, say $\Omega\tilde{X}$. Under these conditions the instantaneous rotation with respect to $\mathscr{R}$ of an arbitrary gimbal can be represented by:

$$n\vec{Z}_1 + \Omega\vec{\tilde{X}} + \psi'_1\vec{\xi} = \omega_1\vec{\xi} + \omega_2\vec{y} + \omega_3\vec{\zeta}$$

with

$$\omega_1 = \psi'_1 + \Omega\cos(nt+\alpha)$$

(3.62)
$$\omega_2 = n\psi_1 - \Omega\sin(nt+\alpha)$$

$$\omega_3 = n$$

after linearisation, and considering Ω as a first order quantity. We deduce from this that the angular momentum at O of the gimbal is $a\omega_1\vec{\xi} + b\omega_2\vec{y} + c\omega_3\vec{\zeta}$ with a, b, c being its principal moments of inertia.

We write T_1 for the moment about $O\xi$ of the forces exerted by the rotor on the outer bearing Oy of the gimbal. We shall assume also that the inner bearing includes a restoring mechanism opposing the motion described by the co-ordinate ψ_1, which has the effect of exerting on the gimbal a couple of moment $-k_1\psi_1$ about the axis, where k_1 is a constant.

The Euler equation for the axis $O\xi$ can be written, for any one of the gimbals

(3.63)
$$a\frac{d\omega_1}{dt} + (c-b)\omega_2\omega_3 = -k_1\psi_1 + T_1$$

that is

(3.64)
$$a\psi''_1 + (k_1 + n^2(c-b))\psi_1 + n(b-a-c)\Omega\sin(nt+\alpha) = T_1.$$

We now consider the rotor whose instantaneous angular velocity is

(3.65)
$$\begin{aligned}&\theta'_1\vec{X}_1 + \theta'_2\vec{Y} + n\vec{Z}_1 + \Omega\vec{X}\\ &= (\theta'_1 + \Omega\cos nt)\vec{X}_1 - \Omega\sin nt\,\vec{Y}_1 + \theta'_2\vec{Y} + n\vec{Z}_1 = n_1\vec{X} + n_2\vec{Y} + n_3\vec{Z}.\end{aligned}$$

We can easily work out the components n_1, n_2, n_3, noting that

(3.66)
$$\begin{aligned}\vec{X}_1 &= \vec{X}\cos\theta_2 + \vec{Z}\sin\theta_2\\ \vec{Y}_1 &= \vec{Y}\cos\theta_1 - (\vec{Z}\cos\theta_2 - \vec{X}\sin\theta_2)\sin\theta_1\\ \vec{Z}_1 &= \vec{Y}\sin\theta_1 + (\vec{Z}\cos\theta_2 - \vec{X}\sin\theta_2)\cos\theta_1.\end{aligned}$$

Thus, after linearisation, we find:

(3.67)
$$\begin{aligned}n_1 &= \theta'_1 - n\theta_2 + \Omega\cos nt\\ n_2 &= \theta'_2 + n\theta_1 - \Omega\sin nt\\ n_3 &= n\end{aligned}$$

and the angular momentum at O of the rotor

(3.68)
$$\vec{H} = An_1\vec{X} + Bn_2\vec{Y} + Cn_3\vec{Z}$$

A, B, C being the principal moments of inertia ($A = B$).

On the other hand, we can represent by $-T_1\vec{\xi} - T_2\vec{y} - T_3\vec{\zeta}$ the moment at O of the forces exerted on the rotor by any particular one of the gimbals, T_1 having the meaning already mentioned and

(3.69)
$$T_2 = k_2\psi_2$$

if we assume that by means of some suitable device this gimbal exerts on the rotor a restoring couple of moment $-k_2\psi_2$ about Oy where k_2 is a positive constant.

To pass from the axes $O\xi y\zeta$ to the system $OXYZ$, we can form the product of the two transformations (3.55), (3.66), which leads to the equation

$$(3.70)\qquad -T_1\vec{\xi}-T_2\vec{y}-T_3\vec{\zeta}=-T_1(\vec{X}\cos\alpha+\vec{Y}\sin\alpha) \\ -T_2(-\vec{X}\sin\alpha+\vec{Y}\cos\alpha)-T_3\vec{Z}.$$

Applying the angular momentum theorem to the rotor at O, in terms of the projections on the axes OX, OY, we obtain, having regard to (3.67), (3.68), (3.70):

$$(3.71)\qquad A\frac{dn_1}{dt}+(C-B)n_2n_3=-\Sigma(T_1\cos\alpha-T_2\sin\alpha)$$

$$B\frac{dn_2}{dt}+(A-C)n_3n_1=-\Sigma(T_1\sin\alpha+T_2\cos\alpha)$$

the equations of motion in a concrete form. The summation signs here refer to the sum over the various gimbals which constitute the complete system. To avoid overloading the formulae we have not thought it necessary to indicate explicitly the suffixes relating to each gimbal $\alpha, a, b, c, k_1, k_2, \psi_1, \psi_2, T_1, T_2$ but it is to be understood that each Σ sign implies a summation of the corresponding summands for all gimbals.

Expressing T_1, T_2 with the help of (3.64), (3.69) we can write (3.71) having regard to (3.67)

$$(3.72)\qquad A(\theta_1''-n\theta_2'-n\Omega\sin\Omega t)+n(C-B)(\theta_2'+n\theta_1-\Omega\sin nt) \\ =-\Sigma[a\psi_1''+(k_1+n^2(c-b))\psi_1 \\ +n(b-a-c)\Omega\sin(nt+\alpha)]\cos\alpha+\Sigma k_2\psi_2\sin\alpha$$

$$(3.73)\qquad B(\theta_2''+n\theta_1'-n\Omega\cos\Omega t)+n(A-C)(\theta_1'-n\theta_2+\Omega\cos nt) \\ =-\Sigma[a\psi_1''+(k_1+n^2(c-b))\psi_1 \\ +n(b-a-c)\Omega\sin(nt+\alpha)]\sin\alpha-\Sigma k_2\psi_2\cos\alpha,$$

and lastly expressing ψ_1, ψ_2 in terms of θ_1, θ_2 by (3.61), integrating (3.61):

$$\psi_1=\theta_1\cos\alpha+\theta_2\sin\alpha \\ \psi_2=-\theta_1\sin\alpha+\theta_2\cos\alpha$$

we can write (3.72) as:

$$(3.74)\qquad (A+\Sigma a\cos^2\alpha)\theta_1''+(\Sigma a\sin\alpha\cos\alpha)\theta_2''-n(A+B-C)\theta_2' \\ +\{n^2(C-B)+\Sigma[k_1\cos^2\alpha+k_2\sin^2\alpha+n^2(c-b)\cos^2\alpha]\}\theta_1 \\ +\{\Sigma[k_1-k_2+n^2(c-b)]\sin\alpha\cos\alpha\}\theta_2 \\ +n\Omega[B-C-A+\Sigma(b-c-a)\cos^2\alpha]\sin nt \\ +n\Omega\Sigma(b-a-c)\sin\alpha\cos\alpha\cdot\cos nt=0.$$

Operating in a similar fashion for (3.73) we obtain:

$$(3.75)\qquad (\Sigma a \sin\alpha\cos\alpha)\theta_1'' + (B + \Sigma a \sin^2\alpha)\theta_2'' + n(A + B - C)\theta_1' \\ + \{\Sigma[k_1 + n^2(c-b) - k_2]\sin\alpha\cos\alpha\}\theta_1 \\ + \{n^2(C-A) + \Sigma[k_1\sin^2\alpha + k_2\cos^2\alpha + n^2(c-b)\sin^2\alpha]\}\theta_2 \\ + n\Omega\Sigma(b-a-c)\sin\alpha\cos\alpha\cdot\sin nt \\ + n\Omega[A - C - B + \Sigma(b-a-c)\sin^2\alpha]\cos nt = 0$$

We can combine (3.74), (3.75) in the matrix form

$$(3.76)\qquad F\theta'' + n(A+B-C)G\theta' + E\theta + n\Omega(L - FG)U = 0$$

where

$$(3.77)\qquad F = \begin{pmatrix} A + \Sigma a\cos^2\alpha & \Sigma a\sin\alpha\cos\alpha \\ \Sigma a\sin\alpha\cos\alpha & B + \Sigma a\sin^2\alpha \end{pmatrix},$$

$$(3.78)\qquad G = \begin{pmatrix} 0 & -1 \\ 1 & 0 \end{pmatrix}$$

$$E = \begin{pmatrix} \Sigma(k_1\cos^2\alpha + k_2\sin^2\alpha) + n^2[C - B + \Sigma(c-b)\cos^2\alpha], & \Sigma[k_1 - k_2 + n^2(c-b)]\sin\alpha\cos\alpha \\ \Sigma[k_1 - k_2 + n^2(c-b)]\sin\alpha\cos\alpha, & \Sigma(k_1\sin^2\alpha + k_2\cos^2\alpha) + n^2[C - A + \Sigma(c-b)\sin^2\alpha] \end{pmatrix}$$

$$(3.79)\qquad L = \begin{pmatrix} -\Sigma(c-b)\sin\alpha\cos\alpha & C - B + \Sigma(c-b)\cos^2\alpha \\ A - C - \Sigma(c-b)\sin^2\alpha & \Sigma(c-b)\sin\alpha\cos\alpha \end{pmatrix},$$

$$U = \begin{pmatrix} \cos nt \\ -\sin nt \end{pmatrix},\quad \theta = \begin{pmatrix} \theta_1 \\ \theta_2 \end{pmatrix}.$$

Inclusion of Damping Terms in the Equations of Motion

No matter how carefully they are designed, the rotor and gimbals, because of their high rotational speeds, are subject to the effect of frictional forces which, however attenuated they may be, can still cause instability. These forces have a moment at O which to a first approximation is proportional to the rotation with respect to the body of the instrument. In addition to this viscosity plays a part in the restoring mechanisms.

To take account of these effects a term $-f_1\cdot\psi_1'$ has to be added to the right-hand side of (3.63) and similarly a term $f_2\cdot\psi_2'$ to the right-hand side of (3.69), f_1 and f_2 being positive dissipation coefficients.

The modifications required in the final equations will be to add terms of the same nature as those involving the stiffness, but with $f_1\dfrac{\mathrm{d}}{\mathrm{d}t}$, $f_2\dfrac{\mathrm{d}}{\mathrm{d}t}$ playing the same role as the operators k_1, k_2 respectively.

The moment at O of the dissipative forces acting on the rotor has components along the axes OX, OY which can be represented by $-D(\theta_1' - n\theta_2)$, $-D(\theta_2' + n\theta_1)$, where $D > 0$.

Thus the additional terms which have to be added to the left-hand sides of the equations (3.74), (3.75) may be written:

$$D(\theta_1' - n\theta_2) + \Sigma(f_1 \cos^2\alpha + f_2 \sin^2\alpha)\cdot\theta_1' + \Sigma(f_1 - f_2)\sin\alpha\cos\alpha\cdot\theta_2'$$
$$D(\theta_2' + n\theta_1) + \Sigma(f_1 - f_2)\sin\alpha\cos\alpha\cdot\theta_1' + \Sigma(f_1 \sin^2\alpha + f_2\cos^2\alpha)\cdot\theta_2'.$$

Finally, we may possibly take account of the derivative $\Omega' = \frac{d\Omega}{dt}$, (where this is assumed to be known) of the rotation imposed on the system. We simply have to add to the left-hand side of (3.64) $a\Omega'\cos(nt+\alpha)$ and of (3.71) $A\Omega'\cos nt$, $-B\Omega'\sin nt$ respectively. This operation brings about the appearance in (3.74), (3.75) of the expressions

$$\Omega'\cdot\Sigma a\cos(nt+\alpha)\cos\alpha + A\Omega'\cos nt$$
$$\Omega'\cdot\Sigma a\cos(nt+\alpha)\sin\alpha - B\Omega'\sin nt$$

which are the components of a vector which can be written

$$\Omega'\cdot\begin{pmatrix} A+\Sigma a\cos^2\alpha & \Sigma a\sin\alpha\cos\alpha \\ \Sigma a\sin\alpha\cos\alpha & B+\Sigma a\sin^2\alpha \end{pmatrix}\begin{pmatrix} \cos nt \\ -\sin nt \end{pmatrix} = \Omega' F U.$$

Finally, with

$$\Delta = \begin{pmatrix} \Sigma(f_1\cos^2\alpha + f_2\sin^2\alpha) & \Sigma(f_1 - f_2)\sin\alpha\cos\alpha \\ \Sigma(f_1 - f_2)\sin\alpha\cos\alpha & \Sigma(f_1\sin^2\alpha + f_2\cos^2\alpha) \end{pmatrix},$$
$$\mathscr{I} = A + B - C,$$

we obtain the equation of motion in the matrix form

$$F\theta'' + (n\mathscr{I}G + DI + \Delta)\theta' + (nDG + E)\theta + [\Omega'F + n\Omega(L - FG)]U = 0. \tag{3.80}$$

Dynamic Stability. Undamped System

Dynamic stability of the system, that is to say stability of the solution $\theta = 0$ of

$$F\theta'' + n(A + B - C)G\theta' + E\theta = 0 \tag{3.81}$$

is achieved if the symmetric matrix E is positive definite. With $E = H + n^2K$

$$H = \begin{pmatrix} \Sigma(k_1\cos^2\alpha + k_2\sin^2\alpha) & \Sigma(k_1 - k_2)\sin\alpha\cos\alpha \\ \Sigma(k_1 - k_2)\sin\alpha\cos\alpha & \Sigma(k_1\sin^2\alpha + k_2\cos^2\alpha) \end{pmatrix}$$

$$K = \begin{pmatrix} C - B + \Sigma(c-b)\cos^2\alpha & \Sigma(c-b)\sin\alpha\cos\alpha \\ \Sigma(c-b)\sin\alpha\cos\alpha & C - A + \Sigma(c-b)\sin^2\alpha \end{pmatrix}$$

it will be noted that H is positive definite because its diagonal elements as well as k_1 and k_2 are positive, and a simple calculation shows that

$$\det H = \sum_{j<1} k_1^{(j)}k_1^{(l)}\sin^2(\alpha^{(l)} - \alpha^{(j)}) + \sum_{j<l} k_2^{(j)}k_2^{(l)}\sin^2(\alpha^{(l)} - \alpha^{(j)}) + \sum_{j,l} k_1^{(j)}k_2^{(l)}\cos^2(\alpha^{(j)} - \alpha^{(l)}) > 0 \tag{3.82}$$

the upper indices referring to the number of the gimbal. It suffices therefore that K be positive definite for there to be stability at all speeds of rotation.

In general we shall impose the conditions

$$C - B + \Sigma(c-b)\cos^2\alpha > 0$$
$$C - A + \Sigma(c-b)\sin^2\alpha > 0 \qquad (3.83)$$

which are sufficient, for example, in the case of a two-gimbal system with $\alpha^{(1)} = 0$, $\alpha^{(2)} = \frac{\pi}{2}$.

However in the case where there is damping defined by (3.80) instability may appear for high values of n, and it will be necessary to introduce a servomechanism to obtain stable operation.

Frequencies of Vibrations of the Free Rotor

Let us consider the motion of the free rotor, i.e. the motion in the absence of gimbals and of restoring springs.

The eigenfrequencies σ are the (real) roots of the equation

$$\det(-F\sigma^2 + in(A+B-C)G\sigma + E) = 0$$

where

$$F = \begin{pmatrix} A & 0 \\ 0 & B \end{pmatrix} \quad E = \begin{pmatrix} n^2(C-B) & 0 \\ 0 & n^2(C-A) \end{pmatrix}$$

and one thus finds, under the stability hypotheses $C-B>0$, $C-A>0$:

$$\sigma_1 = n \quad \sigma_2 = n\sqrt{\frac{(C-A)(C-B)}{AB}}.$$

Noting that $A+B>C$, we find that

$$\sigma_1 = n < \sigma_2 = n\sqrt{\frac{(C-A)(C-B)}{AB}}.$$

Motion of the Free Rotor

Retaining the previous hypothesis, we examine the motion of the rotor which is governed by the equations:

$$A\theta_1'' - n(A+B-C)\theta_2' + n^2(C-B)\theta_1 - n\Omega(A+C-B)\sin nt = 0$$
$$B\theta_2'' + n(A+B-C)\theta_1' + n^2(C-A)\theta_2 + n\Omega(A-B-C)\cos nt = 0. \qquad (3.84)$$

Since n is a frequency of natural vibrations, we have here a case of resonance and it is easily verified that

$$\theta_1 = -\Omega t\cos nt, \quad \theta_2 = \Omega t\sin nt$$

represents the forced vibration.

If we define the axis of the rotor in relation to the frame $O\tilde{X}\tilde{Y}Z_1$ attached to the body by the angles φ_1, φ_2, in the same way as we introduced the angles θ_1, θ_2, relative to $OX_1Y_1Z_1$, we see that the corresponding unit vector may be written

$$\cos\theta_2\cos\theta_1\vec{Z}_1+\sin\theta_2\vec{X}_1-\cos\theta_2\sin\theta_1\vec{Y}_1$$

i.e. after linearisation $\theta_2\vec{X}_1-\theta_1\vec{Y}_1+\vec{Z}_1$, or equivalently $\varphi_2\vec{X}-\varphi_1\vec{Y}+\vec{Z}_1$. Consequently, since

$$\vec{X}_1=\vec{\tilde{X}}\cos nt+\vec{\tilde{Y}}\sin nt$$
$$\vec{Y}_1=-\vec{\tilde{X}}\sin nt+\vec{\tilde{Y}}\cos nt$$

we obtain

$$\varphi_1=\theta_1\cos nt-\theta_2\sin nt,\quad \varphi_2=\theta_1\sin nt+\theta_2\cos nt.$$

In the case of the free rotor, the forced motion is therefore described, relatively to the axes $O\tilde{X}\tilde{Y}Z_1$ fixed to the body by

$$\varphi_1=-\Omega t,\quad \varphi_2=0.$$

Hence the rotation undergone by the rotor relative to the body is the reverse of the rotation imposed by the latter on the former.

Case of a Multigimbal System Without Damping. The Tune Condition

To hope to have an analogous situation in a real case it would be desirable to find ourselves in the resonance case for the differential system (3.74), (3.75). In other words n should be a natural frequency of the free oscillations. We are therefore led to write

$$\det(-Fn^2+in^2(A+B-C)G+E)=0$$

with F, G, E the matrices defined by (3.77), (3.78), i.e.

$$(3.85)\quad \det\begin{pmatrix}\Sigma(k_1\cos^2\alpha+k_2\sin^2\alpha)+n^2[C-B-A+\Sigma(c-b-a)\cos^2\alpha], \\ \Sigma[k_1-k_2+n^2(c-b-a)]\sin\alpha\cos\alpha-in^2(A+B-C), \\ \Sigma[k_1-k_2+n^2(c-b-a)]\sin\alpha\cos\alpha+in^2(A+B-C), \\ \Sigma(k_1\sin^2\alpha+k_2\cos^2\alpha)+n^2[C-B-A+\Sigma(c-b-a)\sin^2\alpha]\end{pmatrix}=0.$$

We write $\mu=a+b-c$; $\mathscr{I}=A+B-C$

$$\lambda_1=\Sigma(k_1+k_2-\mu n^2),\ \lambda_2=\Sigma(k_1-k_2-\mu n^2)\cos 2\alpha,$$
$$\lambda_3=\Sigma(k_1-k_2-\mu n^2)\sin 2\alpha$$

so that we can write (3.85) as:

$$(3.86)\qquad \det\begin{pmatrix}\lambda_1+\lambda_2-2n^2\mathscr{I} & \lambda_3-2in^2\mathscr{I} \\ \lambda_3+2in^2\mathscr{I} & \lambda_1-\lambda_2-2n^2\mathscr{I}\end{pmatrix}=0$$

or

$$4n^2\mathscr{I}\lambda_1-\lambda_1^2+\lambda_2^2+\lambda_3^2=0.$$

Examination of the Two-Gimbal System

We define the values α, $\alpha^{(1)}=0, \alpha^{(2)}=\frac{\pi}{2}$ and to satisfy the tuning condition (3.86), we choose the rigidities and rotational speed n so that

$$\lambda_1=\lambda_2=\lambda_3=0.$$

Explicitly, this means choosing the parameters so that

$$\begin{aligned}k_1^{(1)}+k_2^{(1)}+k_1^{(2)}+k_2^{(2)}&=n^2(\mu^{(1)}+\mu^{(2)})\\ k_1^{(1)}-k_2^{(1)}-k_1^{(2)}+k_2^{(2)}&=n^2(\mu^{(1)}-\mu^{(2)})\end{aligned} \tag{3.87}$$

or

$$n^2=\frac{k_1^{(1)}+k_2^{(2)}}{\mu^{(1)}}=\frac{k_1^{(2)}+k_2^{(1)}}{\mu^{(2)}}$$

which expresses a condition which needs to be satisfied between the inertia and rigidities of the restoring springs of the two gimbals, and also gives the speed of rotation of the tuned system. The equations of motion in this case can be written as:

$$\begin{aligned}&(A+a^{(1)})\theta_1''-n\mathscr{I}\theta_2'+[-n^2\mathscr{I}+n^2(A+a^{(1)})]\theta_1\\ &\quad+n\Omega[B-C-A+b^{(1)}-c^{(1)}-a^{(1)}]\sin nt=0\\ &(B+a^{(2)})\theta_2''+n\mathscr{I}\theta_2'+[-n^2\mathscr{I}+n^2(B+a^{(2)})]\theta_2\\ &\quad+n\Omega[A-C-B+b^{(2)}-c^{(2)}-a^{(2)}]\cos nt=0\end{aligned}$$

and we can look for a forced vibration in the form:

$$\begin{aligned}\theta_1&=pt\cos nt-q\sin nt\\ \theta_2&=-pt\sin nt-q\cos nt\end{aligned}$$

which leads to

$$\begin{aligned}(A+C-B+2a^{(1)})p-2nq\mathscr{I}+\Omega(A+C-B+a^{(1)}+c^{(1)}-b^{(1)})&=0\\ (B+C-A+2a^{(2)})p+2qn\mathscr{I}+\Omega(B+C-A+a^{(2)}+c^{(2)}-b^{(2)})&=0\end{aligned}$$

which has the solution:

$$\begin{aligned}p&=-\Omega\left(1-\frac{\mu^{(1)}+\mu^{(2)}}{2C+2(a^{(1)}+a^{(2)})}\right)\\ q&=\frac{\Omega}{4n}\,\frac{\mu^{(2)}(A+C-B+2a^{(1)})-\mu^{(1)}(B+C-A+2a^{(2)})}{A+a^{(1)}+a^{(2)}}\end{aligned}$$

Now we can suppose the system to be designed in such a way that the supplementary condition:

$$\mu^{(2)}(A+C-B+2a^{(1)})-\mu^{(1)}(B+C-A+2a^{(2)})=0 \tag{3.88}$$

is satisfied. It should be noted that this condition involves only the inertias.

Thus, under the conditions (3.87), (3.88), the forced motion relative to the frame $O\tilde{X}\tilde{Y}Z_1$ is given by:

$$\varphi_1 = -\Omega\left(1 - \frac{\mu^{(1)} + \mu^{(2)}}{2C + 2a^{(1)} + 2a^{(2)}}\right)t \tag{3.89}$$

$$\varphi_2 = 0$$

The motion of the rotor axis in relation to the platform on which it is mounted thus depends on Ω in a simple manner.

Chapter IV. Stability of Systems Governed by the Linear Approximation

In the context of the linearised theory, investigation of the stability of equilibrium of a holonomic mechanical system having n degrees of freedom reduces to an examination of the asymptotic behaviour, as $t \to +\infty$, of the vector solution of a matrix differential equation of the second order.

Thus it is natural to introduce an inertia matrix which operates on the acceleration vector; dissipation and gyroscopic effect matrices applied to the velocity vector; and finally a stiffness matrix, whose symmetrical part corresponds to the conservative forces.

We consider a mechanical system having n degrees of freedom whose configuration is described by the vector co-ordinate $q = \{q_1, \ldots, q_n\}$ and whose linearised equation of motion in the neighbourhood of $q = 0$, an assumed equilibrium position, is, written in matrix form:

$$Aq'' + (\xi\Gamma + D)q' + (K + E)q = 0 \tag{4.1}$$

where A, Γ, D, K, E are real $n \times n$ matrices with constant elements and $\xi \in \mathbb{R}$ is a parameter. The scalar product of two vectors u, v is defined in the usual way as $(u, v) = \sum_{j=1}^{n} u_j \bar{v}_j$, where $\bar{v}_j$ is the value conjugate to v_j.

The mechanical interpretation of the matrices in (4.1) is clear enough: A is the inertia matrix which we shall suppose to be symmetric and positive-definite, D the matrix of the dissipative forces is symmetric and positive, K and E which are respectively symmetric and skew-symmetric are obtained as the result of decomposing the elastic forces acting on the system firstly into forces derived from a potential or conservative forces and secondly into other forces. Lastly the skew-symmetric matrix Γ is associated with the gyroscopic forces, and the parameter ξ depends on and is indicative of the rapid rotation of the gyroscope(s) in the system.

Thus

$$\begin{aligned} &A = A^T \quad (Au, v) = (u, Av), \forall u, v; (Au, u) > 0, \forall u \neq 0 \\ &D = D^T \quad (Du, u) \geqslant 0 \\ &\Gamma = -\Gamma^T \quad (\Gamma u, v) = -(u, \Gamma v) \\ &E = -E^T \\ &K = K^T. \end{aligned} \tag{4.2}$$

We propose to discuss the stability of the solution $q = 0$ of (4.1) under various hypotheses in regard to the matrices A, Γ, D, K, E.

Discussion of the Equation

(4.3)
$$Aq'' + \xi \Gamma q' = 0.$$

Theorem 1. The solution $q = q_0$, (q_0 constant) is stable if and only if $\det \Gamma \neq 0$.

Let us first observe that by (4.2)

$$\det \Gamma = \det(-\Gamma^T) = (-1)^n \det \Gamma^T = (-1)^n \det \Gamma;$$

the condition $\det \Gamma \neq 0$ implies that n must be even.

The condition $\det \Gamma \neq 0$ is necessary for stability because otherwise there would be a $v_0 \neq 0$ such that $\Gamma v_0 = 0$ and then $q = v_0 t + \tilde{q}_0$ would be a solution of (4.3) for which, no matter how small were the norms of $q(0) - q_0 = \tilde{q}_0 - q_0$ and $q'(0) = v_0$, q would not remain in a prescribed neighbourhood of $(q_0, 0)$ for all $t > 0$.

The condition $\det \Gamma \neq 0$ is sufficient; setting $q' = v$ in (4.3) that is $Av' + \xi \Gamma v = 0$, we can write down the characteristic equation:

(4.4)
$$\det(\omega A + \xi \Gamma) = 0.$$

The roots ω of (4.4) are non-zero since $\xi \det \Gamma \neq 0$, and we may associate with each ω an eigenvector v satisfying $\omega A v + \xi \Gamma v = 0$, so that, after scalar multiplication by v, we have

$$\omega(Av, v) + \xi(\Gamma v, v) = 0.$$

As ω may have a complex value, the same may be true of v but at all events (Av, v) is real, and furthermore non-zero since $v \neq 0$, while

$$\overline{(\Gamma v, v)} = -(v, \Gamma v) = -(\Gamma v, v) \neq 0 \quad \text{since } \omega \neq 0.$$

Thus ω is a pure imaginary. Finally, as every real solution of (4.3) satisfies $(Aq'', q') = 0$ or, by integration $(Aq', q') = \text{const.}$ we deduce (since A is positive definite) that q' is bounded in time. This property, taken in conjunction with the fact that all the roots of the characteristic equation have purely imaginary values, implies that the equilibrium $q = 0$ is stable.

Discussion of the Equation

(4.5)
$$Aq'' + \xi \Gamma q' + Kq = 0.$$

Theorem 2.

1. If $\det K < 0$, $q = 0$ is unstable.

2. If K is positive definite or negative definite, the solution $q = 0$ of

(4.6)
$$\xi \Gamma q' + Kq = 0 \text{ is stable.}$$

3. If $\det \Gamma \neq 0$, if the solution $q = 0$ of (4.6) is stable, and if the characteristic roots of (4.6) and of

(4.7)
$$Aq' + \xi \Gamma q = 0$$

are distinct, then for large enough $|\xi|$, the solution $q=0$ of (4.5) is stable, and the characteristic roots of that equation, all pure imaginaries, fall into the following two groups: there are n roots of the order of ξ^{-1} whose products by ξ tend towards the roots of $\det(\sigma\Gamma+K)=0$ as $|\xi|\to\infty$, and there are n other roots of the order of ξ whose quotients by ξ tend towards the roots of $\det(\sigma A+\Gamma)=0$.

In particular it will be seen from 2. and 3. that under the hypotheses K negative definite, $\det\Gamma\neq 0$, the characteristic roots of (4.6) and (4.7) are simple, then for large enough $|\xi|$ the solution $q=0$ of (4.5) is stable (gyroscopic stabilisation).

1. The characteristic equation associated with (4.5) is

$$F(\omega)=\det(\omega^2 A+\xi\omega\Gamma+K)=0 \tag{4.8}$$

If $\det K<0$, we have $F(0)<0$ and since $F(\omega)\sim\omega^{2n}\det A>0$ when $\omega\to+\infty$ the equation (4.8) has at least one positive real root, which implies the instability of $q=0$ and justifies conclusion 1.

2. For every real solution $q(t)$ of (4.6), we have $(\Gamma q', q')=0$ and hence $(Kq,q')=0$, that is after integration $(Kq,q)=$ const. which implies, since K is positive definite or negative definite, that $q=0$ is a stable solution of (4.6).

3. The characteristic polynomial of (4.6)

$$\det(\xi\omega\Gamma+K)=\det(\xi\omega\Gamma^T+K^T)=\det(-\xi\omega\Gamma+K)$$

is an even function of ω. Since, by hypothesis, its roots are simple $\omega=0$ cannot be a root, which implies that $\det K\neq 0$. Furthermore as the roots occur in pairs (of opposite signs) the hypothesis of stability entails that they must be pure imaginaries.

Writing $\sigma=\xi\omega$, where ω is a characteristic root of (4.6) we have, since σ is simple

$$\det(\sigma\Gamma+K)=0 \quad\text{and}\quad \frac{d}{d\sigma}\det(\sigma\Gamma+K)\neq 0,$$

or,

$$\sum_{ij}\gamma_{ij}(\sigma\Gamma+K)^{ij}\neq 0, \tag{4.9}$$

denoting by $(\sigma\Gamma+K)^{ij}$ the cofactor of the (i,j)th element of the determinant of $\sigma\Gamma+K$.

The inequality (4.9) proves that at least one of these cofactors is not null and that consequently the equation

$$(\sigma\Gamma+K)u=0$$

defines the vector u to within a homothetic transformation, or uniquely if the condition be imposed that it should have unit norm.

Moreover, because of the equation

$$\sum_j(\sigma\Gamma+K)_{ij}(\sigma\Gamma+K)^{lj}=0$$

it is clear that $(\sigma\Gamma+K)^{lj}=v_l\cdot u_j$ and similarly, we can deduce from

$$\sum_i(\sigma\Gamma+K)_{ij}(\sigma\Gamma+K)^{ik}=\sum_i\overline{(\sigma\Gamma+K)_{ji}}(\sigma\Gamma+K)^{ik}=0,$$

remembering that σ is purely imaginary and Γ skew-symmetric,

$$(\sigma\Gamma+K)^{ik}=w_k\cdot\bar{u}_i$$

from which it follows that $v_l u_j=w_j\bar{u}_l$.

Now at least one of the coefficients $u_j, j=1,2,\dots,n$, must be non-zero, and we may suppose it to be u_1 without loss of generality. This enables us to write:

$$v_l=\bar{u}_l\frac{w_1}{u_1}.$$

Accordingly we can define a number $\varkappa$ such that

$$v_l=\varkappa\bar{u}_l \quad\text{and}\quad (\sigma\Gamma+K)^{lj}=\varkappa\bar{u}_l u_j,$$

and the inequality (4.9) becomes equivalent to

$$\varkappa\sum_{ij}\gamma_{ij}u_j\bar{u}_i=\varkappa(\Gamma u,u)\neq 0. \tag{4.10}$$

Let us return to (4.5) whose characteristic equation is

$$\det(\omega^2A+\xi\omega\Gamma+K)=0. \tag{4.11}$$

Because of the invariance of this equation under the change from ω to $-\omega$, due to the skew-symmetry of Γ and the symmetry of A and K, it is clear that a necessary condition for stability is that the roots of (4.11) should all be pure imaginaries.

With $\xi\omega=\sigma$, $\xi=\varepsilon^{-1}$ we can rewrite (4.11) as:

$$\det((\varepsilon\sigma)^2A+\sigma\Gamma+K)=0 \tag{4.12}$$

and we see that when $\varepsilon\to 0$, n roots of this equation become infinite whereas the other n tend towards the roots, all distinct, of

$$\det(\sigma\Gamma+K)=0. \tag{4.13}$$

It is on these latter that we shall first focus our attention. If σ_0 is a (simple) root of (4.13) ($\sigma_0\neq 0$ since $\det K\neq 0$), there exists a root σ of (4.12), simple for small enough $|\varepsilon|$, whose limit is σ_0 as $\varepsilon\to 0$. As the cofactors of the determinant (4.12) cannot all simultaneously be zero, because σ is simple, we can introduce the eigenvector w, which is unique after normalisation, and is the solution of

$$((\varepsilon\sigma)^2A+\sigma\Gamma+K)w=0 \tag{4.14}$$

which, when $\varepsilon\to 0$, tends to the eigenvector w_0 associated with σ_0:

$$(\sigma_0\Gamma+K)w_0=0$$

for which we know, from (4.10) ($u=w_0$) that $(\Gamma w_0,w_0)\neq 0$.

By scalar multiplication of (4.14) by w, we can write:

$$\varepsilon^2(i\sigma)^2(Aw,w)+i\sigma(i\Gamma w,w)-(Kw,w)=0 \tag{4.15}$$

and since (Aw, w), $(i\Gamma w, w)$, (Kw, w) are real, $i\sigma$ can be interpreted as the root of an algebraic equation with real coefficients. As the discriminant of this equation, namely $(i\Gamma w, w)^2 + 4\varepsilon^2(Aw, w)(Kw, w)$ tends to the strictly positive quantity $(i\Gamma w_0, w_0)^2$ when $\varepsilon \to 0$, it can be seen that, by making $|\varepsilon|$ small enough, or $|\xi|$ large enough, one can always ensure that $i\sigma$ is real, i.e. that σ is a pure imaginary. The solution corresponding to this simple mode is $we^{\omega t}$ with $\omega = \frac{\sigma}{\xi} \sim \frac{\sigma_0}{\xi}$ and we obtain $\frac{n}{2}$ modes[1] of this type which correspond to motions with a long period of oscillation (slow precession). We now revert to (4.11) and discuss the roots which become infinite when $|\xi| \to \infty$. We introduce $\omega = \sigma\xi$ and it becomes apparent that the equation

$$\det(\sigma^2 A + \sigma\Gamma + \xi^{-2}K) = 0 \tag{4.16}$$

has, for $|\xi| \to \infty$, n roots which tend to zero, corresponding to the motions already studied and n others which tend towards the roots of $\det(\sigma A + \Gamma) = 0$, all of which are non-zero by reason of $\det \Gamma \neq 0$. Let σ_0 be one of these, w_0 an associated normalised eigenvector, so that $(\sigma_0 A + \Gamma)w_0 = 0$

$$\sigma_0(Aw_0, w_0) + (\Gamma w_0, w_0) = 0$$

and since (Aw_0, w_0) is positive, $(\Gamma w_0, w_0)$ is a pure imaginary, and $\sigma_0 \neq 0$, we conclude on the one hand that σ_0 is a pure imaginary, and on the other that $(\Gamma w_0, w_0) \neq 0$.

We consider the root σ of (4.16), simple for large enough $|\xi|$, which tends to σ_0 when $|\xi| \to \infty$, and the corresponding eigenvector w which we may suppose to tend towards w_0 under these same conditions.

Arguing from

$$(i\sigma)^2(Aw, w) + (i\Gamma w, w) - \xi^{-2}(Kw, w) = 0 \tag{4.17}$$

as we did for (4.15) and bearing in mind that the real quantity $(i\Gamma w, w)$ tends to $(i\Gamma w_0, w_0) \neq 0$, we can see that for $|\xi|$ large enough, $i\sigma$ considered as a root of (4.17) is real and consequently σ is a pure imaginary.

The corresponding mode, which like σ is simple, will be described by $q = we^{\sigma\xi t}$, that is to say we obtain in this way $\frac{n}{2}$ very short period modes[1] when $|\xi|$ is large enough (rapid mutation effect).

Systems Comprising Both Gyroscopic Forces and Dissipative Forces

These are described by the matrix equation

$$Aq'' + (D + \xi\Gamma)q' + (K + E)q = 0. \tag{4.18}$$

[1] The assumption $\det \Gamma \neq 0$ implies n even.

1. Case $E = 0$

Theorem 3. Let D be positive definite: if K is positive definite the solution $q = 0$ is asymptotically stable, but if K is negative definite this solution is unstable. The results are valid for any ξ whatsoever.

1. Suppose K to be positive definite and let $q(t)$ be a real solution of (4.18) ($E = 0$). We deduce from the equation, after scalar multiplication by q', taking account of $(\Gamma q', q') = 0$ and integrating with respect to t:

$$(Aq', q') + (Kq, q) = C - 2\int_{t_0}^{t} (Dq', q')\,\mathrm{d}t$$

where C is a constant. Since the left-hand side is positive and since there is a $\delta > 0$ such that $\delta(q', q') \leqslant (Dq', q')$, it is clear that $\int_{t_0}^{t} (q', q')\,\mathrm{d}t < C\delta^{-1}$, $\forall t > t_0$ and

(4.19) $$\int^{+\infty} (q', q')\,\mathrm{d}t < +\infty.$$

If $\omega = \alpha + i\beta$ is a characteristic root, there exists a solution of (4.18):

$$q = \operatorname{Re} q_0 e^{\omega t} = e^{\alpha t}(a \cos \beta t + b \sin \beta t)$$

and (4.19) asserts that $\int^{+\infty} e^{2\alpha t} f(t) < +\infty$ where $f(t)$ is periodic with period $2\pi/\beta$, and is positive or zero. This inequality entails $\alpha < 0$ because $f(t)$ cannot vanish everywhere as this would necessarily imply $q' = 0$ or $q = q_0$, that is $Kq_0 = 0$ or $q_0 = 0$ since K, being positive definite, is invertible. Thus all the characteristic roots lie in the half-plane $\operatorname{Re}\omega < 0$ and there is asymptotic stability.

2. Suppose K negative definite; if ω is a characteristic root with a pure imaginary or zero value, and u an associated eigenvector, we can write

(4.20) $$\omega^2(Au, u) + \omega(Du, u) + \xi\omega(\Gamma u, u) + (Ku, u) = 0$$

and on separating the real and imaginary parts, we obtain $\omega(Du, u) = 0$ that is $(Du, u) = 0$ if $\omega \neq 0$, or $(Ku, u) = 0$ in the case $\omega = 0$.

But as the matrices D and K are positive definite and negative definite respectively, we deduce from this that $u = 0$ which is impossible. Hence there cannot be any characteristic root on the imaginary axis. Accordingly, if we can show that the characteristic roots of

(4.21) $$Aq'' + Dq' + Kq = 0$$

are distributed half and half in the semiplanes $\operatorname{Re}\omega < 0$ and $\operatorname{Re}\omega > 0$, we can be certain that the same property will hold for the roots of the equation $\det(\omega^2 A + \omega(\xi\Gamma + D) + K) = 0$, for any real ξ, by virtue of the fact that the zeros of a polynomial depend in a continuous manner on its coefficients. We modify (4.21) replacing D by θD, where θ is a parameter taking its values in $[0, 1]$:

(4.22) $$Aq'' + \theta Dq' + Kq = 0.$$

It is obvious for $\theta \neq 0$ that (4.22) can have no pure imaginary root; for $\theta = 0$ the characteristic equation is $\det(\omega^2 A + K) = 0$ and denoting by ω a root, and by u an associated eigenvector, we see from $\omega^2(Au, u) + (Ku, u) = 0$ that ω^2 is positive and real; thus for $\theta = 0$ the characteristic roots are real and occur in pairs of opposite sign.

The continuity argument applied when θ runs through the interval [0, 1] shows that the characteristic roots of (4.22) are distributed half in the half-plane $\operatorname{Re}\omega < 0$ and half in the complementary half-plane, and in particular this is true for $\theta = 1$. This shows finally that $q = 0$ is an unstable solution of (4.18) ($E = 0$) irrespective of the value of ξ.

A Modified Approach in the Case of Instability

It is not without interest to resume the analysis by a different method to find out what are the unstable modes when $|\xi|$ is large. To simplify we make the additional assumption that $\det \Gamma \neq 0$ and that characteristic roots of $\xi \Gamma q' + Kq = 0$ are simple. With $\varepsilon = \xi^{-1}$, $\omega\xi = \sigma$ the characteristic equation may be written as:

$$\det(\varepsilon^2 \sigma^2 A + \sigma(\Gamma + \varepsilon D) + K) = 0 \tag{4.23}$$

n of whose roots become infinite, while n others tend towards the roots of

$$\det(\sigma\Gamma + K) = 0 \tag{4.24}$$

as $\varepsilon \to 0$. Let σ_0 be a root of (4.24), this root is finite since $\det \Gamma \neq 0$ and non-zero since $\det K \neq 0$, and is necessarily purely imaginary. We denote by w_0 the associated normalised eigenvector

$$(\sigma_0 \Gamma + K) w_0 = 0$$

which is unique since σ_0 is simple, and which satisfies $(\Gamma w_0, w_0) \neq 0$ since $(K w_0, w_0) \neq 0$.

Considering now, for $\varepsilon \to 0$, the root σ of (4.23) which tends to σ_0 and the associated normalised eigenvector w which tends to w_0, we can write:

$$(\varepsilon^2 \sigma^2 A + \sigma(\Gamma + \varepsilon D) + K) w = 0$$

or after scalar multiplication by w:

$$\varepsilon^2 a \sigma^2 + (i\gamma + \varepsilon d)\sigma + \varkappa = 0 \tag{4.25}$$

where $a = (Aw, w)$, $\gamma = -i(\Gamma w, w)$, $d = (Dw, w)$, $\varkappa = (Kw, w)$, are all real functions of ε which tend, respectively, to

$$a_0 = (A w_0, w_0), \quad \gamma_0 = -i(\Gamma w_0, w_0) \neq 0, \quad d_0 = (D w_0, w_0), \quad \varkappa_0 = (K w_0, w_0).$$

as $\varepsilon \to 0$.

We can find an explicit expression for the root of (4.25) which has a finite limit when $\varepsilon \to 0$ and calculate the first terms of the power series expansion in ε. A simple calculation, whose details we omit, leads to

$$\operatorname{Re}\sigma = -\frac{\varkappa_0 d_0}{\gamma_0^2}\varepsilon\cdot(1+o(1))$$

or

$$\operatorname{Re}\omega = -\frac{\varkappa_0 d_0}{\gamma_0^2}\varepsilon^2(1+o(1))$$

and as $\varkappa_0 < 0$, $d_0 > 0$ we see that the principal part of $\operatorname{Re}\omega$ is positive and of order 2 with respect to ξ^{-1}. The unstable character of the zero solution has thus been confirmed, established it is true by this argument for large $|\xi|$: the unstable modes are those associated with the slow precession movements.

We can complete the investigation by looking for a representation of eigenvalues which become infinite when $|\xi| \to \infty$; we put $\omega = \xi\sigma$ and we are led to discuss the n roots of

(4.26) $$\det(\sigma^2 A + \sigma(\Gamma + \varepsilon D) + \varepsilon^2 K) = 0$$

which tend towards those

(4.27) $$\det(\sigma A + \Gamma) = 0 \quad \text{when} \quad \varepsilon \to 0.$$

We shall suppose that these latter are simple and, by an analysis similar to the one above, it can be shown that $\operatorname{Re}\omega = -\dfrac{d_0}{a_0} + O(\varepsilon)$ where $d_0 = (Dw_0, w_0)$, $a_0 = (Aw_0, w_0)$, and w_0 is the normalised eigenvector associated with the characteristic root σ_0 of (4.27) to which the root σ of (4.26) tends when $\varepsilon \to 0$. The numbers d_0, a_0 are real and positive and consequently $\operatorname{Re}\omega < 0$; the corresponding very high frequency modes are those of the rapid nutations and are stable.

We could have discussed Theorem 3 using a different argument which allows us to weaken slightly the hypotheses, and for that reason our further developments will be given under the heading of

Theorem 4. If D is positive definite, if at least one of the eigenvalues of K is negative, and if $\det K \neq 0$ then $q = 0$ is an unstable solution of 4.1 ($E = 0$) for any ξ.

We establish, as above, that the characteristic equation

(4.28) $$\det(\omega^2 A + \omega(D + \xi\Gamma) + K) = 0$$

has no pure imaginary root, a zero root being excluded by $\det K \neq 0$. To prove the theorem it will be enough to show that the equation has roots in the half-plane $\operatorname{Re}\omega > 0$ or that such is the case for

(4.29) $$\det(\omega^2 A + \omega D + K) = 0$$

by the continuity argument already evoked.

Replacing D by θD in (4.29), we can easily check that the modified equation (4.29) has no roots on the imaginary axis. If one can show that the equation

(4.30) $$\det(\omega^2 A + K) = 0,$$

corresponding to $\theta = 0$ has at least one root ω_0, with $\operatorname{Re}\omega_0 > 0$, then when θ

describes the segment $[0,1]$, the root ω of the modified equation (4.29), varies continuously from ω_0, for $\theta=0$, to ω_1 for $\theta=1$, along a continuous path which never crosses the imaginary axis and consequently $\operatorname{Re}\omega_1>0$. It is sufficient therefore to discuss (4.30); since its roots are in pairs of opposite sign it will be enough to show that it is impossible that all the roots of $\det(-\lambda A+K)$ should be real and positive. Suppose however this were the case. We can write $-\lambda A+K=A^{1/2}(-\lambda I+A^{-1/2}KA^{-1/2})A^{1/2}$, where the matrices $A^{1/2}$, $A^{-1/2}$, are symmetric positive definite matrices, whose precise definition we shall give later. This leads to the supposition that the eigenvalues of the matrix $A^{-1/2}KA^{-1/2}$ are all positive real numbers and hence this symmetric matrix would itself be positive definite, i.e.

$$(A^{-1/2}KA^{-1/2}u,u)>0, \quad \forall u\in\mathbb{R}^n, \quad u\neq 0$$

or with

$$v=A^{-1/2}u, \quad u=A^{1/2}v, \quad (Kv,v)>0, \quad \forall v\in\mathbb{R}^n, \quad v\neq 0,$$

and the matrix K would be positive definite, in contradiction to the hypothesis that it has at least one negative eigenvalue.

2. *Case* $E\neq 0$

We return to the general case described by (4.18) where we have made the change of variable $x=A^{1/2}q$:

$$x''+(\tilde{D}+\xi\tilde{\Gamma})x'+(\tilde{K}+\tilde{E})x=0,$$

$\tilde{D}=A^{-1/2}DA^{-1/2}\ldots$, so that the matrices $\tilde{D},\tilde{\Gamma},\tilde{K},\tilde{E}$ have the same properties of symmetry, skew-symmetry, positivity or not, as the corresponding matrices D, Γ, K, E. To avoid complicating the notation we shall from now on deal with the reduced form:

$$x''+(D+\xi\Gamma)x'+(K+E)x=0. \tag{4.31}$$

Theorem 5. If $\operatorname{tr} D<0$, $x=0$ is an unstable solution of (4.31).

The characteristic equation is:

$$F(\omega)=\det(\omega^2 I+\omega(D+\xi\Gamma)+K+E)=0$$

and the polynomial $F(\omega)$ can be written $F(\omega)=\omega^{2n}+(\operatorname{tr} D)\omega^{2n-1}+\cdots$ so that the sum of its roots $-\operatorname{tr} D$ is positive which implies that at least one of them is in $\operatorname{Re}\omega>0$.

Theorem 6. If $\det(K+E)<0$, $x=0$ is an unstable solution of (4.31).

We have $F(\omega)=\omega^{2n}+\cdots+\det(K+E)$ and hence $F(0)<0$, $F(+\infty)>0$ so that $F(\omega)$ has at least one positive real zero.

Theorem 7. If $K=\Gamma=0$, $x=0$ cannot be an asymptotically stable solution of (4.31): furthermore if $\det E\neq 0$, $x=0$ is unstable.

The proposition is obviously true when $\det E=0$ for there exists an $x_0\neq 0$ such that $Ex_0=0$ and $x=x_0$ is a constant solution. From now on therefore we assume that $\det E\neq 0$. On replacing D by θD, $\theta\in[0,1]$, in (4.31) the characteristic equation becomes

$$\det(\omega^2 I + \theta\omega D + E) = 0. \tag{4.32}$$

Since E is skew symmetric (Eu, u) is purely imaginary or zero for every vector u, and a classical argument shows that (4.32) has no root which is zero or purely imaginary.

For $\theta = 0$ it is easy to see that the characteristic roots are the points in the complex plane which are the vertices of squares centred at the origin and whose sides are parallel to the axis. Hence it can be seen by the continuity argument that the characteristic roots are equally divided between the two half-planes $\mathrm{Re}\,\omega > 0$, $\mathrm{Re}\,\omega < 0$, and it follows from this that the solution $x = 0$ is unstable.

Theorem 8. If K is negative definite and $\Gamma = 0$, then $x = 0$ is an unstable solution of (4.31).

By arguments closely resembling those already used, it can be shown that $\det(K + E) \neq 0$, and then that the characteristic equation has no roots on the imaginary axis, and finally omitting D, that the roots are equally divided between the half-planes $\mathrm{Re}\,\omega < 0$, $\mathrm{Re}\,\omega > 0$.

We shall conclude this analysis with

Theorem 9 [31]. Let

$$Aq'' + (\Gamma + D)q' + (K + E)q = 0, \tag{4.33}$$

where A, D, K, E, Γ are all real matrices of order n: A, D, K being symmetric, E, Γ skew-symmetric, and A, D positive definite and suppose furthermore that $ED^{-1}A$, $ED^{-1}K$ are skew-symmetric and $ED^{-1}\Gamma$ is symmetric.

Then $q = 0$ is an asymptotically stable solution of (4.33) if the eigenvalues of

$$M = ED^{-1}AD^{-1}E - \Gamma D^{-1}E + K$$

are all positive, while $q = 0$ is an unstable solution if at least one of the eigenvalues of M is negative.

Let us first interpret the assumptions of skew-symmetry for $ED^{-1}A$, $ED^{-1}K$ and symmetry for $ED^{-1}\Gamma$.

$$ED^{-1}A = -(ED^{-1}A)^T = -AD^{-1}E^T = AD^{-1}E$$
$$ED^{-1}K = -(ED^{-1}K)^T = -KD^{-1}E^T = KD^{-1}E$$
$$ED^{-1}\Gamma = (ED^{-1}\Gamma)^T = \Gamma^T D^{-1}E^T = \Gamma D^{-1}E.$$

The matrix M is symmetric and has real eigenvalues. Following [31] we introduce Liapounoff's function

$$V = (q' + D^{-1}Eq, A(q' + D^{-1}Eq)) + (q, Mq)$$

which is real-valued. Calculating its derivative $\frac{dV}{dt}$ we obtain, bearing in mind that q can take complex values:

$$\frac{1}{2}\frac{dV}{dt} = \mathrm{Re}\{(q' + D^{-1}Eq, Aq'' + AD^{-1}Eq') + (q, Mq')\}$$

i.e. using (4.33):

$$\frac{1}{2}\frac{\mathrm{d}V}{\mathrm{d}t} = \mathrm{Re}\{(q' + D^{-1}Eq, AD^{-1}Eq' - Dq' - \Gamma q' - Kq - Eq) + (q, ED^{-1}AD^{-1}Eq' - \Gamma D^{-1}Eq' + Kq')\}.$$

But since Γ, $AD^{-1}E$, $ED^{-1}K$ are skew-symmetric, the terms

$$(q', \Gamma q'), \quad (q', AD^{-1}Eq'), \quad (D^{-1}Eq, Kq) = -(q, ED^{-1}Kq)$$

are pure imaginaries and bring no contribution. Hence

$$\frac{1}{2}\frac{\mathrm{d}V}{\mathrm{d}t} = \mathrm{Re}\{-(q', Dq') - (q', Kq) - (q', Eq) - (D^{-1}Eq, Dq') - (D^{-1}Eq, \Gamma q') - (D^{-1}Eq, Eq) + (D^{-1}Eq, AD^{-1}Eq') + (q, ED^{-1}AD^{-1}Eq') - (q, \Gamma D^{-1}Eq') + (q, Kq')\}$$

and noting that

$$(D^{-1}Eq, AD^{-1}Eq') = (q, E^T D^{-1}AD^{-1}Eq') = -(q, ED^{-1}AD^{-1}Eq')$$
$$(D^{-1}Eq, \Gamma q') = (q, E^T D^{-1}\Gamma q') = -(q, ED^{-1}\Gamma q') = -(q, \Gamma D^{-1}Eq')$$

there remains

$$\frac{1}{2}\frac{\mathrm{d}V}{\mathrm{d}t} = \mathrm{Re}\{-(q', Dq') - (q', Eq) - (D^{-1}Eq, Dq') - (D^{-1}Eq, Eq)\}$$
$$= -(D^{-1}(Eq + Dq'), Eq + Dq') \leqslant 0.$$

If M is positive definite than clearly V is positive and vanishes only when $q = q' = 0$; since $\frac{\mathrm{d}V}{\mathrm{d}t} \leqslant 0$, the solution $q = 0$ is therefore stable.

It remains to be shown that it is asymptotically stable, or what amounts to the same thing, that (4.33) has no purely imaginary characteristic root. Now suppose $i\omega$, with ω real, to be a characteristic root and $q = q_0 e^{i\omega t}$, where q_0 is a non-zero constant vector to be an associated solution of (4.33). For such a solution

$$\frac{1}{2}\frac{\mathrm{d}V}{\mathrm{d}t} = -(D^{-1}(Eq_0 + i\omega Dq_0), Eq_0 + i\omega Dq_0)$$

is a non-positive constant: on the other hand if $\frac{\mathrm{d}V}{\mathrm{d}t}$ were non-zero we should have $\lim_{t \to +\infty} V = -\infty$, which is impossible because $V \geqslant 0$.

Thus $\frac{\mathrm{d}V}{\mathrm{d}t} = 0$ and

$$Eq_0 + i\omega Dq_0 = 0 \quad \text{or} \quad \text{or} \quad D^{-1}Eq_0 + i\omega q_0 = 0$$

and by iteration

$$D^{-1}ED^{-1}Eq_0 = -\omega^2 q_0$$

or, in view of $(-A\omega^2 + i\omega(\Gamma + D) + K + E)q_0 = 0$,

$$(AD^{-1}ED^{-1}E - \Gamma D^{-1}E + K)q_0 = 0 \qquad \text{that is} \qquad Mq_0 = 0$$

and hence $q_0 = 0$ since M is invertible. We thus end up with a contradiction, which shows that the characteristic equation of (4.33) cannot have a purely imaginary root.

We now come to the result on instability: suppose that M has at least one negative eigenvalue. Then there exists a vector q_* such that $(q_*, Mq_*) < 0$ and a velocity vector denoted by q'_* such that $V_* < 0$ (we have only to choose q'_* so that the norm of $q'_* + D^{-1}Eq_*$ is small enough). With this choice we shall have, for the motion defined by the initial conditions $q = q_*$, $q' = q'_*$ at $t = 0$: $V \leqslant V_* < 0$, $\forall t \geqslant 0$.

We shall show that the hypothesis that $q = 0$ is a stable solution of (4.33) leads to a contradiction. On this hypothesis the characteristic roots have a negative or zero real part; the solutions of the latter kind of the form $i\omega_j$, $1 \leqslant j \leqslant l$, with ω_j real are necessarily of the type $c_j e^{i\omega_j t}$, where c_j is a constant vector, by reason of the hypothesis of stability, and consequently any solution can be represented by

$$q = \sum_{j=1}^{l} c_j e^{i\omega_j t} + \varepsilon(t), \qquad \lim_{t \to +\infty} \varepsilon(t) = 0.$$

For every solution $ce^{i\omega t}$, it is known that $\dfrac{dV}{dt}$ is constant and non-positive, and thus since V is bounded, we must have $\dfrac{dV}{dt} = 0$, that is $q' + D^{-1}Eq = 0$, whence we deduce as above, using

$$(-A\omega^2 + (D + \Gamma)\omega + K + E)c = 0, \quad \text{and}$$
$$(i\omega + D^{-1}E)c = 0 \text{ that } Mq = 0 \text{ and consequently } V = 0.$$

Finally, for the solution derived from $q = q_*$, $q' = q'_*$ at time $t = 0$, it is clear that $\lim_{t \to +\infty} V = 0$, in contradiction to the inequality $V \leqslant V_* < 0$ which is known to hold for $t \geqslant 0$.

Note that Theorems 3 and 4 are consequences of Theorem 9. Theorem 4 has been improved, as the hypothesis $\det K \neq 0$ is no longer essential.

Eigenmodes

Consider the system governed by the equation

$$Aq'' + Bq = 0, \tag{4.34}$$

where A, B are real symmetric matrices of order n, and A is positive definite.

The matrix A being positive, it is possible to define its positive square root, denoted by $A^{1/2}$ in the following way. It is well-known that there exists a real unitary matrix S, (i.e. a matrix satisfying $S^{-1} = S^T$) such that $S^{-1}AS = \tilde{A}$ is diagonal, the diagonal elements being the eigenvalues of A, which are all positive since A is positive.

Accordingly $\tilde{A} = \begin{pmatrix} \lambda_1 & & 0 \\ & \ddots & \\ 0 & & \lambda_n \end{pmatrix}$ and we can define $\tilde{A}^{1/2} = \begin{pmatrix} \sqrt{\lambda_1} & & 0 \\ & \ddots & \\ 0 & & \sqrt{\lambda_n} \end{pmatrix}$ and then $A^{1/2} = S\tilde{A}^{1/2}S^{-1}$; it is clear that $A^{1/2}$ is symmetric, positive and satisfies $A^{1/2} \cdot A^{1/2} = A$, $\det A^{1/2} = (\det A)^{1/2}$.

Let us now make a change of variable $q \to u : u = A^{1/2} q$, in (4.34) so that it becomes:

$$u'' + A^{-1/2} B A^{-1/2} u = 0.$$

The matrix $C = A^{-1/2} B A^{-1/2}$ is symmetric and can therefore be diagonalised, that is, there exists a real unitary matrix H, $(H^{-1} = H^T)$ such that

$$HCH^{-1} = \begin{pmatrix} \mu_1 & & 0 \\ & \ddots & \\ 0 & & \mu_n \end{pmatrix}.$$

Setting $Hu = v$ or $u = H^{-1} v$ we obtain the equation

$$v'' + \begin{pmatrix} \mu_1 & & 0 \\ & \ddots & \\ 0 & & \mu_n \end{pmatrix} v = 0$$

or in other words

$$v_j'' + \mu_j v_j = 0, \quad 1 \leqslant j \leqslant n. \tag{4.35}$$

Thus by the change of variable $v = HA^{1/2} q$ we have transformed (4.34) into a completely 'uncoupled' system of equations with 'uncoupled' variables. The new variables $v_1, v_2, \ldots, v_n$ are also referred to as modal co-ordinates.

In the more general system $Aq'' + (\xi \Gamma + D)q' + Kq = 0$, this change of variable (with $B = K$) would lead to:

$$v'' + (\xi \tilde{\Gamma} + \tilde{D})v' + \begin{pmatrix} \mu_1 & & 0 \\ & \ddots & \\ 0 & & \mu_n \end{pmatrix} v = 0,$$

where $\tilde{\Gamma}$ is skew-symmetric and $\tilde{D}$ symmetric. Obviously one cannot expect $\tilde{\Gamma}$ to be in diagonal form, but there are numerous cases where $\tilde{D}$ can be regarded as reduced to its diagonal elements, at least to a first approximation, and this greatly simplifies calculations if $\xi = 0$.

Rayleigh's Method

It is interesting, particularly in the more complex situations which will be discussed later, to consider afresh the problem of finding the characteristic eigenmodes of vibration for the equation (4.34) by taking another approach due in principle to

Rayleigh. Let

$$\mu_1 = \operatorname*{Inf}_{q\in\mathbb{R}^n} \frac{(Bq,q)}{(Aq,q)} = \operatorname*{Inf}_{(Aq,q)=1} (Bq,q). \tag{4.36}$$

Since $(Aq,q)=1$ defines a compact set in $\mathbb{R}^n$ (because A is positive definite) and (Bq,q) is continuous, it is safe to assume that $\operatorname{Inf}_{(Aq,q)=1}(Bq,q)$ exists, is finite, and that the infimum is attained for some element v_1, so that we can write:

$$\mu_1 = \frac{(Bv_1,v_1)}{(Av_1,v_1)}. \tag{4.37}$$

The inequality

$$\frac{(B(v_1+\delta q), v_1+\delta q)}{(A(v_1+\delta q), v_1+\delta q)} \geqslant \mu_1$$

and (4.37) together imply

$$(Bv_1 - \mu_1 Av_1, \delta q) + O(\delta q)^2 \geqslant 0, \qquad \forall \delta q,$$

whence $(Bv_1 - \mu_1 Av_1, \delta q) = 0, \forall \delta q$ so that we have

$$Bv_1 - \mu_1 Av_1 = 0. \tag{4.38}$$

Clearly v_1 defines an eigenmode because, with $q = f(t)v_1$, we deduce from (4.34) and (4.38):

$$f'' + \mu_1 f = 0 \quad \text{or} \quad f = (\alpha \cos\sqrt{\mu_1}t + \beta \sin\sqrt{\mu_1}t), \quad \text{if} \quad \mu_1 > 0.$$

We now define

$$\mu_2 = \operatorname*{Inf}_{(q,Av_1)=0} \frac{(Bq,q)}{(Aq,q)} = \operatorname*{Inf}_{\substack{(Aq,q)=1\\(q,Av_1)=0}} (Bq,q);$$

μ_2 is finite, $\mu_2 \geqslant \mu_1$ and there exists a $v_2 \in \mathbb{R}^n$ such that

$$\mu_2 = \frac{(Bv_2,v_2)}{(Av_2,v_2)} \quad \text{and} \quad (v_2, Av_1) = 0.$$

With $\delta q \in \mathbb{R}^n$ satisfying $(\delta q, Av_1) = 0$, we can write, having regard to the definition of μ_2

$$(B(v_2+\delta q), v_2+\delta q) \geqslant \mu_2 (A(v_2+\delta q), v_2+\delta q)$$

or

$$(Bv_2 - \mu_2 Av_2, \delta q) + O(\delta q)^2 \geqslant 0$$

that is

$$(Bv_2 - \mu_2 Av_2, \delta q) = 0, \tag{4.39}$$

for all $\delta q \in \mathbb{R}^n$ such that $(\delta q, Av_1) = 0$. But (4.39) remains true when we replace δq by v_1 because

$$(Bv_2 - \mu_2 Av_2, v_1) = (v_2, Bv_1 - \mu_2 Av_1) = (\mu_1 - \mu_2)(v_2, Av_1) = 0$$

and moreover any element $\delta q \in \mathbb{R}^n$ can be written in the form $\delta q = cv_1 + w$ with $c \in \mathbb{R}$ and $(w, Av_1) = 0$; it suffices to take $c = \dfrac{(\delta q, Av_1)}{(v_1, Av_1)}$.

Thus (4.39) is true for all $\delta q \in \mathbb{R}^n$ and we obtain

$$Bv_2 = \mu_2 Av_2.$$

We can continue this process in the obvious way and find in succession the vectors $v_1, v_2, \ldots, v_n$ and the frequencies $\mu_1 \leqslant \mu_2 \leqslant \cdots \leqslant \mu_n$ such that

$$Bv_j = \mu_j Av_j, \quad (v_j, Av_i) = 0, \quad j \neq i. \tag{4.40}$$

These vectors v_j must be linearly independent because otherwise there would be numbers α_j such that $\sum_1^n \alpha_j v_j = 0$ and hence $\sum_{j=1}^n \alpha_j Av_j = 0$ so that we should have, after scalar multiplication by v_i and taking account of (4.40):

$$\alpha_i(v_i, Av_i) = 0 \quad \text{that is} \quad \alpha_i = 0.$$

If the μ_j are all positive they represent the squares of the eigenfrequencies and the vectors v_j enable us to define the associated eigenmodes.

From now on we define, for any system $w_1, w_2, \ldots, w_{n-1}$ of elements of $\mathbb{R}^n$

$$\nu_j(w_1, \ldots, w_{j-1}) = \operatorname*{Inf}_{\substack{q \in \mathbb{R}^n \\ (q, Aw_l) = 0 \\ 1 \leqslant l \leqslant j-1}} \frac{(Bq, q)}{(Aq, q)}, \quad 2 \leqslant j \leqslant n \tag{4.41}$$

and

$$\nu_1 = \operatorname*{Inf}_{q \in \mathbb{R}^n} \frac{(Bq, q)}{(Aq, q)}$$

With $q = \sum_{i=1}^j c_i v_i \in \mathbb{R}^n$ where v_i are the eigenvectors defined earlier and the c_i are j real coefficients such that $(q, Aw_l) = 0$, $1 \leqslant l \leqslant j-1$ that is to say a non-trivial solution of the set of $j-1$ linear homogeneous equations

$$\sum_{i=1}^{j} c_i(v_i, Aw_l) = 0, \quad 1 \leqslant l \leqslant j-1$$

we can write

$$\frac{(Bq, q)}{(Aq, q)} = \frac{\left(\sum\limits_{i=1}^{j} c_i Bv_i, \sum\limits_{s=1}^{j} c_s v_s\right)}{\left(\sum\limits_{i=1}^{j} c_i Av_i, \sum\limits_{s=1}^{j} c_s v_s\right)} = \frac{\sum\limits_{i=1}^{j} c_i^2 \mu_i (Av_i, v_i)}{\sum\limits_{i=1}^{j} c_i^2 (Av_i, v_i)} \leqslant \mu_j,$$

since $\mu_i \leqslant \mu_{i+1}$ so that we have $\nu_j(w_1, \ldots, w_{j-1}) \leqslant \mu_j$.

On the other hand, remembering that $\nu_j(v_1, \ldots, v_{j-1}) = \mu_j$ we have

$$\operatorname*{Sup}_{w_1, \ldots, w_{j-1}} \nu_j(w_1, \ldots, w_{j-1}) = \mu_j \quad \text{and} \quad \nu_1 = \mu_1 \tag{4.42}$$

Let us now define, again for an arbitrary set of $n-1$ elements $w_1, \ldots, w_{n-1}$

of $\mathbb{R}^n$

(4.43)
$$\zeta_j(w_1,\dots,w_{n-j}) = \operatorname*{Sup}_{\substack{q\in\mathbb{R}^n\\(q,Aw_l)=0\\1\leqslant l\leqslant n-j}} \frac{(Bq,q)}{(Aq,q)}, \quad 1\leqslant j\leqslant n-1$$

and
$$\zeta_n = \operatorname*{Sup}_{q\in\mathbb{R}^n}\frac{(Bq,q)}{(Aq,q)}$$

Since $n-j+1$ real numbers c_i, not all zero, can always be defined so that $q=\sum_{i=j}^n c_i v_i$ satisfies the $n-j$ homogeneous linear equations $(q,Aw_l)=0$ or $\sum_{i=j}^n c_i(v_i,Aw_l)=0$, $1\leqslant l\leqslant n-j$ we see that for such an element q:

$$\frac{(Bq,q)}{(Aq,q)} = \frac{\left(\sum_{i=j}^n c_i B v_i, \sum_{s=j}^n c_s v_s\right)}{\left(\sum_{i=j}^n c_i A v_i, \sum_{s=j}^n c_s v_s\right)} = \frac{\sum_{i=j}^n c_i^2 \mu_i (Av_i,v_i)}{\sum_{i=j}^n c_i^2 (Av_i,v_i)} \geqslant \mu_j$$

from which we deduce:

(4.44)
$$\zeta_j(w_1,\dots,w_{n-j}) \geqslant \mu_j$$

It follows from $(v_k, Av_l)=0$, $l\neq k$, that the subspace defined by the elements q of $\mathbb{R}$ which satisfy $(q,Av_l)=0$ for $j+1\leqslant l\leqslant n$ is generated by the system $(v_1,\dots,v_j)$ and that:

$$\zeta_j(v_n,v_{n-1},\dots,v_{j+1}) = \operatorname*{Sup}_{q\in\{v_1,\dots,v_j\}}\frac{(Bq,q)}{(Aq,q)} = \operatorname{Sup}\frac{\sum_{i=1}^j c_i^2\mu_i(Av_i,v_i)}{\sum_{i=1}^j c_i^2(Av_i,v_i)} = \mu_j$$

whence with (4.44):

$$\operatorname*{Inf}_{w_1,\dots,w_{n-j}} \zeta_j(w_1,\dots,w_{n-j}) = \mu_j, \quad \zeta_n = \mu_n.$$

Collecting these results we can therefore write, for $2\leqslant j\leqslant n-1$,

$$\mu_j = \operatorname*{Sup}_{w_l\in\mathbb{R}^n}\ \operatorname*{Inf}_{\substack{q\in\mathbb{R}^n\\(q,Aw_l)=0\\1\leqslant l\leqslant j-1}}\frac{(Bq,q)}{(Aq,q)} = \operatorname*{Inf}_{w_l\in\mathbb{R}^n}\ \operatorname*{Sup}_{\substack{q\in\mathbb{R}^n\\(q,Aw_l)=0\\1\leqslant l\leqslant n-j}}\frac{(Bq,q)}{(Aq,q)}$$

$$\mu_1 = \operatorname*{Inf}_{q\in\mathbb{R}^n}\frac{(Bq,q)}{(Aq,q)}, \quad \mu_n = \operatorname*{Sup}_{q\in\mathbb{R}^n}\frac{(Bq,q)}{(Aq,q)}$$

Effect on the Eigenvalues of Changes in Structure

1. If the stiffness of the system is increased, with everything else remaining unchanged, i.e. if the symmetric matrix B is replaced by a symmetric matrix $\tilde{B}\geqslant B$, or in other words $(\tilde{B}q,q)\geqslant(Bq,q)$, $\forall q\in\mathbb{R}$, then we have for each j, $\mu_j\leqslant\tilde{\mu}_j$ and so each frequency can only increase.

2. Suppose that we impose an additional constraint on the system compatible with the equilibrium $q=0$ and described by $(q,\alpha)=0$, where $\alpha\in\mathbb{R}^n$ is given. If the constraint is perfect the Lagrange equations contain a multiplier σ and can be written:

$$Aq''+Bq=\sigma\alpha,\quad (q,\alpha)=0,\quad \sigma\in\mathbb{R}. \tag{4.45}$$

The eigenfrequencies can be calculated by Rayleigh's method with minor modifications.

Thus, writing $\tilde{\mu}_1=\inf\limits_{\substack{q\in\mathbb{R}^n\\(q,\alpha)=0}}\dfrac{(Bq,q)}{(Aq,q)}$ for the lower bound attained at $q=\tilde{v}_1\neq 0$, it can be verified that $((B-\tilde{\mu}_1 A)\tilde{v}_1,\delta q)=0$, $\forall\delta q$ such that $(\delta q,\alpha)=0$, from which it follows that $\tilde{\mu}_1$ and $\tilde{v}_1$ are associated eigenfrequency and eigenmode.

The $n-2$ other eigenfrequencies and eigenmodes are defined successively in the natural way, $\tilde{\mu}_2=\inf\limits_{\substack{q\in\mathbb{R}^n\\(q,\alpha)=0\\(q,A\tilde{v}_1)=0}}\dfrac{(Bq,q)}{(Aq,q)}$ is the attained lower bound at an element $\tilde{v}_2\neq 0$, and more generally, with $a=A^{-1}\alpha$ and for $j\leqslant n-1$:

$$\tilde{\mu}_j=\inf_{\substack{q\in\mathbb{R}^n\\(q,Aa)=0,(q,A\tilde{v}_l)=0,1\leqslant l\leqslant j-1}}\frac{(Bq,q)}{(Aq,q)}=v_{j+1}(\tilde{v}_1,\tilde{v}_2,\dots,\tilde{v}_{j-1},a) \tag{4.46}$$

is the attained lower bound at $\tilde{v}_j\in\mathbb{R}^n$, and there is a $\sigma_j\in\mathbb{R}$ such that

$$B\tilde{v}_j=\tilde{\mu}_j A\tilde{v}_j+\sigma_j Aa.$$

It should be noted incidentally that $\tilde{\mu}_i\leqslant\tilde{\mu}_{i+1}$ and

$$(\tilde{v}_i,A\tilde{v}_j)=0,\ i\neq j,\ (\tilde{v}_i,Aa)=0,\ 1\leqslant i\leqslant n-1,\ 1\leqslant j\leqslant n-1, \tag{4.47}$$

the set $a,v_1,\dots,v_{n-1}$ constituting a base of $\mathbb{R}^n$.

Observing that for any given $w_1,\dots,w_{j-1}$ in $\mathbb{R}^n$, we can find real numbers c_i, not all zero, $1\leqslant i\leqslant j$, such that $(q,Aw_l)=0$, $1\leqslant l\leqslant j-1$, with $q=\sum_{i=1}^{j}c_i\tilde{v}_i$, we write

$$v_{j+1}(w_1,\dots,w_{j-1},a)\leqslant\frac{(Bq,q)}{(Aq,q)}=\frac{\sum_{i=1}^{j}c_i^2\tilde{\mu}_i(A\tilde{v}_i,\tilde{v}_i)}{\sum_{i=1}^{j}c_i^2(A\tilde{v}_i,\tilde{v}_i)}\leqslant\tilde{\mu}_j$$

from which we see, by (4.46) that

$$\tilde{\mu}_j=\sup_{w_1,\dots,w_{j-1}}v_{j+1}(w_1,\dots,w_{j-1},a) \tag{4.48}$$

On the other hand it follows from (4.47) that the subspace of $\mathbb{R}^n$ consisting of the elements q satisfying $(q,A\tilde{v}_l)=0$, $j+1\leqslant l\leqslant n-1$, and $(q,Aa)=0$ is generated by the base $(\tilde{v}_1,\dots,\tilde{v}_j)$, so that

$$\zeta_j(\tilde{v}_{n-1},\tilde{v}_{n-2},\dots,\tilde{v}_{j+1},a)=\sup_{q\in\{\tilde{v}_1,\dots,\tilde{v}_j\}}\frac{(Bq,q)}{(Aq,q)}=\tilde{\mu}_j. \tag{4.49}$$

Finally, by finding real numbers c_i not all zero such that with $q=\sum_{i=j}^{n-1} c_i\tilde{v}_i$ we have $(q, Aw_l)=0$ for the w_l given in $\mathbb{R}^n$ such that $1 \leqslant l \leqslant n-j-1$, we can write

$$\zeta_j(w_1,\dots,w_{n-j-1},a) \geqslant \frac{(Bq,q)}{(Aq,q)} = \frac{\sum_{i=j}^{n-1} c_i^2 \tilde{\mu}_i (A\tilde{v}_i,\tilde{v}_i)}{\sum_{i=j}^{n-1} c_i^2 (A\tilde{v}_i,\tilde{v}_i)} \geqslant \tilde{\mu}_j$$

whence we deduce by (4.49) that:

(4.50) $$\tilde{\mu}_j = \operatorname*{Inf}_{w_1,\dots,w_{n-j-1}} \zeta_j(w_1,\dots,w_{n-j-1},a).$$

From the inequality $v_j(w_1,\dots,w_{j-1}) \leqslant v_{j+1}(w_1,w_2,\dots,w_{j-1},a) \leqslant \tilde{\mu}_j$ we obtain by making $w_1,\dots,w_{j-1}$ vary:

(4.51) $$\mu_j \leqslant \tilde{\mu}_j.$$

Noting on the other hand that

$$\operatorname*{Sup}_{w_1,\dots,w_{j-1}} v_{j+1}(w_1,\dots,w_{j-1},a) \leqslant \operatorname*{Sup}_{w_1,\dots,w_j} v_{j+1}(w_1,\dots,w_j) = \mu_{j+1}$$

and recalling (4.48), it becomes apparent that

(4.52) $$\tilde{\mu}_j \leqslant \mu_{j+1}$$

and therefore, with (4.51) we have $\mu_j \leqslant \tilde{\mu}_j \leqslant \mu_{j+1}$.

Also, we see by (4.48) that $\operatorname*{Sup}_a \tilde{\mu}_j = \mu_{j+1}$, and then

$$\mu_j = \operatorname*{Inf}_{w_1,\dots,w_{n-j}} \zeta_j(w_1,\dots,w_{n-j}) \leqslant \operatorname*{Inf}_{w_1,\dots,w_{n-j-1}} \zeta_j(w_1,\dots,w_{n-j-1},a) = \tilde{\mu}_j$$

by (4.50), whence $\operatorname*{Inf}_a \tilde{\mu}_j = \mu_j$.

In conclusion we can state the following:

Theorem. Whenever an additional frictionless constraint, compatible with the equilibrium, is introduced into a system whose eigenfrequencies are μ_j, $1 \leqslant j \leqslant n$, the eigenfrequencies of the modified system separate those of the original system: $\mu_j \leqslant \tilde{\mu}_j \leqslant \mu_{j+1}$.

In particular every multiple eigenfrequency of order r remains an eigenfrequency of the modified system, but of order $r-1$. When the parameters defining the additional constraint are made to vary, $\tilde{\mu}_j$ can take, for each j, any value in the closed interval $[\mu_j, \mu_{j+1}]$

An Example

We consider a double pendulum consisting of two material points M_1, M_2 of the same mass m, suspended from a fixed point O in such a way that $OM_1 = M_1M_2 = l =$ constant, the configuration of the system being defined by the angles θ_1, θ_2 between the downward vertical and the segments OM_1 and M_1M_2 respectively.

The Lagrange equations, linearised in the neighbourhood of the equilibrium position $\theta_1 = \theta_2 = 0$ are:

$$2\theta_1'' + \theta_2'' + 2\frac{g}{l}\theta_1 = 0$$

$$\theta_1'' + \theta_2'' + \frac{g}{l}\theta_2 = 0$$

and the eigenfrequencies of the oscillation are given by:

$$\omega^2 = (2 \pm \sqrt{2})\frac{g}{l}.$$

We now impose the constraint, which we assume to be frictionless, $\theta_2 = k\theta_1$, k const. The equations of motion become:

$$2\theta_1'' + \theta_2'' + 2\frac{g}{l}\theta_1 = k\sigma$$

$$\theta_1'' + \theta_2'' + \frac{g}{l}\theta_2 = -\sigma$$

$$\theta_2 = k\theta_1$$

where σ is the Lagrange multiplier.

Eliminating σ and θ_2 we obtain:

$$(2 + 2k + k^2)\theta_1'' + (2 + k^2)\frac{g}{l}\theta_1 = 0.$$

The eigenfrequency of the oscillation of the modified system is

$$\tilde{\omega}^2 = \frac{2 + k^2}{2 + 2k + k^2}\frac{g}{l} = f(k)\frac{g}{l}$$

and it can be verified that:

$$2 - \sqrt{2} = \operatorname{Inf} f(k) \leqslant f(k) \leqslant \operatorname{Sup} f(k) = 2 + \sqrt{2}.$$

Chapter V. The Stability of Operation of Non-Conservative Mechanical Systems

The stability of operation of numerous mechanical systems which are kept running at a steady operating state maintained by an external source of energy depends to a large extent on the nature of the constraints imposed. The object of this chapter is to illustrate this point by some examples which we shall treat in the context of a linearised theory. These will include the road stability of articulated lorries and trailers, and of air cushion lifting devices.

1. Rolling Motion and Drift Effect

1. The 'rolling condition' for a rigid disc of centre O, radius a, rolling on the plane Sxy of the orthonormal set of axes $Sxyz$ in such a way that the plane of the disc remains parallel to Sz, may be expressed in the form of the vector equation $\vec{V}(O) + \vec{IO} \wedge \vec{\omega} = 0$, where I is the point of contact, $\vec{\omega} = \psi'\vec{z} + \varphi'\vec{n}$ is the instantaneous rotation of the disc, $\vec{n}$ being the unit normal in the direction of the axis of the disc, ϕ the angle of true rotation and $\psi = (O\vec{x}, \vec{n})$.

Thus $\vec{V}(O) = a\varphi'(\vec{n} \wedge \vec{z})$ is in the plane of the disc. However this description is far from representing what actually happens when a wheel fitted with a tyre rolls over level ground. The true state of affairs is in reality much more complex.

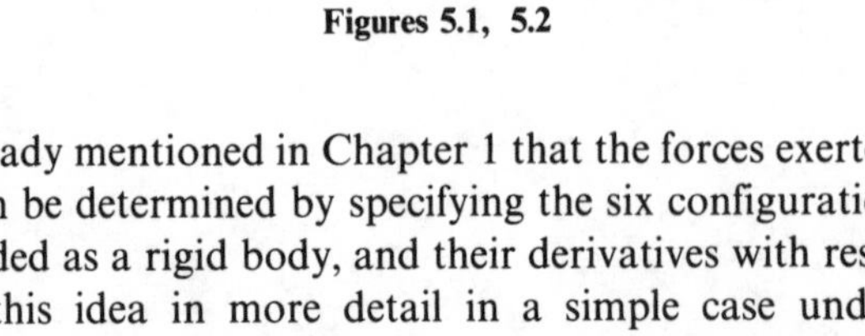

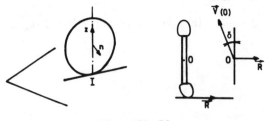

Figures 5.1, 5.2

We have already mentioned in Chapter 1 that the forces exerted by the ground on the wheel can be determined by specifying the six configuration parameters of the wheel, regarded as a rigid body, and their derivatives with respect to time. We shall illustrate this idea in more detail in a simple case under the following assumptions: the plane of the wheel retains a fixed direction normal to the plane of the ground with which it is in contact and the vector $\vec{V}(O)$ representing the velocity of its centre is always parallel to the ground. We denote by δ the angle between $\vec{V}(O)$ and the plane of the wheel, and by $\vec{R}$ the component, normal to this plane, of the force exerted by the ground on the tyre, and we assume that $\vec{R}$ is proportional to δ. Thus

$$|\delta| = \beta|\vec{R}| \quad \text{and} \quad \vec{R}\cdot\vec{V}(O) < 0,$$

β being a positive coefficient which depends, among other things, on the tyre pressure, the height of O, the centre of the wheel, and on the nature of the surfaces in contact.

2. An articulated vehicle can be represented schematically by two wheels O_1, O_2, of equal radius constrained to roll on a plane Sxy, their axles remaining parallel to this plane. The arms OO_1, OO_2 representing the articulated body of the vehicle can pivot at O about an axis Oz, so that both halves remain parallel to the plane Sxy. We shall suppose the whole to have a constant translational speed $V\vec{y}$, $(V > 0)$ and we propose to study the stability of the steady motion in which the arms O_1O and O_2O are parallel to the axis $\vec{y}$, on the basis of the linear equations governing the deviations from the steady state, on the assumption that the deviations remain small [38].

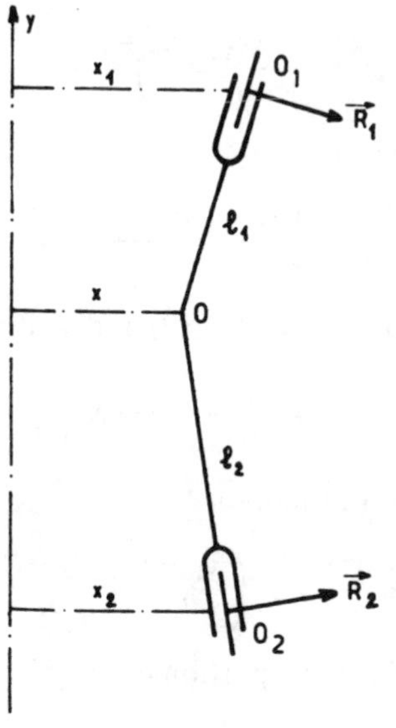

Figure 5.3

a) We write x_1, x_2, x for the x co-ordinates of the centres O_1, O_2, O and R_1, R_2 for the reactions exerted by the ground on the wheels. We define $\overline{OO_1} = l_1$, $\overline{O_2O} = l_2$ as algebraic quantities so as to be able to deal with several configurations as one, and we shall initially assume the total mass m to be concentrated at O. By the theorem on the motion of the centre of inertia applied to the system as a whole we have:

$$m\frac{d^2x}{dt^2} = R_1 + R_2. \tag{5.1}$$

We shall assume that an elastic restoring mechanism at O tends to bring the arms O_1O and O_2O into alignment. The vanishing of the moment at O of the forces exerted on the arm OO_1 and the wheel O_1 give us:

$$-l_1R_1 + k\left(\frac{x_1 - x}{l_1} + \frac{x_2 - x}{l_2}\right) = 0 \tag{5.2}$$

and similarly for the other arm:

$$l_2R_2 - k\left(\frac{x_1 - x}{l_1} + \frac{x_2 - x}{l_2}\right) = 0 \tag{5.3}$$

where k is the stiffness of the restoring system at O. In fact if we assume that the components along Sy of the ground reactions on the wheels are of the same order of magnitude as R_1 and R_2, their possible contribution to (5.2) and (5.3) would be of a higher order than that of R_1, R_2 and it is superfluous to take account of them.

We now have to interpret the kinematic conditions. Let us first ignore the drift effect; the speed of O_1 is borne by OO_1 so that

$$\frac{1}{V}\frac{dx_1}{dt} = \frac{x_1 - x}{l_1} \tag{5.4}$$

and similarly for the other wheel:

$$\frac{1}{V}\frac{dx_2}{dt} = \frac{x - x_2}{l_2}. \tag{5.5}$$

Eliminating R_1, R_2 from (5.1), (5.2), (5.3) and writing

$$k\left(\frac{1}{l_1} + \frac{1}{l_2}\right) = K \tag{5.6}$$

we obtain the third equation of motion

$$m\frac{d^2x}{dt^2} = K\left(\frac{x_1 - x}{l_1} + \frac{x_2 - x}{l_2}\right) \tag{5.7}$$

whence, on seeking solutions proportional to $e^{\omega t}$, we have the characteristic equation:

$$\omega^2 + V\left(\frac{1}{l_2} - \frac{1}{l_1}\right)\omega + \omega_0^2 - \frac{V^2}{l_1 l_2} = 0 \tag{5.8}$$

where

$$\omega_0^2 = \frac{K}{m}\left(\frac{1}{l_1} + \frac{1}{l_2}\right) = \frac{k}{m}\left(\frac{1}{l_1} + \frac{1}{l_2}\right)^2.$$

If V is small, the correction $\frac{V^2}{l_1 l_2}$ can be ignored and the vehicle oscillates at its damped eigenfrequency ω_0 if

$$\frac{l_1 - l_2}{l_1 l_2} > 0, \tag{5.9}$$

or in other words, in the case $l_1 > 0$, $l_2 > 0$, l_2 should be less than l_1 for stability.

However even under these conditions, instability may arise if the speed V is large enough for $V^2 > \omega_0^2 l_1 l_2$ to be satisfied.

Let us now examine how these conclusions would be affected by drift considerations. The only equations to change are (5.4) and (5.5) which become:

$$\frac{1}{V} \cdot \frac{\mathrm{d}x_1}{\mathrm{d}t} = \frac{x_1 - x}{l_1} - \beta_1 R_1 \tag{5.10}$$

$$\frac{1}{V} \frac{\mathrm{d}x_2}{\mathrm{d}t} = \frac{x - x_2}{l_2} - \beta_2 R_2, \tag{5.11}$$

where β_1, β_2 are the drift coefficients of the wheels O_1, O_2, or after eliminating R_1, R_2 from (5.2), (5.10) and (5.3) and (5.11)

$$\frac{1}{V} \frac{\mathrm{d}x_1}{\mathrm{d}t} = \left(1 - \frac{\beta_1 k}{l_1}\right) \cdot \frac{x_1 - x}{l_1} - \frac{\beta_1 k}{l_1} \cdot \frac{x_2 - x}{l_2}$$

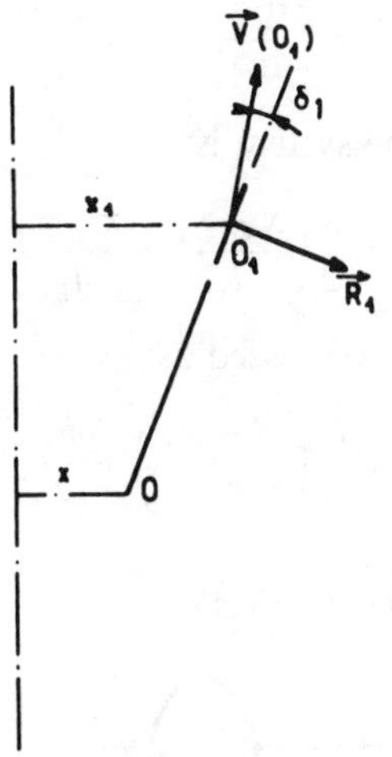

Figure 5.4

$$\frac{1}{V} \frac{\mathrm{d}x_2}{\mathrm{d}t} = -\frac{\beta_2 k}{l_2} \cdot \frac{x_1 - x}{l_1} - \left(1 + \frac{\beta_2 k}{l_2}\right) \cdot \frac{x_2 - x}{l_2} \tag{5.12}$$

which, with (5.7) lead to the characteristic equation

$$\det\begin{pmatrix} m\omega^2+K\left(\dfrac{1}{l_1}+\dfrac{1}{l_2}\right) & -\dfrac{K}{l_1} & -\dfrac{K}{l_2} \\ \dfrac{1}{l_1}\cdot\left(1-\dfrac{\beta_1 K}{l_1}\right) & \dfrac{\omega}{V}-\dfrac{1}{l_1}\left(1-\dfrac{\beta_1 k}{l_1}\right) & \dfrac{\beta_1 k}{l_1 l_2} \\ -\dfrac{1}{l_2}\cdot\left(1+\beta_2\dfrac{K}{l_2}\right) & \dfrac{\beta_2 k}{l_1 l_2} & \dfrac{\omega}{V}+\dfrac{1}{l_2}\left(1+\dfrac{\beta_2 k}{l_2}\right)\end{pmatrix}=0$$

or, having regard to (5.6)

$$m\omega^4+V\left[m\left(\frac{1}{l_2}-\frac{1}{l_1}\right)+mk\left(\frac{\beta_1}{l_1^2}+\frac{\beta_2}{l_2^2}\right)\right]\omega^3$$
$$+\left[k\left(\frac{1}{l_1}+\frac{1}{l_2}\right)^2-\frac{mV^2}{l_1 l_2}+\frac{mV^2 k}{l_1 l_2}\left(\frac{\beta_1}{l_1}-\frac{\beta_2}{l_2}\right)\right]\omega^2=0.$$

If we assume that $\dfrac{\beta_1}{l_1}\sim\dfrac{\beta_2}{l_2}$, we see that the effect of drift is favourable to stability and increases damping.

b) This problem can also be discussed on the hypothesis that the mass is distributed, which is obviously closer to reality. We shall denote by G_1, G_2 the centres of inertia of the arm/wheel systems O_1 and O_2. The corresponding masses are m_1, m_2, the moments of inertia with respect to the axes $G_1 z$, $G_2 z$ are I_1, I_2, and we write

$$\lambda_1=\frac{\overline{OG_1}}{l_1},\quad \lambda_2=\frac{\overline{G_2 O}}{l_2}.$$

The potential energy of the system is:

$$W=\frac{k}{2}\left(\frac{x-x_1}{l_1}+\frac{x-x_2}{l_2}\right)^2 \tag{5.13}$$

and the kinetic energy can be expressed as:

$$T=\frac{I_1}{2}\left(\frac{x'-x_1'}{l_1}\right)^2+\frac{I_2}{2}\left(\frac{x'-x_2'}{l_2}\right)+\frac{m_1}{2}[(1-\lambda_1)x'+\lambda_1 x_1']^2 \tag{5.14}$$
$$+\frac{m_2}{2}[(1-\lambda_2)x'+\lambda_2 x_2']^2$$

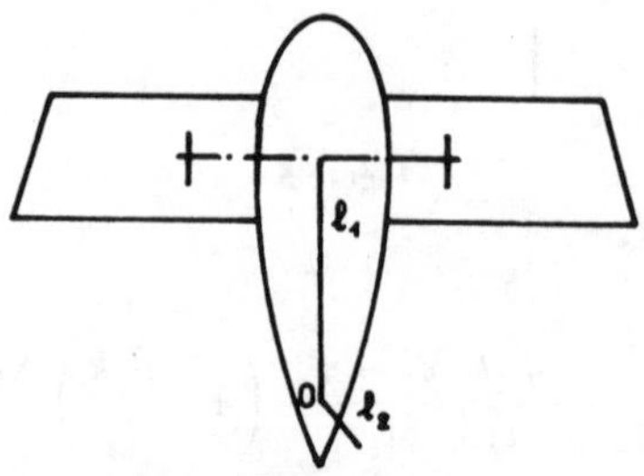

Figure 5.5

if we omit the terms associated with the motion in the $\vec{y}$ direction and the terms arising out of the proper rotations of the wheels which, having regard to our hypotheses play no part in the equations governing x, x_1, x_2. In the absence of drift effect, we have to take account of the non-holonomic constraints (5.4), (5.5) and we obtain the dynamic equation:

$$\frac{\mathrm{d}}{\mathrm{d}t}\left(\frac{\partial T}{\partial x'}\right)-\frac{\partial T}{\partial x}=-\frac{\partial W}{\partial x}$$

or, writing

$$M=\frac{I_1}{l_1^2}+\frac{I_2}{l_2^2}+m_1(1-\lambda_1)^2+m_2(1-\lambda_2)^2,$$

$$\mu_1=m_1(1-\lambda_1)\lambda_1-\frac{I_1}{l_1^2},\quad \mu_2=m_2(1-\lambda_2)\lambda_2-\frac{I_2}{l_2^2},$$

(5.15) $$M\frac{\mathrm{d}^2x}{\mathrm{d}t^2}+\mu_1\frac{\mathrm{d}^2x_1}{\mathrm{d}t^2}+\mu_2\frac{\mathrm{d}^2x_2}{\mathrm{d}t^2}+K\left(\frac{x-x_1}{l_1}+\frac{x-x_2}{l_2}\right)=0.$$

We deduce from (5.4), (5.5), (5.15) the characteristic equation in the form:

(5.16)

$$M\omega^4+MV\left(\frac{1+\frac{\mu_2}{M}}{l_2}-\frac{1+\frac{\mu_1}{M}}{l_1}\right)\omega^3+\left(k\left(\frac{1}{l_1}+\frac{1}{l_2}\right)^2-\frac{M+\mu_1+\mu_2}{l_1l_2}V^2\right)\omega^2=0$$

from which we can easily discuss the stability conditions.

To take account of drift effect, we make use of the constraint relations (5.10) and (5.11), and write down the virtual work equation in the form

$$R_1\delta x_1+R_2\delta x_2-\delta W=\Sigma\left(\frac{\mathrm{d}}{\mathrm{d}t}\left(\frac{\partial T}{\partial x'}\right)-\frac{\partial T}{\partial x}\right)\delta x,$$

where the summation is over the three terms in x, x_1, x_2.

Thus, in addition to the equations (5.10), (5.11), (5.15) which are still valid, we shall have:

(5.17) $$R_1+\frac{k}{l_1}\left(\frac{x-x_1}{l_1}+\frac{x-x_2}{l_2}\right)=\frac{I_1}{l_1}\cdot\frac{x_1''-x''}{l_1}+m_1\lambda_1[(1-\lambda_1)x''+\lambda_1x_1'']$$

(5.18) $$R_2+\frac{k}{l_2}\left(\frac{x-x_1}{l_1}+\frac{x-x_2}{l_2}\right)=\frac{I_2}{l_2}\cdot\frac{x_2''-x''}{l_2}+m_2\lambda_2[(1-\lambda_2)x''+\lambda_2x_2''].$$

R_1 and R_2 can be eliminated from (5.10), (5.17) and (5.11), (5.18), and we thus obtain the three equations of motion. The characteristic equation can be derived from these, enabling stability to be discussed.

Let us however go back to (5.16) to discuss the stability of a taxying aircraft. We see that in the case illustrated in Figure 5.5, with a tail-wheel assembly, we have $l_1>0$, $l_2>0$ and very small, μ_2 is negligible, $k=0$, and the rolling movement is unstable.

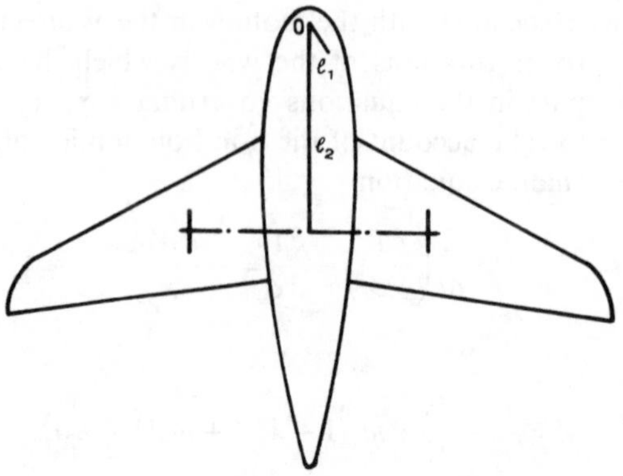

Figure 5.6

In the case illustrated in Figure 5.6, with a nose wheel placed in front under the nose of the aircraft, we have $l_1 < 0$, $l_2 > 0$, μ_1 is negligible, $k = 0$, the movement is stable and the damping factor can be high if $|l_1|$ is small enough.

2. Yawing of Road Trailers

We consider a road trailer hitched to a tractor which we assume to be travelling in uniform rectilinear motion at a speed V. The trailer is articulated at the point P and is in contact with the ground through the wheel-and-axle assembly.

The body of the trailer can become displaced with respect to the centre of the axle and it can also lean over during the course of the motion of the tractor and trailer. All these various effects are illustrated diagrammatically in Figure 5.8.

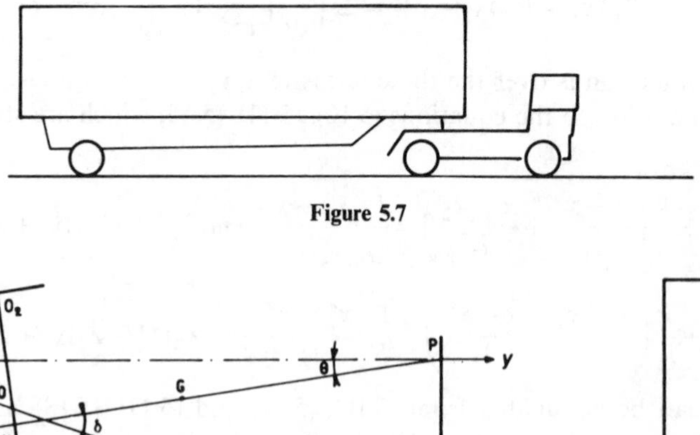

Figure 5.7

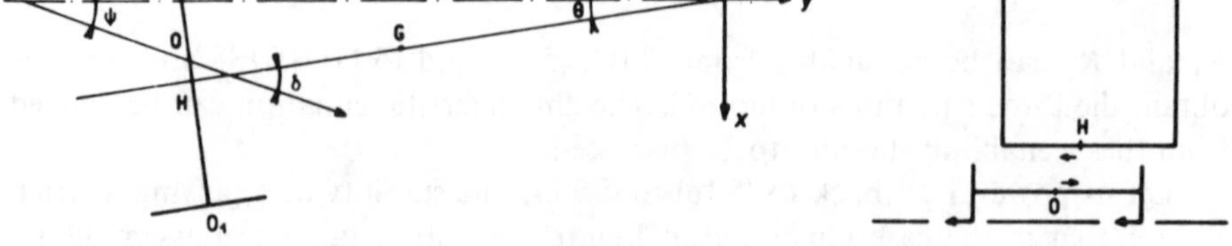

Figure 5.8

Denoting by G the projection of the centre of inertia of the body of the trailer on to the horizontal plane of the axle, we shall assume that PG remains orthogonal to the axle O_1O_2 at H, which may possibly be different from O, the centre of O_1O_2.

Let x be the lateral displacement of the point H, and ξ that of O; we can describe the elastic interaction between the body and axle by a relation of the form

$$x-\xi=kF, \quad k>0, \tag{5.19}$$

where F is the force exerted by the body on the axle in the direction normal to the plane of the road-wheels, the reaction $-F$ acting on the body. On the assumption that the mass of the axles can be ignored, the reactions of the ground in contact with the wheels in the direction normal to their plane can be taken to be $-F/2$ for each wheel; consequently the equation describing the 'drift' effect is:

$$\frac{1}{V}\frac{\mathrm{d}\xi}{\mathrm{d}t}+\frac{x}{a}=\beta F, \quad a=PH, \quad \left(\psi=\frac{1}{V}\frac{\mathrm{d}\xi}{\mathrm{d}t}, \quad \theta=\frac{x}{a}\right). \tag{5.20}$$

We shall write down the dynamic equation with respect to the Galilean axes Pxy, by noting that, to a first approximation, the motion of the trailer-body can be treated as though it were a rotation about the point P, so that

$$I\frac{\mathrm{d}^2\dfrac{x}{a}}{\mathrm{d}t^2}=-Fa.$$

I is the moment of inertia about the vertical axis through P, $-Fa$ is the moment at P of the forces exerted by the axle on the body. On putting

$$I=\mu a^2, \tag{5.21}$$

we have:

$$\mu\frac{\mathrm{d}^2x}{\mathrm{d}t^2}=-F. \tag{5.22}$$

Eliminating ξ, F from (5.19), (5.20), (5.22) we have

$$k\mu\frac{\mathrm{d}^3x}{\mathrm{d}t^3}+\beta\mu V\frac{\mathrm{d}^2x}{\mathrm{d}t^2}+\frac{\mathrm{d}x}{\mathrm{d}t}+\frac{V}{a}x=0$$

and the stability condition can be written:

$$a>\frac{k}{\beta}. \tag{5.23}$$

Suppose the interaction between body and axle includes damping, i.e. that we substitute for (5.19):

$$F=\frac{x-\xi}{k}+f\cdot\left(\frac{\mathrm{d}x}{\mathrm{d}t}-\frac{\mathrm{d}\xi}{\mathrm{d}t}\right).$$

On this hypothesis the equations (5.20) and (5.22) can be written:

$$\frac{1}{V}\frac{\mathrm{d}\xi}{\mathrm{d}t}+\frac{x}{a}-\beta\cdot\frac{x-\xi}{k}-\beta f\cdot\left(\frac{\mathrm{d}x}{\mathrm{d}t}-\frac{\mathrm{d}\xi}{\mathrm{d}t}\right)=0$$

$$\mu\frac{\mathrm{d}^2x}{\mathrm{d}t^2}+f\cdot\left(\frac{\mathrm{d}x}{\mathrm{d}t}-\frac{\mathrm{d}\xi}{\mathrm{d}t}\right)+\frac{x-\xi}{k}=0$$

from which the characteristic equation

$$\mu(1+\beta f V)\omega^3+\left(f+\frac{\mu\beta V}{k}\right)\omega^2+\left(\frac{1}{k}+\frac{fV}{a}\right)\omega+\frac{V}{ka}=0$$

and the stability condition

$$fka > V(\mu k-\mu\beta a-f^2k^2).$$

can be easily deduced.

If $f^2>\frac{\mu}{k}\left(1-\beta\frac{a}{k}\right)$, which is certainly satisfied if (5.23) is true, there is stability no matter what V may be.

If $f^2<\frac{\mu}{k}\left(1-\beta\frac{a}{k}\right)$, there is a critical speed $V_0=\frac{fka}{\mu k-\mu\beta a-f^2k^2}$, above which there is instability.

We may also imagine that there is friction at the point of articulation P; assuming for simplicity friction of a viscous type, we replace (5.22) by:

$$\mu\frac{\mathrm{d}^2x}{\mathrm{d}t^2}=-F-K\frac{\mathrm{d}x}{\mathrm{d}t}$$

and associate with it the equations (5.19) and (5.20) whence, by eliminating F we have:

$$\mu\frac{\mathrm{d}^2x}{\mathrm{d}t^2}+K\frac{\mathrm{d}x}{\mathrm{d}t}+\frac{x-\xi}{k}=0$$

$$\frac{1}{V}\cdot\frac{\mathrm{d}\xi}{\mathrm{d}t}+\frac{x}{a}=\frac{\beta}{k}(x-\xi)$$

and the characteristic equation:

$$\mu\omega^3+\left(K+\beta\frac{\mu}{k}V\right)\omega^2+(K\beta V+1)k^{-1}\omega+\frac{V}{ka}=0.$$

In this case the stability condition is:

$$a>\frac{k}{\left(\beta+\frac{Kk}{\mu V}\right)(1+\beta K V)},$$

which is more favourable than (5.23) [60].

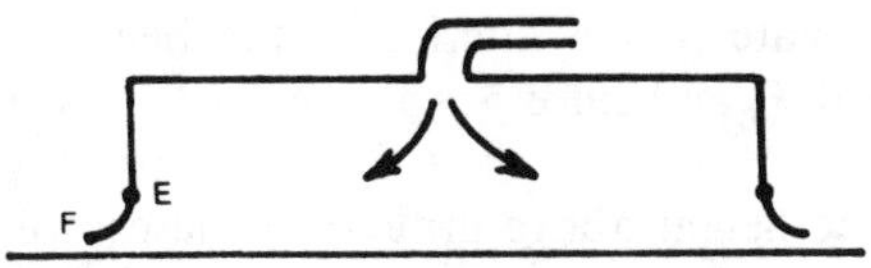

Figure 5.9

3. Lifting by Air-Cushion

The air-cushion or hovercraft principle was designed to allow heavy loads to be moved or vehicles to travel with an almost complete absence of frictional resistance [61]. The cushion of air is contained in a bell-shaped enclosure equipped at its basis with a skirt of some elastic material; the compressed air injected into the enclosure from a compressor escapes below through the annular space between the ground and the bottom edge of the skirt which is blown outwards by the pressure of the air, thus ensuring that the whole assembly is supported without any direct contact with the ground.

Another version is represented in figure 5.10 where the skirt is replaced by an elastic envelope shaped roughly like the lower half of a torus, part of whose outer surface forms the inner wall of the bell-shaped cavity. This envelope is kept under pressure by a by-pass from the compressor outlet.

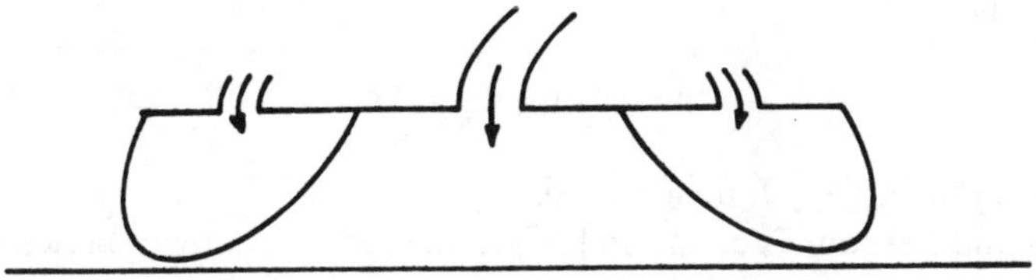

Figure 5.10

We shall describe in more detail below the operation and stability of the system illustrated in Figure 5.9.

We shall write p, ρ, m, v respectively for the pressure, density, mass and volume of the air contained in the cushion, so that:

$$m = v\rho. \tag{5.24}$$

Let q_1 be the inlet (mass) flow rate and q_2 the outlet flow rate, so that the continuity equation is

$$q_1 = q_2 + \frac{\mathrm{d}m}{\mathrm{d}t}. \tag{5.25}$$

The air which can be considered as being stationary inside the bell begins to put itself into a state of motion in the annular space between the skirt and the ground, and is ejected at a velocity V at the trailing edge F. The pressure varies continuously along the curved path EF of the line of flow, from the value p

corresponding to the state in the interior of the bell, to the value p_0 of the atmospheric pressure at F, and since $p > p_0$ the skirt billows outwards as shown in the diagram.

We denote by e the height above ground of the escaping air, and by h the height of E, the points at which the skirt is attached to the bell. We assume that the law governing the behaviour of the material of the skirt is given by:

(5.26) $$h - e = -k(p - p_0) + b,$$

where $k > 0$ is a coefficient of elasticity, and b is the height of the skirt above ground in the absence of deformation.

Now the pressure p depends on the impact flow rate q_1:

(5.27) $$p - p_0 = f(q_1),$$

where $f(q_1)$ is the characteristic function of the compressor, which always has a negative slope

(5.28) $$f_q < 0,$$

while the temperature T of the air in the cushion can be estimated once the operating characteristics of the compressor and the type of flow from compressor to cushion are known. The state of the air is governed by the law $\frac{p}{\rho} = RT$, $R = C_p - C_v$, where C_p, C_v are the specific heats at constant pressure and constant volume, or again by

(5.29) $$p = \rho^\gamma A(s), \quad \gamma = \frac{C_p}{C_v},$$

where $A(s)$ is the entropy function [23].

By applying Bernoulli's equation [36] to the radial jetstream between the skirt and the ground it is known that:

(5.30) $$V \mathrm{d}V = -\frac{\mathrm{d}p}{\rho},$$

the inlet and outlet airspeeds being 0 and V respectively and the inlet and outlet air pressures being p and p_0. In the case of a low-pressure cushion, the mass per unit volume remains constant in the jet so that we obtain, by integration of (5.30):

(5.31) $$p = p_0 + \tfrac{1}{2}\rho V^2.$$

Finally the mass discharged is:

(5.32) $$q_2 = \rho l e V,$$

where l is the length of the periphery of the air cushion.

The Stationary Regime

If M is the total mass supported including that of the cushion, the equilibrium equations can be written:

$$Mg = S(p - p_0), \quad p - p_0 = f(q_1), \quad q_1 = q_2 = \rho leV, \quad p = p_0 + \tfrac{1}{2}\rho V^2,$$

where g is the acceleration of gravity and S the ground surface area covered by the cushion. These enable us to calculate in turn $p - p_0, q_1$ then ρ and V if the leakage height e is specified, or e and V if the temperature T in the enclosure is fixed, so that ρ can be derived from the equation $p/\rho = RT$.

Let us take, for example, $p_0 \sim 10^5$ Pa, $p - p_0 \sim 2500$ Pa, $S = 1.5\,\text{m}^2$, $H = 0.1$ m, $v = SH = 0.15\,\text{m}^3$.

With $g \sim (10\,\text{m/s}^2)$, we obtain for the value of the mass M which can be supported

$$Mg = (1.5) \times 2500 = 3750\,\text{N} \quad \text{and} \quad M \sim 375\,\text{kg}.$$

Assuming $\rho \sim \rho_0 \sim 1.2\,\text{kg/m}^3$ we can deduce from (5.31) an ejection speed of $V \sim 63$ m/s.

Taking $l = 7$ m, $e = 0.006$ m, we can calculate the output $q_1 = q_2$ which specifies with $p - p_0$ a required point on the characteristic curve of the compressor.

Case of an Isentropic Expansion

If the pressure variation $p - p_0$ is considerable, Bernoulli's equation can no longer be integrated by ignoring the variation in density inside the jet.

It is however reasonable to assume that the expansion is isentropic, i.e. that

$$\frac{\mathrm{d}p}{\rho} = \gamma\rho^{\gamma-2}A(s)\mathrm{d}\rho = \frac{\gamma A(s)}{\gamma - 1}\mathrm{d}\rho^{\gamma-1} = \mathrm{d}\frac{c^2}{\gamma - 1}$$

where c is the velocity of sound:

(5.33) $$c^2 = \left(\frac{\partial p}{\partial \rho}\right)_s = \gamma\frac{p}{\rho} = \gamma p^{(\gamma-1)/\gamma}(A(s))^{1/\gamma},$$

whence by integrating (5.30):

(5.34) $$V^2 = \frac{2}{\gamma - 1}(c^2 - c_f^2),$$

c_f being the velocity of sound at the trailing edge,

(5.35) $$c_f^2 = \gamma p_0^{(\gamma-1)/\gamma}(A(s))^{1/\gamma}.$$

Lastly one can write (5.34) as

(5.36) $$\rho\frac{V^2}{2} = \frac{\gamma}{\gamma - 1}p\left(1 - \left(\frac{p_0}{p}\right)^{(\gamma-1)/\gamma}\right),$$

a formula which incidentally agrees with (5.31) in the case of a low-pressure air cushion with $\dfrac{p - p_0}{p_0} \ll 1$.

Finally the delivery output q_2 is given by $q_2 = \rho_f le V$ where ρ_f is the density at the trailing edge, which can be calculated by (5.29): $\rho_f = \rho\left(\dfrac{p_0}{p}\right)^{1/\gamma}$.

It should be noted that $\frac{V}{c_f}$ the Mach number at the output is easily obtained from (5.34):

$$\frac{V^2}{c_f^2} = \frac{2}{\gamma - 1}\left(\frac{c^2}{c_f^2} - 1\right) = \frac{2}{\gamma - 1}\left(\left(\frac{p}{p_0}\right)^{(\gamma-1)/\gamma} - 1\right).$$

Thus the flow is subsonic provided that

$$\frac{p}{p_0} < \left(\frac{\gamma + 1}{2}\right)^{\gamma/(\gamma - 1)},$$

i.e. for $\gamma = 1.4$, $\frac{p}{p_0} < 1.92$, and this sets a limit which it is advisable not to exceed for the air pressure in the cushion.

Dynamic Stability

We shall denote δq_1, δq_2, $\delta \rho, \ldots,$ for the small variations from the corresponding equilibrium values of the co-ordinates of the system when it is subjected to a small perturbation and we shall try to write down the differential equations which they satisfy in the context of a linear theory.

We can write the dynamic equation as:

$$M\frac{\mathrm{d}^2 \delta h}{\mathrm{d}t^2} = S\delta p \tag{5.37}$$

and we shall assume that the perturbation is of an isentropic character

$$\frac{\delta p}{p} = \gamma \frac{\delta \rho}{\rho} \tag{5.38}$$

and that the equations (5.26), (5.27), (5.31) still hold, so that

$$\delta h - \delta e = -k\delta p. \tag{5.39}$$

$$\delta p = f_q \delta q_1. \tag{5.40}$$

$$f\delta\rho + \rho\delta p = \rho V\delta(\rho V). \tag{5.41}$$

To make any progress we have to interpret, in terms of perturbation, the continuity equation.

Let us first suppose, for simplicity, that $\delta q_1 = \delta q_2$, i.e. having regard to (5.40) and (5.32):

$$\frac{\delta p}{f_q} = l(\rho V\delta e + e\delta(\rho V)). \tag{5.42}$$

Starting from (5.38), (5.39), (5.41), (5.42) we can calculate δp in terms of δh, viz: $\left[\frac{1}{f_q} - \frac{le}{V}\left(1 + \frac{f}{\gamma p}\right) - lV\rho k\right]\delta p = l\rho V\delta h$ and, on substitution in (5.37) we obtain:

$$\frac{d^2\delta h}{dt^2}+\omega_0^2\delta h=0 \quad \text{with} \quad \omega_0^2=\frac{l\rho VS}{M\left[\frac{le}{V}\left(1+\frac{f}{\gamma p}\right)+lV\rho k-\frac{1}{f_q}\right]}$$

which can be transformed, noting that $M=\frac{S(p-p_0)}{g}=\frac{\rho SV^2}{2g}$ and $f=p-p_0$: into

$$\omega_0^2=\frac{2g}{e\left(1+\frac{p-p_0}{\gamma p}\right)-\frac{V}{lf_q}+k\rho V^2}.$$

For a low-pressure cushion, the approximation $\omega_0^2=\frac{2g}{e}$ is good enough; thus with $e=0.006\,\mathrm{m}$, $g=10\,\mathrm{m/s^2}$ we find a frequency $\frac{\omega_0}{2\pi}\sim 10\,\mathrm{Hz}$.

However the preceding calculation is insufficient to give an account of dynamic stability. The perturbations δq_1, δq_2 in fact have distinct values, bound to one another by virtue of (5.25) by the equation:

$$\delta q_1=\delta q_2+\frac{d}{dt}(v\delta\rho+\rho\delta v)$$

or, after linearisation:

$$\delta q_1=\delta q_2+v\frac{d}{dt}\delta\rho+\rho S\frac{d}{dt}\delta h.$$

Accordingly we have to replace (5.42) by:

$$\frac{\delta p}{f_q}=l[\rho V\delta e+e\delta(\rho V)]+v\frac{d}{dt}\delta\rho+\rho S\frac{d}{dt}\delta h$$

i.e. with the help of (5.38) and (5.37):

$$\frac{\delta p}{f_q}=l[\rho V\delta e+e\delta(\rho V)]+\frac{v\rho M}{\gamma pS}\frac{d^3\delta h}{dt^3}+\rho S\frac{d}{dt}\delta h \tag{5.43}$$

and by eliminating δp, $\delta\rho$, δe, $\delta(\rho V)$ from (5.37), (5.38), (5.39) (5.41) and (5.43), we arrive at:

$$\frac{d^3\delta h}{dt^3}+\left(\frac{leV}{2v}+\frac{le\gamma p}{Vv\rho}+k\gamma\frac{plV}{v}-\frac{\gamma p}{v\rho f_q}\right)\frac{d^2\delta h}{dt^2}+\frac{\gamma p}{v}\frac{S^2}{M}\frac{d\delta h}{dt}+\frac{S\gamma plV}{Mv}\delta h=0.$$

The coefficients of this equation are all positive, and it can be seen that Routh's stability criterion is

$$\frac{leV}{2v}+\frac{le\gamma p}{vV\rho}+k\gamma\frac{plV}{v}-\frac{\gamma p}{v\rho f_q}-\frac{lV}{S}>0.$$

Chapter VI. Vibrations of Elastic Solids

The study of vibrations in elastic materials leads us to consider various problems which though diverse in formulation are nevertheless all amenable to solution by a uniform method which belongs to the calculus of variations. These problems appear in a great variety of guises depending on the geometric shape of the objects under study. The objects may for example have one, two or no 'privileged' dimensions in Euclidean space depending on whether they fall into the category of beams, plates, or bodies whose volume is of the same order of magnitude as the cube of any of its three cross-sectional dimensions. We may add to this that the material may be inhomogeneous, that static or dynamic boundary conditions may be imposed, that the latter may be transient or permanent, and that it may be appropriate to distinguish between flexional and torsional vibrations, and between longitudinal and transverse vibrations.

The natural vibrations of beams under flexion will be discussed on the basis of the classical equations and under various assumptions as to the nature of the static or kinematic conditions imposed at one or more points. The existence of an infinite set of eigenfrequencies, some of them multiple but of finite order, follows from Rayleigh's method which at the same time defines an algorithm which can be used to calculate their values. When supplemented by the min-max theorem it provides some useful bounds and asymptotic estimates. Several examples including the vibrations of turbine blades and the coupling between vibrations and flow in a supported pipe are described by way of illustration.

The theorem on the expansion of a kinematically admissible vector displacement in a series of eigenfunctions is a useful tool for solving the problem of finding the forced vibrations induced by a harmonic excitation of frequency ω. The form of the transfer function can be deduced from it, as can the representation of the forced vibration caused by an arbitrary excitation obeying any given temporal law, by using a method due to Fourier. The study of the vibrations of three-dimensional elastic bodies is based on a similar analysis in which Rayleigh's algorithm plays an essential role. This applies to the calculation of eigenfrequencies and of vibrations sustained by prescribed displacements or forces at the boundary, or volume forces which are periodic in time t, and to the structure of the transfer function.

An account of the vibrations of bars and of plates completes this exposition, which includes an analysis of the structure of the vibratory state in periodic elastic media and their representation by Bloch functions.

By adopting the 'functional operator' point of view the presentation of all these developments can be given a profound unity; the method of moments, which is described in the section dealing with the vibrations of bars, but which has a more general range of applicability, leads to an algorithm for the numerical analysis of the set of eigenvalues of a solid body in vibration.

I. Flexional Vibrations of Beams

1. Equations of Beam Theory

We suppose the beam in its rest state to have a plane of symmetry Oxy and that the forces acting upon it are symmetric with respect to this plane. In that case the material elements of the beam initially lying in this plane at a given instant of time remain in it, and the beam flexes about Oz, $Oxyz$ denoting an orthonormal set of axes. We assume that there exists within the beam, and lying in the plane of symmetry, a so-called neutral line, which undergoes no change of length during vibration and which, in the rest position is rectilinear and co-incides with the segment of the Ox axis lying between $x = 0$ and $x = l$.

We shall adopt the usual hypothesis that the particles contained in any slice normal to x before deformation remain rigidly connected in any plane section normal to the neutral line. We shall also assume that the motion of the latter can be represented by:

$$y = v(x, t), \quad z = 0$$

and the object of the discussion which follows is to describe the free oscillations which can arise in such a continuous medium, under a variety of different boundary conditions.

We write P for an arbitrary point of abscissa x on the neutral line, $T\vec{y}$ for the shearing stress and $M\vec{z}$ for the moment at P of the forces exerted through the normal section by the elements of the beam whose abscissae exceed x on those whose abscissae are less than x.

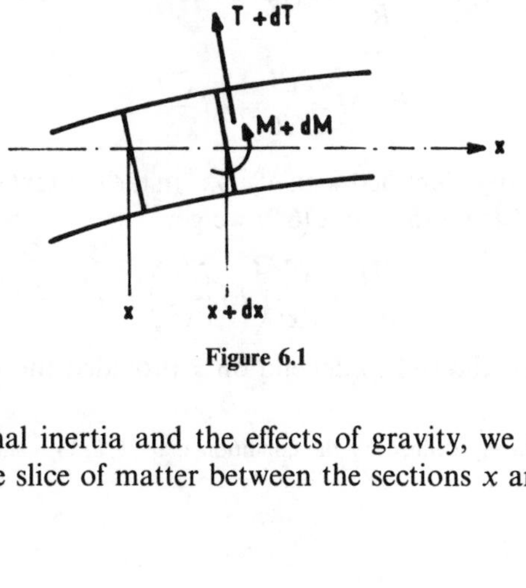

Figure 6.1

Ignoring rotational inertia and the effects of gravity, we obtain by applying Newton's laws to the slice of matter between the sections x and $x + dx$:

(6.1)
$$\rho\frac{\partial^2 v}{\partial t^2}\mathrm{d}x=\mathrm{d}T,\quad \mathrm{d}M+T\mathrm{d}x=0$$

where $\rho(x)$ denotes the linear density [23].

To obtain an expression for the moment M, we take account of the forces of tension or compression exerted along fibres parallel to the neutral line by assuming that these forces are proportional to the relative elongation of the fibre at the point under consideration. Thus, starting from the representation $\vec{P}+y\vec{n}+z\vec{k}$ of an arbitrary point of the normal cross-section at x, where $\vec{k}$ is the unit vector along the axis Oz, $\vec{n}$ the unit vector normal to the neutral line, usually defined by $\vec{n}=R\mathrm{d}\vec{\tau}/\mathrm{d}s$, R being the radius of curvature[1], $\vec{\tau}$ the unit vector tangent at P to the neutral line and s the curvilinear co-ordinate in the direction of increasing x, we have $\vec{\tau}=\mathrm{d}\vec{P}/\mathrm{d}s$, and the relative elongation of the fibre (y,z) described by the point $\vec{P}(s)+y\vec{n}+z\vec{k}$ when s varies and y,z are constants is obtained from the equation $\frac{\mathrm{d}}{\mathrm{d}s}(\vec{P}+y\vec{n}+z\vec{k})=\left(1-\frac{y}{R}\right)\vec{\tau}$, where we have made use of $\frac{\mathrm{d}\vec{n}}{\mathrm{d}s}=-\frac{\vec{\tau}}{R}$.

Denoting by σ the measure of the arc on the fibre (y,z) we have $\frac{\mathrm{d}\sigma-\mathrm{d}s}{\mathrm{d}s}=-\frac{y}{R}$ and since the measure of the arc of this fibre between the sections x and $x+\mathrm{d}x$, before deformation is $\mathrm{d}x=\mathrm{d}s$, we see that the relative elongation sought is:

$$\frac{\mathrm{d}\sigma-\mathrm{d}s}{\mathrm{d}s}=-\frac{y}{R}.$$

Hence the force per unit area exerted on the right section x at the point (x,y,z) by the elements of matter whose abscissae exceed x has the value $-\frac{Ey}{R}\vec{\tau}$, and its moment at P is equal to $\frac{Ey^2}{R}\vec{k}-\frac{Eyz}{R}\vec{n}$. By integration over the right section, we obtain the bending moment along $\vec{k}$ namely

$$M=\frac{E}{R}\int y^2\mathrm{d}y\,\mathrm{d}z=\frac{EI}{R}\qquad \left(\int yz\,\mathrm{d}y\,\mathrm{d}z=0 \text{ by reason of symmetry}\right)$$

that is

(6.2)
$$M=\frac{EI}{R}=EI\frac{\partial^2 v}{\partial x^2}$$

the curvature R^{-1} being identified with $\delta^2 v/\delta x^2$ in the context of a linearised theory.

Eliminating T,M from (6.1) and (6.2) we get

(6.3)
$$\rho\frac{\partial^2 v}{\partial t^2}+\frac{\partial^2}{\partial x^2}\left(EI\frac{\partial^2 v}{\partial x^2}\right)=0$$

where ρ, EI could be allowed to depend on x provided the variation were slow.

---

1 The vector $\vec{n}$ is completely defined by the condition that $(\vec{\tau},\vec{n},\vec{k})$ is an orthogonal system with the same orientation as $Oxyz$.

In studying the free vibrations of the beam, it is natural to look for solutions of (6.3) of the form

$$v = \sin \omega t \cdot \xi$$

with $\xi(x)$ a function of x only which has to satisfy:

$$(EI\xi'')'' - \rho\omega^2\xi = 0, \quad \forall x \in (0, l). \tag{6.4}$$

The boundary conditions now need to be specified; there are several cases to be considered depending on whether the beam is taken to be hinged, fixed (i.e. rigidly attached) or free at one or other of its ends.

1. Beam supported or hinged at both ends:
the kinematic conditions are $\xi(0) = \xi(l) = 0$. The absence of any bending moment at $x = 0$ and $x = l$ imposes the additional conditions:

$$EI\xi''(0) = EI\xi''(l) = 0.$$

2. Beam fixed at both ends:
the boundary conditions, which are all kinematic, are

$$\xi(0) = \xi(l) = 0, \quad \xi'(0) = \xi'(l) = 0.$$

3. Beam hinged at one end, fixed at the other end:
the kinematic conditions are $\xi(0) = \xi'(0) = \xi(l) = 0$. There is no moment at $x = l$, so that there is the additional condition:

$$EI\xi''(l) = 0.$$

4. Beam fixed at one end, free at the other:
the kinematic conditions are $\xi(0) = \xi'(0) = 0$.

The absence of constraint and of moment at $x = l$ imposes the additional conditions:

$$EI\xi''(l) = 0, \quad (EI\xi'')'(l) = 0.$$

5. Beam hinged at one end, free at the other:
the kinematic condition is $\xi(0) = 0$, and in addition we have

$$EI\xi''(0) = 0, \quad EI\xi''(l) = 0, \quad (EI\xi'')'(l) = 0.$$

6. Beam free at both ends:
no kinematic condition but

$$EI\xi''(0) = 0, \quad EI\xi''(l) = 0$$
$$(EI\xi'')'(0) = 0, \quad (EI\xi'')'(l) = 0.$$

2. A Simple Example

Suppose that EI, ρ are independent of x. The equation (6.4) can be written

$$\xi^{IV} = a^4 \xi \quad \text{with} \quad a = \left(\frac{\rho\omega^2}{EI}\right)^{1/4}$$

whose general solution is:

$$\xi = C_1 e^{ax} + C_2 e^{-ax} + C_3 \sin ax + C_4 \cos ax.$$

If we now write down the boundary conditions for any one of the six cases considered above we obtain a system of four linear homogeneous equations whose solutions are the C_i. In order that there should be a non-zero solution, their determinant, which can be expressed as a numerical function $\Delta(\sigma)$ of the variable $\sigma = al$, must vanish; to each root σ of the equation $\Delta(\sigma) = 0$ corresponds an eigenfrequency:

$$\omega = \frac{\sigma^2}{l^2}\sqrt{\frac{EI}{\rho}}. \tag{6.5}$$

We can now calculate the admissible values of σ corresponding to the various situations in §1.

Case 1: Beam Hinged at Either End.
We have

$$\xi = C \sin ax \quad \text{with} \quad \sin al = 0, \quad al = (n+1)\pi$$

and

$$\omega_n = (n+1)^2 \left(\frac{\pi}{l}\right)^2 \sqrt{\frac{EI}{\rho}}, \quad n = 0, 1, 2, \ldots \tag{6.6}$$

For the other cases, the equation $\Delta(\sigma) = 0$ is easily found, and the distribution of its roots is made clear by a graphical discussion.

Case 2: Beam Fixed at Both Ends.
We obtain in this way the equation $\cos\sigma \cosh\sigma = 1$. Examination of the graphs of $y = \cos\sigma$, $z = 1/\cosh\sigma$, aided by the observation that in the neighbourhood of $\sigma = 0$ we have

$$y \sim 1 - \frac{\sigma^2}{2} + \frac{\sigma^4}{24}, \quad z = 1 - \frac{\sigma^2}{2} + \frac{5}{24}\sigma^4$$

implying $y < z$, makes it clear that there is an infinity of solutions $\sigma_n, n = 0, 1, \ldots$ of which the approximate values of the first three are

$$\sigma_0 = 4.730 \quad \sigma_1 = 7.853 \quad \sigma_2 = 10.996$$

and that when n is large σ_n has the asymptotic representation:

$$\sigma_n = n\pi + \frac{3\pi}{2} + o(1/n).$$

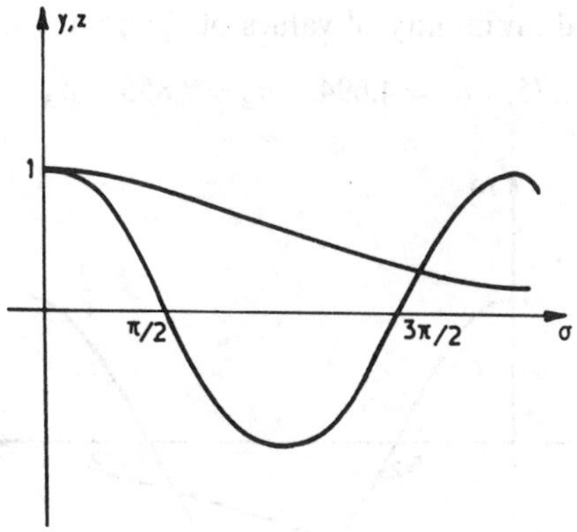

Figure 6.2

Case 3: Beam Hinged at One End, and Fixed at the Other.
The equation $\tan\sigma = \tanh\sigma$, obtained in this case can be discussed with the help of the graphs of $y = \tan\sigma$, $z = \tanh\sigma$.

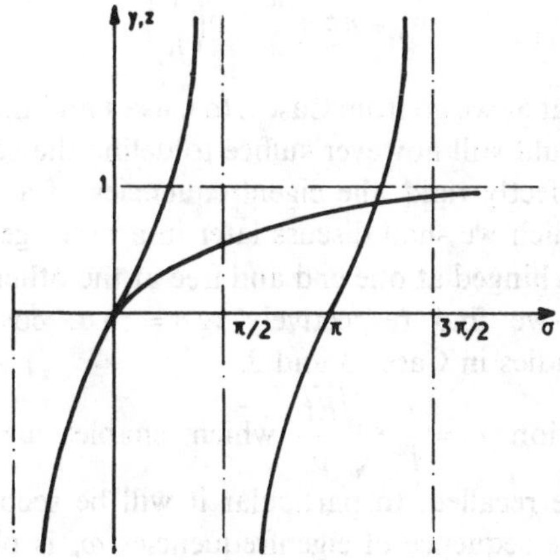

Figure 6.3

We obtain an infinity of solutions σ_n, $n = 0, 1, \ldots$ with

$$\sigma_0 = 3.927, \quad \sigma_1 = 7.069, \quad \sigma_2 = 10.210$$

and the asymptotic representation

$$\sigma_n = n\pi + \frac{5\pi}{4} + o(1/n).$$

Case 4: Beam Fixed at One End, Free at the Other.
We obtain $\cos\sigma\cosh\sigma = -1$ and with the help of the graphs of $y = \cos\sigma$,

$z = -1/\cosh \sigma$, we find an infinity of values of σ_n, $n > 0$ with

$$\sigma_0 = 1{,}875, \quad \sigma_1 = 4{,}694, \quad \sigma_2 = 7{,}855, \quad \sigma_3 = 10{,}996$$

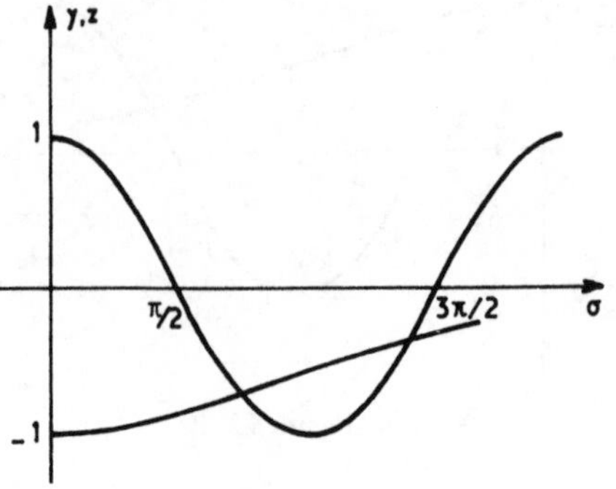

Figure 6.4

and the asymptotic representation

$$\sigma_n = n\pi + \frac{\pi}{2} + o\left(\frac{1}{n}\right).$$

It will be noted that as we go from Case 2 to Case 4 and thus relax the kinematic conditions, which would still however suffice to define the configuration uniquely if the beam were perfectly rigid, the eigenfrequencies of a given order decrease. This is a property which we shall discuss later in a more general context.

In Case 5, a beam hinged at one end and free at the other— or Case 6, a beam free at both ends—we find respectively $\operatorname{tg} \sigma = \operatorname{th} \sigma$, $\cos \sigma \cosh \sigma = 1$, i.e. the equations already studies in Cases 3 and 2.

Finally the relation $\omega = \frac{\sigma^2}{l^2} \cdot \sqrt{\frac{EI}{\rho}}$ which enables us to find the eigenfrequencies should be recalled. In particular it will be seen that in all the cases considered, an infinite sequence of eigenfrequencies ω_n is obtained such that we have, asymptotically

$$\omega_n = O(n^2), \quad \text{as} \quad n \to \infty.$$

3. The Energy Equation

We revert to the general case described by the equation (6.3) and suppose that there exists a solution $v(x, t)$ satisfying the appropriate boundary conditions. Multiplying each side by $\partial v/\partial t$ and integrating with respect to x, we can write:

$$\int_0^l \rho \frac{\partial v}{\partial t} \frac{\partial^2 v}{\partial t^2} \mathrm{d}x + \int_0^l \frac{\partial v}{\partial t} \frac{\partial^2}{\partial x^2}\left(EI \frac{\partial^2 v}{\partial x^2}\right) \mathrm{d}x = 0.$$

On integrating the second integral by parts twice, it is clear that the integrated parts

$$\frac{\partial v}{\partial t}\frac{\partial}{\partial x}\left(EI\frac{\partial^2 v}{\partial x^2}\right)\Bigg|_0^l$$

and

$$\frac{\partial}{\partial t}\left(\frac{\partial v}{\partial x}\right)\cdot EI\frac{\partial^2 v}{\partial x^2}\Bigg|_0^l$$

i.e.

$$-\frac{\partial v}{\partial t}\cdot T\Bigg|_0^l \quad \text{and} \quad \frac{\partial}{\partial t}\left(\frac{\partial v}{\partial x}\right)\cdot M\Bigg|_0^l$$

vanish in all the cases under consideration.

We thus obtain

$$\frac{\partial}{\partial t}\left[\frac{1}{2}\int_0^l \rho\left(\frac{\partial v}{\partial t}\right)^2 \mathrm{d}x + \frac{1}{2}\int_0^l EI\left(\frac{\partial^2 v}{\partial x^2}\right)^2 \mathrm{d}x\right] = 0$$

and by integrating:

(6.7) $$T + V = h,$$

where

$$T = \frac{1}{2}\cdot\int_0^l \rho\left(\frac{\partial v}{\partial t}\right)^2 \mathrm{d}x,$$

is the kinetic energy of the beam, and

$$V = \frac{1}{2}\cdot\int_0^l EI\left(\frac{\partial^2 v}{\partial x^2}\right)^2 \mathrm{d}x,$$

is the elastic energy, h being a constant.

It will be noted that, if $v = \sin\omega t\cdot\xi$ is a representation of harmonic type, (6.7) can be written as:

$$\frac{\omega^2}{2}\int_0^l \rho\xi^2 \mathrm{d}x\cdot\cos^2\omega t + \frac{1}{2}\int_0^l EI\xi''^2 \mathrm{d}x\cdot\sin^2\omega t = h$$

which requires

(6.8) $$\omega^2 = \frac{\int_0^l EI\xi''^2 \mathrm{d}x}{\int_0^l \rho\xi^2 \mathrm{d}x}.$$

We may also observe that as the mean values of the kinetic energy and elastic energy, averaged over time, are

$$\frac{\omega^2}{4}\int_0^l \rho\xi^2 \mathrm{d}x \quad \text{and} \quad \frac{1}{4}\int_0^l EI\xi''^2 \mathrm{d}x,$$

the formula (6.8) expresses the equality of these mean values.

The flexional energy can also be obtained by a direct calculation if we imagine the beam to be subjected, starting from its deformed state at time t, to a virtual displacement

$$\delta v(x), \quad \delta\psi(x) = \delta\frac{\partial v}{\partial x} = \frac{\partial}{\partial x}(\delta v), \quad \text{writing } \psi \text{ for } \frac{\partial v}{\partial x}.$$

The work done by the forces exerted at the ends of the beam to achieve this virtual displacement is

$$T\delta v + M\delta\psi \qquad \text{at } x = l$$

and

$$-T\delta v - M\delta\psi \qquad \text{at } x = 0,$$

with $T = -\partial M/\partial x$, or in all:

$$\delta W = \int_0^l \left[-\frac{\partial M}{\partial x}\delta v + M\delta\left(\frac{\partial v}{\partial x}\right)\right]_x \mathrm{d}x$$

i.e. remembering that

$$\frac{\partial}{\partial x}\delta v = \delta\left(\frac{\partial v}{\partial x}\right):$$

$$\delta W = \int_0^l \left[-\frac{\partial^2 M}{\partial x^2}\delta v + M\delta\left(\frac{\partial^2 v}{\partial x^2}\right)\right]\mathrm{d}x$$

and by (6.2), (6.3)

$$\delta W = \int_0^l \rho\frac{\partial^2 v}{\partial t^2}\delta v\,\mathrm{d}x + \int_0^l EI\frac{\partial^2 v}{\partial x^2}\delta\left(\frac{\partial^2 v}{\partial x^2}\right)\mathrm{d}x$$

or

$$\delta W - \int_0^l \rho\frac{\partial^2 v}{\partial t^2}\delta v\,\mathrm{d}x = \delta\left(\frac{1}{2}\int_0^l EI\left(\frac{\partial^2 v}{\partial x^2}\right)^2 \mathrm{d}x\right)$$

which signifies that the virtual work done by the external forces plus the virtual work done by the inertial forces is equal to the variation in the elastic energy

(6.9)
$$V = \frac{1}{2}\int_0^l EI\left(\frac{\partial^2 v}{\partial x^2}\right)^2 \mathrm{d}x.$$

4. The Modified Equations of Beam Theory; Timoshenko's Model

The effects due to shearing stresses and the inertia of rotation which have been ignored in the above treatment may be taken into account at the cost of a few modifications. We suppose the particles in a cross-section of the beam to constitute an invariable configuration and in particular, to stay in a plane but this plane no longer remains orthogonal to the neutral line, owing to the effects of shear. We shall denote by ψ the angle between Ox and the normal PX to the plane section in its deformed state, and by β the angle of shear, that is to say the angle between

PX and the neutral line, so that

(6.10) $$\psi + \beta = \frac{\partial v}{\partial x}.$$

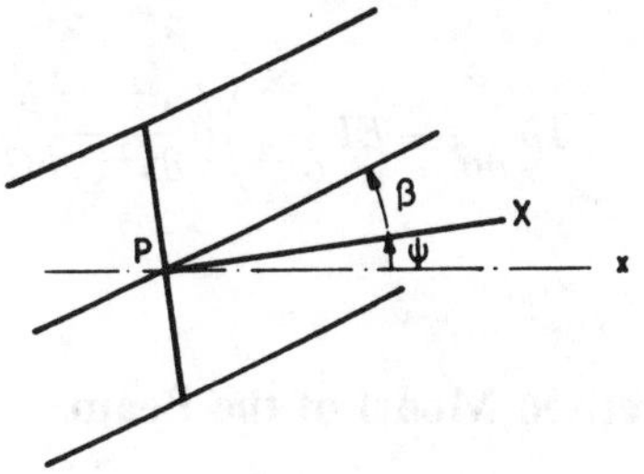

Figure 6.5

It is known that

(6.11) $$T = kAG\beta$$

where G, the modulus of rigidity, depends only on the material, A is the area of the plane section, and k is a geometric constant. The dynamic equation expressing the vanishing of the moment at P of all the forces applied to the slice between the sections x and $x + \mathrm{d}x$ including inertial forces, can be written:

(6.12) $$I\tilde{\rho}\frac{\partial^2 \psi}{\partial t^2} = \frac{\partial M}{\partial x} + T$$

where $\tilde{\rho}$ denotes the mass per unit volume, $\tilde{\rho}A = \rho, I$ being, as above, the geometric moment of inertia of the plane section about the axis Pz. The dynamic equation for the vertical motion of the slice gives

(6.13) $$\frac{\partial T}{\partial x} = \tilde{\rho}A\frac{\partial^2 v}{\partial t^2}$$

and the law (6.11) describing the behaviour of the system is completed by:

(6.14) $$M = EI\frac{\partial \psi}{\partial x}.$$

Eliminating T from (6.11) and (6.13) we arrive at

$$\frac{\partial(kAG\beta)}{\partial x} = \bar{\rho}A\frac{\partial^2 v}{\partial t^2},$$

and then eliminating M and T from (6.11), (6.12), (6.14) we obtain

$$I\tilde{\rho}\frac{\partial^2 \psi}{\partial t^2} = \frac{\partial}{\partial x}\left(EI\frac{\partial \psi}{\partial x}\right) + kAG\beta$$

to which must be adjoined (6.10).

In the simple case where $kAG, I\tilde{\rho}, EI$ are independent of x, we can write, by (6.10):

$$\frac{\partial \psi}{\partial x} = \frac{\partial^2 v}{\partial x^2} - \frac{\partial \beta}{\partial x} = \frac{\partial^2 v}{\partial x^2} - \frac{\tilde{\rho}}{kG}\frac{\partial^2 v}{\partial t^2}$$

and finally:

$$A\tilde{\rho}\frac{\partial^2 v}{\partial t^2} = \left(I\tilde{\rho}\frac{\partial^2}{\partial t^2} - EI\frac{\partial^2}{\partial x^2}\right)\left(\frac{\partial^2}{\partial x^2} - \frac{\tilde{\rho}}{kG}\frac{\partial^2}{\partial t^2}\right)v. \tag{6.15}$$

5. Timoshenko's Discretised Model of the Beam

The neutral axis is represented by a chain of hinged rods of equal length l; this chain moving in the plane Oxy does not shift far from the axis Ox, so that the co-ordinates of each node P_n can be represented by $x = nl$, $y = y_n(t)$. Around each axis $P_n z$ can turn a solid S_n in the form of a parallelepiped of thickness $2h$ ($2h < l$) and height $2d$ and whose base, that is to say the face perpendicular to the plane Oxy is of area A. We write $\psi_n = (Ox, P_n X_n)$ for the angle between the axis Ox and the normal $P_n X_n$ to the base, and

$$\beta_n = (P_n X_n, P_n P_{n+1})$$

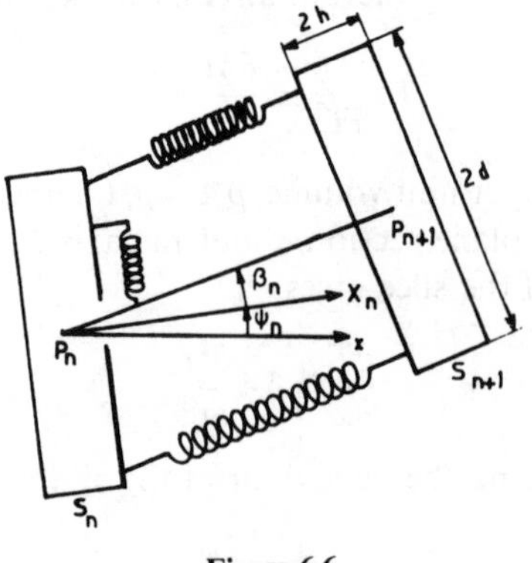

Figure 6.6

so that:

$$\frac{y_{n+1} - y_n}{l} = \psi_n + \beta_n. \tag{6.16}$$

The solids S_n are interconnected by a system of springs whose action can be described as follows: the forces exerted by the springs between S_n and S_{n+1} are zero when $\psi_n = \psi_{n+1}$ and for any arbitrary relative displacement corresponding to $\delta\psi_n$, $\delta\psi_{n+1}$, the work done by these forces can be represented by

$$\delta(-E(\psi_{n+1} - \psi_n)^2 d^2),$$

where E is constant which depends only on the stiffness of the springs, $(\psi_{n+1} - \psi_n)d$ measuring, to a first approximation, their elongation or contraction.

In addition each solid S_n is attached to the thin rod P_nP_{n+1} by a transversely mounted spring, approximately parallel to the base plane through P_n at a distance b from it. The force exerted on the rod P_nP_{n+1} tends to align it with the axis P_nX normal to the base of S_n and can be represented by $-Gb\beta_n$, $b\beta_n$ representing the elongation of the spring and G depending on its rigidity. For a virtual displacement $\delta\beta_n$, the work done by this force will be

$$\delta\left(-\frac{G}{2}b^2\beta_n^2\right).$$

In conclusion we can represent the kinetic energy of the system by:

$$\mathscr{T} = \sum_n \tilde{\rho}Ah\left(\frac{\mathrm{d}y_n}{\mathrm{d}t}\right)^2 + \sum_n \tilde{\rho}Ih\left(\frac{\mathrm{d}\psi_n}{\mathrm{d}t}\right)^2$$

and its elastic energy by:

$$V = \frac{1}{2}\sum_n Gb^2\beta_n^2 + \sum_n E(\psi_{n+1} - \psi_n)^2 d^2,$$

$\delta(-V)$ representing the virtual work by the elastic forces in an arbitrary virtual displacement $\delta\beta_n, \delta\psi_n$.

To write down the equations of motion by Lagrange's method [11], it is more convenient to use the variables y_n, ψ_n with β_n obtained by (6.16) so that:

$$\tilde{\rho}I\frac{\mathrm{d}^2\psi_n}{\mathrm{d}t^2} = \frac{Ed^2}{h}(\psi_{n+1} + \psi_{n-1} - 2\psi_n) + \frac{Gb^2}{2h}\beta_n \tag{6.17}$$

$$\tilde{\rho}A\frac{\mathrm{d}^2y_n}{\mathrm{d}t^2} = \frac{Gb^2}{2hl}(\beta_n - \beta_{n-1}).$$

We now proceed to the limit by allowing l, b and h to tend to zero in such a way that

$$b^2/2h \to kA, \quad l^2/h \to I/d^2,$$

with d remaining fixed, and k being a constant which thus depends on the geometry. We therefore obtain from (6.17), in the limit, the following equations:

$$\tilde{\rho}I\frac{\partial^2\psi}{\partial t^2} = EI\frac{\partial^2\psi}{\partial x^2} + kAG\beta$$

$$\tilde{\rho}A\frac{\partial^2 y}{\partial t^2} = kAG\frac{\partial\beta}{\partial x}$$

$$\frac{\partial y}{\partial x} = \psi + \beta$$

which are the same as those obtained for the continuous model.

6. Rayleigh's Method

We now return to the calculation of the eigenvectors of a beam under flexure. We recall that we have to solve the differential equation

$$(EI\xi'')'' - \rho\omega^2\xi = 0 \tag{6.18}$$

for $\xi(x)$ defined on the interval $(0, l)$ and satisfying various conditions at the ends $0, l$.

This problem has no non-zero solutions unless ω takes on one of certain possible values. These possible values form a discrete sequence of which the smallest is given by:

$$\omega^2 = \operatorname*{Inf}_{\xi \in K} \frac{\int_0^l EI\xi''^2 \, dx}{\int_0^l \rho\xi^2 \, dx}, \tag{6.19}$$

where K is the set of functions $\xi(x)$ with values in $\mathbb{R}$, which belong to $H^2(0, l)$ and which satisfy the kinematic conditions at $x=0$ and $x=l$. We recall that $H^2(0, l)$ is the Sobolev space of order 2 on the open interval $(0, l)$, i.e. the space generated by the functions $\xi(x)$ whose squares are Lebesgue-integrable over $(0, l)$ and whose first and second derivatives, as defined in the theory of distributions, are also square-integrable.

When provided with a scalar product defined by

$$(\xi, \zeta) = \int_0^l (\xi \cdot \zeta + \xi' \cdot \zeta' + \xi'' \cdot \zeta'') \, dx,$$

$H^2(0, l)$ is a Hilbert space.

6.1. Some Elementary Properties of the Spaces $H^1(0, l)$, $H^2(0, l)$

Let $H^1(0, l)$ be the space of functions[2] whose squares are integrable over $(0, l)$ and whose first derivative in the distributional sense is likewise square-integrable over the same interval. When endowed with the scalar product $(\xi, \zeta) = \int_0^l (\xi \cdot \zeta + \xi' \cdot \zeta') \, dx$, $H^1(0, l)$ is a Hilbert space, and it is clear that $H^2(0, l) \subset H^1(0, l)$, both algebraically and topologically.

Let T be a distribution, a continuous linear functional on the space $\mathscr{D}(0, l)$ of indefinitely differentiable functions $\varphi(x)$ with compact support in $(0, l)$ [14]. On the hypotheses that the derivative of T is identifiable with a function $g(x)$ which is a square-integrable function on $(0, l)$ it can be shown that the distribution T itself is a primitive of the function $g(x)$. In fact for any $\varphi \in \mathscr{D}(0, l)$ we have $\langle T', \varphi \rangle = \int_0^l g(s) \cdot \varphi(s) \, ds$, by the definition of $g(x)$, and $\langle T, \varphi \rangle = -\langle T, \varphi' \rangle$, by the rule for calculating the derivative of a distribution. It follows from this, by putting

[2] All functions considered in this section are assumed to be real-valued.

$f(x) = \int_0^x g(s)\,ds$, that, after integration by parts:

$$\langle T, \varphi' \rangle = \int_0^l f(s)\varphi'(s)\,ds.$$

If T_1 is the distribution defined by: $\langle T_1, \varphi \rangle = \int_0^l f(s)\cdot\varphi(s)\,ds$ we see that $T^* = T - T_1$ is a distribution such that

$$\langle T^*, \varphi' \rangle = 0, \quad \forall \varphi \in \mathscr{D}(0, l).$$

Let us now take a reference function φ_0 in $\mathscr{D}(0, l)$ such that $\int_0^l \varphi_0(s)\,ds \neq 0$ and let

$$\theta(x) = \int_0^x \varphi(s)\,ds \cdot \int_0^l \varphi_0(s)\,ds - \int_0^x \varphi_0(s)\,ds \cdot \int_0^l \varphi(s)\,ds$$

be defined for all $\varphi \in \mathscr{D}(0, l)$.

It is easily verified that $\theta \in \mathscr{D}(0, l)$, so that $\langle T^*, \theta' \rangle = 0$ which can be written

$$\langle T^*, \varphi \rangle = a \int_0^l \varphi(s)\,ds, \quad a = \frac{\langle T^*, \varphi_0 \rangle}{\int_0^l \varphi_0(s)\,ds}$$

whence

$$\langle T, \varphi \rangle = \int_0^l (f(s) + a)\varphi(s)\,ds.$$

Let us apply this result to an arbitrary ξ of $H^1(0, l)$ by regarding the function $\xi(x)$ as a distribution whose derivative can be identified with a square integrable function which we denote by ξ', then

$$\int_0^l \xi(x)\varphi(x)\,dx = \int_0^l \left(a + \int_0^x \xi'(s)\,ds \right)\varphi(x)\,dx, \quad \forall \varphi \in \mathscr{D}(0, l)$$

and since $\mathscr{D}(0, l)$ is dense in $L^2(0, l)$, [14], we deduce

$$\xi(x) = a + \int_0^x \xi'(s)\,ds$$

which shows that $\xi(x)$ is a continuous (indeed absolutely continuous [37]) function, which allows us to ascribe a meaning to $\xi(0)$ and to write

$$\xi(x) = \xi(0) + \int_0^x \xi'(s)\,ds.$$

The preceding argument can also be applied to ξ' when ξ is any element of $H^2(0, l)$ because it is obvious that $\xi' \in H^1(0, l)$. Thus $\xi'(x) = \xi'(0) + \int_0^x \xi''(s)\,ds$ is continuous, and the same will clearly be true of $\xi(x)$, so that we can define in a natural manner the values $\xi(0)$, $\xi(l)$, $\xi'(0)$, $\xi'(l)$ which appear in the kinematic conditions and consequently ensure the consistency of the definition of K.

Let us now consider a bounded infinite sequence of elements of $H^2(0, l)$, say ξ_n with $\|\xi_n\|_{H^2} < C$, i.e.

$$\int_0^l (|\xi_n|^2 + |\xi_n'|^2 + |\xi_n''|^2)\,ds < C^2. \tag{6.20}$$

We can extract from this sequence ξ_n a subsequence converging strongly in $H^1(0,l)$ to an element ζ_0 of that space. This property, which one can express by saying that the canonical injection of H^2 into H^1 is compact, is susceptible of important generalisations [14], but can be established very simply in the case which concerns us here.

Starting from

$$\xi_n'(x) = \xi_n'(0) + \int_0^x \xi_n''(s)\,ds \tag{6.21}$$

we deduce $\xi_n'(0)$ and the estimate

$$|\xi_n'(0)|^2 \leqslant 2|\xi_n'(x)|^2 + 2\left|\int_0^x \xi_n''(s)\,ds\right|^2 \leqslant 2|\xi_n'(x)|^2 + 2l\int_0^l |\xi_n''(s)|^2\,ds$$

by Schwarz's inequality. Integrating both sides with respect to x we obtain, using (6.20)

$$l|\xi_n'(0)|^2 \leqslant 2\int_0^l |\xi_n'(x)|^2\,dx + 2l^2\int_0^l |\xi_n''(s)|^2\,ds \leqslant 2(1+l^2)C^2$$

which shows that $\xi_n'(0)$ is a bounded sequence.

It follows from (6.21) that

$$|\xi_n'(x)| \leqslant |\xi_n'(0)| + \int_0^l |\xi_n''(s)|\,ds \leqslant |\xi_n'(0)| + l^{1/2}\cdot\left(\int_0^l |\xi_n''(s)|^2\,ds\right)^{1/2} \leqslant |\xi_n'(0)| + Cl^{1/2}$$

so that the sequence $\xi_n'(x)$ is likewise bounded over the interval $(0,l)$; it is equi-continuous since

$$|\xi_n'(x) - \xi_n'(y)| = \left|\int_y^x \xi_n''(s)\,ds\right| \leqslant |x-y|^{1/2}\cdot\left(\int_0^l |\xi_n''(s)|^2\,ds\right)^{1/2} \leqslant C\cdot|x-y|^{1/2}$$

and consequently, by a theorem of Arzela [17], one can extract from the sequence $\xi_n'(x)$ a subsequence which is uniformly convergent in $(0,l)$ and which tends to a necessarily continuous function $\tilde{\zeta}_0(x)$. There is no inconvenience in retaining the suffix n to index the subsequence and we shall do this to avoid complicating the notation unnecessarily.

Let us now consider:

$$\xi_n(x) = \xi_n(0) + \int_0^x \xi_n'(s)\,ds.$$

By repeating a part of the preceding analysis we see that $\xi_n(0)$ is a bounded sequence, which at the cost of a further extraction, can be taken to be convergent. The sequence $\xi_n(x)$ therefore converges uniformly to a continuous function $\zeta_0(x)$ and $\zeta_0(x) = \zeta_0(0) + \int_0^x \tilde{\zeta}_0(s)\,ds$, from which we deduce, since $\tilde{\zeta}_0(x)$ is continuous, that $\tilde{\zeta}_0(x) = \zeta_0'(x)$ everywhere. Thus $\xi_n(x)$, $\xi_n'(x)$ converge uniformly on $(0,l)$ to $\zeta_0(x)$, $\zeta_0'(x)$ respectively. Clearly $\zeta_0 \in H^1(0,l)$, $\lim_{n\to\infty} \|\xi_n - \zeta_0\|_{H^1} = 0$, and therefore the natural injection $\xi \to i(\xi)$ of $H^2(0,l)$ into $H^1(0,l)$ is compact.

Since $H^2(0,l)$ is a Hilbert space, it has the property that from any bounded infinite sequence of its elements, say ξ_n, $\|\xi_n\|_{H^2} < C$, we can extract a sequence converging weakly to an element ξ_0 of $H^2(0,l)$. We can keep the same indicial notation and suppose that $\xi_n \rightharpoonup \xi_0$. But in view of the preceding argument, by carrying out a fresh extraction if necessary, we can suppose that ξ_n converges strongly to an element ζ_0 in $H^1(0,l)$. We obtain the result $\zeta_0 = i(\xi_0)$, that is ζ_0 is derived from ξ_0 by the canonical injection of H^2 into H^1, which can be justified on the basis of the Banach-Saks approximation theorem [37]: if x_n is a sequence of elements of a Hilbert space $\mathscr{H}$ converging weakly to $x_0, x_n \rightharpoonup x_0$, then, for all $\varepsilon > 0$, there exists a convex linear combination $\sum_{j=1}^m \alpha_j x_j$,

$$\alpha_j \geqslant 0, \quad \sum_{j=1}^m \alpha_j = 1 \quad \text{such that} \quad \left\| \sum_{j=1}^m \alpha_j x_j - x_0 \right\|_{\mathscr{H}} \leqslant \varepsilon.$$

Since the property of convergence is not altered by a translation effected on the indices of the sequence, we can say that, for every positive integer p there exist m non-negative real numbers satisfying $\sum_{j=1}^m \alpha_j = 1$ such that

$$\left\| \sum_{j=1}^m \alpha_j x_{p+j} - x_0 \right\|_{\mathscr{H}} \leqslant \varepsilon.$$

Reverting to the situation described earlier for the sequence ξ_n, weakly converging to ξ_0 in H^2 and strongly converging to ζ_0 in H^1, let p be, for any given $\varepsilon > 0$, a positive integer such that

$$\|\xi_k - \zeta_0\|_{H^1} \leqslant \varepsilon, \quad \forall k > p, \tag{6.22}$$

and $\alpha_1, \ldots, \alpha_m \geqslant 0$, $\sum_{j=1}^m \alpha_j = 1$ a system or real numbers such that

$$\left\| \sum_{j=1}^m \alpha_j \xi_{p+j} - \xi_0 \right\|_{H^2} \leqslant \varepsilon. \tag{6.23}$$

Noting that

$$\|\xi_0 - \zeta_0\|_{H^1} \leqslant \left\| \xi_0 - \sum_{j=1}^m \alpha_j \xi_{p+j} \right\|_{H^1} + \left\| \sum_{j=1}^m \alpha_j \xi_{p+j} - \zeta_0 \right\|_{H^1}$$

or

$$\|\xi_0 - \zeta_0\|_{H^1} \leqslant \left\| \xi_0 - \sum_{j=1}^m \alpha_j \xi_{p+j} \right\|_{H^2} + \left(\sum_{j=1}^m \alpha_j \right) \cdot \operatorname*{Sup}_{1 \leqslant j \leqslant m} \|\xi_{p+j} - \zeta_0\|_{H^1}$$

we deduce from (6.22), (6.23) that:

$$\|\xi_0 - \zeta_0\|_{H^1} \leqq 2\varepsilon \quad \text{that is} \quad \zeta_0 = i(\xi_0).$$

We complete these results by a remark; suppose ξ_n to be an infinite sequence of elements of $H^2(0,l)$, which converge strongly in that space to ξ_0. Then from

$$\xi_n'(x) = \xi_n'(0) + \int_0^x \xi_n''(s)\,\mathrm{d}s$$

and the analogous formula for ξ_0, we easily obtain:

$$l|\xi_n'(0) - \xi_0'(0)|^2 \leqslant 2\int_0^l |\xi_n'(x) - \xi_0'(x)|^2\,\mathrm{d}x + 2l^2 \int_0^l |\xi_n''(x) - \xi_0''(x)|^2\,\mathrm{d}x.$$

i.e.

$$\lim_{n\to\infty} \xi_n'(0) = \xi_0'(0).$$

By

$$\left|\int_0^x \xi_n''(s)\,ds - \int_0^x \xi_0''(s)\,ds\right| \leqslant l^{1/2}\cdot\left(\int_0^l |\xi_n''(s) - \xi_0''(s)|^2\,ds\right)^{1/2}$$

we see that $\xi_n'(x)$ converges uniformly to $\xi_0'(x)$ on $(0, l)$ as $n\to\infty$; under these same conditions $\xi_n(x)$ converges uniformly to $\xi_0(x)$.

6.2. Existence of the Lowest Eigenfrequency

We define $\omega_0 \geqslant 0$ by:

$$\omega_0^2 = \operatorname*{Inf}_{\xi\in K} \frac{\int_0^l EI\xi''^2\,dx}{\int_0^l \rho\xi^2\,dx}, \tag{6.24}$$

where K is the subset of those functions ξ of $H^2(0, l)$ which satisfy the kinematic conditions. We shall for the moment restrict ourselves to the cases of hinged or fixed beams corresponding to the first four situations described in §I.1, reserving for later consideration the modifications needed to deal with the cases where the beams are floating.

Accordingly the conditions imposed are:

$\xi(0) = \xi(l) = 0$, hinged beam

$\xi(0) = \xi'(0) = \xi(l) = \xi'(l) = 0$, fixed beam

$\xi(0) = \xi'(0) = \xi(l) = 0$, beam with one end fixed, the other hinged

$\xi(0) = \xi'(0) = 0$, beam fixed at one end, free at the other.

It is obvious from the concluding remarks of I.6.1 that K is a closed linear variety of $H^2(0, l)$; furthermore by the Banach-Saks theorem, we can add that K is weakly closed in $H^2(0, l)$, because if ξ_n is an infinite sequence of elements of K weakly converging to ξ_0 in $H^2(0, l)$, then one can approximate ξ_0, in terms of the norm in $H^2(0, l)$, as closely as one likes, by a suitable linear combination $\sum_{j=1}^m \alpha_j\xi_j$, with $\alpha_j \geqslant 0$, $\sum_{j=1}^m \alpha_j = 1$, which clearly belongs to K. We suppose ρ, EI to be measurable functions of x, bounded over $[0, l]$, such that

$$a = \operatorname*{Inf}_{x\in[0,l]} \rho, \quad p = \operatorname*{Inf}_{x\in[0,l]} EI \tag{6.25}$$

are strictly positive.

Let ξ_n be an infinite sequence of minimising elements of K, that is to say, such that:

$$\omega_0^2 = \lim_{n\to\infty} \frac{\int_0^l EI\xi_n''^2\,dx}{\int_0^l \rho\xi_n^2\,dx} \tag{6.26}$$

ω_0 being defined by (6.24), and normalised by the condition:

$$\int_0^l \rho \xi_n^2 \, dx = 1. \tag{6.27}$$

From this definition and from (6.25) it follows at once that the sequences

$$\int_0^l \xi_n^2 \, dx \quad \text{and} \quad \int_0^l \xi_n''^2 \, dx$$

are bounded. We shall deduce from this that the sequence $\int_0^l \xi_n'^2 \, dx$ is bounded and that in consequence ξ_n is an infinite bounded sequence in $H^2(0,l)$. Let us suppose that $\varkappa_n^2 = \int_0^l \xi_n'^2 \, dx$ were unbounded, and that a subsequence extracted from it, which we shall suppose indexed in the same way, were to tend to infinity with $n \to \infty$: $\lim_{n\to\infty} \varkappa_n = \infty$. The sequence $\tilde{\xi}_n = \varkappa_n^{-1} \cdot \xi_n$ is bounded in $H^2(0,l)$ so that we can suppose it to converge weakly to $\tilde{\xi}_0$ in $H^2(0,l)$ and strongly to $\tilde{\xi}_0$ in $H^1(0,l)$.

Thus we should have $\lim_{n\to\infty} \int_0^l \tilde{\xi}_n^2 \, dx = \int_0^l \tilde{\xi}_0^2 \, dx$ and since

$$\lim_{n\to\infty} \int_0^l \tilde{\xi}_n^2 \, dx = \lim_{n\to\infty} \varkappa_n^{-2} \cdot \int_0^l \xi_n^2 \, dx = 0, \quad \tilde{\xi}_0 = 0 \quad \text{a.e.} \tag{6.28}$$

But by the definition of $\varkappa_n$, we have

$$\int_0^l \tilde{\xi}_n'^2 \, dx = 1 \quad \text{and} \quad \int_0^l \tilde{\xi}_0'^2 \, dx = \lim_{n\to\infty} \int_0^l \tilde{\xi}_n'^2 \, dx = 1, \tag{6.29}$$

in contradiction to $\tilde{\xi}_0 = 0$. We have thus proved that the minimising sequence ξ_n is bounded in $H^2(0,l)$; we can therefore suppose that it converges weakly to an element ξ_0 in $H^2(0,l)$ and strongly to this same element in $H^1(0,l)$.

In particular

$$\int_0^l \rho |\xi_n - \xi_0|^2 \, dx \leqslant \operatorname{Sup} \rho \cdot \int_0^l |\xi_n - \xi_0|^2 \, dx$$

tends to 0 as $n \to \infty$ whence

$$\lim_{n\to\infty} \int_0^l \rho \xi_n^2 \, dx = \int_0^l \rho \xi_0^2 \, dx$$

and by (6.27)

$$\int_0^l \rho \xi_0^2 \, dx = 1. \tag{6.30}$$

Finally since K is weakly closed in $H^2(0,l)$, it is clear that $\xi_0 \in K$, i.e. that ξ_0 satisfies the kinematic conditions. On the other hand the bilinear form

$$W(\xi, \eta) = \int_0^l EI \xi'' \cdot \eta'' \, dx$$

is positive and continuous on $H^2(0,l)$ and consequently, applying Riesz's theorem [37], we can define a continuous linear operator $\eta \to \mathscr{A}\eta$ from $H^2(0,l)$ into itself, such that $W(\xi, \eta) = (\xi, \mathscr{A}\eta)$ where (,,.) is the scalar product in $H^2(0,l)$.

From $W(\xi_n-\xi_0, \xi_n-\xi_0)\geqslant 0$, i.e.

$$W(\xi_n, \xi_n)\geqslant 2W(\xi_n, \xi_0)-W(\xi_0, \xi_0)$$

and

$$\lim_{n\to\infty} W(\xi_n, \xi_0)=\lim_{n\to\infty}(\xi_n, \mathscr{A}\xi_0)=(\xi_0, \mathscr{A}\xi_0)=W(\xi_0, \xi_0)$$

since ξ_n converges weakly to ξ_0 in $H^2(0,l)$ we conclude that

$$\liminf_{n\to\infty} W(\xi_n, \xi_n)\geqslant W(\xi_0, \xi_0)$$

and with (6.26), (6.27):

$$\omega_0^2=\lim_{n\to\infty} W(\xi_n, \xi_n)=\liminf_{n\to\infty} W(\xi_n, \xi_n)\geqslant W(\xi_0, \xi_0)$$

or

$$\omega_0^2\geqslant\frac{W(\xi_0, \xi_0)}{\int_0^l \rho\xi_0^2\,\mathrm{d}x} \quad \text{by (6.30).}$$

But we have shown that $\xi_0\in K$ and consequently it follows from the definition of ω_0 that we have:

$$\omega_0^2=\frac{\int_0^l EI\xi_0''^2\,\mathrm{d}x}{\int_0^l \rho\xi_0^2\,\mathrm{d}x}.$$

Moreover ω_0 is a true eigenfrequency, i.e. $\omega_0>0$ for $\omega_0=0$ would mean $\xi_0''=0$ almost everywhere and this would imply, since $\xi_0\in H^2(0,l)$, that $\xi_0=\alpha+\gamma x$ with α, γ numerical constants; but in view of the kinematic conditions satisfied by ξ_0 we should have $\alpha=\gamma=0$ or $\xi_0=0$, and this would contradict (6.30).

It remains to be shown that ξ_0 is a solution of the equation (6.4) with $\omega=\omega_0$ and that the remaining boundary conditions are satisfied. Now it follows from the definition of ω_0 that for all $\delta\xi\in K$ and all real λ we have

$$\int_0^l EI(\xi_0+\lambda\delta\xi)''^2\,\mathrm{d}x-\omega_0^2\int_0^l \rho(\xi_0+\lambda\delta\xi)^2\,\mathrm{d}x\geqslant 0$$

which yields

(6.31) $$\int_0^l (EI\xi_0''(\delta\xi)''-\omega_0^2\rho\xi_0\delta\xi)\,\mathrm{d}x=0.$$

Then for all $\delta\xi=\varphi\in\mathscr{D}(0,l)\subset K$, we have

$$\int_0^l (EI\xi_0''\varphi''-\omega_0^2\rho\xi_0\varphi)\,\mathrm{d}x=0$$

which means that

(6.32) $$(EI\xi_0'')''-\omega_0^2\rho\xi_0=0, \quad \forall x\in[0,l]$$

in the distributions sense.

It will be noted that it follows from this equation that $(EI\xi_0'')''$ can be identified with a square-integrable function; consequently $(EI\xi_0'')'$ is a primitive of it and is identifiable with an absolutely continuous function. For a similar reason $EI\xi_0''$ is absolutely continuous and even continuously differentiable.

Let us now return to the equation (6.31) which holds for all $\delta\xi \in K$; using (6.32) we can rewrite it as

$$\int_0^l [EI\xi_0''(\delta\xi)'' - \omega_0^2 \rho\xi_0 \delta\xi - \delta\xi((EI\xi_0'')'' - \omega_0^2 \rho\xi_0)] \,\mathrm{d}x = 0$$

or

$$\int_0^l [EI\xi_0''(\delta\xi)' - \delta\xi(EI\xi_0'')']' \,\mathrm{d}x = 0 \tag{6.33}$$

i.e.

$$(EI\xi_0''\cdot(\delta\xi)' - \delta\xi\cdot(EI\xi_0'')')|_0^l = 0$$

from which it is easily seen that the non-kinematic boundary conditions are satisfied.

Let us now examine how the preceding analysis needs to be modified for the case of beams simply hinged at one end $x = 0$, or free at both ends.

During any actual motion, the moment at O of the momentum is zero which is expressed by the equation $\int_0^l \rho x\xi(x)\,\mathrm{d}x = 0$ and furthermore, in the second case, the total momentum is zero so that $\int_0^l \rho\xi(x)\,\mathrm{d}x = 0$.

Accordingly for a beam free at both ends we shall define ω_0 by (6.24) where K is the set of elements of $H^2(0, l)$ such that

$$\int_0^l \rho\xi \,\mathrm{d}x = 0 \tag{6.34}$$

and

$$\int_0^l \rho x\xi \,\mathrm{d}x = 0. \tag{6.35}$$

It is easily verified that K is weakly closed in $H^2(0, l)$. We can proceed as before, starting from a minimising sequence ξ_n, and reach analogous conclusions.

It will be sufficient to satisfy one's self that ω_0 is not zero and that (6.31) which holds for all $\delta\xi \in K$, remains true for all $\delta\xi \in \mathscr{D}(0, l)$. If ω_0 were zero, we should have $\xi_0'' = 0$, i.e. $\xi_0 = \alpha + \gamma x$ with α, γ real constants; but the conditions (6.34), (6.35) applied to ξ_0 imply $\alpha = \gamma = 0$, or $\xi_0 = 0$, in contradiction to (6.30). As for (6.31) it is obvious that it is satisfied for all $\delta\xi = \alpha + \gamma x$, with α, γ constants because (6.34), (6.35) are satisfied by ξ_0 and since for any $\delta\xi \in \mathscr{D}(0, l)$ one can always find α and γ such that $\delta\xi - (\alpha + \gamma x) \in K$, it follows that (6.31) is true for all $\delta\xi \in \mathscr{D}(0, l)$.

In the case of a beam hinged at one end $x = 0$, and free at the other, we define K by:

$$K = \left\{ \xi : \xi \in H^2(0, l), \int_0^l \rho x\xi \,\mathrm{d}x = 0, \ \xi(0) = 0 \right\}.$$

6.3. Case of a Beam Supporting Additional Concentrated Loads

Let us suppose that additional masses μ_j are fixed to the beam at the points of abscissa x_j, $1 \leqslant j \leqslant p$. We again define

$$\omega_0^2 = \operatorname*{Inf}_{\xi \in K} \frac{\int_0^l EI\xi''^2 \, dx}{\int_0^l \rho\xi^2 \, dx + \sum_{j=1}^p \mu_j \xi^2(x_j)}$$

where K is the set of kinematically admissible elements of $H^2(0, l)$, or in the case of the floating beam, satisfying the zero momentum condition

$$\int_0^l \rho\xi \, dx + \sum_{j=1}^p \mu_j \xi(x_j) = 0, \quad \int_0^l \rho x\xi \, dx + \sum_{j=1}^p \mu_j x_j \xi(x_j) = 0.$$

We can repeat the arguments already developed leading to the conclusion that there exists an element $\xi_0 \in K$, for which the lower bound ω_0^2 is attained. It can be checked that for any element $\delta\xi$ of K, we can write

$$\int_0^l (EI\xi_0''(\delta\xi)'' - \omega_0^2 \rho\xi_0 \delta\xi) \, dx - \sum_{j=1}^p \omega_0^2 \mu_j \xi_0(x_j) \delta\xi(x_j) = 0$$

which holds, a fortiori, for all $\delta\xi \in \mathscr{D}(0, l)$, so that we can write, in the sense of the theory of distributions:

$$(EI\xi_0'')'' - \omega_0^2 \rho\xi_0 - \sum_{j=1}^p \omega_0^2 \mu_j \xi_0(x_j) \delta_j = 0$$

with $\delta_j = \delta(x - x_j)$, being the Dirac measure at the point x_j. This equation is none other than the one for the natural oscillation at frequency ω_0 of a beam carrying concentrated loads.

6.4. Intermediate Conditions Imposed on the Beam

Let us resume the analysis of the general case by supposing, for example, that there is an intermediate support at the point $x = l_1$, where $0 < l_1 < l$. This imposes the additional kinematic condition $\xi(l_1) = 0$.

For the problems defined by the first four sets of boundary conditions, let K_1 be the subset of elements of K which satisfy $\xi(l_1) = 0$ and for which ω_0 is given by

$$\omega_0^2 = \operatorname*{Inf}_{\xi \in K_1} \frac{\int_0^l EI\xi''^2 \, dx}{\int_0^l \rho\xi^2 \, dx}. \tag{6.36}$$

In the case of a beam hinged at $x = 0$ and free at $x = l$, we shall take

$$K_1 = \{\xi : \xi \in H^2(0, l), \xi(0) = 0, \xi(l_1) = 0\}$$

and in that of a beam free at both ends

$$K_1 = \left\{ \xi : \xi \in H^2(0,l), \xi(l_1) = 0, \int_0^l \rho \cdot (x - l_1)\xi(x)\,dx = 0 \right\}.$$

It is easily shown that there is a $\xi_0 \in K_1$ such that

$$\omega_0^2 = \frac{\int_0^l EI\xi_0''^2\,dx}{\int_0^l \rho\xi_0^2\,dx}, \quad \text{and} \quad \omega_0 > 0$$

and hence from (6.36):

$$\int_0^l (EI\xi_0''(\delta\xi)'' - \omega_0^2\rho\xi_0\delta\xi)\,dx = 0 \tag{6.37}$$

for all $\delta\xi \in K_1$, and in particular for all $\delta\xi \in \mathscr{D}(0, l_1)$ or $\delta\xi \in \mathscr{D}(l_1, l)$, which proves that, in the distributional sense:

$$(EI\xi_0'')'' - \omega_0^2\rho\xi_0 = 0 \tag{6.38}$$

on the open intervals $(0, l_1)$ and (l_1, l).

(In the case of a beam free at both ends, it is no longer correct to say that $\mathscr{D}(0, l_1)$, $\mathscr{D}(l_1, l)$ are included in K_1. However (6.37) holds for $\delta\xi = x - l_1$ as ξ_0 satisfies the condition of vanishing moment at $x = l_1$, and it will thus hold for all $\delta\xi$ belonging to $\mathscr{D}(0, l_1)$ or $\mathscr{D}(l_1, l)$ since a constant α can always be found such that $\delta\xi - \alpha(x - l_1) \in K_1$.)

It follows from (6.38) that $EI\xi_0''$, $(EI\xi_0'')'$, being indefinite integrals of a square-integrable function, are uniformly continuous on the open intervals $(0, l_1)$, (l_1, l) and consequently have finite limits when $x \to l_1$, on the right and on the left.

Having regard to (6.38) and for all $\delta\xi \in K_1$ we can rewrite (6.37):

$$\lim_{\varepsilon \to 0} \left(\int_0^{l_1 - \varepsilon} + \int_{l_1 + \varepsilon}^{l} \right) [EI\xi_0''(\delta\xi)'' - \omega_0^2\rho\xi_0\delta\xi - \delta\xi((EI\xi_0'')'' - \omega_0^2\rho\xi_0)]\,dx = 0$$

or

$$\lim_{\varepsilon \to 0} \left(\int_0^{l_1 - \varepsilon} + \int_{l_1 + \varepsilon}^{l} \right) [(EI\xi_0'')(\delta\xi)' - \delta\xi(EI\xi_0'')']'\,dx = 0. \tag{6.39}$$

The expression within the square brackets is absolutely continuous on each of the intervals $(0, l_1 - \varepsilon)$, $(l_1 + \varepsilon, l)$: furthermore it must be remembered that $\delta\xi \in K_1 \subset H^2(0,l)$ and so is continuously differentiable throughout the interval $(0, l)$ and in particular at $x = l_1$.

Lastly, taking into account $\delta\xi(l_1) = 0$ we obtain from (6.39):

$$[EI\xi_0''(\delta\xi)' - \delta\xi(EI\xi_0'')']\Big|_0^l + \lim_{\varepsilon \to 0} EI\xi_0''(\delta\xi)'\Big|_{l_1 + \varepsilon}^{l_1 - \varepsilon} = 0 \tag{6.40}$$

However $(\delta\xi)'$ is continuous at $x = l_1$ and can be given an arbitrary value at this point, which shows that $EI\xi_0''$ is continuous on crossing $x = l_1$. This condition is only natural since it expresses the continuity of the bending moment; the other conditions at the boundaries $x=0$, $x=l$ are deduced in the usual way from (6.40).

The same method could be applied when the prescribed condition at $x=l_1$ corresponds to clamping: $\xi(l_1)=\xi'(l_1)=0$. However it will be found in this case that the two parts of the beam comprised between 0 and l_1 on the one hand, and l_1 and l on the other, can have asynchronous eigenfrequencies. The two parts clearly have no interaction between them, so that their vibrations can be discussed separately.

If ω_0, $\tilde{\omega}_0$ are the lowest eigenfrequencies of the sections of the beams $[0,l_1]$ $[l_1,l]$ respectively, we can imagine the right-hand section to be in a state of flexural vibration at the frequency $\tilde{\omega}_0$ while the left-hand section is at rest, or the converse situation with the right-hand section at rest and the left-hand vibrating at frequency ω_0. The lowest possible vibration frequency of the system as a whole is obviously $\text{Inf}(\omega_0,\tilde{\omega}_0)$. It is not without interest to see what result would be obtained by applying Rayleigh's method to the whole beam. Let us suppose for simplicity that the conditions at $x=0$, $x=l$ correspond to free support. The lowest vibration frequency is then given by

$$\omega^2 = \operatorname*{Inf}_{\xi\in K_1} \frac{\int_0^l EI\xi''^2\,dx}{\int_0^l \rho\xi^2\,dx} \tag{6.41}$$

$$K_1 = \{\xi : \xi\in H^2(0,l), \xi(0)=\xi(l)=0, \xi(l_1)=\xi'(l_1)=0\} \tag{6.42}$$

Now it is clear that the restriction ζ of any function ξ of K_1, to the interval $(0,l_1)$ is a function of $H^2(0,l_1)$ which satisfies $\zeta(0)=\zeta(l_1)=\zeta'(l_1)=0$, and similarly the restriction $\tilde{\zeta}$ of this same function to the interval (l_1,l) is in $H^2(l_1,l)$ and satisfies $\tilde{\zeta}(l_1)=\tilde{\zeta}'(l_1)=\tilde{\zeta}(l)=0$.

Consequently we can write:

$$\frac{\int_0^l EI\xi''^2\,dx}{\int_0^l \rho\xi^2\,dx} = \frac{\int_0^{l_1} EI\zeta''^2\,dx + \int_{l_1}^l EI\tilde{\zeta}''^2\,dx}{\int_0^{l_1} \rho\zeta^2\,dx + \int_{l_1}^l \rho\tilde{\zeta}^2\,dx} \geqslant \text{Inf}\left(\frac{\int_0^{l_1} EI\zeta''^2\,dx}{\int_0^{l_1} \rho\zeta^2\,dx}, \frac{\int_{l_1}^l EI\tilde{\zeta}''^2\,dx}{\int_{l_1}^l \rho\tilde{\zeta}^2\,dx}\right)$$

$$\geqslant \text{Inf}(\omega_0^2,\tilde{\omega}_0^2)$$

However we have seen that equality can be attained for an element $\xi\in K_1$ and therefore the lowest eigenfrequency of the beam is $\omega=\text{Inf}(\omega_0,\tilde{\omega}_0)$. In the practical application of the formulae (6.41), (6.42), it must be remembered that the test functions $\xi(x)$ belonging to $H^2(0,l)$ are in this case continuously differentiable over any interval $(0,l)$ and in particular, at $x=l_1$. If therefore distinct representations of $\xi(x)$ are used in each of the two intervals $(0,l_1)$ and (l_1,l) appropriate boundary conditions will have to be written down at l_1 linking the two.

6.5. Investigation of Higher Frequencies

Let us now return to the general case of a beam subject to boundary conditions at the two ends $x=0$, $x=l$, which correspond to one of the six problems listed earlier. Let ω_0 be the first (lowest) eigenfrequency and $\xi_0(x)$ the corresponding solution or associated mode.

We define K_1 to be the subset of K generated by the elements of K satisfying the orthogonality condition

$$\int_0^l \rho\xi_0\xi\,\mathrm{d}x=0 \tag{6.43}$$

and we define ω_1

$$\omega_1^2=\operatorname*{Inf}_{\xi\in K_1}\frac{\int_0^l EI\xi''^2\,\mathrm{d}x}{\int_0^l \rho\xi^2\,\mathrm{d}x},\quad \omega_1>0. \tag{6.44}$$

It is clear that $\omega_1\geqslant\omega_0$, and that, by using the same arguments as those above, we can establish the existence of an element $\xi_1\in K_1$ which is kinematically admissible, satisfies (6.43), is normalised by $\int_0^l\rho\xi_1^2\,\mathrm{d}x=1$, and such that the lower bound (6.44) is attained for $\xi=\xi_1$.

We deduce from this that

$$\int_0^l (EI\xi_1''(\delta\xi)''-\omega_1^2\rho\xi_1\delta\xi)\,\mathrm{d}x=0 \tag{6.45}$$

for all $\delta\xi\in K_1$. If we can show that (6.45) is true for all $\delta\xi\in K$ or equivalently, for $\delta\xi=\xi_0$ we shall be able to apply the foregoing developments leading to

$$(EI\xi_1'')''-\omega_1^2\rho\xi_1=0,\quad \forall x\in(0,l)$$

and the verification of the remaining boundary conditions at $x=0$ and $x=l$.

Now because of (6.43) it will be sufficient to show that

$$\int_0^l EI\xi_1''\xi_0''\,\mathrm{d}x=\int_0^l\{\xi_1(EI\xi_0'')''+[\xi_1'\cdot EI\xi_0''-\xi_1(EI\xi_0'')']'\}\,\mathrm{d}x$$

vanishes.

But by (6.32) and (6.43) it is plain that $\int_0^l\xi_1(EI\xi_0'')''\,\mathrm{d}x=0$ and the required conclusion follows by applying (6.33) with $\delta\xi=\xi_1\in K_1\subset K$. Clearly this process can be continued, and we can define $\omega_2>0$ by:

$$\omega_2^2=\operatorname*{Inf}_{\xi\in K_2}\frac{\int_0^l EI\xi''^2\,\mathrm{d}x}{\int_0^l \rho\xi^2\,\mathrm{d}x},$$

where K_2 is the subset of elements ξ of K such that

$$\int_0^l \rho\xi_0\xi\,dx = 0 \quad \text{and} \quad \int_0^l \rho\xi_1\xi\,dx = 0.$$

We shall have $\omega_0 \leqslant \omega_1 \leqslant \omega_2$, ξ_2 being the eigenmode associated with ω_2. By iterating in this way we shall obtain an infinity of eigenfrequencies and eigenmodes. Some of the eigenfrequencies may be repeated but the eigenmodes are necessarily all distinct.

7. Examples of Applications

7.1. Beam Fixed at $x = 0$, Free at $x = l$

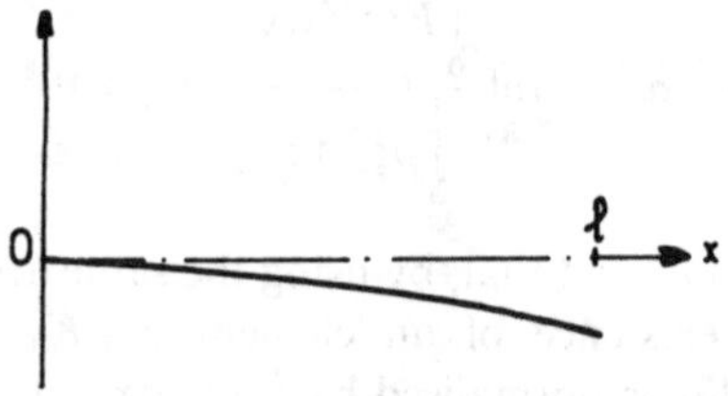

Figure 6.7

We take as test-function $\xi = a(1 - \cos(\pi x/2l))$ which satisfies the kinematic conditions $\xi(0) = \xi'(0) = 0$. The parameter a has a simple interpretation, as it represents the displacement of the free end $x = l$. If we suppose ρ, EI to be independent of x we obtain:

$$\int_0^l EI\xi''^2\,dx = \frac{\pi^4}{32}\cdot\frac{EIa^2}{l^3}, \quad \int_0^l \rho\xi^2\,dx = \rho a^2 l\left(\frac{3}{2} - \frac{4}{\pi}\right)$$

from which we can derive an approximation (in excess of the true value) for the fundamental eigenfrequency

$$\omega = \frac{\pi^2}{8\left(\frac{3}{4} - \frac{2}{\pi}\right)^{1/2}}\sqrt{\frac{EI}{\rho l^4}} = \frac{3.66}{l^2}\sqrt{\frac{EI}{\rho}}.$$

The exact solution has a numerical factor 3.52 instead of 3.66, which is about 4% smaller.

7.2. Beam Fixed at Both Ends

We take as test function $\xi = a(1 - \cos(2\pi x/l))$ which verifies the kinematic conditions $\xi(0) = \xi'(0) = \xi(l) = \xi'(l) = 0$.

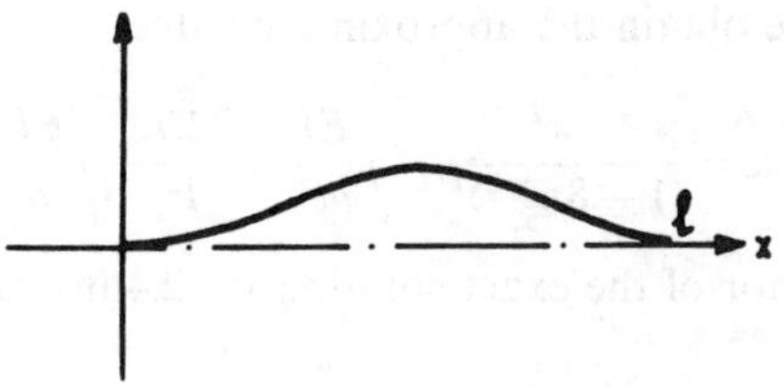

Figure 6.8

On the hypothesis of a homogeneous prismatic beam we obtain the approximate value

$$\omega = \frac{4\pi^2}{\sqrt{3}l^2}\sqrt{\frac{EI}{\rho}} = \frac{22.7}{l^2}\sqrt{\frac{EI}{\rho}}$$

The exact solution contains the factor 22.4, which is 1.3% smaller than the factor 22.7 appearing in the approximation.

7.3. Beam Free at Both Ends

The simplest mode has two nodes; this suggest that we take as test function

$$\xi = b \sin \pi \frac{x}{l} - a, \quad 0 \leqslant x \leqslant l \tag{6.46}$$

with two parameters a and b;

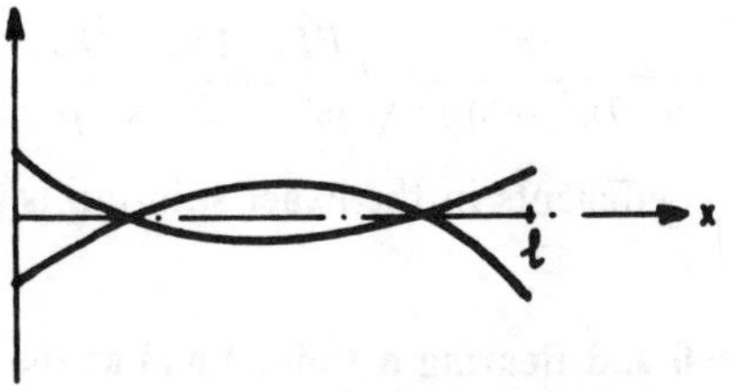

Figure 6.9

but we have to express the conditions (6.34), (6.35), that is to say in the case of a homogeneous beam:

$$\int_0^l \xi \, dx = 0, \quad \int_0^l x\xi \, dx = 0.$$

For ξ given by (6.46), these conditions are compatible and lead to $a = 2b/\pi$, so that we have finally

$$\xi = b\left(\sin \frac{\pi x}{l} - \frac{2}{\pi}\right).$$

For this choice, we obtain the approximate value

$$\omega = \frac{\pi^2}{(1 - 8\pi^{-2})^{1/2}} \sqrt{\frac{EI}{\rho l^4}} = \frac{22.72}{l^2} \sqrt{\frac{EI}{\rho}}$$

and the numerical factor of the exact solution is 22.4 instead of 22.72.

7.4. Beam Hinged at $x = 0$, Free at $x = l$

The test function has to satisfy the kinematic condition $\xi(0) = 0$ and also the zero-moment condition

$$\int_0^l x\xi \, dx = 0 \quad \text{(homogeneous prismatic beam)}. \tag{6.47}$$

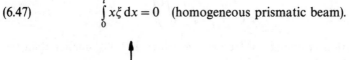

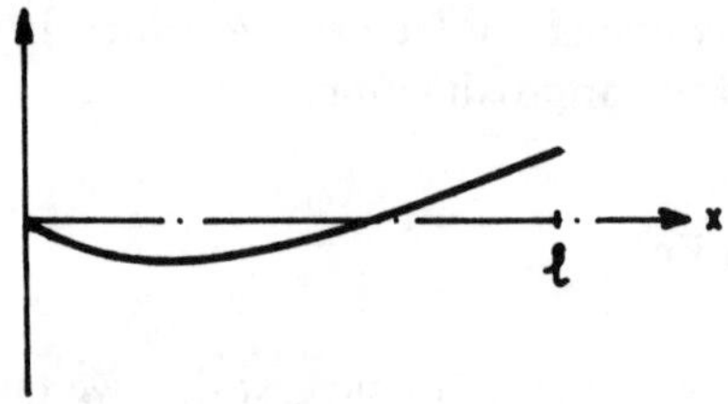

Figure 6.10

Let us take as test function $\xi = a \sin(\pi x/l) + bx/l$ so that by (6.47) $a = -(\pi/3)b$ and an approximate representation of a mode with a node between 0 and l. The upper approximate value of the fundamental frequency is

$$\omega = \frac{\pi^3}{(\pi^2 - 6)^{1/2}} \sqrt{\frac{EI}{\rho l^4}} = \frac{15.8}{l^2} \sqrt{\frac{EI}{\rho}},$$

whereas the numerical coefficients in the exact solution is 15.4.

7.5. Beam Fixed at $x = 0$ and Bearing a Point Load at the Other End

We take $\xi = a(1 - \cos(\pi x/2l))$ which satisfies the kinematic conditions $\xi(0) = \xi'(0) = 0$ and denoting by μ the mass of the load, we calculate

$$\omega^2 = \frac{\int_0^l EI\xi''^2 \, dx}{\int_0^l \rho\xi^2 \, dx + \mu\xi^2(l)}$$

or

$$\omega^2 = \frac{3.03\,EI}{l^3(\mu + 0.23\,m)}$$

for a homogeneous prismatic beam of mass m.

7.6. Beam Supported at Three Points

We shall take the origin at the intermediate point (at which the beam is fixed but free to flex) so that the beam extends from the point $x = -l_2$ to the point $x = l_1$, with $l_1 + l_2 = l$.

We define a test-function by:

$$\xi = a_1 \sin\frac{\pi x}{l_1}, \qquad 0 < x < l_1$$

$$\xi = a_2 \sin\frac{\pi x}{l_2}, \qquad -l_2 < x < 0.$$

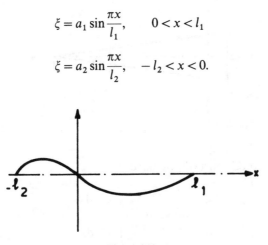

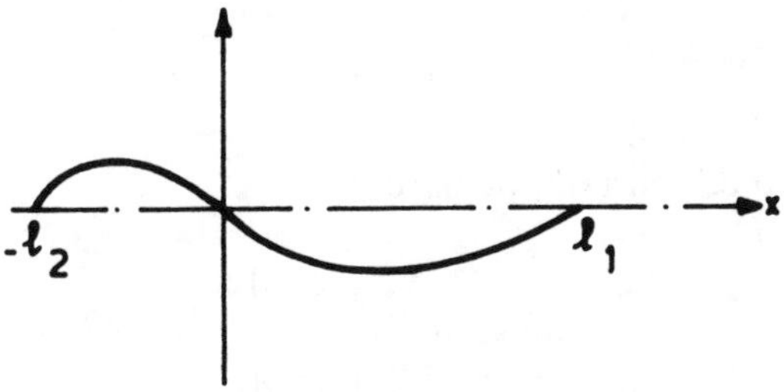

Figure 6.11

The kinematic conditions $\xi(l_1) = \xi(-l_2) = \xi(0) = 0$ are fulfilled; to ensure that ξ belongs to $H^2(-l_2, l_1)$ we have only to stipulate that the derivative $\xi'(x)$ at $x = 0$ should be continuous, which means that $a_1/l_1 = a_2/l_2$.

It will therefore be convenient to represent ξ in the form:

$$\xi = \alpha l_1 \sin\frac{\pi x}{l_1}, \qquad 0 < x < l_1$$

$$\xi = \alpha l_2 \sin\frac{\pi x}{l_2}, \qquad -l_2 < x < 0.$$

and, still on the assumption that the beam is homogeneous and prismatic, we obtain as an approximate value (in excess of the true value) for the fundamental frequency:

$$\omega = \frac{\pi^2}{\sqrt{l_1 l_2 (l_1^2 + l_2^2 - l_1 l_2)}} \cdot \sqrt{\frac{EI}{\rho}}.$$

7.7. Vibration of a Wedge Clamped at $x = 0$. Ritz's Method

In this example ρ and I depend in a simple way on x. Assuming the material to be homogeneous the linear mass is $\rho = \tilde{\rho} A$, A being the cross-sectional area, $\tilde{\rho}$

the constant density, and the fixed end of the wedge having by hypothesis a rectangular cross-section of unit area, we can write

$$A = 2b\left(l - \frac{x}{l}\right) \quad \text{and} \quad I = \frac{2b^3}{3}\left(1 - \frac{x}{l}\right)^3$$

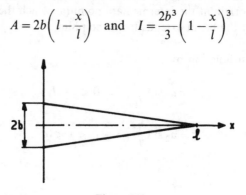

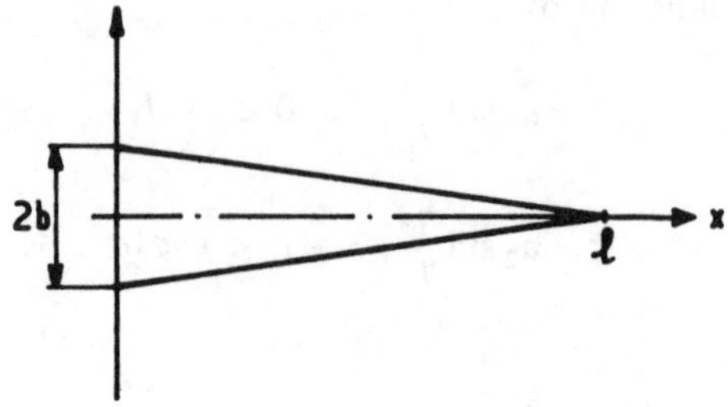

Figure 6.12

Supposing E to be constant, we have

$$\omega_0^2 = \operatorname*{Inf}_{\xi \in K} \frac{E}{\tilde{\rho}} \cdot \frac{\int_0^l I\xi''^2 \, dx}{\int_0^l A\xi^2 \, dx}$$

where K is the subset of those elements ξ of $H^2(0,l)$ for which $\xi(0) = \xi'(0) = 0$.

We can consider taking as test-function, polynomial expressions of the form

$$\xi = a_1\left(\frac{x}{l}\right)^2 + a_2\left(\frac{x}{l}\right)^2\left(1 - \frac{x}{l}\right) + a_3\left(\frac{x}{l}\right)^2\left(1 - \frac{x}{l}\right)^2 + \cdots$$

With $\xi = a_1(x^2/l^2)$ we obtain the approximation

$$\omega = 5.47\frac{b}{l^2}\sqrt{\frac{E}{3\tilde{\rho}}}.$$

We can try to get a better approximation by taking:

$$\xi = a_1\left(\frac{x}{l}\right)^2 + a_2\left(1 - \frac{x}{l}\right)\left(\frac{x}{l}\right)^2 \tag{6.48}$$

where the parameters α_1, α_2 are chosen in such a way as to minimise the ratio

$$\frac{E\int_0^l I\xi''^2 \, dx}{\tilde{\rho}\int_0^l A\xi^2 \, dx} = \frac{W(a_1, a_2)}{T(a_1, a_2)}$$

of the two positive-definite quadratic forms W and T:

$$\omega^2 = \operatorname*{Inf}_{a_1, a_2} \frac{W(a_1, a_2)}{T(a_1, a_2)}.$$

In fact ω^2 will be equal to the minimum of $W(a_1, a_2)$ when a_1 and a_2 vary but are bound by the relation $T(a_1, a_2) = 1$. This minimum must be attained at some point (α_1, α_2) and furthermore $W(\alpha_1, \alpha_2)$ cannot be zero because this would imply $\alpha_1 = \alpha_2 = 0$, contradicting $T(\alpha_1, \alpha_2) = 1$.

Since $W - \omega^2 T$ has a minimum at (α_1, α_2), the partial derivatives of the first order of this function vanish at this point:

$$\frac{\partial W}{\partial \alpha_j} - \omega^2 \frac{\partial T}{\partial \alpha_j} = 0, \quad j = 1, 2$$

or, with

$$S = W - \omega^2 T = \frac{2Eb^3}{3l^3}\left[(a_1 - 2a_2)^2 + \tfrac{24}{5} a_2(a_1 - 2a_2) + 6a_2^2\right]$$

$$- 2b\tilde{\rho}l\omega^2\left(\frac{a_1^2}{30} + \frac{2a_1a_2}{105} + \frac{a_2^2}{280}\right),$$

$$\left(\frac{E}{\tilde{\rho}}\frac{b^2}{3l^4} - \frac{\omega^2}{30}\right)a_1 + \left(\frac{2Eb^2}{15\tilde{\rho}l^4} - \frac{\omega^2}{105}\right)a_2 = 0$$

$$\left(\frac{2Eb^2}{15\tilde{\rho}l^4} - \frac{\omega^2}{105}\right)a_1 + \left(\frac{2Eb^2}{15\tilde{\rho}l^4} - \frac{\omega^2}{280}\right)a_2 = 0.$$

For this system to have a non-null solution (α_1, α_2), the associated determinant must be zero, and this defines the admissible values of ω. We shall denote the smaller by ω which gives

$$\omega = 5.319 \frac{b}{l^2}\sqrt{\frac{E}{3\tilde{\rho}}}.$$

It is interesting to note that an exact solution can be constructed for this problem and yields for the fundamental frequency the value:

$$\omega_0 = 5.315 \frac{b}{l^2}\sqrt{\frac{E}{3\tilde{\rho}}}.$$

The approximation obtained by the Rayleigh-Ritz method is, as can be seen, very good.

7.8. Vibrations of a Supported Pipeline

Let us consider a metal tube inside which flows a liquid of mass μ per unit length at a horizontal velocity V_0. The undeformed tube has a horizontal axis and rests on two supports whose distance apart is l. The equations

$$\frac{\partial M}{\partial x} + T = 0, \quad M = EI\frac{\partial^2 v}{\partial x^2}$$

obtained by regarding the tube as an elastic beam of annular cross-section, still apply (E being Young's modulus and I the geometrical moment of inertia of the annulus). However to write down the equation of motion, the inertia of the fluid has to be taken into account. Thus:

$$\frac{\partial T}{\partial x} = \rho \frac{\partial^2 v}{\partial t^2} + \mu \gamma,$$

where ρ is the linear mass of the tube, γ the vertical acceleration of the fluid. To calculate the latter, we note that we can represent the mean motion of the fluid by the co-ordinates x, y such that

$$y = v(x,t), \quad \frac{dx}{dt} = V_0$$

whence it follows that

$$\gamma = \frac{d^2 y}{dt^2} = \left(V_0 \frac{\partial}{\partial x} + \frac{\partial}{\partial t} \right)^2 v$$

and that the equation of motion is:

$$\frac{\partial^2}{\partial x^2}\left(EI \frac{\partial^2 v}{\partial x^2} \right) + \rho \frac{\partial^2 v}{\partial t^2} + \mu \left(V_0 \frac{\partial}{\partial x} + \frac{\partial}{\partial t} \right)^2 v = 0.$$

Let us assume that the term $2V_0(\partial^2/\partial x \partial t)v$ be neglected, i.e.

$$\frac{\partial^2}{\partial x^2}\left(EI \frac{\partial^2 v}{\partial x^2} \right) + \mu V_0^2 \frac{\partial^2 v}{\partial x^2} + (\rho + \mu) \frac{\partial^2 v}{\partial t^2} = 0.$$

Seeking a solution $v = \xi \sin \omega t$ satisfying the kinematic condition $\xi(0) = \xi(l) = 0$, and the conditions $\xi''(0) = \xi''(l) = 0$, representing the lack of any bending moment at the end of the tube, or in other words flexible connections, we are led to

$$EI\xi^{IV} + \mu V_0^2 \xi'' - (\rho + \mu)\omega^2 \xi = 0$$

and the solution $\xi = C \sin ax$ with

$$EIa^4 - \mu V_0^2 a^2 - (\rho + \mu)\omega^2 = 0 \quad \text{and} \quad a = \frac{n\pi}{l},$$

when n is an integer for which we derive

$$\omega^2 = \frac{EI}{\rho + \mu} \cdot \left(\frac{n\pi}{l} \right)^2 - \frac{\mu V_0^2}{\rho + \mu} \left(\frac{n\pi}{l} \right)^2.$$

The first eigenfrequency is obtained for $n = 1$, namely

$$\omega^2 = \omega_0^2 - \frac{\mu V_0^2 \pi^2}{l^2(\rho + \mu)}, \quad \text{with} \quad \omega_0^2 = \frac{EI}{\rho + \mu} \cdot \frac{\pi^4}{l^4}$$

but it can be observed only if V_0 is not too large, more precisely only if

$$\mu V_0^2 < \frac{EI\pi^2}{l^2}.$$

7.9. Effect of Longitudinal Stress on the Flexural Vibrations of a Beam and Application to Blade Vibrations in Turbomachinery

We shall still assume the existence of an inextensible neutral line, which in the rest position coincides with the axis Ox, but we shall suppose it to be acted upon, at each of the points where it meets the normal cross-section, by a force of tension parallel to Ox, of magnitude $F(x)$ in the positive direction.

We shall ignore deformations due to shearing stresses but we shall take account of rotational inertia; the equations

(6.49) $$\rho\frac{\partial^2 v}{\partial t^2}=\frac{\partial T}{\partial x}, \quad M=EI\frac{\partial^2 v}{\partial x^2}$$

remain valid. On the other hand we have to rewrite the equation derived from the theorem on angular momentum applied to the rotational motion of the slice of the beam between x and $x+\mathrm{d}x$, which now becomes, after division by $\mathrm{d}x$:

(6.50) $$\frac{\partial M}{\partial x}+T-F\frac{\partial v}{\partial x}=I\rho\frac{\partial^2}{\partial t^2}\left(\frac{\partial v}{\partial x}\right).$$

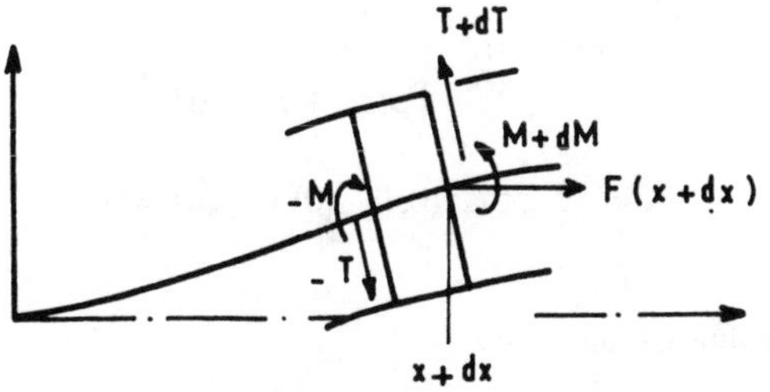

Figure 6.13

Eliminating T, M from (6.49), (6.50) we obtain:

$$\frac{\partial^2}{\partial x^2}\left(EI\frac{\partial^2 v}{\partial x^2}\right)+\rho\frac{\partial^2 v}{\partial t^2}-\frac{\partial}{\partial x}\left(F\frac{\partial v}{\partial x}\right)=\frac{\partial^2}{\partial t^2}\frac{\partial}{\partial x}\left(I\rho\frac{\partial v}{\partial x}\right)$$

and the search for a solution

(6.51) $$v=\sin\omega t\cdot\xi(x)$$

leads to

(6.52) $$(EI\xi'')''-(F\xi')'-\omega^2(\rho\xi-(I\rho\xi')')=0, \quad \forall x\in(0,l).$$

The boundary conditions for a beam fixed at $x=0$ and free at $x=l$ are:

(6.53) $$\xi(0)=\xi'(0)$$

and then $M=T=0$ at $x=l$ so that by (6.49), (6.50) and (6.51):

(6.54) $$\begin{aligned} EI\xi''&=0, \quad x=l\\ (EI\xi'')'-F\xi'+I\rho\omega^2\xi'&=0, \quad x=l. \end{aligned}$$

Multiplying the two sides of (6.52) by ξ and integrating with respect to x from 0 to l we obtain

$$\int_0^l (EI\xi'')''\xi\,\mathrm{d}x - \int_0^l [(F\xi')'\xi + \omega^2(\rho\xi^2 - (I\rho\xi')'\xi)]\,\mathrm{d}x = 0$$

or after some integrating by parts:

$$\int_0^l [EI\xi''^2 + F\xi'^2 - \omega^2(\rho\xi^2 + I\rho\xi'^2)]\,\mathrm{d}x + [((EI\xi'')' - F\xi' + \omega^2 I\rho\xi')\xi - EI\xi''\cdot\xi']|_0^l = 0.$$

The fully integrated term vanishes because of (6.53), (6.54) and consequently

$$\omega^2 = \frac{\int_0^l (EI\xi''^2 + F\xi'^2)\,\mathrm{d}x}{\int_0^l \rho(\xi^2 + I\xi'^2)\,\mathrm{d}x}$$

which suggests developing Rayleigh's method by defining:

$$\omega^2 = \operatorname*{Inf}_{\xi\in K} \frac{\int_0^l (EI\xi''^2 + F\xi'^2)\,\mathrm{d}x}{\int_0^l \rho(\xi^2 + I\xi'^2)\,\mathrm{d}x}$$

with $$K = \{\xi : \xi \in H^2(0,l),\ \xi(0) = \xi'(0) = 0\}.$$

Application to Turbine Blade

The blade attached to the periphery of a wheel of radius r, turning at a speed Ω, can be treated like a straight beam of cross-sectional area $A(x)$ and linear mass $\rho(x)$.

We have to take account, relative to moving axes Oxy carried by the wheel, of the extension force

$$F = \Omega^2 \int_x^l \rho A(s)\cdot(r+s)\,\mathrm{d}s$$

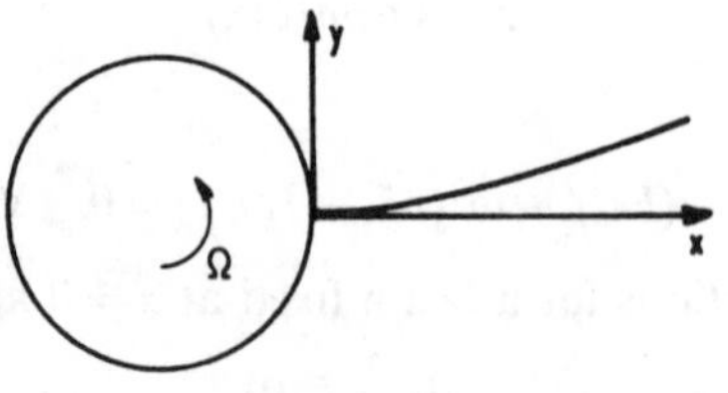

Figure 6.14

which corresponds to the drive-inertia. The Coriolis effects, which are less important will be ignored.

We can also, in a simplified approach, ignore the effect due to the inertia of the relative rotation between the different slices of the beam, and accordingly the fundamental frequency of the natural vibration is given by

$$\omega^2 = \operatorname*{Inf}_{\xi\in K} \frac{\int_0^l (EI\xi''^2 + F\xi'^2)\,dx}{\int_0^l \rho\xi^2\,dx}.$$

It is higher than it would be in the absence of rotation.

7.10. Vibration of Interactive Systems

1. Vibrations of a Beam Resting on an Elastic Support

The first eigenfrequency is given by:

$$\omega^2 = \operatorname*{Inf}_{\xi\in K} \frac{\int_0^l (EI\xi''^2 + k\xi^2)\,dx}{\int_0^l \rho\xi^2\,dx},$$

the rigidity of the suspension being defined by a bounded density function $k(x)$, or by a sum of Dirac distributions $\sum_p k_p\delta(x - x_p)$; we take $K = \{\xi : \xi \in H^2(0, l)$ and satisfying the kinematic conditions$\}$, the kinematic conditions being defined by $\xi(0) = 0$ in the case of a single support at $x = 0$; by $\xi(0) = \xi(l) = 0$ when the beam is supported at each end; by $\xi(0) = \xi'(0) = 0$ when the beam is clamped at $x = 0$; whereas $K = H^2(0, l)$ when the beam is free.

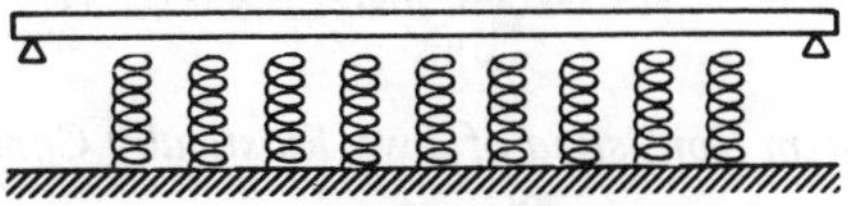

Figure 6.15

2. Vibrations of a System Comprising a Beam with a Mass Hanging from It by an Elastic Suspension

We take as variables $\xi(x)$ and y which are respectively the amplitudes of the deformation of the beam at x and the displacement from its equilibrium position of the mass μ suspended from the point $x = l_1 < l$ by means of a spring of stiffness k.

With $K = \{(\xi, y) : (\xi, y) \in H^2(0, l) \times \mathbb{R} + \text{kin. cond.}\}$, we can define the first eigenfrequency

$$\omega^2 = \operatorname*{Inf}_{(\xi, y)\in K} \frac{\int_0^l EI\xi''^2\,dx + k(\xi(l_1) - y)^2}{\int_0^l \rho\xi^2\,dx + \mu y^2}.$$

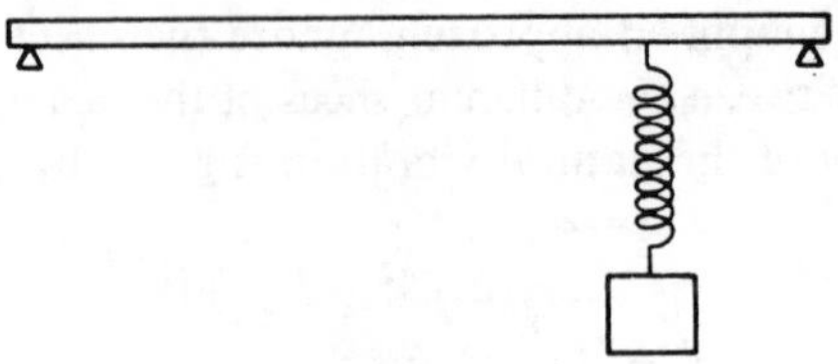

Figure 6.16

3. Vibrations of a Beam Fixed at One End and Suspended Elastically at the Other End with an Added Mass μ Attached

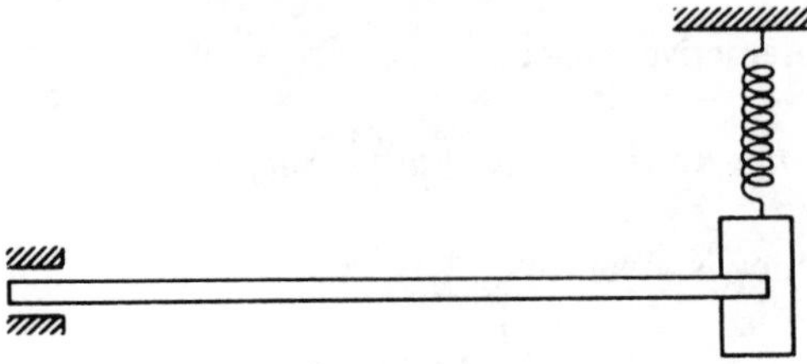

Figure 6.17

The first eigenvalue is given by:

$$\omega^2 = \operatorname*{Inf}_{\xi \in K} \frac{\int_0^l EI\xi''^2\,dx + k\xi^2(l)}{\int_0^l \rho\xi^2\,dx + \mu\xi^2(l)}$$

where k is the modulus of rigidity of the suspension and

$$K = \{\xi, \xi \in H^2(0,l), \xi(0) = \xi'(0) = 0\}\,.$$

4. Vibrations of a System Consisting of Two Elastically Connected Beams

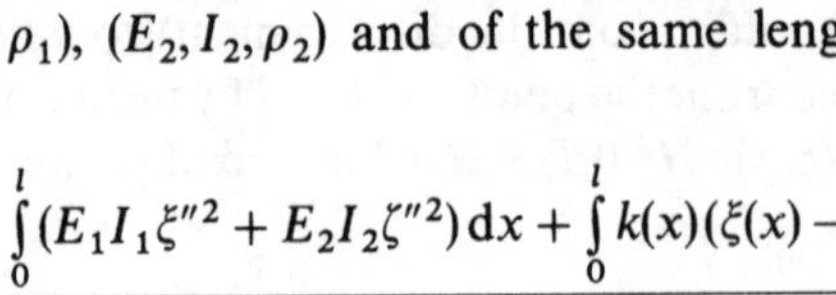

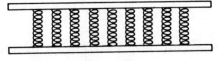

Figure 6.18

Using $\xi(x)$, $\zeta(x)$ to denote the amplitudes of the deformation of the two beams with characteristics (E_1, I_1, ρ_1), (E_2, I_2, ρ_2) and of the same length l, the first eigenfrequency is:

$$\omega^2 = \operatorname*{Inf}_{(\xi,\zeta)\in K} \frac{\int_0^l (E_1I_1\xi''^2 + E_2I_2\zeta''^2)\,dx + \int_0^l k(x)(\xi(x) - \zeta(x))^2\,dx}{\int_0^l (\rho_1\xi^2 + \rho_2\zeta^2)\,dx}$$

with $K = \{(\xi,\zeta):(\xi,\zeta)\in H^2(0,l) \times H^2(0,l) + \text{kin. cond.}\}$ and $k(x)$ representing the density of the rigidity of the elastic connection.

In the case of a free system the appropriate kinematic conditions are:

$$\int_0^l (\rho_1\xi + \rho_2\zeta)\,dx = 0, \quad \int_0^l x(\rho_1\xi + \rho_2\zeta)\,dx = 0$$

which express the fact that the momentum and angular momentum are both zero.

5. Vibrations of a System Consisting of Two Beams and a Mass Elastically Connected

We consider a system comprising two beams (E_1, I_1, ρ_1), (E_2, I_2, ρ_2), of length l, one supported, the other clamped at $x = 0$, and a mass μ attached to the two beams at the point $x = l_1 < l$ by springs of stiffness k_1, k_2.

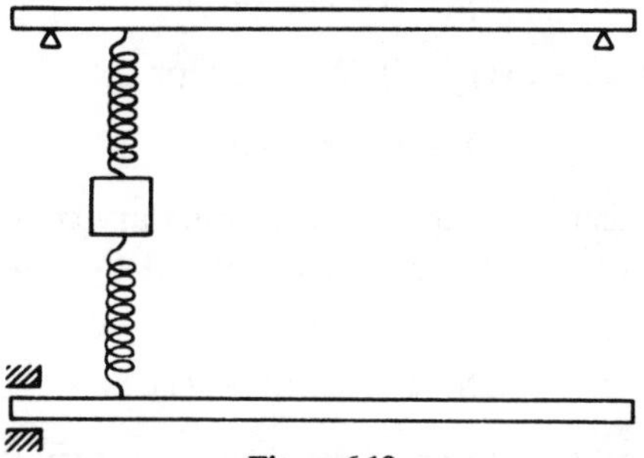

Figure 6.19

Writing $\xi(x)$, $\zeta(x)$, y for the amplitudes of the displacements of the beams and the mass when the system is in harmonic motion, we can define the first eigenfrequency by:

$$\omega^2 = \operatorname*{Inf}_{(\xi,\zeta,y)\in K} \frac{\int_0^l (E_1I_1\xi''^2 + E_2I_2\zeta''^2)\,dx + k_1(\xi(l_1) - y)^2 + k_2(\zeta(l_1) - y)^2}{\int_0^l (\rho_1\xi^2 + \rho_2\zeta^2)\,dx + \mu y^2}$$

with:

$$K = \{(\xi,\zeta,y):(\xi,\zeta,y)\in H^2(0,l) \times H^2(0,l) \times \mathbb{R}, \xi(0) = \xi(l) = 0, \zeta(0) = \zeta'(0) = 0\}$$

A first approximation could be obtained by using the test functions $\xi = a\sin\frac{\pi x}{l}$, $\zeta = b\left(1 - \cos\frac{\pi x}{2l}\right)$, $y = c$ after minimising the Rayleigh quotient with respect to the coefficients a, b, c.

8. Forced Vibrations of Beams Under Flexure

Let us now take up once more the model discussed in Section 1 but this time assuming that vertical forces are exerted on the beam with a linear density of $p(x, t)$.

The equations (6.1), (6.2) remain valid whereas the equation of motion becomes:

$$\rho\frac{\partial^2 v}{\partial t^2} = \frac{\partial T}{\partial x} + p$$

whence, by eliminating T and M:

$$\rho \frac{\partial^2 v}{\partial t^2} + \frac{\partial^2}{\partial x^2}\left(EI \frac{\partial^2 v}{\partial x^2}\right) = p(x, t). \tag{6.55}$$

In the case of a point-force of intensity $f(t)$ exerted at the point of abscissa x', $p(x, t)$ would have to be taken as a generalised function

$$p(x, t) = f(t) \cdot \delta(x - x'),$$

where $\delta(x - x')$ is the Dirac distribution at x'.

Reverting to (6.55) we shall suppose the excitation $p(x, t)$ to be harmonic with respect to time, and of frequency ω, so that:

$$p(x, t) = p(x) \sin \omega t$$

with $p(x)$ square-integrable over $(0, l)$. We can represent

$$v(x, t) = \xi(x) \sin \omega t$$

where $\xi(x) \in H^2(0, l)$ satisfies the appropriate boundary conditions dictated by the assumptions made as to the type of connections at the ends of the beam, and is a solution of:

$$(EI\xi'')'' - \rho\omega^2 \xi = p(x). \tag{6.56}$$

In the case of natural vibrations, $p(x) = 0$, we have seen that there is an infinity of eigenfrequencies $\omega_0 \leqslant \omega_1 \leqslant \cdots$ and of associated eigenmodes $\xi_0, \xi_1, \ldots$ One can therefore try to represent the forced vibration by a series of the form:

$$\xi = \sum_{j=0}^{\infty} c_j \xi_j \tag{6.57}$$

and calculate the coefficients c_j from (6.56), by using the orthogonality relations:

$$\int_0^l \rho \xi_j \xi_k \, dx = 0, \quad j \neq k \tag{6.58}$$

and taking the eigenmodes to be normalised, so that $\int_0^l \rho \xi_j^2 \, dx = 1$.

The calculations which follow and which develop this idea can be justified completely. We shall in the next section below give an outline of the theoretical foundation and then later, on analogous bases, an analysis of the forced vibrations of elastic bodies in three-dimensional space.

Multiplying (6.56) by $\xi_j(x)$ and integrating with respect to x from 0 to l, we have

$$\int_0^l ((EI\xi'')'' - \rho\omega^2 \xi)\xi_j \, dx = \int_0^l p(x)\xi_j(x) \, dx. \tag{6.59}$$

Observing that

$$\int_0^l [(EI\xi'')''\xi_j - (EI\xi_j'')''\xi] \, dx$$

$$= \int_0^l [(EI\xi'')'\xi_j - EI\xi'' \cdot \xi_j' - (EI\xi_j'')'\xi + EI\xi_j'' \cdot \xi'] \, dx = 0$$

by reason of the conditions imposed at the ends, common to all ξ_j and to ξ, and taking account of

$$(EI\xi_j'')'' - \rho\omega_j^2\xi_j = 0$$

and of (6.57), (6.58) we can write (6.59) as:

$$(\omega_j^2 - \omega^2)c_j = \int_0^l p(x)\xi_j(x)\,dx$$

whence

$$\xi(x) = \sum_{j=0}^{\infty} \frac{1}{\omega_j^2 - \omega^2} \cdot \int_0^l p(x)\xi_j(x)\,dx \cdot \xi_j(x)$$

provided that

$$\omega \neq \omega_j, \quad \forall j = 0, 1, \ldots$$

It can be shown that this series converges strongly in $H^2(0, l)$ to the solution ξ of (6.56) which satisfies the boundary conditions of the problem.

In the case of a harmonic excitation by a unit force concentrated at the point x', we have $p(x) = \delta(x - x')$ and the response can be represented by:

$$\xi(x, x', \omega) = \sum_{j=0}^{\infty} \frac{\xi_j(x)\xi_j(x')}{\omega_j^2 - \omega^2}.$$

$\xi(x, x', \omega)$ is said to be the transfer function: this function which is symmetric with respect to x, x' expresses, to within a factor $\sin\omega t$, the response at the point x to a harmonic excitation of unit intensity and of frequency ω exerted at the point x'.

9. The Comparison Method

9.1. The Functional Operator Associated with the Model of a Beam Under Flexure

We consider once again the subspace $K \subset H^2(0, l)$ of displacements $\xi(x)$ which are kinematically admissible, or which satisfy the conditions which replace these in the cases of a beam hinged at one end and free at the other, or completely free at both ends.

Let us first show that $\left(\int_0^l \xi''^2\,dx\right)^{1/2}$ defines, for all $\xi \in K$, a norm equivalent to that of $H^2(0, l)$. On putting

$$\sigma^2 = \operatorname*{Inf}_{\xi \in K} \frac{\int_0^l \xi''^2\,dx}{\int_0^l (\xi^2 + \xi'^2)\,dx} \tag{6.60}$$

we can, by repeating the same argument as that used to establish the existence of the eigenfrequency ω_0 and its associated eigenmode ξ_0 prove that there is a $\zeta \in K$ such that

$$\sigma^2 = \frac{\int_0^l \zeta''^2 \, dx}{\int_0^l (\zeta^2 + \zeta'^2) \, dx}.$$

The possibility that $\sigma = 0$ is thus excluded since it would imply that $\zeta'' = 0$ almost everywhere and this would mean $\zeta = 0$ since $\zeta \in K$.

Accordingly by (6.60) it is clear that for all $\xi \in K$ we have:

$$\int_0^l (\xi^2 + \xi'^2) \, dx \leqslant \sigma^{-2} \int_0^l \xi''^2 \, dx$$

and from

$$\int_0^l \xi''^2 \, dx \leqslant \int_0^l (\xi^2 + \xi'^2 + \xi''^2) \, dx \leqslant (1 + \sigma^{-2}) \int_0^l \xi''^2 \, dx$$

we conclude that

$$\| \xi \|_{H^2} \quad \text{and} \quad \left(\int_0^l \xi''^2 \, dx \right)^{1/2}$$

are equivalent norms on K.

Furthermore

$$[\xi, \eta] = \int_0^l EI \xi'' \eta'' \, dx, \quad \forall (\xi, \eta) \in H^2(0, l) \times H^2(0, l)$$

defines a sesquilinear form, and on the hypothesis that

$$0 < p = \operatorname*{Inf}_x EI \leqslant \operatorname*{Sup}_x EI = q;$$

$$||| \xi ||| = \left(\int_0^l EI \xi''^2 \, dx \right)^{1/2}, \quad \forall \xi \in K$$

is, on K, a norm equivalent to the one induced naturally by $H^2(0, l)$ and denoted by $\| \xi \|_{H^2}$.

The bilinear form $\int_0^l \rho \xi \eta \, dx$ defined on $K \times K$ is bicontinuous and by Riesz's theorem [37] there is a linear operator A from K into K such that

$$\int_0^l \rho \xi \eta \, dx = \int_0^l EI \xi'' (A\eta)'' \, dx = [\xi, A\eta].$$

This operator, which is clearly symmetric is bounded because:

$$\left| \int_0^l \rho \xi \eta \, dx \right| < \left(\int_0^l \rho \xi^2 \, dx \right)^{1/2} \left(\int_0^l \rho \eta^2 \, dx \right)^{1/2} \leqslant \omega_0^{-2} [\xi, \xi]^{1/2} \cdot [\eta, \eta]^{1/2}$$

or

$$|[A\xi, \eta]| \leqslant \omega_0^{-2} [\xi, \xi]^{1/2} \cdot [\eta, \eta]^{1/2}$$

and with $\eta = A\xi$, we have $||| A\xi ||| \leqslant \omega_0^{-2} ||| \xi |||$.

Moreover it is completely continuous; we have seen that given any bounded infinite sequence of elements $\xi_n \in K$, one can always extract a similarly numbered subsequence converging weakly in K to an element $\xi_0 \in K$, and such that it also converges strongly in $H^1(0, l)$ and in particular in $L^2(0, l)$ to ξ_0, so that we have

$$\lim_{n,m \to \infty} \int_0^l \rho(\xi_n - \xi_m)(\xi_n - \xi_m)\,\mathrm{d}x = 0$$

or

$$\lim_{n,m \to \infty} [\xi_n - \xi_m, A(\xi_n - \xi_m)] = 0$$

which ensures complete continuity [37], that is to say the property whereby a subsequence ξ_{n_j}, such that $A\xi_{n_j}$ is strongly convergent in K, can always be extracted from any infinite bounded sequence ξ_n of elements of K.

Observing that

$$\frac{\int_0^l EI\xi''^2\,\mathrm{d}x}{\int_0^l \rho\xi^2\,\mathrm{d}x} = \frac{[\xi, \xi]}{[A\xi, \xi]},$$

$\forall \xi \in K$, we see that the Rayleigh method developed earlier can be identified with the classical method of finding the eigenvalues and eigenvectors of symmetric completely continuous operators in a Hilbert space. Thus we have:

$$\omega_0^{-2} = \operatorname*{Sup}_{\xi \in K} \frac{[A\xi, \xi]}{[\xi, \xi]}$$

the element ξ_0 defined earlier, for which the supremum is attained being an eigenvector:

$$A\xi_0 = \omega_0^{-2}\xi_0 \tag{6.61}$$

which we can take to be normalised by: $[\xi_0, \xi_0] = \omega_0^2$.

Similarly $\omega_1^{-2} = \operatorname*{Sup}_{\xi \in K_1} \dfrac{[A\xi, \xi]}{[\xi, \xi]}$, where K_1 is the subspace of elements $\xi \in K$ such that

$$[A\xi_0, \xi] = 0 \quad \text{or} \quad [\xi_0, \xi] = 0,$$

is the second eigenvalue, ξ_1 the element for which the supremum is attained being the associated eigenvector

$$A\xi_1 = \omega_1^{-2}\xi_1,$$

and ξ_1 being normalised by $[A\xi_1, \xi_1] = 1$ or $[\xi_1, \xi_1] = \omega_1^2$.

More generally, having obtained the first n eigenmodes $0 \leqslant j \leqslant n-1$, satisfying

$$A\xi_j = \omega_j^{-2}\xi_j, \quad [\xi_j, \xi_j] = \omega_j^2, \quad [\xi_j, \xi_k] = 0, \quad j \neq k,$$

we obtain the eigenfrequency of suffix n by,

$$\omega_n^{-2} = \operatorname*{Sup}_{\xi \in K_n} \frac{[A\xi, \xi]}{[\xi, \xi]} \tag{6.62}$$

with $K_n = \{\xi : \xi \in K, \ [\xi_j, \xi] = 0, \ 0 \leqslant j \leqslant n-1\}$ and ξ_n the element for which the supremum is attained, is the associated eigenmode

(6.63) $$A\xi_n = \omega_n^{-2} \xi_n,$$

which we can suppose to be normalised by:

(6.64) $$[\xi_n, \xi_n] = \omega_n^2.$$

It is clear that

(6.65) $$\omega_0 \leqslant \omega_1 \leqslant \cdots \leqslant \omega_n \leqslant \cdots$$

9.2. The Min-Max Principle [37]

Let $\varphi_0, \ldots, \varphi_{n-1}$, be any n elements of $H^2(0, l)$ and let us set

$$v(\varphi_0, \ldots, \varphi_{n-1}) = \sup_{\substack{\xi \in K, [\xi, \varphi_j] = 0 \\ 0 \leqslant j \leqslant n-1}} \frac{[A\xi, \xi]}{[\xi, \xi]},$$

the supremum being thus taken with respect to the set of elements ξ of K orthogonal to φ_j:

$$[\xi, \varphi_j] = 0, \quad 0 \leqslant j \leqslant n-1.$$

We have the result

$$\omega_n^{-2} = \operatorname*{Inf}_{\varphi} v(\varphi_0, \varphi_1, \ldots, \varphi_{n-1}).$$

To prove this we first note that

$$\operatorname*{Inf}_{\varphi} v(\varphi_0, \varphi_1, \ldots, \varphi_{n-1}) \leqslant v(\xi_0, \xi_1, \ldots, \xi_{n-1}) = \omega_n^{-2}.$$

Now let $\xi = \sum_{i=0}^{n} a_i \xi_i$, where the real coefficients a_i are chosen so that

$$[\xi, \varphi_j] = 0, \quad 0 \leqslant j \leqslant n-1,$$

i.e. $$\sum_{i=0}^{n} a_i [\xi_i, \varphi_j] = 0, \quad 0 \leqslant j \leqslant n-1,$$

a set of n homogeneous linear equations in $(n+1)$ unknowns which always has a nonzero solution $a_0 \cdots a_n$ for any given $\varphi_0, \ldots, \varphi_{n-1}$ in $H^2(0, l)$.

Then $\dfrac{[A\xi, \xi]}{[\xi, \xi]} \leqslant v(\varphi_0, \ldots, \varphi_{n-1})$ and on the other hand

$$\frac{[A\xi, \xi]}{[\xi, \xi]} = \frac{\sum_{i,j} a_i a_j [A\xi_i, \xi_j]}{\sum_{i,j} a_i a_j [\xi_i, \xi_j]} = \frac{\sum_{i=0}^{n} a_i^2}{\sum_{i=0}^{n} \omega_i^2 a_i^2} \geqslant \omega_n^{-2}$$

by (6.63), (6.64), (6.65) so that

$$\omega_n^{-2} \leqslant \nu(\varphi_0, \ldots, \varphi_{n-1}), \quad \omega_n^{-2} \leqslant \operatorname*{Inf}_{\varphi} \nu(\varphi_0, \varphi_1, \ldots, \varphi_{n-1})$$

and the conclusion follows.

9.3. Application to Comparison Theorems

1. We shall first apply the Min-Max Theorem to a beam subjected to the conditions enumerated under Cases 2, 3, 4 of §I.1, that is fixed at both ends, hinged at one end and fixed at the other, and fixed at one end but free at the other.

Writing $\tilde{\tilde{K}} \subset \tilde{K} \subset K$ for the subspaces generated by the elements of $H^2(0,l)$ satisfying the corresponding kinematic conditions, $\zeta_0, \ldots, \zeta_{n-1}$, for the first n eigenmodes and $\omega_0, \ldots, \omega_{n-1}$ for the first n eigenmodes corresponding to Case 4 (space K), we know that:

$$\omega_n^{-2} = \operatorname*{Sup}_{\substack{\xi \in K \\ [\xi, \zeta_j] = 0, \\ 0 \leqslant j \leqslant n-1}} \frac{\int_0^l \rho \xi^2 \, dx}{\int_0^l EI \xi''^2 \, dx}.$$

Since $\tilde{K} \subset K$ we have:

$$\operatorname*{Sup}_{\substack{\xi \in \tilde{K} \\ [\xi, \zeta_j] = 0, \\ 0 \leqslant j \leqslant n-1}} \frac{\int_0^l \rho \xi^2 \, dx}{\int_0^l EI \xi''^2 \, dx} = \tilde{\nu}(\zeta_0, \ldots, \zeta_{n-1}) \leqslant \omega_n^{-2}.$$

and by the min-max theorem, we know that $\tilde{\omega}_n^{-2} \leqslant \tilde{\nu}(\zeta_0, \ldots, \zeta_{n-1})$, $\tilde{\omega}_n$ being the eigenfrequency of rank n associated with Case 3. We therefore have $\omega_n \leqslant \tilde{\omega}_n$ and in the same way it could be shown that $\tilde{\omega}_n \leqslant \tilde{\tilde{\omega}}_n$. Qualitatively this expresses the result that imposing additional kinematic constraints has the effect of raising the eigenfrequency at each level. We had already observed this result in the case of a homogeneous prismatic beam (EI, ρ independent of x) but we see here that it applies in a wider context.

2. We now propose to compare the eigenfrequencies of two beams of the same length l, subjected to the same kinematic conditions, or equivalently to the same boundary conditions for the cases 1, 2, 3 of §I.1. The first beam is characterised by specifying $EI(x)$, $\rho(x)$ as given measurable functions of x such that

$$0 < a = \operatorname*{Inf}_{x} \rho \leqslant \rho \leqslant \operatorname*{Sup}_{x} \rho = b < +\infty$$

$$0 < p = \operatorname*{Inf}_{x} EI \leqslant EI \leqslant \operatorname*{Sup}_{x} EI = q < +\infty$$

and the second is a prismatic homogeneous beam for which $EI = p$ and $\rho = b$ are constants.

We introduce the notations

(6.66) $$[\xi,\eta]=\int_0^l EI\xi''\eta''\,dx,\quad \{\xi,\eta\}=\int_0^l p\xi''\eta''\,dx,$$

A is the linear operator from K into K associated with the first beam, B the corresponding operator, likewise from K into K, for the homogeneous prismatic beam. It is clear that:

(6.67) $$\frac{\{B\xi,\xi\}}{\{\xi,\xi\}}=\frac{\int_0^l b\xi^2\,dx}{\int_0^l p\xi''^2\,dx}\geqslant\frac{\int_0^l \rho\xi^2\,dx}{\int_0^l EI\xi''^2\,dx}=\frac{[A\xi,\xi]}{[\xi,\xi]},\quad \forall\xi\in K.$$

Let $\varphi_0,\dots,\varphi_{n-1}$ be an arbitrary system of n elements of $H^2(0,l)$ and let us define

(6.68) $$\nu(\varphi_0,\dots,\varphi_{n-1})=\operatorname*{Sup}_{\substack{\xi\in K\\ [\xi,\varphi_j]=0\\ 0\leqslant j\leqslant n-1}}\frac{[A\xi,\xi]}{[\xi,\xi]}.$$

Since one can always associate with the system φ_j of elements of $H^2(0,l)$ a system ψ_j of elements taken from the same space and such that $p\psi_j''=EI\varphi_j''$ and vice versa, it is clear that

$$\operatorname*{Sup}_{\substack{\xi\in K\\ \{\xi,\psi_j\}=0,\\ 0\leqslant j\leqslant n-1}}\frac{\{B\xi,\xi\}}{\{\xi,\xi\}}=\mu(\psi_0,\dots,\psi_{n-1})\geqslant\nu(\varphi_0,\dots,\varphi_{n-1})$$

so that, if we write $\omega_n,\tilde{\omega}_n$ for the eigenfrequencies of the beams $(EI(x),\rho(x))$ and $(EI=p,\rho=b)$ respectively, we obtain $\omega_n\geqslant\tilde{\omega}_n$.

Similarly with $\tilde{\tilde{\omega}}_n$ as the eigenfrequency of the homogeneous prismatic beam $(EI=q,\rho=a)$ we have $\tilde{\tilde{\omega}}_n\geqslant\omega_n$. In particular, from $\tilde{\omega}_n\leqslant\omega_n\leqslant\tilde{\tilde{\omega}}_n$, it is plain that $\omega_n=O(n^2)$, as $n\to\infty$.

This result remains true for Cases 5 and 6 as the following modified proof shows. Consider, for example, a beam hinged at $x=0$, and free at $x=l$, and let us define:

$$K=\left\{\xi:\xi\in H^2(0,l),\ \xi(0)=0,\int_0^l\rho x\xi\,dx=0\right\},$$

$$\tilde{K}=\left\{\xi:\xi\in H^2(0,l),\ \xi(0)=0,\int_0^l x\xi\,dx=0\right\},$$

retaining the notations A,B for the linear operators from K into K and from $\tilde{K}$ into $\tilde{K}$, associated with the beams defined by $EI(x)$, $\rho(x)$ and $EI=p$, $\rho=b$ respectively. Since $\int_0^l x\xi\,dx$ is a continuous linear functional on K, there is a $\theta\in K$ such that $\int_0^l x\xi\,dx=[\xi,\theta]$, $\forall\xi\in K$ and similarly $\chi\in\tilde{K}$ such that

$$\int_0^l\rho x\xi\,dx=\{\xi,\chi\},\quad\forall\xi\in\tilde{K}.$$

With

$$V = (\xi : \xi \in K, [\xi, \theta] = 0, [\xi, \varphi_j] = 0, 0 \leqslant j \leqslant n-1)$$
$$= (\xi : \xi \in \tilde{K}, \{\xi, \chi\} = 0, \{\xi, \psi_j\} = 0, 0 \leqslant j \leqslant n-1)$$

and

$$\operatorname{Sup}_{\xi \in V} \frac{[A\xi, \xi]}{[\xi, \xi]} = \nu(\theta, \varphi_0, \ldots, \varphi_{n-1}),$$

$$\operatorname{Sup}_{\xi \in V} \frac{\{B\xi, \xi\}}{\{\xi, \xi\}} = \mu(\chi, \psi_0, \ldots, \psi_{n-1})$$

we can write

$$\mu(\psi_0, \ldots, \psi_{n-1}) \geqslant \mu(\chi, \psi_0, \ldots, \psi_{n-1}) \geqslant \nu(\theta, \varphi_0, \ldots, \varphi_{n-1}) \geqslant \omega_{n+1}^{-2}$$

whence $\tilde{\omega}_n^{-2} \geqslant \omega_{n+1}^{-2}$, and therefore $\tilde{\omega}_n \leqslant \omega_{n+1}$.

Similarly we can establish that $\omega_n \leqslant \tilde{\tilde{\omega}}_{n+1}$ and deduce therefrom the asymptotic estimate $\omega_n = O(n^2)$, as $n \to \infty$,

3. We now return to the problem of the eigenfrequencies of turbine blades. Using the notation already used in §I.7.9., we define the form

$$[\xi, \eta] = \int_0^l (EI\xi''\eta'' + F\xi'\eta')\,\mathrm{d}x, \quad \forall (\xi, \eta) \in H^2(0,l) \times H^2(0,l)$$

and the operator A from K into K, $K = \{\xi : \xi \in H^2(0,l), \xi(0) = \xi'(0) = 0\}$ such that

$$[\xi, A\eta] = \int_0^l \rho \xi \eta \,\mathrm{d}x, \quad \forall (\xi, \eta) \in K \times K.$$

We know that

$$\omega_n^{-2} = \operatorname{Inf}_{\varphi} \nu(\varphi_0, \ldots, \varphi_{n-1})$$

with

$$\nu(\varphi_0, \ldots, \varphi_{n-1}) = \operatorname{Sup}_{\substack{\xi \in K \\ [\xi, \varphi_j] = 0, \\ 0 \leqslant j \leqslant n-1}} \frac{\int_0^l \rho \xi^2 \,\mathrm{d}x}{\int_0^l (EI\xi''^2 + F\xi'^2)\,\mathrm{d}x}$$

where $(\varphi_0, \ldots, \varphi_{n-1})$ is a system of n arbitrary elements of $H^2(0,l)$. If with every $\varphi \in H^2(0,l)$ we can associate a $\psi \in H^2(0,l)$ such that

$$\int_0^l (EI\varphi''\xi'' + F\varphi'\xi')\,\mathrm{d}x = \int_0^l EI\psi''\xi''\,\mathrm{d}x, \quad \forall \xi \in K \tag{6.69}$$

and conversely, then it follows, by an obvious inequality, that

$$\nu(\varphi_0, \ldots, \varphi_{n-1}) \leqslant \mu(\psi_0, \ldots, \psi_{n-1})$$

where μ is the functional analogous to ν associated with the same problem in the absence of rotation ($F = 0$). Hence

$$\operatorname{Inf}_{\varphi} \nu(\varphi) \leqslant \operatorname{Inf}_{\psi} \mu(\psi),$$

that is $\tilde{\omega}_n \leqslant \omega_n$ where $\omega_n, \tilde{\omega}_n$ are the eigenfrequencies of rank n of the blades, with or without rotation.

We must now return to the question of finding functions satisfying (6.69). Now for any $\xi \in K$ and $\varphi \in H^2(0,l)$ we can write, on integrating by parts

$$\int_0^l (EI\varphi''\xi'' + F\varphi'\xi')\,dx = \int_0^l \left(EI\varphi'' + \int_x^l F\varphi'\,ds \right)\xi''\,dx - \xi' \int_x^l F\varphi'\,ds \Big|_0^l$$

and therefore, since $\xi'(0) = 0$, (6.69) leads to

$$EI\varphi'' + \int_x^l F\varphi'\,ds = EI\psi''. \tag{6.70}$$

If $\varphi \in H^2(0,l)$, this equation allows us to define ψ in the same space, modulo an affine function of x (assuming $0 < \mathrm{Inf}_x EI < \mathrm{Sup}_x EI < +\infty$). Conversely suppose $\psi \in H^2(0,l)$ to be given, then to obtain φ all we need do is to solve the integral equation

$$\psi' = \varphi' + \int_0^x \frac{d\sigma}{EI(\sigma)} \int_\sigma^l F(s)\varphi'(s)\,ds \tag{6.71}$$

with respect to $\varphi' = \theta$, which can be done by successive approximation using the iterative relation

$$\theta_n = \psi' - \int_0^x \frac{d\sigma}{EI(\sigma)} \int_\sigma^l F(s)\theta_{n-1}(s)\,ds.$$

It can be shown by classical arguments [42] that θ exists, that $\theta - \psi'$ is absolutely continuous, θ' is square-integrable and that φ, the primitive of θ, belongs to $H^2(0,l)$.

10. Forced Excitation of a Beam

10.1. Fourier's Method

Let us again consider a beam of length l, subjected to the conditions represented by one of the Cases 1 to 4, and let us suppose a vertical force to be exerted for a finite time, the linear density of the force being $f(t)\cdot p(x)$, with $f(t) = 0$, for $t \notin [0, t_0]$, where $f(t)$ is bounded and $p(x)$ is square-integrable, and the beam being initially at rest. The displacement $v(x,t)$ satisfies $v(x,0) = 0$, $(\partial v/\partial t)(x,0) = 0$ and is furthermore continuous, as also is the velocity $(\partial v/\partial t)(x,t)$, for $(x,t) \in [0,l] \times \mathbb{R}$.

We now introduce the Fourier transform

$$\hat{v}(x,\omega) = \frac{1}{\sqrt{2\pi}} \int_{-\infty}^{+\infty} e^{-i\omega t} v(x,t)\,dt$$

with $\mathrm{Im}\,\omega = -\varepsilon$, $\varepsilon > 0$. This ensures that the integral converges since

$$e^{-i\omega t} v(x,t) = e^{-\varepsilon t} \cdot v(x,t) \cdot e^{-i\sigma t}, \quad \sigma = \mathrm{Re}\,\omega$$

and $v(x,t) = 0$ for $t \leqslant 0$.

In view of the assumptions regarding the continuity of displacement and velocity with respect to time, we obtain after integrating by parts

$$-\omega^2\hat{v}(x,\omega)=\frac{1}{\sqrt{2\pi}}\int_{-\infty}^{+\infty} e^{-i\omega t}v_{tt}(x,t)\,dt$$

and for the transform of the equation (6.55)

$$\frac{\partial^2}{\partial x^2}\left(EI\frac{\partial^2\hat{v}}{\partial x^2}(x,\omega)\right)-\rho\omega^2\hat{v}(x,\omega)=\hat{f}(\omega)p(x) \tag{6.72}$$

where $\hat{f}(\omega)$ is the Fourier transform of $f(t)$.

The boundary conditions at $x=0$ and $x=l$ can be similarly expressed for $v(x,t)$ and $\hat{v}(x,\omega)$ so that the solution $\hat{v}$ of (6.72) can be obtained by the calculations in Section I.8:

$$\hat{v}(x,\omega)=\sum_{j=0}^{\infty}\hat{f}(\omega)c_j(\omega)\xi_j(x)$$

with

$$c_j(\omega)=\frac{\int_0^l p(x)\xi_j(x)\,dx}{\omega_j^2-\omega^2}$$

and by inversion

$$v(x,t)=\frac{1}{\sqrt{2\pi}}\int_{-\infty-i\varepsilon}^{+\infty-i\varepsilon} e^{i\omega t}\sum_{j=0}^{\infty}\hat{f}(\omega)c_j(\omega)\xi_j(x)\,d\omega. \tag{6.73}$$

Since $f(t)$ has compact support, $\hat{f}(\omega)$ is a holomorphic function of ω throughout the complex plane, and is bounded in any strip parallel to the x-axis.

We shall assume that the order of integration and summation in (6.73) can be interchanged, and therefore

$$v(x,t)=\sum_{j=0}^{\infty}v_j(x,t),$$

$$v_j(x,t)=\frac{1}{\sqrt{2\pi}}\int_{-\infty-i\varepsilon}^{+\infty-i\varepsilon} e^{i\omega t}\hat{f}(\omega)c_j(\omega)\,d\omega\cdot\xi_j(x).$$

To calculate v_j, we can evaluate the integral along the axis $-\infty+i\alpha$, $+\infty+i\alpha$, $\alpha>0$ by using Cauchy's theorem applied to a suitable contour, taking account of the residues at the poles $\pm i\omega_j$ and obtaining:

$$\begin{aligned} v_j(x,t)=\Bigg[&\frac{1}{\sqrt{2\pi}}\int_{-\infty+i\alpha}^{+\infty+i\alpha} e^{i\omega t}\hat{f}(\omega)c_j(\omega)\,d\omega\\ &+i\sqrt{\frac{\pi}{2}}\cdot\omega_j^{-1}\cdot(e^{-i\omega_j t}\hat{f}(-\omega_j)-e^{i\omega_j t}\hat{f}(\omega_j))\cdot\int_0^l p(s)\xi_j(s)\,ds\Bigg]\xi_j(x). \end{aligned} \tag{6.74}$$

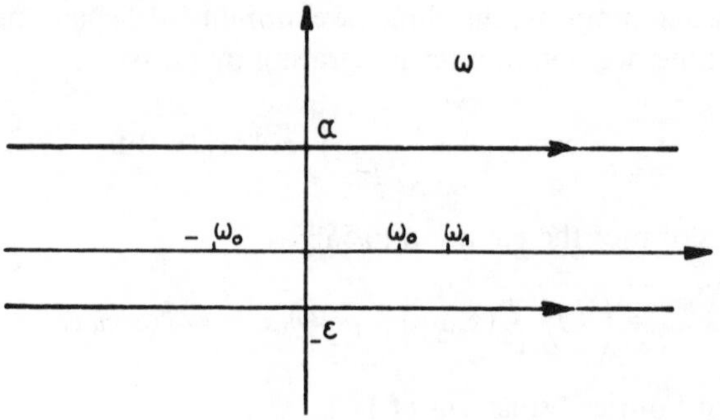

Figure 6.20

The integral is independent of α and can be majorised by:

$$\left|\int_{-\infty+i\alpha}^{+\infty+i\alpha} c_j(\omega)\int_0^{t_0} e^{i\omega(t-\tau)}f(\tau)\,\mathrm{d}\tau\,\mathrm{d}\omega\right| \leqslant e^{-\alpha(t-t_0)}\left|\int_{-\infty+i\alpha}^{+\infty+i\alpha}|c_j(\omega)|\,\mathrm{d}\omega\cdot\int_0^{t_0}|f(\tau)|\,\mathrm{d}\tau\right|$$

whence we conclude that for $t > t_0$ its value is zero.

It therefore becomes apparent that when the excitation has ceased at $t > t_0$ the free oscillations are described by:

$$v(x,t) = i\sqrt{2\pi}\sum_{j=0}^{\infty}\frac{e^{-i\omega_j t}\hat{f}(-\omega_j) - e^{i\omega_j t}\hat{f}(\omega_j)}{2\omega_j}\cdot\int_0^l p(s)\xi_j(s)\,\mathrm{d}s\cdot\xi_j(x). \tag{6.75}$$

Take, for example, the case of a homogeneous prismatic beam of length l, resting on supports and with excitation given by $p(x)f(t)$ where $p(x) = 1$ for

$$|x - l/2| < \delta l/2, \quad 0 < \delta < 1, \quad p(x) = 0 \quad \text{elsewhere}.$$

The eigenfrequencies are

$$\omega_{j-1} = j^2\frac{\pi^2}{l^2}\sqrt{\frac{EI}{\rho}}, \quad j = 1, 2, \ldots$$

and the associated eigenmodes, normalised by

$$\int_0^l \rho\xi_{j-1}^2\,\mathrm{d}x = 1$$

are

$$\xi_{j-1}(x) = \sqrt{\frac{2}{\rho l}}\sin j\frac{\pi x}{l}.$$

We have

$$\int_0^l p(s)\xi_{j-1}(s)\,\mathrm{d}s = \frac{2}{j\pi}\sqrt{\frac{2l}{\rho}}\cdot\sin j\frac{\pi\delta}{2}\cdot\sin j\frac{\pi}{2}$$

whence, by (6.75)

$$v(x,t)=2i\sqrt{\frac{l}{\rho\pi}}\sum_{j=0}^{\infty}(-1)^j\cdot\frac{e^{-i\omega_{2j}t}\hat{f}(-\omega_{2j})-e^{i\omega_{2j}t}\hat{f}(\omega_{2j})}{(2j+1)\omega_{2j}}\sin\left(\frac{2j+1}{2}\pi\delta\right)\xi_{2j}(x).$$

10.2. Boundary Conditions with Elasticity Terms

To illustrate this situation let us consider the case of a beam which is clamped at $x=0$, but whose free end at $x=l$ is connected to a spring which is unextended when the beam is horizontal in its rest position. (A different approach has already been given in §7.10 n°3.)

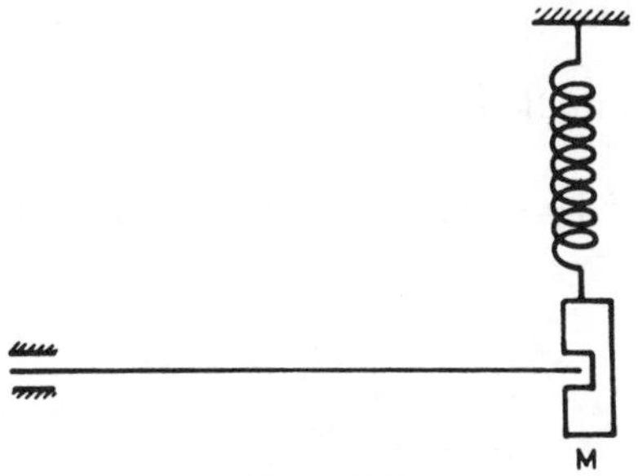

Figure 6.21

To study the motion and determine the eigenfrequencies we can imagine that, because of the spring, a vertical force of intensity f is exerted on the beam at the point $x=l$.

This force is not known beforehand but it is easily expressed in terms of the stiffness of the spring, the displacement of the beam at the point $x=l$, and possibly the inertia of a mass M attached there through which the excitation is transmitted. Writing $v(x,t)$ for the vertical displacement of the beam, so that $y=v(l,t)$ is the displacement at the end $x=l$, we have:

$$\frac{\partial^2}{\partial x^2}\left(EI\frac{\partial^2 v}{\partial x^2}\right)+\rho\frac{\partial^2 v}{\partial t^2}=0,\quad x\in[0,l[$$

and

$$My''=-ky+f$$

$$f=\frac{\partial}{\partial x}\left(EI\frac{\partial^2 v}{\partial x^2}\right)\bigg|_{x=l}.$$

We look for a solution of the form $v(x,t)=\xi(x)\sin\omega t$, where ω is an eigenfrequency of the system, as yet unknown. The equations of motion, expressed in terms of $\xi(x)$ are:

$$(EI\xi'')''-\rho\omega^2\xi=0,\quad x\in[0,l[$$

$$\xi(0)=\xi'(0)=0$$

$$EI\xi''(l)=0,\quad (M\omega^2-k)\xi(l)+(EI\xi'')'(l)=0.$$

It is natural to look for ξ in the subspace K of $H^2(0,l)$ of kinematically admissible displacements, or, by the expansion theorem, in the form of a series

$$\xi(x)=\sum_{j=0}^{\infty} c_j\xi_j(x)$$

strongly convergent in K, where ω_j, ξ_j are the eigenfrequencies and eigenmodes of vibration of a beam fixed at $x=0$ and free at $x=l$.

We have $\omega_j^2 c_j=[\xi,\xi_j]$ and since $A\xi_j=\omega_j^{-2}\xi_j$ we can write:

$$c_j=\omega_j^{-2}[\xi,\xi_j]=[\xi,A\xi_j].$$

But we also know that:

$$[\xi,\xi_j]=\int_0^l EI\xi''\xi_j''\,\mathrm{d}x=\int_0^l (EI\xi'')''\xi_j\,\mathrm{d}x+\int_0^l [EI\xi''\xi_j'-(EI\xi'')'\xi_j]'\,\mathrm{d}x$$

and by the equations of motion

$$[\xi,\xi_j]=\omega^2\int_0^l \rho\xi\xi_j\,\mathrm{d}x+(M\omega^2-k)\xi_j(l)\xi(l)$$

whence

$$c_j=\frac{M\omega^2-k}{\omega_j^2-\omega^2}\xi_j(l)\xi(l)$$

and

$$\xi(x)=(M\omega^2-k)\sum_{j=0}^{\infty}\frac{\xi_j(l)\xi_j(x)\xi(l)}{\omega_j^2-\omega^2}.$$

On setting $x=l$ we obtain the equation which defines the admissible values of ω

$$1=(M\omega^2-k)\sum_{j=0}^{\infty}\frac{\xi_j^2(l)}{\omega_j^2-\omega^2}.$$

The case of an elastic body vibrating within a compressible fluid with a coupling effect is a generalisation of this elementary model which has been studied, with particular reference to the case where the compressibility is slight in [13, 48, 19].

10.3. Forced Vibrations of a Beam Clamped at One End, Bearing a Point Load at the Other End, and Excited at the Clamped End by an Imposed Transverse Motion of Frequency ω

We write $\rho(x)$, $EI(x)$, under the usual hypotheses, for the mechanical characteristics of the beam, fixed at $x=0$, free at $x=l$, and bearing at that end a point load of mass μ. The excitation is produced by a transverse displacement imposed at the point $x=0$, obeying the law $v(0,t)=a\sin\omega t$ where a,ω are given constants. Looking for a representation of the motion of the form

$$v(x,t)=\zeta(x)\sin\omega t$$

we are led to

$$(EI\zeta'')'' - \rho\omega^2\zeta = 0, \quad x \in [0, l[$$
$$\zeta(0) = a, \quad \zeta'(0) = 0$$
$$EI\zeta''(l) = 0, \quad (EI\zeta'')'(l) + \mu\omega^2\zeta(l) = 0$$

this last equation expressing the fact that the shearing constraint at $x = l$ balances the rate of change of momentum due to the load μ. Putting $\zeta = \xi + a$ we can rewrite these equations as:

$$(EI\xi'')'' - \rho\omega^2\xi - \rho a\omega^2 = 0$$
$$\xi(0) = 0, \quad \xi'(0) = 0$$
$$EI\xi''(l) = 0, \quad (EI\xi'')'(l) + \mu\omega^2\xi(l) + \mu a\omega^2 = 0.$$

The natural vibrations of the beam ($a = 0$), and the eigenfrequencies and eigenmodes ω_j, ξ_j, can be found by the classical approach, namely minimising the functional

$$\frac{\int_0^l EI\xi''^2 \,\mathrm{d}x}{\int_0^l \rho\xi^2 \,\mathrm{d}x + \mu\xi^2(l)}$$

on the subspaces $K \supset K_1 \supset K_2 \supset \cdots$ with

$$K = \{\xi : \xi \in H^2(0, l), \ \xi(0) = \xi'(0) = 0\}$$

and

$$K_j = \left\{ \xi : \xi \in K_{j-1}, \int_0^l \rho\xi\xi_{j-1} \,\mathrm{d}x + \mu\xi(l)\xi_{j-1}(l) = 0 \right\}$$

the minimum ω_j^2 of this functional on K_j being attained at the element $\xi_j \in K_j$, which can always be assumed to be normalised by

$$\int_0^l \rho\xi_j^2(x) \,\mathrm{d}x + \mu\xi_j^2(l) = 1,$$

ω_j being always an increasing sequence tending to $+\infty$ with j.

The bilinear form $[\xi, \eta] = \int_0^l EI\xi''\eta'' \,\mathrm{d}x$, defined on $K \times K$, allows us to define a norm $|||\xi||| = [\xi, \xi]^{1/2}$ equivalent to $\|\xi\|_{H^2(0,l)}$ on K. Lastly, applying Riesz's theorem, there exists a positive symmetric linear operator $\xi \to A\xi$ from K into K:

$$\int_0^l \rho\xi\eta \,\mathrm{d}x + \mu\xi(l)\eta(l) = [A\xi, \eta], \quad \forall (\xi, \eta) \in K \times K,$$

and it is clear that this operator is completely continuous, its eigenvalues and eigenvectors being precisely ω_j^{-2}, ξ_j:

$$A\xi_j = \omega_j^{-2}\xi_j, \quad [A\xi_j, \xi_j] = 1,$$
$$[\xi_j, \xi_j] = \omega_j^2, \quad [\xi_j, \xi_k] = 0, \quad j \neq k.$$

Coming back to the problem of the forced vibrations, it seems natural to look for the solution ξ in the space K, or in accordance with the expansion theorem, in the form of a series

$$\xi = \sum_{j=0}^{\infty} c_j \xi_j,$$

strongly convergent in K, from which we deduce:

$$[\xi, \xi_j] = c_j[\xi_j, \xi_j]$$

or

$$c_j = \omega_j^{-2}[\xi, \xi_j] = [\xi, A\xi_j] = \int_0^l \rho\xi\xi_j \mathrm{d}x + \mu\xi(l)\xi_j(l).$$

To calculate these coefficients we multiply the differential equation for ξ by ξ_j and integrate with respect to x from 0 to $l-\varepsilon$, $\varepsilon \uparrow 0$

$$\lim_{\varepsilon\uparrow 0} \int_0^{l-\varepsilon} \xi_j[(EI\xi'')'' - \rho\omega^2\xi - \rho a\omega^2]\mathrm{d}x = 0.$$

Note that because of the equation satisfied by ξ itself, $(EI\xi'')'$, $EI\xi''$ are both continuous on $[0,l]$ and that

$$\xi_j(EI\xi'')'' = (EI\xi_j'')''\xi + [(EI\xi'')'\xi_j - EI\xi''\xi_j' + EI\xi_j''\xi' - (EI\xi_j'')'\xi]'.$$

By integration taking account of the boundary conditions for ξ, ξ_j and in particular $(EI\xi_j'')'(l) + \mu\omega_j^2\xi_j(l) = 0$: we obtain

$$\int_0^l [(EI\xi_j'')''\xi - \rho\omega^2\xi_j\xi - \rho a\omega^2\xi_j]\mathrm{d}x + [-\mu\omega^2(\xi(l)+a)\xi_j(l) + \mu\omega_j^2\xi(l)\xi_j(l)] = 0$$

whence finally if $\omega \neq \omega_j$, $\forall j$ (non-resonance), with $(EI\xi_j'')'' - \rho\omega_j^2\xi_j = 0$:

$$c_j = \frac{a\omega^2}{\omega_j^2 - \omega^2}\left(\int_0^l \rho\xi_j \mathrm{d}x + \mu\xi_j(l)\right), \quad \xi = \sum_{j=0}^{\infty} c_j\xi_j.$$

It remains only to satisfy one's-self of the convergence of the series and to verify that the function which it represents is a solution of the problem. As regards the first point, we note, by an elementary calculation, that the system

$$\begin{gathered} (EI\theta'')'' - \rho = 0, \quad \forall x \in [0,l] \\ \theta(0) = 0, \quad \theta'(0) = 0 \\ EI\theta''(l) = 0, \quad (EI\theta'')'(l) + \mu = 0 \end{gathered}$$

defines uniquely an element $\theta \in K$ and that:

$$\begin{aligned} [\theta, \xi_j] &= \int_0^l EI\theta''\xi_j'' \mathrm{d}x = (EI\theta''\xi_j' - (EI\theta'')'\xi_j)\Big|_0^l + \int_0^l (EI\theta'')''\xi_j \mathrm{d}x \\ &= \mu\xi_j(l) + \int_0^l \rho\xi_j \mathrm{d}x. \end{aligned}$$

However since $\omega_j^{-1} \cdot \xi_j$ is an orthonormal base of K we have

$$|||\theta|||^2 = \sum_{j=0}^{\infty} [\theta, \omega_j^{-1} \xi_j]^2;$$

the convergence of this series, together with $\lim_{j \to \infty} \omega_j = \infty$, implying the convergence of

$$\sum_{j=0}^{\infty} [\theta, \xi_j]^2 \cdot \frac{\omega_j^2}{(\omega_j^2 - \omega^2)^2}$$

and therefore of

$$\sum_{j=0}^{\infty} c_j \xi_j = \xi.$$

Finally we have to verify that ξ is a solution of the problem. We start from

$$[\xi, \xi_j] = \frac{a\omega^2}{\omega_j^2 - \omega^2} \left(\int_0^l \rho \xi_j \mathrm{d}x + \mu \xi_j(l) \right) [\xi_j, \xi_j], \; [\xi_j, \xi_j] = \omega_j^2$$

or

$$[\xi, \xi_j] - \frac{\omega^2}{\omega_j^2} [\xi, \xi_j] = a\omega^2 \left(\int_0^l \rho \xi_j \mathrm{d}x + \mu \xi_j(l) \right)$$

and we represent $[\xi, \xi_j]$, initially by the defining relation

$$[\xi, \xi_j] = \int_0^l EI \xi'' \xi_j'' \mathrm{d}x$$

and then by

$$[\xi, \xi_j] = \omega_j^2 [\xi, A\xi_j] = \left(\int_0^l \rho \xi \xi_j \mathrm{d}x + \mu \xi(l) \xi_j(l) \right) \omega_j^2$$

whence

$$\int_0^l EI \xi'' \xi_j'' \mathrm{d}x - \omega^2 \left(\int_0^l \rho \xi \xi_j \mathrm{d}x + \mu \xi(l) \xi_j(l) \right) = a\omega^2 \left(\int_0^l \rho \xi_j \mathrm{d}x + \mu \xi_j(l) \right).$$

Since $\{\xi_j\}$ is dense in K this relation remains valid if we replace ξ_j by $\varphi \in D(0, l) \subset K$. In other words

$$\int_0^l (EI \xi'' \varphi'' - \omega^2 \rho \xi \varphi - a\omega^2 \rho \varphi) \mathrm{d}x = 0$$

which shows that $(EI\xi'')'' - \rho\omega^2(\xi + a) = 0$ in the distributions sense on $]0, l[$, and in particular that $(EI\xi'')'$, $EI\xi''$ are continuous on $[0, l]$.

Using these results in the above relation, after transformation of

$$\int_0^l EI \xi'' \xi_j'' \mathrm{d}x = (EI\xi'' \cdot \xi_j' - (EI\xi'')' \xi_j) \Big|_0^l + \int_0^l (EI\xi'')'' \xi_j \mathrm{d}x$$

we derive:

$$EI\xi''(l) \cdot \xi_j'(l) - ((EI\xi'')'(l) + \mu\omega^2(\xi(l) + a))\xi_j(l) = 0, \; \forall j$$

i.e. in view of the properties of $\{\xi_j\}$:

$$EI\xi''(l) = 0, \quad (EI\xi'')'(l) + \mu\omega^2(\xi(l) + a) = 0$$

(There is always a function $\chi \in K$ such that $\chi(l)$, $\chi'(l)$ have any prescribed values; χ can be approximated as closely as desired, in terms of the norm in K, by a suitable linear combination of the ξ_j and, with its first derivative χ', pointwise, to the same degree of approximation, by this combination.)

The conditions at $x = 0$ are obviously satisfied by $\xi \in K$.

II. Longitudinal Vibrations of Bars. Torsional Vibrations

1. Equations of the Problem and the Calculation of Eigenvalues

The model we shall consider is one-dimensional, the bar being supposed cylindrical with its generator parallel to the x-axis and the displacement of an arbitrary slice of abscissa x assumed to be parallel to this axis and denoted by $u(x,t)$.

If N is the force exerted normally across the section x by the particles of abscissa greater than x on those of abscissa less than x, and if we assume that external forces of linear density N_e, act on the outside of the bar, the equation of motion is:

$$\rho \frac{\partial^2 u}{\partial t^2} = \frac{\partial N}{\partial x} + N_e$$

where ρ is the linear density [23]. This has to be taken in conjunction with the elasticity equation which describes the behaviour of the bar:

$$\frac{N}{A} = E \frac{\partial u}{\partial x}$$

where A is the cross-sectional area, and E is Young's modulus. We thus obtain the equation of motion:

(6.76) $$\rho \frac{\partial^2 u}{\partial t^2} = \frac{\partial}{\partial x}\left(EA \frac{\partial u}{\partial x}\right) + N_e.$$

Let us, assuming $N_e = 0$, look for harmonic vibrations:

$$u(x,t) = \xi(x) \sin \omega t.$$

We have to solve the equation

(6.77) $$(EA\xi')' + \rho\omega^2 \xi = 0$$

with the following sets of boundary conditions at the ends $x = 0$ and $x = l$ of the bar:

1) Bar fixed at $x = 0$, free at $x = l$
   the kinematic condition $\xi(0) = 0$,
   $N(l) = 0$ or $EA\xi'(l) = 0$

2) Bar free at both ends
$N(0) = N(l) = 0$ or
$EA\xi'(0) = EA\xi'(l) = 0$
3) Bar fixed at both ends
$\xi(0) = \xi(l) = 0$, kinematic conditions.

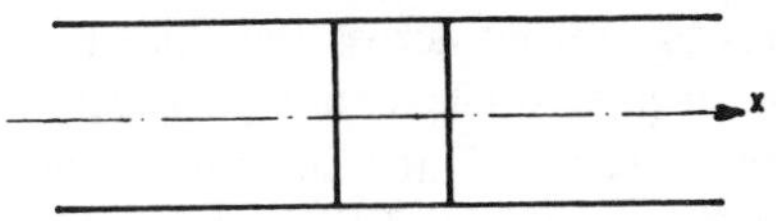

Figure 6.22

It should be noted that equation (6.76) and also (6.77) is still valid even when the bar is not cylindrical and the area $A(x)$ of its cross-section varies slowly with x; similarly ρ may be dependent on x.

However before indicating the general results let us deal with the simple example where EA and ρ are independent of x, and the bar is fixed at $x = 0$, free at $x = l$.

We have to solve

$$EA\xi'' + \rho\omega^2\xi = 0, \quad \forall x \in (0, l), \quad \text{with} \quad \xi(0) = \xi'(l) = 0;$$

and we find:

$$\xi = a \sin\left(\sqrt{\frac{\rho}{EA}}\,\omega x\right), \quad \text{with} \quad \sqrt{\frac{\rho}{EA}}\,\omega l = (2n+1)\frac{\pi}{2}$$

and so the eigenfrequencies are:

$$\omega_n = (2n+1)\frac{\pi}{2l}\sqrt{\frac{EA}{\rho}}, \quad n = 0, 1, 2, \ldots$$

In the general case, we can use Rayleigh's method and show that

$$\omega^2 = \operatorname*{Inf}_{\xi \in K} \frac{\int_0^l EA\xi'^2\,dx}{\int_0^l \rho\xi^2\,dx}$$

where K is the subset of elements ξ of $H^1(0, l)$ which satisfy the kinematic conditions, defines the first eigenfrequency. However in the case of a bar free at both ends, with no kinematic conditions, K will have to be taken as the subset of those elements ξ of $H^1(0, l)$ which satisfy the condition $\int_0^l \rho\xi\,dx = 0$.

The eigenfrequencies of higher order can be defined in the same way as in the flexion problem.

If a point-load μ is placed at the end $x = l$, this end being free, then the lowest eigenfrequency is given by:

$$\omega^2 = \operatorname*{Inf}_{\xi \in K} \frac{\int_0^l EA\xi'^2\,dx}{\int_0^l \rho\xi^2\,dx + \mu\xi^2(l)}$$

with $K=\{\xi:\xi\in H^1(0,l),\ \xi(0)=0\}$ in the case of a bar fixed at $x=0$, and $K=\{\xi:\xi\in H^1(0,l),\ \int_0^l \rho\xi\,dx=0\}$ when it is free at $x=0$.

Torsional vibrations present no new problem. Let Ox be the axis about which a cylindrical bar of generators parallel to Ox is subjected to torsional effects. We denote by θ the angle of torsion of the slice x, i.e. the angle through which this slice turns in relation to some fixed reference direction, and by Γ the moment with respect to Ox of the stresses exerted across this slice by the particles of abscissa greater than x on those of abscissa less than x. Lastly let I be the moment of inertia per unit length about Ox and Γ_e the linear moment with respect to Ox of the forces, if any, acting at the boundaries of the bar, so that we can write:

$$I\frac{\partial^2\theta}{\partial t^2}=\frac{\partial\Gamma}{\partial x}+\Gamma_e$$

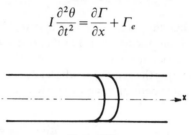

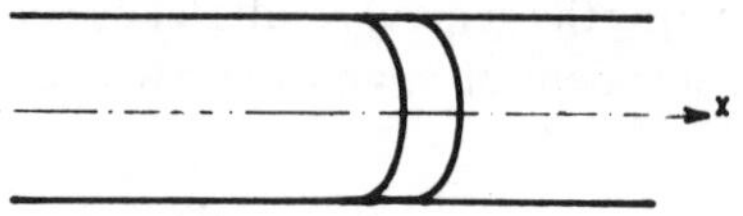

Figure 6.23

In addition we have the equation describing the physical behaviour of the bar $\Gamma=GJ\partial\theta/\partial x$; G the modulus of rigidity, depending only on the material, and J being a constant specified by the geometry. From these we obtain the equation motion [23]:

$$I\frac{\partial^2\theta}{\partial t^2}=\frac{\partial}{\partial x}\left(GJ\frac{\partial\theta}{\partial x}\right)+\Gamma_e.$$

In the case of a bar fixed at $x=0$, and free at $x=l$ we would have as boundary conditions the kinematic condition $\theta=0$, and the condition $GJ(\partial\theta/\partial x)(l)=0$. The problem is formally identical to the preceding one and the same conclusions can be drawn from it.

2. The Associated Functional Operator

The problem of finding the natural frequencies of a shaft under torsion can be reduced to that of finding the eigenvalues of a completely continuous operator in a Hilbert space, as we saw in the case of the flexural vibrations of a beam. On the assumption that GJ and I are measurable functions of $x\in(0,l)$, which are bounded above and below, in this interval, by strictly positive numbers, it suffices to observe that the bilinear form $[\xi,\eta]=\int_0^l GJ\xi'\cdot\eta'\,dx$, defined on $H^1(0,l)\times H^1(0,l)$, induces on the subspace $K=\{\xi:\xi\in H^1(0,l),\ \xi(0)=0\}$, (shaft of length l, with kinematic condition $\theta(0)=0$ or $\xi(0)=0$), a norm $|||\xi|||=[\xi,\xi]^{1/2}$ equivalent to the usual norm of $H^1(0,l)$, then by Riesz's theorem [37], we can define the linear operator $\eta \to A\eta$, from K into K, by:

$$\int_0^l I\xi\eta \,\mathrm{d}x = [\xi, A\eta], \quad \forall \xi \in K.$$

A is a symmetric, positive, completely continuous operator whose eigenvalues are the squares of the reciprocals of the eigenfrequencies of the bar under torsion.

It is incidentally of some interest to show, on this simple example, how we can express $\eta_1 = A\eta$. Writing

$$\int_0^l I\xi\eta \,\mathrm{d}x = \int_0^l GJ\xi'\eta_1' \,\mathrm{d}x = GJ\xi\eta_1'|_0^l - \int_0^l \xi\cdot(GJ\eta_1')' \,\mathrm{d}x$$

we are led to $(GJ\eta_1')' + I\eta = 0$, $\forall x \in (0, l)$

and $$\eta_1(0) = 0, \quad GJ\eta_1'(l) = 0$$

whence

$$\eta_1 = -\int_0^x \frac{\mathrm{d}s}{GJ} \int_s^l I\eta(\sigma) \,\mathrm{d}\sigma \in K. \tag{6.78}$$

This formula suggests that starting from an element $\eta \in K$, we calculate the sequence $\eta_1 = A\eta$, $\eta_2 = A\eta_1, \ldots$ by iterating with the operator A, and this leads us to ask how this can provide information about the approximate values of the eigenvalues. We now prepare to develop this idea more systematically.

3. The Method of Moments

3.1. Introduction

Changing slightly the notation we shall henceforth consider a Hilbert space H, which we shall suppose to be defined over the reals and a positive, completely continuous, linear symmetric operator from H into H, $\varphi \to A\varphi$, $\forall \varphi \in H$.

Starting from an element $\varphi \in H$, we introduce the successive iterates:

$$\varphi_1 = A\varphi, \quad \varphi_2 = A\varphi_1, \ldots, \varphi_{n-1} = A\varphi_{n-2}, \ldots \tag{6.79}$$

and we define the subspace V_n, generated by the elements $\varphi, \varphi_1, \ldots, \varphi_{n-1}$:

$$V_n = \{\varphi, \varphi_1, \ldots, \varphi_{n-1}\}. \tag{6.80}$$

There are only two cases to be considered:

• There exists an integer N such that $\forall n \leqslant N$, V_n is of dimension n and $A\varphi_{N-1} \in V_N$, so that $V_{N+j} = V_N$, for all positive integers j. In this case the subspace V_N is reduced by the operator A, that is

$$\psi \in V_N \to A\psi \in V_N$$

and finding the eigenvalues of A in V_N is equivalent to finding the eigenvalues of a matrix of order N.

• The second case requiring rather more attention is that in which V_n is of dimension n for every natural number n; while V, the closure in H of all the subspaces V_n, which is obviously a subspace in H reduced by the operator A is also a Hilbert space (which may in fact be the same as H). The object of this theory which we shall now develop is to define an algorithm for finding the eigenvalues of A in V.

Writing E_n for the operator which projects V orthogonally on to V_n we introduce the symmetric linear operator, from V into V_n, defined by $A_n = E_n A E_n$, and we propose to examine the relationships which exist between the eigenvalues of A_n and those of A.

3.2. Lanczos's Orthogonalisation Method [57]

Starting from $\psi = \varphi$, we construct $\psi_1 = A\psi - a\psi = \varphi_1 - a\varphi$ orthogonal to φ, which defines

$$a = \frac{[\varphi_1, \varphi]}{[\varphi, \varphi]} = \frac{[A\psi, \psi]}{[\psi, \psi]}$$

where the notation [.,.] is used to denote the scalar product in $V \subset H$. Similarly a_1 and b are determined so that

$$\psi_2 = (A - a_1)\psi_1 - b\psi$$

is orthogonal to ψ and ψ_1 which implies that

$$b = \frac{[A\psi_1, \psi]}{[\psi, \psi]} = \frac{[\psi_1, A\psi]}{[\psi, \psi]} = \frac{[\psi_1, \psi_1]}{[\psi, \psi]}$$

$$a_1 = \frac{[A\psi_1, \psi_1]}{[\psi_1, \psi_1]}.$$

Having constructed $\psi, \psi_1, \ldots, \psi_k$, orthogonal to each other and such that $\{\psi, \psi_1, \ldots, \psi_{j-1}\} = V_j$, $j \leqslant k+1$, we define

(6.81) $$\psi_{k+1} = (A - a_k)\psi_k - b_{k-1}\psi_{k-1}.$$

Now, for $j \leqslant k-2$ we have:

$$[\psi_{k+1}, \psi_j] = [A\psi_k, \psi_j] = [\psi_k, A\psi_j] = 0$$

since $A\psi_j \in V_{j+2} \subset V_k$ and $\psi_k \perp V_k$. To express the fact that ψ_{k+1} is orthogonal to V_{k+1}, it suffices therefore to say that it is orthogonal to ψ_{k-1} and ψ_k, or in other words:

$$b_{k-1} = \frac{[A\psi_k, \psi_{k-1}]}{[\psi_{k-1}, \psi_{k-1}]} = \frac{[\psi_k, A\psi_{k-1}]}{[\psi_{k-1}, \psi_{k-1}]} = \frac{[\psi_k, \psi_k]}{[\psi_{k-1}, \psi_{k-1}]}$$

(6.82) $$a_k = \frac{[A\psi_k, \psi_k]}{[\psi_k, \psi_k]}.$$

It is clear from (6.81) that the orthogonal system $\psi, \psi_1, \ldots, \psi_k, \ldots$ can be represented by:

(6.83)
$$\psi_k = P_k(A)\psi = P_k(A)\varphi$$

where $P_k(\lambda)$ is a polynomial of degree k in λ, defined by the recurrence relation:

(6.84)
$$P_{j+1}(\lambda) = (\lambda - a_j)P_j(\lambda) - b_{j-1}P_{j-1}(\lambda), \quad P_0 = 1, \quad P_1 = \lambda - a.$$

From (6.79) and (6.80) we see that:

(6.85)
$$A^j\varphi = A_n^j\varphi, \quad 0 \leqslant j \leqslant n-1$$

and observing that $P_n(\lambda) = \lambda^n + p_{n-1}(\lambda)$, where $p_{n-1}(\lambda)$ is a polynomial of degree $n-1$, we can write

$$\psi_n = P_n(A)\varphi = A^n\varphi + p_{n-1}(A)\varphi = AA_n^{n-1}\varphi + p_{n-1}(A_n)\varphi$$

whence

$$E_n\psi_n = A_n^n\varphi + p_{n-1}(A_n)\varphi = P_n(A_n)\varphi$$

and

(6.86)
$$P_n(A_n)\varphi = 0$$

because of the orthogonality of ψ_n and V_n.

We can obtain another definition of the polynomial $P_n(\lambda)$ by observing that there exists an unique set of real numbers $\alpha, \alpha_1, \ldots, \alpha_{n-1}$ such that

(6.87)
$$E_n\varphi_n = -\alpha\varphi - \alpha_1\varphi_1 - \cdots - \alpha_{n-1}\varphi_{n-1}$$

or, in view of (6.85) and $E_n\varphi_n = A_n^n\varphi$:

$$(A_n^n + \alpha_{n-1}A_n^{n-1} + \cdots + \alpha I)\varphi = 0.$$

Comparison of this with (6.86), remembering the linear independence of the vectors defined by (6.85), shows that

(6.88)
$$P_n(\lambda) = \lambda^n + \alpha_{n-1}\lambda^{n-1} + \cdots + \alpha$$

3.3. Eigenvalues of A_n

The operator A_n from V_n into V_n is symmetric, hence all its eigenvalues are real and any associated eigenvector can be represented in the form:

$$u = c\varphi + c_1\varphi_1 + \cdots + c_{n-1}\varphi_{n-1}$$

where $A_n u = \lambda u$ and the real numbers c_j are not all zero. However by (6.79) we can write

$$A_n u = c\varphi_1 + c_1\varphi_2 + \cdots + c_{n-2}\varphi_{n-1} + c_{n-1}E_n\varphi_n$$

and using (6.87) we can calculate the c_j and the eigenvalues λ from the set of equations

(6.89)
$$\begin{aligned} -\alpha c_{n-1} &= \lambda c \\ c - \alpha_1 c_{n-1} &= \lambda c_1 \\ c_1 - \alpha_2 c_{n-1} &= \lambda c_2 \\ \cdots & \quad \cdots \\ c_{n-2} - \alpha_{n-1}c_{n-1} &= \lambda c_{n-1}. \end{aligned}$$

The equation which defines the eigenvalues can be derived from these in the form

(6.90)
$$\det\begin{pmatrix} \lambda & \cdots & \cdots & \cdots & \alpha \\ -1 & \lambda & \cdots & \cdots & \alpha_1 \\ 0 & -1 & \lambda & \cdots & \alpha_2 \\ \vdots & & \ddots & \ddots & \vdots \\ \cdots & \cdots & \cdots & -1 & \lambda+\alpha_{n-1} \end{pmatrix} = 0$$

or in the expanded form:

(6.91)
$$\lambda^n + \alpha_{n-1}\lambda^{n-1} + \cdots + \alpha_1\lambda + \alpha = P_n(\lambda) = 0.$$

It also follows from (6.89) that for every root λ of (6.90) or (6.91), the associated eigenvector, normalised by $c_{n-1} = 1$, is unique. The eigenvalues are therefore all simple. We may also add that all these eigenvalues are strictly positive, in consequence of the hypothesis that A is a positive operator, which implies the same for A_n.

A comparison of the eigenvalues of A_{n-1} and A_n brings out an interesting separation property. As a consequence of the recurrence relation (6.84) and the fact that the coefficients b_j defined by (6.82) are strictly positive, it is easily proved that each polynomial $P_j(\lambda)$ has exactly j zeros separating the $j+1$ zeros of $P_{j+1}(\lambda)$.

We shall conclude this section by showing that we can obtain a representation of $P_n(\lambda)$ with the help of the moments $\mu_k = [\varphi, A^k\varphi]$. From (6.86) with $P_n(\lambda)$ defined by (6.91), we obtain, by taking the scalar product with φ_j:

$$\alpha[\varphi, \varphi_j] + \alpha_1[A_n\varphi, \varphi_j] + \cdots + \alpha_{n-1}[A_n^{n-1}\varphi, \varphi_j] + [A_n^n\varphi, \varphi_j] = 0$$

i.e. using $A_n^k\varphi = \varphi_k$, $0 \leqslant k \leqslant n-1$ and

$$[A_n^n\varphi, \varphi_j] = [E_nAA_n^{n-1}\varphi, \varphi_j] = [A\varphi_{n-1}, E_n\varphi_j] = [\varphi_n, \varphi_j], \quad 0 \leqslant j \leqslant n-1,$$

(6.92)
$$\alpha[\varphi, \varphi_j] + \alpha_1[\varphi_1, \varphi_j] + \cdots + \alpha_{n-1}[\varphi_{n-1}, \varphi_j] + [\varphi_n, \varphi_j] = 0, \quad 0 \leqslant j \leqslant n-1.$$

For λ to be an eigenvalue of A_n, it is necessary and sufficient that

(6.93)
$$\alpha + \alpha_1\lambda + \cdots + \alpha_{n-1}\lambda^{n-1} + \lambda^n = 0$$

or in other words that the set of $(n+1)$ equations (6.92) and (6.93) be compatible with respect to $\alpha, \alpha_1, \ldots, \alpha_{n-1}$ so that:

(6.94)
$$\det\begin{pmatrix} \mu_0 & \mu_1 & \mu_2 & \cdots & \mu_n \\ \mu_1 & \mu_2 & \mu_3 & \cdots & \mu_{n+1} \\ \vdots & & & & \\ \mu_{n-1} & \mu_n & \mu_{n+1} & \cdots & \mu_{2n-1} \\ 1 & \lambda & \lambda^2 & \cdots & \lambda^n \end{pmatrix} = 0$$

with $\mu_k = [\varphi, A^k\varphi]$.

The equation (6.94) thus defines the eigenvalues of A_n, directly in terms of the moments μ_k, $0 \leqslant k \leqslant 2n-1$. (The independence of the vectors $\varphi, \varphi_1, \ldots, \varphi_{n-1}$ ensures that the coefficients of λ^n in the expansion of (6.94) is non-zero, or in other words that the polynomial obtained is precisely of degree n as was to be expected.)

We can write $\mu_k = [A^j\varphi, A^{k-j}\varphi]$, since A is symmetric, and consequently $A^j\varphi$ needs to be calculated only up to order $j = n$, in order to work out the equation (6.94).

3.4. Padé's Method

For every real or complex z not belonging to the spectrum of the operator A, which incidentally is discrete spectrum since A is completely continuous [37], we can define the inverse operator $(I - zA)^{-1}$ and

$$R_\varphi(z) = [\varphi, (I - zA)^{-1}\varphi] = \sum_{n=0}^{\infty} [\varphi, A^n\varphi] z^n = \sum_{n=0}^{\infty} \mu_n z^n.$$

Padé's approximation of order n to this series consists in defining a rational fraction $Q(z)/\Delta(z)$, Q and Δ being polynomials of degree $n-1$ and n respectively, such that one can write formally:

$$\sum_{k=0}^{\infty} \mu_k z^k - \frac{Q(z)}{\Delta(z)} = O(z^{2n}). \tag{6.95}$$

Writing

$$\begin{aligned} \Delta(z) &= b_0 + b_1 z + \cdots + b_n z^n \\ Q(z) &= q_0 + q_1 z + \cdots + q_{n-1} z^{n-1}, \end{aligned} \tag{6.96}$$

we thus have the equations

$$\begin{aligned} &\mu_0 b_0 = q_0 \\ &\mu_0 b_1 + \mu_1 b_0 = q_1 \\ &\cdots \\ &\mu_0 b_{n-1} + \mu_1 b_{n-2} + \cdots + \mu_{n-1} b_0 = q_{n-1} \end{aligned} \tag{6.97}$$

and

$$\begin{aligned} &\mu_0 b_n + \mu_1 b_{n-1} + \cdots + \mu_n b_0 = 0 \\ &\mu_1 b_n + \mu_2 b_{n-1} + \cdots + \mu_{n+1} b_0 = 0 \\ &\cdots \\ &\mu_{n-1} b_n + \mu_n b_{n-1} + \cdots + \mu_{2n-1} b_0 = 0. \end{aligned} \tag{6.98}$$

The $n+1$ coefficients b_j are calculated from the set (6.98) of n homogeneous linear equations which, as such, always have at least one solution, and the q_j are then obtained directly using (6.97). Furthermore, by eliminating the b_j from (6.98) and

$$b_0 + b_1 z + \cdots + b_n z^n = 0$$

we can see that the roots of the equation $\Delta(z) = 0$ are the same as those of:

$$\det \begin{pmatrix} \mu_0 & \mu_1 & \cdots & \mu_n \\ \mu_1 & \mu_2 & \cdots & \mu_{n+1} \\ \mu_{n-1} & \mu_n & \cdots & \mu_{2n-1} \\ z^n & z^{n-1} & \cdots & 1 \end{pmatrix} = 0$$

from which it follows, by comparing this with (6.94) that the eigenvalues of A_n correspond, via the transformation $z = 1/\lambda$, to the singularities of the Padé approximation of order $2n$ associated with the full series $\sum_0^\infty \mu_j z^j$ constructed with the help of the moments.

3.5. Approximation of the A Operator

1. It can be shown [57] that the linear operator $A_n = E_n A E_n$ from V into V is an approximation to A in the sense that:

(6.99) $$\lim_{n\to\infty} \| A - A_n \| = 0.$$

If (6.99) were false, there would be a real positive q, an infinite subsequence of A_n, also denoted by A_n, and an infinite sequence of elements $f_n \in V$, with $\| f_n \| = 1$, such that

(6.100) $$\| (A - A_n) f_n \| \geqslant q.$$

Since A is completely continuous, and $\| E_n f_n \| \leqslant 1$, we can by successively extracting suitable subsequences but still retaining the same indicial notation n, assume that the sequences $A f_n$ and $A E_n f_n$ are strongly convergent in V:

$$g = \lim_{n\to\infty} A f_n, \quad h = \lim_{n\to\infty} A E_n f_n.$$

From

$$\| g - h \|^2 = \lim_{n\to\infty} [A f_n - A E_n f_n, g - h]$$
$$= \lim_{n\to\infty} [f_n, (I - E_n) A (g - h)] \leqslant \lim_{n\to\infty} \| (I - E_n) A (g - h) \| = 0,$$

that is $g = h$, and from:

$$(A - A_n) f_n = A f_n - A E_n f_n + (I - E_n)(A E_n f_n - h) + (I - E_n) h$$

we deduce $\lim_{n\to\infty} (A - A_n) f_n = 0$ which contradicts (6.100).

2. We now prove, following [57], that every eigenvalue of A on V is simple[3]. Let ν be the reciprocal of an eigenvalue, with $\nu > 0$, and let n be chosen large enough so that:

(6.101) $$\nu < \| A - A_n \|^{-1}$$

which is always possible by (6.99).

If u is the associated eigenvector we have:

$$u - \nu A u = (I - \nu(A - A_n))(u - \nu(I - \nu(A - A_n))^{-1} A_n u) = 0,$$

the existence of the inverse matrix $(I - \nu(A - A_n))^{-1}$ being ensured by (6.101), and:

[3] This does not conflict with the fact that this same eigenvalue of A may be multiple on H, since $V \subset H$.

(6.102) $$u - \nu(I - \nu(A - A_n))^{-1} A_n u = 0.$$

On the other hand we can represent $A_n u \in V_n$ in the form:

(6.103) $$A_n u = \sum_{k=0}^{n-1} b_k \nu^{-1} \varphi_k$$

with appropriate coefficients b_k and putting

(6.104) $$\xi_k = (I - \nu(A - A_n))^{-1} \varphi_k$$

we can write (6.102) as:

(6.105) $$u = \sum_{k=0}^{n-1} b_k \xi_k.$$

However, from (6.104) written in the form

$$\xi_k - \nu(A - A_n)\xi_k = \varphi_k$$

we see that

(6.106) $$\xi_k = \varphi_k, \quad k = 0, 1, \ldots, n-2,$$

so that, applying A_n to u in (6.105) we obtain:

$$A_n u = \sum_{k=0}^{n-2} b_k \varphi_{k+1} + b_{n-1} A_n \xi_{n-1}$$

and by (6.104):

(6.107) $$(I - \nu(A - A_n))^{-1} A_n u = \sum_{k=0}^{n-2} b_k \xi_{k+1} + b_{n-1}(I - \nu(A - A_n)^{-1}) A_n \xi_{n-1}.$$

On the other hand since $A_n \xi_{n-1} \in V_n$, there exists an unique set of real numbers $\alpha_s(\nu)$ such that:

$$A_n \xi_{n-1} = -\sum_{s=0}^{n-1} \alpha_s(\nu) \varphi_s$$

or
$$(I - \nu(A - A_n))^{-1} A_n \xi_{n-1} = -\sum_{s=0}^{n-1} \alpha_s(\nu) \xi_s$$

and we deduce from (6.107), using (6.102), (6.105) and the linear independence of the ξ_k, $0 \leqslant k \leqslant n-1$ defined by (6.104):

$$b_k = \nu b_{k-1} - \nu \alpha_k(\nu) b_{n-1}, \quad b_{-1} = 0, \quad k = 0, 1, 2, \ldots, n-1.$$

We can normalise the eigenvector by taking $b_{n-1} = 1$ in (6.105) so that:

$$b_k = \nu(b_{k-1} - \alpha_k(\nu)), \quad b_{-1} = 0, \quad k = 0, 1, \ldots, n-2.$$

Thus there is only one eigenvector associated with ν, which proves that the eigenvalue concerned is simple.

3. Denoting by $\lambda_{n+1-j}^{(n)}$ the zeros of the polynomial $P_n(\lambda)$, or in other words the eigenvalues of A_n, arranged in order of decreasing size as j increases from 1 to n,

we know that:

$$0 < \lambda_{n+1-j}^{(n+1)} < \lambda_{n+1-j}^{(n)} < \lambda_{n+2-j}^{(n+1)} \tag{6.108}$$

as the zeros of $P_n(\lambda)$ separate those of $P_{n+1}(\lambda)$. Every eigenvalue λ of A_n satisfies $0 < \lambda < \|A_n\| < \|A\|$ (because $A_n = E_n A E_n$ and $\|E_n\| < 1$), and so we have an upper bound for the whole set of eigenvalues of the operators A_n, $n = 0, 1, \ldots$ For any fixed j, we see from (6.108) that the sequence $\lambda_{n+1-j}^{(n)}$ increases with n, and is bounded and therefore convergent. This allows us to define

$$\lim_{n\to\infty} \lambda_{n+1-j}^{(n)} = \lambda_j > 0. \tag{6.109}$$

We can now show that λ_j is an eigenvalue of A. There exists a $u_n \in V$, with $\|u_n\| = 1$ such that:

$$E_n A E_n u_n = \lambda_{n+1-j}^{(n)} u_n \tag{6.110}$$

and since $\|E_n u_n\| < \|u_n\| = 1$ and A is a completely continuous operator, we can extract a subsequence, which we shall still denote by $AE_n u_n$ which converges strongly in V when n tends to infinity. The sequence $E_n A E_n u_n$ will also converge to the same element, whence it follows by (6.109) and (6.110) that the sequence u_n converges strongly to an element $u \in V$, distinct from 0, since $\|u_n\| = 1$. By proceeding to the limit in (6.110) we obtain $Au = \lambda_j u$, which proves that λ_j is an eigenvalue of A.

The order relation $\lambda_{n+1-j}^{(n)} < \lambda_{n+1-i}^{(n)}$, $1 \leqslant i < j \leqslant n$ implies that $\lambda_j \leqslant \lambda_i$ but more than this is true because it can be shown that the λ_j are necessarily distinct. For suppose that, for some pair of integers i, j with $i \neq j$, we were to have $\lambda_i = \lambda_j = \lambda$; we could then, by successive extractions, construct two sequences of elements of V, with

$$\|u_n\| = \|v_n\| = 1,$$

such that

$$A_n u_n = \lambda_{n+1-i}^{(n)} v_n, \quad A_n v_n = \lambda_{n+1-j}^{(n)} v_n \tag{6.111}$$

which would respectively converge strongly in V to elements u and v belonging to V, which would be of unit norm and be eigenvectors of A for the eigenvalue λ. But it follows from (6.111) that $[u_n, v_n] = 0$, since

$$\lambda_{n+1-i}^{(n)} \neq \lambda_{n+1-j}^{(n)},$$

and also $[u, v] = 0$.

Thus for the eigenvalue λ, the operator A would have two independent eigenvectors, which is impossible since it would contradict the result proved in Section 2.

We conclude therefore that $\lambda_{j+1} < \lambda_j$ for all integers $j \geqslant 1$.

4. We shall now show that if $\lambda > 0$ is an eigenvalue of A, it must be equal to one of the λ_i, unless it is $\tilde{\lambda} = \lim_{j\to\infty} \lambda_j$.

If not there would be an integer j such that $\lambda_{j+1} < \lambda < \lambda_j$, unless $0 < \lambda < \tilde{\lambda}$, a number $\delta > 0$, and an integer n_0 such that for all $n > n_0$ no eigenvalue of A_n could exist in the interval $(\lambda - \delta, \lambda + \delta)$.

Let g by any given element of V and let us try to solve the equation:

(6.112) $$(E_n A E_n - \lambda) f_n = g.$$

We can represent f_n in the form $f_n = -\lambda^{-1}(I - E_n)g + E_n f_n$ and, introducing the orthonormal set of the n eigenvectors of A_n, i.d. $u_1, u_2, \ldots, u_n$ corresponding to the eigenvalues $\lambda_1^{(n)}, \ldots, \lambda_n^{(n)}$ we can write:

$$E_n f_n = a_1 u_1 + \cdots + a_n u_n$$
$$A_n f_n = a_1 \lambda_1^{(n)} u_1 + \cdots + a_n \lambda_n^{(n)} u_n$$
$$E_n g = [g, u_1] u_1 + \cdots + [g, u_n] u_n$$

whence by (6.112) $a_k(\lambda_k^{(n)} - \lambda) = [g, u_k]$ and hence finally

$$f_n = -\lambda^{-1}(I - E_n)g + \sum_{k=1}^{n} \frac{[g, u_k]}{\lambda_k^{(n)} - \lambda} u_k.$$

From $|\lambda_k^{(n)} - \lambda| > \delta$ we deduce:

$$\sum_{k=1}^{n} \frac{[g, u_k]^2}{|\lambda_k^{(n)} - \lambda|^2} < \delta^{-2} \sum_{k=1}^{n} [g, u_k]^2 < \delta^{-2} \|g\|^2$$

and

$$\|f_n\| \leqslant (\lambda^{-1} + \delta^{-1}) \|g\|.$$

The sequence f_n is therefore bounded, and by virtue of the property of complete continuity and (6.112) we can suppose that f_n converges strongly to an element f of V, so that after proceeding to the limit in equation (6.112) we obtain

$$(A - \lambda I) f = g.$$

Thus this equation would have, for every $g \in V$, a solution $f \in V$, which is impossible when λ is an eigenvalue of A.

5. It follows from the preceding considerations that the n eigenvalues of A_n, are approximations from below to the n largest eigenvalues of A on the subspace V. More precisely, if φ be a given element, whose iterates $\varphi_1, \varphi_2, \ldots, \varphi_n$ up to order n, form with φ an independent system, then the roots of the equation of degree n (6.94) form a set of approximations from below to the n largest eigenvalues of A on V.

To illustrate we detail the case $n = 2$: we suppose $\varphi, A\varphi, A^2\varphi$ to be independent and consider the equation:

$$\det \begin{pmatrix} \mu_0 & \mu_1 & \mu_2 \\ \mu_1 & \mu_2 & \mu_3 \\ 1 & \lambda & \lambda^2 \end{pmatrix} = 0$$

which involves calculating the three moments μ_1, μ_2, μ_3.

Setting

$$a = \frac{[\varphi, A\varphi]}{[\varphi, \varphi]} = \frac{\mu_1}{\mu_0}$$

$$\sigma_2 = \frac{[\varphi, (A-a)^2\varphi]}{[\varphi, \varphi]} > 0$$

$$\sigma_3 = \frac{[\varphi, (A-a)^3\varphi]}{[\varphi, \varphi]}$$

we obtain without difficulty:

$$\lambda_1^{(2)} = \frac{[\varphi, A\varphi]}{[\varphi, \varphi]} - \left(\left(\frac{\sigma_3}{2\sigma_2}\right)^2 + \sigma_2\right)^{1/2} + \frac{\sigma_3}{2\sigma_2}$$

$$\lambda_2^{(2)} = \frac{[\varphi, A\varphi]}{[\varphi, \varphi]} + \left(\left(\frac{\sigma_3}{2\sigma_2}\right)^2 + \sigma_2\right)^{1/2} + \frac{\sigma_3}{2\sigma_2}.$$

It will be seen that, whatever the sign of σ_3, we have:

$$\lambda_1^{(2)} < \frac{[\varphi, A\varphi]}{[\varphi, \varphi]} < \lambda_2^{(2)}$$

in accordance with the separation theorem; $\lambda_2^{(2)}$ is an approximation from below to the largest eigenvalue, better than the one provided by Rayleigh's starting from the test function φ; $\lambda_1^{(2)}$ is an approximation from below to the second largest eigenvalue.

III. Vibrations of Elastic Solids

1. Statement of Problem and General Assumptions

We shall suppose that the elastic solid whose vibrational state we wish to study occupies a bounded open domain Ω in $\mathbb{R}^3$, defined by reference to an orthonormal set of axes $Ox_1x_2x_3$, whose boundary $\partial\Omega$ can be regarded as the union of two disjoint parts $\Gamma_1, \Gamma_2 : \partial\Omega = \Gamma_1 \cup \Gamma_2, \Gamma_1 \cap \Gamma_2 = \emptyset$. In one of these parts Γ_1 the displacement of the points is kept at zero, while the points of the other part Γ_2 are not subject to any constraint and may be considered as free to move.

We denote by $u_i(x,t)$, $i = 1, 2, 3$ the components of the displacement at time t of the particle at $x = (x_1, x_2, x_3)$ in Ω, by T_{ij} the strain tensor, by ρ the density, and by ρF_i the vector representing the volume forces which are assumed to be time-independent. The laws of mechanics lead to the set of equations

(6.113) $$\partial_j T_{ij} + \rho F_i = \rho \frac{\partial^2 u_i}{\partial t^2}, \quad \text{where } \partial_j = \frac{\partial}{\partial x_j}.$$

We assume the elastic behaviour to be given by a linear law of the type:

(6.114) $$T_{ij} = a_{ijkh} U_{kh}$$

where U_{kh} is the strain tensor:

(6.115) $$U_{kh} = \tfrac{1}{2}(\partial_k u_h + \partial_h u_k)$$

and $a_{ijkh}(x)$ is a fourth order tensor which is a function of x and is a characteristic of the material under consideration. This tensor satisfies the symmetry conditions:

(6.116) $$a_{ijkh}=a_{jikh}=a_{khij}.$$

We shall also assume that the functions $a_{ijkh}(x)$ are measurable and bounded in Ω, and that the quadratic form $a_{ijkh}(x)\tau_{ij}\tau_{kh}$ is positive-definite, i.e. that there is a positive constant C such that for any symmetric tensor $\tau_{ij}=\tau_{ji}$ and any $x\in\Omega$ the inequality

(6.117) $$a_{ijkh}\tau_{ij}\tau_{kh}\geqslant C\tau_{ij}\tau_{ij}$$

holds (where we adopt the usual summation convention for repeated suffixes).

We denote by u_i^*, T_{ij}^* respectively the displacement-field and the associated constraint field, corresponding to the static problem:

(6.118) $$\partial_j T_{ij}^*+\rho F_i=0$$

with the same constitutive law (6.114), the boundary conditions at the boundary $\partial\Omega$ being zero displacement over Γ_1, no constraints over Γ_2, or in other words $T_{ij}\nu_j=0$ over Γ_2, where ν_j denotes the jth component of the unit vector directed along the outward normal at an arbitrary point of Γ_2.

If u_i, Γ_{ij} are the displacement and constraint fields satisfying (6.113), (6.114) and corresponding to the natural vibrations of the solid, fixed over Γ_1 and free over Γ_2, then it is clear that the field

$$v_i=u_i-u_i^*,\quad \mathscr{T}_{ij}(v)=T_{ij}-T_{ij}^*=a_{ijkh}U_{kh}(v)$$

is a solution of these same equations, with $F_i=0$ and satisfies the same conditions at the boundary.

For this reason we henceforth and without loss of generality, confine our attention to (6.113), (6.114) with $F_i=0$.

Accordingly we can look for solutions of the form

$$u_i=\xi_i(x)\sin\omega t$$

and we are therefore led to the equation:

(6.119) $$\partial_j(a_{ijkh}(x)\varepsilon_{kh})+\rho\omega^2\xi_i=0 \text{ in } \Omega$$

with

(6.120) $$\varepsilon_{kh}=\tfrac{1}{2}(\partial_k\xi_h+\partial_h\xi_k)$$

and with the boundary conditions:

(6.121) $$\xi_i=0$$

on $\Gamma_1, \forall i$, a kinematic constraint, and

(6.122) $$\sigma_{ij}\nu_j=0$$

over $\Gamma_2, \forall i$, with $\sigma_{ij}=a_{ijkh}\varepsilon_{kh}$, where ν_j is the unit vector along the outward normal to Γ_2.

2. The Energy Theorem

Suppose that, for a certain value of ω, the equations (6.119)–(6.122) have a solution. Without bothering for the moment too much about the precise sense in which this solution is defined and proceeding in a purely formal manner, we obtain on integrating both sides of (6.119) after multiplication by ξ_i:

$$\int_\Omega \xi_i \partial_j (a_{ijkh}(x)\varepsilon_{kh})\,\mathrm{d}x + \omega^2 \int_\Omega \rho \xi_i \xi_i \,\mathrm{d}x = 0$$

from which we deduce, after integrating by parts, and taking account of (6.121), (6.122) and the symmetry property (6.116):

$$\int_\Omega a_{ijkh}\varepsilon_{ij}\varepsilon_{kh}\,\mathrm{d}x = \omega^2 \int_\Omega \rho \xi_i \xi_i \,\mathrm{d}x. \tag{6.123}$$

To interpret this equation, let us first of all note that

$$\tfrac{1}{2}\omega^2 \int_\Omega \rho \xi_i \xi_i \,\mathrm{d}x \cdot \cos^2 \omega t = \frac{1}{2}\int_\Omega \rho \frac{\partial u_i}{\partial t}\frac{\partial u_i}{\partial t}\,\mathrm{d}x$$

is the kinetic energy of the solid in vibration. Now let us imagine a virtual displacement δu_i to be impressed, at time t, on each element of the elastic medium. The virtual work which would have to be done by the forces exerted at the boundary in order to bring about this displacement may be written as

$$\int_{\partial\Omega} T_{ij}\nu_j \delta u_i \,\mathrm{d}S = \int_\Omega \partial_j (T_{ij}\delta u_i)\,\mathrm{d}x = \int_\Omega \partial_j T_{ij}\cdot \delta u_i \,\mathrm{d}x + \int_\Omega T_{ij}\partial_j(\delta u_i)\,\mathrm{d}x$$

or, taking into account (6.113) (with $F_i = 0$) and the definition of the U_{ij}:

$$\int_{\partial\Omega} T_{ij}\nu_j \delta u_i \,\mathrm{d}S = \int_\Omega \rho \frac{\partial^2 u_i}{\partial t^2}\delta u_i \,\mathrm{d}x + \int_\Omega a_{ijkh} U_{kh} \delta U_{ij}\,\mathrm{d}x \tag{6.124}$$

which is tantamount to saying that the work done at the boundary plus the work done by the inertial forces (and where applicable the volume forces) is equal to the change in W:

$$W = \frac{1}{2}\int_\Omega a_{ijkh} U_{kh} U_{ij}\,\mathrm{d}x \tag{6.125}$$

which thus represents the potential energy.

In the case of a harmonic motion, this potential energy has the value

$$\tfrac{1}{2}\sin^2 \omega t \cdot \int_\Omega a_{ijkh}\varepsilon_{kh}\varepsilon_{ij}\,\mathrm{d}x$$

and it can be seen that the formula (6.123) expresses the fact that the free oscillations take place at a frequency such that the kinetic and potential energies of the system are equal when averaged over the time of one oscillation.

This remark gives rise to the following formalism. Let $(H^1(\Omega))^3$ be the Sobolev space generated by the vector functions $\xi_i(x)$, $i = 1, 2, 3$, square-integrable on Ω, whose first derivatives in the distributions sense, are also square-integrable on Ω; let K denote the subspace generated by the elements $\xi \in (H^1(\Omega))^3$ satisfying the kinematic constraints $\xi_i = 0$ on Γ_1. At this point, a few additional details are needed to clarify this definition. It is known that if the boundary $\partial\Omega$ is sufficiently regular, more specifically if it is a continuous variety of dimension 2, satisfying a Lipschitz condition, i.e. if it can be represented locally by a suitable choice of co-ordinate system, a_1, a_2, a_3, in the form $a_3 = s(a_1, a_2)$ where s is a continuous Lipschitzian function of the co-ordinates a_1, a_2, then there is a continuous mapping of $H^1(\Omega)$ into $L^2(\partial\Omega)$, denoted by $\xi \to \gamma\xi$, which maps an arbitrary function ξ of $\mathscr{E}(\bar{\Omega})$ the space of continuous and indefinitely differentiable functions in $\bar{\Omega}$ (the closure of Ω), on to the function $\xi|_{\partial\Omega}$ defined by the values taken by ξ at the boundary. As $\mathscr{E}(\bar{\Omega})$ is dense in $H^1(\Omega)$, the trace operator γ allows us to define, in a natural manner, the boundary values of any element $\xi \in H^1(\Omega)$; the continuity of the trace operator ensures moreover that K is a closed linear variety of $(H^1(\Omega))^3$, i.e. a subspace [33].

Accordingly we introduce:

(6.126)
$$\omega_0^2 = \operatorname*{Inf}_{\xi \in K} \frac{\int_\Omega a_{ijkh}(x)\varepsilon_{ij}(\xi)\varepsilon_{kh}(\xi)\,\mathrm{d}x}{\int_\Omega \rho\xi_i\xi_i\,\mathrm{d}x}$$

and our next paragraph will be devoted to showing that this lower bound ω_0^2 is not zero and is attained for an element $\xi \in K$, which satisfies the equations (6.119), (6.120), (6.121), (6.122).

3. Free Vibrations of Elastic Solids

3.1. Existence of the Lowest Eigenfrequency

Let us introduce an infinite sequence $\xi^{(n)}$ of kinematically admissible displacement fields, $\xi^{(n)} \in K$, which are minimising and normalised, so that they satisfy the relations

$$\int_\Omega \rho\xi_i^{(n)}\xi_i^{(n)}\,\mathrm{d}x = 1$$

(6.127)
$$\omega_0^2 = \lim_{n\to\infty} \int_\Omega a_{ijkh}\varepsilon_{ij}^{(n)}\varepsilon_{kh}^{(n)}\,\mathrm{d}x$$

with
$$\varepsilon_{ij}^{(n)} = \tfrac{1}{2}(\partial_j\xi_i^{(n)} + \partial_i\xi_j^{(n)}).$$

1. The sequence $\xi^{(n)}$ is bounded in $(H^1(\Omega))^3$. It is known, from Korn's inequality [18], that there exists a constant C_1 depending only on the domain Ω, such that

(6.128)
$$\int_\Omega \varepsilon_{ij}(\xi)\varepsilon_{ij}(\xi)\,\mathrm{d}x + \int_\Omega \xi_i\xi_i\,\mathrm{d}x \geqslant C_1 \|\xi\|^2$$

for all $\xi \in (H^1(\Omega))^3$, with $\|\xi\|$ defined as usual by:

(6.129)
$$\|\xi\|^2 = \int_\Omega (\xi_i\xi_i + \partial_i\xi_j \cdot \partial_i\xi_j)\,\mathrm{d}x.$$

Since
$$1=\int_\Omega \rho\xi_i^{(n)}\xi_i^{(n)}\,dx \geqslant \operatorname{Inf}_\Omega \rho\cdot\int_\Omega \xi_i^{(n)}\xi_i^{(n)}\,dx$$

we deduce, assuming $\operatorname{Inf}_\Omega \rho = \varkappa^{-1} > 0$, that:

(6.130)
$$\int_\Omega \xi_i^{(n)}\xi_i^{(n)}\,dx \leqslant \varkappa.$$

On the other hand it follows from (6.117) that:

(6.131)
$$\int_\Omega \varepsilon_{ij}^{(n)}\varepsilon_{ij}^{(n)}\,dx \leqslant C^{-1}\cdot\int_\Omega a_{ijkh}\varepsilon_{ij}^{(n)}\varepsilon_{kh}^{(n)}\,dx$$

but the right-hand side of (6.131) tends to $C^{-1}\omega_0^2$ as $n\to\infty$ and in particular defines a bounded sequence.

We deduce from (6.130), (6.131) and (6.128) that $\|\xi^{(n)}\|$ is a bounded sequence.

2. Since $\xi^{(n)}$ is a bounded sequence in the Hilbert space $(H^1(\Omega))^3$ we can extract from it a subsequence, which we may suppose to be indexed in the same way, which converges weakly in $(H^1(\Omega))^3$ to an element $\xi^0\in(H^1(\Omega))^3$, [37]. Furthermore, as the natural injection of $(H^1(\Omega))^3$ in $(L^2(\Omega))^3$ is compact [33], we may suppose that the extracted subsequence converges strongly in $(L^2(\Omega))^3$ to an element $\tilde{\xi}^0\in(L^2(\Omega))^3$. It can easily be proved that $\xi^0=\tilde{\xi}^0$ by an argument similar to the one used in §I.6.1., which we may briefly recall here: if the integer p be chosen so that the $(L^2(\Omega))^3$ norm of $\xi^{(n)}$, for $n>p$, differs from that of $\tilde{\xi}^0$ by less than $\varepsilon>0$, then a finite convex linear combination of the ξ^m, $m>p$, at a distance less than ε from ξ^0, can be found in $(H^1(\Omega))^3$ since ξ^m converges weakly to ξ^0. A fortiori, this convex linear combination is distant less than ε from ξ^0 in $(L^2(\Omega))^3$ and because of the way p has been chosen, and the fact that all its terms are of rank p or greater, it must approximate to within ε of $\tilde{\xi}^0$ in $(L^2(\Omega))^3$. We therefore have

$$\|\xi^0-\tilde{\xi}^0\|_{(L^2(\Omega))^3}\leqslant 2\varepsilon \quad \text{which proves that } \xi^0=\tilde{\xi}^0.$$

3. The strong convergence of $\xi^{(n)}$ to ξ^0 in $(L^2(\Omega))^3$ implies

(6.132)
$$\lim_{n\to\infty}\int_\Omega \rho\xi_i^{(n)}\xi_i^{(n)}\,dx=\int_\Omega \rho\xi_i^0\xi_i^0\,dx=1.$$

On the other hand the bilinear form defined on $(H^1(\Omega))^3$ by

$$W(\xi,\zeta)=\int_\Omega a_{ijkh}\varepsilon_{ij}(\xi)\varepsilon_{kh}(\zeta)\,dx$$

is obviously symmetric, positive, and bicontinuous. It follows (see §I.6.2.) that the mapping $\xi\to W(\xi,\xi)$ of $(H^1(\Omega))^3$ into $\mathbb{R}$ is weakly lower semicontinuous and consequently

$$\liminf_n W(\xi^{(n)},\xi^{(n)})\geqslant W(\xi^0,\xi^0)$$

or, since

$$\omega_0^2=\lim_{n\to+\infty} W(\xi^{(n)},\xi^{(n)}),$$

(6.133)
$$\omega_0^2\geqslant W(\xi^0,\xi^0).$$

4. Let us now show that ξ^0 is a permissible displacement, i.e. that $\gamma\xi^0=0$ on Γ_1.

Since the trace operator from $(H^1(\Omega))^3$ into $L^2(\partial\Omega)$ is continuous, there exists for any given $\varepsilon > 0$, an $\eta > 0$ such that:

$$\| \gamma\xi - \gamma\xi^0 \|_{(L^2(\partial\Omega))^3} < \varepsilon \quad \text{if} \quad \| \xi - \xi^0 \|_{(H^1(\Omega))^3} < \eta.$$

Now $\xi^{(n)}$ converges weakly to ξ^0 in $(H^1(\Omega))^3$, and by the Banach-Saks theorem [37], we can therefore find a convex linear combination

$$\sum_1^n \alpha_j \xi^{(j)}, \; \alpha_j \geqslant 0, \; \sum_1^n \alpha_j = 1,$$

such that

$$\left\| \sum_1^n \alpha_j \xi^{(j)} - \xi^0 \right\|_{(H^1(\Omega))^3} < \eta$$

whence
$$\left\| \gamma\left(\sum_1^n \alpha_j \xi^{(j)} \right) - \gamma\xi^0 \right\|_{(L^2(\partial\Omega))^3} < \varepsilon$$

and a fortiori

$$\int_{\Gamma_1} \left| \sum_1^n \alpha_j \gamma\xi^{(j)} - \gamma\xi^0 \right|^2 \mathrm{d}S < \varepsilon^2$$

so that since $\gamma\xi^{(j)} = 0$ on Γ_1, $\int_{\Gamma_1} |\gamma\xi^0|^2 \mathrm{d}S < \varepsilon^2$ and as ε may be arbitrarily small, this proves that $\gamma\xi^0 = 0$ on Γ_1.

We have thus proved that $\xi^0 \in K$; by (6.132), (6.133) and from the definition of ω_0 we deduce:

$$\omega_0^2 = \frac{\int_\Omega a_{ijkh}\varepsilon_{ij}(\xi^0)\varepsilon_{kh}(\xi^0)\mathrm{d}x}{\int_\Omega \rho\xi_i^0\xi_i^0\mathrm{d}x}, \quad \text{with} \quad \xi^0 \in K.$$

Let us now complete this result by showing that $\omega_0 \neq 0$. If ω_0 were zero, this would imply $\varepsilon_{ij}(\xi^0) = 0$ almost every where in Ω, and since $\xi^0 \in (H^1(\Omega))^3$, we know that it would then follow that $\xi^0 = a + b \wedge x$, with a, b fixed vectors in $\mathbb{R}^3$. This would mean that ξ^0 would be a rigid-body displacement. If Γ_1 contained at least three non-collinear points, which would have to be the case if Γ_1 had non-zero measure, the condition $\gamma\xi^0 = 0$ on Γ_1 could not be satisfied unless a and b were both non-zero, i.e. unless $\xi^0 = 0$. But this is incompatible with the condition $\int_\Omega \rho\xi_i^0\xi_i^0\mathrm{d}x = 1$.

5. It only remains to specify the precise sense in which the pair ω_0, ξ^0 is a solution to the problem posed at the outset. To simplify the notation let us write

$$W_{\text{pot}} = \int_\Omega a_{ijkh}\varepsilon_{ij}(\xi)\varepsilon_{kh}(\xi)\mathrm{d}x, \; W_{\text{kin}} = \int_\Omega \rho\xi_i\xi_i\mathrm{d}x$$

for all $\xi \in K$.

It then follows from the definition of ω_0, that for all $\delta\xi \in K$, $\lambda \in \mathbb{R}$

$$W_{\text{pot}}(\xi^0 + \lambda\delta\xi) \geqslant \omega_0^2 W_{\text{kin}}(\xi^0 + \lambda\delta\xi)$$

i.e. bearing in mind that there is equality for $\lambda = 0$, and that $a_{ijkh} = a_{khij}$:

(6.134)
$$\int_\Omega a_{ijkh}\varepsilon_{kh}(\xi^0)\varepsilon_{ij}(\delta\xi)\,dx = \omega_0^2 \int_\Omega \rho\xi^0\delta\xi\,dx$$

or

$$\int_\Omega \sigma_{ij}^0\varepsilon_{ij}(\delta\xi)\,dx = \omega_0^2 \int_\Omega \rho\xi^0\delta\xi\,dx$$

with
$$\sigma_{ij}^0 = a_{ijkh}\varepsilon_{kh}(\xi^0).$$

However in view of the symmetry of the tensor a_{ijkh} we can equally well write this as

(6.135)
$$\int_\Omega \sigma_{ij}^0\partial_j(\delta\xi_i)\,dx = \omega_0^2 \int_\Omega \rho\xi_i^0\delta\xi_i\,dx$$

and so for all $\delta\xi\in(\mathscr{D}(\Omega))^3 \subset K$ (where $\mathscr{D}(\Omega)$ is the space of indefinitely differentiable functions with compact support in Ω), we at once obtain:

(6.136)
$$\partial_j\sigma_{ij}^0 + \rho\omega_0^2\xi_i^0 = 0$$

in the sense of the theory of distributions. In particular, it follows from this formula that $\partial_j\sigma_{ij}^0$ can be identified with square-integrable function in Ω.

On the other hand, going back to (6.135), which holds for all $\delta\xi\in K$, and using (6.136), we can write:

(6.137)
$$\int_\Omega (\sigma_{ij}^0\partial_j(\delta\xi_i) + \partial_j\sigma_{ij}^0\cdot\delta\xi_i)\,dx = 0$$

which it now remains to interpret. To this end, consider the linear functional defined on $(H^1(\Omega))^3$ by:

(6.138)
$$f(u) = \int_\Omega (\sigma_{ij}^0\partial_j u_i + \partial_j\sigma_{ij}^0\cdot u_i)\,dx.$$

This linear functional is continuous because σ_{ij}^0, $\partial_j\sigma_{ij}^0$ are both square-integrable, in Ω, and it is obvious that $f(u)=0$, $\forall u\in(\mathscr{D}(\Omega))^3$; now since $(\mathscr{D}(\Omega))^3$ is dense in the subspace spanned by those elements u of $(H^1(\Omega))^3$ which vanish on $\partial\Omega$, we conclude that $u\in(H^1(\Omega))^3$, $\gamma u = 0$ together imply that $f(u)=0$.

We have already indicated that the trace operator γ is continuous from $H^1(\Omega)$ into $L^2(\partial\Omega)$; we denote by $H^{1/2}(\partial\Omega)$ the image of $H^1(\Omega)$ under γ, and we have $H^{1/2}(\partial\Omega)\subset L^2(\partial\Omega)$.

We know moreover, cf. [33], that $H^{1/2}(\partial\Omega)$ can be provided with the structure of a topological space in such a way that:

(a) the preceding relation of inclusion remains valid topologically,
(b) the mapping of $H^1(\Omega)$ on to $H^{1/2}(\partial\Omega)$ under the trace operation is continuous, and
(c) there exists a linear and continuous lift R of $H^{1/2}(\partial\Omega)$ into $H^1(\Omega)$.

Accordingly for all $v\in(H^{1/2}(\partial\Omega))^3$, we may define $\varphi(v)=f(Rv)$ as a continuous linear functional on $(H^{1/2}(\partial\Omega))^3$, because $f(Rv)$ is independent of the particular lift; if R^* is another, we have $\gamma(Rv-R^*v)=0$ and consequently $f(Rv-R^*v)=0$, or $f(Rv)=f(R^*v)$. It follows from all this that $f(u)$ defined by (6.138) can be written:

$$f(u)=\varphi(\gamma u), \quad \forall u\in(H^1(\Omega))^3.$$

Since $\varphi(v)$ is a continuous linear functional on $(H^{1/2}(\partial\Omega))^3$, φ is an element of the dual space usually denoted by $(H^{-1/2}(\partial\Omega))^3$. If, as is natural, we identify $L^2(\partial\Omega)$ with its dual, we can write

$$H^{1/2}(\partial\Omega)\subset L^2(\partial\Omega)\subset H^{-1/2}(\partial\Omega).$$

We shall find it convenient to write, formally

$$\varphi(v)=\int_{\partial\Omega}\sigma_{ij}^0 v_j\cdot v_i\,dS$$

the constraint $\sigma_{ij}^0 v_j$ at the boundary being defined as a continuous linear functional on $(H^{1/2}(\partial\Omega))^3$, i.e. as an element of $(H^{-1/2}(\partial\Omega))^3$.

Thus we write

$$f(u)=\int_{\partial\Omega}\sigma_{ij}^0 v_j u_i\,dS$$

and it is clear from (6.137) that

$$\int_{\partial\Omega}\sigma_{ij}^0 v_j\delta\xi_i\,dS=0$$

for all $\delta\xi_i\in H^{1/2}(\partial\Omega)$, vanishing on Γ_1, or equivalently, for all $\delta\xi\in\gamma(K)$, which from a generalised point of view can be interpreted by saying that $\sigma_{ij}^0 v_j=0$ on Γ_2, which is precisely the condition (6.122).

3.2. Higher Eigenfrequencies

Let us consider the subspace K_1 spanned by the elements $\xi\in(H^1(\Omega))^3$ which are kinematically admissible and such that

$$\int_\Omega\rho\xi_i\xi_i^0\,dx=0.$$

We can repeat the preceding analysis, defining ω_1 by

$$\omega_1^2=\operatorname*{Inf}_{\xi\in K_1}\frac{W_{\text{pot}}(\xi)}{W_{\text{kin}}(\xi)}=\operatorname*{Inf}_{\xi\in K_1}\frac{\int_\Omega a_{ijkh}\varepsilon_{ij}(\xi)\varepsilon_{kh}(\xi)\,dx}{\int_\Omega\rho\xi_i\xi_i\,dx} \tag{6.139}$$

and noting that $\omega_0\leqslant\omega_1$.

We can form a minimising sequence and establish the existence of an element $\xi^{(1)}\in K_1$, for which the infimum in (6.139) is attained. We can then easily deduce, by repeating a now familiar argument, that for all $\delta\xi\in K_1$ we have:

$$\int_\Omega\sigma_{ij}^{(1)}\partial_j(\delta\xi_i)\,dx=\omega_1^2\int_\Omega\rho\xi_i^{(1)}\delta\xi_i\,dx, \quad \text{with } \sigma_{ij}^{(1)}=a_{ijkh}\varepsilon_{kh}(\xi^{(1)}). \tag{6.140}$$

It will therefore be sufficient to establish that this formula remains true when $\delta\xi$ is replaced by ξ^0 in order to guarantee that $\partial_j\sigma_{ij}^{(1)} + \rho\omega_1^2\xi_i^{(1)} = 0$, in the sense of the theory of distributions.

Now we can write:

$$\int_\Omega \sigma_{ij}^{(1)}\partial_j\xi_i^0\,dx - \omega_1^2\int_\Omega \rho\xi_i^{(1)}\xi_i^0\,dx \tag{6.141}$$

$$= \int_\Omega (a_{ijkh}\varepsilon_{kh}(\xi^{(1)})\varepsilon_{ij}(\xi^0) - \omega_1^2\rho\xi_i^{(1)}\xi_i^0)\,dx$$

and using (6.134) and the symmetry property $a_{ijkh} = a_{khij}$, with $\delta\xi = \xi^{(1)}\in K_1 \subset K$, we obtain for the right-hand side of (6.141):

$$(\omega_0^2 - \omega_1^2)\int_\Omega \rho\xi_i^{(1)}\xi_i^0\,dx = 0.$$

Thus (6.140) has been established for $\delta\xi = \xi^0$.

Finally it can be verified that $\xi^{(1)}$ satisfies the required boundary conditions on Γ_1, by a reasoning similar to the one developed for ξ^0.

The process which led to the construction of $\xi^{(1)}$ can obviously be continued and by iteration we can define an infinite sequence of eigenfrequencies ω_j and associated eigenmodes $\xi^{(j)}$, orthogonal to each other in the sense that

$$\int_\Omega \rho\xi_i^{(j)}\xi_i^{(k)}\,dx = 0, \quad j \neq k.$$

3.3. Case Where There Are No Kinematic Conditions

Suppose $\Gamma_1 = \emptyset$; clearly the method for finding ξ^0, ω_0 gives $\omega_0 = 0$, $\xi^0 = a + b \wedge x$, where a, b are vectors which are constants, i.e. a solution corresponding to a rigid-body displacement.

On the other hand, for every state of vibration of the solid, it is clear that, since there is no external force acting on the system, its total momentum must remain constant. In other words if the displacement be represented by $u(x,t) = \xi(x)\cdot\sin\omega t$, the condition

$$\int_\Omega \rho\xi\cdot v\,dx = 0 \quad \text{with} \quad v = a + b \wedge x \tag{6.142}$$

where a, b are arbitrary vectors in $\mathbb{R}^3$, is satisfied.

This suggests that to obtain the eigenfrequencies and associated eigenmodes, it will suffice to develop the method described in Section II.3.2, by seeking the minimum of $W_{\text{pot}}(\xi)/W_{\text{kin}}(\xi)$ over the subspace generated by the elements of $(H^1(\Omega))^3$ satisfying (6.142).

3.4. Properties of Eigenmodes and Eigenfrequencies

We resume our discussion of the general case for a Γ_1 of positive measure and with

$$K = \{\xi : \xi \in (H^1(\Omega))^3, \quad \gamma\xi = 0 \text{ on } \Gamma_1\}.$$

The bilinear form

$$[\xi, \eta] = \int_\Omega a_{ijkh}\varepsilon_{ij}(\xi)\varepsilon_{kh}(\eta)\,dx$$

defined on $(H^1(\Omega))^3 \times (H^1(\Omega))^3$, allows us to define, for all $\xi \in K$, a norm $|||\xi||| = [\xi, \xi]^{1/2}$ which is equivalent to the one induced on K by the usual norm in $(H^1(\Omega))^3$, namely:

$$\|\xi\| = \left(\int_\Omega (\xi_i\xi_i + \partial_j\xi_i \cdot \partial_j\xi_i)\,dx\right)^{1/2}.$$

In fact, from

$$\int_\Omega a_{ijkh}\varepsilon_{ij}(\xi)\varepsilon_{kh}(\xi)\,dx \geqslant C\int_\Omega \varepsilon_{ij}(\xi)\varepsilon_{ij}(\xi)\,dx$$

and then

$$\int_\Omega a_{ijkh}\varepsilon_{ij}(\xi)\varepsilon_{kh}(\xi)\,dx \geqslant \omega_0^2\int_\Omega \rho\xi_i\xi_i\,dx \geqslant \omega_0^2 \operatorname*{Inf}_\Omega \rho \cdot \int_\Omega \xi_i\xi_i\,dx$$

and lastly from Korn's inequality (6.128), it follows that:

$$\left(C^{-1} + \omega_0^{-2}\cdot\left(\operatorname*{Inf}_\Omega \rho\right)^{-1}\right)\int_\Omega a_{ijkh}\varepsilon_{ij}(\xi)\varepsilon_{kh}(\xi)\,dx \geqslant C_1\|\xi\|^2.$$

It is moreover evident that:

$$\int_\Omega a_{ijkh}\varepsilon_{ij}(\xi)\varepsilon_{kh}(\xi)\,dx \leqslant C_2\|\xi\|^2$$

where C_2 is a constant depending only on the tensor a_{ijkh} and on Ω.

Having said this, let us now observe that $\int_\Omega \rho\xi_i\eta_i\,dx$ is a continuous bilinear form on $K \times K$, so that by Riesz's theorem [37] we can define a continuous linear mapping A of K into K, $\xi \to A\xi$ by

(6.143) $$\int_\Omega \rho\xi_i\eta_i\,dx = \int_\Omega a_{ijkh}\varepsilon_{ij}(A\xi)\varepsilon_{kh}(\eta)\,dx = [A\xi, \eta].$$

Noting from the definition of ω_0 that

$$\left|\int_\Omega \rho\xi_i\eta_i\,dx\right| \leqslant \left(\int_\Omega \rho\xi_i\xi_i\,dx\right)^{1/2}\left(\int_\Omega \rho\eta_i\eta_i\,dx\right)^{1/2} \leqslant \omega_0^{-2}[\xi, \xi]^{1/2}\cdot[\eta, \eta]^{1/2}$$

we obtain by (6.143):

$$|[A\xi, \eta]| \leqslant \omega_0^{-2}[\xi, \xi]^{1/2}[\eta, \eta]^{1/2}$$

and with $\eta = A\xi$

(6.144) $$|||A\xi||| \leqslant \omega_0^{-2}|||\xi|||$$

which shows that A is a continuous linear operator. It is obviously symmetric and since

$$\frac{W_{\text{pot}}}{W_{\text{kin}}} = \frac{[\xi, \xi]}{[A\xi, \xi]}$$

it is clear that

$$\operatorname*{Sup}_{\xi \in K} \frac{[A\xi, \xi]}{[\xi, \xi]} = \omega_0^{-2}$$

the supremum being attained for $\xi = \xi^0$. Thus ξ^0 is an eigenmode of A, of eigenvalue ω_0^{-2}, that is:

$$A\xi^0 = \omega_0^{-2}\xi^0 \quad \text{and} \quad [A\xi^0, \xi^0] = 1 \quad \text{or} \quad [\xi^0, \xi^0] = \omega_0^2.$$

We can likewise say that ω_1^{-2} is the upper bound of $\frac{[A\xi, \xi]}{[\xi, \xi]}$ in the subspace K_1 spanned by the elements $\xi \in K$ such that $[A\xi^0, \xi] = 0$ or $[\xi^0, \xi] = 0$, this upper bound being attained for $\xi^{(1)}$, and so on. The reader will recognise the classical process for finding the eigenvalues and eigenfunctions of a completely continuous symmetric linear operator (we shall verify later that A is completely continuous).

Thus we have generally

$$A\xi^{(j)} = \omega_j^{-2}\xi^{(j)} \tag{6.145}$$

$$[\xi^{(i)}, \xi^{(j)}] = 0, i \neq j, \quad [A\xi^{(j)}, \xi^{(j)}] = 1 \quad \text{or} \quad [\xi^{(j)}, \xi^{(j)}] = \omega_j^2$$

and

$$\omega_{j+1}^{-2} \leqslant \omega_j^{-2}.$$

Moreover, from the very way in which the ω_j have been constructed it is clear that

$$|||A\xi||| \leqslant \omega_n^{-2} |||\xi|||, \quad \forall \xi \in K_n \tag{6.146}$$

that is to say for all $\xi \in K$ such that $[\xi, \xi^{(j)}] = 0$, $0 \leqslant j \leqslant n-1$.

The proposition that the operator A should be completely continuous in K, i.e. that it should transform every infinite bounded sequence of elements of K into a compact sequence, or into one from which a subsequence can be extracted which is strongly convergent in K, is one which plays an essential role in Rayleigh's algorithm, and which we have already justified in Section II.3.1. We can however give a more concise demonstration here by noting that, from any infinite bounded sequence of elements $\xi^{(n)} \in K$ we can extract a subsequence (numbered in the same way) which converges weakly to $\xi^0 \in K$ (weak convergence in terms of the scalar product in $(H^1(\Omega))^3$; K as a subspace of $(H^1(\Omega))^3$ being weakly closed), and which also converges strongly to ξ^0 in $(L^2(\Omega))^3$ because the canonical injection of $(H^1(\Omega))^3$ into $(L^2(\Omega))^3$ is compact.

Consequently

$$\int_\Omega \rho(\xi^{(m)} - \xi^{(n)})(\xi^{(m)} - \xi^{(n)})\,\mathrm{d}x$$

tends to 0 if n and $m \to \infty$, or

$$\lim_{n\to\infty, m\to\infty} [A(\xi^{(m)} - \xi^{(n)}), \xi^{(m)} - \xi^{(n)}] = 0,$$

which ensures the complete continuity of the operator A on K, [37].

We shall deduce from this that $\lim \omega_j = \infty$, a result which implies, in particular, that the number of eigenmodes corresponding to each multiple eigenfrequency is finite. For if the sequence ω_j were bounded, then $\xi^{(j)}$ would be an infinite bounded sequence in K, since

$$[\xi^{(j)}, \xi^{(j)}] = \omega_j^2.$$

As A is completely continuous we may suppose $A\xi^{(j)}$ to be a strongly convergent sequence in K, and this would result in a contradiction since

$$2 = |[A\xi^{(j)} - A\xi^{(l)}, \xi^{(j)} - \xi^{(l)}]| \leqslant |||A\xi^{(j)} - A\xi^{(l)}||| \cdot (\omega_j + \omega_l)$$

thus proving that ω_j must be unbounded.

We now come to the theorem on the series-expansion of ξ, [37]. Let $\xi \in K$ and

$$\zeta^{(n)} = \xi - \sum_0^n \omega_j^{-2} [\xi, \xi^{(j)}] \xi^{(j)}. \tag{6.147}$$

Clearly $[\zeta^{(n)}, \xi^{(j)}] = 0$, $\forall j = 0, 1, \ldots, n$ and in consequence of (6.146) we have

$$|||A\zeta^{(n)}||| \leqslant \omega_{n+1}^{-2} |||\zeta^{(n)}|||.$$

On the other hand the sequence $\zeta^{(n)}$ is bounded because:

$$|||\zeta^{(n)}|||^2 = [\xi, \xi] - \sum_0^n \omega_j^{-2} [\xi, \xi^{(j)}]^2 < [\xi, \xi]$$

and thus

$$\lim_{n\to\infty} A\zeta^{(n)} = 0 \tag{6.148}$$

whence, applying A to (6.147) and taking (6.145) into account, we obtain

$$A\xi = \sum_{j=0}^{\infty} \omega_j^{-4} [\xi, \xi^{(j)}] \cdot \xi^{(j)}$$

in the sense of strong convergence in K.

We can complete this result with the following remark: from the definition of A by (6.143), it is clear that $A\xi = 0$ implies $\xi = 0$; on the other hand the series

$$\sum_0^{\infty} \omega_j^{-2} [\xi, \xi^{(j)}] \xi^{(j)}$$

converges in K since

$$\left\| \sum_p^n \omega_j^{-2} [\xi, \xi^{(j)}] \xi^{(j)} \right\|^2 = \sum_p^n \omega_j^{-2} [\xi, \xi^{(j)}]^2$$

and

$$\sum_0^\infty \omega_j^{-2}[\xi, \xi^{(j)}]^2 \leqslant [\xi, \xi].$$

Let

$$h = \xi - \sum_0^\infty \omega_j^{-2}[\xi, \xi^{(j)}]\xi^{(j)} = \lim_{n\to\infty} \zeta^{(n)}.$$

We can write:

$$Ah = \lim_{n\to\infty} A\zeta^{(n)} = 0$$

by reason of the continuity of A and by (6.148), and therefore $h = 0$.

We have thus established that for all $\xi \in K$

(6.149) $$\xi = \sum_0^\infty \omega_j^{-2}[\xi, \xi^{(j)}]\xi^{(j)}$$

in the sense of strong convergence in K, or again, remembering that:

(6.150) $$\omega_j^{-2}[\xi, \xi^{(j)}] = [\xi, A\xi^{(j)}] = \int_\Omega \rho \xi_k \xi_k^{(j)}\,\mathrm{d}x$$

(6.151) $$\xi = \sum_0^\infty \left(\int_\Omega \rho \xi_k \xi_k^{(j)}\,\mathrm{d}x \right) \xi^{(j)}.$$

4. Forced Vibrations of Elastic Solids

4.1. Excitation by Periodic Forces Acting on Part of the Boundary

We shall suppose the displacement to be zero over a part Γ_1 of the boundary $\partial\Omega$, where Γ_1 has positive measure, while on the complementary part Γ_2 the constraints imposed obey the law

(6.152) $$T_{ij}\nu_j = p_i(x)\sin\omega t$$

where $p_i(x) \in L^2(\Gamma_2)$ and ω are given.

We shall try to represent the state of vibration of the solid by:

$$u = \xi(x)\sin\omega t, \quad T_{ij} = \sigma_{ij}\sin\omega t$$

so that the equations of the problem can be written:

(6.153) $$\partial_j\sigma_{ij} + \rho\omega^2\xi_i = 0, \quad \sigma_{ij} = a_{ijkh}\varepsilon_{kh}(\xi)$$

with $\xi \in K$ and

(6.154) $$\sigma_{ij}\nu_j = p_i(x), \quad x \in \Gamma_2,$$

ν_j being the outward unit vector normal on Γ_2.

Let us now begin by supposing that there is a solution ξ to this problem. Since

$\xi \in K$, it may be represented, by virtue of the expansion theorem, in the form (6.149)

$$\xi = \sum_{j=0}^{\infty} \omega_j^{-2}[\xi, \xi^{(j)}]\xi^{(j)}$$

Now we know that:

$$[\xi, \xi^{(l)}] = \int_\Omega a_{ijkh}\varepsilon_{ij}(\xi^{(l)})\varepsilon_{kh}(\xi)\,\mathrm{d}x = \int_\Omega \sigma_{ij}(\xi)\partial_j \xi_i^{(l)}\,\mathrm{d}x$$

or, by the interpretation of the functional

$$\int_{\partial\Omega} \sigma_{ij}v_j v_i \mathrm{d}S, \quad v \in (H^{1/2}(\partial\Omega))^3, \quad v = \gamma u, \quad \forall u \in (H^1(\Omega))^3,$$

resulting from the analysis carried out in connection with (6.138):

$$[\xi, \xi^{(l)}] = -\int_\Omega \partial_j \sigma_{ij}\xi_i^{(l)}\,\mathrm{d}x + \int_{\partial\Omega} \sigma_{ij}v_j\xi_i^{(l)}\,\mathrm{d}S,$$

i.e. by (6.153), (6.154):

(6.155)
$$[\xi, \xi^{(l)}] = \omega^2 \int_\Omega \rho\xi_i\xi_i^{(l)}\,\mathrm{d}x + \int_{\partial\Omega} p_i\xi_i^{(l)}\,\mathrm{d}S$$

or by (6.150):

(6.156)
$$\left(1 - \frac{\omega^2}{\omega_l^2}\right)[\xi, \xi^{(l)}] = \int_{\partial\Omega} p_i\xi_i^{(l)}\mathrm{d}S.$$

Let us now assume that $\omega \neq \omega_l$, $\forall l$. We can then deduce from (6.156) that if there is a solution to the problem, it is unique and can be represented by the series

(6.157)
$$\xi = \sum_{l=0}^{\infty} \left(\frac{1}{\omega_l^2 - \omega^2} \int_{\Gamma_2} p_i\xi_i^{(l)}\,\mathrm{d}S\right)\xi^{(l)}$$

which is convergent in the strong sense in K (we can replace $\partial\Omega$ by Γ_2 because $\xi^{(l)} \in K$). If ω is equal to one of the eigenfrequencies of vibration of the solid, say ω_h, then (6.156) shows that there can be no solution unless

(6.158)
$$\int_{\partial\Omega} p_i\xi_i^{(h)}\,\mathrm{d}S = 0$$

for all the eigenmodes $\xi^{(h)}$, finite in number, corresponding to the eigenvalue ω_h.

If this condition is satisfied, the solution, if it exists, will still be represented by (6.157), where the summation is over all integers $l \geqslant 0$, with the exception of $l = h$ but terms containing eigenmodes ξ_h corresponding to the eigenvalue ω_h with arbitrary coefficients c_h, can be added, so that the general solution is:

(6.159)
$$\xi = \sum_{l=0, l\neq h}^{\infty} \left(\frac{1}{\omega_l^2 - \omega^2} \int_{\Gamma_2} p_i\xi_i^{(l)}\,\mathrm{d}S\right)\xi^{(l)} + c_h\xi^{(h)}.$$

We shall now take (6.157) as our point of departure, or in the critical case (6.158) and (6.159) and show that these formulae define the solution of the problem. We first have to establish that the expansions (6.157) or (6.159) are convergent in K, and then that the equations, (6.153) and (6.154) are satisfied. To prove convergence, let $\zeta \in K$, the displacement field, and let τ_{ij} be the associated stress field associated with the static problem:

(6.160)
$$\begin{aligned} \partial_j \tau_{ij} &= 0, \quad \tau_{ij} = a_{ijkh}\varepsilon_{kh}(\zeta) \\ \tau_{ij} v_j &= p_i, \quad \forall x \in \Gamma_2. \end{aligned}$$

It is known (since Γ_1 has a positive measure), [18], that this problem has an unique solution ζ in K, and that we can write:

$$\int_{\partial\Omega} p_i \xi_i^{(l)} \, dS = \int_{\partial\Omega} \tau_{ij} v_j \xi_i^{(l)} \, dS = \int_{\Omega} \partial_j \tau_{ij} \cdot \xi_i^{(l)} \, dx + \int_{\Omega} \tau_{ij} \cdot \partial_j \xi_i^{(l)} \, dx$$

i.e. by (6.160)

$$\int_{\partial\Omega} p_i \xi_i^{(l)} \, dS = \int_{\Omega} a_{ijkh} \varepsilon_{kh}(\zeta) \varepsilon_{ij}(\xi^{(l)}) \, dx = [\zeta, \xi^{(l)}].$$

Remembering that $\xi^{(l)}$ is an orthogonal sequence in K, in the sense that

$$[\xi^{(m)}, \xi^{(l)}] = \delta_{ml} \cdot \omega_l^2, \quad \delta_{ml} = 0, m \neq l$$
$$= 1, m = l$$

we arrive at:

$$\left\| \sum_n^m \left(\frac{1}{\omega_l^2 - \omega^2} \int_{\partial\Omega} p_i \xi_i^{(l)} \, dS \right) \xi^{(l)} \right\|^2 = \sum_n^m \frac{\omega_l^2}{(\omega_l^2 - \omega^2)^2} [\zeta, \xi^{(l)}]^2$$

an expression which tends to zero as n and m tend to infinity, because the series whose general terms are

$$\frac{\omega_l^2}{(\omega_l^2 - \omega^2)^2} [\zeta, \xi^{(l)}]^2 \quad \text{and} \quad \omega_l^{-2} \cdot [\zeta, \xi^{(l)}]^2$$

have the same behaviour as regards convergence since $\lim_{l \to \infty} \omega_l = +\infty$ and, on the other hand

$$\sum_{l=0}^{\infty} \omega_l^{-2} [\zeta, \xi^{(l)}]^2 = [\zeta, \zeta].$$

It remains to check that (6.153), (6.154) are satisfied.

Now we deduce from (6.157), (6.159) that

$$(\omega_l^2 - \omega^2)[\xi, \xi^{(l)}] = \omega_l^2 \int_{\partial\Omega} p_i \xi_i^{(l)} \, dS, \quad \forall l$$

(no matter what c_h is in the second case), which is equivalent to (6.155), i.e. by reason of the definition of $[\xi, \xi^{(l)}]$:

(6.161)
$$\int_{\Omega} (-\sigma_{ij} \partial_j \xi_i^{(l)} + \rho \omega^2 \xi_i \xi_i^{(l)}) \, dx + \int_{\partial\Omega} p_i \xi_i^{(l)} \, dS = 0, \quad \forall l.$$

Since $\xi^{(l)}$ is a base of K, this formula remains valid if we replace $\xi^{(l)}$ by an arbitrary element of K, and in particular by any $\varphi \in (\mathscr{D}(\Omega))^3$ which leads to:

$$\int_\Omega (-\sigma_{ij}\partial_j \varphi_i + \rho\omega^2 \xi_i \varphi_i)\,\mathrm{d}x = 0$$

from which it follows that $\partial_j \sigma_{ij} + \rho\omega^2 \xi_i = 0$ in Ω in the distributions sense.

Reverting to (6.161) which can be written

$$\int_\Omega (\partial_j \sigma_{ij} + \rho\omega^2 \xi_i)\xi_i^{(l)}\,\mathrm{d}x - \int_{\partial\Omega} (\sigma_{ij}(\xi)\nu_j - p_i)\xi_i^{(l)}\,\mathrm{d}S = 0$$

we see that we are left with:

$$\int_{\partial\Omega} (\sigma_{ij}(\xi)\nu_j - p_i)\xi_i^{(l)}\,\mathrm{d}S = 0$$

so that for all $\delta\xi \in \gamma(K)$ we can write:

$$\int_{\partial\Omega} (\sigma_{ij}(\xi)\nu_j - p_i)\delta\xi_i\,\mathrm{d}S = 0$$

whose interpretation leads to (6.154).

In conclusion, let us indicate a method of approximate calculation. Suppose we are given r independent functions $\zeta^{(1)}, \ldots, \zeta^{(r)}$ belonging to K, not necessarily eigenmodes.

We can assume the motion of the system to be described, at least approximately, by

$$u = \sum_1^r q_p(t)\zeta^{(p)}(x)$$

where the coefficients $q_p(t)$ have yet to be determined. To do this we apply the virtual work principle which asserts that, for an arbitrary virtual displacement defined by the changes δq_p from that state at time t, the virtual work done by the forces acting at the boundary, plus the virtual work done by the inertial forces is equal to the change in potential energy.

Denoting the surface density of the forces acting at the boundary Γ_2 by $p_i(x, t)$, the virtual work done by these is:

$$\sum_{l=1}^r \int_{\Gamma_2} \delta q_l \zeta_i^{(l)}(x) p_i(x, t)\,\mathrm{d}S = \sum_{l=1}^r Q_l \delta q_l$$

where

(6.162)
$$Q_l(t) = \int_{\Gamma_2} \zeta_i^{(l)}(x) p_i(x, t)\,\mathrm{d}S.$$

The virtual work of the inertial forces is:

$$-\sum_{m,l} \int_\Omega \rho q_m''(t)\zeta_i^{(m)}(x)\delta q_l \zeta_i^{(l)}(x)\,\mathrm{d}x = -\sum_{l,m} a_{lm} q_m'' \delta q_l$$

(6.163)
$$a_{lm} = \int_\Omega \rho(x)\zeta_i^{(l)}(x)\zeta_i^{(m)}(x)\,\mathrm{d}x.$$

Lastly the potential energy U is given by

$$2U = \int_\Omega a_{ijkh}\varepsilon_{ij}\left(\sum_l q_l \zeta^{(l)}\right)\varepsilon_{kh}\left(\sum_m q_m \zeta^{(m)}\right) dx = \sum_{l,m} b_{lm} q_l q_m$$

where

(6.164) $$b_{lm} = \int_\Omega a_{ijkh}\varepsilon_{ij}(\zeta^{(l)})\varepsilon_{kh}(\zeta^{(m)})\, dx$$

(the Einstein summation convention is here, as usual, used for dummy suffixes which run from 1 to 3; while the summation sign $\sum_l$ is retained for those indices which run from 1 to r).

Finally we obtain:

(6.165) $$\sum_m a_{lm} q''_m + \sum_m b_{lm} q_m = Q_l$$

which can be interpreted as the set of Lagrange's equations for a discrete system with r degrees of freedom whose kinetic energy is $2T = \sum_{l,m} a_{lm} q'_l q'_m$, and for which the virtual work of the applied forces both external and internal is:

$$\sum_l Q_l \delta q_l - \sum_{lm} b_{lm} q_m \delta q_l.$$

In the case where $\zeta^{(j)} = \xi^{(j)}$ (modal co-ordinates) we obtain:

$$2U = \sum_{l,m} [\xi^{(l)}, \xi^{(m)}] q_l q_m = \sum_l \omega_l^2 q_l^2$$

$$2T = \sum_{l,m} [A\xi^{(l)}, \xi^{(m)}] q'_l q'_m = \sum_l q_l'^2$$

and the differential equations which determine the evolution of the q_l are decoupled (i.e. the variables have been separated):

$$q''_l + \omega_l^2 q_l = Q_l.$$

If we assume $p_i(x, t) = p_i(x) \sin \omega t$, with $\omega \neq \omega_l$, $\forall l$; then we have:

$$q''_l + \omega_l^2 q_l = Q_l^* \sin \omega t$$

where

$$Q_l^* = \int_{\Gamma_2} \xi_i^{(l)}(x) p_i(x)\, dS.$$

Thus $q_l(t) = \dfrac{Q_l^* \sin \omega t}{\omega_l^2 - \omega^2}$ represents the forced vibration and the displacement field can be written:

$$u = \sum_{l=1}^{r} q_l(t)\xi^{(l)}(x) = \left(\sum_{l=1}^{r} \frac{Q_l^*}{\omega_l^2 - \omega^2} \xi^{(l)}(x)\right) \sin \omega t$$

which is none other than the sum of the first r terms of the convergent series associated with the exact solutions.

In numerous practical cases, useful information about the state of vibration of the system can be derived by starting from the equations (6.165) after having performed the calculations described by (6.162), (6.163), (6.164).

4.2. Excitation by Periodic Displacements Imposed on Some Part of the Boundary

We shall suppose the displacements at the boundary $\partial\Omega$ to be prescribed in the form:

$$\gamma u = f \sin \omega t, \quad \text{with} \quad f \in (H^{1/2}(\partial\Omega))^3$$

(the case where $f=0$ on Γ_1 being of course included).

We look for a representation of the displacements in Ω, which are harmonic in $t, u = \xi(x) \sin \omega t$, the associated stress tensor being denoted by $T_{ij} = \tilde{\sigma}_{ij} \cdot \sin \omega t$ so that the equations of the problem are:

$$\partial_j \tilde{\sigma}_{ij} + \rho\omega^2 \xi_i = 0, \quad \forall x \in \Omega, \quad \tilde{\sigma}_{ij} = a_{ijkh}\varepsilon_{kh}(\xi)$$

(6.166) $$\gamma\xi = f, \quad \xi \in (H^1(\Omega))^3.$$

We write $\zeta \in (H^1(\omega))^3$ the displacement field, τ_{ij} for the associated stress tensor corresponding to the static problem:

$$\partial_j \tau_{ij} = 0, \quad \forall x \in \Omega, \quad \tau_{ij} = a_{ijkh}\varepsilon_{kh}(\zeta)$$

(6.167) $$\gamma\zeta = f.$$

It is known [18], that ζ exists and is unique; we can therefore seek the solution of (6.166) by putting:

(6.168) $$\xi = \zeta + \chi$$

where the displacement field χ, by reason of (6.166), (6.167), (6.168) has to satisfy:

$$\partial_j \sigma_{ij} + \rho\omega^2(\zeta_i + \chi_i) = 0, \quad \forall x \in \Omega, \quad \sigma_{ij} = a_{ijkh}\varepsilon_{kh}(\chi)$$

(6.169) $$\gamma\chi = 0.$$

It is therefore apparent that χ has to be sought in the subspace

$$G = \{\chi : \chi \in (H^1(\Omega))^3, \gamma\chi = 0\}.$$

We can define the eigenfrequencies μ_l and the associated eigenmodes $\eta^{(l)}$ which correspond to the natural vibrations of an elastic body when the displacements are kept null at every point of the boundary $\partial\Omega$. It follows from the expansion theorem that any $\chi \in G$ can be represented by the series:

(6.170) $$\chi = \sum_l \mu_l^{-2} [\chi, \eta^{(l)}] \eta^{(l)}$$

which converges strongly in $G \subset (H^1(\Omega))^3$.

If we assume that $\chi \in G$ is a solution of (6.169), we can represent it by (6.170) and try to calculate the coefficients of this series. Thus we have:

$$[\chi, \eta^{(l)}] = \int_\Omega a_{ijkh}\varepsilon_{kh}(\chi)\varepsilon_{ij}(\eta^{(l)}) \,dx = \int_\Omega \sigma_{ij}\partial_j \eta_i^{(l)} \,dx = -\int_\Omega \partial_j \sigma_{ij} \eta_i^{(l)} \,dx$$

since $\gamma\eta^{(l)}=0$, and hence, by (6.169)

$$[\chi,\eta^{(l)}]=\omega^2\int_\Omega \rho\zeta_i\eta_i^{(l)}\,dx+\omega^2\int_\Omega \rho\chi_i\eta_i^{(l)}\,dx. \tag{6.171}$$

If we denote by $\mathscr{A}$ the linear operator induced by Riesz's theorem, starting from the continuous bilinear functional

$$\int_\Omega \rho\xi\cdot\zeta\,dx$$

defined on $G\times G$, such that

$$\int_\Omega \rho\xi\cdot\zeta\,dx=\int_\Omega a_{ijkh}\varepsilon_{kh}(\mathscr{A}\xi)\varepsilon_{ij}(\zeta)\,dx=[\mathscr{A}\xi,\zeta],$$

it is known that its eigenfunctions are $\eta^{(l)}$ and its eigenvalues are μ_l^{-2}. Thus it is permissible to write:

$$\int_\Omega \rho\chi_i\eta_i^{(l)}\,dx=[\chi,\mathscr{A}\eta^{(l)}]=\mu_l^{-2}[\chi,\eta^{(l)}]$$

and by (6.171)

$$\left(1-\frac{\omega^2}{\mu_l^2}\right)[\chi,\eta^{(l)}]=\omega^2\int_\Omega \rho\zeta_i\eta_i^{(l)}\,dx. \tag{6.172}$$

It is plain that in the case $\omega\neq\mu_l$, $\forall l$, the solution χ of (6.169), if it exists, is represented by

$$\chi=\sum_l\left(\frac{\omega^2}{\mu_l^2-\omega^2}\int_\Omega \rho\zeta_i\eta_i^{(l)}\,dx\right)\eta^{(l)}. \tag{6.173}$$

If ω is equal to an eigenvalue $\omega=\mu_h$, which may possibly be multiple, then there can be no solution unless:

$$\int_\Omega \rho\zeta_i\eta_i^{(h)}\,dx=0 \tag{6.174}$$

and under these conditions, the solution χ, provided that it exists, admits of the representation:

$$\chi=\sum_{l\neq h}\left(\frac{\omega^2}{\mu_l^2-\omega^2}\int_\Omega \rho\zeta_i\eta_i^{(l)}\,dx\right)\eta^{(l)}+c_h\eta^{(h)}. \tag{6.175}$$

It remains to be shown that the expansions (6.173) or (6.175) converge in $(H^1(\Omega))^3$ and define a solution of (6.169).

Let us first discuss the question of convergence. As ρ is bounded in Ω, $\rho\zeta$ like ζ is square-integrable in Ω; it is therefore known, [18], that there is a $\theta\in(H^1(\Omega))^3$ such that

$$\partial_j\bar{\sigma}_{ij}+\rho\zeta_i=0,\quad \forall x\in\Omega,$$
$$\bar{\sigma}_{ij}=a_{ijkh}\varepsilon_{kh}(\theta),\quad \gamma\theta=0\quad\text{or}\quad\theta\in G$$

and we can write

$$\int_\Omega \rho\zeta_i\eta_i^{(l)}\,dx = -\int_\Omega \partial_j\bar{\sigma}_{ij}\eta_i^{(l)}\,dx = \int_\Omega \bar{\sigma}_{ij}\partial_j\eta_i^{(l)}\,dx = \int_\Omega a_{ijkh}\varepsilon_{kh}(\theta)\varepsilon_{ij}(\eta^{(l)})\,dx = [\theta, \eta^{(l)}].$$

Since $\theta \in G$, we can write, by the expansion theorem

$$\theta = \sum_l \mu_l^{-2}[\theta, \eta^{(l)}]\eta^{(l)}$$

in the sense of strong convergence in $(H^1(\Omega))^3$ and remembering $[\eta^{(l)}, \eta^{(p)}] = \delta_{lp}\mu_l^2$, we get:

$$[\theta, \theta] = \sum_i \mu_l^{-2}[\theta, \eta^{(l)}]^2. \tag{6.176}$$

It is then easy to show that the series which appear in (6.173) or (6.175) are strongly convergent since

$$\left\|\left|\sum_n^m\left(\frac{\omega^2}{\mu_l^2-\omega^2}\int_\Omega \rho\xi_i\eta_i^{(l)}\,dx\right)\eta^{(l)}\right\|\right|^2 = \sum_n^m \frac{\omega^4\mu_l^2}{(\mu_l^2-\omega^2)^2}[\theta, \eta^{(l)}]^2$$

tends to zero as n and m tend to infinity, as the series whose general term is

$$\frac{\omega^4\mu_l^2}{(\mu_l^2-\omega^2)^2}[\theta, \eta^{(l)}]^2$$

behaves like the series of term

$$\mu_l^{-2}[\theta, \eta^{(l)}]^2$$

(because $\mu_l \to \infty$ if $l \to \infty$), which by (6.176) obviously converges.

On the other hand, starting from (6.173) or from (6.174), (6.175) we get back to (6.172) and then (6.171) which we shall write in the form

$$\int_\Omega a_{ijkh}\varepsilon_{kh}(\chi)\varepsilon_{ij}(\eta^{(l)})\,dx = \omega^2\int_\Omega \rho(\zeta_i + \chi_i)\eta_i^{(l)}\,dx.$$

Since $\eta^{(l)}$ is a base of G and $(\mathscr{D}(\Omega))^3 \subset G$, we can say that for all $\varphi \in (\mathscr{D}(\Omega))^3$:

$$\int_\Omega a_{ijkh}\varepsilon_{kh}(\chi)\partial_j\varphi_i\,dx = \omega^2\int_\Omega \rho(\zeta_i + \chi_i)\varphi_i\,dx$$

that is to say $\partial_j(a_{ijkh}\varepsilon_{kh}(\chi)) + \omega^2\rho(\zeta_i + \chi_i) = 0$ in the sense of distributions in Ω, which is precisely the substance of (6.169).

4.3. Excitation by Periodic Volume Forces

Suppose the elastic body to be excited by volume forces of density

$$\rho F_i \sin\omega t, \quad F_i(x) \in L^2(\Omega).$$

Assuming a displacement field and associated constraint field given by $u = \xi\sin\omega t$, $T_{ij} = \sigma_{ij}\sin\omega t$, we are led to the following set of equations

(6.177) $$\partial_j \sigma_{ij} + \rho F_i + \rho \omega^2 \xi_i = 0, \quad \forall x \in \Omega$$

(6.178) $$\gamma \xi = 0 \quad \text{on} \quad \Gamma_1$$

(6.179) $$\sigma_{ij} v_j = 0 \quad \text{on} \quad \Gamma_2$$

on the hypothesis that the displacements on the part Γ_1 of the boundary, and the stresses on the complementary part Γ_2, are both kept at zero.

Denoting by ω_l, $\xi^{(l)}$ the eigenfrequencies and eigenmodes corresponding to the natural vibrations (i.e. those in the absence of the F_i term), we know that every solution $\xi \in (H^1(\Omega))^3$ of this problem can be represented, because of (6.178), by the series

$$\xi = \sum_l \omega_l^{-2} [\xi, \xi^{(l)}] \xi^{(l)}.$$

Furthermore we can write:

$$[\xi, \xi^{(l)}] = \int_\Omega a_{ijkh} \varepsilon_{kh}(\xi) \varepsilon_{ij}(\xi^{(l)}) \mathrm{d}x = \int_\Omega \sigma_{ij} \partial_j \xi_i^{(l)} \mathrm{d}x$$

$$= -\int_\Omega \partial_j \sigma_{ij} \xi_i^{(l)} \mathrm{d}x + \int_{\partial\Omega} \sigma_{ij} v_j \xi_i^{(l)} \mathrm{d}S.$$

But the integral over $\partial\Omega$ vanishes by reason of (6.178), (6.179) and we obtain, after taking account of (6.177):

(6.180) $$[\xi, \xi^{(l)}] = \int_\Omega \rho F_i \xi_i^{(l)} \mathrm{d}x + \omega^2 \int_\Omega \rho \xi_i \xi_i^{(l)} \mathrm{d}x$$

or by (6.150):

(6.181) $$\left(1 - \frac{\omega^2}{\omega_l^2}\right) [\xi, \xi^{(l)}] = \int_\Omega \rho F_i \xi_i^{(l)} \mathrm{d}x.$$

If $\omega \neq \omega_l$, $\forall l$, the solution, if it exists, of the problem is therefore represented by:

(6.182) $$\xi = \sum_l \left(\frac{1}{\omega_l^2 - \omega^2} \int_\Omega \rho F_i \xi_i^{(l)} \mathrm{d}x \right) \xi^{(l)}.$$

If ω is equal to an eigenfrequency ω_h, then it is necessary that

(6.183) $$\int_\Omega \rho F_i \xi_i^{(h)} \mathrm{d}x = 0$$

and we have the representation:

(6.184) $$\xi = \sum_{l \neq h} \left(\frac{1}{\omega_l^2 - \omega^2} \int_\Omega \rho F_i \xi_i^{(l)} \mathrm{d}x \right) \xi^{(l)} + c_h \xi^{(h)}.$$

As in the preceding cases, it remains to be shown that these series converge and that they provide a solution to the problem.

We introduce the field of displacements $\zeta \in (H^1(\Omega))^3$ and the associated field of constraints τ_{ij} defined uniquely [18] by

$$\begin{gathered} \partial_j \tau_{ij} + \rho F_i = 0 \\ \gamma\zeta = 0 \quad \text{on} \quad \Gamma_1, \quad \tau_{ij} v_j = 0 \quad \text{on} \quad \Gamma_2. \end{gathered} \tag{6.185}$$

We can then write

$$\int_\Omega \rho F_i \xi_i^{(l)} \mathrm{d}x = -\int_\Omega \partial_j \tau_{ij} \xi_i^{(l)} \mathrm{d}x = \int_\Omega \tau_{ij} \partial_j \xi_i^{(l)} \mathrm{d}x$$

$$- \int_\Omega a_{ijkh} \varepsilon_{kh}(\zeta) \varepsilon_{ij}(\xi^{(l)}) \mathrm{d}x = [\zeta, \xi^{(l)}]$$

and

$$\left\| \sum_n^m \left(\frac{1}{\omega_l^2 - \omega^2} \int_\Omega \rho F_i \xi_i^{(l)} \mathrm{d}x \right) \xi^{(l)} \right\|^2 = \sum_n^m \frac{\omega_l^2}{(\omega_l^2 - \omega^2)^2} [\zeta, \xi^{(l)}]^2.$$

Noticing that $\sum_0^\infty \omega_l^{-2} [\zeta, \xi^{(l)}]^2 = [\zeta, \zeta]$, since $\zeta \in K$, we deduce that the expansions (6.182) or (6.184) converge strongly in $(H^1(\Omega))^3$.

As these series obviously define an element ξ of K, it remains only to verify that (6.177) and (6.179) are satisfied; but from (6.182) or (6.184), (6.183) we can deduce (6.181) and then the validity for all l of (6.180), which we can write in the form:

$$\int_\Omega (\sigma_{ij} \varepsilon_{ij}(\xi^{(l)}) - \rho F_i \xi_i^{(l)} - \omega^2 \rho \xi_i \xi_i^{(l)}) \mathrm{d}x = 0 \tag{6.186}$$

with

$$\sigma_{ij} = a_{ijkh} \varepsilon_{kh}(\xi).$$

Since $\xi^{(l)}$ is a base of K and $(\mathscr{D}(\Omega))^3 \subset K$, (6.186) remains valid if $\xi^{(l)}$ be replaced by any element $\varphi \in (\mathscr{D}(\Omega))^3$, so that

$$\int_\Omega (\sigma_{ij} \varepsilon_{ij}(\varphi) - \rho F_i \varphi_i - \omega^2 \rho \xi_i \varphi_i) \mathrm{d}x = 0$$

which means that, in the distributions sense, we have:

$$\partial_j \sigma_{ij} + \rho F_i + \omega^2 \rho \xi_i = 0 \tag{6.187}$$

in Ω.

Reverting to (6.186) we can write for all $\delta\xi \in K$:

$$\int_\Omega (\sigma_{ij} \partial_j (\delta\xi_i) - \rho F_i \delta\xi_i - \omega^2 \rho \xi_i \delta\xi_i) \mathrm{d}x = 0$$

or

$$\int_\Omega (\partial_j \sigma_{ij} + \rho F_i + \omega^2 \rho \xi_i) \delta\xi_i \mathrm{d}x - \int_{\partial\Omega} \sigma_{ij} v_j \delta\xi_i \mathrm{d}S = 0$$

that is, by (6.187):

$$\int_{\partial\Omega} \sigma_{ij} v_j \delta\xi_i \mathrm{d}S = 0, \quad \forall \delta\xi \in K$$

or

$$\sigma_{ij} v_j = 0 \quad \text{on} \quad \Gamma_2.$$

5. Vibrations of Non-Linear Elastic Media

The study of large-amplitude vibrations is made difficult by the need to take account of laws involving non-linear behaviour. There are however certain cases amenable to calculation as shown by the example worked out below [1].

For an incompressible hyperelastic material whose deformation is described by $x_i = x_i(X, t)$, where x_1, x_2, x_3 represent the co-ordinates at time t of the particle of matter which at time $t = 0$ was at the point $X_1 X_2 X_3$, it is known [36] that the stress tensor $T = (T_{ij})$ at x can be represented by $T = -pI + \alpha B + \beta B^{-1}$. Here $B = F \cdot F^T$, $F = \left(\dfrac{\partial x_i}{\partial X_j}\right)$ is the tensor gradient, p the pressure term, α and β denote functions, specific to the material, depending on the invariants of B (that is to say of the first two fundamental symmetric functions of the zeros of the polynomial $\det(B - \lambda I)$, the third having the value unity by the hypothesis of incompressibility of the material). If $\rho(x)$ denotes the density at x, the equations of motion can be written:

$$\rho \frac{\partial^2 x_j}{\partial t^2} = \frac{\partial T_{ij}}{\partial x_i}.$$

We consider in particular material occupying the region $(X, Y) \in \mathbb{R}^2$, $0 \leqslant Z \leqslant 1$, of mass $\rho(Z)$ per unit volume and subjected to the deformation:

$$x = X + u(Z, t), \quad y = Y + v(Z, t), \quad z = Z,$$

where we use, from now on, the notation (x, y, z), (X, Y, Z) for the co-ordinates mentioned earlier.

We obtain without difficulty

$$F = \begin{pmatrix} 1 & 0 & u_Z \\ 0 & 1 & v_Z \\ 0 & 0 & 1 \end{pmatrix}, \qquad B = \begin{pmatrix} 1 + u_Z^2 & u_Z v_Z & u_Z \\ u_Z v_Z & 1 + v_Z^2 & v_Z \\ u_Z & v_Z & 1 \end{pmatrix},$$

and the invariants of B:

$$\mathscr{B}_{(1)} = \mathscr{B}_{(2)} = u_Z^2 + v_Z^2 + 3, \quad \mathscr{B}_{(3)} = 1,$$

and lastly the law of behaviour [36]:

$$\begin{aligned} &T_{xx} = -p + \alpha(1 + u_Z^2) + \beta, \quad T_{yy} = -p + \alpha(1 + v_Z^2) + \beta, \\ &T_{zz} = -p + \alpha + \beta(1 + u_Z^2 + v_Z^2) \\ &T_{xy} = T_{yx} = \alpha u_Z v_Z, \quad T_{yz} = T_{zy} = (\alpha - \beta) v_Z, \quad T_{zx} = T_{xz} = (\alpha - \beta) u_Z \end{aligned}$$

$$\left(u_Z, \dots = \frac{\partial u}{\partial Z}, \dots\right)$$

where α and β are taken to depend only on $u_Z^2 + v_Z^2$ and Z. If we write μ for

$\alpha-\beta$, the equations of motion can be written:

$$-p_x+(\mu u_Z)_Z=\rho(Z)u_{tt},\quad p_x=\frac{\partial p}{\partial x},\quad u_{tt}=\frac{\partial^2 u}{\partial t^2},$$

$$-p_y+(\mu v_Z)_Z=\rho(Z)v_{tt}$$

$$\frac{\partial T_{zz}}{\partial z}=[-p+\alpha+\beta(1+u_Z^2+v_Z^2)]_Z=0,$$

from which we see that without loss of generality we may suppose p to be independent of x and y, and thus u, v will be solutions of:

$$(\mu u_Z)_Z=\rho(Z)u_{tt}$$
$$(\mu v_Z)_Z=\rho(Z)v_{tt}$$

in $0\leqslant Z\leqslant 1$, with $\mu=\mu(u_Z^2+v_Z^2,Z)$.

We shall prescribe the following boundary conditions:

$u(0)=v(0)=0$ i.e. no deformation on the lower face;

$u_Z(1)=v_Z(1)=0$ i.e. absence of shear on the upper face;

and we shall seek natural vibrations of the form:

$$u(Z,t)=\xi(Z)\cos\omega t$$
$$v(Z,t)=\xi(Z)\sin\omega t.$$

Accordingly the eigenfrequencies are the values ω for which the boundary problem:

$$(\mu(\xi_Z^2,Z)\cdot\xi_Z)_Z+\rho\omega^2\xi=0,\quad 0\leqslant Z\leqslant 1,\quad \xi(0)=\xi_Z(1)=0$$

has a non-trivial solution $\xi(Z)$.

A simple case is that in which μ is independent of the invariants of B. Under these circumstances and assuming $\mu>0$, we have to solve a problem which is formally identical to the one which we have already met in connection with the longitudinal vibrations of an elastic bar fixed at one end and free at the other.

IV. Vibrations of Plane Elastic Plates

1. Description of Stresses; Equations of Motion

The plate in its undeformed state is symmetrically situated with respect to the plane Ox_1x_2; we denote by Ω the bounded open connected domain which is the set of points of the material lying in this plane, the region of space occupied by the material being then defined, in terms of the orthonormal co-ordinate system $Ox_1x_2x_3$, by $(x_1,x_2)\in\Omega$, $-h/2<x_3<h/2$ where h is the plate thickness, assumed to be small in comparison with the linear dimensions of Ω (thin-plate assumption). Incidentally h may depend on the co-ordinates x_1, x_2 provided thickness varies

slowly,

$$\left|\frac{\partial h}{\partial x_1}\right| + \left|\frac{\partial h}{\partial x_2}\right| \ll 1.$$

To obtain a global picture of the stresses in each section of the material perpendicular to the median plane we consider an element of area of centre $P \in \Omega$, with oriented unit normal $\vec{n}(n_1, n_2, 0)$, which intersects the plane Ox_1x_2 along an arc of length ds, and which extends from the upper part of the plate to the lower part of the plate, or in other words from $x_3 = -h/2$ to $x_3 = h/2$.

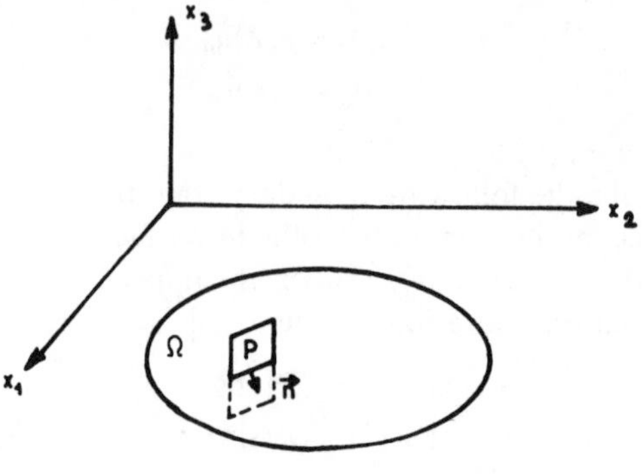

Figure 6.24

The stresses exerted by the part of the material containing $\vec{n}$ on the other part, through this element of area have (x_1, x_2, x_3) components at each point M of the element which are, per unit area:

$$\vec{F} = \begin{matrix} T_{11}n_1 + T_{12}n_2 \\ T_{21}n_1 + T_{22}n_2 \\ T_{31}n_1 + T_{32}n_2 \end{matrix}$$

where, with the usual notation, T_{ij} is the stress tensor at the point M. The value at P of the x_3 component of the resultant of these stresses is $T\mathrm{d}s + O(\mathrm{d}s^2)$ with

$$(6.188) \qquad T = \left(\int_{-h/2}^{h/2} T_{31}\,\mathrm{d}x_3 \right) n_1 + \left(\int_{-h/2}^{h/2} T_{32}\,\mathrm{d}x_3 \right) n_2 = H_1 n_1 + H_2 n_2 .$$

If we write $\vec{\tau}$ to denote the vector derived by rotating the vector $\vec{n}$ through $+\pi/2$ radians about the axis Px_3, we can represent $\overrightarrow{PM} = r\vec{\tau} + x_3\vec{x}_3$, for every point M of the element of area considered, and the resultant moment at P of the stresses acting through this element is:

$$\left(\int_{-h/2}^{h/2} \overrightarrow{PM} \wedge \vec{F}\,\mathrm{d}x_3 \right) \mathrm{d}s + O(\mathrm{d}s^2)$$

or

$$\vec{G} = \int_{-h/2}^{h/2} \overrightarrow{PM} \wedge \vec{F}\,\mathrm{d}x_3 = G_1\vec{x}_1 + G_2\vec{x}_2 + G_3\vec{x}_3 , \qquad \text{with}$$

(6.189)

$$G_1 = -\left(\int_{-h/2}^{h/2} x_3 T_{21} \mathrm{d}x_3\right) n_1 - \left(\int_{-h/2}^{h/2} x_3 T_{22} \mathrm{d}x_3\right) n_2 = M_{21} n_1 + M_{22} n_2$$

$$G_2 = \left(\int_{-h/2}^{h/2} x_3 T_{11} \mathrm{d}x_3\right) n_1 + \left(\int_{-h/2}^{h/2} x_3 T_{12} \mathrm{d}x_3\right) n_2 = -M_{11} n_1 - M_{12} n_2$$

$G_3 = 0$, and with

$$M_{\alpha\beta} = -\int_{-h/2}^{h/2} x_3 T_{\alpha\beta} \mathrm{d}x_3,$$

symmetric in α, β ($= 1, 2$).

Let us now suppose that the plate is bent slightly and that the deformation of the median plane can be represented by

$$(x_1, x_2, 0) \to (x_1, x_2, w(x_1, x_2, t)).$$

Adopting the formulae (6.188), (6.189) for the global stresses per unit length in Ω, acting across a section normal to the plate, and writing the dynamic equations for the part of the material whose projection on the median plane is an arbitrary region $\mathscr{A} \in \Omega$, we have to begin with:

$$-\iint_{\mathscr{A}} q \,\mathrm{d}x_1 \,\mathrm{d}x_2 + \int_{\partial\mathscr{A}} T \,\mathrm{d}s = \int_{\mathscr{A}} \rho \frac{\partial^2 w}{\partial t^2} \mathrm{d}x_1 \,\mathrm{d}x_2,$$

where $-q(x_1, x_2, t)$ denotes the surface density of the loads, and ρ is the mass of the plate per unit area. Expressing T by (6.188) we can write:

$$\iint_{\mathscr{A}} \left(-q + \frac{\partial H_1}{\partial x_1} + \frac{\partial H_2}{\partial x_2} - \rho \frac{\partial^2 w}{\partial t^2}\right) \mathrm{d}x_1 \,\mathrm{d}x_2 = 0$$

from which we obtain the local equation

$$-q + \frac{\partial H_1}{\partial x_1} + \frac{\partial H_2}{\partial x_2} = \rho \frac{\partial^2 w}{\partial t^2}. \tag{6.190}$$

Next we express the fact that the x_1 and x_2 components of the moment at O of the inertial forces and the applied external forces are both zero.

$$\int_{\partial\mathscr{A}} x_2 T \,\mathrm{d}s + \int_{\partial\mathscr{A}} G_1 \,\mathrm{d}s - \iint_{\mathscr{A}} x_2 \left(\rho \frac{\partial^2 w}{\partial t^2} + q\right) \mathrm{d}x_1 \,\mathrm{d}x_2 = 0$$

$$-\int_{\partial\mathscr{A}} x_1 T \,\mathrm{d}s + \int_{\partial\mathscr{A}} G_2 \,\mathrm{d}s + \iint_{\mathscr{A}} x_1 \left(\rho \frac{\partial^2 w}{\partial t^2} + q\right) \mathrm{d}x_1 \,\mathrm{d}x_2 = 0$$

from which we deduce, having regard to (6.188), (6.189), (6.190):

$$H_1 + \frac{\partial M_{11}}{\partial x_1} + \frac{\partial M_{12}}{\partial x_2} = 0$$

$$H_2 + \frac{\partial M_{21}}{\partial x_1} + \frac{\partial M_{22}}{\partial x_2} = 0. \tag{6.191}$$

Eliminating H_1, H_2 from (6.190) and (6.191) we find

$$\frac{\partial^2 M_{\alpha\beta}}{\partial x_\alpha \partial x_\beta} + \rho \frac{\partial^2 w}{\partial t^2} + q = 0 \tag{6.192}$$

using the summation convention and with the indices α, β running from 1 to 2.

2. Potential Energy of a Plate

Suppose that, starting from its deformed state, defined by $w(x_1, x_2, t)$ at time t, the plate undergoes a virtual displacement $\delta w(x_1, x_2)$. If the unit normal at an arbitrary point of the deformed median plane be denoted by $\vec{n} = \left(-\frac{\partial w}{\partial x_1}, -\frac{\partial w}{\partial x_2}, 1\right)$ the vector rotation associated with the virtual displacement δw is $\vec{n} \wedge \delta\vec{n}$, and the virtual work done by the stresses at the edge of the plate is given by

$$T\delta w + (\vec{n} \wedge \delta\vec{n})\vec{G} = T\delta w + G_1 \frac{\partial \delta w}{\partial x_2} - G_2 \frac{\partial \delta w}{\partial x_1}$$

per unit length of the boundary $\partial\Omega$, or in total:

$$\int_{\partial\Omega} \left(T\delta w + G_1 \frac{\partial \delta w}{\partial x_2} - G_2 \frac{\partial \delta w}{\partial x_1} \right) \mathrm{d}s,$$

an expression which can be transformed, using (6.188) and (6.189), into:

$$\begin{aligned}
&\int_{\partial\Omega} \left\{ \left(H_1 \delta w + M_{21} \frac{\partial \delta w}{\partial x_2} + M_{11} \frac{\partial \delta w}{\partial x_1} \right) n_1 \right. \\
&\qquad \left. + \left(H_2 \delta w + M_{22} \frac{\partial \delta w}{\partial x_2} + M_{12} \frac{\partial \delta w}{\partial x_1} \right) n_2 \right\} \mathrm{d}s \\
&= \int_{\Omega} \left\{ \left(\frac{\partial H_1}{\partial x_1} + \frac{\partial H_2}{\partial x_2} \right) \delta w + \left(H_1 + \frac{\partial M_{11}}{\partial x_1} + \frac{\partial M_{12}}{\partial x_2} \right) \frac{\partial \delta w}{\partial x_1} \right. \\
&\qquad \left. + \left(H_2 + \frac{\partial M_{21}}{\partial x_1} + \frac{\partial M_{22}}{\partial x_2} \right) \frac{\partial \delta w}{\partial x_2} \right\} \mathrm{d}x_1\, \mathrm{d}x_2 + \int_{\Omega} M_{\alpha\beta} \frac{\partial^2 \delta w}{\partial x_\alpha \partial x_\beta} \mathrm{d}x_1\, \mathrm{d}x_2 \\
&= \int_{\Omega} \left(q + \rho \frac{\partial^2 w}{\partial t^2} \right) \delta w\, \mathrm{d}x_1\, \mathrm{d}x_2 + \int_{\Omega} M_{\alpha\beta} \frac{\partial^2 \delta w}{\partial x_\alpha \partial x_\beta} \mathrm{d}x_1\, \mathrm{d}x_2
\end{aligned}$$

by (6.190) and (6.191).

If we postulate that the behaviour of the plate obeys the law:

$$M_{\alpha\beta} = A_{\alpha\beta\gamma\delta} \frac{\partial^2 w}{\partial x_\gamma \partial x_\delta} \tag{6.193}$$

where the 4th order tensor $A_{\alpha\beta\gamma\delta}(x_1, x_2)$ is symmetric:

$$A_{\alpha\beta\gamma\delta} = A_{\beta\alpha\gamma\delta} = A_{\gamma\delta\alpha\beta},$$

we see that the virtual work done by the forces acting at the edges plus that done by the surface forces and inertial forces is equal to the change in value of the functional

(6.194) $$W=\frac{1}{2}\int_{\Omega} A_{\alpha\beta\gamma\delta}(x_1,x_2)\frac{\partial^2 w}{\partial x_\alpha \partial x_\beta}\cdot\frac{\partial^2 w}{\partial x_\gamma \partial x_\delta}\,\mathrm{d}x_1\,\mathrm{d}x_2$$

which thus represents the potential energy.

3. Determination of the Law of Behaviour

We shall assume the validity of Hooke's law in the interior of the material which we shall regard as a three-dimensional medium, one dimension of which, namely the thickness, is small. This assumption implies that the so-called volume forces must be defined in the present case by means of a surface density, as has already been indicated and, in this respect are representative of the stresses which may possibly be imposed on one or other of the faces of the plate.

The classical theory is based on the following approximations formulated either in terms of the field of displacements $u_i(x_1,x_2,x_3)$, $i=1,2,3$, of the point (x_1,x_2,x_3) in the interior of the plate, or the stress field associated with it by Hooke's law, in the form

$$T_{ij}=\lambda\theta\delta_{ij}+2\mu U_{ij},\quad \theta=U_{ii},\quad U_{ij}=\tfrac{1}{2}(\partial_j u_i+\partial_i u_j),$$

where λ,μ are the Lamé coefficients.

1) The displacement of any point in the median plane is orthogonal to this plane:

(6.195) $$\begin{aligned} u_1(x_1,x_2,0)&=0\\ u_2(x_1,x_2,0)&=0\\ u_3(x_1,x_2,0)&=w(x_1,x_2). \end{aligned}$$

2) The internal stresses on the median plane, i.e. T_{13}, T_{23}, T_{33} are nil for $x_3=0$, that is to say:

(6.196) $$\begin{aligned} \partial_3 u_1+\partial_1 u_3&=0\\ \partial_3 u_2+\partial_2 u_3&=0 \qquad \text{for } x_3=0\\ (\lambda+2\mu)\partial_3 u_3+\lambda(\partial_1 u_1+\partial_2 u_2)&=0. \end{aligned}$$

3) T_{33} in the interior of the plate is of the order of magnitude of x_3^2:

(6.197) $$T_{33}=O(x_3^2).$$

We deduce from (6.195), (6.196) the expressions:

$$u_1=-x_3\frac{\partial w}{\partial x_1}+O(x_3^2),\quad u_2=-x_3\frac{\partial w}{\partial x_2}+O(x_3^2)$$

and by (6.197) and the third equation of (6.196), we have

$$(\lambda+2\mu)\partial_3 u_3+\lambda(\partial_1 u_1+\partial_2 u_2)=O(x_3^2)$$

inside the plate, whence:

$$u_3 = w(x_1, x_2) + \frac{\lambda}{2(\lambda + 2\mu)} x_3^2 \Delta w + O(x_3^3). \tag{6.198}$$

Thus

$$\theta = -\frac{2\mu}{\lambda + 2\mu} x_3 \Delta w + O(x_3^2)$$

and

$$T_{11} = -\frac{2\mu\lambda}{\lambda + 2\mu} x_3 \Delta w - 2\mu x_3 \frac{\partial^2 w}{\partial x_1^2} + O(x_3^2)$$

$$T_{21} = -2\mu x_3 \frac{\partial^2 w}{\partial x_1 \partial x_2} + O(x_3^2)$$

$$T_{22} = -\frac{2\mu\lambda}{\lambda + 2\mu} x_3 \Delta w - 2\mu x_3 \frac{\partial^2 w}{\partial x_2^2} + O(x_3^2)$$

and lastly, by (6.189)

$$M_{11} = \frac{h^3}{12}\left(\frac{2\mu\lambda}{\lambda + 2\mu} \Delta w + 2\mu \frac{\partial^2 w}{\partial x_1^2}\right)$$

$$M_{12} = \frac{\mu h^3}{6} \frac{\partial^2 w}{\partial x_1 \partial x_2}$$

$$M_{22} = \frac{h^3}{12}\left(\frac{2\mu\lambda}{\lambda + 2\mu} \Delta w + 2\mu \frac{\partial^2 w}{\partial x_2^2}\right),$$

these formulae remaining valid even if h is slowly variable, i.e. in the case of a plate of non-uniform thickness. Introducing Young's modulus E and Poisson's ratio σ:

$$E = \mu \frac{3\lambda + 2\mu}{\lambda + \mu}, \qquad \sigma = \frac{\lambda}{2(\lambda + \mu)}$$

or

$$\mu = \frac{E}{2(1 + \sigma)}, \qquad \lambda = \frac{E\sigma}{(1 + \sigma)(1 - 2\sigma)}$$

we can finally write:

$$\begin{aligned} M_{11} &= D\left(\frac{\partial^2 w}{\partial x_1^2} + \sigma \frac{\partial^2 w}{\partial x_2^2}\right), \\ M_{12} = M_{21} &= D(1 - \sigma) \frac{\partial^2 w}{\partial x_1 \partial x_2}, \\ M_{22} &= D\left(\frac{\partial^2 w}{\partial x_2^2} + \sigma \frac{\partial^2 w}{\partial x_1^2}\right), \end{aligned} \tag{6.199}$$

with $D = (Eh^3/12)(1 - \sigma^2)^{-1}$, the coefficient of flexural rigidity, and the potential energy of the plate is:

$$W = \frac{1}{2}\int_{\Omega} D\left\{\left(\frac{\partial^2 w}{\partial x_1^2}\right)^2 + \left(\frac{\partial^2 w}{\partial x_2^2}\right)^2 + 2\sigma\frac{\partial^2 w}{\partial x_1^2}\cdot\frac{\partial^2 w}{\partial x_2^2} + 2(1-\sigma)\left(\frac{\partial^2 w}{\partial x_1 \partial x_2}\right)^2\right\} dx_1\, dx_2. \tag{6.200}$$

4. Eigenfrequencies and Eigenmodes

We consider the general case where the behaviour of the plate is described by the equation (6.193) and the potential energy by (6.194). We recall that the tensor $A_{\alpha\beta\gamma\delta}$ satisfies the usual symmetry conditions $A_{\alpha\beta\gamma\delta} = A_{\beta\alpha\gamma\delta} = A_{\gamma\delta\alpha\beta}$, and is a measurable function of x in Ω, and that there exists a constant C such that

$$A_{\alpha\beta\gamma\delta}\varepsilon_{\alpha\beta}\varepsilon_{\gamma\delta} \geqslant C\varepsilon_{\alpha\beta}\varepsilon_{\alpha\beta} \tag{6.201}$$

for all $x \in \Omega$ and every symmetric tensor $\varepsilon_{\alpha\beta}$. All these conditions are certainly satisfied in the isotropic, homogeneous case described by (6.199).

To study the natural vibrations of the plate, we shall suppose it to be clamped over some part Γ_1 of the boundary $\partial\Omega$, so that

$$\left.\begin{array}{l} w = 0 \\ \dfrac{dw}{dn} = 0 \end{array}\right., \quad \forall x \in \Gamma_1, \quad \Gamma_1 \subset \partial\Omega, \tag{6.202}$$

where Γ_1 has non-zero measure, $\vec{n}$ is the unit vector in the plane Ox_1x_2 directed along the outward normal to $\partial\Omega$. The plate is free over the complementary portion Γ_2 of its edge ($\Gamma_1 \cup \Gamma_2 = \partial\Omega$, $\Gamma_1 \cap \Gamma_2 = \emptyset$). We shall explain below the precise sense in which the conditions (6.202) should be interpreted.

We shall be looking for a solution, harmonic with respect to time, and of frequency ω, of the form $w = w(x_1, x_2)\sin \omega t$ and hence, assuming that there is no surface load, we deduce from (6.192) and (6.193), the equation of motion in the form:

$$\frac{\partial^2}{\partial x_\alpha \partial x_\beta}\left(A_{\alpha\beta\gamma\delta}\frac{\partial^2 w}{\partial x_\gamma \partial x_\delta}\right) - \rho\omega^2 w = 0 \tag{6.203}$$

with w subject to the kinematic constraints (6.202).

We look for w in the space $H^2(\Omega)$ of functions $w(x_1, x_2)$, square-integrable in Ω, whose partial derivatives of the first and second order, in the distributions sense, can be identified with functions which are square-integrable on Ω.[1] If $\partial\Omega$ is continuous and satisfies a Lipschitz condition, which we shall assume to be the case, it is known [33] that it is possible to define a continuous linear mapping $u \to \gamma u$ from $H^2(\Omega)$ into $H^1(\partial\Omega)$ such that for $u \in \mathscr{E}(\bar{\Omega})$, the space of infinitely

[4] According to Sobolev's theorem it is known that $w \in H^2(\Omega)$, $\Omega \subset \mathbb{R}^2$ imply $w \subset c(\Omega)$, that is w is continuous in the open set Ω.

differentiable functions in $\bar{\Omega}$, we have:

$$\gamma u = u, \quad \gamma \frac{\partial u}{\partial x_\alpha} = \frac{\partial u}{\partial x_\alpha},$$

where the terms on the right represent, for $u \in \mathscr{E}(\bar{\Omega})$ the natural values at the boundary.

Since $\partial\Omega$ is continuous and satisfies a Lipschitz condition, n_1 and n_2 the components of the unit normal vector exist almost everywhere, are locally measurable, and obviously bounded, and thus one can define $\frac{\mathrm{d}u}{\mathrm{d}n} = n_\alpha \frac{\partial u}{\partial x_\alpha}$ as a square-integrable function on $\partial\Omega$.

We can therefore introduce the subset $K \subset H^2(\Omega)$ of kinematically admissible elements:

(6.204) $$K = \left\{ w : w \in H^2(\Omega), \gamma w = 0, \gamma \frac{\mathrm{d}w}{\mathrm{d}n} = 0 \quad \text{on} \quad \Gamma_1 \right\}$$

Since the trace operator is continuous, it is clear that the linear variety K is strongly closed in $H^2(\Omega)$ and is consequently a subspace; by reason of convexity, K is also weakly closed.

With $W(\xi)$ defined by (6.194) (with ξ instead of w) and

$$T(\xi) = \int_\Omega \rho \xi^2 \, \mathrm{d}x_1 \, \mathrm{d}x_2$$

we define

(6.205) $$\omega_0^2 = \operatorname*{Inf}_{\xi \in K} \frac{W(\xi)}{T(\xi)}$$

and we shall suppose the surface density $\rho(x_1, x_2)$ to be such that:

(6.206) $$0 < \rho_0 < \rho(x_1, x_2) < \rho_1, \quad \forall x \in \Omega,$$

where ρ_0 and ρ_1 are constants.

Clearly the lower bound of W/T on K is non-negative and we can define a minimising sequence $\xi^{(n)}$, normalised by:

(6.207) $$\int_\Omega \rho \xi^{(n)2} \, \mathrm{d}x_1 \, \mathrm{d}x_2 = 1$$

such therefore that:

(6.208) $$\lim_{n \to \infty} W(\xi^{(n)}) = \omega_0^2.$$

Let us first show that this sequence is bounded in $H^2(\Omega)$; it is clear from (6.206) and (6.207) that:

$$\int_\Omega \xi^{(n)2} \, \mathrm{d}x_1 \, \mathrm{d}x_2 \leqslant \rho_0^{-1}$$

and by (6.208), (6.201) that

$$\int_\Omega \xi_{\alpha\beta}^{(n)} \xi_{\alpha\beta}^{(n)} \, dx_1 \, dx_2$$

is also a bounded infinite sequence.

It remains to be shown that $\int_\Omega \xi_\alpha^{(n)} \xi_\alpha^{(n)} \, dx_1 \, dx_2$ is a bounded sequence; now if this were not so, we should have (after replacing it, if necessary, by a suitable subsequence indexed in the same way).

$$\lim_{n \to \infty} \varkappa_n^2 = \infty \quad \text{with} \quad \varkappa_n^2 = \int_\Omega \xi_\alpha^{(n)} \xi_\alpha^{(n)} \, dx_1 \, dx_2 .$$

Introducing $\zeta^{(n)} = \varkappa_n^{-1} \xi^{(n)}$ we see clearly that $\zeta^{(n)}$ is a bounded sequence in $H^2(\Omega)$, which moreover is such that:

(6.209) $$\lim_{n \to \infty} \int_\Omega \zeta^{(n)} \zeta^{(n)} \, dx_1 \, dx_2 = 0,$$

(6.210) $$\int_\Omega \zeta_\alpha^{(n)} \zeta_\alpha^{(n)} \, dx_1 \, dx_2 = 1.$$

Since the canonical injection of $H^2(\Omega)$ into $H^1(\Omega)$ is compact [33] we may suppose that the sequence $\zeta^{(n)}$ converges weakly to $\zeta^{(0)}$ in $H^2(\Omega)$ and strongly to the same element in $H^1(\Omega)$. We could then deduce from (6.209) that

$$\int_\Omega \zeta^{(0)} \zeta^{(0)} \, dx_1 \, dx_2 = 0 \quad \text{or} \quad \zeta^{(0)} = 0 \text{ p.p.}$$

and from (6.210) that

$$\int_\Omega \zeta_\alpha^{(0)} \zeta_\alpha^{(0)} \, dx_1 \, dx_2 = 1,$$

which would be a contradiction.

Thus the minimising sequence $\zeta^{(n)}$ must be bounded in $H^2(\Omega)$; we can therefore, at the cost of extractions if need be, suppose that it converges weakly in $H^2(\Omega)$ to an element $\xi^{(0)}$ and strongly in $H^1(\Omega)$ to the same element. Since K is weakly closed in $H^2(\Omega)$, it can be seen that $\xi^{(0)} \in K$, i.e. that it satisfies the kinematic conditions. The fact that the sequence converges strongly to $\xi^{(0)}$ in $H^1(\Omega)$ ensures moreover that

(6.211) $$\int_\Omega \rho \xi^{(0)} \xi^{(0)} \, dx_1 \, dx_2 = 1$$

and consequently $\omega_0^2 \leqslant W(\xi^{(0)})$.

The bilinear form

$$W(\xi, \eta) = \int_\Omega A_{\alpha\beta\gamma\delta} \xi_{\alpha\beta} \eta_{\gamma\delta} \, dx_1 \, dx_2$$

is continuous on $H^2(\Omega)$; the associated quadratic form $W(\xi, \xi)$ is positive and consequently weakly lower semi-continuous, so that:

$$\liminf_n W(\xi^{(n)}, \xi^{(n)}) \geqslant W(\xi^{(0)}, \xi^{(0)}) = W(\xi^{(0)})$$

since $\xi^{(0)}$ is the weak limit of $\xi^{(n)}$ in $H^2(\Omega)$. We deduce from this that $\omega_0^2 \geqslant W(\xi^{(0)})$, and since $\xi^{(0)} \in K$:

(6.212)
$$\omega_0^2 = \frac{W(\xi^{(0)})}{T(\xi^{(0)})}.$$

We shall now show that $w = \xi^{(0)}$ is a solution of (6.203). For all $\delta\xi \in K$, $\lambda \in \mathbb{R}$, we have:

$$W(\xi^{(0)} + \lambda\delta\xi) - \omega_0^2 T(\xi^{(0)} + \lambda\delta\xi) \geqslant 0$$

whence, in view of (6.211):

(6.213)
$$\int_\Omega (A_{\alpha\beta\gamma\delta}\xi_{\gamma\delta}^{(0)}(\delta\xi)_{\alpha\beta} - \omega_0^2\rho\xi^{(0)}\delta\xi)\,\mathrm{d}x_1\,\mathrm{d}x_2 = 0.$$

Taking $\delta\xi \in \mathscr{D}(\Omega) \subset K$, we arrive at:

(6.214)
$$(A_{\alpha\beta\gamma\delta}\xi_{\gamma\delta}^{(0)})_{\alpha\beta} - \omega_0^2\rho\xi^{(0)} = 0$$

in the distributions sense, or in other words $\xi^{(0)}$ satisfies the elastic plate partial differential equation. Note that $\omega_0 > 0$, because $\omega_0 = 0$ would imply $W(\xi^{(0)}) = 0$, i.e. $\xi_{\alpha\beta}^{(0)} = 0$ almost everywhere in Ω. This would mean that $\xi^{(0)}$ would be an affine function of the co-ordinates and could not satisfy the kinematic conditions without being zero which would be incompatible with (6.211).

We now have to examine in what sense the boundary conditions at the edge Γ_2 free from constraint are satisfied. We first note that, using (6.214) the relation (6.213) can be rewritten:

(6.215)
$$\int_\Omega [A_{\alpha\beta\gamma\delta}\xi_{\gamma\delta}^{(0)}(\delta\xi)_{\alpha\beta} - (A_{\alpha\beta\gamma\delta}\xi_{\gamma\delta}^{(0)})_{\alpha\beta}\delta\xi]\,\mathrm{d}x_1\,\mathrm{d}x_2 = 0$$

for all $\delta\xi \in K$, because $(A_{\alpha\beta\gamma\delta}\xi_{\gamma\delta}^{(0)})_{\alpha\beta}$ can be identified with the square-integrable function $\omega_0^2\rho\xi^{(0)}$ in Ω.

The formula (6.215) suggests that we should introduce the linear functional $f(u)$, defined for all $u \in H^2(\Omega)$ by

$$f(u) = \int_\Omega [A_{\alpha\beta\gamma\delta}\xi_{\gamma\delta}^{(0)}u_{\alpha\beta} - (A_{\alpha\beta\gamma\delta}\xi_{\gamma\delta}^{(0)})_{\alpha\beta}u]\,\mathrm{d}x_1\,\mathrm{d}x_2$$

which is continuous on $H^2(\Omega)$. It also follows from (6.215) that $f(\varphi) = 0$ for all $\varphi \in \mathscr{D}(\Omega)$ and also on the closure of $\mathscr{D}(\Omega)$ in $H^2(\Omega)$, which in fact is the same as the set of elements u of $H^2(\Omega)$ satisfying $u = \dfrac{\mathrm{d}u}{\mathrm{d}n} = 0$ on $\partial\Omega$.

The trace mapping of $H^2(\Omega)$ into $H^1(\delta\Omega)$ defined earlier is not surjective; but it is well-known that one can define a space $H^{3/2}(\partial\Omega)$, such that, both algebraically and topologically $H^{3/2}(\partial\Omega) \subset H^1(\partial\Omega)$ and such that the trace operator γ is a continuous operator from $H^2(\Omega)$ onto $H^{3/2}(\partial\Omega)$. Furthermore, provided one accepts more restrictive boundary conditions (local maps having continuous Holderian second derivatives) we can make use of a lifting theorem [33] whereby there exists a continuous linear operator from $H^{3/2}(\partial\Omega) \times H^{1/2}(\partial\Omega)$ into $H^2(\Omega)$,

$$(h_0, h_1) \to u = R(h_0, h_1) \in H^2(\Omega),$$

such that

$$h_0 = \gamma u, \quad h_1 = \gamma \frac{du}{dn}.$$

Using these ideas we see that we can define a linear functional on $H^{3/2}(\partial\Omega) \times H^{1/2}(\partial\Omega)$ by putting

$$\varphi(h_0, h_1) = f(R(h_0, h_1)).$$

This functional is independent of R, because if R and $\tilde{R}$ are two distinct liftings, and $u, \tilde{u}$ the corresponding images, so that $u = R(h_0, h_1)$, $\tilde{u} = \tilde{R}(h_0, h_1)$ we have $\gamma(u - \tilde{u}) = 0$, $\gamma \dfrac{d(u-\tilde{u})}{dn} = 0$ and consequently $f(u - \tilde{u}) = 0$. The functional $\varphi(h_0, h_1)$ obtained by composition of two continuous operations is itself continuous on $H^{3/2}(\partial\Omega) \times H^{1/2}(\partial\Omega)$ and by (6.215) it is clear that this functional vanishes for any element (h_0, h_1) such that $(h_0, h_1) = 0$ on Γ_1.

We shall now try to obtain an interpretation of this functional and to enable this to be done with greater convenience we shall make such assumptions regarding regularity as are needed to justify the steps of our calculations.

Starting from:

$$f(\delta\xi) = \int_\Omega [A_{\alpha\beta\gamma\delta}\xi^{(0)}_{\gamma\delta}(\delta\xi)_{\alpha\beta} - (A_{\alpha\beta\gamma\delta}\xi^{(0)}_{\gamma\delta})_{\alpha\beta}\delta\xi]\, dx_1\, dx_2$$

$$= \int_\Omega [(A_{\alpha\beta\gamma\delta}\xi^{(0)}_{\gamma\delta}(\delta\xi)_\alpha)_\beta - ((A_{\alpha\beta\gamma\delta}\xi^{(0)}_{\gamma\delta})_\beta \delta\xi)_\alpha]\, dx_1\, dx_2$$

$$f(\delta\xi) = \int_{\partial\Omega} [n_\beta A_{\alpha\beta\gamma\delta}\xi^{(0)}_{\gamma\delta}(\delta\xi)_\alpha - n_\alpha(A_{\alpha\beta\gamma\delta}\xi^{(0)}_{\gamma\delta})_\beta\delta\xi]\, ds \tag{6.216}$$

or, having regard to:

$$(\delta\xi)_1 = \frac{d\delta\xi}{dn} n_1 - \frac{d\delta\xi}{ds} n_2$$

$$(\delta\xi)_2 = \frac{d\delta\xi}{dn} n_2 + \frac{d\delta\xi}{ds} n_1,$$

from the definition of $M_{\alpha\beta}$ by (6.193), with $\xi^{(0)}$ in place of w, and (6.191), we get:

$$f(\delta\xi) = \int_{\partial\Omega} \left[n_\beta M_{1\beta}\left(\frac{d\delta\xi}{dn} n_1 - \frac{d\delta\xi}{ds} n_2\right) + n_\beta M_{2\beta}\left(\frac{d\delta\xi}{dn} n_2 + \frac{d\delta\xi}{ds} n_1\right) + n_\alpha H_\alpha \delta\xi \right] ds$$

$$= \int_{\partial\Omega} \left[n_\alpha n_\beta M_{\alpha\beta} \frac{d\delta\xi}{dn} + (n_\beta M_{2\beta} n_1 - n_\beta M_{1\beta} n_2)\frac{d\delta\xi}{ds} + H_\alpha n_\alpha \delta\xi \right] ds$$

and using (6.189) we obtain after integration by parts:

$$f(\delta\xi) = \int_{\partial\Omega} \left[n_\alpha n_\beta M_{\alpha\beta} \frac{d\delta\xi}{dn} + \left(H_\alpha n_\alpha - \frac{d}{ds}(G_\alpha n_\alpha)\right)\delta\xi \right] ds.$$

The condition $f(\delta\xi)=0$ for all $\delta\xi\in K$ can thus be interpreted by:

$$n_\alpha n_\beta M_{\alpha\beta}=0 \quad \text{on} \quad \Gamma_2$$

$$H_\alpha n_\alpha-\frac{\mathrm{d}}{\mathrm{d}s}(G_\alpha n_\alpha)=0 \quad \text{on} \quad \Gamma_2;$$

$n_\alpha n_\beta M_{\alpha\beta}$ is the moment of the forces acting at the edge, with respect to the tangent there.

The higher order frequencies can be determined by using the classical methods; we introduce the subspace K_1 of elements $\xi\in H^2(\Omega)$ that are kinematically admissible and such that:

(6.217)
$$\int_\Omega \rho\xi\xi^{(0)}\,\mathrm{d}x_1\,\mathrm{d}x_2=0,$$

where $K_1\subset K\subset H^2(\Omega)$. K_1 is strongly closed in $H^2(\Omega)$ and we define $\omega_1>0$ by:

$$\omega_1^2=\operatorname*{Inf}_{\xi\in K_1}\frac{W(\xi)}{T(\xi)},$$

which obviously satisfies $\omega_0\leqslant\omega_1$.

We shall show, as was done for $\xi^{(0)}$, that there exists a $\xi^{(1)}\in K_1$ such that

(6.218)
$$\omega_1^2=\frac{W(\xi^{(1)})}{T(\xi^{(1)})}$$

and satisfying

(6.219)
$$\int_\Omega [A_{\alpha\beta\gamma\delta}\xi^{(1)}_{\gamma\delta}(\delta\xi)_{\alpha\beta}-\omega_1^2\rho\xi^{(1)}\delta\xi]\,\mathrm{d}x_1\,\mathrm{d}x_2=0$$

for all $\delta\xi\in K_1$. We can extend the validity of (6.219) to $\delta\xi=\xi^{(0)}$ and consequently to all $\delta\xi\in K$. For

$$\int_\Omega (A_{\alpha\beta\gamma\delta}\xi^{(1)}_{\gamma\delta}\xi^{(0)}_{\alpha\beta}-\omega_1^2\rho\xi^{(1)}\xi^{(0)})\,\mathrm{d}x_1\,\mathrm{d}x_2$$

$$=\int_\Omega (A_{\alpha\beta\gamma\delta}\xi^{(0)}_{\gamma\delta}\xi^{(1)}_{\alpha\beta}-\omega_0^2\rho\xi^{(1)}\xi^{(0)})\,\mathrm{d}x_1\,\mathrm{d}x_2=0$$

because of the symmetry of the tensor $A_{\alpha\beta\gamma\delta}$, of (6.217) applied to $\xi=\xi^{(1)}$, and finally because of (6.213) with $\delta\xi=\xi^{(1)}\in K$.

Thus we can guarantee that (6.219) holds for all $\delta\xi\in K$ and deduce the same consequences for $\xi^{(1)}$ as for $\xi^{(0)}$. In other words we can prove that $\xi^{(1)}$ satisfies throughout Ω the plate-vibration equation (6.203) with $w=\xi^{(1)}$, $\omega=\omega_1$ as well as the required boundary conditions at the free edge. We can therefore find, by an obvious iterative procedure, an infinity of eigenfrequencies $\omega_0\leqslant\omega_1\leqslant\omega_2\leqslant\cdots\leqslant\omega_n\leqslant\cdots$ and of associated eigenmodes $\xi^{(n)}$ forming an orthogonal sequence

$$\int_\Omega \rho\xi^{(j)}\xi^{(k)}\,\mathrm{d}x_1\,\mathrm{d}x_2=0,\quad j\neq k$$

normalised by:

(6.220)
$$\int_\Omega \rho \xi^{(j)2} \, dx_1 \, dx_2 = 1.$$

Also we can verify that the sequence ω_n tends to $+\infty$, from which it follows in particular that the number of eigenmodes corresponding to a given eigenfrequency is finite. If indeed the sequence ω_n were bounded it would follow from (6.218) with index n, (6.220) and (6.201) that

$$\int_\Omega (\xi^{(n)} \xi^{(n)} + \xi^{(n)}_{\alpha\beta} \xi^{(n)}_{\alpha\beta}) \, dx_1 \, dx_2$$

would be bounded as n tends to infinity. By a familiar argument one could then deduce that $\xi^{(n)}$ would be bounded in $H^2(\Omega)$ and there is no restriction in supposing it to converge weakly to $\zeta^{(0)}$ in $H^2(\Omega)$ and strongly to the same element in $H^1(\Omega)$; but this would entail

$$\lim_{n,m \to \infty} \int_\Omega \rho (\xi^{(n)} - \xi^{(m)})^2 \, dx_1 \, dx_2 = 0$$

which is impossible since

$$\int_\Omega \rho (\xi^{(n)} - \xi^{(m)})^2 \, dx_1 \, dx_2 = 2$$

for $n \neq m$.

If the plate carries additional loads consisting of masses μ_j, concentrated at certain points P_j, $1 \leqslant j \leqslant r$ the first eigenfrequency can be defined by:

$$\omega_0^2 = \operatorname*{Inf}_{\xi \in K} \frac{\int_\Omega A_{\alpha\beta\gamma\delta} \xi_{\alpha\beta} \xi_{\gamma\delta} \, dx_1 \, dx_2}{\int_\Omega \rho \xi^2 \, dx_1 \, dx_2 + \sum_{j=1}^{r} \mu_j \xi^2(P_j)}$$

and the higher-order frequencies can be obtained in an analogous manner by minimising the functional successively on the subspaces $K_1 K_2, \ldots$

5. Forced Vibrations

Suppose the plate to be subjected to loading of surface density

$$q = q(x_1, x_2) \sin \omega t,$$

where ω is a given frequency, q is square-integrable on Ω, and the plate is held fixed along the edge Γ_1, and free along Γ_2. With a displacement

$$w = w(x_1, x_2) \sin \omega t,$$

it can be seen by (6.192) and (6.193) that the amplitude of the forced vibration

$w(x_1, x_2)$ has to satisfy the equation

(6.221) $$\frac{\partial^2}{\partial x_\alpha \partial x_\beta}\left(A_{\alpha\beta\gamma\delta}\frac{\partial^2 w}{\partial x_\gamma \partial x_\delta}\right) + q(x_1, x_2) - \rho\omega^2 w = 0, \; (x_1, x_2) \in \Omega.$$

But any element $w \in K$, that is any kinematically admissible displacement, can be represented by a series of eigenmodes, converging strongly in $H^2(\Omega)$:

$$w = \sum_j c_j \xi^{(j)}$$

whose coefficients c_j are given by:

$$c_j = \int_\Omega \rho w \xi^{(j)} \mathrm{d}x_1 \, \mathrm{d}x_2$$

or with (6.221), by

(6.222) $$\omega^2 c_j = \int_\Omega \rho q \xi^{(j)} \mathrm{d}x_1 \, \mathrm{d}x_2 + \int_\Omega (A_{\alpha\beta\gamma\delta} w_{\gamma\delta})_{\alpha\beta} \xi^{(j)} \mathrm{d}x_1 \, \mathrm{d}x_2.$$

Furthermore we have seen above that the interpretation of the functional

$$f(\delta\xi) = \int_\Omega (A_{\alpha\beta\gamma\delta} w_{\gamma\delta} (\delta\xi)_{\alpha\beta} - (A_{\alpha\beta\gamma\delta} w_{\gamma\delta})_{\alpha\beta} \delta\xi) \mathrm{d}x_1 \, \mathrm{d}x_2,$$

$\forall \delta\xi \in H^2(\Omega)$, with $w \in H^2(\Omega)$ and $(A_{\alpha\beta\gamma\delta} w_{\gamma\delta})_{\alpha\beta} \in L^2(\Omega)$, was closely bound up with the boundary conditions imposed on w at the free edge Γ_2. (The relevant calculation was done for $\xi^{(0)}$ but is valid for any $w \in K$). More precisely we can say that $f(\delta\xi) = 0$, $\forall \delta\xi \in K$, expresses for $w \in K$ the condition that the plate should be free along the edge Γ_2, and consequently:

$$\int_\Omega (A_{\alpha\beta\gamma\delta} w_{\gamma\delta})_{\alpha\beta} \xi^{(j)} \mathrm{d}x_1 \, \mathrm{d}x_2 = \int_\Omega A_{\alpha\beta\gamma\delta} w_{\gamma\delta} \cdot \xi^{(j)}_{\alpha\beta} \mathrm{d}x_1 \, \mathrm{d}x_2 = \int_\Omega A_{\alpha\beta\gamma\delta} \xi^{(j)}_{\gamma\delta} w_{\alpha\beta} \mathrm{d}x_1 \, \mathrm{d}x_2$$

$$= \omega_j^2 \int_\Omega \rho \xi^{(j)} w \, \mathrm{d}x_1 \, \mathrm{d}x_2$$

by virtue of the symmetry of $A_{\alpha\beta\gamma\delta}$ and the extension of (6.219) to $\xi^{(j)}$ valid for all $\delta\xi \in K$.

Coming back to (6.222) we obtain, for calculating the c_j:

$$(\omega_j^2 - \omega^2) c_j + \int_\Omega \rho q \xi^{(j)} \mathrm{d}x_1 \, \mathrm{d}x_2 = 0$$

and if $\omega \neq \omega_j$, $\forall j$, we have the series expansion:

$$w = \sum_j \left(\frac{1}{\omega^2 - \omega_j^2} \int_\Omega \rho q \xi^{(j)} \mathrm{d}x_1 \, \mathrm{d}x_2 \right) \xi^{(j)}.$$

One can fully justify this calculation, showing that the series converges strongly in $H^2(\Omega)$ and defines the solution to the problem.

6. Eigenfrequencies and Eigenmodes of Vibration of Complex Systems

6.1. Free Vibrations of a Plate Supported Elastically over a Part U of Its Area, U Open and $\bar{U} \subset \Omega$

If $k(x, y)$ is the surface density of the rigidity of the elastic suspension we can write

$$\omega^2 = \operatorname*{Inf}_{w \in K} \frac{\int_\Omega A_{\alpha\beta\gamma\delta} w_{,\alpha\beta} \cdot w_{,\gamma\delta} \, dx_\alpha \, dx_\beta + \int_U k w^2 \, dx_1 \, dx_2}{\int_\Omega \rho w^2 \, dx_1 \, dx_2}$$

where $K = \{w : w \in H^2(\Omega)$ and w satisfies the kinematic conditions$\}$.

In the case where the plate is free at its edge and U is of positive measure, we can take $K = H^2(\Omega)$.

6.2. Eigenfrequencies and Eigenmodes of a Rectangular Plate Reinforced by Regularly Spaced Stiffeners

The plate extends over the domain $\Omega : 0 < x < l$, $0 < y < L$; it is assumed to be homogeneous with D being the coefficient of flexural regidity, σ Poisson's ratio, ρ the mass per unit area. The stiffeners are regarded as beams under flexure, of characteristics E, I and linear (mass) density v, whose neutral axes are parallel to the axis Ox, and regularly spaced at a distance of d apart: $y = jd$, $j = 0, 1, \ldots, p$, $L = pd$ (so that there are $p + 1$ stiffeners).

We denote by $w(x, y)$ the amplitude of the displacement of the plate, and by $\xi_j(x)$ that of the jth stiffener, at the point x, y.

The fundamental eigenfrequency is

$$w^2 = \operatorname*{Inf}_{(w,\xi) \in K} \frac{\int_\Omega D\{(w_{,xx})^2 + (w_{,yy})^2 + 2\sigma w_{,xx} w_{,yy} + 2(1-\sigma)(w_{,xy})^2\} \, dx \, dy + \sum_j \int_0^l EI \xi_j'^2(x) \, dx}{\int_\Omega \rho w^2 \, dx \, dy + \sum_j \int_0^l v \xi_j^2 \, dx}$$

$$K = \{(w, \xi) : w \in H^2(\Omega),\ \xi_j \in H^2(0, l),\ w(x, jd) = \xi_j(x),\ \forall j + \text{kinematic conditions}\}$$

In the case of a free plate, the admissible displacements will have to satisfy the conditions

$$\int_\Omega \rho w \, dx \, dy + \sum_j \int_0^l v \xi_j \, dx = 0$$

$$\int_\Omega \rho x w \, dx \, dy + \sum_j \int_0^l x v \xi_j \, dx = 0$$

$$\int_\Omega \rho y w \, dx \, dy + \sum_j j\alpha \int_0^l v \xi_j \, dx = 0$$

which express the fact that the total momentum of the system is equal to zero.

V. Vibrations in Periodic Media

We consider an elastic medium of indefinite extent having a periodic structure; the property of spatial periodicity being defined by reference to three given linearly independent vectors b_1, b_2, b_3 of $\mathbb{R}^3$. A function defined on $\mathbb{R}^3$ is said to be periodic if

$$f(x+\gamma_1 b_1+\gamma_2 b_2+\gamma_3 b_3)=f(x),$$

$\forall x \in \mathbb{R}^3$ and $\forall(\gamma_1,\gamma_2,\gamma_3)\in\mathbb{Z}^3$. We shall suppose that the elastic properties of the material, the tensor $a_{mnpq}(x)$ and the density $\rho(x)$ are periodic with respect to x, and we propose to describe the states of elastic vibration which can appear in such a medium.

1. Formulation of the Problem and Some Consequences of Korn's Inequality

Let $\mathscr{E}$ be the vector space of periodic, infinitely differentiable, complex-valued functions $f(x)$ defined on $\mathbb{R}^3$; Ω the open bounded domain defined by the parallelepiped whose sides are the vectors b_1, b_2, b_3 issuing from the origin of the co-ordinate system of $\mathbb{R}^3$. We denote by $H^1(\Omega)$ the completion, relative to the norm defined by

$$\|u\| = \left(\int_\Omega (u\cdot\bar{u} + \partial_j u \partial_j \bar{u})\,dx\right)^{1/2},$$

($\bar{u}$ being the complex conjugate of u), of the space $\mathscr{E}(\Omega)$ generated by the elements of $\mathscr{E}$, with restriction of their definition to Ω, the extension of the elements of $H^1(\Omega)$ in $\mathbb{R}^3$ being obtained by periodicity. It is useful to notice that $\mathscr{D}(\Omega)\subset\mathscr{E}(\Omega)$.

As regards the tensor $a_{mnpq}(x)$ we shall make the usual symmetry assumptions in connection with the elastic behaviour of the material:

$$a_{mnpq} = a_{nmpq} = a_{pqmn}$$

and we shall assume that the real-valued functions $a_{mnpq}(x)$, are measurable and bounded on Ω and satisfy:

$$a_{mnpq}(x)\varepsilon_{mn}\bar{\varepsilon}_{pq} \geqslant C\varepsilon_{pq}\bar{\varepsilon}_{pq} \tag{6.223}$$

for all $x\in\Omega$ and all complex-valued symmetric tensors ε_{pq} where C is a positive constant. The density $\rho(x)$ is assumed to be measurable, bounded and to have a strictly positive lower bound.

For any pair u, v of elements of $(H^1(\Omega))^3$ we introduce the continuous bilinear form

$$[u,v] = \int_\Omega a_{mnpq}(x)(\partial_n u_m + ik_n u_m)\overline{(\partial_q v_p + ik_q v_p)}\,dx \tag{6.224}$$

where i denotes the complex unity ($i^2=-1$), $k\in\Omega$ is a given real vector whose components are k_1, k_2, k_3. We note that $[u,u]\geqslant 0$.

We shall however establish that there exist positive constants α, β such that for all $u \in (H^1(\Omega))^3$ we have:

$$(6.225) \qquad [u,u] + \alpha \| u \|^2_{(L^2(\Omega))^3} \geqslant \beta \| u \|^2_{(H^1(\Omega))^3}.$$

We begin by recalling Korn's inequality, for the domain Ω, and in fact if we take account of the periodicity properties, we can give a very simple proof. We first establish it for every field $u \in \mathscr{E}^3$. Setting

$$\varepsilon_{pq} = \tfrac{1}{2}(\partial_p u_q + \partial_q u_p), \quad r_{pq} = \tfrac{1}{2}(\partial_p u_q - \partial_q u_p)$$

we can write on the one hand:

$$(6.226) \qquad \partial_p u_q \overline{\partial_p u_q} = \varepsilon_{pq} \bar{\varepsilon}_{pq} + r_{pq} \bar{r}_{pq}$$

then

$$\varepsilon_{pq} \bar{\varepsilon}_{pq} - r_{pq} \bar{r}_{pq} = \tfrac{1}{2}(\partial_p u_q \overline{\partial_q u_p} + \partial_q u_p \overline{\partial_p u_q})$$

and

$$\varepsilon_{pq} \bar{\varepsilon}_{pq} - r_{pq} \bar{r}_{pq} - \varepsilon_{pp} \bar{\varepsilon}_{qq} = \tfrac{1}{2}[(u_q \overline{\partial_q u_p})_p - (u_q \overline{\partial_p u_p})_q + (u_p \overline{\partial_p u_q})_q - (u_p \overline{\partial_q u_q})_p].$$

Because of the periodicity property, we deduce from this result, by Green's formula

$$\int_\Omega (\varepsilon_{pq} \bar{\varepsilon}_{pq} - r_{pq} \bar{r}_{pq} - \varepsilon_{pp} \bar{\varepsilon}_{qq}) \mathrm{d}x = 0$$

and consequently:

$$\int_\Omega r_{pq} \bar{r}_{pq} \mathrm{d}x \leqslant \int_\Omega \varepsilon_{pq} \bar{\varepsilon}_{pq} \mathrm{d}x,$$

an inequality which remains valid by extension for every element $u \in (H^1(\Omega))^3$. Having regard to (6.226) we obtain:

$$(6.227) \qquad \int_\Omega \partial_p u_q \overline{\partial_p u_q} \mathrm{d}x \leqslant 2 \int_\Omega \varepsilon_{pq} \bar{\varepsilon}_{pq} \mathrm{d}x.$$

Reverting to the definition (6.224) we write:

$$(6.228) \qquad [u,u] = \int_\Omega a_{mnpq} \partial_n u_m \overline{\partial_q u_p} \mathrm{d}x + i \int_\Omega a_{mnpq}(k_n u_m \overline{\partial_q u_p} - k_q \bar{u}_p \partial_n u_m) \mathrm{d}x + \int_\Omega a_{mnpq} k_n k_q u_m \bar{u}_p \mathrm{d}x$$

a formula whose individual terms we now propose to estimate.

By (6.223), (6.227) we have:

$$(6.229) \qquad \int_\Omega a_{mnpq} \partial_n u_m \overline{\partial_q u_p} \mathrm{d}x = \int_\Omega a_{mnpq} \varepsilon_{mn} \bar{\varepsilon}_{pq} \mathrm{d}x \geqslant C \int_\Omega \varepsilon_{mn} \bar{\varepsilon}_{mn} \mathrm{d}x \geqslant \frac{C}{2} \int_\Omega \partial_m u_n \overline{\partial_m u_n} \mathrm{d}x$$

With μ, b positive constants such that

$$|a_{mnpq}(x)| \leqslant \mu, \quad \forall x \in \Omega, \quad |k_j| \leqslant b, \quad \forall k \in \Omega,$$

we can estimate the second integral of (6.228) by:

$$\text{(6.230)} \quad \left| \int_\Omega a_{mnpq}(k_n u_m \overline{\partial_q u_p} - k_q \bar{u}_p \partial_n u_m) \mathrm{d}x \right| \leqslant 6\mu b \int_\Omega \sum_{mpq} |u_m| |\partial_q u_p| \mathrm{d}x$$

$$\leqslant 18\sqrt{3}\mu b \left(\int_\Omega u_m \bar{u}_m \mathrm{d}x \right)^{1/2} \cdot \left(\int_\Omega \partial_q u_p \overline{\partial_q u_p} \mathrm{d}x \right)^{1/2}$$

$$\leqslant 18\sqrt{3}\mu b C_1 \int_\Omega u_m \bar{u}_m \mathrm{d}x + \frac{18\sqrt{3}\mu b}{C_1} \int_\Omega \partial_q u_p \overline{\partial_q u_p} \mathrm{d}x$$

for any positive C_1.

The third integral of (6.228) can lastly be majorised by:

$$\text{(6.231)} \quad \left| \int_\Omega a_{mnpq} k_n k_q u_m \bar{u}_p \mathrm{d}x \right| \leqslant 9\mu b^2 \int_\Omega \left(\sum_m |u_m| \right)^2 \mathrm{d}x \leqslant 27\mu b^2 \int_\Omega u_m \bar{u}_m \mathrm{d}x.$$

Choosing C_1 so that $18\dfrac{\sqrt{3}\mu b}{C_1} = \dfrac{C}{4}$ we obtain from (6.228) and the estimates (6.229), (6.230), (6.231):

$$[u, u] \geqslant \frac{C}{4} \int_\Omega \partial_m u_n \overline{\partial_m u_n} \mathrm{d}x - 4\frac{(18\sqrt{3}\mu b)^2}{C} \int_\Omega u_m \bar{u}_m \mathrm{d}x - 27\mu b^2 \int_\Omega u_m \bar{u}_m \mathrm{d}x$$

or

$$[u, u] \geqslant \frac{C}{4} \int_\Omega (u_m \bar{u}_m + \partial_m u_n \overline{\partial_m u_n}) \mathrm{d}x - \left(\frac{C}{4} + 27\mu b^2 + \frac{4(18\sqrt{3}\mu b)^2}{C} \right) \int_\Omega u_m \bar{u}_m \mathrm{d}x$$

(i.e. (6.225)).

2. Bloch Waves [7]

We define

$$\text{(6.232)} \quad \omega_0^2 = \operatorname*{Inf}_{\xi \in (H^1(\Omega))^3} \frac{[\xi, \xi]}{\int_\Omega \rho \xi_p \bar{\xi}_p \mathrm{d}x}.$$

Then every infinite minimising sequence $\xi^{(n)}$, normalised by

$$\int_\Omega \rho \xi_p^{(n)} \cdot \bar{\xi}_p^{(n)} \mathrm{d}x = 1,$$

is, by (6.225), bounded in $(H^1(\Omega))^3$ and by repeating the arguments already used in Sections I, III, IV we can establish the existence of an element $\xi^{(0)} \in (H^1(\Omega))^3$ for which the lower bound in (6.232) is attained.

For all $\delta\xi \in (H^1(\Omega))^3$ and all complex λ we have:

$$[\xi^{(0)} + \lambda\delta\xi, \xi^{(0)} + \lambda\delta\xi] - \omega_0^2 \int_\Omega \rho(\xi_p^{(0)} + \lambda\delta\xi_p)\overline{(\xi_p^{(0)} + \lambda\delta\xi_p)}\,\mathrm{d}x \geqslant 0$$

or

$$\mathrm{Re}\{\bar{\lambda}([\xi^{(0)}, \delta\xi] - \omega_0^2 \int_\Omega \rho\xi_p^{(0)}\overline{\delta\xi_p}\,\mathrm{d}x)\} + O(\lambda^2) \geqslant 0$$

so that:

$$[\xi^{(0)}, \delta\xi] - \omega_0^2 \int_\Omega \rho\xi_p^{(0)}\overline{\delta\xi_p}\,\mathrm{d}x = 0, \quad \forall\delta\xi \in (H^1(\Omega))^3 \tag{6.233}$$

Let us introduce

$$\zeta^{(0)} = e^{ikx}\xi^{(0)}, \quad \delta\zeta = e^{ikx}\delta\xi \tag{6.234}$$

with $kx = k_1x_1 + k_2x_2 + k_3x_3.$

Noting that

$$[\xi^{(0)}, \delta\xi] = \int_\Omega a_{mnpq}\partial_n\zeta_m^{(0)}\overline{\partial_q\delta\zeta_p}\,\mathrm{d}x = \int_\Omega a_{mnpq}\varepsilon_{mn}(\zeta^{(0)})\cdot\overline{\partial_q\delta\zeta_p}\,\mathrm{d}x$$

we can write (6.233) in the form

$$\int_\Omega (a_{pqmn}\varepsilon_{mn}(\zeta^{(0)})\overline{\partial_q\delta\zeta_p} - \omega_0^2\rho\zeta_p^{(0)}\overline{\delta\zeta_p})\mathrm{d}x = 0$$

and taking for $\delta\zeta$ any arbitrary element of $(\mathscr{D}(\Omega))^3$, which is legitimate since by (6.234) there is an element $\delta\xi$ of $(\mathscr{D}(\Omega))^3$, corresponding to it, we see that:

$$\partial_q(a_{pqmn}\varepsilon_{mn}(\zeta^{(0)})) + \omega_0^2\rho\zeta_p^{(0)} = 0$$

in the distributions sense.

Thus for a given k in Ω, $\omega_0(k)$ exists and there is a displacement field u given by:

$$u(x,t) = \zeta^{(0)}e^{-i\omega_0(k)t} = e^{i(k\cdot x - \omega_0(k)t)}\cdot\xi^{(0)}(x,k) \tag{6.235}$$

where $\xi^{(0)}(x,k)$ is periodic in x, representing vibrations which can occur in a medium of indefinite extent.

We observe also that $\omega_0(k) = 0$ if and only if

$$\varepsilon_{mn}(\zeta^{(0)}) = 0 \quad \text{that is} \quad e^{ikx}\xi^{(0)}(x) = a + f \wedge x,$$

a and f being fixed elements of $\mathbb{R}^3$.

Now the periodicity of $\xi^{(0)}(x)$, implies that this identity can be satisfied only if $f = 0$ and $\xi^{(0)}(x) = ae^{-ikx}$ with

$$k\cdot b_j = 0 \bmod 2\pi.$$

If we introduce the conjugate lattice of vectors of base c_1, c_2, c_3 defined by $c_i\cdot b_j = \delta_{ij}$, then (6.236) is equivalent to:

$$k = \gamma_1c_1 + \gamma_2c_2 + \gamma_3c_3, \quad (\gamma_1, \gamma_2, \gamma_3) \in \mathbb{Z}^3$$

and $k \in \Omega$; except for these finitely many particular values of k, $\omega_0(k)$ is non-zero.

We can then define the eigenvalues of higher order, for example:

$$\omega_1^2 = \operatorname*{Inf}_{\xi \in K_1} \frac{[\xi, \xi]}{\int \rho \xi \bar{\xi} \mathrm{d}x}$$

where K_1 is the subspace spanned by the elements ξ of $(H^1(\Omega))^3$ which are orthogonal to $\xi^{(0)}$ in the sense that

$$\int_\Omega \rho \xi^{(0)} \bar{\xi} \mathrm{d}x = 0.$$

We thus obtain the second eigenfrequency $\omega_1 \geqslant \omega_0$ and the associated eigenmode $\xi^{(1)}$; continuing in this way we can define an infinite sequence of eigenfrequencies

$$\omega_0 \leqslant \omega_1 \leqslant \cdots \leqslant \omega_s \leqslant \cdots$$

and corresponding associated eigenmodes $e^{ikx} \xi^{(s)}$ with $\xi^{(s)}$ periodic in x.

It can be shown that the eigenmodes corresponding to a multiple eigenfrequency are finite in number.

Chapter VII. Modal Analysis and Vibrations of Structures

In complex structures certain excitative forces, for example those due to variations in flow in pipelines, or to the action of wind, often exhibit a random behaviour. The response characteristics of a system acted upon by such forces can be defined schematically by a modal description based on the displacements intercorrelation spectrum which can be calculated from the intercorrelation spectrum of the excitative forces, by means of the transfer function.

The vibrations of a suspension bridge provide an interesting illustration of the coupling effects between the aerodynamic stresses acting on a structure in oscillatory motion and the elastic response of the structure to such stresses. The stability conditions can be assessed on the basis of a theoretical approach supplemented by experimental data.

I. Vibrations of Structures

Free Vibrations

The vibratory state of a complex system comprising an assemblage of beams, plates, pipelines through which fluids are passing, and so on, may be studied by taking as a starting point for discussion a discretised model of the structure consisting of the set Ω of the solid parts of the system. This means that we seek, at the outset, to represent the field of kinematically admissible displacements of the x of Ω by

$$x \to x + u, \; u_j = \sum_{k=1}^{r} \zeta_{jk}(x) q_k, \quad j = 1, 2, 3 \tag{7.1}$$

where $\zeta = \{\zeta_{jk}(x)\}$ is an explicitly known $3 \times r$ matrix resulting from a decomposition of the structure into finite elements, and where the independent variables q_k, describe the r degrees of freedom allowed to the system. By applying the method of virtual work, we can easily obtain the set of differential equations obeyed by the q_k. Any virtual displacement can be defined, in matrix notation, by $\delta u = \zeta \cdot \delta q$ and by first calculating the virtual work of the rate of change of momentum we have

$$\int_{\Omega} (\rho(x) \zeta q'', \zeta \delta q) \, \mathrm{d}x = \left(\int_{\Omega} \zeta^T \rho \zeta \, \mathrm{d}x \cdot q'', \delta q \right) = (a q'', \delta q),$$

where $a = \int_\Omega \zeta^T \rho \zeta \,\mathrm{d}x$ is an $r \times r$ matrix and the notation $(.,.)$ is used to denote the scalar product, ζ^T is the transpose of the matrix ζ, and $\rho(x)$ is the mass per unit volume at x.

We shall assume that, following a displacement u, the stresses acting on a volume element $\mathrm{d}x$ surrounding x, due to the neighbouring elements, can be represented by the linear approximation

$$(-\nu\dot{u} - \varkappa u)\mathrm{d}x \quad \text{where} \quad \nu = \{\nu_{ij}(x)\}, \quad \varkappa = \{\varkappa_{ij}(x)\}$$

are symmetric, positive 3×3 matrices associated with the dissipation and the elastic rigidity properties respectively of the structure, and $\dot{u}$ is the partial derivative with respect to time of the displacement. On these assumptions the virtual work of these forces is

$$\int_\Omega (-\nu\zeta q' - \varkappa\zeta q, \zeta\delta q)\,\mathrm{d}x = -(bq' + cq, \delta q)$$

with

$$b = \int_\Omega \zeta^T \nu \zeta \,\mathrm{d}x, \quad c = \int_\Omega \zeta^T \varkappa \zeta \,\mathrm{d}x$$

from which we obtain the equation for the free vibrations

$$aq'' + bq' + cq = 0, \tag{7.2}$$

a, b, c being symmetric $r \times r$ matrices with constant coefficients.

The symmetric, positive-definite matrix a, is invertible and the matrix $a^{-1/2} \cdot c \cdot a^{-1/2}$ can be diagonalised; that is to say, there exists a real unitary matrix σ, with $\sigma^T = \sigma^{-1}$ such that

$$\sigma^T a^{-1/2} c a^{-1/2} \sigma = \begin{pmatrix} \omega_1^2 & & 0 \\ & \ddots & \\ 0 & & \omega_r^2 \end{pmatrix} = \gamma, \tag{7.3}$$

the eigenvalues ω_j^2 of $a^{-1/2} c a^{-1/2}$ defining the squares of the eigenfrequencies of the structure.

By the change of variable $q = a^{-1/2}\sigma\tilde{q}$, $\tilde{q} = \sigma^T a^{1/2} q$, we obtain the equation

$$\tilde{q}'' + \beta\tilde{q}' + \gamma\tilde{q} = 0 \tag{7.4}$$

where

$$\beta = \sigma^T a^{-1/2} b a^{-1/2} \sigma.$$

The damping matrix β is symmetric but not in diagonal form; however it is usual to retain only the diagonal terms which are generally preponderant. On this hypothesis the equations (7.4) are decoupled and can be written

$$\tilde{q}_k'' + 2\varepsilon_k \omega_k \tilde{q}_k' + \omega_k^2 \tilde{q}_k = 0$$

the parameter $1/2\varepsilon_k$ is the overtension and the co-ordinate $\tilde{q}_k$ is said to be modal.

Forced Vibrations

Take as an example the case of the interaction between a fluid and pipeline and let us suppose the excitation to be produced by a pressure fluctuation $p(x, t)$ at a part of the surface denoted by Γ. The corresponding virtual work can be expressed by:

$$\int_\Gamma (p(x,t)n, \delta u)\,\mathrm{d}x = \left(\int_\Gamma \zeta^T pn\,\mathrm{d}x, \delta q \right) = \sum_{k=1}^{r} Q_k \delta q_k$$

where $\mathrm{d}x$ is the element of area on Γ,

$$Q_k(t) = \int_\Gamma \sum_{j=1}^{3} p(x,t) n_j(x) \zeta_{jk}(x)\,\mathrm{d}x, \tag{7.5}$$

with $n = (n_1, n_2, n_3)$ being the unit vector at x normal to Γ away from the fluid. The equations for the forced vibrations of the system can then be written, assuming we have chosen modal co-ordinates:

$$q_k'' + 2\varepsilon_k \omega_k q_k' + \omega_k^2 q_k = \mu_k^{-1} Q_k(t), \quad 1 \leqslant k \leqslant r \tag{7.6}$$

with

$$\mu_k = \sum_{j=1}^{3} \int_\Omega \rho \zeta_{jk} \zeta_{jk}\,\mathrm{d}x$$

which we can assume to be equal to 1, by an appropriate choice of modal co-ordinates.

If we suppose the structure to have been excited by $p(x,t) = \delta(x - x_0) e^{i\omega t}$, $x_0 \in \Gamma$ where δ is the Dirac surface-distribution, we obtain the transfer function by calculating the response at the current point x of Γ. Now for the excitation under consideration we have:

$$Q_k = \sum_{j=1}^{3} n_j(x_0) \zeta_{jk}(x_0) e^{i\omega t} \tag{7.7}$$

and it only remains to express the periodic solution of frequency ω of the equations (7.6), (7.7) with $\mu_k = 1$, i.e.:

$$q_k(t) = \frac{\sum_{j=1}^{3} n_j(x_0) \zeta_{jk}(x_0)}{\omega_k^2 - \omega^2 + 2i\varepsilon_k \omega_k \omega} \cdot e^{i\omega t}$$

from which we obtain the complex displacement at $x \in \Gamma$ in the direction normal to the wall

$$y(x,t) = \sum_{j,k} n_j(x) \zeta_{jk}(x) q_k(t) = G(x, x_0, \omega) e^{i\omega t} \tag{7.8}$$

or with

$$\chi_k(x) = \sum_{j=1}^{3} n_j(x) \zeta_{jk}(x):$$

$$G(x, x_0, \omega) = \sum_{k=1}^{r} \frac{\chi_k(x) \chi_k(x_0)}{\omega_k^2 - \omega^2 + 2i\varepsilon_k \omega_k}. \tag{7.9}$$

We obtain similarly for $\frac{\partial y}{\partial t}$, the velocity of displacement normal to the wall, the transfer function:

$$H(x, x_0, \omega) = i\omega G(x, x_0, \omega).$$

Random Excitation of Structures

1. Most often the pressure fluctuations are generated locally by singularities in the flow, which appear for example in elbow-bends and are random functions of time and space. For this reason they are generally characterised by their intercorrelation spectrum, which we shall now define. Spectral analysis of the pressure fluctuations shows that the range of frequencies is bounded and that the pressure at x_0, t, can be represented by the real scalar:

$$p(x_0, t) = \frac{1}{\sqrt{2\pi}} \int_{-\infty}^{+\infty} \hat{p}(x_0, \omega) e^{i\omega t} \, d\omega, \tag{7.10}$$

where $\hat{p}(x_0, \omega)$ is the complex amplitude of a vibration of frequency ω; this representation is valid in the case of a continuous spectrum, but the modifications required to deal with the case of a discrete spectrum will be explained later on.

For a harmonic excitation described by a pressure term $\hat{p}(x_0, \omega) e^{i\omega t}$, $x_0 \in \Gamma$, we obtain, by (7.5)

$$Q_k(t) = \int_\Gamma \sum_{j=1}^{3} \hat{p}(x_0, \omega) n_j(x_0) \zeta_{jk}(x_0) \, dx_0 \cdot e^{i\omega t}$$

and on comparing (7.7) with the response associated with it by (7.8), we can write down, after an additional integration with respect to ω, the response which corresponds to the pressure-field (7.10), namely,

$$y(x, t) = \frac{1}{\sqrt{2\pi}} \int_{-\infty}^{+\infty} \int_\Gamma G(x, x_0, \omega) \hat{p}(x_0, \omega) e^{i\omega t} \, dx_0 \, d\omega. \tag{7.11}$$

Returning to (7.10) we apply Parseval's formula to $p(x_0, t)$ and $p(x'_0, t + \tau)$ whose spectrum is $\hat{p}(x'_0, \omega) e^{i\omega\tau}$ (assuming the pressure fluctuations are square-integrable with respect to time) and this leads to:

$$\int_{-\infty}^{+\infty} p(x'_0, t + \tau) \bar{p}(x_0, t) \, dt = \int_{-\infty}^{+\infty} \hat{p}(x'_0, \omega) \bar{\hat{p}}(x_0, \omega) e^{i\omega\tau} \, d\omega \tag{7.12}$$

suggesting the following notations:

$$\begin{aligned} R_p(x_0, x'_0, \tau) &= \int_{-\infty}^{+\infty} p(x'_0, t + \tau) \bar{p}(x_0, t) \, dt \\ R_y(x_0, x'_0, \tau) &= \int_{-\infty}^{+\infty} y(x'_0, t + \tau) \bar{y}(x_0, t) \, dt \\ R_v(x_0, x'_0, \tau) &= \int_{-\infty}^{+\infty} v(x'_0, t + \tau) \bar{v}(x_0, t) \, dt \end{aligned} \tag{7.13}$$

which are respectively the intercorrelation functions of pressure, displacement and velocity respectively. The conjugate values $\bar{p}, \bar{y}, \bar{v}$ have been introduced in anticipation of the case when these quantities are complex.

Defining the pressure intercorrelation spectrum by:

$$S_p(x_0, x'_0, \omega) = \hat{p}(x'_0, \omega)\bar{\hat{p}}(x_0, \omega) \tag{7.14}$$

we interpret (7.12) as:

$$R_p(x_0, x'_0, \tau) = \int_{-\infty}^{+\infty} S_p(x_0, x'_0, \omega) e^{i\omega\tau} \, d\omega. \tag{7.15}$$

We can now once more apply Parseval's theorem to the formula (7.11) which expresses the displacement by means of a Fourier transform, and obtain:

$$\int_{-\infty}^{+\infty} y(x', t + \tau)\bar{y}(x, t) \, dt$$
$$= \int_{-\infty}^{+\infty} e^{i\omega\tau} \int_{\Gamma} G(x', x'_0, \omega)\hat{p}(x'_0, \omega) \, dx'_0 \int_{\Gamma} \bar{G}(x, x_0, \omega)\bar{\hat{p}}(x_0, \omega) \, dx_0 \, d\omega$$

or with

$$S_y(x, x', \omega) = \int_{\Gamma}\int_{\Gamma} \bar{G}(x, x_0, \omega) \cdot G(x', x'_0, \omega)\bar{\hat{p}}(x_0, \omega)\hat{p}(x'_0, \omega) \, dx_0 \, dx'_0, \tag{7.16}$$

the displacement intercorrelation spectrum:

$$R_y(x, x', \tau) = \int_{-\infty}^{+\infty} e^{i\omega\tau} S_y(x, x', \omega) \, d\omega. \tag{7.17}$$

Substituting (7.14) in (7.16) we obtain:

$$S_y(x, x', \omega) = \int_{\Gamma}\int_{\Gamma} \bar{G}(x, x_0, \omega) G(x', x'_0, \omega) S_p(x_0, x'_0, \omega) \, dx_0 \, dx'_0 \tag{7.18}$$

and similarly

$$S_v(x, x', \omega) = \int_{\Gamma}\int_{\Gamma} \bar{H}(x, x_0, \omega) H(x', x'_0, \omega) S_p(x_0, x'_0, \omega) \, dx_0 \, dx'_0 \tag{7.19}$$

for the velocity intercorrelation spectrum, the displacement velocity $\frac{\partial y}{\partial t}$ being given by a formula similar to (7.11), obtained by substituting $H = i\omega G$ for G.

2. In the case of a discrete spectrum, we represent the pressure fluctuations by

$$p(x_0, t) = \sum_l \hat{p}_l(x_0) e^{i\omega_l t}$$

and the displacement at $x \in \Gamma$ by

$$y(x, t) = \sum_l G(x, x_0, \omega_l)\hat{p}_l(x_0) e^{i\omega_l t}.$$

We can then define the intercorrelation functions

$$R_p(x_0, x'_0, \tau) = \lim_{T \to \infty} \frac{1}{2T} \int_{-T}^{+T} p(x'_0, t + \tau)\bar{p}(x_0, t) \, dt$$

$$R_y(x_0, x_0', \tau) = \lim_{T\to\infty} \frac{1}{2T} \int_{-T}^{+T} y(x_0', t+\tau)\bar{y}(x_0, t)\,\mathrm{d}t$$

and we shall then obtain between the associated intercorrelation spectra S_p, S_y the same relation as in (7.18).

3. Suppose the structure to be put into vibration by some excitation source at the point $\xi \in \Gamma$, in such a way that

$$p(x_0, t) = p(t)\cdot\delta(x_0 - \xi).$$

Writing $\hat{p}(\omega) = \frac{1}{\sqrt{2\pi}} \int_{-\infty}^{+\infty} p(t)e^{-i\omega t}\,\mathrm{d}t$, we obtain by (7.10) and (7.14) the pressure intercorrelation spectrum

$$S_p(x_0, x_0', \omega) = |\hat{p}(\omega)|^2 \delta(x_0 - \xi)\delta(x_0' - \xi)$$

and by (7.18), (7.19):

$$S_y(x, x', \omega) = \bar{G}(x, \xi, \omega)G(x', \xi, \omega)|\hat{p}(\omega)|^2$$
$$S_v(x, x', \omega) = \bar{H}(x, \xi, \omega)H(x', \xi, \omega)|\hat{p}(\omega)|^2.$$

If the random pressure at ξ is white noise, $|\hat{p}(\omega)| = 1$, we see that the velocity intercorrelation spectrum corresponding to such excitation, in the range of frequencies under consideration, is given by:

$$S_v(x, x', \omega) = \phi_\xi(x, x', \omega) = \bar{H}(x, \xi, \omega)H(x', \xi, \omega). \tag{7.20}$$

It is interesting to note that, by (7.19), $S_v(x, x, \omega)$, the power spectrum at x, for any given excitation, can be represented by:

$$S_v(x, x, \omega) = \int_\Gamma\int_\Gamma \bar{H}(x, x_0, \omega)H(x, x_0', \omega)S_p(x_0, x_0', \omega)\,\mathrm{d}x_0\,\mathrm{d}x_0'$$

or, in view of the symmetry properties of the transfer function, and of the definition (7.20):

$$S_v(x, x, \omega) = \int_\Gamma\int_\Gamma \phi_x(x_0, x_0', \omega)S_p(x_0, x_0', \omega)\,\mathrm{d}x_0\,\mathrm{d}x_0'. \tag{7.21}$$

In practice the natural sources of excitation of the structure (turbulence, instabilities of flow, ...) cover a range of frequencies which can go beyond 2,000 Hz. The determination of the vibration characteristics has to cover a wide spectrum and one comes up against several difficulties. Modelling by finite elements is not well suited to the lower frequency modes. Furthermore in a 2,000 Hz frequency range there can well be a very large number of eigenmodes, perhaps 300 or more, and the eigenfrequencies corresponding to these modes become closer and closer as the frequency increases.

It is precisely in such cases where there is a large number of modes that one can usefully introduce [6] the concept of a mean power spectrum in a band of frequencies of width $\Delta\omega$ around ω

$$\tilde{S}_v(x, x, \omega) = \frac{1}{\Delta\omega} \int_{\omega-\Delta\omega/2}^{\omega+\Delta\omega/2} S_v(x, x, \beta)\,\mathrm{d}\beta$$

and if one assumes that $S_v(x_0, x'_0, \omega)$ varies slowly in the neighbourhood of ω, we can write, after integrating (7.21):

$$\tilde{S}_v(x, x, \omega) = \int_\Gamma\int_\Gamma \tilde{\phi}_x(x_0, x'_0, \omega) S_p(x_0, x'_0, \omega)\, dx_0\, dx'_0$$

where

$$\tilde{\phi}_x(x_0, x'_0, \omega) = \frac{1}{\Delta\omega} \int_{\omega-\Delta\omega/2}^{\omega+\Delta\omega/2} \phi_x(x_0, x'_0, \beta)\, d\beta$$

is the mean, with respect to frequency, of the function $\phi_x(x_0, x'_0, \omega)$. This mean which can be directly measured experimentally, represents all the properties of the structure in the frequency-band $\Delta\omega$ which may include a large number of modes of vibration.

Experience shows that it is often possible to represent the pressure intercorrelation spectrum by:

$$S_p(x, x', \omega) = S_p(\omega) \exp\left\{ -\frac{|x - x'|}{\lambda} + i\frac{(x - x', u)}{V_f}\omega \right\}$$

where $S_p(\omega)$ is a mean spectral density, λ a correlation length, V_f a velocity which is a characteristic of the mean flow, known as the fluctuation transport velocity and u the unit vector along the direction of mean flow.

It is interesting to relate these spectral quantities to certain physical characteristics of the steady flow which are essentially:

- an average pressure ΔP often taken to be equal to the static pressure jump from one side to the other of the singularity;
- an average flow velocity V, generally taken as equal to the maximum velocity in the singularity;
- one or more characteristic dimensions, denoted by D.

We can define a dimensionless spectral density

$$F_p(s) = \frac{S_p(\omega)}{(\Delta P)^2} \cdot \frac{V}{D}$$

with $s = \frac{\omega D}{V}$, being Strouhal's number.

The dimensionless mean-square value of the pressure fluctuations is, having regard to (7.15) with $\tau = 0$, defined by:

$$\left(\frac{\sigma}{\Delta P}\right)^2 = \int_0^\infty F_p(s)\, ds.$$

The dimensionless spectral density is generally a decreasing function which can be fairly well represented by:

$$F_p(s) = \frac{K}{1 + \left(\frac{s}{s_c}\right)^2}$$

where s_c is Strouhal's cut-off number.

The set of dimensionless quantities defined in this way appear to have rather slight, if any, connection with the Reynolds and Mach numbers of the flow and to vary little from one singularity to another.

II. Vibrations in Suspension Bridges

Suspension bridges are structures which, under the action of wind, can undergo dangerous oscillations.

The accident to the Tacoma Narrows bridge at Puget Sound doubtless one of the most striking examples of this, was due to lateral and torsional vibrations of the deck caused by the action of a wind blowing at a speed of 18 m/s (42 m.p.h.). The event was filmed by a witness, from one of the banks of the sound, and thanks to this unique document we have a very moving record of the disaster.

An understanding of the mechanism which generates these vibrations was acquired shortly afterwards. The explanation involves taking account of the coupling between the aerodynamic forces acting on the oscillating structure and the elastic forces generated in response to them.

It is therefore essentially a problem in aero-elasticity and it is well-known that these questions are also of great importance in aeronautics.

The account which follows, which is greatly indebted to [39], sets out to analyse the flexural-torsional modes of vibration of a suspension bridge structure and to explain the process of excitation under incompressible flow in order to determine the critical conditions.

The Equilibrium Configuration

The bridge is represented schematically by an uniform beam of length $L + 2L_1$, suspended, at regular intervals, from an inextensible cable, supported at the tops of pylons at $x = \pm\frac{L}{2}$ and anchored to piers at $x = \pm\left(\frac{L}{2} + L_1\right)$ on each bank. The actual structure obviously comprises two such cables, but for the purpose of studying equilibrium and analysing purely flexural movements, the geometrical configurations are identical and it is possible to argue as though there were only one cable.

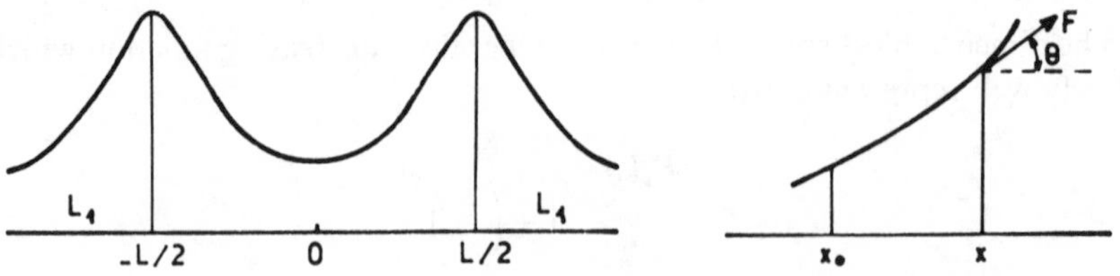

Figure 7.1

Supposing the structure to be in equilibrium, we denote by $F(x)$ the tension exerted by the parts of the cable of abscissa greater than x on those of abscissa less than x, by θ the angle of inclination of the cable to the horizontal and by x, y the co-ordinates of an arbitrary point on the cable, where the y-axis coincides with the upward vertical.

By applying the equations of mechanics to the part of the cable between a reference point x_0 and the point x, the cable being connected to the roadway in this interval only through the suspending rods, we can write:

(7.22) $$F\cos\theta = F_0\cos\theta_0 = Q, \quad Q \text{ being a constant independent of } x$$

(7.23) $$F\sin\theta - F_0\sin\theta_0 = mg(x - x_0)$$

m denoting the mass per unit length of the bridge, that of the cable and of the suspending rods being negligible in comparison and therefore ignored.

Since $\frac{dy}{dx} = \tan\theta$, we obtain on eliminating F, θ from (7.22), (7.23):

$$Q\frac{dy}{dx} - F_0\sin\theta_0 = mg(x - x_0)$$

and, as Q does not depend on x:

(7.24) $$\frac{d^2y}{dx^2} = \frac{mg}{Q}$$

so that the equilibrium configuration is given by:

(7.25) $$y = -f\left(1 - \left(\frac{2x}{L}\right)^2\right) + h$$

where h is the height of the pylons above the roadway, f is the sag and

(7.26) $$Q = \frac{mgL^2}{8f}.$$

The Flexure Equation Assuming Small Disturbances

We shall assume that the point (x, y) of the suspension cable goes to $(x + \xi, y + \eta)$ under the effect of disturbance, with $y(x)$ given by (7.25), while the corresponding point on the deck of the bridge at the bottom end of the suspending rod hanging from the cable at the point (x, y) goes to (x, η). This is compatible with the assumption that the suspending rods are inextensible if we ignore small quantities of the second order, since

$$(\xi^2 + y^2)^{1/2} = y\left(1 + \frac{\xi^2}{y^2}\right)^{1/2} = y\left(1 + O\left(\frac{\xi^2}{y^2}\right)\right).$$

The length of the portion of cable between the points 0 and x being

$$\int_0^x [(1+\xi')^2 + (y' + \eta')^2]^{1/2}\,dx = \int_0^x (1 + y'^2)^{1/2}\cdot\left[1 + \frac{\xi' + \eta' y'}{1 + y'^2} + O(\xi'^2 + \eta'^2)\right]dx$$

the condition that the suspension cable is inextensible is satisfied to the second order if we impose the relation

(7.27)
$$\xi' + \eta' y' = 0$$

on the perturbations ξ, η with $y' = \frac{dy}{dx}$ being calculated from (7.25).

The bridge itself (i.e. the deck) can be treated like a beam. At the point x the shearing stress T and the bending moment M are calculated in the usual way from the formulae of Chapter 7:

$$T + \frac{\partial M}{\partial x} = 0, \quad M = EI\frac{\partial^2 \eta}{\partial x^2},$$

on the supposition that EI is independent of x, so that

$$T = -EI\frac{\partial^3 \eta}{\partial x^3}.$$

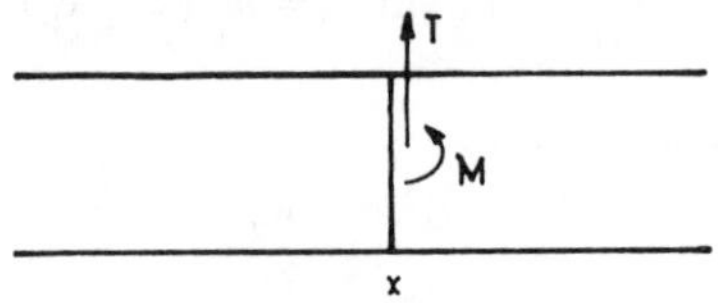

Figure 7.2

To write down the equations of motion, holding at time t, we denote by $F + \Delta F$ the tension of the suspension cable at the point x, F being the value found for the equilibrium configuration; by θ the inclination to the horizontal of the cable at the point x in its deformed state, so that by applying the equations of mechanics to the part of the cable, and to the deck comprised between the points x_0 and x, we obtain (ignoring the mass of the cable and suspending rods):

(7.28)
$$(F + \Delta F)\cos\theta = (F_0 + \Delta F_0)\cos\theta_0 = Q + \Delta Q$$

(7.29)
$$(F + \Delta F)\sin\theta - (F_0 + \Delta F_0)\sin\theta_0 - mg(x - x_0) + T - T_0 = \int_{x_0}^{x} m\frac{\partial^2 \eta}{\partial t^2}\mathrm{d}x.$$

where Q is given by (7.26), ΔQ is independent of x (having the same value on either side of each pylon, because the suspension cable is supported at the top on rollers).

Now

$$\tan\theta = \frac{y' + \eta'}{1 + \xi'} \sim y' + \eta'$$

and we deduce from (7.28), (7.29):

$$(Q + \Delta Q)\left(\frac{dy}{dx} + \frac{\partial \eta}{\partial x}\right) + (F_0 - \Delta F_0)\sin\theta_0 - mg(x - x_0) - EI\frac{\partial^3 \eta}{\partial x^3} - T_0 = \int_{x_0}^{x} m\frac{\partial^2 \eta}{\partial t^2}\mathrm{d}x$$

and by differentiation with respect to x:

$$(Q+\varDelta Q)\left(\frac{\mathrm{d}^2 y}{\mathrm{d}x^2}+\frac{\partial^2\eta}{\partial x^2}\right)-mg-EI\frac{\partial^4\eta}{\partial x^4}-m\frac{\partial^2\eta}{\partial t^2}=0. \tag{7.30}$$

Neglecting the second order term $\varDelta Q\cdot\frac{\partial^2\eta}{\partial x^2}$ and using (7.24), (7.26) we get:

$$\frac{8f}{L^2}\varDelta Q+\frac{mgL^2}{8f}\frac{\partial^2\eta}{\partial x^2}-EI\frac{\partial^4\eta}{\partial x^4}-m\frac{\partial^2\eta}{\partial t^2}=0 \tag{7.31}$$

with the obvious boundary conditions:

$$\eta\left(\frac{L}{2}\right)=\eta\left(-\frac{L}{2}\right)=0 \tag{7.32}$$

$$\eta\left(\frac{L}{2}+L_1\right)=\eta\left(-\frac{L}{2}-L_1\right)=0. \tag{7.33}$$

In addition, by integrating (7.27) with respect to x from $-\frac{L}{2}-L_1$ to $\frac{L}{2}+L_1$, and using the expression (7.25) for $y(x)$ we obtain:

$$0=\int_{-L_1-L/2}^{L_1+L/2}(\xi'+\eta' y')\,\mathrm{d}x=(\xi+\eta y')\Big|_{-L_1-L/2}^{L_1+L/2}-\int_{-L_1-L/2}^{L_1+L/2}\eta y''\,\mathrm{d}x$$

i.e.

$$\int_{-L_1-L/2}^{L_1+L/2}\eta\,\mathrm{d}x=0. \tag{7.34}$$

Free Flexural Vibrations in the Absence of Stiffness

Let us consider initially the case where the stiffness term in (7.31) can be neglected. We have $EI=0$ so that:

$$mg\frac{\varDelta Q}{Q}+Q\frac{\partial^2\eta}{\partial x^2}=m\frac{\partial^2\eta}{\partial t^2}. \tag{7.35}$$

We look for solutions of the form $\eta=\eta(x)\cdot\sin\omega t$ arriving at:

$$mg\frac{\varDelta Q}{Q}+(Q\eta''+m\omega^2\eta)\sin\omega t=0$$

and since $\frac{\varDelta Q}{Q}$ depends only on t, and Q is constant, we are led to put

$$\varDelta Q=-h\sin\omega t, \tag{7.36}$$

so that η is the solution of the differential equation

$$\eta''+\left(\frac{2\sigma}{L}\right)^2\eta=\frac{8fh}{L^2 Q} \tag{7.37}$$

where we have put

$$\sigma=\frac{\omega L}{2}\sqrt{\frac{m}{Q}}=\sqrt{\frac{2f}{g}}\cdot\omega. \tag{7.38}$$

a) Symmetric Modes: $\eta(x)=\eta(-x)$

In the bay of the bridge $0\leqslant x\leqslant\frac{L}{2}$, $\eta(x)$ is defined by (7.37) with the boundary conditions $\eta'(0)=0$, $\eta\left(\frac{L}{2}\right)=0$ whence

$$\eta(x)=\frac{2fh}{Q\sigma^2}\left(1-\frac{\cos\left(\frac{2\sigma x}{L}\right)}{\cos\sigma}\right).$$

For $\frac{L}{2}\leqslant x\leqslant\frac{L}{2}+L_1$, we calculate $\eta(x)$ the solution of (7.37) which satisfies $\eta\left(\frac{L}{2}=\eta\left(\frac{L}{2}+L_1\right)=0\right.$, that is:

$$\eta(x)=\frac{2fh}{Q\sigma^2}\left(1-\frac{\cos\left[\sigma(1+\alpha)-\frac{2\sigma x}{L}\right]}{\cos\alpha\sigma}\right)$$

with $\alpha=\frac{L_1}{L}<\frac{1}{2}$.

To obtain σ or in other words the frequencies ω given by (7.38) we use (7.34) or:

$$\tan\sigma+2\tan(\sigma\alpha)=(1+2\alpha)\sigma$$

with $\sigma=\sqrt{\frac{2f}{g}}\cdot\omega$.

For $\alpha=\frac{1}{3}$, we obtain a first eigenfrequency whose value is approximately $\omega=1.5\pi\sqrt{\frac{g}{2f}}$.

b) Skew-Symmetric Modes: $\eta(x)=-\eta(-x)$

Two cases need to be distinguished depending on whether the cable and deck are independent or not.

1. Cable Independent of Deck

The skew-symmetry condition requires us to take $h=0$ in (7.37); and we then obtain for the main span, in view of

$$\eta\left(-\frac{L}{2}\right)=\eta\left(\frac{L}{2}\right)=0:$$

$$\eta(x)=\eta_0\sin\left(2\sigma\frac{x}{L}\right),\quad \sigma=n\pi,\quad n\text{ integer}$$

and for the lateral bays, for example the right-hand bay in the diagram

$$\frac{L}{2} < x < \frac{L}{2} + L_1 \quad \eta(x) = \eta_0 \sin\left[\frac{2\sigma}{L}\left(x - \frac{L}{2}\right)\right]$$

with $\sigma = \frac{m\pi}{2\alpha}$, where m is an integer. It follows from this that the integers m, n have to satisfy the relation $2\alpha = \frac{n}{m}$, the corresponding frequency being defined by

$$\omega = \sqrt{\frac{g}{2f}} \cdot n\pi. \tag{7.39}$$

Lastly the condition (7.34) is certainly satisfied.

2. The Cable is Tied to the Deck at Its Midpoint

In this case the tension in the cable has a discontinuity at the point $x = 0$. We shall denote the tension by $Q + \Delta Q_1$ in $x > 0$ and $Q + \Delta Q_2$ in $x < 0$ and on the assumption that there is a vibratory mode of frequency ω, we are again led by equation (7.35) to write

$$\Delta Q_1 = -h_1 \sin \omega t, \quad \Delta Q_2 = -h_2 \sin \omega t,$$

where h_1 and h_2 are constants.

If the deck is treated as a rigid structure as regards horizontal movement, the latter being described by the perturbation of the midpoint of the cable (i.e. the change in its horizontal co-ordinate), $\xi = \xi(0) \sin \omega t$, then the laws of mechanics applied to the portion of the deck between the two pylons (provided this part is structurally separated from the approach bays and simply rests on supports at each end of the main span), yield the equation

$$mL \frac{\mathrm{d}^2 \xi}{\mathrm{d}t^2} = \Delta Q_1 - \Delta Q_2 = (h_2 - h_1) \sin \omega t$$

or

$$mL\omega^2 \xi(0) = h_1 - h_2. \tag{7.40}$$

For $0 < x < \frac{L}{2}$, we calculate the solution of (7.37) taking $h = h_1$ and the boundary conditions $\eta\left(\frac{L}{2}\right) = 0$, $\eta(0) = 0$, the latter being a consequence of the assumption of skew-symmetry, from which we have:

$$\eta(x) = \frac{2fh_1}{Q\sigma^2}\left(1 - \frac{\cos \sigma\left(\frac{1}{2} - \frac{2x}{L}\right)}{\cos \frac{\sigma}{2}}\right). \tag{7.41}$$

For the bay $\frac{L}{2} < x < \frac{L}{2} + L_1$, still with $h = h_1$ and with the boundary conditions

$\eta\left(\frac{L}{2}\right)=\eta\left(\frac{L}{2}+L_1\right)=0$ we find

$$\eta(x)=\frac{2fh_1}{Q\sigma^2}\left(1-\frac{\cos\sigma\left(1+\alpha-\dfrac{2x}{L}\right)}{\cos\alpha\sigma}\right). \tag{7.42}$$

Similarly, with $h=h_2$ for $x<0$, we get

$$\begin{aligned}
\eta(x)&=\frac{2fh_2}{Q\sigma^2}\left(1-\frac{\cos\sigma\left(\dfrac{1}{2}+\dfrac{2x}{L}\right)}{\cos\dfrac{\sigma}{2}}\right), \quad -\frac{L}{2}<x<0\\
\eta(x)&=\frac{2fh_2}{Q\sigma^2}\left(1-\frac{\cos\sigma\left(1+\alpha+\dfrac{2x}{L}\right)}{\cos\alpha\sigma}\right), \quad -\frac{L}{2}-L_1<x<-\frac{L}{2}
\end{aligned} \tag{7.43}$$

and to satisfy the skew-symmetry condition, we must have

$$h_1=-h_2=h. \tag{7.44}$$

It remains to take account of the inextensible nature of the cable between the anchorage points $x=0$ and $x=\frac{L}{2}+L_1$. Going back to (7.27) and integrating with respect to x over this interval, we obtain:

$$\int_0^{L_1+L/2}\xi'\,dx=[-\eta y']_0^{L_1+L/2}+\int_0^{L_1+L/2}\eta y''\,dx$$

i.e.

$$-\xi(0)=\frac{8f}{L^2}\int_0^{L_1+L/2}\eta\,dx \tag{7.45}$$

or by (7.40), (7.44)

$$h=-\frac{4fm\omega^2}{L}\int_0^{L_1+L/2}\eta\,dx$$

and lastly with the aid of the representations (7.41), (7.42):

$$\tan\frac{\sigma}{2}+\tan\sigma\alpha=\left(\tfrac{1}{2}+\alpha+\frac{L^2}{32f^2}\right)\sigma \tag{7.46}$$

which, via $\sigma=\sqrt{\frac{2f}{g}}\,\omega$, defines the eigenfrequencies.

Torsional Vibrations of a Suspension Bridge

We shall suppose the structure of the bridge to consist of two parallel girders of the same stiffness to which the deck is attached as indicated on the figure below which represents a cross-section.

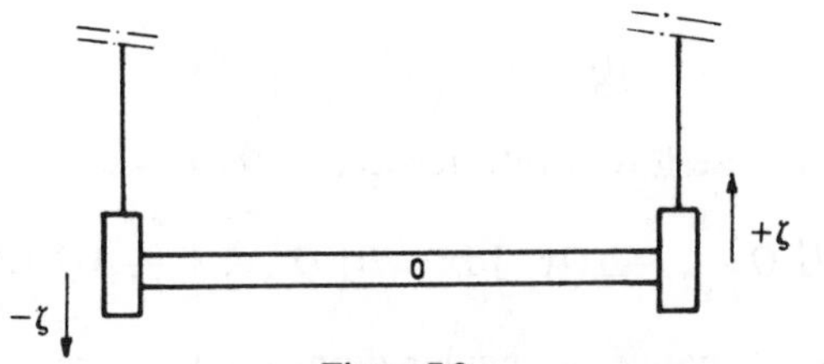

Figure 7.3

We propose to study torsional movements in which the longitudinal axis of the bridge remains undeformed and coincides with the axis Ox.

Accordingly the vertical displacements of the centres of the cross-sections of abscissa x of the stiffness girders are denoted by $\zeta(x,t)$, $-\zeta(x,t)$ and the torsion angle i.e. the angle through which the deck is twisted in this cross-section is $\theta = \frac{\zeta}{l}$, where $2l$ is the width of the bridge (deck).

We apply the theorem on angular momentum to the slice of the bridge between the sections x and $x + \mathrm{d}x$ and with this aim we begin by calculating the moment with respect to Ox of the stresses exerted.

1. The moment of the shearing stresses in the girders is

$$2l\frac{\partial T}{\partial x}\mathrm{d}x.$$

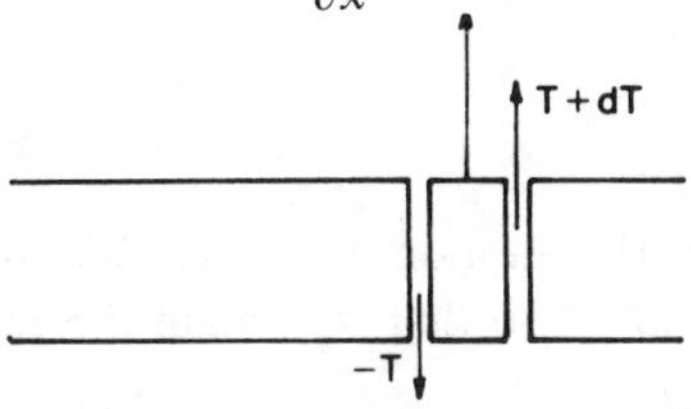

Figure 7.4

2. Moment of the torsional stresses in the deck. The moment with respect to Ox of the forces exerted by the elements of the deck of abscissa greater than x on those of abscissa less than x may be represented by $KG\frac{\partial \theta}{\partial x}$, where K depends on the geometry and G on the constitutive material; for the slice $\mathrm{d}x$ the moment concerned has the value $KG\frac{\partial^2 \theta}{\partial x^2}\mathrm{d}x$.

3. Moment of the forces exerted by the suspending rods. Denoting by $Q + \Delta Q$ the horizontal component of the tension in the cable carrying the girder whose movement is described by ζ, the vertical component of the force exerted by the suspending rods on the section $\mathrm{d}x$ is:

$$\frac{\partial}{\partial x}\left((Q+\Delta Q)\frac{\partial(y+\zeta)}{\partial x}\right)\mathrm{d}x = (Q+\Delta Q)\left(y'' + \frac{\partial^2 \zeta}{\partial x^2}\right)\mathrm{d}x.$$

The data for the other girder are $Q - \Delta Q$ and $-\zeta$ so that the lifting force is:

$$(Q-\Delta Q)\left(y''-\frac{\partial^2\zeta}{\partial x^2}\right)\mathrm{d}x$$

and the moment of these forces with respect to Ox is

$$2l\left(Q\frac{\partial^2\zeta}{\partial x^2}+\Delta Q y''\right)\mathrm{d}x=2l\left(Q\frac{\partial^2\zeta}{\partial x^2}+\frac{8f}{L^2}\Delta Q\right)\mathrm{d}x.$$

4. Finally the rate of change of angular momentum in the rotational motion of the section is $\mathscr{I}\frac{\partial^2\theta}{\partial t^2}\mathrm{d}x$, where $\mathscr{I}$ is the moment of inertia of the bridge, per unit length, about the longitudinal axis Ox.

Finally therefore we obtain the equation

$$\mathscr{I}\frac{\partial^2\theta}{\partial t^2}=2l\frac{\partial T}{\partial x}+KG\frac{\partial^2\theta}{\partial x^2}+2l\left(Q\frac{\partial^2\zeta}{\partial x^2}+\frac{8f}{L^2}\Delta Q\right)$$

or, with $\zeta=\theta l$, $T=-EI\frac{\partial^3\zeta}{\partial x^3}$ and putting $\mathscr{I}=2mr^2$, where m and EI represent respectively the mass per unit length and the rigidity of each of the two bridge girders:

$$\frac{mr^2}{l^2}\frac{\partial^2\zeta}{\partial t^2}+EI\frac{\partial^4\zeta}{\partial x^4}-\left(\frac{mgL^2}{8f}+\frac{KG}{2l^2}\right)\frac{\partial^2\zeta}{\partial x^2}-\frac{8f}{L^2}\Delta Q=0. \tag{7.47}$$

Symmetric Modes

a) Flexure

The equation describing the motion is (7.31). We represent ΔQ and η by $\Delta Q=-h\sin\omega t$, $\eta=\eta(x)\sin\omega t$ so that we obtain the equation for $\eta(x)$:

$$\frac{mgL^2}{8f}\eta''-EI\eta^{IV}+m\omega^2\eta-\frac{8fh}{L^2}=0. \tag{7.48}$$

Having regard to the boundary conditions

$$\eta\left(\pm\frac{L}{2}\right)=0,\quad \eta\left(\pm\left(\frac{L}{2}+L_1\right)\right)=0$$

and the symmetry assumption, we can adopt the representations

$$\begin{aligned}
&\eta=A\left[1-\frac{\cos\left(2\sigma\frac{x}{L}\right)}{\cos\sigma}\right], && -\frac{L}{2}<x<\frac{L}{2}\\
&\eta=A\left[1-\frac{\cos\sigma\left(1+\alpha-\frac{2x}{L}\right)}{\cos\alpha\sigma}\right], && \frac{L}{2}<x<\frac{L}{2}+L_1\\
&\eta(x)=\eta(-x), && -\left(\frac{L}{2}+L_1\right)<x<-\frac{L}{2}
\end{aligned} \tag{7.49}$$

which are continuous[5] at $x = \pm\frac{L}{2}$ and by substitution in (7.48) we obtain the conditions

$$m\omega^2 A = \frac{8fh}{L^2} \quad m\omega^2 - \frac{16\sigma^4 EI}{L^4} - \frac{mg\sigma^2}{2f} = 0. \tag{7.50}$$

Lastly we can interpret the condition (7.34) i.e. with (7.49):

$$\tan\sigma + 2\tan\alpha\sigma = (1+2\alpha)\sigma, \quad \alpha = \frac{L_1}{L} < 1/2 \tag{7.51}$$

which allows us to obtain the admissible values σ_n; the corresponding frequencies are obtained using (7.50):

$$\omega_{f_n}^2 = \frac{16EI}{mL^4}\sigma_n^4 + \frac{g}{2f}\sigma_n^2. \tag{7.52}$$

b) Torsion

Again adopting the representation $\Delta Q = -h\sin\omega t$, we obtain from (7.47) with $\zeta = \zeta(x)\sin\omega t$:

$$EI\zeta^{IV} - \left(\frac{mgL^2}{8f} + \frac{KG}{2l^2}\right)\zeta'' - \frac{mr^2}{l^2}\omega^2\zeta + \frac{8f}{L^2}h = 0 \tag{7.53}$$

which has the same structure as (7.48).

Accordingly we can write:

$$\zeta = B\left(1 - \frac{\cos\left(2\sigma\frac{x}{L}\right)}{\cos\sigma}\right), \qquad -\frac{L}{2} < x < \frac{L}{2}$$

$$\zeta = B\left(1 - \frac{\cos\sigma\left(1+\alpha-\frac{2x}{L}\right)}{\cos\sigma\alpha}\right), \qquad \frac{L}{2} < x < \frac{L}{2} + L_1$$

$$\zeta(x) = \zeta(-x), \qquad -\frac{L}{2} - L_1 < x < -\frac{L}{2}$$

[5] The representation (7.49) which is an exact solution of (7.48) does not ensure that the derivatives η', η'' at $x = \frac{L}{2}$, are continuous. We should, strictly speaking, have included this (continuity of the slope of the deformed deck and of the bending moment) as a necessary condition to be satisfied but, on the basis of the result obtained above in the case of zero rigidity, the representation (7.49) constitutes an acceptable approximation, if we assume that the effect of the stiffening of the deck is only secondary in the case of a suspension bridge. This is moreover the conclusion to which one would be led, if one were to calculate an exact solution of (7.48) satisfying all the boundary conditions, on the assumption of low rigidity.

and by substitution in (7.53):

$$\omega^2 \cdot \frac{mr^2}{l^2} B = \frac{8fh}{L^2}$$

$$-\frac{mr^2}{l^2}\omega^2 + \frac{16EI\sigma^4}{L^4} + \left(\frac{mg}{2f} + \frac{2KG}{l^2L^2}\right)\sigma^2 = 0$$

whence

$$\omega_{T_n}^2 = \frac{l^2}{r^2}\left(\frac{16EI}{mL^4}\sigma_n^4 + \frac{g}{2f}\sigma_n^2\right) + \frac{2KG}{mr^2L^2}\sigma_n^2 \tag{7.54}$$

the values σ_n again being defined by the equation (7.51).

For the other modes it is clear that the relations (7.52), (7.54) tying ω_f, ω_T to σ continue to hold; it is only the determination of σ which may change, depending on the parity of the mode, and on the structure of the bridge and in particular whether there is a central attachment or not. To conclude it is interesting to note that:

$$\frac{\omega_T^2 - \omega_f^2}{\omega_f^2} = \left(\frac{l^2}{r^2} - 1\right) + \frac{\dfrac{4fKG}{gr^2L^2m}}{1 + \dfrac{32fEI}{gmL^4}\sigma^2}. \tag{7.55}$$

Vibrations Induced by Wind

The flow of an incompressible fluid around a cylindrical obstacle of diameter b, in a direction normal to the generators, presents features which vary depending particularly on the Reynolds number $\mathscr{R} = \rho\dfrac{Vb}{\mu}$, where ρ is the density and μ the viscosity of the fluid, and V is the flow velocity upstream.

For an obstacle which is symmetrical with respect to the axis $x'x$ of the upstream flow, a symmetric wake is observed which is stationary at low speeds and more particularly for $\mathscr{R} < 50$. At this limit symmetric vortices may appear.

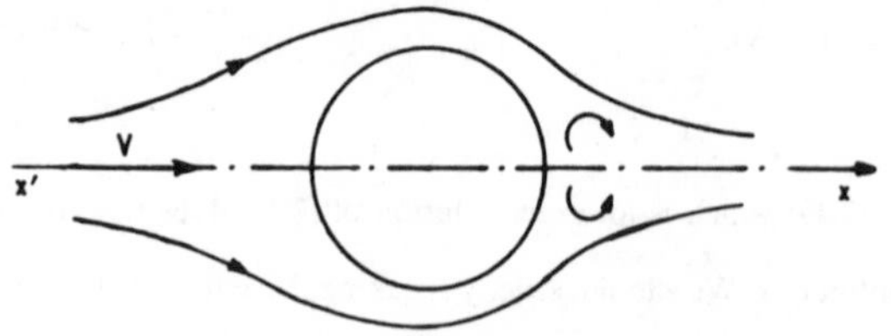

Figure 7.5

For Reynolds numbers above 50 the appearance of the phenomenon changes. From the downstream edge of the obstacle eddies detach themselves alternatively from each side forming two streams of vortices known as vortex streets and this configuration assumes a regular pattern for some distance past the obstacle so that one can define the distance a separating two consecutive vortices in the same street, as well as the distance h between the two streets. The two streams of vortices move away from the solid at a velocity V_s different from V. The theory of alternating vortex streets due to Benard and Karman [56] leads on an analysis of stability, to the relation $\cosh\frac{\pi h}{a}=\sqrt{2}$, that is $\frac{h}{a}\sim 0.281$. The ratio a/b is not given by the theory but for an obstacle in the form of a thin-wing profile, experiments indicate that

$$\frac{b}{h}\sim 0.66 \tag{7.56}$$

The resistance offered by an obstacle to the flow can very generally be represented by:

$$R=c_x\rho\frac{bV^2}{2} \tag{7.57}$$

where c_x is a dimensionless constant (for the flow parallel to a plane, and with R denoting the resistance per unit length of obstacle measured in the direction perpendicular to the plane of flow). The drag coefficient c_x has a value of approximately 1.8 for the Reynolds numbers under consideration.

On the other hand the von Karman [56] vortex street theory based on streams of alternate vortices leads to the expression

$$R=\rho\sqrt{8}V_s(V-2V_s)h+\frac{4\rho}{\pi}\left(\frac{h}{0.281}\right)V_s^2. \tag{7.58}$$

By equating this to the expression given by (7.57) and taking (7.56) into account and $c_x\sim 1.8$, we find $\frac{V_s}{V}=0.24$ or $\frac{V-V_s}{V}=0.76$.

In any case experience enables us to determine the frequency of appearance of vortices in the wake of an obstacle, expressed by $N=\frac{V-V_s}{a}$. For a cylinder we have an empirical formula expressing the Strouhal's number

$$s=\frac{Nb}{V}=0.198\cdot\left(1-\frac{19.7}{\mathscr{R}}\right)$$

The results which we obtain, taken as a whole, appear to be quite consistent and, in every case, the frequency of appearance of vortices in the slipstream of a bridge caused by wind is proportional to the wind-velocity, because $\mathscr{R}$ is very large due to the low viscosity of air. It should be pointed out that this description is acceptable only on the assumption of a fixed bridge.

It is reasonable to think that owing to the periodic phenomenon associated with the generation of alternating streams of vortices, an alternating lateral force (that is to say a force perpendicular to the direction of the wind, and alternating in direction) will be exerted on the obstacle, and that this force can be represented by:

$$F = \beta c_x \rho \frac{V^2 b}{2} \sin(2\pi N t),$$

where N is the frequency of the vortices, β a very small numerical factor of magnitude around 0.05, so that clearly F is very weak compared with R.

Under the action of a horizontal wind against the bridge, this force would be vertical and consequently could induce flexural vibrations. As there is no reason to think that the resultant of the vertical forces would always pass through the central axis of the bridge deck, one can see that a transverse couple would also be exerted on the deck which would give rise to torsional oscillations as well.

Aerodynamic Forces Exerted on the Deck of the Bridge

It is essential to be able to calculate the forces exerted by the wind on the deck of the bridge as soon as the latter begins to vibrate, if one wishes to be in a position to forecast the critical conditions, that is to say the conditions under which dynamically unstable oscillations can appear.

If one likens the deck to a plate of width $2l$, placed at an angle of incidence θ in an incompressible fluid flow of velocity V, the lift per unit length perpendicular to the plane of flow is, to a first approximation, given by:

$$F = \frac{\partial c_x}{\partial \theta} \cdot \theta \cdot \frac{\rho V^2}{2} \cdot 2l$$

But if the centre of the deck is in vertical motion described by the co-ordinate η, then the flow velocity with respect to axes tied to that point is $V\vec{z} - \frac{\partial \eta}{\partial t}\vec{y}$, so that the apparent angle of incidence is $\theta - \frac{1}{V} \cdot \frac{\partial \eta}{\partial t}$ and the lift can be taken to be equal to:

$$F = \rho V^2 l \frac{\partial c_x}{\partial \theta} \cdot \left(\theta - \frac{1}{V} \cdot \frac{\partial \eta}{\partial t}\right). \tag{7.59}$$

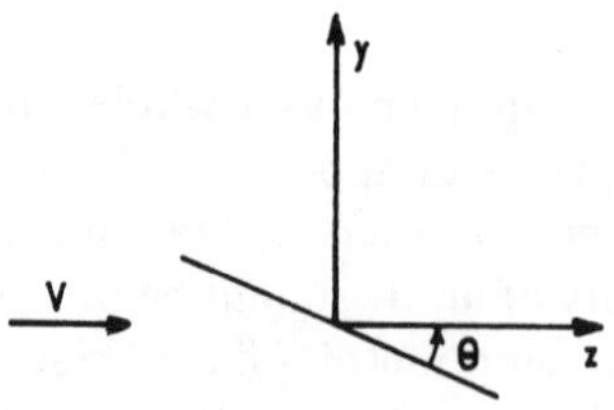

Figure 7.6

We can assume that $\frac{\partial c_x}{\partial \theta}$ is constant; for the deck of a bridge, experience suggests taking $\frac{\partial c_x}{\partial \theta} = 4$, and that the lift is applied at a point distant $\frac{l}{2}$ from the leading edge.

However this representation is rather crude and takes no account of the effects of lag in the formation of the stresses on the deck in motion.

A more accurate theory has been developed for calculating the stresses around wings in motion oscillating at a frequency ω in an incompressible-fluid flow of velocity V which is open-ended upstream [4]. If θ is the angle of incidence, y the vertical co-ordinate at the centre of the chord of the profile, $2l$ its width, and if we assume that y and θ vary with time while remaining proportional to $e^{i\omega t}$, then F the lift per unit length of profile, and M the moment of the forces about the midpoint O of the profile counted positively in the sense of increasing θ, are given by:

$$F = -\pi\rho l^2 \ddot{y} + 2\pi\rho l V^2 \left[C\left(\frac{\omega l}{V}\right)\cdot\left(\theta - \frac{\dot{y}}{V}\right) + \left(1 + C\left(\frac{\omega l}{V}\right)\right)\frac{l}{2}\cdot\frac{\dot{\theta}}{V}\right]$$

$$(7.60) \quad M = -\frac{\pi\rho l^4}{8}\ddot{\theta} + \pi\rho l^2 V^2 \left[C\left(\frac{\omega l}{V}\right)\cdot\left(\theta - \frac{\dot{y}}{V}\right) - \left(1 - C\left(\frac{\omega l}{V}\right)\right)\frac{l}{2}\cdot\frac{\dot{\theta}}{V}\right]$$

where

$$C(k) = H(k) - iG(k) = \frac{H_1^{(2)}(k)}{H_1^{(2)}(k) + iH_0^{(2)}(k)}$$

with $H_n^{(2)}(k) = J_n(k) - iY_n(k)$, being Hankel's function, and $\dot{y} = \frac{\partial y}{\partial t}, \ldots$. Observing that the multiplier $C\left(\frac{\omega l}{V}\right)$ in the formulae (7.60) acts on the expression $\theta - \frac{\dot{y}}{V} + \frac{l\dot{\theta}}{2V} = -\frac{w}{V}$, we see that in the case where the motion of the profile is described by some law, the stresses can be calculated by introducing the Fourier transform of w:

$$w = \frac{1}{\sqrt{2\pi}} \int_{-\infty}^{+\infty} f(\omega) e^{i\omega t} \mathrm{d}\omega$$

so that we have:

$$(7.61) \quad F = -\pi\rho l^2 \ddot{y} + 2\pi\rho l V^2 \left[\frac{l\dot{\theta}}{2V} - \frac{1}{\sqrt{2\pi} V} \int_{-\infty}^{+\infty} C\left(\frac{\omega l}{V}\right) f(\omega) e^{i\omega t} \mathrm{d}\omega\right]$$

$$M = -\pi\rho\frac{l^4}{8}\ddot{\theta} + \pi\rho l^2 V^2 \left[-\frac{l\dot{\theta}}{2V} - \frac{1}{\sqrt{2\pi}\, V} \int_{-\infty}^{+\infty} C\left(\frac{\omega l}{V}\right) f(\omega) e^{i\omega t} \mathrm{d}\omega\right].$$

Coming back to the formulae (7.60) we can ignore the inertial terms in $\ddot{y}, \ddot{\theta}$, which, involving as they do the density ρ of the air, will be small in comparison with the corresponding terms relating to the bridge in the equations of motion of the latter.

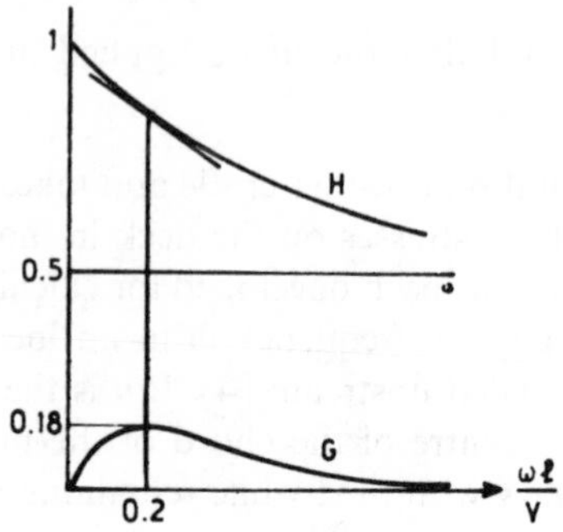

Figure 7.7

The graphs of the functions H and G shown in the figure suggest that with the crude approximation $H=1$, $G=0$, that is $C=1$, and neglecting the inertial terms and the rotational speed $\dot{\theta}$, the formulae (7.60) reduce to the result indicated by (7.59) with $\frac{\partial c_x}{\partial \theta}=2\pi$, the moment corresponding to a force applied at the point distant $\frac{l}{2}$ from the leading edge. One can see therefore to what extent this first representation is very crude, because in reality $C=1$ corresponds to $\frac{\omega l}{V}=0$. If we assume that $\frac{\omega l}{V}$ is in the neighbourhood of 0.2, the value for which G has a maximum of 0.18, we can approximate to $C\left(\frac{\omega l}{V}\right)$ by the expression:

$$C\left(\frac{\omega l}{V}\right)=0.93-\frac{\omega l}{V}-i\cdot 0.18$$

or introduce in the representations of F and M the equivalent operator:

$$C=0.93-\frac{\omega l}{V}-\frac{0.18}{\frac{\omega l}{V}}\frac{1}{V}\frac{\mathrm{d}}{\mathrm{d}t}$$

from which we deduce by (7.60) after leaving out the inertial terms

$$F=2\pi\rho l V^2\left[\left(0.93-\frac{\omega l}{V}\right)\left(\theta-\frac{\dot{y}}{V}\right)+\left(\frac{1+0.93}{2}-\frac{\omega l}{2V}-\frac{0.18}{\frac{\omega l}{V}}\right)\frac{l}{V}\dot{\theta}\right]$$

and since

(7.62)
$$\frac{1+0.93}{2}-\frac{\omega l}{2V}-\frac{0.18}{\frac{\omega l}{V}}\sim 0$$

when $\frac{\omega l}{V} \sim 0.2$ we can simply write:

$$F = 2\pi\rho l V^2\left(0.93 - \frac{\omega l}{V}\right)\left(\theta - \frac{\dot{y}}{V}\right). \tag{7.63}$$

In the same way, starting from (7.60) we can calculate

$$M = \pi\rho l^2 V^2\left[\left(0.93 - \frac{\omega l}{V}\right)\left(\theta - \frac{\dot{y}}{V}\right) - \left(\frac{0.18}{\frac{\omega l}{V}} + \frac{1 - 0.93}{2} + \frac{\omega l}{2V}\right)\frac{l}{V}\dot{\theta}\right]$$

or having regard to (7.62):

$$M = \pi\rho l^2 V^2\left[\left(0.93 - \frac{\omega l}{V}\right)\left(\theta - \frac{\dot{y}}{V}\right) - \frac{l}{V}\dot{\theta}\right]. \tag{7.64}$$

Discussion Based on a Simplified Model

Consider the rudimentary model of a suspension bridge consisting of a rectangular plate of great length, and of width of $2l$, suspended at its edges by identical springs of stiffness $\frac{K}{2}$ per unit length, and which is horizontal in its position of rest.

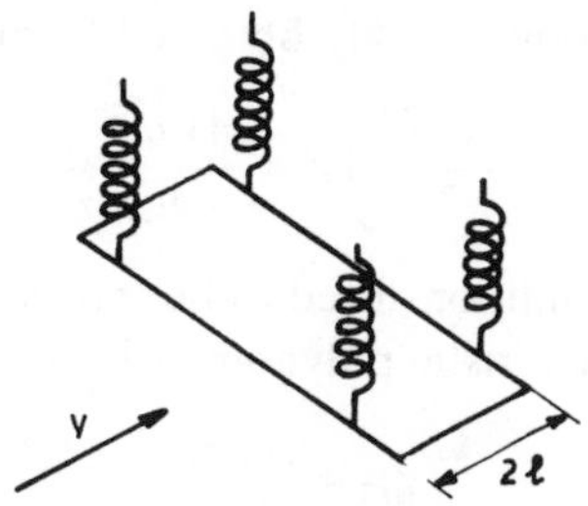

Figure 7.8

We shall suppose it to be acted upon by a cross-wind of velocity V and denote by y the co-ordinate describing the vertical movement and by θ the angle of incidence with respect to the direction of the wind. We shall assume that the effect of the wind on the plate resolves itself into a force F given by (7.59) applied at a distance $\frac{l}{2}$ from the leading edge.

The equations of motion can be written:

$$m\frac{d^2y}{dt^2} + Ky = F \tag{7.65}$$

$$I\frac{d^2\theta}{dt^2} + Kl^2\theta = \frac{Fl}{2} \tag{7.66}$$

where m is the mass, I the moment of inertia with respect to the longitudinal axis of the plate, per unit length, and

$$F = AV^2\theta - AV\frac{\mathrm{d}y}{\mathrm{d}t}, \tag{7.67}$$

where $A = \rho l \dfrac{\partial c_x}{\partial \theta}$ is a positive constant.

If the co-ordinate θ is kept constant, the motion is described by:

$$m\frac{\mathrm{d}^2 y}{\mathrm{d}t^2} + AV\frac{\mathrm{d}y}{\mathrm{d}t} + Ky = 0$$

and is stable; on the other hand if y is kept constant, the angle of rotation θ is given by:

$$I\frac{\mathrm{d}^2\theta}{\mathrm{d}t^2} + \left(Kl^2 - \frac{AV^2 l}{2}\right)\theta = 0$$

and for the motion to be stable we must have $V < \sqrt{\dfrac{2Kl}{A}}$.

Suppose now that these variables are coupled and put:

$$I = mr^2, \quad \omega_1^2 = \frac{K}{m}, \quad \omega_2^2 = \frac{Kl^2}{I} = \frac{l^2}{r^2}\omega_1^2.$$

The characteristic equation is easily obtained from (7.65), (7.66), (7.67) and is:

$$s^4 + \frac{AV}{m}s^3 + \left(\omega_1^2 + \omega_2^2 - \frac{AV^2 l}{2I}\right)s^2 + \frac{AV\omega_2^2}{m}s + \omega_1^2\left(\omega_2^2 - \frac{AV^2 l}{2I}\right) = 0.$$

For stability, Routh's conditions need to be satisfied, so that, on the one hand, the coefficients of the characteristic polynomial have to be positive, or

$$\omega_2^2 > \frac{AV^2 l}{2I} \tag{7.68}$$

and on the other hand:

$$\frac{AV}{m}\left(\omega_1^2 + \omega_2^2 - \frac{AV^2 l}{2I}\right) - \frac{AV\omega_2^2}{m} > 0,$$

$$\left(\frac{AV}{m}\right)^2 \omega_2^2\left(\omega_1^2 + \omega_2^2 - \frac{AV^2 l}{2I}\right) - \left(\frac{AV}{m}\right)^2 \omega_2^4 - \left(\frac{AV}{m}\right)^2 \omega_1^2\left(\omega_2^2 - \frac{AV^2 l}{2I}\right) > 0,$$

which reduce to

$$\omega_1^2 > \frac{AV^2 l}{2I} \tag{7.69}$$

$$\omega_1^2 - \omega_2^2 > 0 \tag{7.70}$$

respectively.

There is therefore a condition which limits the velocity V, but there is also a structural condition (7.70) which can be expressed in the form $\frac{l}{r} < 1$; now a condition of this type is unnatural, since for example $r^2 = \frac{l^2}{3}$ in the case of a homogeneous plate. We can transfer these results to the suspension bridge case by taking for ω_1, ω_2 the values $(\omega_f)_n$, $(\omega_T)_n$ of the flexural and torsional eigenfrequencies in any vibration mode.

However in any traditional design of these bridges, one always finds that $(\omega_T)_n > (\omega_f)_n$ because of the way the mass is distributed, which means that the condition (7.70) can never be satisfied. It may well be however that this situation might be improved if the effects of damping, which are particularly manifest in torsion, were taken into account.

A More Realistic Approach

It seems appropriate to take up the previous analysis again using the expressions (7.63), (7.64) to represent the aerodynamic stresses. The equations of motion with the variables θ, $\varphi = \frac{y}{l}$ are:

$$m(\varphi'' + \omega_1^2 \varphi) = \frac{F}{l}$$

$$I(\theta'' + \omega_2^2 \theta) = M \tag{7.71}$$

where ω_1, ω_2 are the flexural and torsional eigenfrequencies; m and I, the mass and moment of inertia with respect to the longitudinal axis, in each case of a section of unit length of the bridge.

If we write, in the expressions (7.63), (7.64):

$$m_0 = 2\pi\rho l^2\left(0.93 - \frac{\omega l}{V}\right), \quad m_1 = 2\pi\rho l^2, \tag{7.72}$$

we have for $\frac{\omega l}{V} \sim 0.2$:

$$m_0 \sim 4.5\rho l^2, \quad m_1 \sim 6.3\,\rho l^2$$

and

$$\frac{m_0}{m} \sim \frac{m_1}{m} \sim \frac{1}{50}. \tag{7.73}$$

With

$$I = mr^2, \frac{V}{l} = \omega_0,$$

we can write (7.71):

$$\varphi'' + \frac{m_0}{m}\omega_0\varphi' + \omega_1^2\varphi - \frac{m_0}{m}\omega_0^2\theta = 0.$$

$$\theta'' + \frac{m_1 l^2}{2mr^2}\omega_0\theta' + \left(\omega_2^2 - \frac{m_0 l^2}{2mr^2}\omega_0^2\right)\theta + \frac{m_0 l^2}{2mr^2}\omega_0\varphi' = 0,$$

from which we deduce with

$$\frac{\omega_1^2}{\omega_0^2} = \Omega_1^2, \quad \frac{\omega_2^2}{\omega_0^2} = \Omega_2^2, \quad \frac{l^2}{2r^2} = \gamma,$$

$\frac{m_0}{m} = \varepsilon_0$, $\frac{m_1}{m} = \varepsilon_1$, and $s = x\omega_0$ as the characteristic variable, the characteristic equation:

(7.74)
$$x^4 + (\varepsilon_0 + \varepsilon_1\gamma)x^3 + (\Omega_1^2 + \Omega_2^2 + \varepsilon_0\varepsilon_1\gamma - \varepsilon_0\gamma)x^2 + (\varepsilon_0\Omega_2^2 + \varepsilon_1\gamma\Omega_1^2)x + \Omega_1^2(\Omega_2^2 - \varepsilon_0\gamma) = 0$$

for which we now have to write down Routh's conditions.

Let us first give a few numerical details, taking as our example the Tacoma bridge, for which we have $l = 6\,\text{m}$, $f = 70\,\text{m}$, $\gamma = \frac{l^2}{2r^2} \sim 1$, a wind velocity $V \sim 18\,\text{m/s}$, $g = 10\,\text{m/s}^2$.

We know that $\omega_1 = \sqrt{\frac{g}{2f}}\sigma$ with $\sigma = \pi$ or 1.5π or 2π. With $\sigma = \pi$ we obtain a value ω_1 which corresponds approximately to 8 oscillations a minute and we have

$$\Omega_1 = \frac{l\omega_1}{V} = \frac{\pi l}{V}\sqrt{\frac{g}{2f}} \quad \text{ou} \quad \Omega_1^2 \sim 0.08.$$

By (7.55), with $\frac{l^2}{2r^2} = 1$ and L very large we can justify: $\Omega_2^2 \sim 2\Omega_1^2 = 0.16$, so that

$$\Omega_2^2 - \varepsilon_0\gamma \sim 0.16 - \frac{1}{50} \sim 0.14 > 0$$

and it will be seen that all the coefficients of (7.74) are positive.

There remain the two conditions

$$(\varepsilon_0 + \varepsilon_1\gamma)(\Omega_2^2 + \Omega_1^2 + \varepsilon_0\varepsilon_1\gamma - \varepsilon_0\gamma)(\varepsilon_0\Omega_2^2 + \varepsilon_1\gamma\Omega_1^2) - (\varepsilon_0\Omega_2^2 + \varepsilon_1\gamma\Omega_1^2)^2 - (\varepsilon_0 + \varepsilon_1\gamma)^2\Omega_1^2(\Omega_2^2 - \varepsilon_0\gamma) > 0$$

or omitting $\varepsilon_0\varepsilon_1\gamma$ which is small compared with $\varepsilon_0\gamma$:

(7.75)
$$\varepsilon_0\gamma(\Omega_2^2 - \Omega_1^2)[\varepsilon_1(\Omega_2^2 - \Omega_1^2) - \varepsilon_0(\varepsilon_0 + \varepsilon_1\gamma)] > 0$$

and

$$(\varepsilon_0 + \varepsilon_1\gamma)(\Omega_2^2 + \Omega_1^2 + \varepsilon_0\varepsilon_1\gamma - \varepsilon_0\gamma) - (\varepsilon_0\Omega_2^2 + \varepsilon_1\gamma\Omega_1^2) > 0$$

or
$$\varepsilon_1\gamma\Omega_2^2 + \varepsilon_0\Omega_1^2 - \varepsilon_0\gamma(\varepsilon_0 + \varepsilon_1\gamma)(1 - \varepsilon_1) > 0;$$

i.e. since $\gamma \sim 1$, $\varepsilon_0 \sim \varepsilon_1 \sim 1/50$:

$$\Omega_1^2 + \Omega_2^2 - 0.04 > 0$$

which is satisfied because $\Omega_1^2 + \Omega_2^2 \sim 0.24$.

To sum up the vital condition to be retained is the one derived from (7.75)

$$\Omega_2^2 - \Omega_1^2 - \varepsilon_0(1+\gamma) > 0$$

or, more explicitly:

$$\frac{l^2}{V^2}(\omega_T^2 - \omega_f^2) - \frac{m_0}{m}\left(1 + \frac{l^2}{2r^2}\right) > 0$$

which can be written, after (7.72):

$$0.93\left(\frac{V}{\omega l}\right)^2 - \frac{V}{\omega l} < \frac{m}{2\pi\rho l^2}\,\frac{\omega_T^2 - \omega_f^2}{\left(1 + \dfrac{l^2}{2r^2}\right)\omega^2} \tag{7.76}$$

where we assume that ω is the frequency of the mode of oscillation which establishes itself when the wind reaches the critical velocity.

We see that (7.76) is certainly satisfied if

$$0.93\left(\frac{V}{l}\right)^2 < \frac{m}{2\pi\rho l^2}\,\frac{\omega_T^2 - \omega_f^2}{1 + \dfrac{l^2}{2r^2}}$$

where ω does not appear, and we therefore have an approximation by defect to the critical velocity.

Chapter VIII. Synchronisation Theory

Whenever a system capable of entering into a state of natural vibration at a frequency ω_0 is acted upon by a periodic force of frequency ω, which is non-linear but of low amplitude, one generally finds that the system is kept in a state of forced vibration at the frequency ω, that is at a frequency synchronous with that of the excitation. We shall find it convenient to use the customary O-notation after introducing a dimensionless parameter μ which is small compared with 1, so that for example in the situation described above we shall say that the amplitude of the periodic force with the angular frequency ω is $O(\mu)$. If, however, $\frac{\omega-\omega_0}{\omega_0}=O(1)$ the amplitude of the forced vibration is $O(\mu)$, while in the case of resonance $\frac{\omega-\omega_0}{\omega_0}=O(\mu)$ the oscillations have an appreciable amplitude $O(1)$ and instabilities can appear.

A result of an asymptotic nature, and of general applicability, allows us to analyse weakly non-linear systems, in particular in the case of resonance, and to obtain an analytic representation of the synchronised motion with the conditions for stability. The response curve which expresses the amplitude of the forced vibration as a function of the relative frequency difference or detuning $\frac{\omega-\omega_0}{\omega_0}$ provides a graphic illustration of the physically admissible situations because it is relatively easy to distinguish the arcs corresponding to stable solutions from those corresponding to unstable solutions. This approach will be applied to several examples, and will provide, among other things, a means of calculating the conditions favourable to the occurrence of oscillations sustained by friction.

In certain systems a phenomenon of subharmonic synchronisation can be observed – the forced vibration becomes established at a frequency which is an integral submultiple of the excitation frequency. The excitation may be due to an external agent or may, on the contrary, result from some internal element in the system (parametric excitation); in each case the asymptotic theorem enables us to calculate the subharmonic synchronised regimes and to determine their stability.

As is only natural, vibration in machines are usually generated by periodic excitations acting on an elastic structure. It is often essential to take account of the non-linear coupling which can exist between the source of the excitatory forces and the vibrating system. One is therefore led to investigate several classes of strongly non-linear periodic systems, with regular or singular perturbation, for which appropriate methods have been developed which enable the stability or instability of the vibratory regimes which can arise to be investigated. Among the applications may be cited the generalisation of van der Pol's model with amplitude delay effect, the study of a certain type of regulator, and the analysis of stability of rotating machinery on an elastic foundation with damping when an imperfectly balanced shaft is driven by a motor with a steep characteristic.

The synchronisation of the rotation of the shaft of a machine resting on an elastic support and acted upon by alternating vertical forces, with or without resonance, leads to the development of an asymptotic theory involving multi-scale parameters, whose general results we have formulated and applied to the mechanical problem to which it owes its origin.

1. Non-Linear Interactions in Vibrating Systems

In many cases the modelling of vibratory mechanisms by linear differential systems provides a satisfactory approach; however non-linear interactions, even of slight amplitude, may have a considerable effect on the behaviour of systems, and for that reason should not be systematically ignored.

Let us begin by considering the very simple case of the pendulum. It is generally assumed that the oscillations of small amplitude are governed by the equations:

(8.1)
$$\theta'' + \frac{g}{l}\theta = 0, \quad \theta'' = \frac{d^2\theta}{dt^2}$$

where g is the accelaration due to gravity, l the length of the pendulum, θ the angle of deflection, and therefore have a period $T = 2\pi\sqrt{l/g}$, which is independent of the amplitude.

The equation (8.1) is a substitute for the true equation

(8.2)
$$\theta'' + \frac{g}{l}\sin\theta = 0$$

which is non-linear, and whose periodic solutions can be studied without difficulty. The period of oscillations of amplitude θ_0 is:

$$T = \sqrt{\frac{2l}{g}} \cdot \int_{-\theta_0}^{\theta_0} \frac{d\theta}{(\cos\theta - \cos\theta_0)^{1/2}}$$

or, writing k for $\sin\frac{\theta_0}{2}$

$$T = 4\sqrt{\frac{l}{g}} \int_0^l \frac{du}{\sqrt{1-k^2u^2} \cdot \sqrt{1-u^2}}$$

a result which leads to the expansion

$$T = 4\sqrt{\frac{l}{g}}\cdot\left(\int_0^1 \frac{du}{\sqrt{1-u^2}} + \frac{k^2}{2}\int_0^1 \frac{u^2 du}{\sqrt{1-u^2}} + \cdots\right)$$

$$= 2\pi\sqrt{\frac{l}{g}}\left(1 + \frac{k^2}{4} + \cdots\right).$$

The period is a function of the amplitude, but the correction due to the non-linear effect is of the second order in k or θ_0 for small oscillations, which explains the so-called isochronous property of pendulums. The vibratory motion of a system with one degree of freedom described by an equation of the type $x'' + F(x) = 0$ can be discussed with equal facility. Such an equation describes for example the motion of a mass subjected to a restoring force $-F(x)$, depending non-linearly on the displacement x. Non-linear characteristics $x \to F(x)$ appear in most physical systems, even the most simple. For example an arrangement of springs as shown in Figure 8.1 leads to the piece-wise linear characteristics illustrated.

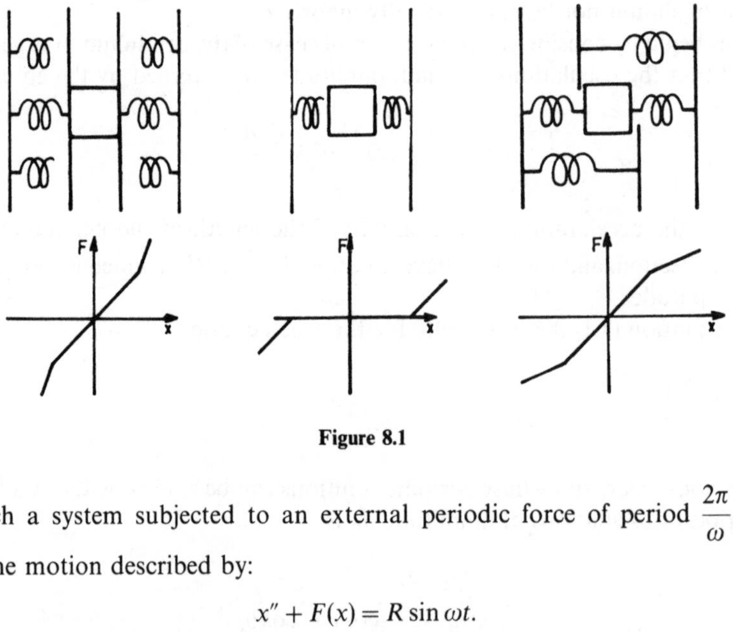

Figure 8.1

Such a system subjected to an external periodic force of period $\frac{2\pi}{\omega}$ would have the motion described by:

$$x'' + F(x) = R \sin \omega t.$$

The equation thus obtained could take account of the forced vibrations but it becomes much more difficult to investigate. It often happens however that the terms representing the non-linear effects can be multiplied by a dimensionless numerical factor μ, small compared with unity, and that one can then develop an asymptotic expansion which remains valid when $\mu \to 0$. Take for example the case of a system with two degrees of freedom with co-ordinates x_1, x_2 governed by the equations:

(8.3)
$$x_1'' + \omega_1^2 x_1 = h_1 \cos\omega t + \mu f(x_1, x_2, x_1', x_2')$$
$$x_2'' + \omega_2^2 x_2 = h_2 \cos\omega t + \mu g(x_1, x_2, x_1', x_2')\,.$$

If ω is sensibly different from ω_1 or ω_2 there is no resonance and (8.3) has a periodic solution of period $\frac{2\pi}{\omega}$ which differs little from the forced vibration obtained when $\mu = 0$. However the situation changes when $\omega \sim \omega_1$ or $\omega \sim \omega_2$: suppose, for example, that

(8.4)
$$\omega_2^2 = \omega^2(1 + \mu\eta)$$

where η is a parameter of fixed value.

As the forced vibration derived from $x_2'' + \omega_2^2 x_2 = h_2 \cos\omega t$ is $x = \frac{h_2}{\mu\eta\omega^2}\cos\omega t$, it seems reasonable to assume that the amplitude h_2 should be reduced in the proportion μ, that is to say to consider instead of (8.3):

$$x_1'' + \omega_1^2 x_1 = h_1 \cos\omega t + \mu f(x_1, x_2, x_1', x_2')$$
$$x_2'' + \omega_2^2 x_2 = \mu h_2 \cos\omega t + \mu g(x_1, x_2, x_1', x_2')$$

which suggests the introduction of the variables y_1, y_2:

$$x_1 = \frac{h_1}{\omega_1^2 - \omega^2}\cos\omega t + y_1, \quad x_2 = \frac{\mu h_2}{\omega_2^2 - \omega^2}\cos\omega t + y_2$$

whence, in view of (8.4):

$$y_1'' + \omega_1^2 y_1 = \mu F$$
$$y_2'' + \omega^2 y_2 = \mu(-\eta\omega^2 y_2 + G).$$

F and G depending on y_1, y_2, y_1', y_2', t, being periodic in t with period $\frac{2\pi}{\omega}$.

We next introduce the variables u_1, v_1, u_2, v_2 defined by:

$$y_1 = u_1$$
$$y_1' = v_1$$
$$y_2 = u_2 \cos\omega t + v_2 \sin\omega t$$
$$y_2' = -\omega u_2 \sin\omega t + \omega v_2 \cos\omega t$$

and by a simple calculation we arrive at the set of differential equations of the first order:

(8.5)
$$\frac{du_1}{dt} = v_1$$
$$\frac{dv_1}{dt} = -\omega_1^2 u_1 + \mu F$$
$$\frac{du_2}{dt} = -\mu\frac{\sin\omega t}{\omega}[-\eta\omega^2(u_2\cos\omega t + v_2\sin\omega t) + G]$$
$$\frac{dv_2}{dt} = \mu\frac{\cos\omega t}{\omega}[-\eta\omega^2(u_2\cos\omega t + v_2\sin\omega t) + G]$$

whose right-hand sides are functions of u_1, v_1, u_2, v_2, t and periodic in t with period $\frac{2\pi}{\omega}$.

More generally one can see the merit of developing the theory of the standard system:

$$\begin{aligned} \frac{dx}{dt} &= Ax + \mu f(x, y, t, \mu) \\ \frac{dy}{dt} &= \mu g(x, y, t, \mu) \end{aligned} \tag{8.6}$$

where x, y are vector variables of dimension n and p, and A is an $n \times n$ matrix with real constant coefficients. We suppose f and g to have values in $\mathbb{R}^n$ and $\mathbb{R}^p$ respectively, to be continuous, t-periodic with period T, and continuously differentiable in x, y, μ for $(x, y, t, \mu) \in \mathscr{U} \times \mathscr{V} \times \mathbb{R} \times \mathscr{I}$ where $\mathscr{U}$ and $\mathscr{V}$ are open subsets of $\mathbb{R}^n$ and $\mathbb{R}^p$ respectively containing 0 and y_0, y_0 being such that $G(0, y_0) = 0$ with

$$G(x, y) = \frac{1}{T}\int_0^T g(x, y, t, 0)\, dt, \quad \mathscr{I} = [0, \mu_0].$$

Then, if we define

$$S = \frac{1}{T}\int_0^T P(t)\, dt, \quad P(t) = \frac{\partial g}{\partial y}(0, y_0, t, 0) \tag{8.7}$$

we can state:

Theorem 1

If S is non-singular, and if $\frac{dx}{dt} = Ax$ has no periodic solution of period T, the system (8.6) has for small enough μ a periodic solution of period T which, as $\mu \to 0$, tends to the solution $x = 0$, $y = y_0$. If the eigenvalues of S and A all have a negative real part, this periodic solution is asymptotically stable for $t \to +\infty$, $\mu > 0$.

Assuming $y_0 = 0$, which is no restriction, we put:

$$g(0, 0, t, 0) = r(t), \quad g_x(0, 0, t, 0) = Q(t), \quad g_y(0, 0, t, 0) = P(t)$$

and write the system (8.6) as

$$\begin{aligned} \frac{dx}{dt} &= Ax + \mu f(x, y, t, \mu) \\ \frac{dy}{dt} &= \mu P(t) y + \mu Q(t) x + \mu r(t) + \mu \delta(x, y, t, \mu) \end{aligned} \tag{8.8}$$

with

$$\delta(x, y, t, \mu) = g(x, y, t, \mu) - g(0, 0, t, 0) - g_x(0, 0, t, 0) \cdot x - g_y(0, 0, t, 0) \cdot y$$

so that

$$|\delta| \leqslant O(\mu) + (|x| + |y|)\varepsilon(|x| + |y| + |\mu|), \quad \lim_{\lambda \downarrow 0} \varepsilon(\lambda) = 0,$$

$$\lim_{\substack{x, x^*, y, y^*, \\ \mu \to 0}} \operatorname*{Sup}_{t} \frac{|\delta(x, y, t, \mu) - \delta(x^*, y^*, t, \mu)|}{|x - x^*| + |y - y^*|} = 0.$$

Since by (8.7) the mean value of $P(t) - S$ is zero, we can define a $p \times p$ matrix $H(t)$, periodic with period T, such that $\frac{dH}{dt} = P(t) - S$.

We now make the change of variable

$$y \to z = (I - \mu H(t))y = Ky$$

in (8.8), and after some calculations arrive at

$$\frac{dz}{dt} = \mu Sz + \mu Qx + \mu r - \mu^2 Hr + \mu^2(SH - HP)K^{-1}z - \mu^2 HQx + \mu K\delta(x, K^{-1}z, t, \mu).$$

It is known [42] that the equation

$$\frac{dz}{dt} = \mu Sz + \mu r(t)$$

with $\int_0^T r(t)\,dt = 0$ has a unique t-periodic solution of period T namely $z_0(t, \mu)$ such that $\lim_{\mu \to 0} \operatorname*{Sup}_{t} |z_0(t, \mu)| = 0$.

Putting $z = z_0(t, \mu) + \zeta$ we can rewrite (8.8) in terms of the variables x, ζ as:

$$\frac{dx}{dt} = Ax + \mu F(x, \zeta, t, \mu)$$

$$\frac{d\zeta}{dt} = \mu S\zeta + \mu Qx + \mu k(t, \mu) + \mu L(x, \zeta, t, \mu)$$

with

$$\lim_{\mu \to 0} \operatorname*{Sup}_{t} |k(t, \mu)| = 0, \quad L(0, 0, t, \mu) = 0$$

and

$$\lim_{\substack{x, x^*, \zeta, \zeta^*, \\ \mu \to 0}} \operatorname*{Sup}_{t} \frac{|L(x, \zeta, t, \mu) - L(x^*, \zeta^*, t, \mu)|}{|x - x^*| + |\zeta - \zeta^*|} = 0.$$

These estimates are essential to establish the validity and convergence of the process of successive approximation which can be described as follows: having obtained the $(m-1)$th periodic approximation of period T $x_{m-1}(t)$, $\zeta_{m-1}(t)$, the mth approximation $x_m(t)$, $\zeta_m(t)$ is obtained by defining it to be the periodic solution of the same period of the linear system

$$\frac{dx_m}{dt} = Ax_m + \mu F(x_{m-1}(t), \zeta_{m-1}(t), t, \mu)$$

$$\frac{d\zeta_m}{dt} = \mu S\zeta_m + \mu Qx_m + \mu k(t, \mu) + \mu L(x_{m-1}(t), \zeta_{m-1}(t), t, \mu).$$

The reference [42] may be consulted for the details.

2. Non-Linear Oscillations of a System with One Degree of Freedom

2.1. Reduction to Standard Form

We shall apply Theorem 1 to the equation:

$$x'' + \omega_1^2 x = \mu f(x, x', \omega t) \tag{8.9}$$

x denoting a real scalar variable, $f(x, x', \omega t)$ a continuous function of the variables $x, x', \omega t$, periodic in t with period $\frac{2\pi}{\omega}$, and continuously differentiable with respect to x, x'. From the physical point of view, the equation (8.9) is representative of a rather wide class of oscillators—in fact it represents a harmonic oscillator disturbed by a non-linear periodic force of frequency ω whose amplitude is of the order of the small factor μ. We are interested in finding out what happens in the neighbourhood of resonance, that is to say when $\omega - \omega_1$ is small, or more generally when we assume that ω differs little from a specified fraction of ω_1:

$$\omega_1 = \omega \frac{m}{N} (1 + \mu\eta)^{1/2}. \tag{8.10}$$

Here m/N denotes an irreducible fraction, η is a given number which may depend on μ, but which we shall for simplicity take to be a constant.

Let us introduce the variables:

$$\varphi = \omega \frac{t}{N} \text{ and} \tag{8.11}$$

$$x = y \cos(z + m\varphi) \tag{8.12}$$

so that

$$\frac{dx}{d\varphi} = -my \sin(z + m\varphi) \tag{8.13}$$

provided we impose the condition

$$\frac{dy}{d\varphi} \cos(z + m\varphi) - y \frac{dz}{d\varphi} \sin(z + m\varphi) = 0. \tag{8.14}$$

We can moreover rewrite (8.9) in the form:

$$\frac{\omega^2}{N^2} \frac{d^2x}{d\varphi^2} + \omega_1^2 x = \mu f\left(x, \frac{\omega}{N} \frac{dx}{d\varphi}, N\varphi\right). \tag{8.15}$$

Differentiating (8.13) with respect to φ and taking account of (8.9), (8.10), (8.12) we get:

$$\frac{dy}{d\varphi} \sin(z + m\varphi) + y \frac{dz}{d\varphi} \cos(z + m\varphi)$$

$$= \mu \left[m\eta y \cos(z + m\varphi) - \frac{N^2}{m\omega^2} f\left(y \cos(z + m\varphi), \frac{-m\omega}{N} y \sin(z + m\varphi), N\varphi\right) \right]$$

whence we deduce, with the help of (8.14):

$$\frac{dy}{d\varphi} = \mu g \sin(z + m\varphi) \tag{8.16}$$

$$\frac{dz}{d\varphi} = \mu \frac{g}{y} \cos(z + m\varphi)$$

with

$$g(y, z, \varphi, \mu) = m\eta y \cos(z + m\varphi) - \frac{N^2}{m\omega^2} f\left(y\cos(z + m\varphi), \frac{-m\omega}{N} y \sin(z + m\varphi), N\varphi \right). \tag{8.17}$$

Clearly g and the right-hand sides of (8.16) are periodic in φ and of period 2π. To every solution 2π-periodic in φ of the system in the standard form (8.16) corresponds by (8.12) and (8.11) a solution (8.9) periodic in t, and of period $N\frac{2\pi}{\omega}$; $N = 1$ corresponds to the normal synchronisation case and $N > 1$ to the subharmonic synchronisation case. Lastly it is clear that a system which has been put into the form (8.16) lends itself to the use of Theorem 1.

2.2. The Associated Functions

We introduce the mean values of the right-hand sides of (8.16) for $\mu = 0$, regarding ω as a function of μ defined by (8.10) with ω_1 as a given constant.

$$Y(y, z) = -\frac{m}{2\pi\omega_1^2} \int_0^{2\pi} f(y\cos(z + m\varphi), - \omega_1 y \sin(z + m\varphi), N\varphi) \cdot \sin(z + m\varphi)\, d\varphi \tag{8.18}$$

$$Z(y, z) = \frac{m\eta}{2} + \frac{\bar{Z}}{y} \tag{8.19}$$

with

$$\bar{Z}(y, z) = -\frac{m}{2\pi\omega_1^2} \int_0^{2\pi} f(y\cos(z + m\varphi), - \omega_1 y \sin(z + m\varphi), N\varphi) \cdot \cos(z + m\varphi)\, d\varphi \tag{8.20}$$

Accordingly the associated system, in the general case, is:

$$\frac{dy}{d\varphi} = \mu Y(y, z), \quad \frac{dz}{d\varphi} = \mu Z(y, z) \tag{8.21}$$

and the synchronisation equations are

$$Y(y, z) = 0, \quad \frac{m\eta}{2} + \frac{\tilde{Z}(y, z)}{y} = 0. \tag{8.22}$$

By Theorem 1, if there is a solution (y_0, z_0) of these equations, if f is continuously differentiable in x, x' in a neighbourhood of $x = y_0 \cos(z_0 + m\varphi)$, $x' = -\omega_1 y_0 \sin(z_0 + m\varphi)$ and if the Jacobian matrix S associated with (8.22) at the point y_0, z_0 is non-singular, then the set of equations (8.16) has a periodic solution of period 2π which tends to (y_0, z_0) as $\mu \to 0$. Furthermore if the eigenvalues of S have a negative real part the periodic solution will be asymptotically stable for $t \to +\infty$, $\mu > 0$.

2.3. Choice of the Numbers m and N

In practice ω and ω_1 are given and the ratio ω_1/ω, which is thus determined, may be approximated as closely as is desired by fractions such as m/N. It might be thought therefore that synchronisation is always possible in an infinite number of ways as long as the system is stable.

However on looking at the reduced equations and the proof of Theorem 1 we perceive that the permitted upper limit for μ, in order that synchronisation should be possible, diminishes as m increases.

2.4. Case of an Autonomous System

Let us again consider the equation (8.9), but this time assuming that f does not depend on t. We can introduce the variables φ, y, z defined by (8.11), (8.12), ω being defined by (8.10), with m/N given but η becoming an unknown.

The equations (8.16) remain valid with

$$g = m\eta y \cos(z + m\varphi) - \frac{N^2}{m\omega^2} f\left(y \cos(z + m\varphi), \frac{-m\omega}{N} y \sin(z + m\varphi)\right)$$

and (8.10) still holds, so that the synchronisation equations can be written:

$$Y(y) = 0 \tag{8.23}$$

$$\frac{m\eta}{2} + \frac{\tilde{Z}(y)}{y} = 0.$$

The first determines y i.e. the mean amplitude of the self-sustained motion, while the second determines η by:

$$\eta = \frac{1}{\pi\omega_1^2 y} \int_0^{2\pi} f(y \cos\psi, -\omega_1 y \sin\psi) \cos\psi \, d\psi. \tag{8.24}$$

This formula, and also incidentally the equation defining y is independent of m and N. We can take $m = N = 1$ and the frequency of the self-induced oscillation will be:

$$\omega = \omega_1 (1 + \mu\eta)^{-1/2} \sim \omega_1 \left(1 - \frac{\mu\eta}{2}\right). \tag{8.25}$$

We could have operated more simply, from (8.16), by taking $m = N = 1$ and noting that g depends only on $z + \varphi = \psi$ and y. Since

$$d\psi = dz + d\varphi = \left(1 + \mu \frac{g}{y} \cos \psi\right) d\varphi$$

we can write:

$$\frac{dy}{d\psi} = \mu \frac{y g \sin \psi}{y + \mu g \cos \psi}$$

and we can apply the synchronisation theorem directly to this equation. We find once more that the admissible values of the amplitude are obtained by solving:

$$\int_0^{2\pi} g \sin \psi \, d\psi = 0 \quad \text{or} \quad Y = 0$$

with

$$Y = -\frac{1}{2\pi\omega_1^2} \int_0^{2\pi} f(y \cos \psi, -\omega_1 y \sin \psi) \cdot \sin \psi \, d\psi. \tag{8.26}$$

The associated equation is $\dfrac{dy}{d\psi} = \mu Y(y)$ and the stability condition can be written as $Y'(y) < 0$, since $\varphi \to +\infty$ as $\psi \to +\infty$ because z is periodic in φ and therefore bounded.

Taking for example the van der Pol equation:

$$x'' + \mu(x^2 - 1)x' + x = 0$$

we have

$$\omega_1 = 1, \quad f = (1 - x^2)x', \quad Y = \frac{y}{2}\left(1 - \frac{y^2}{4}\right)$$

whence $y = 2$, $\dfrac{dY}{dy}(2) < 0$, $\eta = 0$ so that $\omega \sim \omega_1$ to the second order in μ.

3. Synchronisation of a Non-Linear Oscillator Sustained by a Periodic Couple. Response Curve. Stability

Consider the oscillator described by the differential equation

$$x'' + \omega_1^2 x = \mu k(x, x') + \mu H \cos \omega t, \quad H > 0, \quad \mu > 0. \tag{8.27}$$

We can construct without difficulty the associated functions, which for $m = N = 1$ are:

$$Y = -\frac{H}{2\omega_1^2} \sin z + P(y), \quad \frac{\tilde{Z}}{y} = -\frac{H}{2\omega_1^2 y} \cos z + Q(y)$$

with

$$P(y) = -\frac{1}{2\pi\omega_1^2}\int_0^{2\pi} k(y\cos\psi, -\omega_1 y\sin\psi)\sin\psi\, d\psi$$

(8.28)
$$Q(y) = -\frac{1}{2\pi\omega_1^2 y}\int_0^{2\pi} k(y\cos\psi, -\omega_1 y\sin\psi)\cos\psi\, d\psi$$

so that the synchronisation equations are:

(8.29)
$$-\frac{H}{2\omega_1^2}\sin z + P(y) = 0, \quad \frac{\eta}{2} - \frac{H}{2y\omega_1^2}\cos z + Q(y) = 0.$$

Eliminating z we obtain:

(8.30)
$$y^2\left(Q + \frac{\eta}{2}\right)^2 + P^2 = \frac{H^2}{4\omega_1^4}$$

an equation which defines the admissible values of y as a function of the parameter η, which measures the discrepancy between the excitation frequency and the eigenfrequency of the system. Thus for every value of η, i.e. of the frequency $\omega = \omega_1(1+\mu\eta)^{-1/2}$ of the excitatory force, we obtain, by (8.30) the amplitudes of the possible synchronised regimes. The tracing of the response curve is facilitated by the following remarks: Writing

(8.31)
$$\frac{\eta}{2} = -Q(y) \pm \frac{1}{y}\sqrt{\frac{H^2}{4\omega_1^4} - P^2(y)}$$

and observing from (8.28) that $P \to 0$, and Q tends to a finite limit as $y \to 0$, we deduce that for each $y > 0$ satisfying $P^2(y) < \frac{H^2}{4\omega_1^4}$, there are two values of η symmetric with respect to the line $\eta = -2Q(y)$ and that these values become $\pm\infty$ when $y \to 0$.

These indications though somewhat fragmentary allow us to predict response curves of the types illustrated in the following diagrams.

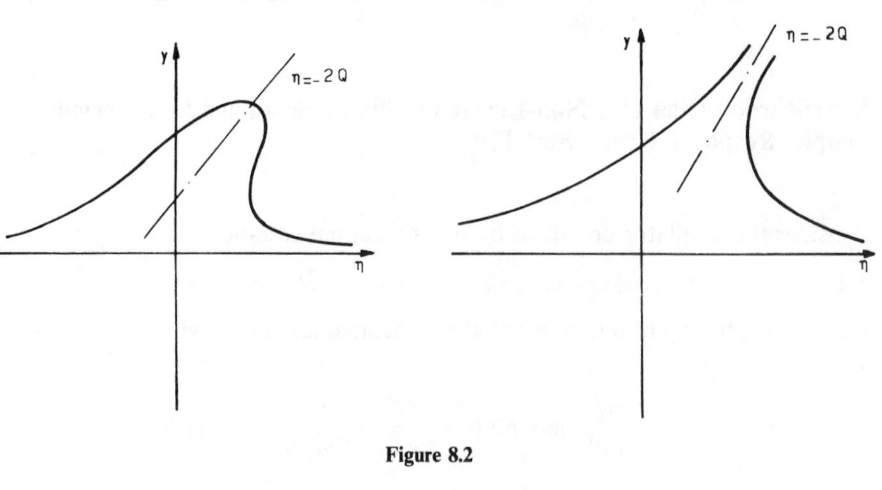

Figure 8.2

Let us now tackle the problem of the stability of the possible synchronised regimes. To do this we study the eigenvalues of the matrix

$$\begin{pmatrix} \dfrac{\partial Y}{\partial y} & \dfrac{\partial Y}{\partial z} \\ \dfrac{\partial}{\partial y}\left(\dfrac{\tilde{Z}}{y}\right) & \dfrac{\partial}{\partial z}\left(\dfrac{\tilde{Z}}{y}\right) \end{pmatrix} = \begin{pmatrix} P' & -\dfrac{H}{2\omega_1^2}\cos z \\ Q' + \dfrac{H}{2\omega_1^2 y^2}\cos z & \dfrac{H}{2\omega_1^2 y}\sin z \end{pmatrix}$$

with
$$P' = \frac{\mathrm{d}P}{\mathrm{d}y}, \quad Q' = \frac{\mathrm{d}Q}{\mathrm{d}y}.$$

The equation defining these eigenvalues is

$$\lambda^2 - \left(P' + \frac{H}{2\omega_1^2 y}\sin z\right)\lambda + \frac{HP'\sin z}{2\omega_1^2 y} + \frac{H\cos z}{2\omega_1^2}\left(Q' + \frac{H}{2\omega_1^2 y^2}\cos z\right) = 0$$

from which we obtain the conditions for asymptotic stability:

$$P' + \frac{H}{2\omega_1^2 y}\sin z < 0 \tag{8.32}$$

$$\frac{P'\sin z}{y} + \cos z \cdot \left(Q' + \frac{H}{2\omega_1^2 y^2}\cos z\right) > 0. \tag{8.33}$$

The condition (8.32) can be written, having regard to (8.29) and $y > 0$ (we can always confine our attention to positive amplitudes):

$$(yP)' < 0. \tag{8.34}$$

We transform (8.33) on the basis of analogous considerations:

$$PP' + y\left(Q + \frac{\eta}{2}\right)\cdot\left[y\left(Q + \frac{\eta}{2}\right)\right]' > 0. \tag{8.35}$$

On the other hand, by differentiating (8.30) we have:

$$2y\left(Q + \frac{\eta}{2}\right)\left[y\left(Q + \frac{\eta}{2}\right)\right]' + 2PP' + y^2\left(Q + \frac{\eta}{2}\right)\frac{\mathrm{d}\eta}{\mathrm{d}y} = 0$$

which allows us to interpret (8.35) as

$$\left(Q + \frac{\eta}{2}\right)\frac{\mathrm{d}\eta}{\mathrm{d}y} < 0 \tag{8.36}$$

where $\dfrac{\mathrm{d}\eta}{\mathrm{d}y}$ is the differential coefficient of the response curve. The conditions (8.34) and (8.36) are in general easy to interpret.

Let us take for example $k = -a\omega_1 x' + c\omega_1^2 x^3$, $a > 0$ (Duffing's equation). We find $P = -a(y/2)$, $Q = -(3/8)cy^2$ and the equation of the response curve is:

$$\frac{\eta}{2} = \frac{3}{8}cy^2 \pm \frac{1}{2y}\sqrt{\frac{H^2}{\omega_1^4} - a^2 y^2}.$$

We note that the condition $(yP)' < 0$ is always satisfied. The condition (8.36) shows that the arcs MAB, DCN, correspond to stable synchronised regimes.

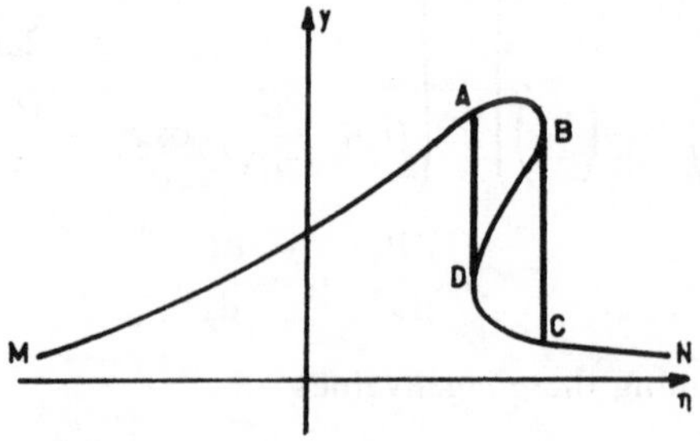

Figure 8.3

This diagram gives us a simple explanation of the physical phenomenon of the jump in amplitude observed when the excitation frequency is varied, sufficiently slowly however to allow us to assume that the steady state is reached at each point in time. If η decreases from $+\infty$ to $-\infty$ the image point describes the arc NCD, jumps at A, and then describes the arc AM. If, on the contrary, η increases from $-\infty$ to $+\infty$, the image point describes the arc MAB, jumps at C and then describes the arc CN.

4. Oscillations Sustained by Friction

Consider a mass m suspended from the lower end of a fixed vertical rod and in contact with a disc which is rotating uniformly with a tangential velocity v_0. We denote by x the horizontal displacement of the mass m, assumed to be in permanent contact with the disc.

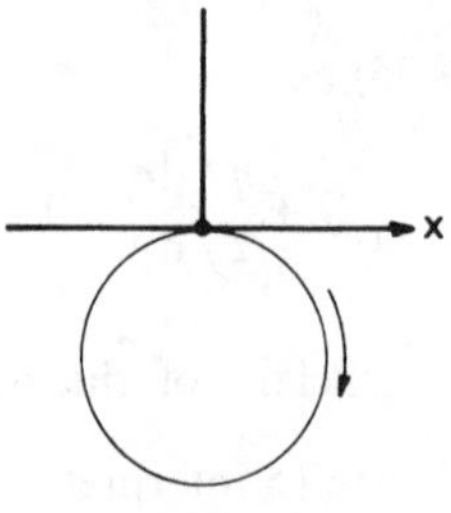

Figure 8.4

The relative velocity of the disc with respect to the mass is $v = v_0 - x'$ and the horizontal force which it exerts on the mass is a function of v which can be represented by:

$$R = R_0 - \mu R_*\left(\frac{v}{v_*} - \frac{v^3}{3v_*^3}\right), \qquad v > 0$$

$$R = -R_0 - \mu R_*\left(\frac{v}{v_*} - \frac{v^3}{3v_*^3}\right), \quad v < 0$$

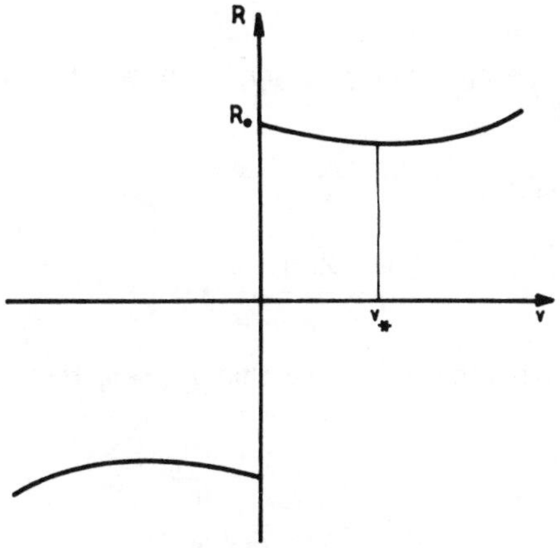

Figure 8.5

where μ is a small positive parameter, R_0, R_* are positive constants, and the minimum of R for $v > 0$ is attained when $v = v_* > 0$.

Assuming that $v_0 - x'$ remains positive during the course of the motion (we shall come back to this later) and denoting the restoring force exerted by the rod by

$$-cx - \mu\frac{R_*}{v_*}(1 - \alpha^2)x'$$

where c represents the rigidity coefficient and the second term represents an $O(\mu)$ damping effect, which depends on the parameter α, $0 < \alpha < 1$, we can write down the equation of motion:

$$mx'' + \mu\frac{R_*}{v_*}(1 - \alpha^2)x' + cx = R$$

or, with the variable:

$$\xi = x - \frac{R_0}{c} + \mu\frac{R_* v_0}{c v_*}\left(1 - \frac{v_0^2}{3v_*^2}\right),$$

$$m\xi'' - \mu R_*\left[v_*^{-1}\left(\alpha^2 - \frac{v_0^2}{v_*^2}\right)\xi' + \frac{v_0}{v_*^3}\xi'^2 - \frac{\xi'^3}{3v_*^3}\right] + c\xi = 0.$$

Putting:

$$\omega_1^2 = \frac{c}{m},\ r_1 = \frac{R_*}{mv_*}\left(\alpha^2 - \frac{v_0^2}{v_*^2}\right),\ r_2 = \frac{R_* v_0}{mv_*^3},\ r_3 = \frac{R_*}{3mv_*^3},$$

we get:

$$\xi'' + \omega_1^2 \xi = \mu(r_1 \xi' + r_2 \xi'^2 - r_3 \xi'^3)$$

an 'autonomous' equation to which we can apply the equations (8.23) or (8.24) and (8.26) which lead to:

$$Y(y) = -\frac{1}{2\pi\omega_1^2} \cdot \int_0^{2\pi} (-r_1\omega_1 y \sin\psi + r_2\omega_1^2 y^2 \sin^2\psi + r_3\omega_1^3 y^3 \sin^3\psi) \sin\psi \, d\psi$$

$$= \frac{y}{2\omega_1}(r_1 - \tfrac{3}{4} r_3 \omega_1^2 y^2) = 0$$

$$\frac{\eta}{2} + \frac{\tilde{Z}(y)}{y} = \frac{\eta}{2} = 0.$$

Thus, with the proviso that $r_1 > 0$, i.e. that $\alpha v_* > v_0$ there exists a synchronised regime of amplitude

$$y_0 = \frac{2}{\omega_1\sqrt{3}}\sqrt{\frac{r_1}{r_3}},$$

with $\eta = 0$. However we need to make sure that $x' < v_0$ throughout the motion, or in other words that $\omega_1 y_0 < v_0$, which requires that $\frac{2}{\sqrt{5}} av_* < v_0$ so that the tangential velocity of the disc has to satisfy the condition:

$$\frac{2}{\sqrt{5}} \alpha v_* < v_0 < \alpha v_*.$$

The synchronised oscillatory motion will then be able to establish itself and it will be stable because $Y'(y_0) < 0$; its period will be equal to $\frac{2\pi}{\omega_1}$, to the second order in μ. A model of this kind can account for the way in which the vibrations of the strings of a musical instrument are sustained by the player's bow.

5. Parametric Excitation of a Non-Linear System

Consider the second-order differential equation

$$x'' + \mu a\omega_1 x' + \omega_1^2(1 - \mu h \cos\omega t)x - \mu c\omega_1^2 x^3 = 0 \tag{8.37}$$

with a, h, c positive constants, which represents the motion of a system having one degree of freedom, subjected to a damping force of viscous type and an elastic restoring force whose linear component is of a strength modulated by the term

$-\mu h \cos \omega t$. We shall assume that we can represent ω by $\omega_1 = \omega \frac{m}{N}(1+\mu\eta)^{1/2}$ with $m=1$, $N=2$.

It is clear that $x=0$ is a solution of (8.37), but as we shall show later, it may happen that this solution is unstable and that the system begins to oscillate spontaneously. As the modulation term in the expression for the stiffness (i.e. the expression representing the elastic restoring force) is responsible for the excitation, we shall say that there is parametric excitation.

A simple calculation leads to the associated functions and the synchronisation equations:

(8.38)
$$Y = -\frac{ay}{2} - \frac{hy}{4}\sin 2z = 0$$
$$Z = \frac{\eta}{2} - \frac{3}{8}cy^2 - \frac{h}{4}\cos 2z = 0$$

and hence by elimination of z, to the response curve equation:

(8.39)
$$\eta = \tfrac{3}{4}cy^2 \pm \sqrt{\frac{h^2}{4} - a^2}$$

which makes it clear that the condition $h > 2a$ has to be satisfied before there can be any parametric excitation. The response curve consists of two parabolic arcs derived from one another by a translation parallel to the η-axis of amplitude $2\sqrt{\frac{h^2}{4} - a^2}$.

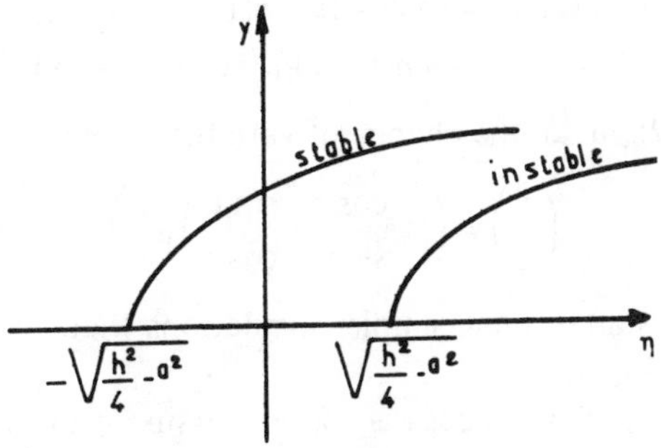

Figure 8.6

Let us now discuss the stability of the various possible synchronised oscillatory motions. The equation which defines the eigenvalues λ of the stability matrix is:

$$\lambda^2 + \left(\frac{a}{2} - \frac{h}{4}\sin 2z\right)\lambda - \left(\frac{a}{2} + \frac{h}{4}\sin 2z\right)\frac{h}{2}\sin 2z - \frac{3h}{8}cy^2\cos 2z = 0$$

from which we deduce the stability conditions:

(8.40)
$$a - \frac{h}{2}\sin 2z > 0$$

(8.41) $$(2a + h\sin 2z)\sin 2z + 3cy^2\cos 2z < 0.$$

Having regard to (8.38) the inequality (8.40) becomes simply $a > 0$ and is satisfied, whereas (8.41) can be written $\cos 2z < 0$, or:

$$\eta < \frac{3c}{4} y^2.$$

This conclusion shows that one of the two parabolic arcs, namely the one on the left in Figure 8.6, corresponds to stable motions, while that on the right corresponds to unstable ones.

However the question of whether the equilibrium configurations $y = 0$ are stable or not remains open, and so to complete this analysis, we shall approach the problem in a slightly different way.

We return to the equation (8.37) with $\omega_1 = \frac{\omega}{2}(1 + \mu\eta)^{1/2}$ and substitute for t the variable τ defined by $\omega t = 2\tau$, so that we can write

$$\frac{d^2x}{d\tau^2} + \mu a(1 + \mu\eta)^{1/2}\frac{dx}{d\tau} + (1 + \mu\eta)(1 - \mu h\cos 2\tau)x - \mu c(1 + \mu\eta)x^3 = 0$$

or

(8.42) $$\frac{dx}{d\tau} = y \qquad \frac{dy}{d\tau} = -x + \mu k(x, y, \tau, \mu)$$

(8.43) $$k(x, y, \tau, \mu) = hx\cos 2\tau - \eta(1 - \mu h\cos 2\tau)x - a(1 + \mu\eta)^{1/2}y + c(1 + \mu\eta)x^3.$$

We now, in (8.42), make the change of variable defined by:

$$\begin{pmatrix} x \\ y \end{pmatrix} = \begin{pmatrix} \cos\tau & \sin\tau \\ -\sin\tau & \cos\tau \end{pmatrix}\begin{pmatrix} u \\ v \end{pmatrix}$$

from which it follows after a few simple calculations that:

(8.44) $$\frac{du}{d\tau} = -\mu\sin\tau\cdot k(u\cos\tau + v\sin\tau, -u\sin\tau + v\cos\tau, \tau, \mu)$$
$$\frac{dv}{d\tau} = \mu\cos\tau\cdot k(u\cos\tau + v\sin\tau, -u\sin\tau + v\cos\tau, \tau, \mu)$$

which is a set of equations in standard form for which the associated equations are:

(8.45) $$-2au + (h + 2\eta)v - \frac{3c}{2}v(u^2 + v^2) = 0$$
$$(h - 2\eta)u - 2av + \frac{3c}{2}u(u^2 + v^2) = 0$$

which admit of the solution $u = v = 0$. The equation which defines the associated eigenvalues is:

$$\lambda^2 + 4a\lambda + 4a^2 - h^2 + 4\eta^2 = 0$$

so that $u = v = 0$ is a stable solution if $\eta^2 > \frac{h^2}{4} - a^2$, and an unstable solution otherwise. We could, starting from (8.45), look for the other possible synchronised oscillatory motions but we should simply rediscover the results already obtained.

When $h < 2a$, the equilibrium position is always stable and there are no periodic solutions.

If $h > 2a$, we can distinguish on the response curve the stable arcs from those which are unstable. If the frequency ω varies slowly in the neighbourhood of $2\omega_1$, we can foretell, by looking in Figure 8.7, when the parametric excitation will give rise to the characteristic stable oscillations, starting from the rest position.

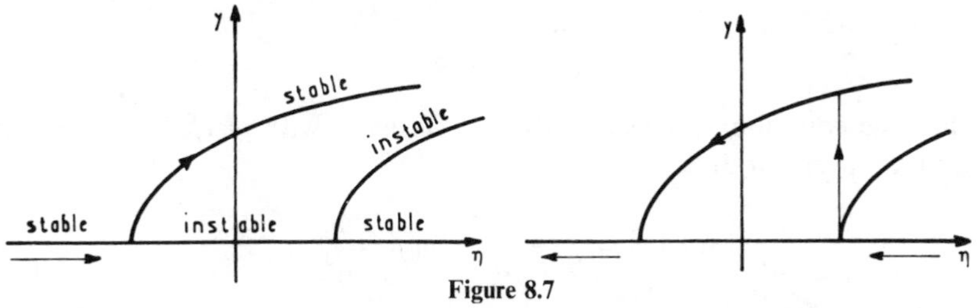

Figure 8.7

It is interesting to repeat the preceding analysis in the case $N = 1$, i.e. when $\omega_1 = \omega(1 + \mu\eta)^{1/2}$. We can in fact show that the equilibrium position $x = 0$ is always stable in this case so that a periodic motion due to parametric excitation is never observed.

We again return to the equation (8.37) with $\omega t = \tau$, $\omega_1 = \omega(1 + \mu\eta)^{1/2}$:

$$\frac{d^2x}{d\tau^2} + x = \mu\left[hx\cos\tau - \eta(1 - \mu h\cos\tau)x - a(1 + \mu\eta)^{1/2}\frac{dx}{d\tau} + c(1 + \mu\eta)x^3\right]$$

and obtain for the synchronisation equations:

$$-2au + 2\eta v - \frac{3c}{2}v(u^2 + v^2) = 0$$

$$-2\eta u - 2av + \frac{3c}{2}u(u^2 + v^2) = 0$$

in which h no longer appears. It is easily verified that $u = v = 0$ is the only real solution of these equations and that the stability conditions are satisfied.

6. Subharmonic Synchronisation

It is an interesting property of a non-linear oscillator that under the effect of an excitatory force of period $\frac{2\pi}{\omega}$, it may under certain conditions respond by oscillating with a period N times as great. In these circumstances we say that there is subharmonic synchronisation of order N and we have already seen in §5 an example of this situation with $N=2$.

The effect is closely connected with non-linearity as can be appreciated from the following simple model [47]. Let us consider the differential equation:

$$x''+g(x)=h\cos\omega t \tag{8.46}$$

and try to find out what the function $g(x)$ would have to be to allow the existence of a solution of the form $x=a\cos\frac{\omega t}{3}$. By substitution in (8.46) we have at once:

$$g(x)=g\left(a\cos\frac{\omega t}{3}\right)=h\cos\omega t+\frac{a\omega^2}{9}\cos\frac{\omega t}{3}=4h\cos^3\frac{\omega t}{3}+\left(\frac{a\omega^2}{9}-3h\right)\cos\frac{\omega t}{3}$$

or

$$g(x)=\frac{4h}{a^3}x^3+\left(\frac{\omega^2}{9}-\frac{3h}{a}\right)x.$$

Thus the equation $x''+\omega_0^2(1+\beta x^2)x=h\cos\omega t$, with ω_0,β,h,ω given, has a solution $x=a\cos(\omega t/3)$ if

$$\omega_0^2=\frac{\omega^2}{9}-\frac{3h}{a},\quad \omega_0^2\beta=\frac{4h}{a^3}.$$

The amplitude a is defined by $a=\dfrac{1}{3h((\omega^2/9)-\omega_0^2)}$ and a compatibility condition is:

$$27h^2\beta\omega_0^2=4\left(\frac{\omega^2}{9}-\omega_0^2\right)^3.$$

In the case of an oscillator which is only slightly non-linear the general theory allows a detailed analysis to be carried out, and we shall do this on the model represented by:

$$x''+\mu a\omega_1 x'+\omega_1^2x=\mu c\omega_1^2x^3+8\omega_1^2G\cos\omega t. \tag{8.47}$$

For $\mu=0$ the forced-vibration regime is described by:

$$x=\frac{8\omega_1^2G}{\omega_1^2-\omega^2}\cos\omega t;$$

we suppose ω_1,a,G to be given positive quantities and assume that ω is represented by:

$$\omega_1=\frac{\omega}{3}(1+\mu\eta)^{1/2} \tag{8.48}$$

with η given. We substitute for x, the unknown u defined by:

$$x=\frac{8\omega_1^2G}{\omega_1^2-\omega^2}(\cos\omega t+u) \tag{8.49}$$

so that (8.47) becomes

(8.50) $$u'' + \omega_1^2 u = \mu\left[c\omega_1^2\left(\frac{8\omega_1^2 G}{\omega_1^2 - \omega^2}\right)^2 (\cos\omega t + u)^3 - a\omega_1(u' - \omega\sin\omega t)\right]$$

for which we shall now look for a periodic solution which has a period three times as large as the period $\frac{2\pi}{\omega}$. To this end we introduce the new variable $\varphi = (\omega t/3)$ and put

$$u = y\cos(z + \varphi),$$

where the amplitude y and the phase z, which are periodic functions of φ of period 2π, have to be determined from the equation (8.50).

We require y and z to satisfy:

$$\frac{dy}{d\varphi}\cos(z+\varphi) - y\frac{dz}{d\varphi}\sin(z+\varphi) = 0$$

so that $u' = -\frac{\omega}{3} y\sin(z+\varphi)$ and by (8.48):

$$u'' + \omega_1^2 u = -\frac{\omega^2}{9}\left[\frac{dy}{d\varphi}\sin(z+\varphi) + y\frac{dz}{d\varphi}\cos(z+\varphi)\right] + \mu\frac{\omega^2}{9}\eta y\cos(z+\varphi)$$

whence by (8.50):

$$\frac{dy}{d\varphi} = \mu g\sin(z+\varphi)$$

$$\frac{dz}{d\varphi} = \mu\frac{g}{y}\cos(z+\varphi)$$

with, for $\mu = 0$:

$$g = \eta y\cos(z+\varphi) - cG^2(\cos 3\varphi + y\cos(z+\varphi))^3 - a(3\sin 3\varphi + y\sin(z+\varphi)).$$

Calculation of the associated functions leads to:

$$Y = -\frac{3cG^2}{8}y^2\sin 3z - \frac{ay}{2}$$

$$Z = \frac{\eta}{2} - cG^2(\tfrac{3}{4} + \tfrac{3}{8}y\cos 3z + \tfrac{3}{8}y^2)$$

and with $h = \frac{3cG^2}{4}$, we obtain the synchronisation points from the equations

(8.51) $$\sin 3z = -\frac{a}{hy}$$

$$\cos 3z = \frac{\eta - h(2 + y^2)}{hy}$$

or by eliminating z from the equation defining the amplitudes:

$$\eta = h(2 + y^2) \pm \sqrt{h^2 y^2 - a^2}\,. \tag{8.52}$$

The stability matrix can be written

$$\begin{pmatrix} -2hy\sin 3z - a & -3hy^2\cos 3z \\ -h(\cos 3z + 2y) & 3hy\sin 3z \end{pmatrix}$$

whose eigenvalues are the solutions of:

$$\lambda^2 + (a - hy\sin 3z)\lambda - (2hy\sin 3z + a)(3hy\sin 3z) - 3h^2y^2\cos 3z(\cos 3z + 2y) = 0$$

or, in view of (8.51), (8.52):

$$\lambda^2 + 2a\lambda + p = 0$$
$$p = -3hy^2\left(h \pm 2\sqrt{h^2y^2 - a^2}\right)$$

the sign $\pm$ being assigned to agree with the one appearing in (8.52).

There is stability if and only if $p > 0$. To interpret this condition, we may suppose $c > 0$, i.e. $h > 0$, the analysis being analogous on the contrary supposition. For $h > 0$, we thus see that the stable motions can correspond only to the points of the branch of (8.52) considered with the negative sign, for the values of y such that

$$h - 2\sqrt{h^2y^2 - a^2} < 0 \quad \text{or} \quad y > \frac{\sqrt{4a^2 + h^2}}{2h},$$

that is to say situated on the arc of the response curve lying on the far side, in the sense of increasing y, of the point where the tangent is parallel to the y-axis.

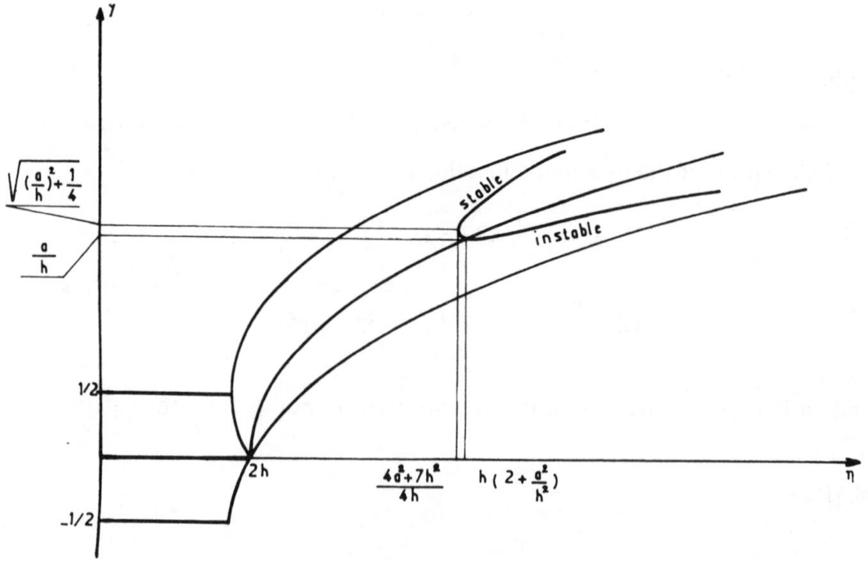

Figure 8.8

To draw the graph of the response curve (8.52) it can be helpful to use the asymptotic parabolas whose equations are

$$\eta = h(2 + y^2) \pm hy.$$

In conclusion for $\eta > \dfrac{4a^2 + 7h^2}{4h}$ there exists only one stable subharmonic regime of order 3. It can be represented by (8.49) with

$$u = y\cos\left(z + \frac{\omega t}{3}\right) \tag{8.53}$$

the asymptotic values of y and z when $\mu \to 0$ being obtained by means of (8.51), and (8.52) restricted to its stable section. By (8.51) three admissible values of the phase z are obtained for each value of the amplitude y. These phases are $2\pi/3$ apart but the expression for u in (8.53) shows that we pass from one to another by a translation of the time-scale of $2\pi/3$, which has no effect on the solution.

7. Non-Linear Excitation of Vibrating Systems. Some Model Equations

The system in the standard form (8.6), which has been the subject-matter of the preceding discussions could be handled conveniently because its integration was easy for $\mu = 0$. In the same spirit of perturbation theory we can put forward a more general model of the form:

$$\frac{\mathrm{d}x}{\mathrm{d}t} = Ax + f(y, t, \mu) + \mu h(x, y, t, \mu) \tag{8.54}$$

$$\frac{\mathrm{d}y}{\mathrm{d}t} = \mu g(x, y, t, \mu)$$

with x, y having values in $\mathbb{R}^n, \mathbb{R}^p$; f, g, h periodic in t and of period T, and A a matrix with constant entries.

We can reduce to the form (8.54) the study of the motion of a rotating machine, whose configuration is described by a vector variable $x \in \mathbb{R}^m$ as regards the supporting structure, and an angular variable φ as regards the rotor, leading to the equations:

$$\frac{\mathrm{d}x}{\mathrm{d}t} = Ax + f\left(\frac{\mathrm{d}\varphi}{\mathrm{d}t}, \varphi, t, \mu\right) + \mu h\left(x, \frac{\mathrm{d}\varphi}{\mathrm{d}t}, \varphi, t, \mu\right)$$

$$\frac{\mathrm{d}^2\varphi}{\mathrm{d}t^2} = \mu g\left(x, \frac{\mathrm{d}\varphi}{\mathrm{d}t}, \varphi, t, \mu\right)$$

f, g, h being periodic in φ of period 2π, and in t of period T.

Putting $\varphi = \dfrac{2\pi t}{T} + \psi$, and seeking x, ψ as t-periodic functions of period T, we arrive, after introducing a new variable $\tilde{\omega}$ and the parameter $v = \sqrt{\mu}$ at the set of equations:

$$\frac{dx}{dt} = Ax + f\left(\frac{2\pi}{T} + \nu\tilde{\omega}, \frac{2\pi t}{T} + \psi, t, \nu^2\right) + \nu^2 h\left(x, \frac{2\pi}{T} + \nu\tilde{\omega}, \frac{2\pi t}{T} + \psi, t, \nu^2\right)$$

$$\frac{d\psi}{dt} = \nu\tilde{\omega}$$

$$\frac{d\tilde{\omega}}{dt} = \nu g\left(x, \frac{2\pi}{T} + \nu\tilde{\omega}, \frac{2\pi t}{T} + \psi, t, \nu^2\right)$$

which have the same structure as those of (8.54) since the terms on the right are t-periodic of period T.

The autonomous system

(8.55)
$$\frac{dx}{dt} = Ax + f\left(\frac{d\varphi}{dt}, \varphi\right)$$

$$\frac{d^2\varphi}{dt^2} = \mu g\left(x, \frac{d\varphi}{dt}, \varphi\right)$$

where f and g are φ-periodic of period 2π is not reducible to (8.54); however the study of the vibratory regime associated with it, when μ is a small parameter, may also be developed by a perturbation method, as we shall see later, when we apply the theory to a mechanical regulator.

Lastly the techniques of singular perturbation may prove to be necessary in the analysis of certain systems, for example those described by:

(8.56)
$$\mu\frac{dx}{dt} = A(t)x + h(t) + \mu g(x, y, t)$$

$$\frac{dy}{dt} = By + k(x, t) + \mu l(x, y, t)$$

with $(x, y) \in \mathbb{R}^n \times \mathbb{R}^p$, and where all the given functions are t-periodic of period T.

8. On a Class of Strongly Non-Linear Systems

8.1. Periodic Regimes and Stability

We revert to the system (8.54) for which the functions f, g, h, which are continuous with respect to all their arguments, and periodic in t of period T are assumed to be continuously differentiable twice in x, y, μ in suitably chosen open sets. A is assumed to be a matrix with constant entries such that $\frac{dx}{dt} = Ax$ has no solution periodic in t of period T. For $\mu = 0$ the integration of (8.54) reduces to that of

$$\frac{\mathrm{d}x}{\mathrm{d}t} = Ax + f(y_0, t, 0) \tag{8.57}$$

$$y = y_0$$

with $y_0 \in \mathbb{R}^p$ constant. The first equation (8.57) has an unique periodic solution of period T [42], which can be represented by $x = \xi(t, y_0)$. Let us introduce the new unknowns x_1, y_1 defined by:

$$\begin{aligned} x &= \xi(t, y_0) + \mu x_1 \\ y &= y_0 + \mu y_1 \end{aligned} \tag{8.58}$$

whence by (8.54):

$$\frac{\mathrm{d}x_1}{\mathrm{d}t} = Ax_1 + \mu^{-1}\cdot(f(y_0 + \mu y_1, t, \mu) - f(y_0, t, 0)) + h(\xi + \mu x_1, y_0 + \mu y_1, t, \mu)$$

$$\begin{aligned} \frac{\mathrm{d}y_1}{\mathrm{d}t} = g(\xi, y_0, t, 0) + \mu[g_x(\xi, y_0, t, 0)\cdot x_1 + g_y(\xi, y_0, t, 0)\cdot y_1 \\ + g_\mu(\xi, y_0, t, 0)] + O(\mu^2) \end{aligned}$$

or with

$$\begin{aligned} k(x_1, y_1, t) = g_x(\xi(t, y_0), y_0, t, 0)\cdot x_1 + g_y(\xi(t, y_0), y_0, t, 0)\cdot y_1 \\ + g_\mu(\xi(t, y_0), y_0, t, 0) \end{aligned} \tag{8.59}$$

$$\frac{\mathrm{d}x_1}{\mathrm{d}t} = Ax_1 + f_y(y_0, t, 0)\cdot y_1 + f_\mu(y_0, t, 0) + h(\xi(t, y_0), y_0, t, 0) + O(\mu) \tag{8.60}$$

$$\frac{\mathrm{d}y_1}{\mathrm{d}t} = g(\xi(t, y_0), y_0, t, 0) + \mu k(x_1, y_1, t) + O(\mu^2).$$

Observing that by partial differentiation of (8.57) with respect to y_0 we can write:

$$\frac{\mathrm{d}}{\mathrm{d}t}\left(\frac{\partial \xi}{\partial y_0}\right) = A\frac{\partial \xi}{\partial y_0} + f_y(y_0, t, 0) \tag{8.61}$$

and substituting for x_1, y_1 the variables y_1, z_1 with z_1 defined by:

$$x_1 = \frac{\partial \xi}{\partial y_0}\cdot y_1 + z_1 \tag{8.62}$$

we can write (8.60) in the form:

(8.63)

$$\frac{\mathrm{d}z_1}{\mathrm{d}t} = Az_1 + f_\mu(y_0, t, 0) + h(\xi(t, y_0), y_0, t, 0) - \frac{\partial \xi}{\partial y_0}\cdot g(\xi(t, y_0), y_0, t, 0) + O(\mu)$$

$$\frac{\mathrm{d}y_1}{\mathrm{d}t} = g(\xi(t, y_0), y_0, t, 0) + \mu k\left(\frac{\partial \xi}{\partial y_0}\cdot y_1 + z_1, y_1, t\right) + O(\mu^2).$$

Let $\zeta(t, y_0)$ be the unique periodic solution of period T

(8.64) $$\frac{d\zeta}{dt} = A\zeta + f_\mu(y_0, t, 0) + h(\xi(t, y_0), y_0, t, 0) - \frac{\partial \xi}{\partial y_0} \cdot g(\xi(t, y_0), y_0, t, 0)$$

and let us from now on suppose y_0 to be chosen so that

(8.65) $$\int_0^T g(\xi(t, y_0), y_0, t, 0)\, dt = 0.$$

With y_2, z_2 new variables introduced by:

(8.66) $$y_1 = \int_0^t g(\xi(s, y_0), y_0, s, 0)\, ds + y_2$$

$$z_1 = \zeta(t, y_0) + z_2$$

the condition (8.65) ensures that the property of periodicity persists in changing from (y_1, z_1) to (y_2, z_2) and vice-versa. In view of this we can transform the equations (8.63) into:

$$\frac{dz_2}{dt} = Az_2 + O(\mu)$$

$$\frac{dy_2}{dt} = \mu k\left(\frac{\partial \xi}{\partial y_0} \cdot \left(\int_0^t g(\xi(s, y_0), y_0, s, 0)\, ds + y_2\right) + \zeta(t, y_0) + z_2,\right.$$
$$\left.\int_0^t g(\xi(s, y_0), y_0, s, 0)\, ds + y_2, t\right) + O(\mu^2)$$

which is a periodic system in standard form to which Theorem 1 may be applied.

The synchronisation point (y_2, z_2) is defined by $z_2 = 0$, y_2 the solution of the linear equation

(8.67) $$Sy_2 = b$$

with

(8.68) $$S = \int_0^T \left[g_x(\xi(t, y_0), y_0, t, 0) \cdot \frac{\partial \xi}{\partial y_0} + g_y(\xi(t, y_0), y_0, t, 0)\right] dt$$

$$= \frac{\partial}{\partial y_0} \int_0^T g(\xi(t, y_0), y_0, t, 0)\, dt$$

$$b = -\int_0^T k\left(\frac{\partial \xi}{\partial y_0} \cdot \int_0^t g(\xi(s, y_0), y_0, s, 0)\, ds + \zeta(t, y_0),\right.$$
$$\left.\int_0^t g(\xi(s, y_0), y_0, s, 0)\, ds, t\right) dt.$$

We can therefore enunciate the following:

Theorem 2

If the t-periodic functions f, g, h of period T are continuously differentiable twice with respect to x, y, μ in a neighbourhood of $x = \xi(t, y_0)$, $y = y_0$, $\mu = 0$, where $\xi(t, y_0)$ denotes the t-periodic solution of (8.57) and y_0 satisfies (8.65), if the equation $\frac{dx}{dt} = Ax$ has no periodic solution of period T and if the matrix S defined by (8.68) is invertible, then the set of equations (8.54) has, for small enough μ, a periodic solution of period T whose asymptotic representation, to the first order in μ, can be obtained from the formulae (8.58), (8.62), (8.66) with $z_2 = 0$ and y_2 the solution of (8.67).

Moreover if all the real parts of the eigenvalues of A and S are negative, the periodic solution so obtained is asymptotically stable when $t \to +\infty$ provided $\mu > 0$ is small enough.

8.2. Van der Pol's Equation with Amplitude Delay Effect

Consider the oscillator described by:

$$x'' + \omega^2 x = \mu[(1-y)x]' + \mu p \cos t \tag{8.69}$$

$$ky' + y = x^2 \tag{8.70}$$

where x, y are scalar variables, p, k positive constants and μ a small positive parameter. For $k = 0$ we have the classical van der Pol equation. However supposing $\omega^2 = 1 + \mu\eta$, with η given, we can rewrite (8.69) as:

$$x'' + x = \mu[-\eta x + p\cos t + (1-y)x' - k^{-1}(x^2 - y)x] = \mu h(x, x', y, t) \tag{8.71}$$

which suggests taking for x a representation of the type:

$$x = u\cos(v + t) \tag{8.72}$$

such that

$$u'\cos(v+t) - uv'\sin(v+t) = 0 \tag{8.73}$$

so that

$$x' = -u\sin(v+t) \tag{8.74}$$

and by (8.71)

$$u'\sin(v+t) + uv'\cos(v+t) = -\mu h(u\cos(v+t), -u\sin(v+t), y, t). \tag{8.75}$$

Calculating u', v' from (8.73), (8.75) and remembering (8.70) we can finally write:

$$\begin{aligned} u' &= -\mu h(u\cos(v+t), -u\sin(v+t), y, t)\cdot\sin(v+t) \\ v' &= -\mu u^{-1} h(u\cos(v+t), -u\sin(v+t), y, t)\cdot\cos(v+t) \end{aligned} \tag{8.76}$$

$$ky' + y = u^2 \cos^2 (v + t) \tag{8.77}$$

a system which can be discussed with the help of Theorem 2.

We begin by writing down the periodic solution of period 2π of (8.77) when u and v are constants, namely

$$y(t, u, v) = \frac{u^2}{2}\left(1 + \frac{2k}{4k^2+1}\sin(2v + 2t) + (4k^2+1)^{-1}\cos(2v + 2t)\right). \tag{8.78}$$

The synchronisation equation (8.65) can thus be written in this case:

$$-\int_0^{2\pi} h(u\cos(v+t), -u\sin(v+t), y(t,u,v), t)\cdot\sin(v+t)\,dt = 0 \tag{8.79}$$

$$-\int_0^{2\pi} u^{-1} h(u\cos(v+t), -u\sin(v+t), y(t,u,v), t)\cdot\cos(v+t)\,dt = 0$$

or after elementary calculation:

$$-p\sin v + u - \frac{8k^2+3}{4(4k^2+1)}u^3 = 0 \tag{8.80}$$

$$-pu^{-1}\cos v + \eta + \frac{ku^2}{2(4k^2+1)} = 0.$$

For a given η, every solution (u, v) of (8.80) for which the associated Jacobian matrix is invertible, defines a synchronisation point and an asymptotic representation of the periodic solution can be obtained by inserting this solution in (8.72), (8.74), (8.78).

Let us return to the discussion of (8.80). Putting

$$P(u) = u - \frac{8k^2+3}{4(4k^2+1)}u^3, \quad Q(u) = \frac{ku^2}{2(4k^2+1)} \tag{8.81}$$

and eliminating v we obtain:

$$u^2(\eta + Q(u))^2 + P^2(u) = p^2 \tag{8.82}$$

which defines the response curve, that is to say the amplitude of the synchronised oscillation as a function of the frequency detuning η.

We may also note incidentally that (8.80) would equally well give us the synchronisation point corresponding to non-sustained oscillations, i.e. to the case when $p = 0$. The first equation would give the amplitude of the free oscillations

$$u = 2\sqrt{\frac{4k^2+1}{8k^2+3}} = u_l < 2$$

and the second the value of η, i.e. the frequency:

$$\omega = (1 + \mu\eta)^{1/2} = 1 - \frac{\mu k}{8k^2+3}.$$

Let us now come back to the case of forced oscillations ($p \neq 0$) and note that the stability conditions can be expressed by:

$$(8.83)\qquad (Q+\eta)\frac{d\eta}{du} < 0$$

$$(uP)' < 0$$

with $\frac{d\eta}{du}$ calculated on the response curve.

Now $(uP)' = 2u\left(1 - \frac{8k^2+3}{2(4k^2+1)}u^2\right) < 0$ is equivalent, bearing in mind that $u > 0$, to:

$$u^2 > \frac{u_l^2}{2}.$$

It therefore seems appropriate to introduce the variable $\sigma = (u^2/u_l^2)$ by means of which the response curve is represented by:

$$(8.84)\qquad \eta + \frac{2k}{8k^2+3}\sigma = \pm\sqrt{\frac{q^2}{\sigma} - (1-\sigma)^2}, \quad q^2 = \frac{p^2}{u_l^2}.$$

It is then a matter of determining the arcs on this curve for which we have $\sigma > 1/2$ and at the same time

$$(8.85)\qquad (Q+\eta)\frac{d\eta}{d\sigma} < 0.$$

An elementary discussion leads us to distinguish the three cases:

Case 1: $0 < q^2 < 1/8$

If $\sigma_1 < \sigma_2 < \sigma_3$ are the distinct roots of $q^2 - \sigma(1-\sigma)^2 = 0$, the admissible intervals for σ are

$$0 < \sigma < \sigma_1, \quad \sigma_2 < \sigma < \sigma_3, \quad \text{with} \quad \sigma_1 < 1/3 < 1/2 < \sigma_2 < 1 < \sigma_3.$$

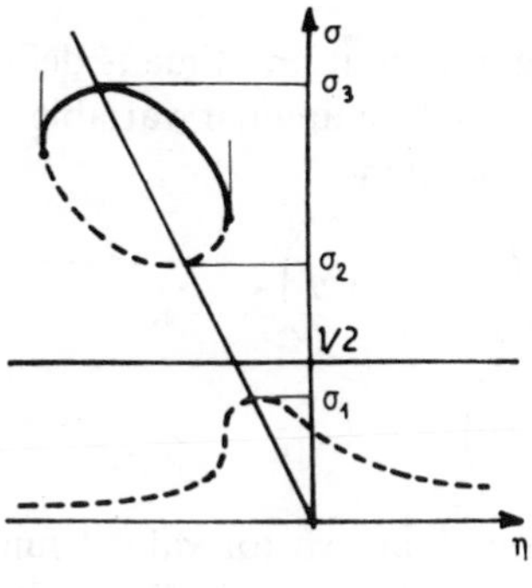

Figure 8.9

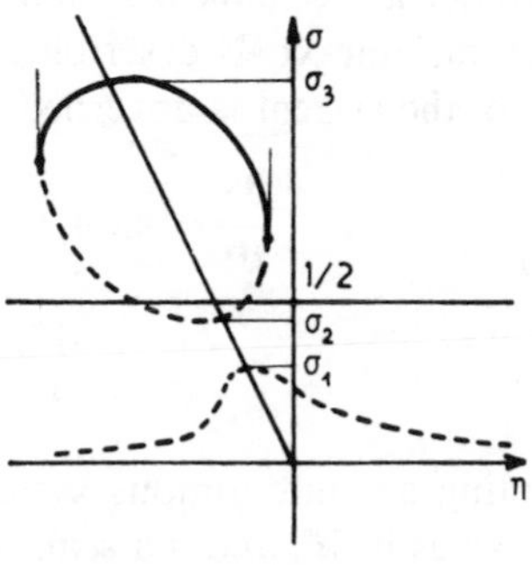

Figure 8.10

Case 2: $1/8 < q^2 < 4/27$

The equation $q^2 - \sigma(1-\sigma)^2 = 0$ again has three distinct roots

$$\sigma_1 < \sigma_2 < \sigma_3, \quad \sigma_1 < 1/3, \quad \sigma_2 < 1/2 < 1 < \sigma_3$$

and the admissible intervals for σ are $0 < \sigma < \sigma_1$, $\sigma_2 < \sigma < \sigma_3$.

Case 3: $4/27 < q^2$

σ_3 is the only real root of $q^2 - \sigma(1-\sigma)^2 = 0$.

The admissible interval is $0 < \sigma < \sigma_3$.

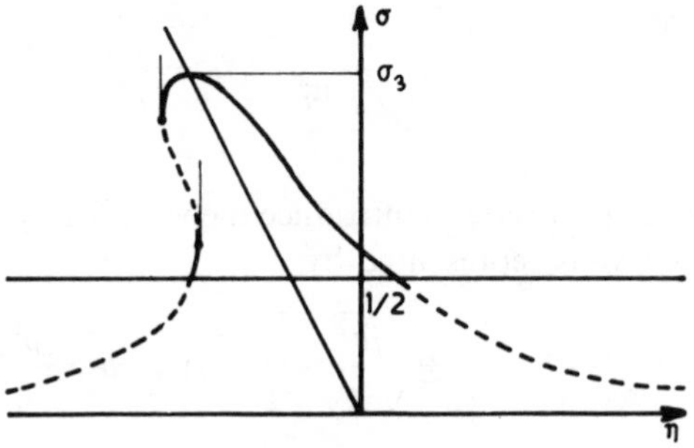

Figure 8.11

To interpret the stability condition (8.85) we note that $\dfrac{d\eta}{du} = \dfrac{2u}{u_l^2}\dfrac{d\eta}{d\sigma}$ and that the sign of $Q + \eta$ on the response curve is the one associated with the radical in (8.84). These remarks enable us immediately to sketch in the arcs of the response curve which correspond to stable regimes. These are shown by a continuous as opposed to a dotted line in the diagram in Figure 8.11.

9. Non-Linear Coupling Between the Excitation Forces and the Elastic Reactions of the Structure on Which They Are Exerted

We consider a mechanical system whose configuration at any time is defined by a vector variable $x \in \mathbb{R}^n$ describing the structure, and an angular variable φ. The motion of the system is governed by the set of equations

$$\frac{dx}{dt} = Ax + f\left(\frac{d\varphi}{dt}, \varphi\right) + \mu h\left(x, \frac{d\varphi}{dt}, \varphi, \mu\right) \tag{8.86}$$

$$\frac{d^2\varphi}{dt^2} = \mu g\left(x, \frac{d\varphi}{dt}, \varphi\right) \tag{8.87}$$

constituting an autonomous system. Here f and h are vector-valued functions taking values in $\mathbb{R}^n$, and g a scalar-valued function with values in $\mathbb{R}$. All three are periodic in φ with period 2π and continuously differentiable twice with respect to

$x, \frac{d\varphi}{dt}, \mu$; while μ is a small positive parameter. A is an $n \times n$ matrix with constant elements, and we shall assume it to be 'stable', i.e. the real parts of its eigenvalues are all negative. If $\mu = 0$, the angular velocity $\frac{d\varphi}{dt}$ has a constant value ω which can be chosen arbitrarily, and the two equations become a single linear differential equation representing the motion of a mechanical system under the action of the periodic forces associated with the term $f(\omega, \omega t)$.

On the other hand if $\mu \neq 0$ there is a non-linear coupling effect and there exist in general one or more periodic motions for which $\frac{d\varphi}{dt} = \omega_0 + O(\mu)$, where ω_0 has a well-determined value. We now propose to clarify this result by a method leading to the calculation of the frequencies ω_0 and the stability criterion.

Following [45], we change the form of expression of the system (8.86), (8.87) by regarding $\omega = \frac{d\varphi}{dt}$ as an unknown. We note that x and ω, henceforth to be considered as functions of φ, satisfy the equations

$$\frac{dx}{d\varphi} = \omega^{-1} A x + \omega^{-1} f(\omega, \varphi) + \mu \omega^{-1} h(x, \omega, \varphi, \mu), \tag{8.88}$$

$$\frac{d\omega}{d\varphi} = \mu \omega^{-1} g(x, \omega, \varphi), \tag{8.89}$$

equations for which it is natural to seek solutions x, ω periodic in φ and of period 2π. If any such solutions exist, the representation of φ as a function of t will be obtainable from $dt = \frac{d\varphi}{\omega(\varphi)}$, provided that $\omega(\varphi)$ never vanishes, a hypothesis whose truth will be verified later. The solution expressed as a function of t will be periodic of period $T = \int_0^{2\pi} \frac{d\varphi}{\omega(\varphi)}$.

Reverting to (8.88), (8.89), it is clear that when $\mu = 0$ we have $\omega = \omega_0$ an arbitrary constant and that we can define $x = \xi(\varphi, \omega_0)$ to be the unique periodic solution of period 2π of:

$$\frac{d\xi}{d\varphi} = \omega_0^{-1} A \xi + \omega_0^{-1} f(\omega_0, \varphi). \tag{8.90}$$

To discuss the case $\mu \neq 0$, we introduce the unknowns x_1, ω_1 defined by:

$$\begin{aligned} x &= \xi(\varphi, \omega_0) + \mu x_1 \\ \omega &= \omega_0 + \mu \omega_1 \end{aligned} \tag{8.91}$$

where ω_0 is a parameter which will be defined precisely later on. A simple calculation then leads from (8.88), (8.89) to:

$$\begin{aligned} \frac{dx_1}{d\varphi} = {} & \omega_0^{-1} A x_1 - \omega_1 \omega_0^{-2} A \xi - \omega_1 \omega_0^{-2} f(\omega_0, \varphi) \\ & + \omega_0^{-1} f_\omega(\omega_0, \varphi) \cdot \omega_1 + \omega_0^{-1} h(\xi, \omega_0, \varphi, 0) + O(\mu) \end{aligned} \tag{8.92}$$

(8.93)
$$\frac{d\omega_1}{d\varphi} = \omega_0^{-1} g(\xi, \omega_0, \varphi) + \mu[\omega_0^{-1} g_x(\xi, \omega_0, \varphi) \cdot x_1 + \omega_0^{-1} g_\omega(\xi, \omega_0, \varphi) \cdot \omega_1 - \omega_0^{-2} g(\xi, \omega_0, \varphi)\omega_1] + O(\mu^2).$$

Observing that by differentiating both sides of (8.90) with respect to ω_0 we have:

(8.94)
$$\frac{d}{d\varphi}\left(\frac{\partial \xi}{\partial \omega_0}\right) = \omega_0^{-1} A \frac{\partial \xi}{\partial \omega_0} - \omega_0^{-2} A\xi - \omega_0^{-2} f(\omega_0, \varphi) + \omega_0^{-1} f_\omega(\omega_0, \varphi),$$

we introduce a new variable $u_1 \in \mathbb{R}^n$ defined by:

(8.95)
$$x_1 = \omega_1 \frac{\partial \xi}{\partial \omega_0} + u_1$$

so that by (8.92) and (8.94) we arrive at:

$$\frac{du_1}{d\varphi} + \frac{d\omega_1}{d\varphi} \cdot \frac{\partial \xi}{\partial \omega_0} = \omega_0^{-1} A u_1 + \omega_0^{-1} h(\xi, \omega_0, \varphi, 0) + O(\mu)$$

and by (8.93):

(8.96)
$$\frac{du_1}{d\varphi} = \omega_0^{-1} A u_1 - \omega_0^{-1} g(\xi, \omega_0, \varphi) \cdot \frac{\partial \xi}{\partial \omega_0} + \omega_0^{-1} h(\xi, \omega_0, \varphi, 0) + O(\mu).$$

From now on we assume that $\omega_0 \neq 0$ and that it satisfies the condition

(8.97)
$$\int_0^{2\pi} g(\xi(\varphi, \omega_0), \omega_0, \varphi)\, d\varphi = 0$$

so that with the new variables u_2, ω_2 defined by:

(8.98)
$$u_1 = \zeta(\varphi, \omega_0) + u_2$$
$$\omega_1 = \int_0^{\varphi} \omega_0^{-1} g(\xi(\varphi, \omega_0), \omega_0, \varphi)\, d\varphi + \omega_2 = \theta(\varphi, \omega_0) + \omega_2$$

where $\zeta(\varphi, \omega_0)$ is the unique φ-periodic solution of period 2π of:

(8.99)
$$\frac{d\zeta}{d\varphi} = \omega_0^{-1} A\zeta - \omega_0^{-1} g(\xi(\varphi, \omega_0), \omega_0, \varphi) \cdot \frac{\partial \xi}{\partial \omega_0} + \omega_0^{-1} h(\xi(\varphi, \omega_0), \omega_0, \varphi, 0)$$

we are led via (8.93), (8.95), (8.96) to seek the periodic solution of period 2π of the system:

(8.100)
$$\frac{du_2}{d\varphi} = \omega_0^{-1} A u_2 + O(\mu)$$

(8.101)
$$\frac{d\omega_2}{d\varphi} = \mu[\omega_0^{-1} g_x(\xi(\varphi, \omega_0), \omega_0, \varphi) \cdot \left((\theta(\varphi, \omega_0) + \omega_2)\frac{\partial \xi}{\partial \omega_0} + \zeta(\varphi, \omega_0) + u_2\right) + (\omega_0^{-1} g_\omega(\xi(\varphi, \omega_0), \omega_0, \varphi) + \omega_0^{-2} g(\xi(\varphi, \omega_0), \omega_0, \varphi))(\theta(\varphi, \omega_0) + \omega_2)] + O(\mu^2)$$

which appears in standard form, since it is clear that (8.101) can be written in the form:

$$\frac{d\omega_2}{d\varphi} = \mu k(\omega_2, u_2, \varphi, \mu)$$

where k is φ-periodic of period 2π. Since the matrix A is stable we can apply Theorem 1 which leads to our having to solve for ω_2 the equation:

$$\int_0^{2\pi} k(\omega_2, 0, \varphi, 0)\, d\varphi = 0.$$

After removal of the factor ω_0^{-1} this equation can be written:

$$\sigma\omega_2 + a = 0$$

with

$$\sigma = \int_0^{2\pi} \left[g_x(\xi(\varphi, \omega_0), \omega_0, \varphi) \cdot \frac{\partial \xi}{\partial \omega_0} + g_\omega(\xi(\varphi, \omega_0), \omega_0, \varphi) - \omega_0^{-1} g(\xi(\varphi, \omega_0), \omega_0, \varphi) \right] d\varphi \tag{8.102}$$

and

$$a = \int_0^{2\pi} \left[g_x(\xi(\varphi, \omega_0), \omega_0, \varphi) \left(\theta(\varphi, \omega_0) \cdot \frac{\partial \xi}{\partial \omega_0} + \zeta(\varphi, \omega_0) \right) + (g_\omega(\xi(\varphi, \omega_0), \omega_0, \varphi) - \omega_0^{-1} g(\xi(\varphi, \omega_0), \omega_0, \varphi)) \cdot \theta(\varphi, \omega_0) \right] d\varphi. \tag{8.103}$$

In the case where $\sigma \neq 0$ we can ensure that the system (8.100), (8.101) has, for small enough μ, a periodic solution of period 2π represented by:

$$u_2 = O(\mu), \quad \omega_2 = -\frac{a}{\sigma} + O(\mu). \tag{8.104}$$

Furthermore this solution is asymptotically stable for $\omega_0 > 0$, $\mu > 0$ and $\varphi \to +\infty$, if $\sigma < 0$; moreover this conclusion would still hold good for $\omega_0 < 0$, as long as we take into account that as t has to increase, φ must be made to tend to $-\infty$. Bearing in mind the formulae (8.91), (8.95), (8.98), (8.104), the asymptotic representation of the solution can be written:

$$x = \xi(\varphi, \omega_0) + \mu \left[\left(\theta(\varphi, \omega_0) - \frac{a}{\sigma} \right) \frac{\partial \xi}{\partial \omega_0} + \zeta(\varphi, \omega_0) \right] + O(\mu^2), \tag{8.105}$$

$$\omega = \omega_0 + \mu \left(\theta(\varphi, \omega_0) - \frac{a}{\sigma} \right) + O(\mu^2).$$

Finally it is useful to observe that, having regard to (8.97) we can rewrite the expression for σ in the form:

(8.106)
$$\sigma = \frac{d}{d\omega} \int_0^{2\pi} g(\xi(\varphi, \omega), \omega, \varphi) d\varphi \big|_{\omega=\omega_0}$$

and that we can enunciate the following theorem

Theorem 3

On the same regularity assumptions for f, g, h and A as those of Theorem 2, the autonomous system (8.86), (8.87) has, for small enough μ, a solution $x(t)$, t-periodic of period $T = \int_0^{2\pi} \frac{d\varphi}{\omega(\varphi)}$, represented asymptotically by the formulae (8.105) with $t = \int^{\varphi} \frac{d\varphi}{\omega(\varphi)}$, provided that $\omega_0 \neq 0$ and satisfies

(8.107)
$$\int_0^{2\pi} g(\xi(\varphi, \omega_0), \omega_0, \varphi) d\varphi = 0$$

where $\xi(\varphi, \omega_0)$ is the unique φ-periodic solution of period 2π of (8.90) and provided σ, defined by (8.106), is non-zero.

Furthermore there is asymptotic stability of the solution thus obtained for $\mu > 0$ and $t \to +\infty$, if $\sigma < 0$.

Application to Bouasse and Sarda's Regulator

A drum of radius R is fixed to a horizontal shaft. Around the circumference of the drum is wrapped a thread which is held taut by a weight of mass M attached to the vertically hanging free end. The drum is made to revolve under the action of gravity. To regulate the motion the shaft is provided with a crank to which is

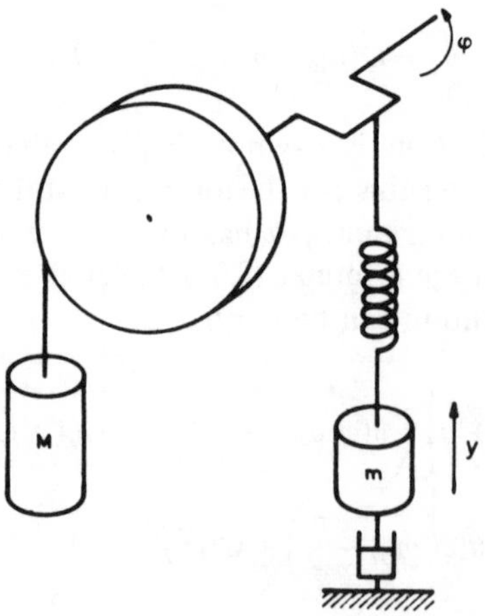

Figure 8.12

attached a device consisting essentially of a weight of mass m suspended from a spring of stiffness k with viscous type damping. We denote the radius of the crank arm by r, the upward vertical displacement of the mass m by y, and the angle between the crank-arm and the horizontal by φ.

Under certain conditions it is found that the shaft can turn with an almost constant angular velocity $\frac{d\varphi}{dt} = \omega$. A heuristic approach to this problem was first presented in [38] and then completed to take account of damping effect on the co-ordinate φ in [16, 35].

However the preceding theory has the advantage of providing a sound basis for calculation and of showing that we have here an example of a phenomenon of asymptotic character whose analysis requires a small parameter condition to be taken into account. The equations of motion can be written

$$m\frac{d^2y}{dt^2} + b\frac{dy}{dt} + ky = kr\sin\varphi - mg$$

$$(J + MR^2)\frac{d^2\varphi}{dt^2} = MgR + k(y - r\sin\varphi)r\cos\varphi - f\frac{d\varphi}{dt}$$

where J is the moment of inertia of the drum, $b\frac{dy}{dt}$, $f\frac{d\varphi}{dt}$ the terms corresponding to the viscous dissipation of energy which hinders the movement of the mass m and the rotation of the drum.

With $p = \sqrt{\frac{k}{m}}$, $2\varepsilon p = \frac{b}{m}$ we can rewrite the equations in the form:

$$\frac{d^2y}{dt^2} + 2\varepsilon p\frac{dy}{dt} + p^2 y = k\frac{r}{m}\sin\varphi - g$$

$$\frac{d^2\varphi}{dt^2} = \mu\left[\frac{g}{R} + \frac{kr}{MR^2}(y - r\sin\varphi)\cos\varphi - \frac{f}{MR^2}\frac{d\varphi}{dt}\right]$$

where $\mu = \frac{MR^2}{J + MR^2}$, and we can apply Theorem 3, if we can assume μ to be a small parameter, i.e. that J, the moment of inertia of the drum is large in comparison with MR^2.

To be able to express the condition (8.107) we first have to calculate the periodic solution $y(\varphi, \omega_0)$, of period 2π, of the equation

$$\omega_0^2\frac{d^2y}{d\varphi^2} + 2\varepsilon p\omega_0\frac{dy}{d\varphi} + p^2 y = \frac{kr}{m}\sin\varphi - g$$

namely

$$y = -\frac{g}{p^2} + \alpha\cos\varphi + \beta\sin\varphi$$

with

$$\alpha = \frac{2p\varepsilon\omega_0 kr}{m[(p^2 - \omega_0^2)^2 + 4p^2\varepsilon^2\omega_0^2]}, \quad \beta = \frac{(p^2 - \omega_0^2)kr}{m[(p^2 - \omega_0^2)^2 + 4p^2\varepsilon^2\omega_0^2]}.$$

Accordingly condition (8.107) can be written:

$$\int_0^{2\pi}\left[MgR + kr\left(\alpha\cos\varphi + \beta\sin\varphi - \frac{g}{p^2} - r\sin\varphi\right)\cos\varphi - f\omega_0\right]d\varphi = 0$$

that is:

$$\frac{MgR}{kr^2} - \frac{f\omega_0}{kr^2} - \frac{\varepsilon\omega_0 p^3}{(p^2-\omega_0^2)^2 + 4p^2\varepsilon^2\omega_0^2} = 0$$

an equation which defines the admissible values of ω_0. In a plane referred to the two axes z, ω_0 we can represent the graphs of:

$$z = \frac{\varepsilon\omega_0 p^3}{(p^2-\omega_0^2)^2 + 4p^2\varepsilon^2\omega_0^2} \quad \text{and} \quad z = \frac{MgR}{kr^2} - \frac{f\omega_0}{kr^2}.$$

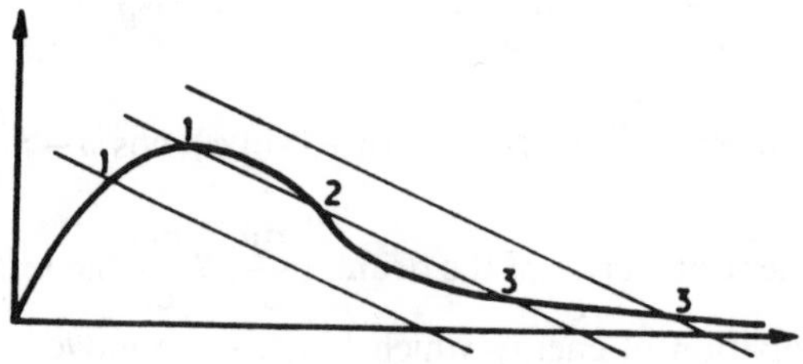

Figure 8.13

It is clear from the diagram that, depending on the particular case, there will be either 1 or 3 points of intersection. The extreme points 1 and 3 correspond to stable regimes.

10. Stability of Rotation of a Machine Mounted on an Elastic Base and Driven by a Motor with a Steep Characteristic Curve

We shall suppose that the horizontal shaft is rigid and that it rotates in bearings integral with the machine which is mounted on an elastic base. We shall assume that the motion can be described by two co-ordinates, φ the angle of rotation of the shaft, and y which measures the vertical displacement of the whole. We shall also suppose the shaft to be imperfectly balanced and that M, the torque of the motor, depends on the speed of rotation in accordance with the law:

$M = M_0 - h\varphi'$, where M_0 and h are positive constants. If we further suppose that the torque of the resisting couple can be represented by $k\varphi'$, where k is a positive constant, and we denote by m_1 the mass representing the want of balance and r its distance from the shaft, m_2 the mass of the machine including shaft, $m = m_1 + m_2$ the total mass, J_1 the moment of inertia of the shaft, $J = J_1 + m_1 r^2$ the moment of inertia of the shaft plus imbalance, and lastly by c the stiffness of the suspension, we can write the equations of motion in the form:

$$\begin{aligned} &my'' - m_1 r(\varphi'' \sin\varphi + \varphi'^2 \cos\varphi) + cy = 0 \\ &J\varphi'' - m_1 r y'' \sin\varphi = M_0 - (h+k)\varphi', \end{aligned} \qquad y'' = \frac{d^2 y}{dt^2} \tag{8.108}$$

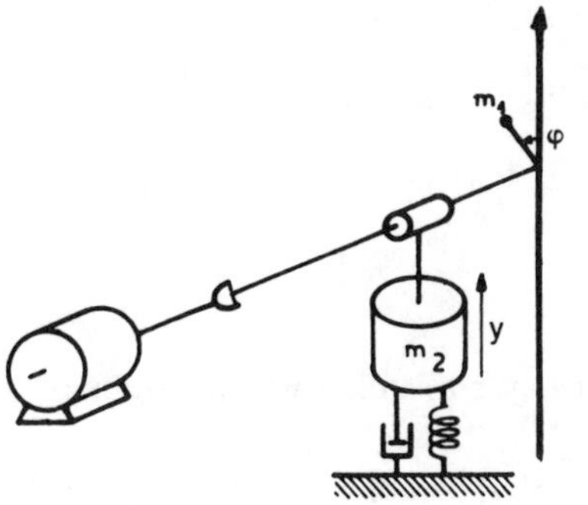

Figure 8.14

An experimental study of this problem, conducted initially by Sommerfeld, reveals that under certain conditions a zone of instability appears. The problem has been discussed on a theoretical basis by several authors [38, 35] but the method used is open to objection on the grounds of its heuristic nature and leads to results that are incorrect. Besides which these authors have not appreciated that we are here in the presence of an example of a singular perturbation, and that it is only by recognising this context that we can hope or claim to build up a satisfactory theory.

We shall, in point of fact, suppose that the driving motor has a sluggish characteristic, or in other words the characteristic curve has a steep slope so that a large change of motor torque M is needed to produce a small change in angular speed φ'. The reduced equations, obtained from (8.108) by adding a damping term for the suspension can be written:

$$\begin{aligned} &y'' - ar\varphi'' \sin\varphi = -\omega_0^2 y - f\omega_0 y' + ar\varphi'^2 \cos\varphi \\ &\varphi'' - \frac{b}{r} y'' \sin\varphi = \mu^{-1}\omega_0(\omega_1 - \varphi') \end{aligned} \tag{8.109}$$

where

$$a = \frac{m_1}{m_1 + m_2} < 1, \quad b = \frac{m_1 r^2}{J_1 + m_1 r^2} < 1, \quad \omega_0^2 = \frac{c}{m}, \quad \omega_1 = \frac{M_0}{h+k} \tag{8.110}$$

$$\mu = \frac{J\omega_0}{h+k}, \quad f > 0, \quad \omega_0 > 0,$$

and the assumption of a steep characteristic means that μ is a small parameter.

We solve the equations (8.109) with respect to y'', φ'':

(8.111)

$$(1-ab\sin^2\varphi)(y''+f\omega_0 y'+\omega_0^2 y)=-ab(f\omega_0 y'+\omega_0^2 y)\sin^2\varphi+ar\varphi'^2\cos\varphi$$
$$+\mu^{-1}ar(\omega_1-\varphi')\omega_0\sin\varphi$$

$$(1-ab\sin^2\varphi)\varphi''=-\frac{b}{r}\sin\varphi\cdot(f\omega_0 y'+\omega_0^2 y)+ab\varphi'^2\sin\varphi\cos\varphi$$
$$+\mu^{-1}\omega_0(\omega_1-\varphi')$$

and setting

$$\varphi'=\omega_1+\mu\omega$$
$$y'=z$$

we henceforth consider ω, y, z as unknown functions of the variable φ so that:

$$y'=\frac{dy}{d\varphi}\cdot(\omega_1+\mu\omega)=z \quad \text{or} \quad \frac{dy}{d\varphi}=\omega_1^{-1}z+O(\mu)$$

and then

$$y''=\frac{dz}{dt}=\frac{dz}{d\varphi}\cdot(\omega_1+\mu\omega)$$

$$\varphi''=\mu\frac{d\omega}{dt}=\mu\frac{d\omega}{d\varphi}\cdot(\omega_1+\mu\omega)$$

and by (8.111):

$$\frac{dz}{d\varphi}=-\frac{\omega_0^2}{\omega_1}y-f\frac{\omega_0}{\omega_1}z+\frac{-ab(f\omega_0 z+\omega_0^2 y)\sin^2\varphi+ar\omega_1^2\cos\varphi-ar\omega_0\omega\sin\varphi}{\omega_1(1-ab\sin^2\varphi)}$$
$$+O(\mu)$$

$$\mu\frac{d\omega}{d\varphi}=\frac{-br^{-1}\sin\varphi\cdot(\omega_0^2 y+f\omega_0 z)+ab\omega_1^2\sin\varphi\cos\varphi-\omega_0\omega}{\omega_1(1-ab\sin^2\varphi)}+O(\mu)$$

It is convenient to substitute for ω:

$$\tilde{\omega}=\omega+br^{-1}\sin\varphi\cdot(\omega_0 y+fz)$$

so that finally we obtain, in terms of $y, z, \tilde{\omega}$

$$\mu\frac{d\tilde{\omega}}{d\varphi}=\frac{ab\omega_1^2\sin\varphi\cos\varphi-\omega_0\tilde{\omega}}{\omega_1(1-ab\sin^2\varphi)}+O(\mu)$$

(8.112)
$$\frac{dy}{d\varphi}=\omega_1^{-1}\cdot z+O(\mu)$$

$$\frac{dz}{d\varphi}=-\frac{\omega_0^2}{\omega_1}y-f\frac{\omega_0}{\omega_1}z+ar\frac{\omega_1^2\cos\varphi-\omega_0\tilde{\omega}\sin\varphi}{\omega_1(1-ab\sin^2\varphi)}+O(\mu).$$

11. Periodic Differential Equations with Singular Perturbation

We have thus arrived at a system of equations whose general structure may be formulated as follows:

$$\text{(8.113)} \qquad \mu\frac{dx}{dt} = A(t)x + h(t) + \mu g(x, y, t, \mu)$$

$$\frac{dy}{dt} = By + k(x, t) + \mu l(x, y, t, \mu)$$

with $(x, y) \in \mathbb{R}^n \times \mathbb{R}^p$, and $A(t)$, $h(t)$, $k(x, t)$, $g(x, y, t, \mu)$, $l(x, y, t, \mu)$ functions with real values depending continuously on their arguments, periodic in t of period T and with k, g, l continuously differentiable in x, y. In addition we shall suppose that the real parts of the eigenvalues of the matrix $A(t)$ are all negative, for every t. Because of the continuity assumption it follows that there is a real $\alpha > 0$ such that $\operatorname{Re}\lambda(t) < -2\alpha$, for every eigenvalue $\lambda(t)$ of the matrix $A(t)$. Lastly we shall suppose that $\frac{dy}{dt} = By$ has no periodic solution of period T.

The equations (8.112) obviously have this structure as is seen by taking $n = 1$, $p = 2$ and changing x, y into $\tilde{\omega}$, $\begin{pmatrix} y \\ z \end{pmatrix}$ and t into φ.

We shall first show that, for small enough μ, the equations (8.113) have a periodic solution of period T and then go on to discuss its stability.

11.1. Study of a Linear System with Singular Perturbation

$$\text{(8.114)} \qquad \mu\frac{dx}{dt} = A(t)x + h(t).$$

We write $X(t, \mu)$ for the solution matrix defined by:

$$\mu\frac{dX}{dt} = A(t)X, \quad X(0, \mu) = I$$

Since the $n \times n$ matrix $A(t)$ is periodic and continuous it is uniformly continuous and recalling that $\lambda(t) < -2\alpha < 0$ for every eigenvalue $\lambda(t)$ of $A(t)$, it is known, by a lemma of Flatto Levinson [42], that there exist positive constants K and μ_0, depending only on $A(t)$, such that:

$$\text{(8.115)} \qquad |X(t, \mu)X^{-1}(s, \mu)| < K\exp\left[-\frac{\alpha}{\mu}(t - s)\right],$$

$$s < t, \quad \mu \in]0, \mu_0].$$

The equation (8.114) has a unique periodic solution which can be represented by:

$$\text{(8.116)} \qquad \xi(t, \mu) = \mu^{-1}\int_{-\infty}^{t} X(t, \mu)X^{-1}(s, \mu)h(s)\,ds.$$

Indeed differentiation with respect to t gives:

$$\frac{d\xi}{dt} = \mu^{-1}h(t) + \mu^{-1} \int_{-\infty}^{t} \mu^{-1}A(t)X(t,\mu)X^{-1}(s,\mu)h(s)\,ds$$

which shows at once that (8.114) is satisfied. Furthermore by Floquet's theory [42]. It is known that $X(t,\mu) = Q(t,\mu)e^{tL(\mu)}$, with $Q(t,\mu)$ an $n \times n$ matrix, t-periodic of period T, and $L(\mu)$ an $n \times n$ matrix not depending on t, so that

$$\xi(t,\mu) = \mu^{-1} \int_{0}^{+\infty} Q(t,\mu)e^{\tau L(\mu)}Q^{-1}(t-\tau,\mu)h(t-\tau)\,d\tau$$

which ensures that $\xi(t,\mu)$ is t-periodic of period T.

Using the bound given by (8.115) we obtain from (8.116):

$$|\xi(t,\mu)| \leqslant \mu^{-1} \int_{-\infty}^{t} K\exp\left[-\frac{\alpha}{\mu}(t-s)\right]\cdot|h(s)|\,ds \leqslant \frac{K}{\alpha}\|h\|, \text{ or}$$

(8.117)
$$\|\xi\| \leqslant \frac{K}{\alpha}\|h\|, \text{ with}$$

$$\|\xi\| = \operatorname*{Sup}_t|\xi(t,\mu)|, \quad \|h\| = \operatorname*{Sup}_t|h(t)|$$

K/α only depending on $A(t)$.

Let us now examine the behaviour of $\xi(t,\mu)$ as $\mu \to 0$.

With $p(s) = A^{-1}(s)h(s)$, s-periodic of period T and

(8.118)
$$\xi_0(t) = \mu^{-1} \int_{-\infty}^{t} X(t,\mu)X^{-1}(s,\mu)A(s)A^{-1}(t)h(t)\,ds$$

which, as will be seen later, does not depend on μ, we can write:

$$\xi(t,\mu) - \xi_0(t) = \mu^{-1} \int_{-\infty}^{t} X(t,\mu)X^{-1}(s,\mu)A(s)(p(s)-p(t))\,ds$$

whence we deduce from (8.115) and the change of variable $s \to \tau$, $t = s + \mu\tau$:

$$|\xi(t,\mu) - \xi_0(t)| < K^* \int_{0}^{+\infty} e^{-\alpha\tau}|p(t-\mu\tau) - p(t)|\,d\tau$$

with
$$\operatorname*{Sup}_s|A(s)| = \frac{K^*}{K}.$$

Since $p(s)$ which is continuous and periodic in s with period T, is ipso facto, uniformly continuous on the real line, it is easily proved that:

$$\|\xi - \xi_0\| = \operatorname*{Sup}_t|\xi(t,\mu) - \xi_0(t)|$$

tends to 0 as $\mu \to 0$, or in other words:

(8.119)
$$\xi(t,\mu) = \xi_0(t) + \varepsilon(t,\mu)$$

$$\lim_{\mu\to 0}\operatorname*{Sup}_t|\varepsilon(t,\mu)| = 0.$$

To evaluate $\xi_0(t)$ defined by (8.118) it suffices to calculate

$$\mu^{-1} \int_{-\infty}^{t} X^{-1}(s, \mu) A(s) \, ds.$$

Now we can verify in the usual way that the solution matrix satisfies

$$\frac{dX^{-1}(s, \mu)}{ds} = -\mu^{-1} X^{-1}(s, \mu) A(s)$$

whence

$$\mu^{-1} \int_{-\infty}^{t} X^{-1}(s, \mu) A(s) \, ds = -\int_{-\infty}^{t} \frac{dX^{-1}}{ds} ds = -X^{-1}(t, \mu)$$

so that by (8.118), we have

$$\xi_0(t) = -A^{-1}(t) h(t). \tag{8.120}$$

Lastly from (8.114), of which $\xi(t, \mu)$ is a periodic solution, and from (8.119), (8.120) it is clear that:

$$\lim_{\mu \to 0} \left\| \mu \frac{d\xi}{dt} \right\| = \lim_{\mu \to 0} \operatorname{Sup}_{t} \left| \mu \frac{d\xi}{dt} \right| = 0. \tag{8.121}$$

11.2. The Non-Linear System

We denote by $x_0(t, \mu)$, $y_0(t, \mu)$ the unique periodic solution of period T of the equations:

$$\mu \frac{dx_0}{dt} = A(t) x_0 + h(t) \tag{8.122}$$

$$\frac{dy_0}{dt} = B y_0 + k(x_0, t)$$

which, as $\mu \to 0$, tends uniformly with respect to t to ξ_0 defined by (8.120) and by ζ_0 the unique periodic solution of:

$$\frac{d\zeta_0}{dt} = B\zeta_0 + k(-A^{-1}(t) h(t), t). \tag{8.123}$$

We can now describe an iterative scheme of successive approximation defined, for $m \geqslant 1$, by:

$$\mu \frac{dx_m}{dt} = A(t) x_m + h(t) + \mu g(x_{m-1}, y_{m-1}, t) \tag{8.124}$$

$$\frac{dy_m}{dt} = B y_m + k(x_m, t) + \mu l(x_{m-1}, y_{m-1}, t),$$

x_m, y_m being t-periodic solutions of period T, subject to suitable regularity conditions; with $\mathscr{U}, \mathscr{V}$ bounded open neighbourhoods of $\xi_0(t)$, $\zeta_0(t)$ in $\mathbb{R}^n, \mathbb{R}^p$

respectively, we assume g, k, l to be continuously differentiable in (x, y) within $\mathscr{U} \times \mathscr{V}$ and we denote by c an upper bound in $\mathscr{U} \times \mathscr{V} \times \mathbb{R}$ of the norms $|g(x, y, t)|$, $|l(x, y, t)|$, $|g_x(x, y, t)|$, $|g_y(x, y, t)|$, $|l_x(x, y, t)|$, $|l_y(x, y, t)|$, $|k_x(x, t)|$[1].

We shall suppose μ to be small enough to ensure that all approximations up to the $(m-1)$th are contained in the neighbourhood $\mathscr{U} \times \mathscr{V}$. From the estimates we shall obtain, it will then follow that the process of constructing approximations can be pursued indefinitely and that the sequence converges to a solution of the problem.

We have seen that the periodic solution of a linear differential equation of the type (8.114) satisfies an estimate

$$\|\xi\| \leqslant K_1 \|h\|, \quad \text{with} \quad K_1 = K/\alpha \tag{8.125}$$

depending only on $A(t)$. We know also that the unique periodic solution of the equation $\frac{dy}{dt} = By + q(t)$, where $q(t)$ is t-periodic of period T, and where the associated homogeneous equation has no periodic solution of period T, satisfies the estimate

$$\|y\| \leqslant K_2 \|q\|, \tag{8.126}$$

the constant K_2 depending only on B.

In view of all this we obtain from (8.122) and (8.124) written with $m = 1$:

$$\mu \frac{d(x_1 - x_0)}{dt} = A(t) \cdot (x_1 - x_0) + \mu g(x_0, y_0, t)$$

$$\frac{d(y_1 - y_0)}{dt} = B(y_1 - y_0) + k(x_1, t) - k(x_0, t) + \mu l(x_0, y_0, t)$$

whence

$$\|x_1 - x_0\| \leqslant \mu K_1 c$$

$$\|y_1 - y_0\| \leqslant K_2[c\|x_1 - x_0\| + \mu c] \leqslant \mu(K_1 K_2 c + K_2)c$$

and

$$\|x_1 - x_0\| + \|y_1 - y_0\| \leqslant \mu(K_1 K_2 c + K_1 + K_2)c$$

and then by forming the difference between the sets of equations (8.124) written for two consecutive values of m

$$\mu \frac{d(x_m - x_{m-1})}{dt} = A(t)(x_m - x_{m-1}) + \mu(g(x_{m-1}, y_{m-1}, t) - g(x_{m-2}, y_{m-2}, t))$$

$$\frac{d(y_m - y_{m-1})}{dt} = B(y_m - y_{m-1}) + k(x_m, t) - k(x_{m-1}, t) + \mu(l(x_{m-1}, y_{m-1}, t) - l(x_{m-2}, y_{m-2}, t))$$

[1] We have here implicitly assumed that g and l do not depend on μ, but this restriction could be lifted without difficulty.

we deduce the estimates

$$\|x_m - x_{m-1}\| \leqslant \mu K_1 c(\|x_{m-1} - x_{m-2}\| + \|y_{m-1} - y_{m-2}\|)$$
$$\|y_m - y_{m-1}\| \leqslant K_2 c(\|x_m - x_{m-1}\| + \mu(\|x_{m-1} - x_{m-2}\| + \|y_{m-1} + y_{m-2}\|))$$

which imply

$$\|x_m - x_{m-1}\| + \|y_m - y_{m-1}\| \leqslant \mu[K_1 K_2 c^2 + K_1 c + K_2 c] \cdot(\|x_{m-1} - x_{m-2}\| + \|y_{m-1} - y_{m-2}\|)$$

or with
$$\varkappa = K_1 K_2 c^2 + (K_1 + K_2)c,$$

$$\|x_m - x_{m-1}\| + \|y_m - y_{m-1}\| \leqslant (\mu\varkappa)^{m-1}(\|x_1 - x_0\| + \|y_1 - y_0\|) \leqslant (\mu\varkappa)^m$$

by using the analogous estimates of lower rank, and finally by repeated use of the triangle inequality:

$$\|x_m - x_0\| + \|y_m - y_0\| \leqslant \frac{\mu\varkappa}{1 - \mu\varkappa}$$

which legitimises the approximation process if μ is chosen small enough, and in any event such that $|\mu\varkappa| < 1$. On this hypotheses the convergence of the process is assured and it has thus been proved that the set of equations (8.113) has, for small enough μ, a periodic solution $x(t, \mu)$, $y(t, \mu)$ of period T, which tends uniformly with respect to t to $\xi_0(t)$, $\zeta_0(t)$ as $\mu \to 0$.

11.3. Stability of the Periodic Solution

The variational linear system associated with (8.113) and its periodic solution can be written [40]:

(8.127)
$$\mu \frac{du}{dt} = A(t)u + \mu P(t, \mu)u + \mu Q(t, \mu)v$$

$$\frac{dv}{dt} = Bv + H(t, \mu)u + \mu R(t, \mu)v$$

where the matrices P, Q, H, R, all of which are periodic in t of period T, have finite limits when $\mu \to 0$.

We shall assume that the matrix B is stable, i.e. the real parts of its eigenvalues are all negative:

(8.128)
$$\operatorname{Re} \lambda_j < -2\beta < 0$$

and that $A(t)$ is a symmetric matrix.

We can always assume that B has been brought into triangular form by a suitable linear transformation on v, with constant coefficients, if necessary complex, which are independent of μ.

$$B = \begin{pmatrix} \lambda_1 & & b_{ij} \\ & \ddots & \\ 0 & & \lambda_p \end{pmatrix}, \quad b_{ij} = 0 \text{ if } i > j$$

and such that

$$\sum_{i<j} b_{ij}\bar{v}_i v_j < \beta |v|^2,$$

with

$$|v|^2 = v_i \bar{v}_i = (v, v)$$

where the bar indicates the complex conjugate. (It should be understood that the assumption of a reduction to triangular form implies that the new co-ordinates which have been introduced for v may assume complex values.)

Scalar multiplication of the equations (8.127) by u and v respectively yields:

$$\mu\left[\left(u, \frac{\mathrm{d}u}{\mathrm{d}t}\right) + \left(\frac{\mathrm{d}u}{\mathrm{d}t}, u\right)\right] = 2\,\mathrm{Re}\,[(u, A(t)u) + \mu(u, P(t,\mu)u) + \mu(u, Q(t,\mu)v)]$$

whence

$$\frac{\mu}{2}\frac{\mathrm{d}|u|^2}{\mathrm{d}t} \leqslant -2\alpha|u|^2 + \mu c|u|^2 + \mu c|u|\cdot|v| \leqslant -2\alpha|u|^2 + 2\mu c|u|^2 + \mu c|v|^2$$

using the facts that, because of the symmetry of $A(t)$:$(u, A(t)u) < -2\alpha|u|^2$, and

$$\frac{1}{2}\frac{\mathrm{d}|v|^2}{\mathrm{d}t} \leqslant -2\beta|v|^2 + \beta|v|^2 + c|v|\,|u| + \mu c|v|^2$$

$$\leqslant -\beta|v|^2 + \frac{c}{\rho}|v|^2 + c\rho|u|^2 + \mu c|v|^2$$

with c a generic constant connected with the bounds of $|P(t,\mu)|$, $|Q(t,\mu)|$, $|H(t,\mu)|$, $|R(t,\mu)|$, and ρ a positive number which, provisionally may have any value. By adding the preceding estimates, after multiplying the first of them by a number $\chi > 0$, we obtain:

$$\frac{1}{2}\frac{\mathrm{d}}{\mathrm{d}t}(\mu\chi|u|^2 + |v|^2) \leqslant -(2\chi\alpha - 2\mu c\chi - c\rho)|u|^2 - \left(\beta - \mu\chi c - \frac{c}{\rho} - \mu c\right)|v|^2.$$

Let us first choose $\rho > 0$ so that $\beta - \frac{c}{\rho} > \frac{3\beta}{4}$, and then $\chi > 0$ so that $2\chi\alpha - c\rho > \alpha$, and lastly $\mu > 0$ small enough so that

$$\alpha - 2\mu\chi c > \frac{\alpha}{2} \quad \text{and} \quad \beta - \frac{c}{\rho} - \mu(\chi c + c) > \frac{\beta}{2}.$$

Under these conditions we obtain

$$\frac{\mathrm{d}}{\mathrm{d}t}(\mu\chi|u|^2 + |v|^2) < -(\alpha|u|^2 + \beta|v|^2) \tag{8.129}$$

from which it follows that u and v tend to 0 as $t \to +\infty$. Thus there is asymptotic stability of the periodic solution for $\mu > 0$ small enough as $t \to +\infty$ and we can state the following theorem:

Theorem 4

In the set of differential equations with singular perturbation described by (8.113) let $A(t)$, $h(t)$ be continuous and t-periodic of period T, and let the real parts of the eigenvalues of $A(t)$ be strictly negative. Let $k(x, t)$ be t-periodic of period T and continuous, and continuously differentiable in x in $\mathscr{U} \times \mathbb{R}$, where $\mathscr{U}$ is a neighbourhood of $\xi_0(t) = -A^{-1}(t)h(t)$ in $\mathbb{R}^n$. With B a constant $p \times p$ matrix such that $\frac{dy}{dt} = By$ has no periodic solution of period T, we introduce $\zeta_0(t)$ defined as the periodic solution of period T of the equation $\frac{dy}{dt} = By + k(\xi_0(t), t)$ and $\mathscr{V}$ a neighbourhood of $\zeta_0(t)$ in $\mathbb{R}^p$. Assuming g, l to be t-periodic, continuous and continuously differentiable in x, y in $\mathscr{U} \times \mathscr{V} \times \mathbb{R}$, then the set of equations (8.113) has, for small enough μ a periodic solution $x(t, \mu)$, $y(t, \mu)$ of period T, which, when $\mu \to 0$ tends, uniformly with respect to t, to $\xi_0(t)$, $\zeta_0(t)$. If $A(t)$ is a symmetric matrix and if the real parts of the eigenvalues of B are all negative, then this solution is, for small enough $\mu > 0$, asymptotically stable as $t \to +\infty$.

12. Application to the Study of the Stability of a Rotating Machine Mounted on an Elastic Suspension and Driven by Motor with a Steep Characteristic Curve

We return to the equations (8.112) to which we can now apply the theory which we have just developed. The matrix A is reduced to a single element

$$A(\varphi) < -\frac{\omega_0}{\omega_1} < 0$$

and the matrix B is stable. The existence and stability of the periodic solution follow from this. With the object of obtaining an asymptotic representation, we first calculate the term equivalent to $\xi_0(t)$, $\zeta_0(t)$ of the general theory. We thus find:

$$\tilde{\omega} = ab\frac{\omega_1^2}{\omega_0}\sin\varphi\cos\varphi \tag{8.130}$$

and to obtain the analogue of ζ_0, we have to calculate the 2π-periodic solution of the linear approximation obtained from the last two equations of (8.112), after taking account of (8.130):

$$\frac{dy}{d\varphi} = \omega_1^{-1} z$$

$$\frac{dz}{d\varphi} = -\frac{\omega_0^2}{\omega_1} y - f\frac{\omega_0}{\omega_1} z + ar\omega_1 \cos\varphi$$

or, in other words:

$$\begin{aligned} y &= C\sin\varphi + D\cos\varphi \\ z &= \omega_1(C\cos\varphi - D\sin\varphi) \end{aligned} \tag{8.131}$$

with

$$C=\frac{arf\omega_0\omega_1}{\delta},\quad D=\frac{ar}{\delta}\cdot(\omega_0^2-\omega_1^2),\tag{8.132}$$

$$\delta=f^2\omega_0^2+\left(\frac{\omega_0^2-\omega_1^2}{\omega_1}\right)^2.$$

Remembering that

$$\varphi'=\omega_1+\mu\omega,$$
$$\tilde{\omega}=\omega+br^{-1}\sin\varphi\cdot(\omega_0 y+fz)$$

and with the help of (8.130), (8.131), (8.132) we obtain an asymptotic representation of φ':

$$\varphi'=\omega_1-\frac{\mu ab\omega_1^2}{2\delta}\left[f\omega_1+\frac{\omega_0^2-\omega_1^2}{\omega_0}\sin 2\varphi-f\omega_1\cos 2\varphi\right]+\mu\varepsilon(\mu)\tag{8.133}$$

The conclusion which we have just reached on the existence and stability of rotation at the speed ω_1 does not seem to agree with the experimental results which, under certain conditions, provide evidence of a range of values of ω_1 giving rise to instability.

However the theory can account for these unstable speeds as soon as we assume that the coefficient of viscous dissipation associated with the suspension is small, and of the same order of magnitude as μ.

To analyse the situation we consider once more the equations (8.109) of the problem, but substituting μf for f:

$$\begin{aligned}y''-ar\,\varphi''\sin\varphi&=-\omega_0^2 y-\mu f\omega_0 y'+ar\varphi'^2\cos\varphi\\ \varphi''-br^{-1}y''\sin\varphi&=\mu^{-1}\omega_0(\omega_1-\varphi').\end{aligned}\tag{8.134}$$

As will become apparent later on, it will be necessary to express explicitly in these equations all the terms of the first order in μ and that is why we cannot be content simply to rewrite (8.112) with f replaced by μf.

Solving (8.134) for y'', φ'' we find:

$$\begin{aligned}(y''+\omega_0^2 y)(1-ab\sin^2\varphi)&=-ab\omega_0^2 y\sin^2\varphi+ar\varphi'^2\cos\varphi+\mu^{-1}ar\omega_0\\&\quad\cdot(\omega_1-\varphi')\sin\varphi-\mu f\omega_0 y'\\(1-ab\sin^2\varphi)\varphi''&=-br^{-1}\omega_0^2 y\sin\varphi+ab\varphi'^2\sin\varphi\cos\varphi\\&\quad+\mu^{-1}\omega_0(\omega_1-\varphi')-\mu br^{-1}f\omega_0 y'\sin\varphi.\end{aligned}\tag{8.135}$$

With z,ω defined by:

$$\begin{aligned}y'&=z\\ \varphi'&=\omega_1+\mu\omega\end{aligned}\tag{8.136}$$

and φ as variable, this becomes:

$$y'=(\omega_1+\mu\omega)\frac{dy}{d\varphi}=z$$

whence

$$\frac{dy}{d\varphi}=\omega_1^{-1}z-\mu\frac{\omega}{\omega_1^2}z+O(\mu^2)$$

and then
$$y'' = (\omega_1 + \mu\omega)\frac{dz}{d\varphi},$$

i.e. with (8.135) and after a few calculations:

$$\frac{dz}{d\varphi} = -\frac{\omega_0^2}{\omega_1}y + \frac{-ab\omega_0^2 y\sin^2\varphi + ar(\omega_1^2\cos\varphi - \omega_0\omega\sin\varphi)}{\omega_1(1 - ab\sin^2\varphi)}$$
$$+\mu\left\{\frac{\omega_0^2\omega}{\omega_1^2}y + (1 - ab\sin^2\varphi)^{-1}\cdot\left[2ar\omega\cos\varphi - f\frac{\omega_0}{\omega_1}z - \frac{\omega}{\omega_1^2}(ar(\omega_1^2\cos\varphi\right.\right.$$
$$\left.\left. - \omega_0\omega\sin\varphi) - ab\omega_0^2 y\sin^2\varphi)\right]\right\} + O(\mu^2).$$

To calculate $\frac{d\omega}{d\varphi}$ we use (8.136), then (8.135), from which we obtain:

$$\varphi'' = \mu(\omega_1 + \mu\omega)\frac{d\omega}{d\varphi}$$

and

$$\mu(1 - ab\sin^2\varphi)(\omega_1 + \mu\omega)\frac{d\omega}{d\varphi}$$
$$= -br^{-1}\omega_0^2 y\sin\varphi + ab(\omega_1^2 + 2\mu\omega_1\omega)\sin\varphi\cos\varphi - \omega_0\omega$$
$$-\mu br^{-1}f\omega_0 z\sin\varphi + O(\mu^2).$$

Introducing $\tilde{\omega} = \omega + br^{-1}\omega_0\sin\varphi\cdot y$ we finally obtain:

(8.137)
$$\mu\frac{d\tilde{\omega}}{d\varphi} = \frac{ab\omega_1^2\sin\varphi\cos\varphi - \omega_0\tilde{\omega}}{\omega_1(1 - ab\sin^2\varphi)} - \mu\left[br^{-1}\frac{\omega_0}{\omega_1}z(f - 1 + ab\sin^2\varphi)\sin\varphi\right.$$
$$+ b\omega_0 r^{-1}y(-1 + 2ab\sin^2\varphi)\cos\varphi + b\frac{\omega_0^2}{\omega_1^2}r^{-1}y\sin\varphi\cdot\tilde{\omega}$$
$$\left. - \frac{\tilde{\omega}}{\omega_1^2}(ab\omega_1^2\sin\varphi\cos\varphi + \omega_0\tilde{\omega})\right]\cdot(1 - ab\sin^2\varphi)^{-1} + O(\mu^2)$$

an equation with which should be associated those obtained for y and z which can be written:

(8.138)
$$\frac{dy}{d\varphi} = \omega_1^{-1}z - \mu\left(\frac{\tilde{\omega}}{\omega_1^2}z - \frac{b\omega_0 r^{-1}\sin\varphi}{\omega_1^2}yz\right) + O(\mu^2)$$
$$\frac{dz}{d\varphi} = -\frac{\omega_0^2}{\omega_1}y + \frac{ar\omega_1^2\cos\varphi - ar\omega_0\tilde{\omega}\sin\varphi}{\omega_1(1 - ab\sin^2\varphi)}$$
$$+\mu\left\{\frac{\omega_0^2\tilde{\omega}}{\omega_1^2}y - \frac{b\omega_0^3 r^{-1}}{\omega_1^2}\sin\varphi\cdot y^2 + (1 - ab\sin^2\varphi)^{-1}\left[2ar\tilde{\omega}\cos\varphi\right.\right.$$
$$-f\frac{\omega_0}{\omega_1}z - 2ab\omega_0\sin\varphi\cos\varphi\cdot y - (\tilde{\omega} - b\omega_0 r^{-1}\sin\varphi\cdot y)$$
$$\left.\left.\cdot(ar\cos\varphi - ar\omega_0\omega_1^{-2}\tilde{\omega}\sin\varphi)\right]\right\} + O(\mu^2).$$

Again we obtain a system having the structure of (8.113), the matrix A being reduced to a single element:

$$A(\varphi) = -\omega_0\omega_1^{-1}(1 - ab\sin^2\varphi)^{-1} \leqslant -\frac{\omega_0}{\omega_1} < 0$$

and the matrix B being defined by

$$B = \begin{pmatrix} 0 & \omega_1^{-1} \\ -\dfrac{\omega_0^2}{\omega_1} & 0 \end{pmatrix}$$

i.e. having as eigenvalues the numbers $\pm i\dfrac{\omega_0}{\omega_1}$, ($i$ being the complex unit).

We shall assume that $\dfrac{\mathrm{d}}{\mathrm{d}\varphi}\begin{pmatrix} y \\ z \end{pmatrix} = B\begin{pmatrix} y \\ z \end{pmatrix}$ has no periodic solution of period 2π, i.e. that $\dfrac{\omega_0}{\omega_1}$ is not an integer; on this assumption Theorem 4 ensures that the set of equations (8.137), (8.138) has a periodic solution of period 2π which can be represented asymptotically by

$$\tilde{\omega} = ab\frac{\omega_1^2}{\omega_0}\sin\varphi\cos\varphi + \varepsilon(\mu)$$

with y and z, the 2π-periodic solution of:

$$\frac{\mathrm{d}y}{\mathrm{d}\varphi} = \omega_1^{-1}z$$

$$\frac{\mathrm{d}z}{\mathrm{d}\varphi} = -\frac{\omega_0^2}{\omega_1}y + ar\omega_1\cos\varphi$$

or finally:

$$\omega = -\frac{ab\omega_1^4}{2\omega_0(\omega_0^2 - \omega_1^2)}\sin 2\varphi + \varepsilon(\mu),$$

or

$$\varphi' = \omega_1 - \frac{\mu ab\omega_1^4}{2\omega_0(\omega_0^2 - \omega_1^2)}\sin 2\varphi + \mu\varepsilon(\mu)$$

(8.139)

$$y = \frac{ar\omega_1^2}{\omega_0^2 - \omega_1^2}\cos\varphi + \varepsilon(\mu)$$

$$z = -\frac{ar\omega_1^3}{\omega_0^2 - \omega_1^2}\sin\varphi + \varepsilon(\mu).$$

However Theorem 4 does not enable us to say whether the solution so obtained is stable because the eigenvalues of B lie on the imaginary axis.

13. Analysis of Stability

We now have to reconsider the problem starting from the equations (8.113), on the assumption that the eigenvalues of B are the pure imaginaries $\lambda_j = i\nu_j$, where

ν_j is real and $\nu_j \neq \nu_{j'}$ if $j \neq j'$, and are such that $\frac{dy}{dt} = By$ has no T-periodic solution, and retaining the previous hypotheses regarding $A(t)$. The linear variational system can again be written:

$$\mu \frac{du}{dt} = A(t)u + \mu P(t, \mu)u + \mu Q(t, \mu)v \tag{8.140}$$
$$\frac{dv}{dt} = Bv + H(t, \mu)u + \mu R(t, \mu)v$$

with

$$P(t, \mu) = g_x(x(t, \mu), y(t, \mu), t, \mu),$$
$$Q(t, \mu) = g_y(x(t, \mu), y(t, \mu), t, \mu)$$
$$H(t, \mu) = k_x(x(t, \mu), t) + \mu l_x(x(t, \mu), y(t, \mu), t, \mu),$$
$$R(t, \mu) = l_y(x(t, \mu), y(t, \mu), t, \mu)$$

where $x(t, \mu)$, $y(t, \mu)$ is the T-periodic solution which as $\mu \to 0$ tends uniformly with respect to t to $\xi_0(t)$, $\zeta_0(t)$ the T-periodic solution of:

$$A(t)\xi(t) + h(t) = 0$$
$$\frac{d\zeta}{dt} = B\zeta + k(\xi(t), t).$$

It is clear that the matrices $P(t, \mu), \ldots, R(t, \mu)$ have limits $P(t), \ldots, R(t)$ which are finite, T-periodic and continuous as $\mu \to 0$, for example, $\lim_{\mu \to 0} H(t, \mu) = k_x(\xi_0(t), t) = H(t)$; but some care needs to be taken in examining the behaviour of the derivative $\frac{dH(t, \mu)}{dt}$. Thus provided k and l are continuously differentiable twice:

$$\mu \frac{dH}{dt} = \mu k_{xx}(x(t, \mu), t) \frac{dx}{dt} + \mu k_{xt}(x(t, \mu), t) + \mu^2 l_{xx}(x(t, \mu), \ldots) \frac{dx}{dt}$$
$$+ \mu^2 l_{xy}(x(t, \mu), \ldots) \frac{dy}{dt} + \mu^2 l_{xt}(x(t, \mu), \ldots)$$

and since $\mu \frac{dx}{dt}(t, \mu)$ tends to 0, while $\frac{dy}{dt}(t, \mu)$ has a finite limit when $\mu \to 0$, it is clear that:

$$\lim_{\mu \to 0} \operatorname{Sup}_t \left| \mu \frac{dH(t, \mu)}{dt} \right| = 0.$$

We know that the periodic solution $x(t, \mu)$, $y(t, \mu)$ will be stable if every solution u, v of the linear variational system tends to 0 as $t \to +\infty$.

To examine this point, let us first note that:

$$H(t, \mu)u = H(t, \mu)A^{-1}(t)A(t)u$$
$$= H(t, \mu)A^{-1}(t)\left[\mu \frac{du}{dt} - \mu P(t, \mu)u - \mu Q(t, \mu)v\right]$$
$$= \mu \frac{d}{dt}(H(t, \mu)A^{-1}(t)u) - \mu (H(t, \mu)A^{-1}(t))'u$$
$$- \mu H(t, \mu)A^{-1}(t)(P(t, \mu)u + Q(t, \mu)v)$$

provided that $A(t)$ is differentiable. Accordingly we can rewrite the second equation (8.140) in the form:

$$\frac{dv}{dt} = (B + \mu R(t))v + \mu \frac{d}{dt}(H(t,\mu)A^{-1}(t)u) - \mu(H(t,\mu)A^{-1}(t))'u$$
$$- \mu H(t,\mu)A^{-1}(t)(P(t,\mu)u + Q(t,\mu)v) + \mu\varepsilon(\mu)v$$

or with

(8.141) $$w = v - \mu H(t,\mu)A^{-1}(t)u$$

$$\frac{dw}{dt} = (B + \mu R(t) - \mu H(t,\mu)A^{-1}(t)Q(t) + \mu\varepsilon(\mu))v + \varepsilon(\mu)u$$

(we use the notation $\varepsilon(\mu)$ to denote any arbitrary matrix which vanishes with μ, uniformly with respect to t), or again with

(8.142) $$G(t) = R(t) - H(t)A^{-1}(t)Q(t)$$

$$\frac{dw}{dt} = (B + \mu G(t) + \mu\varepsilon(\mu))w + \varepsilon(\mu)u.$$

Let us now, in the equation above, make the change of variable:

(8.143) $$w = (I - \mu K(t))q$$

where $K(t)$ is a T-periodic matrix to be defined later, so that the equation becomes:

$$(I - \mu K)\frac{dq}{dt} - \mu \frac{dK}{dt} q = (B + \mu G + \mu\varepsilon(\mu))(I - \mu K)q + \varepsilon(\mu)u$$

or

$$\frac{dq}{dt} = (I + \mu K + \mu\varepsilon(\mu))\left(\mu \frac{dK}{dt} + B + \mu(G - BK) + \mu\varepsilon(\mu)\right)q + \varepsilon(\mu)u$$

that is

(8.144) $$\frac{dq}{dt} = (B + \mu S + \mu\varepsilon(\mu))q + \varepsilon(\mu)u$$

if one has taken care to ensure that the T-periodic $K(t)$, and the constant matrix S are such that they satisfy:

(8.145) $$\frac{dK}{dt} = BK - KB + S - G.$$

It is known [40] that the pair $K(t), S$ exist if $\lambda_j - \lambda_{j'} \not\equiv 0 \bmod \frac{2\pi i}{T} \forall j, j', j \neq j'$. It is also clear that if $K(t)$, S is a particular solution of this equation, then $K(t) + C$, $S - BC + CB$ is another solution whatever the constant matrix C. For this reason we can normalise the solution by stipulating that the mean value of $K(t)$ should be zero. On this hypothesis however, we have, after integrating both sides of (8.145) with respect to t over a period T:

$$S=\frac{1}{T}\int_0^T G(t)\,\mathrm{d}t. \tag{8.146}$$

Having regard to the transformations (8.141), (8.143), we need to know whether every solution u, q of the equations:

$$\begin{aligned}\mu\frac{\mathrm{d}u}{\mathrm{d}t}&=(A(t)+\varepsilon(\mu))u+\mu L(t,\mu)q\\ \frac{\mathrm{d}q}{\mathrm{d}t}&=(B+\mu S+\mu\varepsilon(\mu))q+\varepsilon(\mu)u\end{aligned} \tag{8.147}$$

tends to 0 as $t\to+\infty$ for small enough μ. (The matrix $L(t,\mu)$ is T-periodic, and regular in the neighbourhood of $\mu=0$.) The eigenvalues of $B+\mu S$ are solutions of:

$$\det(B+\mu S-\tilde{\lambda}I)=f_0(\tilde{\lambda})+\mu f_1(\tilde{\lambda})+\cdots+\mu^p f_p(\tilde{\lambda})=F(\tilde{\lambda},\mu)=0$$

and we shall suppose that those of B, defined by $f_0(\tilde{\lambda})=0$ are all distinct. If λ is one of them, we have $f_0(\lambda)=0$, $\frac{\mathrm{d}f_0}{\mathrm{d}\lambda}(\lambda)\neq 0$ and it is easily seen from the theorem on implicit functions that the equation $F(\tilde{\lambda},\mu)=0$ defines, for small enough μ, $\tilde{\lambda}(\mu)$ as an analytic function of μ taking the value λ for $\mu=0$. Moreover all the eigenvalues $\tilde{\lambda}(\mu)$ are distinct and we can define an invertible matrix $J(\mu)$, analytic in a neighbourhood of $\mu=0$ which is such that:

$$J^{-1}(\mu)(B+\mu S)J(\mu)=\begin{pmatrix}\tilde{\lambda}_1 & & 0\\ & \ddots & \\ 0 & & \tilde{\lambda}_p\end{pmatrix}. \tag{8.148}$$

Provided we make a suitable change of variable $q\to J(\mu)q$, we see that we can without any inconvenience suppose the matrix $B+\mu S$ on the right-hand side of (8.147) to have been put into the diagonal form (8.148).

Note that $\tilde{\lambda}_j(\mu)$, the eigenvalue of $B+\mu S$, can be written:

$$\tilde{\lambda}_j(\mu)=\lambda_j+O(\mu),\quad \lambda_j=i\nu_j,\quad \nu_j\text{ real}$$

or with

$$\operatorname{Re}\tilde{\lambda}_j(\mu)=-\sigma_j(\mu)\cdot\mu,$$

$$\tilde{\lambda}_j(\mu)=-\sigma_j(\mu)\cdot\mu+i(\nu_j+\operatorname{Im}O(\mu)).$$

We shall now make the assumption that there exists a number $\sigma>0$ such that

$$\sigma_j(\mu)\geqslant\sigma,\quad \forall j\in[1,2,\ldots,p],\quad \forall\mu\in]0,\mu_0] \tag{8.149}$$

an assumption designed, as we shall show, to ensure stability.

In consequence of the successive reductions we are thus led to examine the behaviour, as $t\to+\infty$, of the solutions of the linear system:

$$\begin{aligned}\mu\frac{\mathrm{d}u}{\mathrm{d}t}&=(A(t)+\varepsilon(\mu))u+\mu L(t,\mu)q\\ \frac{\mathrm{d}q}{\mathrm{d}t}&=\left(\begin{pmatrix}\tilde{\lambda}_1 & 0\\ 0 & \tilde{\lambda}_p\end{pmatrix}+\mu\varepsilon(\mu)\right)q+\varepsilon(\mu)u\end{aligned} \tag{8.150}$$

with $$\operatorname{Re}\tilde{\lambda}_j = -\sigma_j(\mu)\cdot\mu < -\sigma\mu.$$

Retaining the hypothesis that $A(t)$ is symmetric, and that the real parts of its eigenvalues are all less than -2α, we can by using the classical procedure, deduce from (8.150) the inequalities:

$$\frac{\chi\mu}{2}\frac{\mathrm{d}|u|^2}{dt} \leqslant -\chi\alpha|u|^2 + \chi\mu\delta c|u|^2 + \chi\mu c\delta^{-1}|q|^2$$

$$\frac{1}{2}\frac{\mathrm{d}|q|^2}{dt} \leqslant -\sigma\mu|q|^2 + \mu\varepsilon(\mu)|q|^2 + \varkappa\varepsilon(\mu)|u|^2 + \varkappa^{-1}\varepsilon(\mu)|q|^2 \tag{8.151}$$

where $\chi, \delta, \varkappa$ are positive numbers, for the time being arbitrary, c is an upper bound of $|L(t,\mu)|$, $\varepsilon(\mu)$ a non-decreasing positive-valued function for $\mu > 0$, and such that $\lim_{\mu\downarrow 0}\varepsilon(\mu) = 0$.

To prove that every solution of (8.150) tends to 0 as $t \to +\infty$, it will be sufficient to show that we can satisfy the inequalities

$$\begin{aligned} -\chi\alpha + \chi\mu\delta c + \varkappa\varepsilon_1(\mu) &< 0 \\ -\mu\sigma + \mu\varepsilon_3(\mu) + \varkappa^{-1}\varepsilon_2(\mu) + \chi\mu c\delta^{-1} &< 0 \end{aligned} \tag{8.152}$$

i.e. since α and σ are fixed positive numbers, to show that we can define χ, δ, κ as functions of μ satisfying the relations:

$$\begin{aligned} \mu\delta \ll 1, &\qquad \varkappa\chi^{-1}\varepsilon_1(\mu) \ll 1 \\ \chi\delta^{-1} \ll 1, &\qquad (\varkappa\mu)^{-1}\varepsilon_2(\mu) \ll 1 \end{aligned}$$

(where the sign $\ll$ is to be interpreted as meaning "small compared with", for all μ in a suitable neighbourhood of 0).

Let us put $\varkappa\chi^{-1}\varepsilon_1(\mu) = \eta_1(\mu)$, $(\varkappa\mu)^{-1}\varepsilon_2(\mu) = \eta_2(\mu)$ with η_1, η_2 given positive functions tending to zero with μ.

We then have

$$\varkappa = \frac{\varepsilon_2(\mu)}{\mu\eta_2(\mu)}, \quad \chi = \frac{\varepsilon_1(\mu)\varepsilon_2(\mu)}{\mu\eta_1(\mu)\eta_2(\mu)}.$$

so that we have only to satisfy $\chi\delta^{-1} \ll 1$, $\mu\delta \ll 1$.

We put $\mu\delta = \eta_3(\mu)$ with η_3 a given positive function tending to 0 with μ, which thus defines δ and leads to

$$\chi\delta^{-1} = \frac{\varepsilon_1(\mu)\varepsilon_2(\mu)}{\eta_1(\mu)\eta_2(\mu)\eta_3(\mu)}.$$

It remains that η_1, η_2, η_3 have to be such that:

$\lim_{\mu\to 0}\frac{\varepsilon_1(\mu)\varepsilon_2(\mu)}{\eta_1(\mu)\eta_2(\mu)\eta_3(\mu)} = 0$ and we can take for example $\eta_1 = \eta_2 = \eta_3 = (\varepsilon_1\cdot\varepsilon_2)^{1/4}$.

We can complete this analysis by a result on instability. Let us suppose the eigenvalues of B to be distinct, pure imaginaries, satisfying $\lambda_j - \lambda_{j'} \not\equiv 0 \bmod \frac{2\pi i}{T}$,

$j \neq j'$, and divided into two groups such that

$$\tilde{\lambda}_j(\mu) = \sigma_j(\mu)\cdot\mu + i(v_j + O(\mu)), \quad O(\mu) \text{ real}$$

with $\sigma_j(\mu) \geqslant \sigma > 0$, for $\begin{matrix}\forall j: 1 \leqslant j \leqslant r \\ \forall \mu \in]0, \mu_0]\end{matrix}$ (for the first group)

and $\operatorname{Re}\tilde{\lambda}_j(\mu) \leqslant 0$, for $\forall j, r+1 \leqslant j \leqslant p$ (for the second group).

From the equations (8.150) we can deduce the inequalities:

$$\frac{\chi\mu}{2}\frac{\mathrm{d}}{\mathrm{d}t}|u|^2 \leqslant -\chi\alpha|u|^2 + \chi\mu\delta c|u|^2 + \chi\mu\delta^{-1}c|q|^2$$

$$\frac{1}{2}\frac{\mathrm{d}}{\mathrm{d}t}\left(\sum_1^r |q_j|^2 - \sum_{r+1}^p |q_j|^2\right) \geqslant \mu\sigma\sum_1^r |q_j|^2 - \mu\varepsilon(\mu)|q|^2 - \varkappa\varepsilon(\mu)|u|^2 - \varkappa^{-1}\varepsilon(\mu)|q|^2$$

whence, by combining them:

$$\begin{aligned}\frac{1}{2}\frac{\mathrm{d}}{\mathrm{d}t}&\left(\sum_1^r |q_j|^2 - \sum_{r+1}^p |q_j|^2 - \chi\mu|u|^2\right)\\ &\geqslant \mu\sigma\sum_1^r |q_j|^2 + (\chi\alpha - \chi\delta\mu c - \chi\varepsilon(\mu))|u|^2\\ &\quad - (\chi\mu\delta^{-1}c + \mu\varepsilon(\mu) + \varkappa^{-1}\varepsilon(\mu))|q|^2\end{aligned}$$

and we impose on the choice of $\chi, \delta, \varkappa$ the conditions:

$$\chi\alpha - \chi\delta\mu c - \varkappa\varepsilon(\mu) > 0$$

$$\chi\mu\delta^{-1}c + \mu\varepsilon(\mu) + \varkappa^{-1}\varepsilon(\mu) < \frac{\sigma\mu}{2}$$

which are the same as those of (8.152) apart from changing σ to $\sigma/2$.

We therefore obtain, for small enough μ, the inequality

$$\frac{\mathrm{d}}{\mathrm{d}t}\left(\sum_1^r |q_j|^2 - \sum_{r+1}^p |q_j|^2 - \chi\mu|u|^2\right) \geqslant \mu\sigma\left(\sum_1^r |q_j|^2 - \sum_{r+1}^p |q_j|^2 - \chi\mu|u|^2\right)$$

from which it follows that certain solutions of (8.150) can become infinite when $t \to +\infty$, and consequently the periodic solution provided by Theorem 4 is unstable.

Lastly we can enunciate the following theorem:

Theorem 5

In addition to the general hypotheses of Theorem 4 we assume that $A(t)$ is symmetric and differentiable and that λ_j the eigenvalues of B, are all distinct, purely imaginary and such that $\lambda_j - \lambda_{j'} \not\equiv 0 \bmod \frac{2\pi i}{T}$, for $j \neq j'$, and lastly that k, g, l are continuously differentiable twice. Using the notation

$$H(t) = k_x(\xi_0(t), t), \quad Q(t) = g_y(\xi_0(t), \zeta_0(t), t, 0),$$
$$R(t) = l_y(\xi_0(t), \zeta_0(t), t, 0), \quad \text{then} \quad G(t) = R(t) - H(t)A^{-1}(t)Q(t)$$

and $S=(1/T)\cdot\int_0^T G(t)\,dt$, we denote the real parts of the eigenvalues of $B+\mu S$, for small μ, by $\sigma_j(\mu)\cdot\mu$, $1\leqslant j\leqslant p$.

If there exists a number $\sigma>0$ such that $\sigma_j(\mu)<-\sigma<0$, $\forall j\in[1,\dots,p]$, $\forall\mu\in]0,\mu_0]$, then the periodic solution of the system (8.113) is asymptotically stable for $t\to+\infty$ and $\mu>0$ small enough.

If, for at least one index $j\in[1,\dots,p]$, there exists a number $\sigma>0$ such that $\sigma_j(\mu)>\sigma$, $\forall\mu\in]0,\mu_0]$, then the periodic solution of the system (8.113) is unstable for $t\to+\infty$ and small enough μ.

Application to the Dynamic Stability of a Rotating Shaft Mounted on an Elastic Suspension, When the Driving Motor Has a Steep Characteristic

In this case B is a 2×2 matrix with eigenvalues $\pm\dfrac{i\omega_0}{\omega_1}$ and to express the stability condition it suffices to write down the condition that the trace of the matrix S is negative, with the further restriction that $2\dfrac{\omega_0}{\omega_1}\not\equiv 0 \bmod 1$.

We have recourse to the equations (8.137), (8.138) to evaluate the terms g, l, k and to the representations (8.138) for the calculation of $H(t)$, $Q(t)$, $R(t)$ appearing in the definition of $G(t)$.

Thus we have:

$$\operatorname{trace} R(\varphi)=b\omega_0\omega_1^{-2}r^{-1}\sin\varphi\cdot z(\varphi)-f\omega_0\omega_1^{-1}(1-ab\sin^2\varphi)^{-1}$$

with

$$z(\varphi)=-\frac{ar\omega_1^3}{\omega_0^2-\omega_1^2}\sin\varphi$$

then

$$H(\varphi)=\begin{pmatrix}h_{11}\\ h_{21}\end{pmatrix},\quad h_{11}=0,\quad h_{21}=-\frac{ar\omega_0\sin\varphi}{\omega_1(1-ab\sin^2\varphi)}$$

$$Q(\varphi)=(q_{11},q_{12})$$

with

$$q_{12}=-br^{-1}\frac{\omega_0}{\omega}(f-1+ab\sin^2\varphi)(1-ab\sin^2\varphi)^{-1}\sin\varphi$$

$$A^{-1}(\varphi)=-\frac{\omega_1}{\omega_0}(1-ab\sin^2\varphi)$$

whence

$$\begin{aligned}\operatorname{trace}(HA^{-1}Q)&=-(h_{11}q_{11}+h_{21}q_{12})\omega_1\omega_0^{-1}(1-ab\sin^2\varphi)\\&=-ab\omega_0\omega_1^{-1}(f-1+ab\sin^2\varphi)(1-ab\sin^2\varphi)^{-1}\sin^2\varphi\end{aligned}$$

and finally the stability condition

$$\frac{1}{2\pi}\int_0^{2\pi}\operatorname{trace}(R(\varphi)-H(\varphi)A^{-1}(\varphi)Q(\varphi))\,d\varphi=-\frac{ab\omega_0^3}{2\omega_1(\omega_0^2-\omega_1^2)}-f\frac{\omega_0}{\omega_1}<0$$

or

$$2f+ab\frac{\omega_0^2}{\omega_0^2-\omega_1^2}>0.$$

Accordingly the rotation ω_1 is stable if $\omega_1 < \omega_0$ or

$$\omega_1 > \omega_0 \sqrt{1 + \frac{ab}{2f}}$$

and unstable for ω_1 in the interval $\omega_0 < \omega_1 < \omega_0 \sqrt{1 + \frac{ab}{2f}}$.

14. Rotation of an Unbalanced Shaft Sustained by Alternating Vertical Displacements

We consider a system consisting of a frame of mass m_1 mounted in an elastic suspension which allows it only one degree of freedom corresponding to a movement of vertical translation, represented by the co-ordinate Y. An imperfectly balanced horizontal shaft can turn freely in bearings integral with the frame, the amount of rotation being indicated by the angular co-ordinate φ. The mass of the rotating shaft is m, the distance of its centre of inertia G from the geometrical axis of rotation is r, and I is its moment of inertia about this axis.

We shall suppose the movement imposed on the frame to be described by $Y = a \sin \omega t$, where a, ω are given positive quantities. Under certain conditions it may happen that the rotation of the shaft is synchronised to the frequency ω, that is to say that the shaft rotates at an almost constant angular velocity equal to ω.

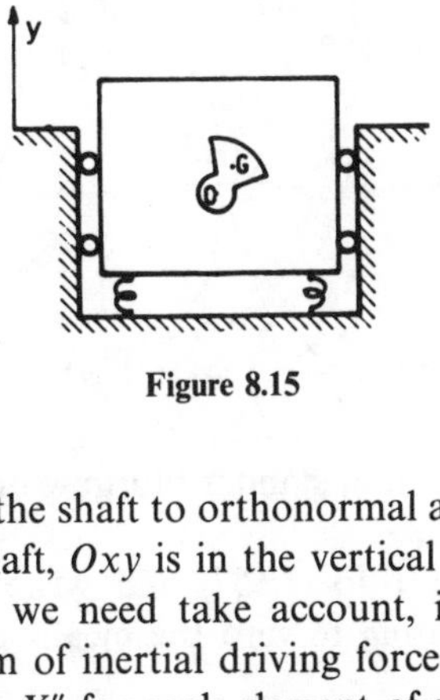

Figure 8.15

Referring the motion of the shaft to orthonormal axes $Oxyz$, where Oz is along the geometric axis of the shaft, Oxy is in the vertical plane containing G and Oy along the upward vertical, we need take account, in relation to this frame of reference, only of the system of inertial driving forces which, parallel to the axis Oy and of magnitude $-dm \cdot Y''$ for each element of mass dm, are equivalent, for the shaft as a whole, to a single force $ma\omega^2 \sin \omega t \cdot \vec{y}$ passing through G. Their moment at O is therefore:

$$\begin{aligned}\overrightarrow{OG} \wedge ma\omega^2 \sin \omega t \vec{y} &= mar\omega^2 \sin \omega t(\vec{x} \cos \varphi + \vec{y} \sin \varphi) \wedge \vec{y} \\ &= mar\omega^2 \sin \omega t \cdot \cos \varphi \vec{z}.\end{aligned}$$

Taking account of the moment of the gravitational forces $-mgr \cos \varphi \vec{z}$ and of the moment of the frictional forces opposing the rotation of the shaft, and represented by $-k\varphi'\cdot\vec{z}$, with k a positive constant, we can write the equation of motion as:

$$I\varphi'' = -mgr\cos\varphi - k\varphi' + mar\omega^2 \sin\omega t\cdot\cos\varphi. \tag{8.153}$$

Writing $I = m\rho^2$, ρ being the radius of gyration and recalling that $I = m\rho^2 = mr^2 + I_G$, where I_G is the moment of inertia about an axis parallel to Oz passing through G, it can be seen that $r < \rho$ and, with $\mu = \frac{mra}{I} = \frac{ra}{\rho^2}$, we can rewrite (8.153) as:

$$\varphi'' = \mu\left[\left(\omega^2 \sin\omega t - \frac{g}{a}\right)\cos\varphi - \frac{k}{mra}\varphi'\right]. \tag{8.154}$$

We shall suppose that $\mu = \frac{ra}{\rho^2}$ can be considered as a small parameter, while $\sqrt{\frac{g}{a}}$ and $\frac{k}{mra} = \omega_0$ are of the order of ω.

Taking $\tau = \omega t$ as a new variable and writing

$$\varphi = \tau + \psi \tag{8.155}$$

we obtain from (8.154):

$$\ddot{\psi} = \mu\left[\left(\sin\tau - \frac{g}{a\omega^2}\right)\cos(\tau+\psi) - \frac{\omega_0}{\omega}(1+\dot{\psi})\right], \tag{8.156}$$

$$\dot{\psi} = \frac{d\psi}{d\tau}, \quad \ddot{\psi} = \frac{d^2\psi}{d\tau^2}$$

which we shall write, with $\nu = \mu^{1/2} = \frac{\sqrt{ra}}{\rho}$, in the form:

$$\begin{aligned} &\dot{\psi} = \nu\xi \\ &\dot{\xi} = \nu\left[\left(\sin\tau - \frac{g}{a\omega^2}\right)\cos(\tau+\psi) - \frac{\omega_0}{\omega}(1+\nu\xi)\right] \end{aligned} \tag{8.157}$$

which now appears as a system in standard form, with right-hand sides which are periodic in τ with period 2π.

We can apply Theorem 1 to (8.157) and write down the synchronisation equations, obtained by equating to zero the means of the right-hand sides with respect to τ, for $\nu = 0$:

$$\xi = 0, \quad -\frac{\sin\psi}{2} - \frac{\omega_0}{\omega} = 0$$

and it will be seen that the associated Jacobian matrix is

$$\begin{pmatrix} 0 & 1 \\ -\frac{\cos\psi}{2} & 0 \end{pmatrix}.$$

Thus a synchronised regime can establish itself, which corresponds to the values

(8.158) $$\xi = 0, \quad \sin\psi_0 = -\frac{2\omega_0}{\omega}$$

provided that $\omega > 2\omega_0$, and assuming as we may that $\cos\psi_0 > 0$, since the equation for the eigenvalues of the stability matrix is

$$\lambda^2 + \frac{\cos\psi_0}{2} = 0$$

(there would certainly be instability if we took the solution ψ_0 for which $\cos\psi_0 < 0$).

The periodic solution $\psi(\tau, \nu)$ of period 2π, whose existence has thus been proved, can be represented by:

$$\psi(\tau, \nu) = \psi_0 + O(\nu) \quad \text{or} \quad \varphi = \omega t + \psi_0 + O(\nu)$$

but as the eigenvalues of the stability matrix are pure imaginaries, we are unable, at this stage of the analysis, to pronounce on the stability of the solution.

Note that we could have carried out similar calculations by introducing instead of (8.155), $\varphi = -\tau + \psi$; we would then have obtained a synchronised regime corresponding to a quasi-uniform rotation $-\omega$ of the shaft, with the same question-mark in regard to stability.

To go into this question more deeply we shall have to calculate the higher-order terms in the asymptotic expansion and it will be sufficient to consider the case described by the equations (8.157).

To do this we put

(8.159) $$\begin{aligned}\psi &= \psi_0 + \nu\psi_1\\ \xi &= \nu\xi_1\end{aligned}$$

so that ψ_1, ξ_1 are solution of:

(8.160) $$\begin{aligned}\dot{\psi}_1 &= \nu\xi_1\\ \dot{\xi}_1 &= \left[\left(\sin\tau - \frac{g}{a\omega^2}\right)\cos(\tau + \psi_0 + \nu\psi_1) - \frac{\omega_0}{\omega}(1 + \nu^2\xi_1)\right].\end{aligned}$$

We can write the second of these equations in the form:

$$\begin{aligned}\dot{\xi}_1 = \left(\sin\tau - \frac{g}{a\omega^2}\right)\cos(\tau + \psi_0) - \frac{\omega_0}{\omega} + \nu\Bigg[\left(\frac{g}{a\omega^2} - \sin\tau\right)\sin(\tau + \psi_0)\psi_1\\ - \frac{\nu\psi_1^2}{2}\left(\sin\tau - \frac{g}{a\omega^2}\right)\cos(\tau + \psi_0) - \frac{\nu\omega_0}{\omega}\xi_1 + O(\nu^2\psi_1^3)\Bigg]\end{aligned}$$

which suggests substituting for ξ_1 the co-ordinate ζ_1 defined by:

(8.161) $$\xi_1 = f(\tau) + \zeta_1$$

with

(8.162) $$f(\tau) = \int_0^\tau \left[\left(\sin\tau - \frac{g}{a\omega^2}\right)\cos(\tau + \psi_0) - \frac{\omega_0}{\omega}\right]\mathrm{d}\tau$$

periodic in τ of period 2π, by virtue of (8.158).

Thus, by means of the co-ordinates ψ_1, ζ_1 the system (8.160) can be written:

(8.163)
$$\dot{\psi}_1 = \nu(\zeta_1 + f(\tau))$$
$$\dot{\zeta}_1 = \nu\left[\left(\frac{g}{a\omega^2} - \sin\tau\right)\sin(\tau+\psi_0)\cdot\psi_1 - \frac{\nu\psi_1^2}{2}\left(\sin\tau - \frac{g}{a\omega^2}\right)\cos(\tau+\psi_0)\right.$$
$$\left. - \frac{\nu\omega_0}{\omega}(f(\tau)+\zeta_1) + O(\nu^2\psi_1^3)\right]$$

to which the synchronisation theorem can again be applied. We thus arrive at

(8.164)
$$\zeta_1 + \bar{f} = 0, \quad \bar{f} = \frac{1}{2\pi}\int_0^{2\pi} f(\tau)\,\mathrm{d}\tau$$
$$-\frac{\cos\psi_0}{2}\cdot\psi_1 = 0.$$

It will also be observed that the set of equations obtained for calculating the synchronisation point, corresponding to the approximation of order 1 in ν, is linear and that the associated determinant is precisely the same as that of the stability matrix, which is indeed only a particular illustration of a general situation [40].

Thus for small enough ν, the periodic solution can be represented by

(8.165)
$$\zeta_1(\tau,\nu) = -\bar{f} + O(\nu)$$
$$\psi_1(\tau,\nu) = O(\nu)$$

or coming back to (8.159)

$$\psi = \psi_0 + O(\nu^2)$$
$$\xi = \nu(f(\tau) - \bar{f}) + O(\nu^2).$$

To go further, that is to say to evaluate the $O(\nu^2)$ terms, we introduce ψ_2 and ζ_2 by:

(8.166)
$$\psi_1 = \nu\psi_2$$
$$\zeta_1 = -\bar{f} + \nu\zeta_2$$

and going back to (8.163) we obtain:

(8.167)
$$\dot{\psi}_2 = f(\tau) - \bar{f} + \nu\zeta_2$$
$$\dot{\zeta}_2 = \nu\left[\left(\frac{g}{a\omega^2} - \sin\tau\right)\sin(\tau+\psi_0)\cdot\psi_2 - \frac{\omega_0}{\omega}(f(\tau) - \bar{f} + \nu\zeta_2) + O(\nu^2)\right]$$

which we shall write in terms of the variables ζ_2 and χ_2, the latter defined by:

(8.168)
$$\psi_2 = \chi_2 + h(\tau)$$

with

(8.169)
$$h(\tau) = \int_0^{\tau} (f(\tau) - \bar{f})\,\mathrm{d}\tau,$$

τ-periodic of period 2π:

$$\dot{\chi}_2 = \nu\zeta_2$$

(8.170)
$$\dot{\zeta}_2 = \nu\left[\left(\frac{g}{a\omega^2} - \sin\tau\right)\sin(\tau + \psi_0)\cdot(\chi_2 + h(\tau)) - \frac{\omega_0}{\omega}(f(\tau) - \bar{f} + \nu\zeta_2) + O(\nu^2)\right]$$

a system to which we can again apply the synchronisation technique, which leads to the associated equations:

$$\zeta_2 = 0$$

(8.171)
$$-\frac{\cos\psi_0}{2}\cdot\chi_2 + \frac{1}{2\pi}\int_0^{2\pi}\left(\frac{g}{a\omega^2} - \sin\tau\right)\sin(\tau + \psi_0)h(\tau)\,d\tau = 0$$

i.e. by (8.162) and (8.169)

$$\zeta_2 = 0, \quad \chi_2 = \frac{g}{a\omega^2}\cos\psi_0 - \frac{\sin\psi_0}{8}.$$

Finally by (8.159), (8.161), (8.166) and (8.168) we obtain the asymptotic representation:

(8.172)
$$\psi = \psi_0 + \nu^2\left(\frac{g}{a\omega^2}\cos(\tau + \psi_0) - \tfrac{1}{8}\sin(2\tau + \psi_0)\right) + O(\nu^3)$$

with $\quad \varphi = \psi + \tau, \quad \tau = \omega t.$

15. Stability of Rotation of the Shaft

We have seen that it is not possible to answer the question of stability from a knowledge of the approximation of order 0. In point of fact to find the answer we have to analyse the behaviour for $t \to +\infty$ of the solutions of the set of linear variational equations derived from (8.157) in the neighbourhood of the periodic solution $\xi(\tau, \nu)$, $\psi(\tau, \nu)$ whose existence has been established. We shall henceforth reserve the notation ξ, ψ for these variations, so that the system can be written [40]:

$$\dot{\psi} = \nu\xi$$
$$\dot{\xi} = \nu\left[\left(\frac{g}{a\omega^2} - \sin\tau\right)\sin(\tau + \psi(\tau, \nu))\cdot\psi - \frac{\omega_0}{\omega}\nu\xi\right]$$

or with $\psi(\tau, \nu) = \psi_0 + O(\nu^2)$

$$\dot{\psi} = \nu\xi$$
$$\dot{\xi} = \nu\left[\left(\frac{g}{a\omega^2} - \sin\tau\right)\sin(\tau + \psi_0)\cdot\psi - \frac{\omega_0}{\omega}\nu\xi + O(\nu^2)\psi\right]$$

or again in matrix form:

$$\begin{pmatrix} \dot{\psi} \\ \dot{\xi} \end{pmatrix} = \nu[A(\tau) + \nu A_1(\tau) + O(\nu^2)]\begin{pmatrix} \psi \\ \xi \end{pmatrix} \tag{8.173}$$

with

$$A(\tau) = \begin{pmatrix} 0 & 1 \\ \left(\frac{g}{a\omega^2} - \sin\tau\right)\sin(\tau + \psi_0) & 0 \end{pmatrix} \tag{8.174}$$

$$A_1(\tau) = \begin{pmatrix} 0 & 0 \\ 0 & -\omega_0\omega^{-1} \end{pmatrix} \tag{8.175}$$

and the solution $\psi(\tau, \nu)$ will be stable if the null solution of (8.173) is itself asymptotically stable when $t \to +\infty$. This poses a problem which can obviously be discussed in the more general context of the following question. If $x \in \mathbb{R}^n$, and $A(t)$, $A_1(t)$ are continuous, T-periodic, $n \times n$ matrices, how does one recognise, for $t \to +\infty$, whether the null solution of

$$\frac{dx}{dt} = \mu[A(t) + \mu A_1(t) + O(\mu^2)]x \tag{8.176}$$

when $\mu > 0$ is a small parameter, and $O(\mu^2)$ is independent of x, is stable or not?

Let us, in (8.176), make the change of variable:

$$x = (I + \mu U + \mu^2 U_1)y \tag{8.177}$$

where U, U_1 are $n \times n$ matrices, continuously differentiable and periodic in t with period T, which have yet to be fully defined.

By substituting in (8.176) and expanding up to order 2 in μ, we find:

$$\frac{dy}{dt} = \mu\left[A - \frac{dU}{dt} + \mu\left(AU - UA + A_1 - \frac{dU_1}{dt} + U\frac{dU}{dt}\right) + O(\mu^2)\right]y.$$

We define $U(t)$ to be t-periodic of period T and such that $A - \frac{dU}{dt}$ is a constant matrix, which leads to putting:

$$S = \frac{1}{T}\int_0^T A(t)\,dt \tag{8.178}$$

$$U = \int_0^t (A(t) - S)\,dt.$$

Similarly we shall define $U_1(t)$ to be t-periodic of period T, and such that

$$A_1 + AU - UA + U\frac{dU}{dt} - \frac{dU_1}{dt} = S_1,$$

where S_1 is a constant matrix. This equation can be simplified, using (8.178), to

$$A_1 + AU - US - \frac{dU_1}{dt} = S_1$$

and is solved by:

$$S_1 = \frac{1}{T}\int_0^T (A_1(t) + A(t)U(t) - U(t)S)\,dt$$

and

$$U_1(t) = \int_0^t (A_1(t) + A(t)U(t) - U(t)S - S_1)\,dt.$$

Finally the stability problem relating to the equation (8.176) is transformed into an analogous problem for:

$$\frac{dy}{dt} = \mu(S + \mu S_1 + O(\mu^2))y$$

assuming $\mu > 0$ is small enough. This problem is simpler in form because S and S_1 are now matrices with constant elements. If the real parts of the eigenvalues of S are distinct and negative or zero, and if, for $\mu \in]0, \mu_0]$, the eigenvalues of $S + \mu S_1$ which are in the neighbourhood of the pure imaginary eigenvalues of S, have a negative real part of order 1 in μ, then the solution $y = 0$ will be asymptotically stable for $t \to +\infty$ and small enough μ.

Let us apply this result to the set of equations (8.173), (8.174), (8.175).

We have

$$S = \frac{1}{2\pi}\int_0^{2\pi} A(\tau)\,d\tau = \begin{pmatrix} 0 & 1 \\ -\dfrac{\cos\psi_0}{2} & 0 \end{pmatrix}$$

$$U(\tau) = \begin{pmatrix} 0 & 0 \\ \chi(\tau) & 0 \end{pmatrix}$$

with $$\chi(\tau) = \int_0^\tau \left(\left(\frac{g}{a\omega^2} - \sin\tau\right)\sin(\tau + \psi_0) + \frac{\cos\psi_0}{2}\right)d\tau$$

$$= \frac{g}{a\omega^2}(\cos\psi_0 - \cos(\tau + \psi_0)) + \tfrac{1}{4}(\sin(2\tau + \psi_0) - \sin\psi_0).$$

We then calculate:

$$AU + A_1 - US = \begin{pmatrix} \chi & 0 \\ 0 & -\dfrac{\omega_0}{\omega} - \chi \end{pmatrix}$$

then $$S_1 = \begin{pmatrix} \bar{\chi} & 0 \\ 0 & -\dfrac{\omega_0}{\omega} - \bar{\chi} \end{pmatrix}, \quad S + \nu S_1 = \begin{pmatrix} \nu\bar{\chi} & 1 \\ -\dfrac{\cos\psi_0}{2} & -\nu\left(\dfrac{\omega_0}{\omega} + \bar{\chi}\right) \end{pmatrix}$$

The eigenvalues of $S + \nu S_1$ are

$$-\frac{\nu\omega_0}{2\omega} \pm i\sqrt{\frac{\cos\psi_0}{2} - \nu^2\left(\bar{\chi}\left(\bar{\chi} + \frac{\omega_0}{\omega}\right) + \frac{\omega_0^2}{4\omega^2}\right)}$$

whose real part $-\nu\dfrac{\omega_0}{2\omega}$ is negative for sufficiently small positive ν; the periodic solution $\psi(\tau, \nu)$ is therefore asymptotically stable, at least if $\nu > 0$ is small enough.

16. Synchronisation of the Rotation of an Unbalanced Shaft Sustained by Alternating Vertical Forces

We consider afresh the problem examined in §14, but this time under the hypothesis that vertical forces of intensity $H \sin \omega t$ act on the frame. Denoting henceforth by y the co-ordinate associated with the vertical displacement of the frame, the equations of motion are:

$$(8.179)\qquad \begin{aligned} I\varphi'' &= -m(y''+g)r\cos\varphi - k\varphi' \\ (m+m_1)y'' + \sigma y' + cy + mr(\varphi''\cos\varphi - \varphi'^2\sin\varphi) &= H\sin\omega t \end{aligned}$$

where c represents the rigidity and σ the damping of the frame mounting, and the co-ordinates φ and y being from now on coupled.

We define the positive parameters ω_0, a, f, κ by:

$$\omega_0^2 = \frac{c}{m+m_1}, \quad H = a(m+m_1)\omega_0^2, \quad \sigma = \left(\frac{r}{\rho}\right)^2 f\omega_0(m+m_1), \quad k = \left(\frac{r}{\rho}\right)^2 I\varkappa\omega_0$$

and we assume that $\mu = (r/\rho)^2$ is a small parameter.

We can thus write the equations (8.179) in the form:

$$(8.180)\qquad \varphi'' = -\mu[(y''+g)r^{-1}\cos\varphi + \varkappa\omega_0\varphi']$$

$$(8.181)\qquad y'' + \omega_0^2 y = -\frac{mr}{m+m_1}(\varphi''\cos\varphi - \varphi'^2\sin\varphi) + a\omega_0^2\sin\omega t - \mu f\omega_0 y'.$$

We shall suppose that $m(m+m_1)^{-1}$ is of the same order of magnitude as μ, that is $m/(m+m_1) = \mu p$, $p = O(1)$, so that we are led to consider the system consisting of (8.180) and:

$$(8.182)\qquad y'' + \omega_0^2 y = -\mu[f\omega_0 y' + pr(\varphi''\cos\varphi - \varphi'^2\sin\varphi)] + a\omega_0^2\sin\omega t$$

and the discussion of this calls for a distinction to be made between the non-resonant case $\omega \neq \omega_0$, and the case of resonance $\omega \sim \omega_0$. Note that we could also have dealt with the case where $m/(m+m_1) = O(1)$ by the methods which follow, but we shall not do so.

16.1. The Non-Resonant Case

Isolating the forced vibration, we describe the vertical displacement with the help of the variable z:

$$y = \frac{a\omega_0^2}{\omega_0^2 - \omega^2}\sin\omega t + z,$$

and for time variable we take $\tau = \omega t$, and introduce the angular variable ψ defined through the equation:

$$\varphi = \omega t + \psi = \tau + \psi.$$

The equations of motion then become:

$$\ddot{\psi} = -\mu\left[\left(\ddot{z} + \frac{g}{\omega^2} - \frac{a\omega_0^2}{\omega_0^2 - \omega^2}\sin\tau\right)r^{-1}\cos(\tau+\psi) + \varkappa\frac{\omega_0}{\omega}(1+\dot{\psi})\right]$$

$$\ddot{z} + \left(\frac{\omega_0}{\omega}\right)^2 z = -\mu\left[f\frac{\omega_0}{\omega}\dot{z} + \frac{fa\omega_0^2}{\omega_0^2-\omega^2}\cdot\frac{\omega_0}{\omega}\cos\tau\right.$$
$$\left. + pr(\ddot{\psi}\cos(\tau+\psi) - (1+\dot{\psi})^2\sin(\tau+\psi))\right]$$

with $\quad \dot{z} = \dfrac{\mathrm{d}z}{\mathrm{d}\tau}\cdots$

Putting $\dot{\psi} = \nu\chi$, with $\nu = \mu^{1/2}$, and then $\dot{z} = \xi$, we finally obtain a set of first-order differential equations in standard form:

(8.183)
$$\dot{\psi} = \nu\chi$$
$$\dot{\chi} = \nu\left[\left(\left(\frac{\omega_0}{\omega}\right)^2 z - \frac{g}{\omega^2} + \frac{a\omega_0^2}{\omega_0^2-\omega^2}\sin\tau\right)r^{-1}\cos(\tau+\psi)\right.$$
$$\left. - \varkappa\frac{\omega_0}{\omega}(1+\nu\chi) + O(\nu^2)\right]$$
$$\dot{z} = \xi$$

(8.184)
$$\dot{\xi} = -\left(\frac{\omega_0}{\omega}\right)^2 z + \nu^2\left[-f\frac{\omega_0}{\omega}\xi - \frac{f\omega_0^3}{\omega(\omega_0^2-\omega^2)}\cos\tau\right.$$
$$\left. + pr(1+\nu\chi)^2\sin(\tau+\psi) + O(\nu^2)\right]$$

to which we can apply Theorem 1, provided that the linear system obtained when $\nu = 0$ has no periodic solution of period 2π, or in other words, providing $\omega_0/\omega \not\equiv 0 \bmod 1$, a condition which we shall always assume to be satisfied.

We can write down the synchronisation equations by taking the means with respect to τ of the right-hand sides of (8.183) with $z = 0$ and $\nu = 0$. This gives us: $\chi = 0$

(8.185)
$$-\frac{a\omega_0^2}{2(\omega_0^2-\omega^2)r}\sin\psi - \varkappa\frac{\omega_0}{\omega} = 0,$$

which shows that

(8.186)
$$2a^{-1}r\varkappa\left|\frac{\omega_0^2-\omega^2}{\omega_0\omega}\right| < 1$$

is a necessary condition.

The stability matrix associated with (8.185) is:

$$\begin{pmatrix} 0 & 1 \\ -\dfrac{a\omega_0^2\cos\psi}{2(\omega_0^2-\omega^2)r} & 0 \end{pmatrix}$$

whose eigenvalues are given by:

$$\lambda^2 + \frac{a\omega_0^2}{2(\omega_0^2-\omega^2)r}\cos\psi = 0.$$

Of the two solutions provided by the equation (8.185) the only one which can correspond to a stable motion is the ψ_0 for which:

$$\sin\psi_0 = -\frac{2r\varkappa a^{-1}}{\omega_0\omega}(\omega_0^2-\omega^2) \quad \text{and} \quad (\omega_0^2-\omega^2)\cos\psi_0 > 0.$$

Strictly speaking Theorem 1 while guaranteeing for sufficiently small positive v, the existence of a solution of (8.183), (8.184) which is periodic in τ and of period 2π, leaves open the question of its stability. Before tackling this question, we shall first give the details of the asymptotic representation of the solution.

First of all it is clear, by (8.184) that z and ξ are $O(v^2)$. Setting

$$\psi = \psi_0 + v\psi_1, \quad \chi = v\chi_1, \tag{8.187}$$

we can write (8.183) in the form:

$$\dot\psi_1 = v\chi_1$$
$$\dot\chi_1 = \left[\left(\frac{a\omega_0^2\sin\tau}{\omega_0^2-\omega^2} - \frac{g}{\omega^2}\right)r^{-1}(\cos(\tau+\psi_0) - v\psi_1\sin(\tau+\psi_0)) - \varkappa\frac{\omega_0}{\omega} + O(v^2)\right]$$

and putting

$$h(\tau) = \int_0^\tau \left[r^{-1}\left(\frac{a\omega_0^2\sin\tau}{\omega_0^2-\omega^2} - \frac{g}{\omega^2}\right)\cos(\tau+\psi_0) - \frac{\varkappa\omega_0}{\omega}\right]d\tau$$

which is τ-periodic of period 2π, because of the choice of ψ_0, and then

$$\chi_1 = h(\tau) + \zeta_1 \tag{8.188}$$

we obtain

$$\dot\psi_1 = v(\zeta_1 + h(\tau))$$
$$\dot\zeta_1 = v\sin(\tau+\psi_0)\cdot\left[\frac{g}{\omega^2} - \frac{a\omega_0^2}{\omega_0^2-\omega^2}\sin\tau\right]r^{-1}\psi_1 + O(v^2)$$

to which we can once more apply the synchronisation theorem, leading to the associated equations:

$$\zeta_1 + \bar h = 0$$
$$\psi_1 = 0$$

and by (8.187), (8.188):

$$\begin{aligned} \psi(\tau, v) &= \psi_0 + O(v^2) \\ \chi(\tau, v) &= v[h(\tau) - \bar{h} + O(v)] \end{aligned} \tag{8.189}$$

with $\xi(\tau, v) = O(v^2)$, $z = O(v^2)$, the asymptotic representation to the first order in v.

16.2. Analysis of Stability

We have to study the variational system formed from the set of equations (8.183), (8.184) in the neighbourhood of their periodic solution. We shall denote these variations from now on by ψ, χ, z, ξ and make use of the representations (8.189). Accordingly we obtain:

$$\begin{aligned} \dot{\psi} &= v\chi \\ \dot{\chi} &= v\left[\left(\frac{\omega_0}{\omega}\right)^2 r^{-1}\cos(\tau+\psi_0)\cdot z + \left(\frac{g}{\omega^2} - \frac{a\omega_0^2}{\omega_0^2-\omega^2}\sin\tau\right) r^{-1}\sin(\tau+\psi_0)\cdot\psi \right. \\ &\qquad \left. - \frac{v\varkappa\omega_0}{\omega}\chi + O(v^2)\right] \\ \dot{z} &= \xi \\ \dot{\xi} &= -\left(\frac{\omega_0}{\omega}\right)^2 z + v^2\left[-f\frac{\omega_0}{\omega}\xi + pr\cos(\tau+\psi_0)\cdot\psi + O(v)\right] \end{aligned} \tag{8.190}$$

where $O(v)$, $O(v^2)$ are representations, linear in ψ, χ, z, ξ and of order v, v^2 respectively.

If, for small enough v, every solution of the linear system (8.190) tends to 0 as $\tau \to +\infty$, then we shall be able to guarantee, under the same conditions, the asymptotic stability of the periodic solution of the non-linear system (8.183), (8.184).

We shall begin by transforming the system (8.190), by substituting for z, ξ the variables u, v:

$$\xi + i\frac{\omega_0}{\omega}z = u\exp(i\omega_0\tau/\omega)$$

$$\xi - i\frac{\omega_0}{\omega}z = v\exp(-i\omega_0\tau/\omega)$$

giving:

$$\begin{aligned} \dot{\psi} &= v\chi \\ \dot{\chi} &= v\left[\frac{i\omega_0}{2\omega r}(v\exp(-i\omega_0\tau/\omega) - u\exp(i\omega_0\tau/\omega))\cos(\tau+\psi_0)\right. \\ &\qquad \left. + \left(\frac{g}{\omega^2} - \frac{a\omega_0^2}{\omega_0^2-\omega^2}\sin\tau\right) r^{-1}\sin(\tau+\psi_0)\cdot\psi - \frac{v\varkappa\omega_0}{\omega}\chi + O(v^2)\right]. \end{aligned}$$

$$(8.191)\quad \dot{u} = v^2 \exp(-i\omega_0\tau/\omega)\left[-f\frac{\omega_0}{2\omega}(u\exp(i\omega_0\tau/\omega) + v\exp(-i\omega_0\tau/\omega)) + pr\cos(\tau+\psi_0)\cdot\psi + O(v)\right]$$

$$\dot{v} = v^2 \exp(i\omega_0\tau/\omega)\left[-f\frac{\omega_0}{2\omega}(u\exp(i\omega_0\tau/\omega) + v\exp(-i\omega_0\tau/\omega)) + pr\cos(\tau+\psi_0)\cdot\psi + O(v)\right].$$

It will be seen that we have obtained a linear system which after a self-evident change of notation can be written in matrix form:

$$(8.192)\qquad \dot{x} = v[A(\theta) + vA_1(\theta) + O(v^2)]x, \quad x\in\mathbb{C}^n$$

with

$$(8.193)\qquad \theta = \{\theta_1,\ldots,\theta_p\},\quad \theta_j = \omega_j\tau,\quad \dot{x} = \frac{dx}{d\tau}$$

the matrices $A(\theta)$, $A_1(\theta)$ being polynomials in $\exp(i\theta_j)$, $1 \leqslant j \leqslant p$, and[2] therefore periodic of period 2π with respect to each variable θ_j.

In the case (8.191) we have $n=4$, $p=2$, so that there are two variables $\theta_1 = \frac{\omega_0}{\omega}\tau$, $\theta_2 = \tau$ and the matrices $A(\theta)$, $A_2(\theta)$ are polynomial functions of $e^{i\theta_1}$, $e^{i\theta_2}$ which can be written, in explicit form, as

(8.194)

$$A(\theta) = \begin{pmatrix} 0 & 1 & 0 & 0 \\ r^{-1}\left(\frac{g}{\omega^2} - \frac{a\omega_0^2\sin\tau}{\omega_0^2-\omega^2}\right)\sin(\tau+\psi_0) & 0 & -\frac{i\omega_0}{2\omega r}\exp(i\omega_0\tau/\omega)\cos(\tau+\psi_0) & \frac{i\omega_0}{2\omega r}\exp(-i\omega_0\tau/\omega)\cos(\tau+\psi_0) \\ 0 & 0 & 0 & 0 \\ 0 & 0 & 0 & 0 \end{pmatrix}$$

(8.195)

$$A_1(\theta) = \begin{pmatrix} 0 & 0 & 0 & 0 \\ 0 & -\varkappa\frac{\omega_0}{\omega} & 0 & 0 \\ pr\exp(-i\omega_0\tau/\omega)\cos(\tau+\psi_0) & 0 & -f\frac{\omega_0}{2\omega} & -f\frac{\omega_0}{2\omega}\exp(-2i\omega_0\tau/\omega) \\ pr\exp(i\omega_0\tau/\omega)\cos(\tau+\psi_0) & 0 & -f\frac{\omega_0}{2\omega}\exp(2i\omega_0\tau/\omega) & -f\frac{\omega_0}{2\omega} \end{pmatrix}$$

[2] The number p introduced here to represent the number of suffixes must not be confused with the dimensionless coefficient written with the same letter in (8.182).

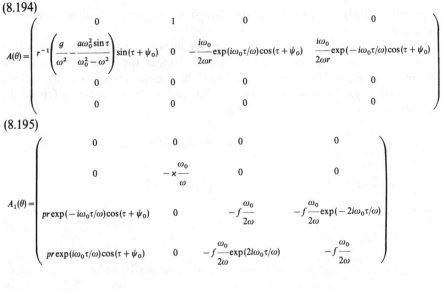

Coming back to the general system (8.192), (8.193) it is clear that we can develop an analysis closely similar to the one already presented for the periodic system (8.176).

We carry out on (8.192) the change of variable $x \to y$ defined by (8.177), U, U_1 now denoting matrices whose elements are polynomials in $e^{i\theta_j}$, $j \in [1, 2, \ldots, p]$. Noting that

$$d/d\tau = \sum_{j=1}^{p} \omega_j(\partial/\partial\theta_j),$$

we are led to the following pair of matrix equations for the determination of the matrix pairs (U, S), (U_1, S_1):

$$A(\theta) - S = \sum_{j=1}^{p} \omega_j \frac{\partial U}{\partial \theta_j} \tag{8.196}$$

$$A_1(\theta) + A(\theta)U(\theta) - U(\theta)S - S_1 = \sum_{j=1}^{p} \omega_j \frac{\partial U_1}{\partial \theta_j}. \tag{8.197}$$

Provided that the frequencies ω_j are such that

$$\sum_{j=1}^{p} k_j \omega_j \neq 0, \quad \forall (k_1, \ldots, k_p) \in \mathbb{Z}^p - 0,$$

one can easily calculate the constant matrix S and the associated matrix $U(\theta)$ satisfying (8.196) and then determine S_1 and $U_1(\theta)$ by (8.197).

One ends up, after the change of variables (8.177), with an equation of the type

$$\dot{y} = v(S + vS_1 + O(v^2))y$$

where S and S_1 are constant matrices, this equation being of a form which makes it easy to analyse the asymptotic stability of the solution $y = 0$, for sufficiently small positive v.

Applying this method to the case (8.194), (8.195), we obtain:

$$S = \begin{pmatrix} 0 & 1 & 0 & 0 \\ -\dfrac{a\omega_0^2 \cos\psi_0}{2(\omega_0^2 - \omega^2)r} & 0 & 0 & 0 \\ 0 & 0 & 0 & 0 \\ 0 & 0 & 0 & 0 \end{pmatrix}$$

$$U = \begin{pmatrix} 0 & 0 & 0 & 0 \\ w_{21} & 0 & w_{23} & w_{24} \\ 0 & 0 & 0 & 0 \\ 0 & 0 & 0 & 0 \end{pmatrix}$$

$$w_{21} = -\frac{g}{r\omega^2}\cos(\tau + \psi_0) + \frac{a\omega_0^2}{2r(\omega_0^2 - \omega^2)}\sin(2\tau + \psi_0)$$

$$w_{23} = -\frac{\omega_0}{2r(\omega_0^2 - \omega^2)}(\omega_0 \cos(\tau + \psi_0) - i\omega \sin(\tau + \psi_0))\exp(+i\omega_0\tau/\omega)$$

$$w_{24} = -\frac{\omega_0}{2r(\omega_0^2-\omega^2)}(\omega_0\cos(\tau+\psi_0)+i\omega\sin(\tau+\psi_0))\exp(-i\omega_0\tau/\omega)$$

whence

$$AU = \begin{pmatrix} w_{21} & 0 & w_{23} & w_{24} \\ 0 & 0 & 0 & 0 \\ 0 & 0 & 0 & 0 \\ 0 & 0 & 0 & 0 \end{pmatrix}$$

$$US = \begin{pmatrix} 0 & 0 & 0 & 0 \\ 0 & w_{21} & 0 & 0 \\ 0 & 0 & 0 & 0 \\ 0 & 0 & 0 & 0 \end{pmatrix}$$

and
$$P(\theta) = A_1(\theta) + A(\theta)U(\theta) - U(\theta)S$$

$$P(\theta) = \begin{pmatrix} w_{21} & 0 & w_{23} & w_{24} \\ 0 & -\varkappa\frac{\omega_0}{\omega} - w_{21} & 0 & 0 \\ pr\exp(-i\omega_0\tau/\omega)\cos(\tau+\psi_0) & 0 & -f\frac{\omega_0}{2\omega} & -f\frac{\omega_0}{\omega}\exp(-2i\omega_0\tau/\omega) \\ pr\exp(i\omega_0\tau/\omega)\cos(\tau+\psi_0) & 0 & -f\frac{\omega_0}{2\omega}\exp(2i\omega_0\tau/\omega) & -f\frac{\omega_0}{2\omega} \end{pmatrix}$$

then

$$S_1 = \begin{pmatrix} 0 & 0 & 0 & 0 \\ 0 & -\varkappa\frac{\omega_0}{\omega} & 0 & 0 \\ 0 & 0 & -f\frac{\omega_0}{2\omega} & 0 \\ 0 & 0 & 0 & -f\frac{\omega_0}{2\omega} \end{pmatrix}$$

$$S + vS_1 = \begin{pmatrix} 0 & 1 & 0 & 0 \\ -\frac{a\omega_0^2\cos\psi_0}{2(\omega_0^2-\omega^2)r} & -v\varkappa\frac{\omega_0}{\omega} & 0 & 0 \\ 0 & 0 & -vf\frac{\omega_0}{2\omega} & 0 \\ 0 & 0 & 0 & -vf\frac{\omega_0}{2\omega} \end{pmatrix}$$

and as $f, \varkappa, (\omega_0^2-\omega^2)\cos\psi_0$ are positive, the eigenvalues of $S + vS_1$ have a real part of the order of v which is negative if $v > 0$. We deduce from this that the periodic solution is stable for τ or $t \to +\infty$.

17. Synchronisation of the Rotation of an Unbalanced Shaft Sustained by Alternating Forces in the Case of Resonance

The equations are those already written in (8.180), (8.182); however we now assume that ω is near ω_0 and express this by:

(8.198) $$\omega_0 = \omega(1+\mu\eta)^{1/2}$$

with ω_0, η given, and $\omega_0 > 0$.

The amplitude of the excitatory force must be of the order of μ, and for that reason we represent it by μH. Thus the equations of the problem are:

(8.199) $$\begin{aligned} \varphi'' &= -\mu[(y''+g)r^{-1}\cos\varphi + \varkappa\omega_0\varphi'] \\ y''+\omega_0^2 y &= -\mu[f\omega_0 y' + pr(\varphi''\cos\varphi - \varphi'^2\sin\varphi) - a\omega_0^2\sin\omega t]. \end{aligned}$$

We introduce $\omega t = \tau$, $\varphi = \tau + \psi$ and by transforming the equations (8.199) obtain, taking (8.198) into account:

(8.200) $$\begin{aligned} \ddot{\psi} &= -\mu[(\ddot{y} + g\omega_0^{-2})r^{-1}\cos(\tau+\psi) + \varkappa(1+\dot{\psi}) + O(\mu)] \\ \ddot{y}+y &= -\mu[\eta y + f(1+\mu\eta)^{1/2}\dot{y} + pr(\ddot{\psi}\cos(\tau+\psi) \\ &\quad - (1+\dot{\psi})^2\sin(\tau+\psi)) - a(1+\mu\eta)\sin\tau]. \end{aligned}$$

We introduce at this stage the new variables u, v defined through:

(8.201) $$y = u\cos(\tau+v)$$

imposing on them the condition

(8.202) $$\dot{u}\cos(\tau+v) - u\dot{v}\sin(\tau+v) = 0$$

so that

(8.203) $$\dot{y} = -u\sin(\tau+v)$$

and

(8.204) $$\ddot{y} = -\dot{u}\sin(\tau+v) - u\dot{v}\cos(\tau+v) - y.$$

The variable χ having been defined, as before, by $\dot{\psi} = \nu\chi$, with $\nu = \mu^{1/2}$ we deduce from (8.200) with the help of (8.202), (8.204):

(8.205) $$\begin{aligned} \dot{\psi} &= \nu\chi \\ \dot{\chi} &= \nu\left[\left(u\cos(\tau+v) - \frac{g}{\omega_0^2}\right)r^{-1}\cos(\tau+\psi) - \varkappa(1+\nu\chi) + O(\nu^2)\right] \\ \dot{u} &= \nu^2[\eta u\cos(\tau+v) - fu\sin(\tau+v) - pr(1+\nu\chi)^2\sin(\tau+\psi) \\ &\quad - a\sin\tau + O(\nu^2)]\sin(\tau+v) \\ \dot{v} &= \nu^2[\eta u\cos(\tau+v) - fu\sin(\tau+v) - pr(1+\nu\chi)^2\sin(\tau+\psi) \\ &\quad - a\sin\tau + O(\nu^2)]\frac{\cos(\tau+v)}{u} \end{aligned}$$

which is a set of equations in which the terms on the right-hand side, including those not explicitly developed and represented by $O(v^2)$ are periodic in τ with period 2π.

17.1. The Modified Standard System

The preceding calculation suggests that it might be useful to have available a theory which would enable one to settle existence and stability questions for the periodic solution of a system of differential equations of the type:

$$(8.206)\qquad \begin{aligned} \frac{dx}{dt} &= vf(x, y, t, v) \\ \frac{dy}{dt} &= v^2 g(x, y, t, v) \end{aligned}$$

where $x \in \mathbb{R}^n$, $y \in \mathbb{R}^p$, f and g are periodic in t with period T, and v is a small parameter.

We begin by analysing the linear system:

$$(8.207)\qquad \begin{aligned} \frac{dx}{dt} &= v(S_{11}x + S_{12}y + f(t)) \\ \frac{dy}{dt} &= v^2(S_{21}x + S_{22}y + g(t)) \end{aligned}$$

where S_{ij} are matrices with constant elements, f and g are t-periodic mappings of period T sending $\mathbb{R}$ into $\mathbb{R}^n$, $\mathbb{R}^p$ respectively.

Let $Z(t, v)$ be the solution matrix associated with (8.207), or in other words the $(n+p) \times (n+p)$ matrix defined by:

$$(8.208)\qquad \frac{dZ}{dt} = vSZ, \quad Z(0, v) = I$$

with $S = \begin{pmatrix} S_{11} & S_{12} \\ vS_{21} & vS_{22} \end{pmatrix}$ and I the identity matrix [42].

It is known that Z can be represented by:

$$Z(t, v) = I + vtS + \frac{(vt)^2}{2!}S^2 + \cdots + \frac{(vt)^q}{q!}S^q + \cdots$$

the series normally converging for bounded v and t, and since it is clear, by recurrence, that

$$S^q = \begin{pmatrix} S_{11}^{(q)}(v) & S_{12}^{(q)}(v) \\ vS_{21}^{(q)}(v) & vS_{22}^{(q)}(v) \end{pmatrix}$$

where $S_{ij}^{(q)}(v)$ are matrices whose elements are polynomials in v, it will be seen that we can write:

$$(8.209)\qquad Z(t, v) = I + vt\begin{pmatrix} \Theta_{11}(t, v) & \Theta_{12}(t, v) \\ v\Theta_{21}(t, v) & v\Theta_{22}(t, v) \end{pmatrix}$$

the matrices $\Theta_{ij}(t, \nu)$ being holomorphic functions of t and ν, such that

$$\Theta_{ij}(t, \nu) = S_{ij} + O(\nu) \tag{8.210}$$

the $O(\nu)$ term being of the same order of magnitude as ν, uniformly with respect to t over any finite interval.

The solution of (8.207), emanating from the point $\begin{pmatrix} c_1 \\ c_2 \end{pmatrix}$ at $t = 0$, is described by the classical formula [42]:

$$\begin{pmatrix} x \\ y \end{pmatrix} = Z(t, \nu)\begin{pmatrix} c_1 \\ c_2 \end{pmatrix} + \int_0^t Z(t - \tau, \nu)\begin{pmatrix} \nu f(\tau) \\ \nu^2 g(\tau) \end{pmatrix} \mathrm{d}\tau \tag{8.211}$$

and the stipulation that this solution is to be t-periodic with period T requires that $\begin{pmatrix} c_1 \\ c_2 \end{pmatrix}$ should satisfy the condition:

$$(I - Z(T, \nu))\begin{pmatrix} c_1 \\ c_2 \end{pmatrix} = \int_0^T Z(T - \tau, \nu)\begin{pmatrix} \nu f(\tau) \\ \nu^2 g(\tau) \end{pmatrix} \mathrm{d}\tau \tag{8.212}$$

which can be written, having regard to (8.209) and (8.210),

$$\begin{pmatrix} S_{11} + O(\nu) & S_{12} + O(\nu) \\ \nu(S_{21} + O(\nu)) & \nu(S_{22} + O(\nu)) \end{pmatrix}\begin{pmatrix} c_1 \\ c_2 \end{pmatrix} = -\frac{1}{\nu T}\int_0^T \left(\begin{pmatrix} I_1 & 0 \\ 0 & I_2 \end{pmatrix} + O(\nu)\right)\begin{pmatrix} \nu f(\tau) \\ \nu^2 g(\tau) \end{pmatrix} \mathrm{d}\tau$$

or

$$(S_{11} + O(\nu))c_1 + (S_{12} + O(\nu))c_2 = -\frac{1}{T}\left[\int_0^T f(\tau)\mathrm{d}\tau + O(\nu f) + O(\nu^2 g)\right]$$

$$(S_{21} + O(\nu))c_1 + (S_{22} + O(\nu))c_2 = -\frac{1}{T}\left[\int_0^T g(\tau)\mathrm{d}\tau + O(f) + O(\nu g)\right].$$

On the assumption that $\begin{pmatrix} S_{11} & S_{12} \\ S_{21} & S_{22} \end{pmatrix}$ is invertible, it is clear that for sufficiently small ν, $\begin{pmatrix} c_1 \\ c_2 \end{pmatrix}$ can be made to satisfy the periodicity condition and furthermore that there exists a real α depending only on the S_{ij}, such that

$$|c_1| \leqslant \alpha(\|f\| + \|g\|)$$
$$|c_2| \leqslant \alpha(\|f\| + \|g\|),$$

$$\|f\| = \operatorname*{Sup}_\tau |f(\tau)|, \quad \|g\| = \operatorname*{Sup}_\tau |g(\tau)|.$$

Reverting to (8.211), we can therefore ensure that there is a $\nu_0 > 0$ and a number $\tilde{\alpha}$ depending only on S_{ij}, such that for $|\nu| < \nu_0$ the system (8.207) has an unique t-periodic solution $x(t, \nu)$, $y(t, \nu)$ of period T satisfying the estimate:

$$\begin{aligned} \|x\| &\leqslant \tilde{\alpha}(\|f\| + \|g\|) \\ \|y\| &\leqslant \tilde{\alpha}(\|f\| + \|g\|) \end{aligned} \quad \text{with} \quad \|x\| = \operatorname*{Sup}_t |x(t, \nu)|. \tag{8.213}$$

This estimate can be improved, if in addition to the assumptions already made, we assume that f and g have zero means, i.e.

$$\int_0^T f(\tau)\,\mathrm{d}\tau = 0, \quad \int_0^T g(\tau)\,\mathrm{d}\tau = 0. \tag{8.214}$$

In fact, starting from the formula (8.212), integrating by parts its right-hand side and using (8.208), (8.214), we get

$$\begin{aligned}
&-\nu T[(S_{11}+O(\nu))c_1+(S_{12}+O(\nu))c_2] \\
&-\nu^2 T[(S_{21}+O(\nu))c_1+(S_{22}+O(\nu))c_2] \\
&=\int_0^T \begin{pmatrix} \nu S_{11} & \nu S_{12} \\ \nu^2 S_{21} & \nu^2 S_{22} \end{pmatrix} Z(T-\tau,\nu) \begin{pmatrix} \nu \int_0^\tau f(s)\,\mathrm{d}s \\ \nu^2 \int_0^\tau g(s)\,\mathrm{d}s \end{pmatrix} \mathrm{d}\tau
\end{aligned}$$

and by (8.209) or $Z(T-\tau,\nu) = \begin{pmatrix} I_1 & 0 \\ 0 & I_2 \end{pmatrix} + O(\nu)$:

$$\begin{aligned}
&(S_{11}+O(\nu))c_1+(S_{12}+O(\nu))c_2 \\
&= -\frac{1}{T}\left[\int_0^T \left(S_{11}\left(\nu\int_0^\tau f(s)\,\mathrm{d}s\right) + S_{12}\left(\nu^2\int_0^\tau g(s)\,\mathrm{d}s\right)\right)\mathrm{d}\tau + O(\nu^2 f) + O(\nu^3 g)\right] \\
&(S_{21}+O(\nu))c_1+(S_{22}+O(\nu))c_2 \\
&= -\frac{1}{T}\left[\int_0^T \left(S_{21}\left(\nu\int_0^\tau f(s)\,\mathrm{d}s\right) + S_{22}\left(\nu^2\int_0^\tau g(s)\,\mathrm{d}s\right)\right)\mathrm{d}\tau + O(\nu^2 f) + O(\nu^3 g)\right]
\end{aligned}$$

from which we see that there exists a $\nu_0 > 0$, $\alpha > 0$ depending only on the S_{ij}, such that for $|\nu| < \nu_0$ we have:

$$|c_1| \leqslant \alpha(|\nu|\,\|f\| + \nu^2\|g\|)$$
$$|c_2| \leqslant \alpha(|\nu|\,\|f\| + \nu^2\|g\|)$$

and similarly it is apparent that there is an $\tilde{\alpha}$ depending only on the S_{ij} and such that the periodic solution of (8.207), represented by (8.211), satisfies, for small enough ν:

$$\begin{aligned}
\|x\| &\leqslant \tilde{\alpha}(|\nu|\,\|f\| + \nu^2\|g\|) \\
\|y\| &\leqslant \tilde{\alpha}(|\nu|\,\|f\| + \nu^2\|g\|).
\end{aligned} \tag{8.215}$$

17.2. Synchronisation of Non-Linear System

If the non-linear system (8.206) has a T-periodic solution which reduces to (x_0, y_0) when $\nu \to 0$, it is clear, by integrating the two equations of (8.206) that:

$$\int_0^T f(x_0, y_0, t, 0)\,\mathrm{d}t = 0, \quad \int_0^T g(x_0, y_0, t, 0)\,\mathrm{d}t = 0. \tag{8.216}$$

From now on we shall suppose that x_0, y_0 satisfy these equations and that, in the neighbourhood of this point, for all real t and $|v| < v_0$, the functions f and g, with values in $\mathbb{R}^n$, $\mathbb{R}^p$ respectively, are continuous, t-periodic with period T, and continuously differentiable twice in x, y. We may add that there is no real restriction in assuming $x_0 = y_0 = 0$, as this is tantamount to a translation of the co-ordinate axes.

Accordingly we assume

$$\int_0^T f(0,0,t,0)\mathrm{d}t = 0, \quad \int_0^T g(0,0,t,0)\mathrm{d}t = 0 \tag{8.217}$$

and put

$$\begin{aligned} r_1(t) &= f(0,0,t,0), \quad r_2(t) = g(0,0,t,0) \\ P_{11}(t) &= f_x(0,0,t,0), \quad P_{12}(t) = f_y(0,0,t,0) \\ P_{21}(t) &= g_x(0,0,t,0), \quad P_{22}(t) = g_y(0,0,t,0) \end{aligned} \tag{8.218}$$

so that we can represent f, g in the form:

$$\begin{aligned} f(x,y,t,v) &= r_1(t) + P_{11}(t)x + P_{12}(t)y + \delta_1(x,y,t,v) \\ g(x,y,t,v) &= r_2(t) + P_{21}(t)x + P_{22}(t)y + \delta_2(x,y,t,v) \end{aligned}$$

where

$$|\delta_j| = O(v) + O(|x|^2 + |y|^2), \quad j = 1,2 \tag{8.219}$$

$$\lim_{\substack{x,x^*,y,y^*,\\ v\to 0}} \frac{|\delta_j(x,y,t,v) - \delta_j(x^*,y^*,t,v)|}{|x-x^*| + |y-y^*|} = 0 \tag{8.220}$$

and rewrite (8.206) as:

$$\begin{aligned} \frac{\mathrm{d}x}{\mathrm{d}t} &= v[r_1(t) + P_{11}(t)x + P_{12}(t)y + \delta_1(x,y,t,v)] \\ \frac{\mathrm{d}y}{\mathrm{d}t} &= v^2[r_2(t) + P_{21}(t)x + P_{22}(t)y + \delta_2(x,y,t,v)]. \end{aligned} \tag{8.221}$$

We now introduce the constant matrices:

$$S_{ij} = \frac{1}{T}\int_0^T P_{ij}(t)\mathrm{d}t \tag{8.222}$$

and the t-periodic matrices $U_{ij}(t)$ of period T:

$$U_{ij}(t) = \int_0^t (P_{ij}(\tau) - S_{ij})\mathrm{d}\tau \tag{8.223}$$

and, in (8.221), make the change of variable:

$$\begin{pmatrix}\xi\\ \eta\end{pmatrix} = K\cdot\begin{pmatrix}x\\ y\end{pmatrix} \tag{8.224}$$

K being the T-periodic matrix:

(8.225)
$$K = I - \nu\begin{pmatrix} U_{11}(t) & U_{12}(t) \\ \nu U_{21}(t) & \nu U_{22}(t) \end{pmatrix}$$

obviously invertible if $|\nu|$ is sufficiently small.

We easily obtain the transform of (8.221) by (8.224) with the help of (8.225), (8.223) in the form:

(8.226)
$$\begin{pmatrix} \xi' \\ \eta' \end{pmatrix} = \nu\begin{pmatrix} S_{11} & S_{12} \\ \nu S_{21} & \nu S_{22} \end{pmatrix}\begin{pmatrix} \xi \\ \eta \end{pmatrix} + \begin{pmatrix} \nu r_1 \\ \nu^2 r_2 \end{pmatrix} - \nu\begin{pmatrix} U_{11} & U_{12} \\ \nu U_{21} & \nu U_{22} \end{pmatrix}\begin{pmatrix} \nu r_1 \\ \nu^2 r_2 \end{pmatrix}$$
$$+ \nu^2\left[\begin{pmatrix} S_{11} & S_{12} \\ \nu S_{21} & \nu S_{22} \end{pmatrix}\begin{pmatrix} U_{11} & U_{12} \\ \nu U_{21} & \nu U_{22} \end{pmatrix} - \begin{pmatrix} U_{11} & U_{12} \\ \nu U_{21} & \nu U_{22} \end{pmatrix}\right.$$
$$\left.\cdot\begin{pmatrix} P_{11} & P_{12} \\ \nu P_{21} & \nu P_{22} \end{pmatrix}\right]\cdot K^{-1}\begin{pmatrix} \xi \\ \eta \end{pmatrix} + K\begin{pmatrix} \nu\delta_1 \\ \nu^2\delta_2 \end{pmatrix}.$$

We now introduce $\xi_0(t, \nu)$, $\eta_0(t, \nu)$, the t-periodic solution of period T of the linear system:

$$\begin{pmatrix} \xi_0' \\ \eta_0' \end{pmatrix} = \nu\begin{pmatrix} S_{11} & S_{12} \\ \nu S_{21} & \nu S_{22} \end{pmatrix}\begin{pmatrix} \xi_0 \\ \eta_0 \end{pmatrix} + \nu\begin{pmatrix} r_1 \\ \nu r_2 \end{pmatrix}$$

The existence of this solution is assured, for small enough $|\nu|$, if as will always be assumed to be so in what follows, the matrix $\begin{pmatrix} S_{11} & S_{12} \\ S_{21} & S_{22} \end{pmatrix}$ is invertible. By (8.215) which can be applied here because of (8.217), we can add that ξ_0, η_0 are of order ν:

(8.227)
$$\|\xi_0\| = O(\nu), \quad \|\eta_0\| = O(\nu).$$

To simplify the notation of (8.226) let

$$\begin{pmatrix} M_{11} & M_{12} \\ \nu M_{21} & \nu M_{22} \end{pmatrix} := \begin{pmatrix} S_{11} & S_{12} \\ \nu S_{21} & \nu S_{22} \end{pmatrix}\begin{pmatrix} U_{11} & U_{12} \\ \nu U_{21}, & \nu U_{22} \end{pmatrix} - \begin{pmatrix} U_{11} & U_{12} \\ \nu U_{21} & \nu U_{22} \end{pmatrix}\begin{pmatrix} P_{11} & P_{12} \\ \nu P_{21} & \nu P_{22} \end{pmatrix}$$

the $M_{ij}(t, \nu)$ thus being T-periodic matrices which depend linearly on ν. Substituting in (8.226) for ξ, η the variable $\tilde{\xi}, \tilde{\eta}$ defined by:

(8.228)
$$\xi = \xi_0(t, \nu) + \tilde{\xi}$$
$$\eta = \eta_0(t, \nu) + \tilde{\eta}$$

we obtain the equation:

(8.229)
$$\begin{pmatrix} \tilde{\xi}' \\ \tilde{\eta}' \end{pmatrix} = \nu\begin{pmatrix} S_{11} & S_{12} \\ \nu S_{21} & \nu S_{22} \end{pmatrix}\begin{pmatrix} \tilde{\xi} \\ \tilde{\eta} \end{pmatrix} - \nu\begin{pmatrix} U_{11} & U_{12} \\ \nu U_{21} & \nu U_{22} \end{pmatrix}\begin{pmatrix} \nu r_1 \\ \nu^2 r_2 \end{pmatrix}$$
$$+ \nu^2\begin{pmatrix} M_{11} & M_{12} \\ \nu M_{21} & \nu M_{22} \end{pmatrix}K^{-1}\cdot\begin{pmatrix} \xi_0 \\ \eta_0 \end{pmatrix} + K\cdot\begin{pmatrix} \nu\delta_1\left(K^{-1}\cdot\begin{pmatrix} \xi_0 \\ \eta_0 \end{pmatrix}, t, \nu\right) \\ \nu^2\delta_2\left(K^{-1}\cdot\begin{pmatrix} \xi_0 \\ \eta_0 \end{pmatrix}, t, \nu\right) \end{pmatrix}$$
$$+ \nu^2\begin{pmatrix} M_{11} & M_{12} \\ \nu M_{21} & \nu M_{22} \end{pmatrix}K^{-1}\cdot\begin{pmatrix} \tilde{\xi} \\ \tilde{\eta} \end{pmatrix} + K\cdot\begin{pmatrix} \nu(\delta_1 - \delta_1^0) \\ \nu^2(\delta_2 - \delta_2^0) \end{pmatrix}$$

with

$$\delta_j^0 = \delta_j\left(K^{-1}\cdot\begin{pmatrix}\xi_0\\ \eta_0\end{pmatrix}, t, v\right), \quad j=1,2$$

$$\delta_j = \delta_j\left(K^{-1}\cdot\begin{pmatrix}\xi_0+\tilde{\xi}\\ \eta_0+\tilde{\eta}\end{pmatrix}, t, v\right)$$

and we can write, with an obvious notation:

$$\begin{aligned}\tilde{\xi}' &= v(S_{11}\tilde{\xi}+S_{12}\tilde{\eta}) + vk_1(t,v) + vL_1(\tilde{\xi},\tilde{\eta},t,v)\\ \tilde{\eta}' &= v^2(S_{21}\tilde{\xi}+S_{22}\tilde{\eta}) + v^2k_2(t,v) + v^2L_2(\tilde{\xi},\tilde{\eta},t,v)\end{aligned} \tag{8.230}$$

where $\|k_j\| = \operatorname{Sup}_t |k_j(t,v)| = O(v)$

$$\lim_{\tilde{\xi},\tilde{\eta}\to 0} \|L_j\| = \lim_{\tilde{\xi},\tilde{\eta}\to 0} \operatorname{Sup}_t |L_j(\tilde{\xi},\tilde{\eta},t,v)| = 0 \tag{8.231}$$

and

$$\lim_{\substack{\tilde{\xi},\tilde{\xi}^*,\tilde{\eta},\tilde{\eta}^*,\\ v\to 0}} \sup_t \frac{|L_j(\tilde{\xi},\tilde{\eta},t,v) - L_j(\tilde{\xi}^*,\tilde{\eta}^*,t,v)|}{|\tilde{\xi}-\tilde{\xi}^*| + |\tilde{\eta}-\tilde{\eta}^*|} = 0.$$

These conditions are sufficient to enable us to define, with (8.230) as starting point, a process of successive approximation which starting from the initial approximation $\tilde{\xi}_0 = \tilde{\eta}_0 = 0$ and having resulted in the approximation $\tilde{\xi}_{m-1}$, $\tilde{\eta}_{m-1}$, calculates $\tilde{\xi}_m, \tilde{\eta}_m$ by means of the linear system

$$\begin{aligned}\tilde{\xi}'_m &= v(S_{11}\tilde{\xi}_m + S_{12}\tilde{\eta}_m) + vk_1(t,v) + vL_1(\tilde{\xi}_{m-1}(t), \tilde{\eta}_{m-1}(t), t, v)\\ \tilde{\eta}'_m &= v^2(S_{21}\tilde{\xi}_m + S_{22}\tilde{\eta}_m) + v^2k_2(t,v) + v^2L_2(\tilde{\xi}_{m-1}(t), \tilde{\eta}_{m-1}(t), t, v).\end{aligned} \tag{8.232}$$

The T-periodic approximation $\tilde{\xi}_{m-1}(t), \tilde{\eta}_{m-1}(t)$ having been obtained, we shall be able to calculate the unique T-periodic solution $\tilde{\xi}_m(t)$, $\tilde{\eta}_m(t)$ of (8.232) for $|v| < v_1$ and show, thanks to the estimates of the type (8.213) and the behaviour described by (8.231), that, for a suitably chosen v_1, this approximation converges, as $m\to\infty$, to a T-periodic solution of the equations (8.230). We can thus state the following theorem.

Theorem 6

Let $f(x,y,t,v)$, $g(x,y,t,v)$ be continuous mappings of $(x,y,t,v)\in\mathscr{U}\times\mathscr{V}\times\mathbb{R}\times\mathscr{I}$ into $\mathbb{R}^n, \mathbb{R}^p$ respectively, where $\mathscr{U},\mathscr{V}$ are open subsets of $\mathbb{R}^n, \mathbb{R}^p$, respectively, and $\mathscr{I}$ is a neighbourhood of zero on the axis of reals. Let f, g be t-periodic of period T, and continuously differentiable twice with respect to x, y, v. Assuming that there exists an $(x_0, y_0)\in\mathscr{U}\times\mathscr{V}$ such that:

$$\begin{aligned}&\int_0^T f(x_0,y_0,t,0)\,dt = 0,\\ &\int_0^T g(x_0,y_0,t,0)\,dt = 0\end{aligned} \tag{8.233}$$

and that the matrix $\begin{pmatrix} S_{11} & S_{12} \\ S_{21} & S_{22} \end{pmatrix}$ where

$$S_{11} = \frac{1}{T}\int_0^T f_x(x_0, y_0, t, 0)\,dt,$$

$$S_{12} = \frac{1}{T}\int_0^T f_y(x_0, y_0, t, 0)\,dt,$$

$$S_{21} = \frac{1}{T}\int_0^T g_x(x_0, y_0, t, 0)\,dt,$$

$$S_{22} = \frac{1}{T}\int_0^T g_y(x_0, y_0, t, 0)\,dt$$

is invertible, then there exists a real $v_1 > 0$ such that for all v with $|v| < v_1$, the set of equations (8.206) has an unique periodic solution of period T which, when $v \to 0$, tends uniformly to the point (x_0, y_0).

17.3. Stability Criterion for Periodic Solution

We take up once more the assumptions of Theorem 6, defining for the periodic system (8.206) the synchronisation point (x_0, y_0) which satisfies (8.233).

With the object of giving a precise expression to the asymptotic representation of the periodic solution, we introduce the variables ξ, η defined by:

(8.234)
$$x = x_0 + v\xi$$
$$y = y_0 + v\eta$$

so that the transformed equations can be written:

$$\xi' = f(x_0, y_0, t, 0) + v[P_{11}(t)\xi + P_{12}(t)\eta + f_v + O(v)]$$
$$\eta' = vg(x_0, y_0, t, 0) + v^2[P_{21}(t)\xi + P_{22}(t)\eta + g_v + O(v)]$$

with
$$f_v = \frac{\partial f}{\partial v}(x_0, y_0, t, 0), \quad g_v = \frac{\partial g}{\partial v}(x_0, y_0, t, 0),$$
$$P_{11}(t) = f_x(x_0, y_0, t, 0), \ldots$$

or with ζ, χ defined by:

(8.235)
$$\xi = \int_0^t f(x_0, y_0, t, 0)\,dt + \zeta$$

$$\eta = v\int_0^t g(x_0, y_0, t, 0)\,dt + \chi$$

$$\zeta' = v\left[P_{11}\cdot\zeta + P_{12}\cdot\chi + P_{11}\cdot\int_0^t f(x_0, y_0, t, 0)\,dt + f_v + O(v)\right]$$

$$\chi' = v^2\left[P_{21}\cdot\zeta + P_{22}\cdot\chi + P_{21}\cdot\int_0^t f(x_0, y_0, t, 0)\,dt + g_v + O(v)\right]$$

to which the synchronisation theorem, Theorem 6, can again be applied, leading one to define the point ζ_0, χ_0 by:

$$(8.236)\qquad \begin{aligned} &S_{11}\cdot\zeta_0 + S_{12}\cdot\chi_0 + \overline{P_{11}(t)\cdot\int_0^t f(x_0, y_0, t, 0)\,dt} + \bar{f}_\nu = 0 \\ &S_{21}\cdot\zeta_0 + S_{22}\cdot\chi_0 + \overline{P_{21}(t)\cdot\int_0^t f(x_0, y_0, t, 0)\,dt} + \bar{g}_\nu = 0 \end{aligned}$$

(where the horizontal line denotes the operation of averaging the t-periodic functions over the period T). This system has an unique solution because the matrix $\begin{pmatrix} S_{11} & S_{12} \\ S_{21} & S_{22} \end{pmatrix}$ is invertible.

By (8.234), (8.235) and (8.236) we are therefore led to the asymptotic representation:

$$(8.237)\qquad \begin{aligned} x(t,\nu) &= x_0 + \nu\left(\zeta_0 + \int_0^t f(x_0, y_0, t, 0)\,dt\right) + O(\nu^2) \\ y(t,\nu) &= y_0 + \nu\chi_0 + O(\nu^2). \end{aligned}$$

To investigate the stability of this periodic solution, it is necessary to discuss the behaviour for $t \to +\infty$ of the solution of the linear periodic variational system derived from (8.206) in the neighbourhood of its periodic solution. Reserving henceforth the notation (x, y) to denote this variation, the system can be written in the form:

$$\begin{aligned} x' &= \nu[f_x(x(t,\nu), y(t,\nu), t, \nu)\cdot x + f_y(x(t,\nu), y(t,\nu), t, \nu)\cdot y] \\ y' &= \nu^2[g_x(x(t,\nu), y(t,\nu), t, \nu)\cdot x + g_y(x(t,\nu), y(t,\nu), t, \nu)\cdot y] \end{aligned}$$

or, using the expansion (8.237):

$$(8.238)\qquad \begin{aligned} x' &= \nu[(P_{11}(t) + \nu Q_{11}(t) + O(\nu^2))\cdot x + (P_{12}(t) + \nu Q_{12}(t) + O(\nu^2))\cdot y] \\ y' &= \nu^2[(P_{21}(t) + \nu Q_{21}(t) + O(\nu^2))\cdot x + (P_{22}(t) + \nu Q_{22}(t) + O(\nu^2))\cdot y] \end{aligned}$$

the matrices $P_{ij}(t)$, $Q_{ij}(t)$ being easily made explicit after insertion in $f_x, f_y, \ldots$ of the representations (8.237) and expansion to the first order in ν; thus, for example:

$$\begin{aligned} P_{11}(t) &= f_x(x_0, y_0, t, 0) \\ Q_{11}(t) &= f_{xx}(x_0, y_0, t, 0)\cdot\left(\zeta_0 + \int_0^t f(x_0, y_0, t, 0)\,dt\right) + f_{xy}(x_0, y_0, t, 0)\cdot\chi_0 \\ &\quad + f_{x\nu}(x_0, y_0, t, 0), \text{ etc.} \end{aligned}$$

We now, in (8.238), make the change of variable $\begin{pmatrix} x \\ y \end{pmatrix} \to \begin{pmatrix} u \\ v \end{pmatrix}$ defined by:

$$(8.239)\qquad \begin{pmatrix} u \\ v \end{pmatrix} = \left[I - \nu\begin{pmatrix} U_{11} & U_{12} \\ \nu U_{21} & \nu U_{22} \end{pmatrix} - \nu^2\begin{pmatrix} V_{11} & V_{12} \\ \nu V_{21} & \nu V_{22} \end{pmatrix}\right]\cdot\begin{pmatrix} x \\ y \end{pmatrix}$$

where the matrices $U_{ij}(t)$, $V_{ij}(t)$ are T-periodic but are otherwise as yet unspecified, and obtain:

$$\begin{pmatrix} u' \\ v' \end{pmatrix} = \left[I - v \begin{pmatrix} U_{11} & U_{12} \\ vU_{21} & vU_{22} \end{pmatrix} - v^2 \begin{pmatrix} V_{11} & V_{12} \\ vV_{21} & vV_{22} \end{pmatrix} \right]$$

$$\cdot \left[v \begin{pmatrix} P_{11} & P_{12} \\ vP_{21} & vP_{22} \end{pmatrix} + v^2 \begin{pmatrix} Q_{11} & Q_{12} \\ vQ_{21} & vQ_{22} \end{pmatrix} + v^3 \begin{pmatrix} O(1) & O(1) \\ O(v) & O(v) \end{pmatrix} \right] \cdot \begin{pmatrix} x \\ y \end{pmatrix}$$

$$- \left[v \begin{pmatrix} U'_{11} & U'_{12} \\ vU'_{21} & vU'_{22} \end{pmatrix} + v^2 \begin{pmatrix} V'_{11} & V'_{12} \\ vV'_{21} & vV'_{22} \end{pmatrix} \right] \cdot \begin{pmatrix} x \\ y \end{pmatrix}$$

where the terms $O(1)$, $O(v)$ are matrices whose elements are of the same order of magnitude as 1, v respectively. Taking

$$U_{ij}(t) = \int_0^t (P_{ij}(\tau) - S_{ij})\,\mathrm{d}\tau$$

with

$$S_{ij} = \frac{1}{T} \int_0^T P_{ij}(t)\,\mathrm{d}t$$

and then

$$V_{ij}(t) = \int_0^t (R_{ij}(\tau) - \Sigma_{ij})\,\mathrm{d}\tau \tag{8.240}$$

where the Σ_{ij} are the mean values of the periodic matrices $R_{ij}(t)$ which remain to be defined, we can write:

$$\begin{pmatrix} u' \\ v' \end{pmatrix} = \left[v \begin{pmatrix} S_{11} & S_{12} \\ vS_{21} & vS_{22} \end{pmatrix} + v^2 \begin{pmatrix} Q_{11} & Q_{12} \\ vQ_{21} & vQ_{22} \end{pmatrix} \right.$$

$$- v^2 \begin{pmatrix} U_{11} & U_{12} \\ vU_{21} & vU_{22} \end{pmatrix} \begin{pmatrix} P_{11} & P_{12} \\ vP_{21} & vP_{22} \end{pmatrix} - v^2 \begin{pmatrix} V'_{11} & V'_{12} \\ vV'_{21} & vV'_{22} \end{pmatrix}$$

$$\left. + v^3 \begin{pmatrix} O(1) & O(1) \\ O(v) & O(v) \end{pmatrix} \right] \cdot \left[I + v \begin{pmatrix} U_{11} & U_{12} \\ vU_{21} & vU_{22} \end{pmatrix} + v^2 \begin{pmatrix} O(1) & O(1) \\ O(v) & O(v) \end{pmatrix} \right] \begin{pmatrix} u \\ v \end{pmatrix}$$

$$\begin{pmatrix} u' \\ v' \end{pmatrix} = \left\{ v \begin{pmatrix} S_{11} & S_{12} \\ vS_{21} & vS_{22} \end{pmatrix} + v^2 \left[\begin{pmatrix} Q_{11} & Q_{12} \\ vQ_{21} & vQ_{22} \end{pmatrix} \right. \right.$$

$$\left. + \begin{pmatrix} S_{11} & S_{12} \\ vS_{21} & vS_{22} \end{pmatrix} \begin{pmatrix} U_{11} & U_{12} \\ vU_{21} & vU_{22} \end{pmatrix} - \begin{pmatrix} U_{11} & U_{12} \\ vU_{21} & vU_{22} \end{pmatrix} \begin{pmatrix} P_{11} & P_{12} \\ vP_{21} & vP_{22} \end{pmatrix} \right]$$

$$\left. - v^2 \begin{pmatrix} V'_{11} & V'_{12} \\ vV'_{21} & vV'_{22} \end{pmatrix} + v^3 \begin{pmatrix} O(1) & O(1) \\ O(v) & O(v) \end{pmatrix} \right\} \begin{pmatrix} u \\ v \end{pmatrix}$$

or with

$$\begin{pmatrix} R_{11} & R_{12} \\ vR_{21} & vR_{22} \end{pmatrix} = \begin{pmatrix} S_{11}U_{11} - U_{11}P_{11} + Q_{11} & S_{11}U_{12} - U_{11}P_{12} + Q_{12} \\ v(S_{21}U_{11} - U_{21}P_{11} + Q_{21}) & v(S_{21}U_{12} - U_{21}P_{12} + Q_{22}) \end{pmatrix}$$

and $V_{ij}(t)$ defined by (8.240):

$$\begin{pmatrix} u' \\ v' \end{pmatrix} = v \begin{pmatrix} S_{11} + v\Sigma_{11} + O(v^2) & S_{12} + v\Sigma_{12} + O(v^2) \\ v(S_{21} + v\Sigma_{21} + O(v^2)) & v(S_{22} + v\Sigma_{22} + O(v^2)) \end{pmatrix} \cdot \begin{pmatrix} u \\ v \end{pmatrix} \tag{8.241}$$

the matrix elements $O(v^2)$ being t-periodic with period T and of order v^2 uniformly with respect to t.

The question which has to be answered is whether, for small $v > 0$, every solution (u, v) of (8.241) tends to 0 as $t \to +\infty$. The answer depends essentially on the properties of the numerical matrix:

(8.242) $$\begin{pmatrix} S_{11} + v\Sigma_{11} & S_{12} + v\Sigma_{12} \\ vS_{21} & vS_{22} \end{pmatrix}$$

whose eigenvalues tend, as $v \to 0$ to the roots of the equation

$$\det(S_{11} - \lambda)\cdot(-\lambda)^p = 0$$

which has a zero of order p.

We first investigate the behaviour of the eigenvalues which tend to 0 and to this end we put $\lambda = v\sigma$ in the expression for the characteristic polynomial. Thus from:

$$\det\begin{pmatrix} S_{11} + v(\Sigma_{11} - \sigma I) & S_{12} + v\Sigma_{12} \\ vS_{21} & v(S_{22} - \sigma I) \end{pmatrix} = v^p \det\begin{pmatrix} S_{11} + v(\Sigma_{11} - \sigma I) & S_{12} + v\Sigma_{12} \\ S_{21} & S_{22} - \sigma I \end{pmatrix}$$

it is clear that the limiting values of σ, when $v \to 0$, are the roots of:

(8.243) $$\det\begin{pmatrix} S_{11} & S_{12} \\ S_{21} & S_{22} - \sigma I \end{pmatrix} = 0.$$

We shall make the hypotheses that the roots of (8.243) are all simple and have their real parts negative. We can then represent the eigenvalues in the neighbourhood of zero by:

(8.244) $$\lambda_j = v(\sigma_j + O(v)), \quad \operatorname{Re}\sigma_j < 0, \\ \sigma_j \neq \sigma_{j'}, \quad j \neq j', \quad j \in [1, 2, \ldots, p].$$

The other n eigenvalues of (8.242) tend to the roots of

(8.245) $$\det(S_{11} - \lambda I) = 0, \quad \text{as} \quad v \to 0.$$

We shall suppose the roots of this equation to be simple, non-zero, and with negative or zero real parts. In actual fact, it is by no means exceptional to obtain purely imaginary roots. If λ_0 is one of these, we shall need to know, for the corresponding eigenvalue of (8.242) the correction term to the first order in v. We obtain it without difficulty in the form

$$\lambda = \lambda_0 + v(\sigma + O(v)),$$

as the solution of

$$\theta(\lambda, v) = \det\begin{pmatrix} S_{11} + v\Sigma_{11} - \lambda I & S_{12} + v\Sigma_{12} \\ vS_{21} & vS_{22} - \lambda I \end{pmatrix} = 0$$

with

(8.246) $$\theta(\lambda_0, 0) = 0 \quad \text{and} \quad \sigma = -\left.\frac{\theta_v}{\theta_\lambda}\right|_{\substack{v=0 \\ \lambda=\lambda_0}}$$

We suppose the numbers σ associated in this way with the purely imaginary eigenvalues of (8.245) all to have a negative real part.

It can be proved that if all these conditions are fulfilled, then, for small enough positive ν, every solution of (8.241) tends to zero as $t \to +\infty$, or in other words the zero solution is asymptotically stable.

We note in fact that the eigenvectors of the matrix (8.242), which we suppose to be of unit norm, have as limit when $\nu \to 0$, a set of independent vectors which can be defined as follows: we associate with each root λ of (8.245) the normalised vector of $\mathbb{R}^{n+p}$ such that

$$(S_{11} - \lambda I)u = 0, \quad v = 0$$

and with every root σ of (8.243) the vector $\begin{pmatrix} u \\ v \end{pmatrix}$ suitably normalised which is a solution of:

$$S_{11}u + S_{12}v = 0$$
$$S_{21}u + (S_{22} - \sigma)v = 0.$$

We note that these equations are equivalent to

$$u = -S_{11}^{-1}S_{12}v, \quad (S_{22} - S_{21}S_{11}^{-1}S_{12} - \sigma I)v = 0$$

and consequently that (8.243) is equivalent to:

$$\det(S_{22} - S_{21}S_{11}^{-1}S_{12} - \sigma I) = 0$$

so that σ is a simple eigenvalue of the matrix $S_{22} - S_{21}S_{11}^{-1}S_{12}$ from which we deduce the uniqueness of v and of u.

The $n+p$ eigenvectors $\begin{pmatrix} u \\ v \end{pmatrix}$ defined in this way through taking the limit as $\nu \to 0$, form a linearly independent set as can easily be shown by direct verification. It follows from this that the matrix $\mathscr{R}(\nu)$ which allows diagonalisation of (8.242) has a limit $\mathscr{R}(0)$ which is regular, or in other words invertible, when $\nu \to 0$. We can therefore analyse the stability of the null solution of (8.241) by making the change of variable

$$\begin{pmatrix} u \\ v \end{pmatrix} = \mathscr{R}(\nu)\begin{pmatrix} \tilde{u} \\ \tilde{v} \end{pmatrix}$$

which leads to

$$\begin{pmatrix} \tilde{u}' \\ \tilde{v}' \end{pmatrix} = (\nu(\diagdown) + O(\nu^3))\begin{pmatrix} \tilde{u} \\ \tilde{v} \end{pmatrix}$$

the matrix $(\diagdown)$ having non-zero elements only on its leading diagonal, these being the eigenvalues of (8.242). From the properties of these eigenvalues the conclusion of stability can easily be drawn.

We may therefore state the following theorem.

Theorem 7

With the same hypotheses as those of Theorem 6 implying the existence of a t-periodic solution of the set of equations (8.206) of period T, tending toward the synchronisation point x_0, y_0 when $\nu \to 0$, we define the t-periodic matrices

$$U_{ij}(t) = \int_0^t (P_{ij}(\tau) - S_{ij})\,d\tau$$

and the matrices $Q_{11}(t)$, $Q_{12}(t)$ which are the derivatives with respect to ν, taken for $\nu = 0$, of

$$f_x(x(t,\nu), y(t,\nu), t, \nu),\ f_y(x(t,\nu), y(t,\nu), t, \nu)$$

with $x(t,\nu)$, $y(t,\nu)$ defined by the expansions (8.237). We introduce the mean values

$$\Sigma_{11} = \frac{1}{T}\int_0^T (S_{11}U_{11}(t) - U_{11}(t)P_{11}(t) + Q_{11}(t))\,dt$$

$$\Sigma_{12} = \frac{1}{T}\int_0^T (S_{11}U_{12}(t) - U_{11}(t)P_{12}(t) + Q_{12}(t))\,dt$$

and make the following assumptions additional to those of Theorem 6:

1. The roots of

$$\det\begin{pmatrix} S_{11} & S_{12} \\ S_{21} & S_{22} - \sigma \end{pmatrix} = 0$$

are simple and their real parts are strictly negative.

2. The roots of $\det(S_{11} - \lambda) = 0$ are simple, non-zero and their real parts are non-positive.

For all roots λ_0 with zero real part, the number σ defined by (8.246) has a strictly negative real part.

Under these assumptions the periodic solution of period T of the system (8.206), is for small enough $\nu > 0$, asymptotically stable when $t \to +\infty$. The conditions stated are merely sufficient; some of them could be weakened.

17.4. Application

Let us apply Theorem 6 to the set of equations (8.205) to which we have reduced the problem of the synchronisation of the rotary motion of an unbalanced shaft sustained by alternating vertical forces of a frequency in the neighbourhood of resonance.

The synchronisation equations can be written:

(8.247)
$$\chi = 0$$

$$\frac{ur^{-1}}{2}\cos(v - \psi) - \varkappa = 0$$

$$-\frac{fu}{2} - \frac{a\cos v}{2} - \frac{pr}{2}\cos(v - \psi) = 0$$

$$\frac{\eta}{2} - \frac{pr}{2u}\sin(\psi - v) + \frac{a}{2u}\sin v = 0.$$

From the two last, namely

$$(8.248)\qquad \begin{aligned} a\cos v &= -(fu + pr\cos(v-\psi)) \\ a\sin v &= -\eta u + pr\sin(\psi - v) \end{aligned}$$

which will later on serve for the calculation of v, we obtain:

$$a^2 = (f^2+\eta^2)u^2 + 2fpru\cos(v-\psi) - 2\eta pru\sin(\psi-v) + p^2r^2$$

whence, bearing in mind the second equation (8.247) recalled here

$$(8.249)\qquad \begin{aligned} u\cos(v-\psi) &= 2\varkappa r \\ u\sin(v-\psi) &= \frac{a^2}{2\eta pr} - (2\eta pr)^{-1}[(f^2+\eta^2)u^2 + 4fp\varkappa r^2 + p^2r^2] \end{aligned}$$

and by eliminating $v-\psi$, we obtain the equation which enables us to calculate the amplitude u

$$4\eta^2p^2r^2u^2 = 16\eta^2p^2r^4\varkappa^2 + [(a^2 - 4fp\varkappa r^2 - p^2r^2) - (f^2+\eta^2)u^2]^2$$

or

$$(8.250)\qquad \begin{aligned} &(f^2+\eta^2)^2u^4 - 2[(f^2+\eta^2)(a^2-4fp\varkappa r^2-p^2r^2) + 2\eta^2p^2r^2]u^2 \\ &\quad + (a^2-4fp\varkappa r^2-p^2r^2)^2 + 16\eta^2p^2r^4\varkappa^2 = 0. \end{aligned}$$

A simple calculation shows that this equation has a real positive root only if

$$(8.251)\qquad \left(\frac{a}{r}\right)^2 > p(p+4f\varkappa) - \frac{\eta^2p^2}{f^2+\eta^2} + 4(f^2+\eta^2)\varkappa^2.$$

Thus for a given value of the detuning η between the frequencies ω, ω_0 we can obtain by (8.250), (8.249), (8.248) and $\chi = 0$ two possible synchronised motions, provided that the condition (8.251) is fulfilled. This expresses the physical condition that the amplitude of the excitatory force represented by a has to exceed a certain threshold value.

It could easily be verified that for such values the Jacobian matrix associated with (8.247) is invertible. It would then remain only to analyse the stability of the motion on the basis of Theorem 7.

Chapter IX. Stability of a Column Under Compression – Mathieu's Equation

We discuss the problem of the buckling of an elastic column under a vertical load in the two cases where its lower end is jointed (i.e. free to swivel), or fixed (firmly embedded), and where the load acts vertically through the point of swivel in the former case. The critical static and dynamic loads can be calculated for the continuous model or the discretised models. The difference between these two concepts becomes clearly apparent if the loading is arranged in such a way that the load stresses at the point of application always have the same direction as the column.

An analysis of the forced vibrations of the column caused by a vertical loading whose magnitude is a periodic function of time leads to the discussion of the regions of instability of a Mathieu equation. These can be determined by an analysis based on evaluating determinants of infinite order. An introduction to the basic theory of these determinants is given in this chapter.

Buckling of a Column

We consider a column on which a time-dependent vertical load $P(t)$ is imposed at its top end A. The column may be treated as a vertical beam under flexure, and we denote by $v(x, t)$ the horizontal displacement of a point Q at the point x of the neutral axis, by T the shear, and M the moment at Q of the stresses exerted across the horizontal cross-section at Q on the material of the column in the part between A and Q.

We have seen (Chapter 6) that it is possible to express, in the context of the theory of beams, M and T in the form:

(9.1) $$M = EI\frac{\partial^2 v}{\partial x^2}, \quad T + \frac{\partial M}{\partial x} = 0.$$

In the static case P is independent of t, and the equilibrium of the part AQ gives:

(9.2) $$Pv + M = 0$$

or

(9.3) $$Pv + EI\frac{\partial^2 v}{\partial x^2} = 0.$$

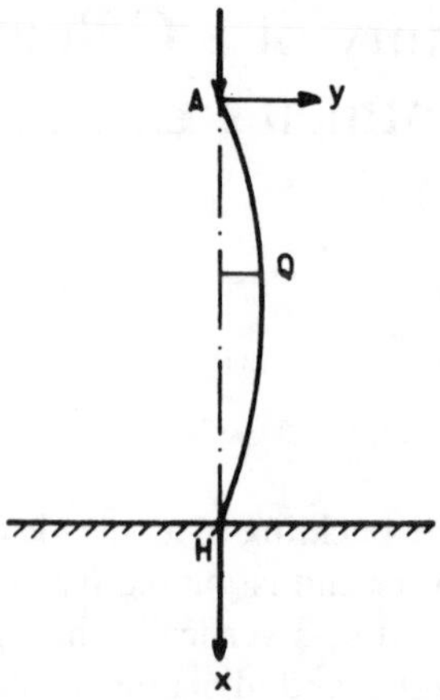

Figure 9.1

To this equation must be added the boundary condition

(9.4) $$v = 0 \quad \text{at} \quad x = 0, x = l$$

on the hypothesis, implicit in what has been said, that A remains vertically above H, which is a legitimate assumption if the column is not embedded at H, but is simply hinged at that point.

However it is clear that the system (9.3), (9.4) where P and EI are constants, has no solution other than $v = 0$, except when $l\sqrt{\frac{P}{EI}} = k\pi$, $k = 1, 2, \ldots$ Accordingly the first critical load corresponds to $k = 1$, and has the value:

(9.5) $$P_1^* = \pi^2 \frac{EI}{l^2}.$$

In the dynamic case, we have to take account of the effects of inertia; the external forces applied to the part AQ together with the inertia forces have zero moment at Q so that:

(9.6) $$Pv + M + \int_0^x \rho \frac{\partial^2 v}{\partial t^2}(\xi, t) \cdot (x - \xi)\, d\xi = 0$$

where ρ is the mass per unit length.

We deduce from (9.6) after differentiating twice with respect to x:

$$P\frac{\partial^2 v}{\partial x^2} + \frac{\partial^2 M}{\partial x^2} + \rho \frac{\partial^2 v}{\partial t^2} = 0$$

and taking account of (9.1):

(9.7) $$EI\frac{\partial^4 v}{\partial x^4} + P(t)\frac{\partial^2 v}{\partial x^2} + \rho \frac{\partial^2 v}{\partial t^2} = 0$$

on the assumption that EI is independent of x. The boundary conditions (9.4) suggest that the solutions can be represented by:

$$v(x,t)=f_k(t)\sin\frac{k\pi x}{l} \tag{9.8}$$

with $f_k(t)$ satisfying the differential equation:

$$\frac{d^2 f_k}{dt^2}+\left(\frac{k\pi}{l}\right)^4\frac{EI}{\rho}\left(1-\frac{P(t)}{P_k^*}\right)f_k=0 \tag{9.9}$$

$$P_k^*=\frac{k^2\pi^2 EI}{l^2}. \tag{9.10}$$

Suppose the excitation to be periodic:

$$P(t)=P_0+P_1\phi(t), \tag{9.11}$$

where $\phi(t)$ is periodic in t with period T, and P_0, P_1 are constants.

We can put (9.9) in the form

$$\frac{d^2 f}{dt^2}+\Omega^2(1-2\mu\phi(t))f=0 \tag{9.12}$$

with the pair (Ω,μ) being defined for every integer k by:

$$\Omega_k=\frac{k\pi}{l}\sqrt{\frac{EI}{\rho}\left(1-\frac{P_0}{P_k^*}\right)},\quad \mu_k=\frac{P_1}{2(P_k^*-P_0)}. \tag{9.13}$$

In the particular case where $\phi(t)=\cos\omega t$ we obtain from (9.12) a Mathieu equation, or in the general case a Hill equation and it becomes essential to be able to tell whether or not the null solution of (9.12) is stable.

Analysis of Stability

Consider the equation

$$\frac{d^2 y}{dt^2}+Q(t)y=0 \tag{9.14}$$

in which $Q(t)$ is a continuous periodic function of period T.

Introducing the solutions $y_1(t)$, $y_2(t)$ of (9.14) satisfying the initial conditions:

$$\begin{aligned} y_1(0)&=1 \quad y_1'(0)=0\\ y_2(0)&=0 \quad y_2'(0)=1, \end{aligned} \tag{9.15}$$

and noting that $y_1(t+T)$, $y_2(t+T)$ is also a solution of (9.14), it can clearly be seen that there exists a matrix $A=(a_{ij})$ such that:

$$\left.\begin{aligned} y_1(t+T)&=a_{11}y_1(t)+a_{12}y_2(t)\\ y_2(t+T)&=a_{21}y_1(t)+a_{22}y_2(t) \end{aligned}\right\},\ \forall t \tag{9.16}$$

and taking (9.15) into account it is clear that:

$$a_{11} = y_1(T) \quad a_{12} = y_1'(T)$$
$$a_{21} = y_2(T) \quad a_{22} = y_2'(T). \tag{9.17}$$

We now seek to diagonalise A, and to this end, let us calculate its eigenvalues ρ, which are the roots of

$$\det \begin{vmatrix} y_1(T) - \rho & y_1'(T) \\ y_2(T) & y_2'(T) - \rho \end{vmatrix} = 0$$

i.e.

$$\rho^2 - 2\Delta\rho + 1 = 0 \tag{9.18}$$

because $(y_1 y_2' - y_1' y_2)(t)$, the Wronskian of (9.14) is independent of t and its value for $t = 0$ is 1.

In addition we have:

$$\Delta = \tfrac{1}{2}(y_1(T) + y_2'(T)). \tag{9.19}$$

If the roots ρ_1, ρ_2 of (9.18) are distinct we can define a matrix S, such that $SAS^{-1} = \begin{pmatrix} \rho_1 & 0 \\ 0 & \rho_2 \end{pmatrix}$ and writing $\begin{pmatrix} \tilde{y}_1 \\ \tilde{y}_2 \end{pmatrix} = S \begin{pmatrix} y_1 \\ y_2 \end{pmatrix}$, we can verify using (9.16) that

$$\tilde{y}_1(t + T) = \rho_1 \tilde{y}_1(t), \quad \tilde{y}_2(t + T) = \rho_2 \tilde{y}_2(t)$$

whence
$$\tilde{y}_j(t) = \exp\left(\frac{t}{T} \log \rho_j\right) \cdot p_j(t), \quad j = 1, 2,$$

with $p_j(t)$ periodic of period T.

In view of the foregoing, if $|\Delta| > 1$, then ρ_1 and ρ_2 are real, and $\rho_1 \rho_2 = 1$; one of the roots is of modulus greater than 1 and the corresponding solution $\tilde{y}(t)$ is unbounded as $t \to +\infty$. This implies that $y = 0$ is an unstable solution of (9.14).

If $|\Delta| < 1$, ρ_1 and ρ_2 have modulus 1 and the solutions $\tilde{y}_j(t)$ are bounded, so that the null solution of (9.14) is stable.

If $\Delta = \pm 1$, (9.18) has a double root $\rho = +1$ or $\rho = -1$. Even if it is no longer certain that A can be diagonalised, it can at least be put into triangular form, i.e. there is a matrix S such that $SAS^{-1} = \begin{pmatrix} \rho & 0 \\ & \rho \end{pmatrix}$ and the solution $\tilde{y}_1(t)$ satisfies $\tilde{y}_1(t + T) = \rho \tilde{y}_1(t)$ with $\rho = +1$ or $\rho = -1$. In the case $\Delta = +1$ we have a periodic solution of period T, and in the other case $\Delta = -1$ a periodic solution of period $2T$.

It is tempting to apply these results to the equation (9.12), basing the discussion of the various possibilities on the position in the plane of the point of co-ordinates Ω, μ. The determinant Δ appears as a continuous function of μ and Ω and the regions of stability correspond to $|\Delta| < 1$, while those of instability are defined by $|\Delta| > 1$. The boundaries between them define the sets of points for which there exist periodic solutions of period T or $2T$.

Incidentally we can define more precisely what happens near the Ω axis, that is when μ is close to zero.

When $\mu \to 0$ the fundamental solutions $y_1(t)$, $y_2(t)$ tend respectively to the limits $y_1 = \cos \Omega t$, $y_2 = \dfrac{\sin \Omega t}{\Omega}$ so that Δ defined by (9.19) tends to $\Delta_0 = \cos \Omega T$.

It is therefore clear that $|\Delta| < 1$ when $\dfrac{\Omega T}{\pi}$ is not an integer; for $|\mu|$ sufficiently small, $|\Delta| < 1$ and the null solution is stable. An important consequence is that the separating boundary curves of which we spoke earlier, drawn in the half-plane $\mu > 0$ (or in the half-plane $\mu < 0$) can issue only from points on the axis $\mu = 0$ corresponding to values $\Omega = \dfrac{n\pi}{T}$, for which n is an integer.

A Discretised Model of the Loaded Column

One could analyse the conditions under which buckling of a loaded column will occur under various somewhat different hypotheses, for example assuming the column to be embedded at H and that its top end A is subjected to lateral displacements. To illustrate the variety of the available methods we shall argue on a discretised model. We consider a system formed of two bars of equal length l, hinged at the point K, the whole being able to turn about a fixed horizontal axis through H. The position of the bars is defined relatively to the upward vertical by their inclinations θ_1, θ_2 and it is assumed that there are elastic restoring couples at H and at K which cause $\theta_1 = \theta_2 = 0$ to be an equilibrium position. We shall also assume that the bars behave, for inertial purposes, as point masses of value m_1 and m_2 concentrated at points distant a_1 and a_2 from H and K respectively ($a_1 < l, a_2 < l$). Lastly at the top of the upper bar is a vertical load of magnitude P acting downwards.

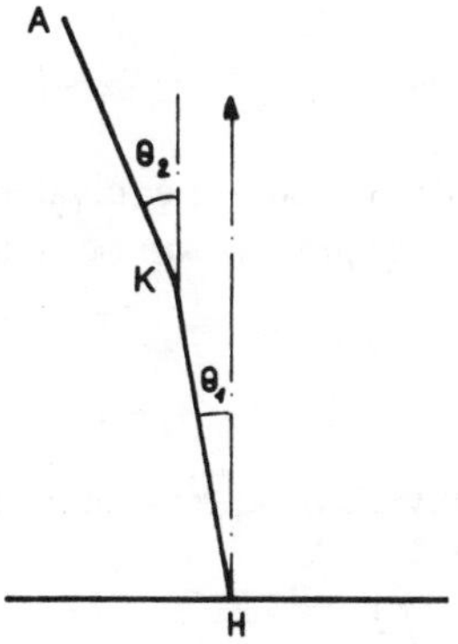

Figure 9.2

Denoting by $-c\theta_1$, $-c(\theta_2 - \theta_1)$ the moments of the elastic couples exerted at H on the bar HK, and at K on the bar KA, c being a positive constant, we can write down, within the context of a linear theory, with the angles θ_1, θ_2 assumed to be small, the following expression for the kinetic energy:

$$T = \tfrac{1}{2}(m_1 a_1^2 \theta_1'^2 + m_2(l\theta_1' + a_2\theta_2')^2)$$

while the virtual work done by the forces acting on the system is

$$\begin{aligned} &-c\theta_1\delta\theta_1 - c(\theta_2-\theta_1)\delta(\theta_2-\theta_1) + Pl(\theta_1\delta\theta_1 + \theta_2\delta\theta_2) \\ &= \delta\left[-\frac{c}{2}(\theta_1^2 + (\theta_2-\theta_1)^2) + \frac{Pl}{2}(\theta_1^2+\theta_2^2)\right] \end{aligned}$$

from which we deduce, by Lagrange's method, that

$$\begin{aligned} &(m_1 a_1^2 + m_2 l^2)\theta_1'' + m_2 l a_2 \theta_2'' + (2c - Pl)\theta_1 - c\theta_2 = 0 \\ &m_2 l a_2 \theta_1'' + m_2 a_2^2 \theta_2'' - c\theta_1 + (c - Pl)\theta_2 = 0 \end{aligned}$$

and we now propose, starting from these equations, to discuss the stability of the solution $\theta_1 = \theta_2 = 0$.

The characteristic equation is:

$$p_0\omega^4 + p_2\omega^2 + p_4 = 0 \tag{9.20}$$

with

$$\begin{aligned} &p_0 = m_1 m_2 a_1^2 a_2^2 \\ &p_2 = [m_1 a_1^2 + m_2 a_2^2 + m_2(l + a_2)^2]c - [m_1 a_1^2 + m_2(l^2 + a_2^2)]Pl \\ &p_4 = c^2 - 3cPl + P^2 l^2. \end{aligned} \tag{9.21}$$

The fact that we have chosen the same constant c at both joints means that we have modelled a column with uniform flexural rigidity. We could approximate to the case of uniform mass distribution by taking

$$m_1 = m_2 = \frac{m}{2}, \quad a_1 = a_2 = \frac{l}{2}$$

or alternatively

$$m_1 = \frac{m}{2}, \quad m_2 = \frac{m}{4}, \quad a_1 = a_2 = l.$$

The conditions for stability, which are equivalent to the condition that every solution ω of (9.20) is strictly imaginary, can be written as:

$$p_2 > 0, \quad p_4 > 0 \tag{9.22}$$

$$p_2^2 - 4p_0 p_4 > 0. \tag{9.23}$$

To interpret (9.22) it is convenient to represent in a plane of co-ordinates $\frac{Pl}{c}, p$ the graphs of

$$p_2 = p_2\left(\frac{Pl}{c}\right) \quad \text{and} \quad p_4 = p_4\left(\frac{Pl}{c}\right),$$

which are respectively a half-line, and a parabolic arc.

Now the abscissae of the points R and S are $\frac{3-\sqrt{5}}{2}, \frac{3+\sqrt{5}}{2}$, respectively, while that of U is:

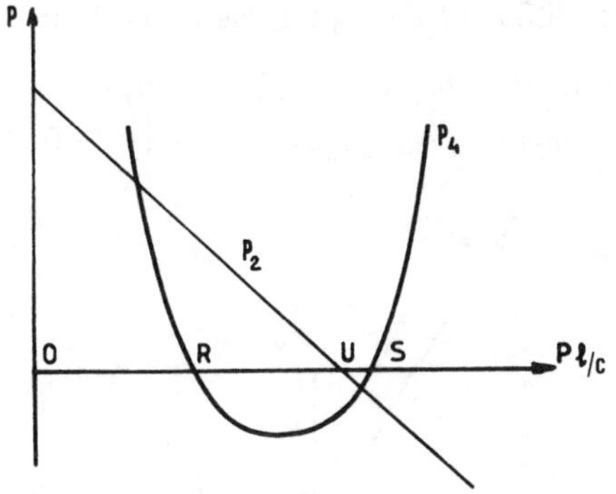

Figure 9.3

$$1 < \frac{Pl}{c} = \frac{m_1 a_1^2 + m_2 a_2^2 + m_2(l+a_2)^2}{m_1 a_1^2 + m_2(l^2 + a_2^2)} < \frac{l^2 + 2la_2 + 2a_2^2}{l^2 + a_2^2} < \frac{5}{2}$$

which proves that U lies between R and S. The conditions (9.22) are therefore equivalent to:

(9.24) $$P < P_1 = \frac{c}{l} \cdot \frac{3-\sqrt{5}}{2}.$$

We must now examine the condition (9.23). For $P=0$ we have:

$$p_2^2 - 4p_0 p_4 = c^2[(\sqrt{m_1}\,a_1 + \sqrt{m_2}\,a_2)^2 + m_2(l+a_2)^2] \cdot [(\sqrt{m_1}\,a_1 - \sqrt{m_2}\,a_2)^2 + m_2(l+a_2)^2] > 0$$

and similarly $p_2^2 - 4p_0 p_4$ takes a positive value at R. Since this expression is of the second degree in P and is positive at O and at R, either the polynomial stays positive on the segment OR, in which case the stability condition reduces to (9.24), or else it has two zeros there and in that case the half-sum of these zeros must lie between 0 and $\frac{3-\sqrt{5}}{2}$. It can easily be shown that the latter situation cannot occur (a simple calculation shows in fact that the half-sum exceeds 1), which confirms that P_1, as given by (9.24), must be the critical load from the dynamic viewpoint. Furthermore this is also the critical load from the static viewpoint, as the equations for static equilibrium obtained by putting $\theta_1'' = \theta_2'' = 0$ in the dynamic equations, are satisfied for values of θ_1, θ_2 not both zero when $P = P_1$.

The Discretised Model with Slave Load [59]

Let us consider the same problem but from now on assuming that, with the help of a suitable servo-device, the load P always has the same direction as the upper bar, its magnitude remaining constant.

We obtain without difficulty the equations of motion:

$$(m_1 a_1^2 + m_2 l^2)\theta_1'' + m_2 l a_2 \theta_2'' + (2c - Pl)\theta_1 - (c - Pl)\theta_2 = 0$$
$$m_2 l a_2 \theta_1'' + m_2 a_2^2 \theta_2'' - c\theta_1 + c\theta_2 = 0$$

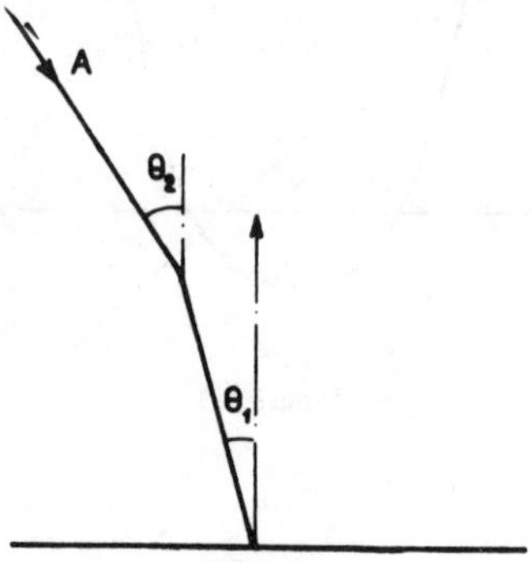

Figure 9.4

The characteristic equation can be written:

$$p_0\omega^4 + p_2\omega^2 + p_4 = 0$$

with

$$p_0 = m_1 m_2 a_1^2 a_2^2$$
$$p_2 = (m_1 a_1^2 + m_2 a_2^2 + m_2(l + a_2)^2)c - m_2 a_2(l + a_2)Pl$$
$$p_4 = c^2$$

and we note already that, since $p_4 > 0$, static instability cannot occur, i.e. the only equilibrium configuration is the one given by $\theta_1 = \theta_2 = 0$ no matter what the value of P. To analyse dynamic stability let us take two examples:

1. $m_1 = m_2 = \dfrac{m}{2}, \quad a_1 = a_2 = \dfrac{l}{2}$

whence $\qquad p_0 = \dfrac{m^2 l^4}{64}, \quad p_2 = \dfrac{ml^2}{8}(11c - 3Pl), \quad p_4 = c^2;$

we obtain: $\qquad p_2^2 - 4p_0p_4 = \frac{3}{64} m^2 l^4 (39c^2 - 22cPl + 3P^2l^2)$

so that the condition for stability is reduced to $P < \dfrac{3c}{l}$.

2. $m_1 = \dfrac{m}{2}$, $m_2 = \dfrac{m}{4}$, $a_1 = a_2 = l$ (this situation meaning that each bar of mass $\dfrac{m}{2}$ is represented schematically by two point masses $\dfrac{m}{4}$, one at each of its ends).

We have in this case

$$p_0 = \frac{m^2 l^4}{8}, \quad p_2 = \frac{ml^2}{4}(7c - 2Pl), \quad p_4 = c^2$$

and

$$p_2^2 - 4p_0p_4 = \frac{m^2l^4}{16}(41c^2 - 28cPl + 4P^2l^2)$$

so that the condition for stability is

$$P < \left(\frac{7}{2} - \sqrt{2}\right)\frac{c}{l} \sim 2.09 \cdot \frac{c}{l}.$$

We now see appear an essential difference from the case with vertical loading. Under slave loading the critical value depends on the mass distribution and has no connection with the critical load of the static problem.

Description of the Asymptotic Nature of the Zones of Instability for the Mathieu Equation

Considering yet again the differential equation

(9.25) $$y'' + \Omega^2(1 - 2\mu\cos\omega t)y = 0$$

we have seen that, if a certain condition between μ and Ω is satisfied, there may exist periodic solutions of period $T = \frac{2\pi}{\omega}$ or of period $2T$.

Let us begin by looking for a periodic solution of period $2T$ which may, a priori, be represented by its Fourier series:

(9.26) $$y = b_0 + \sum_{k \geqslant 1}\left[a_k \sin\left(\frac{k\omega t}{2}\right) + b_k \cos\left(\frac{k\omega t}{2}\right)\right].$$

However, as this series may equally well represent a periodic solution of period T, when k takes only even values, we can kill two birds with one stone by starting from (9.26) which we shall substitute for y in the equation (9.25).

We obtain in this way four separate sets of linear equations, two for the coefficients a_k, the first containing only the odd-numbered coefficients and the second the even-numbered coefficients, and similarly for the coefficients b_k:

(9.27) $$\begin{gathered}\left(1 + \mu - \frac{\omega^2}{4\Omega^2}\right)a_1 - \mu a_3 = 0\\ \dots\\ \left(1 - k^2\frac{\omega^2}{4\Omega^2}\right)a_k - \mu(a_{k-2} + a_{k+2}) = 0, \quad k = 3, 5, 7\dots\end{gathered}$$

(9.28) $$\begin{gathered}\left(1 - \mu - \frac{\omega^2}{4\Omega^2}\right)b_1 - \mu b_3 = 0\\ \dots\\ \left(1 - k^2\frac{\omega^2}{4\Omega^2}\right)b_k - \mu(b_{k-2} + b_{k+2}) = 0, \quad k = 3, 5, 7\dots\end{gathered}$$

$$\left(1-\frac{\omega^2}{\Omega^2}\right)a_2-\mu a_4=0$$

(9.29) $$\cdots$$

$$\left(1-k^2\frac{\omega^2}{4\Omega^2}\right)a_k-\mu(a_{k-2}+a_{k+2})=0, \quad k=4,6\ldots$$

$$b_0-\mu b_2=0$$

$$\left(1-\frac{\omega^2}{\Omega^2}\right)b_2-\mu(2b_0+b_4)=0$$

(9.30) $$\cdots$$

$$\left(1-\frac{k^2\omega^2}{4\Omega^2}\right)b_k-\mu(b_{k-2}+b_{k+2})=0, \quad k=4,6\ldots$$

These sets of homogeneous linear equations contain an infinity of equations and unknowns. For each set there will be no non-trivial solution unless the associated infinite determinant has the value zero. The vanishing of the determinant expresses a certain dependence between μ and Ω. We shall return later to this concept of an infinite determinant, and we shall be able to give a precise meaning to it in the case which concerns us here. For the moment however we note that if μ, Ω are such that (9.27) has a non-trivial solution a_k, then by taking the even-numbered a_k, and taking all b_k to be zero, the series (9.26) will give us a periodic solution of period $2T$ to the equation (9.25).

We can make an analogous remark if we associate, with a non-trivial solution of (9.28), the null solutions of the complementary systems. We thus see that the existence of a periodic solution of period $2T$ is conditional on the possibility of solving the two equations:

(9.31) $$\det\begin{pmatrix} 1\pm\mu-\dfrac{\omega^2}{4\Omega^2} & -\mu & 0\cdots \\ -\mu & 1-\dfrac{9\omega^2}{4\Omega^2} & -\mu\cdots \\ 0 & -\mu & 1-\dfrac{25\omega^2}{4\Omega^2} \\ & & & \ddots \end{pmatrix}=0$$

Similarly, to obtain the periodic solutions of period T, we have only to solve:

(9.32) $$\det\begin{pmatrix} 1-\dfrac{\omega^2}{\Omega^2} & -\mu & 0\cdots & \\ -\mu & 1-\dfrac{4\omega^2}{\Omega^2} & -\mu & 0\cdots \\ 0 & -\mu & 1-\dfrac{9\omega^2}{\Omega^2} & -\mu \\ & & & \ddots \end{pmatrix}=0$$

associated with (9.29), or

$$\det\begin{pmatrix} 1 & -\mu & 0\cdots & & \\ -2\mu & 1-\dfrac{\omega^2}{\Omega^2} & -\mu & 0\cdots & \\ 0 & -\mu & 1-\dfrac{4\omega^2}{\Omega^2} & -\mu & \\ 0 & 0 & -\mu & 1-\dfrac{9\omega^2}{\Omega^2} & \\ & & & & \ddots \end{pmatrix}=0 \tag{9.33}$$

associated with (9.30).

An infinite determinant whose elements are $a_{ij}=\delta_{ij}+c_{ij}$, $\delta_{ij}=0$, $i\neq j$, $\delta_{ii}=1$, is said to be in normal form if $\sum_{i,j}|c_{ij}|<+\infty$; we shall explain in more detail the implications of this hypothesis, but let us observe that those we have considered so far can all be reduced to normal form.

We consider the set of equations (9.27) or (9.28). It suffices in fact to multiply the pth equation by $-\dfrac{4\Omega^2}{(2p-1)^2\omega^2}$; the associate determinant has as its elements $\delta_{ij}+c_{ij}$ with:

$$c_{11}=-\frac{4\Omega^2}{\omega^2}(1\pm\mu),\quad c_{pp}=-\frac{4\Omega^2}{(2p-1)^2\omega^2},\quad p\neq 1$$

$$\begin{aligned} c_{pq}&=\frac{4\Omega^2}{(2p-1)^2\omega^2}\mu, && q=p\pm 1\\ &=0, && q\neq p\pm 1 \end{aligned}$$

and the series $\sum_{i,j}|c_{ij}|$ is convergent. Analogous remarks could be made for the sets (9.29) and (9.30).

Lastly it is not without interest for numerical applications to note that the determinants so far considered can be reduced to the form:

$$\delta_1=\det\begin{pmatrix} a_1 & 1 & 0 & 0\cdots\\ 1 & a_2 & 1 & 0\\ 0 & 1 & a_3 & 1\\ \vdots & & & \end{pmatrix}$$

which can be expanded in the form of a continued fraction.

If we denote by δ_j the infinite determinant obtained from δ_1 by suppressing the first $j-1$ lines and the first $j-1$ columns, we can write, by using the usual rules for expanding a determinant, and anticipating somewhat the existence results to be given later,

$$\begin{aligned} \delta_1&=a_1\delta_2-\delta_3\\ \delta_j&=a_j\delta_{j+1}-\delta_{j+2}. \end{aligned}$$

Expressing the condition $\delta_1 = 0$ we are therefore led to:

$$a_1 = \frac{\delta_3}{\delta_2} = \cfrac{1}{\cfrac{\delta_2}{\delta_3}}, \quad \frac{\delta_2}{\delta_3} = a_2 - \cfrac{1}{\cfrac{\delta_3}{\delta_4}}, \ldots, \frac{\delta_j}{\delta_{j+1}} = a_j - \cfrac{1}{\cfrac{\delta_{j+1}}{\delta_{j+2}}}$$

from which we derive the expression in the form of the continued fraction:

$$a_1 - \cfrac{1}{a_2 - \cfrac{1}{a_3 - \ddots}} = 0 \tag{9.34}$$

Let us take for example the equation (9.31), where we introduce:

$$a_1 = -\mu^{-1}\left(1 \pm \mu - \frac{\omega^2}{4\Omega^2}\right), \quad a_p = -\mu^{-1}\left(1 - \frac{(2p-1)^2}{4\Omega^2}\omega^2\right).$$

We can write:

$$1 \pm \mu - \frac{\omega^2}{4\Omega^2} - \cfrac{\mu^2}{1 - \cfrac{9\omega^2}{4\Omega^2} - \cfrac{\mu^2}{1 - \cfrac{25\omega^2}{4\Omega^2} - \cfrac{\mu^2}{\ldots}}} = 0$$

which can be solved with respect to ω by successive approximation.

The first approximation leads us to take:

$$\frac{\omega^2}{4\Omega^2} = 1 \pm \mu \quad \text{or} \quad \omega = \omega^* = 2\Omega(1 \pm \mu)^{1/2};$$

from the second we get

$$\frac{\omega^2}{4\Omega^2} = 1 \pm \mu - \cfrac{\mu^2}{1 - \cfrac{9\omega^{*2}}{4\Omega^2}}$$

i.e. $\omega = 2\Omega\left(1 \pm \mu + \dfrac{\mu^2}{8 \pm 9\mu}\right)^{1/2}$ which constitutes an asymptotic representation of the boundary of the first zone of instability, the one emanating from the point $\mu = 0, \dfrac{\omega}{2\Omega} = 1$.

We have already seen, in fact, that the zones of instability in the (μ, Ω) plane emanate from the points $\mu = 0, \Omega = n\dfrac{\pi}{T}$, where T is the period of the equation, i.e. $\dfrac{\omega}{2\Omega} = \dfrac{1}{n}$, with n an integer.

To define the second zone of instability, corresponding to $n = 2$, we need to examine the equations (9.32) and (9.33) and more precisely their solutions which take the value $\dfrac{\omega}{2\Omega} = \dfrac{1}{2}$ for $\mu = 0$.

By limiting ourselves to the determinants of order 2 we write

$$\det\begin{pmatrix} 1-\dfrac{\omega^2}{\Omega^2} & -\mu \\ -\mu & 1-\dfrac{4\omega^2}{\Omega^2} \end{pmatrix}=0 \quad\text{and}\quad \det\begin{pmatrix} 1 & -\mu \\ -2\mu & 1-\dfrac{\omega^2}{\Omega^3} \end{pmatrix}=0$$

from which we derive the approximations

$$\omega=\Omega\left(1+\frac{\mu^2}{3}\right)^{1/2}, \quad \omega=\Omega(1-2\mu^2)^{1/2}.$$

To find the third zone of instability we come back to (9.31), and consider the solutions which, for $\mu=0$, are such that $\dfrac{\omega}{2\Omega}=\dfrac{1}{3}$. Again limiting ourselves to a determinant of order 2:

$$\det\begin{pmatrix} 1\pm\mu-\dfrac{\omega^2}{4\Omega^2} & -\mu \\ -\mu & 1-\dfrac{9\omega^2}{4\Omega^2} \end{pmatrix}=0$$

we obtain the approximations

$$\omega=\tfrac{2}{3}\Omega\left(1-\frac{9\mu^2}{8\pm 9\mu}\right)^{1/2}$$

Figure 9.5

Normal Form of Infinite Determinant. Analysis of Convergence [30]

Suppose that we are given the matrix of complex-valued elements a_{mn}, the suffixes m, n running through all the integers from $-\infty$ to $+\infty$. We define c_{mn} such that:

(9.35) $$a_{mn}=\delta_{mn}+c_{mn}$$

and we shall assume that the double series

(9.36) $$\sum_{-\infty}^{+\infty}\sum_{-\infty}^{+\infty}|c_{mn}|<+\infty$$

is convergent.

We write $\Delta_p^q = \| a_{mn} \|_{-p}^q$, for the determinant of order $p+q+1$ whose elements are the a_{mn}, with $-p \leqslant m \leqslant q$, $-p \leqslant n \leqslant q$. We shall prove [30] that Δ_p^q has a limit Δ when p and q both tend to infinity, independently of each other, and this limit will by definition be the value of the infinite determinant $\| a_{mn} \| = \Delta$.

1. We first recall Hadamard's lemma [37] which states that for any determinant of order n, with complex elements u_{ij}, we have

$$\| u_{ij} \| \leqslant \prod_{i=1}^{n} \left(\sum_{j=1}^{n} |u_{ij}|^2 \right)^{1/2}.$$

Thus for any infinite determinant $\| a_{mn} \|$, of normal form, i.e. satisfying (9.35), (9.36), the quantity

$$H = \prod_{m=-\infty}^{+\infty} \left(\sum_{n=-\infty}^{+\infty} |a'_{mn}|^2 \right)^{1/2} \tag{9.37}$$

where

$$a'_{mn} = a_{mn} \quad m \neq n$$

$$\begin{aligned} a'_{mm} &= a_{mm} \quad \text{if} \quad |a_{mm}| > 1 \\ &= 1 \quad \text{if} \quad |a_{mm}| \leqslant 1 \end{aligned} \tag{9.38}$$

is finite.

Noting first of all that:

$$0 < |a'_{mm}| - 1 < |c_{mm}|$$

it follows from (9.36) that:

$$\sum_m \alpha_m < +\infty, \quad \text{with} \quad \alpha_m = |a'_{mm}| - 1$$

and

$$\sum_m \varepsilon_m < +\infty, \quad \text{with} \quad \varepsilon_m = \sum_{\substack{n \\ n \neq m}} |a_{mn}|.$$

From the obvious inequality:

$$\sum_{\substack{n \\ n \neq m}} |a'_{mn}|^2 < \left(\sum_{\substack{n \\ n \neq m}} |a_{mn}| \right)^2 = \varepsilon_m^2$$

it will be seen that we can define:

$$p_m = \left(\sum_n |a'_{mn}|^2 - 1 \right)^{1/2}$$

and that

$$p_m^2 < |a'_{mm}|^2 - 1 + \sum_{\substack{n \\ n \neq m}} |a'_{mn}|^2 < (|a'_{mm}| - 1)(|a'_{mm}| + 1) + \varepsilon_m^2,$$

or with

$$A = \operatorname{Sup}_m |a_{mm}| + 1,$$

$$p_m^2 < A|c_{mm}| + \varepsilon_m^2$$

and

$$\sum_m p_m^2 < A \sum_m |c_{mm}| + \sum_m \varepsilon_m^2,$$

bearing in mind that $\sum_m \varepsilon_m^2 < +\infty$, since $\sum_m \varepsilon_m < +\infty$ by hypothesis.

The series $\sum_m p_m^2$ is therefore convergent and the same is true of the infinite product $H = \prod_{m=-\infty}^{+\infty} (1 + p_m^2)^{1/2}$.

It is useful to note that, in consequence of Hadamard's lemma, every subdeterminant extracted from the infinite determinant $\| a_{mn} \|$ is of modulus H or less.

2. By expanding the determinant in terms of the elements of the $(q+1)$th row, we have

$$\Delta_p^{q+1} = a_{q+1,q+1} \cdot \Delta_p^q + \theta_p^q \cdot H$$

with $$|\theta_p^q| \leqslant |a_{q+1,-p}| + |a_{q+1,-p+1}| + \cdots + |a_{q+1,q}|$$

and consequently:

$$|\Delta_p^{q+1} - \Delta_p^q| \leqslant H \sum_{j=-p}^{q+1} |c_{q+1,j}|.$$

Similarly

$$|\Delta_{p+1}^r - \Delta_p^r| \leqslant H \sum_{s=-p-1}^{r} |c_{-p-1,s}|$$

and with the help of the decomposition

$$\Delta_{p+\alpha}^{q+\beta} - \Delta_p^q = \sum_{j=1}^{\alpha} (\Delta_{p+j}^{q+\beta} - \Delta_{p+j-1}^{q+\beta}) + \sum_{l=1}^{\beta} (\Delta_p^{q+l} - \Delta_p^{q+l-1}).$$

we easily arrive at:

$$|\Delta_{p+\alpha}^{q+\beta} - \Delta_p^q| < H\left(\sum_{i \leqslant -p-1} \sum_{j=-\infty}^{+\infty} |c_{ij}| + \sum_{i \geqslant q+1} \sum_{j=-\infty}^{+\infty} |c_{ij}| \right)$$

which shows, since the double series $\sum\sum |c_{ij}|$ is convergent, that $|\Delta_{p+\alpha}^{q+\beta} - \Delta_p^q|$ can be made arbitrarily small if p, q are large enough, no matter what the positive integers α, β may be, and this proves the existence of the limit $\Delta = \lim_{\substack{p\to\infty \\ q\to\infty}} \Delta_p^q$.

Note, for use later on, that the preceding estimate applied to $\Delta_p^q - \Delta_1^1$, shows that Δ tends to 1 as $\sum_{ij} |c_{ij}|$ tends to 0.

3. With $\| a_{mn} \|$ an infinite determinant in normal form, let us now suppose that there exists an infinite bounded sequence of numbers x_n, not all zero, where $-\infty < n < +\infty$, and with $|x_n| < M$, such that

(9.39) $$\sum_{n=-\infty}^{+\infty} a_{mn} x_n = 0, \quad \forall m.$$

Then we have $\Delta = 0$.

For i any integer between $-p$ and q, we can write, in view of (9.39):

$$\sum_{j=-p}^{q} a_{ij}x_j = b_i \tag{9.40}$$

with

$$b_i = -\sum_{\substack{s>q\\ s<-p}} a_{is}x_s \quad \text{and} \quad |b_i| < M \sum_{\substack{s>q\\ s<-p}} |a_{is}|. \tag{9.41}$$

By solving the set (9.40) of $p+q+1$ linear equations, we obtain the estimates:

$$|\Delta_p^q x_j| < H \sum_{i=-p}^{q} |b_i| < MH \sum_{i=-p}^{q} \sum_{\substack{s>q\\ s<-p}} |a_{is}| = MH \sum_{i=-p}^{q} \sum_{\substack{s>q\\ s<-p}} |c_{is}|.$$

The expression on the right tends to 0 if p and q tend to $+\infty$, while Δ_p^q tends to Δ. By taking limits we have $\Delta x_j = 0$ and since by hypothesis the x_j are not all zero we deduce that $\Delta = 0$.

Hill's Equation

Consider the differential equation

$$y'' + Q(t)y = 0 \tag{9.42}$$

where $Q(t)$ is a real-valued periodic function of t. We may suppose without loss of generality that its period is equal to π and, subject to suitable assumptions regarding regularity, that $Q(t)$ can be represented by its Fourier series:

$$Q(t) = \sum_{-\infty}^{+\infty} q_n e^{2int}, \quad \bar{q}_n = q_{-n} \tag{9.43}$$

with

$$\sum_{-\infty}^{+\infty} |q_n| < +\infty \tag{9.44}$$

(a sufficient condition would be for $Q(t)$ to be continuously twice-differentiable).

It is known that (9.42) has solutions of the type

$$y = e^{i\alpha t} p(t) \tag{9.45}$$

where $p(t)$ is periodic with period π, $i\alpha = \dfrac{\log \rho}{\pi}$ with ρ a root of

$$\rho^2 - 2\Delta\rho + 1 = 0, \quad 2\Delta = y_1(\pi) + y_2'(\pi),$$

$y_1(t)$, $y_2(t)$ fundamental solutions of (9.42), and $\Delta = \cos \pi\alpha$.

If $Q(t)$ can be expanded in a series (9.43) satisfying the condition (9.44) then it is clear that the solution $y(t)$ represented by (9.45) can be expanded in an absolutely convergent Fourier series

$$y(t) = \sum_{-\infty}^{+\infty} p_n e^{i(\alpha+2n)t} \tag{9.46}$$

and substituting in (9.42) we obtain:

(9.47)
$$\sum_{n=-\infty}^{+\infty} [q_{m-n} - (\alpha + 2m)^2 \delta_{mn}] p_n = 0$$

or denoting q_0 from now on by the real number λ, and after multiplication by the factor $(\lambda - (\alpha + 2m)^2)^{-1}$:

(9.48)
$$p_m + \sum_{\substack{n=-\infty \\ n \neq m}}^{+\infty} \frac{q_{m-n} \cdot p_n}{\lambda - (\alpha + 2m)^2} = 0.$$

Since it follows from (9.44) that:

$$\sum_m \sum_{\substack{n=-\infty \\ n \neq m}}^{+\infty} \left| \frac{q_{m-n}}{\lambda - (\alpha + 2m)^2} \right| < +\infty$$

provided $\lambda - (\alpha + 2m)^2 \neq 0$, for all integral m, it will be seen that the infinite determinant with elements $d_{mn} = \delta_{mn} + \dfrac{q_{m-n}}{\lambda - (\alpha + 2m)^2}$ (where we assume from now on that $q_0 = 0$) is in normal form, and consequently it is possible to define:

$$D(\alpha, \lambda) = \left\| \delta_{mn} + \frac{q_{m-n}}{\lambda - (\alpha + 2m)^2} \right\|_{-\infty}^{+\infty}$$

which has a finite value, except for the pairs α, λ such that

$$\lambda - (\alpha + 2m)^2 = 0, \quad \text{for some integer } m.$$

$D(\alpha, \lambda)$ has the following properties [30]:

1. $D(\alpha, \lambda)$ is an analytic function of α, for all α in the complex plane, except for $\alpha = \pm\sqrt{\lambda} - 2m$, $m \in \mathbb{Z}$. For it is clear that for any α belonging to a compact set not containing any point $\pm\sqrt{\lambda} - 2m$, the sequence D_r^s of determinants formed from the elements d_{mn}, with indices between $-r$ and s, converges uniformly with respect to α to D, when r and s become infinite.

2. $D(\alpha, \lambda)$ is periodic with respect to α with period 2 and tends to 1 as $\alpha \to \pm i\infty$.

The periodicity is a consequence of $d_{mn}(\alpha + 2) = d_{m-1,n-1}(\alpha)$ and of the fact that the value of the infinite determinant is obviously not altered by a displacement of the rows and columns by one step in a direction parallel to the leading diagonal $m = n$.

It follows from the property

$$\lim_{\alpha \to \pm i\infty} \sum_m \sum_{\substack{n=-\infty \\ n \neq m}}^{+\infty} \left| \frac{q_{m-n}}{\lambda - (\alpha + 2m)^2} \right| = 0, \quad \text{that} \quad \lim_{\alpha \to \pm i\infty} D(\alpha, \lambda) = 1.$$

3. If $\lambda \neq 0$, the points $\alpha = \pm\sqrt{\lambda} - 2m$, $m \in \mathbb{Z}$ are at most simple poles of $D(\alpha, \lambda)$. Let us fix $m = m_0$ and let r, s be integers such that $-r < m_0 < s$; expanding D_r^s

in terms of the elements of the m_0th row, we can write

$$D_r^s = \tilde{D}_r^s + (\lambda - (\alpha + 2m_0)^2)^{-1} \cdot \sum_{\substack{n=-r \\ n \neq m_0}}^{s} q_{m_0 - n} \cdot H_n$$

with $|H_n| \leqslant H$, for α in the neighbourhood of $\pm\sqrt{\lambda} - 2m_0$, and $\tilde{D}_r^s$ a determinant of normal type obtained by deleting from D the m_0th row and column. Making r and s tend to ∞, we obtain:

$$D(\alpha, \lambda) = \tilde{D}(\alpha, \lambda) + (\lambda - (\alpha + 2m_0)^2)^{-1} \cdot \sum_{\substack{n=-\infty \\ n \neq m_0}}^{+\infty} q_{m_0 - n} \cdot H_n$$

where $\tilde{D}(\alpha, \lambda)$ is the infinite determinant obtained by simply removing the m_0th row and column from $\| d_{mn} \|$. Consequently $\tilde{D}(\alpha, \lambda)$ is analytic in the neighbourhood of $\pm\sqrt{\lambda} - 2m_0$ and it follows from the estimate:

$$|(\lambda - (\alpha - 2m_0)^2) D(\alpha, \lambda)| < |(\lambda - (\alpha + 2m_0)^2) \tilde{D}(\alpha, \lambda)| + H \sum_{-\infty}^{+\infty} |q_n|$$

that $\pm\sqrt{\lambda} - 2m_0$ is at most a simple pole of $D(\alpha, \lambda)$.

4. $D(\alpha, \lambda)$ is real for α real and $D(\alpha, \lambda) = \overline{D(\bar{\alpha}, \lambda)}$. We consider the determinant D_s^s and the term in its expansion which corresponds to the permutations:

$$(9.49) \qquad \begin{array}{ll} m_1, m_2, \ldots, m_{2s+1} & -s \leqslant m_j \leqslant s \\ n_1, n_2, \ldots, n_{2s+1} & -s \leqslant n_j \leqslant s \end{array}$$

namely

$$(9.50) \qquad (-1)^{I+J} \frac{q_{m_1 - n_1} \cdot q_{m_2 - n_2} \cdot \ldots \cdot q_{m_{2s+1} - n_{2s+1}}}{\prod_{j=1}^{2s+1}{}' (\lambda - (\alpha + 2m_j)^2)}$$

where the notation $\prod'$ denotes the omission in the product of the factors of index j for which $m_j = n_j$.

If we interchange the two permutations (9.49) the numerator of (9.50) is replaced by its conjugate and, in the denominator, only the order of the factors is altered, because the same set of integers is involved.

Thus D_s^s is real for real α and the same is true of $D(\alpha, \lambda)$.

5. $D(\alpha, \lambda) = \overline{D(-\bar{\alpha}, \lambda)}$.

For the element (m, n) of $D_s^s(\alpha, \lambda)$ is:

$$\delta_{mn} + \frac{q_{m-n}}{\lambda - (\alpha + 2m)^2} = \delta_{mn} + \overline{\left[\frac{q_{n-m}}{\lambda - (-\bar{\alpha} - 2m)^2} \right]}$$

and is consequently equal to the conjugate of the element $(-m, -n)$ of $D_s^s(-\bar{\alpha}, \lambda)$; thus $D_s^s(\alpha, \lambda) = \overline{D_s^s(-\bar{\alpha}, \lambda)}$ and $D(\alpha, \lambda) = \overline{D(-\bar{\alpha}, \lambda)}$; and combining 4 and 5, we see that $D(\alpha, \lambda) = D(-\alpha, \lambda)$.

6. Let ρ be the residue of $D(\alpha, \lambda)$ at $\alpha = \sqrt{\lambda}$; it follows from 4 or 5, depending on whether λ is negative or positive, that $-\rho$ is the residue at $\alpha = -\sqrt{\lambda}$, and that, by reason of periodicity, ρ, $-\rho$ are the residue at the poles

$$\alpha = \sqrt{\lambda} - 2m, \alpha = -\sqrt{\lambda} - 2m, \forall m \in \mathbb{Z}.$$

The function $D(\alpha, \lambda) - \frac{\pi\rho}{2}\left[\cot\frac{\pi}{2}(\alpha - \sqrt{\lambda}) - \cot\frac{\pi}{2}(\alpha + \sqrt{\lambda})\right]$ is holomorphic in the complex plane of α and is periodic with period 2. As it is bounded in the strip $|\operatorname{Re}\alpha| \leqslant 1$ and hence also in the whole plane, it follows from Liouville's theorem that it reduces to a constant, whose value can be determined for $\alpha \to i\infty$. Thus

$$D(\alpha, \lambda) = 1 + \frac{\pi\rho}{2}\left[\cot\frac{\pi}{2}(\alpha - \sqrt{\lambda}) - \cot\frac{\pi}{2}(\alpha + \sqrt{\lambda})\right] \tag{9.51}$$

where ρ, which is independent of α can be expressed by:

$$\rho = \pi^{-1}\cdot(1 - D(0, \lambda))\tan\left(\frac{\pi}{2}\sqrt{\lambda}\right). \tag{9.52}$$

7. It follows from the equations (9.48) that the equation $D(\alpha, \lambda) = 0$ defines, in terms of λ regarded as a real parameter, the admissible values of α, the characteristic coefficient of (9.45), provided that $\alpha \neq \pm\sqrt{\lambda} - 2m$, $\forall m \in \mathbb{Z}$.

To include this eventuality, for example $\alpha = \sqrt{\lambda} - 2m_0$, let us note that the coefficients p_n would not cease to satisfy (9.48) for all $m \neq m_0$, while the m_0th equation would be obtained from (9.47) in the form:

$$\sum_{\substack{n=-\infty \\ n \neq m_0}}^{+\infty} q_{m_0 - n}\cdot p_n = 0.$$

The associated determinant would still be of normal type and having regard to our notation would have the value:

$$\lim_{\alpha \to \sqrt{\lambda} - 2m_0} D(\alpha, \lambda)(\lambda - (\alpha + 2m_0)^2).$$

Accordingly the condition which expresses the fact that (9.47) has a non-zero solution $\{p_n\}$ may be generally formulated in the form of the equation:

$$\frac{\pi}{4}(\alpha^2 - \lambda^2)\prod_{\substack{m=-\infty \\ m \neq 0}}^{+\infty}\left(1 + \frac{\alpha + \sqrt{\lambda}}{2m}\right)\left(1 + \frac{\alpha + \sqrt{\lambda}}{2m}\right)D(\alpha, \lambda) = 0$$

i.e.

$$\sin\frac{\pi}{2}(\alpha - \sqrt{\lambda})\sin\frac{\pi}{2}(\alpha + \sqrt{\lambda})\cdot D(\alpha, \lambda) = 0 \tag{9.53}$$

which defines the basic relation between α and λ which has to hold for there to be a solution $y = e^{i\alpha t}p(t)$ of (9.42), with $p(t)$ periodic of period π.

In view of (9.51) and (9.52) we may write (9.53) in the form

$$1 - \cos\pi\alpha = 2\sin^2\left(\frac{\pi}{2}\sqrt{\lambda}\right)\cdot D(0, \lambda)$$

and we may recall that

$$2 \cos \pi\alpha = y_1(\pi, \lambda) + y_2'(\pi, \lambda)$$

where $y_1(t, \lambda)$, $y_2(t, \lambda)$ is the fundamental solution of (9.42).

8. If there is a damping term, i.e. if (9.42) is replaced by $y'' + 2ay' + Q(t)y = 0$, with $Q(t)$ periodic of period π, the change of variable $z = e^{at}y$, leads to $z'' + (Q(t) - a^2)z = 0$, an equation to which the preceding theory can be applied after replacing Q by $Q - a^2$.

Chapter X. The Method of Amplitude Variation and Its Application to Coupled Oscillators

The behaviour of a system of oscillators coupled through the interaction of non-linear forces, that are either time-independent or quasi-periodic in time, may be investigated by looking for quasi-periodic solutions of the equations of motion, but the conditions required for success by this method are generally-speaking very constrictive.

By restricting one's objectives to looking for solutions which remain bounded for all future time, one can, on the contrary, develop a theory having wide application possibilities. The essential result, which is mainly based on the idea of averaging operations carried out in the phase-space, allows us to predict the existence of such solutions, and to discuss their stability, for an extensive class of coupled oscillators, and to specify the extensions needed to deal with the cases of multiple frequencies, the decay of certain vibrational eigenmodes, and resonance for non-autonomous systems.

Applied to a non-linear gyroscopic system, the theory provides an interesting illustration of the phenomenon known as bifurcation.

Posing the Problem

We consider (in the real field) the set of differential equations:

(10.1) $$\frac{d^2 q_j}{dt^2} + \omega_j^2 q_j = \mu f_j\left(q, \frac{dq}{dt}\right), \quad 1 \leqslant j \leqslant n$$

(10.2) $$q = \{q_1, q_2, \ldots, q_n\} \in \mathbb{R}^n, \frac{dq}{dt} = \left\{\frac{dq_1}{dt}, \ldots, \frac{dq_n}{dt}\right\}$$

representing the motion of n weakly-coupled harmonic oscillators, with

$$\omega = (\omega_1, \omega_2, \ldots, \omega_n) \in \mathbb{R}^n$$

the given set of eigenfrequencies, μ a real parameter, $f_j\left(q, \frac{dq}{dt}\right)$ the given coupling functions.

Associating with each variable q_j its amplitude a_j and phase θ_j, defined by:

(10.3) $$q_j = a_j \sin \theta_j$$

(10.4) $$\frac{dq_j}{dt} = a_j\omega_j \cos\theta_j$$

which necessitates:

(10.5) $$\frac{da_j}{dt}\sin\theta_j + a_j\cos\theta_j \cdot \frac{d\theta_j}{dt} = a_j\omega_j\cos\theta_j,$$

we deduce from (10.1) and (10.4):

$$\frac{d^2q_j}{dt^2} = \frac{da_j}{dt}\omega_j\cos\theta_j - a_j\omega_j\sin\theta_j \cdot \frac{d\theta_j}{dt} = -\omega_j^2 a_j \sin\theta_j + \mu f_j(a\sin\theta, a\omega\cos\theta)$$

and with (10.5)

(10.6) $$\frac{da_j}{dt} = \mu A_j(a,\theta)$$

$$\frac{d\theta_j}{dt} = \omega_j + \mu B_j(a,\theta)$$

where

$$A_j(a,\theta) = \omega_j^{-1} f_j(a\sin\theta, a\omega\cos\theta)\cdot\cos\theta_j$$
$$B_j(a,\theta) = -(\omega_j a_j)^{-1} f_j(a\sin\theta, a\omega\cos\theta)\cdot\sin\theta_j$$

are periodic functions of each θ_ν, $1 \leqslant \nu \leqslant n$, of period 2π. We shall, in what follows, basically keep to the structure of the system (10.6) but we can forget about its origin and consider more generally:

(10.7) $$\frac{da}{dt} = \mu A(a,\theta) + \mu\tilde{A}(a,\theta,\mu)$$

$$\frac{d\theta}{dt} = \omega + \mu B(a,\theta) + \mu\tilde{B}(a,\theta,\mu)$$

where $A(a,\theta)$, $\tilde{A}(a,\theta,\mu)$; $B(a,\theta)$, $\tilde{B}(a,\theta,\mu)$ are continuous functions of

$$(a,\theta,\mu)\in\bar{G}\times T^q\times I,$$

with values in $\mathbb{R}^n$, $\mathbb{R}^q$ respectively, G being an open and bounded subset of $\mathbb{R}^n$, and T^q a q-dimensional torus defined by

$$\theta = (\theta_1, \theta_2, \ldots, \theta_q), \quad 0 \leqslant \theta_j \leqslant 2\pi$$

(any function of the terms on the left of (10.7) is periodic of period 2π with respect to each θ_j),

$$I = [0, \mu_0] \subset \mathbb{R}^+.$$

We shall also assume that $A(a,\theta)$, $B(a,\theta)$ are continuously differentiable and that

$$\lim_{\mu\to 0} \sup_{\substack{a\in\bar{G}\\ \theta\in T^q}} (\|\tilde{A}(a,\theta,\mu)\| + \|\tilde{B}(a,\theta,\mu)\|) = 0.$$

The idea that the system (10.7) could have a quasi-periodic solution having the structure:

$$a = g(\varphi_1, \ldots, \varphi_q) \tag{10.8}$$
$$\theta = \omega t + h(\varphi_1, \ldots, \varphi_q), \quad \varphi_j = \omega_j t$$

where g and h are 2π-periodic functions of each argument φ_j, has given rise to numerous research papers in the recent past [5]; however the results obtained, based for the most part, on a technique of accelerated convergence introduced by Kolmogoroff [2], cannot claim to be generally applicable, for several reasons.

In the first place, the validity of the iterative process used is based largely on the assumption that the right-hand sides of (10.7) are analytic functions of the variables (a, θ). Secondly it is known [51] that for all

$$\omega \in D = \{\omega \in \mathbb{R}^q, 0 \leqslant \omega_j \leqslant 1, \forall j \in [1, 2, \ldots, q]\}$$

with the exception of a subset $E \subset D$ of measure zero, there exists a constant c, depending on ω, such that

$$|(\omega, k)^{-1}| \leqslant c|k|^{q+1}, \quad \forall k \neq 0, \quad \text{with } k = (k_1, \ldots, k_q) \tag{10.9}$$

being a set of integers with arbitrary signs, and

$$(\omega, k) = \sum_{j=1}^{q} \omega_j k_j, \quad |k| = \sum_{j=1}^{q} |k_j|.$$

Now to get over the difficulties associated with the presence of small divisors, the above-mentioned method requires that ω should satisfy an inequality of the type (10.9). This condition may seem quite natural and even fairly harmless from a statistical point of view, in view of the result recalled above. However the frequency vector ω having been given it is usually difficult to tell whether or not it satisfies the condition because we do not know how to specify the subset E of measure zero. Finally it is necessary to modify the frequency vector slightly, admittedly by an amount which vanishes with μ, i.e. a quasi-periodic solution of type (10.8) cannot be constructed unless ω in (10.7) is replaced by $\tilde{\omega} = \omega + \delta(\mu)$, where $\lim_{\mu \to 0} \delta(\mu) = 0$. If the second equation of (10.7) contains, in an additive form, parameters which one is allowed to choose arbitrarily, then one can, in certain cases adjust these terms in such a way as to modify the frequency-vector in the manner mentioned; but this amounts to a theoretical adjustment, which is difficult to achieve in a concrete case, because we do not know how to define $\delta(\mu)$ explicitly. It will even be impossible when, as usually happens, these parameters are absent.

The method based on the calculation of time-averages has proved to be a valuable tool in clarifying the concept of an integral variety [10, 43]; the idea of using spatial means, in the phase-space, will be developed here with a view to enabling a solution to the problem defined earlier to be found by a simple method.

Having defined:

(10.10)
$$A^0(a) = \frac{1}{(2\pi)^q} \int_0^{2\pi} \cdots \int_0^{2\pi} A(a,\theta)\, d\theta_1 \cdots d\theta_q$$
$$B^0(a) = \frac{1}{(2\pi)^q} \int_0^{2\pi} \cdots \int_0^{2\pi} B(a,\theta)\, d\theta_1 \cdots d\theta_q$$

we can establish the following approximation theorem.

Theorem

If $a^0 \in G$ is such that $A^0(a^0) = 0$ and $H = \left.\frac{\partial A^0}{\partial a}\right|_{a^0}$ is a stable matrix and if the frequencies $(\omega_1, \ldots, \omega_q)$ are independent, i.e. are such that

$$(k,\omega) = k_1\omega_1 + \cdots + k_q\omega_q \neq 0$$

for any set of integers $k_1, \ldots, k_q$ of arbitrary sign, and not all zero, then there exist positive numerical constants η, μ_1, c and positive-valued functions $\varepsilon_1(\mu)$, $\varepsilon_2(\mu)$ tending to 0 with μ, such that, subject to the two conditions:

$$\| a(0) - a^0 \| < \eta, \quad 0 < \mu < \operatorname{Inf}(\mu_0, \mu_1),$$

every solution $a(t)$, $\theta(t)$ of (10.7) can be extended indefinitely into the future and satisfies the relations:

$$\| a(t) - a^0 \| \leqslant c \| a(0) - a^0 \| + \varepsilon_1(\mu), \quad \forall t > 0$$
$$\left\| \frac{d\theta}{dt} - \omega \right\| \leqslant \varepsilon_2(\mu).$$

It will be noted that stability of amplitude is implied in this enunciation. The extensions needed to deal with certain situations which are of interest from a practical point of view will be developed later. These include:

—Cases where certain eigenfrequencies coincide.

—Extinction of certain vibrational eigenmodes for the system of coupled harmonic oscillators; applying the approximation theorem to the autonomous system (10.1), i.e. in reality to the equations (10.6), we can ask what conclusions can be reached in those cases when certain components of a^0 are zero. The fact is that the calculation of A^0 and H retains its significance whereas the corresponding components of B_j become infinite for $a = a^0$. Nevertheless the theorem remains valid in such a case.

—The non-autonomous system and resonance.

Lemma

The function

(10.11)
$$u(a,\theta,\mu) = \int_0^{+\infty} e^{-\mu s}(A^0(a) - A(a, \theta + \omega s))\, ds, \quad \mu > 0$$

satisfies the equation:

$$\sum_{j=1}^{q} \omega_j \frac{\partial u}{\partial \theta_j} = \mu u + A(a,\theta) - A^0(a) \tag{10.12}$$

and

$$\lim_{\mu\to 0} \mu \operatorname*{Sup}_{\substack{a\in\bar{G}\\ \theta\in T^q}} \left(\| u(a,\theta,\mu) \| + \left\| \frac{\partial u}{\partial a}(a,\theta,\mu) \right\| + \left\| \frac{\partial u}{\partial \theta}(a,\theta,\mu) \right\| \right) = 0. \tag{10.13}$$

Similar results apply to:

$$v(a,\theta,\mu) = \int_0^{+\infty} e^{-\mu s}(B^0(a) - B(a,\theta+\omega s))\,\mathrm{d}s. \tag{10.14}$$

Proof.

We write:

$$\sum_{j=1}^{q} \omega_j \frac{\partial u}{\partial \theta_j} = \int_0^{+\infty} e^{-\mu s} \frac{d}{\mathrm{d}s}(A^0(a) - A(a,\theta+\omega s))\,\mathrm{d}s$$
$$= \mu u + A(a,\theta) - A^0(a)$$

after integrating by parts.

Because of the continuity of $A(a,\theta)$ and its partial derivatives $\frac{\partial A}{\partial a}, \frac{\partial A}{\partial \theta}$, on the compact set $\bar{G} \times T^q$, to establish (10.13) it suffices to show that

$$\mu u(a,\theta,\mu), \quad \mu\frac{\partial u}{\partial a}(a,\theta,\mu), \quad \mu\frac{\partial u}{\partial \theta}(a,\theta,\mu)$$

all tend to 0 when $\mu \to 0$, for every fixed pair (a,θ) of $\bar{G} \times T^q$.

With $M = \operatorname*{Sup}_{\substack{a\in\bar{G}\\ \theta\in T^q}} \| A(a,\theta) \|$ and

$$\mu u(a,\theta,\mu) = \int_0^{+\infty} e^{-\tau}\left(A^0(a) - A\left(a, \theta + \omega\frac{\tau}{\mu} \right)\right)\mathrm{d}\tau$$

we can define η and l so that

$$\left\| \left(\int_0^{\eta} + \int_l^{+\infty} \right) e^{-\tau}\left(A^0(a) - A\left(a, \theta + \omega\frac{\tau}{\mu} \right)\right)\mathrm{d}\tau \right\| < \varepsilon$$

by, for example, choosing them so that $\eta, l{:}\, 2\eta M < \frac{\varepsilon}{2},\ 2Me^{-l} < \frac{\varepsilon}{2}$.

With $g(\tau) = \int_0^\tau \left(A\left(a, \theta + \frac{\omega\sigma}{\mu} \right) - A^0(a) \right)\mathrm{d}\sigma$, we write:

$$\int_\eta^l e^{-\tau}\left(A^0(a) - A\left(a, \theta + \omega\frac{\tau}{\mu} \right)\right)\mathrm{d}\tau = -\int_\eta^l e^{-\tau}\frac{\mathrm{d}g}{\mathrm{d}\tau}\mathrm{d}\tau$$
$$= e^{-\eta}g(\eta) - e^{-l}g(l) - \int_\eta^l e^{-\tau}g(\tau)\,\mathrm{d}\tau.$$

Noting that $\|e^{-\eta}g(\eta)\| \leqslant 2M\eta e^{-\eta} < \varepsilon$, if η is sufficiently small and $\|e^{-l}g(l)\| \leqslant 2Mle^{-l} < \varepsilon$, if l is sufficiently large, we can see that it remains only to investigate the behaviour of:

$$\int_\eta^l e^{-\tau}g(\tau)\,\mathrm{d}\tau = \int_\eta^l \tau e^{-\tau}\left(\frac{\mu}{\tau}\int_0^{\tau/\mu}(A(a,\theta+\omega t) - A^0(a))\,\mathrm{d}t\right)\mathrm{d}\tau$$

as $\mu \to 0$. We shall here make use of H. Weyl's ergodic theorem [2], which may be stated in the form: if the frequencies $\omega_1 \cdots \omega_q$ are independent and $f(\theta)$ is a continuous function on the torus T^q, then

$$\lim_{T\to\infty}\frac{1}{T}\int_0^T f(\theta+\omega t)\,\mathrm{d}t = \frac{1}{(2\pi)^q}\int_0^{2\pi}\cdots\int_0^{2\pi} f(\varphi)\,\mathrm{d}\varphi_1\,\mathrm{d}\varphi_2\cdots\mathrm{d}\varphi_q.$$

Thus for fixed a, θ, there is a T_0 such that $\frac{\tau}{\mu} > T_0$ implies

$$\left\|\frac{\mu}{\tau}\int_0^{\tau/\mu}(A(a,\theta+\omega t) - A^0(a))\,\mathrm{d}t\right\| < \frac{\varepsilon}{\int_0^{+\infty}\tau e^{-\tau}\,\mathrm{d}\tau}$$

so that, for $0 < \mu < \frac{\eta}{T_0}$, we have

$$\left\|\int_\eta^l e^{-\tau}g(\tau)\,\mathrm{d}\tau\right\| < \varepsilon$$

and this completes the proof of the lemma.

We can develop the same arguments for $\frac{\partial u}{\partial a}, \frac{\partial u}{\partial \theta}$.

We now come to the proof of the theorem.

We first make the following change of variables in (10.7):

$$a = x + \mu u(x,\varphi,\mu)$$
$$\theta = \varphi + \mu v(x,\varphi,\mu) \tag{10.15}$$

where the vector-functions u and v are defined by (10.11) and (10.14) after replacing (a,θ) by (x,φ) in these two equations.

A straightforward calculation using (10.12) and the analogous result for v leads to

$$\left(I + \mu\frac{\partial u}{\partial x}\right)\cdot\frac{\mathrm{d}x}{\mathrm{d}t} + \mu\frac{\partial u}{\partial \varphi}\cdot\left(\frac{\mathrm{d}\varphi}{\mathrm{d}t} - \omega\right)$$
$$= \mu A(x+\mu u, \varphi+\mu v) - \mu A(x,\varphi) + \mu A^0(x) - \mu^2 u + \mu\varepsilon(\mu)$$
$$\mu\frac{\partial v}{\partial x}\cdot\frac{\mathrm{d}x}{\mathrm{d}t} + \left(I + \mu\frac{\partial v}{\partial \varphi}\right)\cdot\left(\frac{\mathrm{d}\varphi}{\mathrm{d}t} - \omega\right) \tag{10.16}$$
$$= \mu B(x+\mu u, \varphi+\mu v) - \mu B(x,\varphi) + \mu B^0(x) - \mu^2 v + \mu\varepsilon(\mu)$$

where the notation $\varepsilon(\mu)$ represents terms tending to 0 with μ, uniformly on $\bar{G} \times T^q$.

In view of the lemma, the matrix

$$\begin{pmatrix} I+\mu\dfrac{\partial u}{\partial x} & \mu\dfrac{\partial u}{\partial \varphi} \\ \mu\dfrac{\partial v}{\partial x} & I+\mu\dfrac{\partial v}{\partial \varphi} \end{pmatrix}$$

is invertible for small enough μ and besides it tends to the unit matrix when $\mu \to 0$, so that we can write (10.16) in the form:

$$\text{(10.17)} \qquad \begin{aligned} \frac{\mathrm{d}x}{\mathrm{d}t} &= \mu A^0(x) + \mu\varepsilon(\mu) \\ \frac{\mathrm{d}\varphi}{\mathrm{d}t} &= \omega + \mu B^0(x) + \mu\varepsilon(\mu). \end{aligned}$$

We suppose that there exists an $a^0 \in G$, such that $A^0(a^0)=0$ and $H = \left.\dfrac{\partial A^0}{\partial a}\right|_{a^0}$ is a stable matrix, i.e. all its eigenvalues have a negative real part.

Putting $x - a^0 = \xi$, we can rewrite (10.17) in the form:

$$\text{(10.18)} \qquad \begin{aligned} \frac{\mathrm{d}\xi}{\mathrm{d}t} &= \mu H\cdot\xi + \mu X \\ \frac{\mathrm{d}\varphi}{\mathrm{d}t} &= \omega + \mu B^0(a^0) + \mu\Phi \end{aligned}$$

with $X(\xi, \varphi, \mu)$, $\Phi(\xi, \varphi, \mu)$ such that

$$\text{(10.19)} \qquad \|X\| \leqslant \|\xi\|\cdot\varepsilon_3(\|\xi\|) + \varepsilon_4(\mu)$$

$$\text{(10.20)} \qquad \|\Phi\| \leqslant c\|\xi\| + \varepsilon_5(\mu)$$

c being a numerical constant, and $\varepsilon_3, \varepsilon_4, \varepsilon_5$ non-decreasing numerical functions of their arguments, tending to 0 with these.

We denote by $\xi(t)$, $\varphi(t)$ a solution of (10.18) corresponding to the initial values: $\xi(0)=\xi^0$, $\varphi(0)=\varphi^0$ and defined in a certain neighbourhood of $t=0$.

Denoting by $Y(t)$ the resolvent matrix associated with H, we can write, by integrating the first equation of (10.18):

$$\text{(10.21)} \qquad \xi(t) = Y(\mu t)\xi^0 + \mu\int_0^t Y(\mu(t-s))\cdot X(\xi(s), \varphi(s), \mu)\,\mathrm{d}s.$$

The hypothesis that the matrix H is stable implies that there are positive constants m and h satisfying:

$$\text{(10.22)} \qquad \|Y(t)\| \leqslant m e^{-ht}, \quad \forall t > 0;$$

let us introduce an $r > 0$, such that the ball in $\mathbb{R}^n$ of centre a^0 and radius r is contained in $\bar{G}$, and

$$\text{(10.23)} \qquad \gamma = \operatorname{Inf}\left(\frac{r}{2}, \gamma_1\right), \quad \text{with } \gamma_1 \text{ chosen so that}$$

(10.24) $$\varepsilon_3(\gamma_1) < \frac{h}{2m}.$$

The function $\sigma(\tau) = \operatorname{Sup}_{t\in[0,\tau]} \|\xi(t)\|$, $\tau > 0$, is continuous and non-decreasing; assuming ξ^0 to be such that

(10.25) $$\|\xi^0\| < \frac{\gamma}{4m}$$

and μ small enough so that:

(10.26) $$\varepsilon_4(\mu) < \frac{\gamma h}{4m},$$

we shall now show that:

(10.27) $$\sigma(\tau) < \gamma, \quad \forall\tau > 0,$$

at least so long as the solution $\xi(t)$, $\varphi(t)$ of (10.18), (10.20) exists. Consequently the point $\xi(t)$ will be inside the ball of centre a^0 and radius $r/2$, as long as it can be defined at all, and this allows the indefinite continuation of the solution which may thus be considered as defined for the indefinite future, i.e. for all $t > 0$, [42].

Since $\sigma(0) = \|\xi^0\| < \gamma$ (as $m \geqslant 1$), if (10.27) were false, there would be, by virtue of the continuity of $\sigma(\tau)$, a positive number T such that:

$$\tau\in[0, T[\to \sigma(\tau) < \gamma \quad \text{and} \quad \sigma(T) = \gamma.$$

We deduce from (10.21), using (10.19) and (10.22),

(10.28) $$\|\xi(t)\| \leqslant m\|\xi^0\| + \frac{m}{h}(\varepsilon_4(\mu) + \sigma(t)\varepsilon_3(\sigma(t))), \quad \forall\tau\in[0, T]$$

whence

(10.29) $$\sigma(\tau) \leqslant m\|\xi^0\| + \frac{m}{h}(\varepsilon_4(\mu) + \sigma(\tau)\varepsilon_3(\sigma(\tau))), \quad \forall\tau\in[0, T];$$

but $\sigma(\tau) \leqslant \gamma$, $\forall\tau\in[0, T]$, implies by (10.23), (10.24):

$$\varepsilon_3(\sigma(\tau)) < \frac{h}{2m}$$

and coming back to (10.29):

$$\sigma(\tau) \leqslant 2m\|\xi^0\| + \frac{2m}{h}\varepsilon_4(\mu) < \gamma, \quad \forall\tau\in[0, T]$$

by (10.25) and (10.26), contrary to $\sigma(T) = \gamma$.

We have thus proved that, subject to the condition (10.25), with γ defined by (10.23), (10.24) and with μ small enough and positive, the solution $\xi(t)$, $\varphi(t)$ is defined for $t > 0$ and satisfies $\sigma(\tau) < \gamma$, $\forall\tau > 0$. From this follow by the same argument:

$$\|\xi(t)\| \leqslant \sigma(t) \leqslant 2m\|\xi^0\| + \frac{2m}{h}\varepsilon_4(\mu), \quad \forall t > 0$$

and the estimates of the approximation theorem.

Cases Where Certain Oscillations Have the Same Frequency

We resume our consideration of the system (10.17) on the same assumptions as before, the only difference being that we shall suppose $\omega_{q-1} = \omega_q$, the $q-1$ frequencies $\omega_1, \omega_2, \ldots, \omega_{q-1}$ forming an independent system.

The approach must be modified as follows. We introduce the variable $\chi = \theta_q - \theta_{q-1}$ and write

$$\bar{\theta} = (\theta_1, \theta_2, \ldots, q_{q-1}), \quad \bar{\omega} = (\omega_1, \omega_2, \ldots, \omega_{q-1}),$$

these being $q-1$ dimensional vectors. We write the set of equations (10.7) with the help of the variables $a, \chi, \bar{\theta}$ using the notations:

$$\begin{aligned} A^*(a, \chi, \bar{\theta}) &= A(a, \theta_1, \ldots, \theta_{q-1}, \theta_{q-1} + \chi) \\ B^*(a, \chi, \bar{\theta}) &= B(a, \theta_1, \ldots, \theta_{q-1}, \theta_{q-1} + \chi) \end{aligned}$$

and similarly $\tilde{A}^*(a, \chi, \bar{\theta}, \mu)$, $\tilde{B}^*(a, \chi, \bar{\theta}, \mu)$:

$$\begin{aligned} \frac{da}{dt} &= \mu A^*(a, \chi, \bar{\theta}) + \mu \tilde{A}^*(a, \chi, \bar{\theta}, \mu) \\ \frac{d\chi}{dt} &= \mu(B_q^*(a, \chi, \bar{\theta}) - B_{q-1}^*(a, \chi, \bar{\theta})) \\ &\quad + \mu(\tilde{B}_q^*(a, \chi, \bar{\theta}, \mu) - \tilde{B}_{q-1}^*(a, \chi, \bar{\theta}, \mu)) \\ \frac{d\bar{\theta}}{dt} &= \bar{\omega} + \mu \bar{B}^*(a, \chi, \bar{\theta}) + \mu \bar{\tilde{B}}^*(a, \chi, \bar{\theta}, \mu) \end{aligned} \tag{10.30}$$

where $\bar{B}^*$, $\bar{\tilde{B}}^*$ are the vectors obtained by retaining the first $q-1$ components of B^* and $\tilde{B}^*$ respectively.

Since the terms appearing on the right-hand side of (10.30) are 2π-periodic with respect to $\theta_1, \theta_2, \ldots, \theta_{q-1}$, we can apply the approximation theorem, keeping separate the "amplitude" variables (a, χ) from the "phase" variable $\bar{\theta}$.

In fact we calculate

$$A^{*0}(a, \chi) = \frac{1}{(2\pi)^{q-1}} \int_0^{2\pi} \cdots \int_0^{2\pi} A^*(a, \chi, \bar{\theta}) d\theta_1 d\theta_2 \cdots d\theta_{q-1}$$

$$\begin{aligned} C^{*0}(a, \chi) = \frac{1}{(2\pi)^{q-1}} \int_0^{2\pi} \cdots \int_0^{2\pi} & (B_q^*(a, \chi, \bar{\theta}) \\ & - B_{q-1}^*(a, \chi, \bar{\theta})) d\theta_1 d\theta_2 \cdots d\theta_{q-1} \end{aligned}$$

and we are led to solve the system

$$A^{*0}(a,\chi)=0$$
$$C^{*0}(a,\chi)=0.$$

If a^0, χ^0 is a solution, $a^0\in G$, such that the square matrix

$$\mathscr{H}=\begin{pmatrix}\dfrac{\partial A^{*0}}{\partial a} & \dfrac{\partial A^{*0}}{\partial \chi}\\[2ex] \dfrac{\partial C^{*0}}{\partial a} & \dfrac{\partial C^{*0}}{\partial \chi}\end{pmatrix}$$

calculated at a^0, χ^0 is stable, we can again apply the approximation theorem. In other words, if the initial values $a(0)$,

$$\chi(0)=\theta_q(0)-\theta_{q-1}(0)$$

are close enough to a^0, χ^0 and μ is small enough and positive, then the solution can be defined for all $t>0$, the amplitude variables are stable and the estimates given by the theorem can be maintained, with a replaced by (a,χ) and θ by $\bar{\theta}$. It is clear that more complicated cases, for example,

$$\omega_{q-3}=\omega_{q-2},\quad \omega_{q-1}=\omega_q,\quad \omega_1,\omega_2,\ldots,\omega_{q-3},\omega_{q-1}$$

forming an independent system of $q-2$ frequencies could be discussed by a suitable modification of this method.

Coupled Oscillators; Non-Autonomous System and Resonance. A Modified Approach

We consider the set of differential equations:

$$\frac{d^2q_s}{dt^2}+\omega_s^2 q_s=\mu f_s\left(q_{s'},\frac{dq_{s'}}{dt},\omega_r t\right), \tag{10.31}$$

$1\leqslant s\leqslant n$, $1\leqslant s'\leqslant n$, $n+1\leqslant r\leqslant m$, where $\omega_1,\omega_2,\ldots,\omega_m$ are independent frequencies and the f_s, are 2π-periodic with respect to each variable $\theta_r=\omega_r t$, are continuous functions with respect to all their arguments and continuously differentiable with respect to q and $\dfrac{dq}{dt}$. We can reduce (10.31) to the form (10.6), with $1\leqslant j\leqslant n$, or better still to the form (10.7) with $\theta=(\theta_1,\ldots,\theta_n,\ldots,\theta_m)$, the variables θ_r, $n+1\leqslant r\leqslant m$ being governed by $\dfrac{d\theta_r}{dt}=\omega_r$. On these assumptions we can apply the foregoing theory.

Certain difficulties however may arise when we try to put this programme into effect which can prevent a successful outcome if one of the frequencies ω_r, $n+1\leqslant r\leqslant m$ is equal to an eigenvalue ω_j, $1\leqslant j\leqslant n$ (resonance), or if certain co-ordinates of the vector a^0 have the value zero (decay of eigenmode).

It is for this reason that we propose, at first in the general case where the frequencies $\omega_1, \ldots, \omega_m$ are independent, to introduce polar variables and Cartesian variables in the form:

$$q_j = \xi_j \sin\theta_j \tag{10.32}$$

$$\frac{dq_j}{dt} = \omega_j \xi_j \cos\theta_j, \quad 1 \leqslant j \leqslant p$$

$$q_k = \xi_k \sin\theta_k + \eta_k \cos\theta_k \tag{10.33}$$

$$\frac{dq_k}{dt} = \omega_k \xi_k \cos\theta_k - \omega_k \eta_k \sin\theta_k,$$

$p+1 \leqslant k \leqslant n$, with $\theta_k = \omega_k t$ with the understanding that we shall provide the justification for this division into two classes later on.

From (10.32) we deduce:

$$\frac{d\xi_j}{dt}\sin\theta_j + \xi_j \cos\theta_j \cdot \frac{d\theta_j}{dt} = \omega_j \xi_j \cos\theta_j \tag{10.34}$$

and then with (10.31)

$$\begin{aligned}\frac{d^2 q_j}{dt^2} + \omega_j^2 q_j &= \omega_j \frac{d\xi_j}{dt}\cos\theta_j - \omega_j \xi_j \sin\theta_j \frac{d\theta_j}{dt} + \omega_j^2 \xi_j \sin\theta_j \\ &= \mu f_j(\xi_{j'} \sin\theta_{j'}, \xi_{k'} \sin\theta_{k'} \\ &\quad + \eta_{k'} \cos\theta_{k'}, \omega_{j'} \xi_{j'} \cos\theta_{j'}, \omega_{k'} \xi_{k'} \cos\theta_{k'} \\ &\quad - \omega_{k'} \eta_{k'} \sin\theta_{k'}, \theta_r) = \mu \tilde{f}_j(\xi_{j'}, \xi_{k'}, \eta_{k'}, \theta_{j'}, \theta_{k'}, \theta_r)\end{aligned} \tag{10.35}$$

and finally from (10.34), (10.35):

$$\frac{d\xi_j}{dt} = \frac{\mu}{\omega_j}\tilde{f}_j \cdot \cos\theta_j = \mu A_j(\xi_{j'}, \xi_{k'}, \eta_{k'}, \theta_{j'}, \theta_{k'}, \theta_r) \tag{10.36}$$

$$\frac{d\theta_j}{dt} = \omega_j - \frac{\mu}{\omega_j \xi_j}\tilde{f}_j \cdot \sin\theta_j = \omega_j + \mu B_j(\xi_{j'}, \xi_{k'}, \eta_{k'}, \theta_{j'}, \theta_{k'}, \theta_r).$$

A similar calculation carried out on (10.33), remembering that $\theta_k = \omega_k t$ leads to:

$$\frac{d\xi_k}{dt} = \frac{\mu}{\omega_k}\tilde{f}_k \cdot \cos\theta_k = \mu A_k(\xi_{j'}, \xi_{k'}, \eta_{k'}, \theta_{j'}, \theta_{k'}, \theta_r) \tag{10.37}$$

$$\frac{d\eta_k}{dt} = -\frac{\mu}{\omega_k}\tilde{f}_k \cdot \sin\theta_k = \mu C_k(\xi_{j'}, \xi_{k'}, \eta_{k'}, \theta_{j'}, \theta_{k'}, \theta_r)$$

We now introduce the vector functions $u(\xi_{j'}, \xi_{k'}, \eta_{k'}, \theta_{j'}, \theta_{k'}, \theta_r)$, v, w depending on the same arguments, 2π-periodic in $\theta_{j'}, \theta_{k'}, \theta_r$ and constructed as in the statement of the lemma in such a way that:

$$(10.38)\quad \sum_{\nu=1}^{m} \omega_\nu \frac{\partial u_j}{\partial \theta_\nu} = \mu u_j + A_j(\xi_{j'}, \xi_{k'}, \eta_{k'}, \theta_{j'}, \theta_{k'}, \theta_r) - A_j^0(\xi_{j'}, \xi_{k'}, \eta_{k'})$$

$$\sum_{\nu=1}^{m} \omega_\nu \frac{\partial u_j}{\partial \theta_\nu} = \mu v_j + B_j(\xi_{j'}, \xi_{k'}, \eta_{k'}, \theta_{j'}, \theta_{k'}, \theta_r) - B_j^0(\xi_{j'}, \xi_{k'}, \eta_{k'}), \quad 1 \leqslant j \leqslant p$$

$$\sum_{\nu=1}^{m} \omega_\nu \frac{\partial u_k}{\partial \theta_\nu} = \mu u_k + A_k(\xi, \eta, \theta) - A_k^0(\xi, \eta)$$

$$\sum_{\nu=1}^{m} \omega_\nu \frac{\partial w_k}{\partial \theta_\nu} = \mu w_k + C_k(\xi, \eta, \theta) - C_k^0(\xi, \eta), \quad p+1 \leqslant k \leqslant n$$

with

$$A_\nu^0 = \frac{1}{(2\pi)^m} \int_0^{2\pi} A_\nu(\xi, \eta, \theta) \mathrm{d}\theta_1 \cdots \mathrm{d}\theta_m, \quad \nu = j \quad \text{or} \quad \nu = k$$

$$(10.39)\quad B_j^0 = \frac{1}{(2\pi)^m} \int_0^{2\pi} B_j(\xi, \eta, \theta) \mathrm{d}\theta_1 \cdots \mathrm{d}\theta_m$$

$$C_k^0 = \frac{1}{(2\pi)^m} \int_0^{2\pi} C_k(\xi, \eta, \theta) \mathrm{d}\theta_1 \cdots \mathrm{d}\theta_m.$$

We then make the following change of variables in (10.36), (10.37)

$$(10.40)\quad \begin{aligned} \xi_j &= x_j + \mu u_j(x_{j'}, x_{k'}, y_{k'}, \varphi_{j'}, \varphi_{k'}, \varphi_r) = x_j + \mu u_j(x, y, \varphi) \\ \theta_j &= \varphi_j + \mu v_j(x_{j'}, x_{k'}, y_{k'}, \varphi_{j'}, \varphi_{k'}, \varphi_r) = \varphi_j + \mu v_j(x, y, \varphi) \\ \xi_k &= x_k + \mu u_k(x, y, \varphi) \\ \eta_k &= y_k + \mu w_k(x, y, \varphi) \end{aligned}$$

where, in the interests of notational uniformity, we have introduced $\varphi_k = \theta_k = \omega_k t$, $\varphi_r = \theta_r = \omega_r t$, and we thus arrive at the equations:

$$\sum_{j'=1}^{p} \left[\left(\delta_{jj'} + \mu \frac{\partial u_j}{\partial x_{j'}} \right) \frac{\mathrm{d}x_{j'}}{\mathrm{d}t} + \mu \frac{\partial u_j}{\partial \varphi_{j'}} \left(\frac{\mathrm{d}\varphi_{j'}}{\mathrm{d}t} - \omega_{j'} \right) \right] + \mu \sum_{k'=p+1}^{n} \left(\frac{\partial u_j}{\partial x_{k'}} \frac{\mathrm{d}x_{k'}}{\mathrm{d}t} + \frac{\partial u_j}{\partial y_{k'}} \frac{\mathrm{d}y_{k'}}{\mathrm{d}t} \right)$$
$$+ \mu \sum_{\nu=1}^{m} \frac{\partial u_j}{\partial \varphi_\nu} \omega_\nu = \mu A_j(x_{j'} + \mu u_{j'}, \ldots, \varphi_{j'} + \mu v_{j'}, \varphi_{k'}, \varphi_r)$$

$$\sum_{j'=1}^{p} \left[\mu \frac{\partial v}{\partial x_{j'}} \frac{\mathrm{d}x_{j'}}{\mathrm{d}t} + \left(\delta_{jj'} + \mu \frac{\partial v_j}{\partial \varphi_{j'}} \right) \left(\frac{\mathrm{d}\varphi_{j'}}{\mathrm{d}t} - \omega_{j'} \right) \right]$$
$$+ \mu \sum_{k'=p+1}^{n} \left(\frac{\partial v_j}{\partial x_{k'}} \frac{\mathrm{d}x_{k'}}{\mathrm{d}t} + \frac{\partial v_j}{\partial y_{k'}} \frac{\mathrm{d}y_{k'}}{\mathrm{d}t} \right) + \mu \sum_{\nu=1}^{m} \frac{\partial v_j}{\partial \varphi_\nu} \omega_\nu = \mu B_j(\cdots)$$

$$\sum_{j'=1}^{p} \mu \left[\frac{\partial u_k}{\partial x_{j'}} \frac{\mathrm{d}x_{j'}}{\mathrm{d}t} + \frac{\partial u_k}{\partial \varphi_{j'}} \left(\frac{\mathrm{d}\varphi_{j'}}{\mathrm{d}t} - \omega_{j'} \right) \right]$$
$$+ \sum_{k'=p+1}^{n} \left[\left(\delta_{kk'} + \mu \frac{\partial u_k}{\partial x_{k'}} \right) \frac{\mathrm{d}x_{k'}}{\mathrm{d}t} + \mu \frac{\partial u_k}{\partial y_{k'}} \frac{\mathrm{d}y_{k'}}{\mathrm{d}t} \right] + \mu \sum_{\nu=1}^{m} \frac{\partial u_k}{\partial \varphi_\nu} \omega_\nu = \mu A_k(\cdots)$$

$$\sum_{j'=1}^{p} u\left[\frac{\partial w_k}{\partial x_{j'}}\frac{dx_{j'}}{dt}+\frac{\partial w_k}{\partial \varphi_{j'}}\left(\frac{d\varphi_{j'}}{dt}-\omega_{j'}\right)\right]$$

$$+\sum_{k'=p+1}^{n}\left[\mu\frac{\partial w_k}{\partial x_{k'}}\frac{dx_{k'}}{dt}+\left(\delta_{kk'}+\mu\frac{\partial w_k}{\partial y_{k'}}\right)\frac{dy_{k'}}{dt}\right]+\mu\sum_{\nu=1}^{m}\frac{\partial w_k}{\partial \varphi_\nu}\omega_\nu=\mu C_k(\cdots)$$

which, in view of (10.38) and the lemma can be written, if $\mu>0$ is small enough:

(10.41)
$$\frac{dx_j}{dt}=\mu A_j^0(x_{j'},x_{k'},y_{k'})+\mu\varepsilon(\mu)$$

$$\frac{d\varphi_j}{dt}=\omega_j+\mu B_j^0(x_{j'},x_{k'},y_{k'})+\mu\varepsilon(\mu),\quad 1\leqslant j\leqslant p$$

$$\frac{dx_k}{dt}=\mu A_k^0(x_{j'},x_{k'},y_{k'})+\mu\varepsilon(\mu)$$

$$\frac{dy_k}{dt}=\mu C_k^0(x_{j'},x_{k'},y_{k'})+\mu\varepsilon(\mu),\quad p+1\leqslant k\leqslant n$$

$$\frac{d\varphi_k}{dt}=\omega_k$$

$$\frac{d\varphi_r}{dt}=\omega_r,\quad n+1\leqslant r\leqslant m$$

where the notation $\varepsilon(\mu)$ is used to denote any functions of the variables of the problem, 2π-periodic in $\varphi_j, \varphi_k, \varphi_r$, and tending uniformly to 0 as $\mu\to 0$.

By applying the approximation theorem to (10.41) we shall be led to solving the set of $2n-p$ equations in $2n-p$ unknowns:

(10.42)
$$A_j^0(x_{j'},x_{k'},y_{k'})=0$$
$$A_k^0(x_{j'},x_{k'},y_{k'})=0$$
$$C_k^0(x_{j'},x_{k'},y_{k'})=0$$

where it should be remembered that the left-hand sides are the means with respect to $\varphi_1,\ldots,\varphi_m$ of:

(10.43) $A_j=\omega_j^{-1}\tilde{f}_j\cos\varphi_j,\quad A_k=\omega_k^{-1}\tilde{f}_k\cos\varphi_k,\quad C_k=-\omega_k^{-1}\tilde{f}_k\sin\varphi_k,$

respectively, with

(10.44)
$$\tilde{f}=f(x_{j'}\sin\varphi_{j'},x_{k'}\sin\varphi_{k'}+y_{k'}\cos\varphi_{k'},\omega_{j'}x_{j'}\cos\varphi_{j'},$$
$$\omega_{k'}x_{k'}\cos\varphi_{k'}-\omega_{k'}y_{k'}\sin\varphi_{k'},\varphi_r).$$

Suppose now that (10.42) has a solution $x_j=a_j$, $x_k=a_k$, $y_k=b_k$, such that the a_j are all non-zero and for at least one value of k, say, k^*, $\rho^*=(a_{k^*}^2+b_{k^*}^2)^{1/2}\neq 0$.

It is easily seen that the equations (10.42) are still satisfied, if while leaving $x_j=a_j$, $x_k=a_k$, $y_k=b_k$ unchanged for $k\neq k^*$, we replace x_{k^*} and y_{k^*} by $\rho^*\cos\chi$, $\rho^*\sin\chi$ respectively, with any χ. Indeed for any equation of the set (10.42), other than the k^*th, this is simply tantamount to adding the phase χ to the variable

φ_{k*}, in the corresponding expression in (10.43), and this obviously has no effect on the calculation of the mean. On the other hand the mean is also unaffected by the substitution affecting the two equations of rank k^*, since the new equations are obtained from the old by multiplying by the matrix $\begin{pmatrix} \cos\chi & \sin\chi \\ -\sin\chi & \cos\chi \end{pmatrix}$. In such a case the Jacobian associated with (10.42) would have the value zero, which would prohibit the use of the approximation theorem.

However recourse to the Cartesian variables is useful in the case of resonance or the case of the decay of certain eigenmodes, which cases we shall now examine.

Case of Resonance

Let us suppose one of the frequencies ω_r of the non-autonomous system (10.31) to be equal to an eigenfrequency ω_s; consider for example the situation where $\omega_n = \omega_{n+1}$, the set $\omega_1, \ldots, \omega_n, \omega_{n+2}, \ldots, \omega_m$ being independent. We introduce the polar and Cartesian variables defined by (10.32) and (10.33) with $p = n-1$, i.e. $n-1$ pairs of polar variables and one pair of Cartesian. To define the change of variables (10.40) we have to identify the variable φ_{n+1} with φ_n, so that the system of frequencies to be considered is from now on $(\omega_1, \ldots, \omega_n, \omega_{n+2}, \ldots, \omega_m)$ with all the means being calculated with respect to $m-1$ variables $\varphi_1, \ldots, \varphi_n, \varphi_{n+2}, \ldots, \varphi_m$ (r varying from $n+2$ to m).

We shall again be led to a system having the structure of (10.41) with $1 \leqslant j \leqslant n-1, k = n, n+2 \leqslant r \leqslant m$ and we shall still have to solve the set of equations of type (10.42), but the difficulty mentioned at the end of the preceding paragraph no longer arises because the variable φ_n, with respect to which it will (among others) be necessary to apply the averaging operator, appears in $\tilde{f}$ as a polar variable but also as an excitation variable since φ_{n+1} has been identified with φ_n.

Case Where Certain Eigenmodes Decay (Degeneracy)

Consider once more the system (10.31) on the hypothesis that the frequencies $(\omega_1, \ldots, \omega_m)$ are independent; we introduce polar and Cartesian variables by (10.32) and (10.33) and making the change of variable (10.40) arrive at the reduced form (10.41) and the associated equations (10.42). Suppose that there exists a vector $a \in \mathbb{R}^p$, whose components $a_1, \ldots, a_p$ are all non-zero, and such that $(a, 0, 0)$ is the solution of the first associated equation (10.42), i.e.

(10.45) $$A_j^0(a_{j'}, 0, 0) = 0, \quad 1 \leqslant j \leqslant p.$$

It is easily verified that we also have in this case:

(10.46) $$A_k^0(a_{j'}, 0, 0) = 0, \quad C_k^0(a_{j'}, 0, 0) = 0, \quad p+1 \leqslant k \leqslant n$$

as can be seen from the representations (10.43), the means being calculated by first integrating with respect to the variable φ_k.

To be able to apply the approximation theorem to the system (10.41), under the conditions (10.45), we have to consider the matrix:

$$S=\begin{pmatrix} \dfrac{\partial A_j^0}{\partial x_{j'}} & \dfrac{\partial A_j^0}{\partial x_{k'}} & \dfrac{\partial A_j^0}{\partial y_{k'}} \\ \dfrac{\partial A_k^0}{\partial x_{j'}} & \dfrac{\partial A_k^0}{\partial x_{k'}} & \dfrac{\partial A_k^0}{\partial y_{k'}} \\ \dfrac{\partial C_k^0}{\partial x_{j'}} & \dfrac{\partial C_k^0}{\partial x_{k'}} & \dfrac{\partial C_k^0}{\partial y_{k'}} \end{pmatrix}$$

calculated at the point $x_{j'}=a_{j'}$, $x_{k'}=y_{k'}=0$ and make sure that it is stable.

Now it is easy to verify, from the expressions (10.43), (10.44), that:

$$\frac{\partial A_k^0}{\partial x_{j'}}=\frac{\partial C_k^0}{\partial x_{j'}}=0; \quad \frac{\partial A_j^0}{\partial x_{k'}}=\frac{\partial A_j^0}{\partial y_{k'}}=0$$

and

$$\frac{\partial A_k^0}{\partial x_{k'}}=\frac{\partial C_k^0}{\partial x_{k'}}=\frac{\partial A_k^0}{\partial y_{k'}}=\frac{\partial C_k^0}{\partial y_{k'}}=0, \quad k\neq k'$$

and lastly that

$$\frac{\partial A_k^0}{\partial x_k}=\frac{\partial C_k^0}{\partial y_k}, \quad \frac{\partial A_k^0}{\partial y_k}=-\frac{\partial C_k^0}{\partial x_k}.$$

The matrix S can therefore be written in the form

$$S=\left(\begin{array}{c|ccc|ccc} \dfrac{\partial A_j^0}{\partial x_{j'}} & & 0 & & & 0 & \\ \hline & \alpha_{p+1} & & 0 & \beta_{p+1} & & 0 \\ 0 & & \ddots & & & \ddots & \\ & 0 & & \alpha_n & 0 & & \beta_n \\ \hline & -\beta_{p+1} & & 0 & \alpha_{p+1} & & 0 \\ 0 & & \ddots & & & \ddots & \\ & 0 & & -\beta_n & 0 & & \alpha_n \end{array}\right)$$

whose eigenvalues are, in addition to those of the matrix $\dfrac{\partial A_j^0}{\partial x_{j'}}$ the numbers $\alpha_k \pm i\beta_k$, $p+1\leqslant k\leqslant n$.

However from another point of view, it is clear that the $n\times n$ matrix

$$H=\begin{pmatrix} \dfrac{\partial A_j^0}{\partial x_{j'}} & 0 \\ 0 & \dfrac{\partial A_k^0}{\partial x_{k'}} \end{pmatrix}$$

is the one we should have obtained if we had used polar variables for all the degrees of freedom of the system. Since the eigenvalues of H are, in addition to those of $\dfrac{\partial A_j^0}{\partial x_j}$, the numbers α_k, $p+1\leqslant k\leqslant n$, it is clear that S and H are alike in respect of stability or instability.

In other words, when the frequencies $(\omega_1,\ldots,\omega_m)$ are independent we can introduce only polar variables into the calculation and it does not then matter, as far as applying the approximation theorem is concerned, whether some of the components of the vector a^0 (i.e. the solution of the associated equation $A_j^0(a^0)=0,\ 1\leqslant j\leqslant n$) are zero. However if a^0 has any zero components the corresponding components of $B_j(x)$ are not continuous in a^0.

Case of Oscillators Coupled Through Linear Terms

Let us first consider the linear system

(10.47) $$\frac{d^2q_s}{dt^2}+\sum_{j=1}^{n} b_{sj}\frac{dq_j}{dt}+\sum_{j=1}^{n} c_{sj}q_j=0,\quad 1\leqslant s\leqslant n$$

where $\mathscr{B}=(b_{ij})$, $\mathscr{C}=(c_{ij})$ are real, constant $(n\times n)$ matrices. Writing p for $p=\dfrac{dq}{dt}$, (10.47) can be put in the form:

(10.48) $$\frac{dq}{dt}=p$$
$$\frac{dp}{dt}=-\mathscr{C}q-\mathscr{B}p$$

or

$$\frac{d}{dt}\begin{pmatrix} q \\ p \end{pmatrix}=\mathscr{L}\cdot\begin{pmatrix} q \\ p \end{pmatrix}$$

where

(10.49) $$\mathscr{L}=\begin{pmatrix} 0 & I \\ -\mathscr{C} & -\mathscr{B} \end{pmatrix}.\quad \text{is a } 2n\times 2n \text{ matrix.}$$

If $q=ve^{i\omega t}$, $p=i\omega ve^{i\omega t}$, is a solution of (10.48) with v a constant vector, we have

$$i\omega\begin{pmatrix} v \\ i\omega v \end{pmatrix}=\mathscr{L}\begin{pmatrix} v \\ i\omega v \end{pmatrix}$$

i.e. with (10.49):

(10.50) $$(-\omega^2+i\omega\mathscr{B}+\mathscr{C})v=0$$

and

$$\det(-\omega^2 I+i\omega\mathscr{B}+\mathscr{C})=0$$

We shall suppose that this last equation has all its roots real and distinct, and hence necessarily in pairs, equal in magnitude but of opposite signs, $\pm\omega_1,\pm\omega_2,\ldots,\pm\omega_n$. We shall denote by

$$\begin{pmatrix} v_{jk} \\ i\omega_k v_{jk} \end{pmatrix},\quad \begin{pmatrix} \bar{v}_{jk} \\ -i\omega_k \bar{v}_{jk} \end{pmatrix},$$

for $k=1,2,\ldots,n$, the $2n$ eigenvectors of $\mathscr{L}$, which moreover form a linearly independent set since the $2n$ eigenvalues are distinct. We thus have

$$\det\begin{pmatrix} v_{jk} & \bar{v}_{jk} \\ i\omega_k v_{jk} & -i\omega_k \bar{v}_{jk} \end{pmatrix} \neq 0$$

or, in terms of real numbers, by taking linear combinations of the columns:

$$\det\begin{pmatrix} \dfrac{v_{jk}-\bar{v}_{jk}}{2i} & \dfrac{v_{jk}+\bar{v}_{jk}}{2} \\ \omega_k \dfrac{v_{jk}+\bar{v}_{jk}}{2} & -\omega_k \dfrac{v_{jk}-\bar{v}_{jk}}{2i} \end{pmatrix} \tag{10.51}$$

$$= \det\begin{pmatrix} \rho_{jk}\sin\chi_{jk} & \rho_{jk}\cos\chi_{jk} \\ \omega_k \rho_{jk}\cos\chi_{jk} & -\omega_k\rho_{jk}\sin\chi_{jk} \end{pmatrix} \neq 0$$

with

$$v_{jk} = \rho_{jk} e^{i\chi_{jk}}, \ \rho, \chi \text{ real.} \tag{10.52}$$

We can represent the general solution of (10.47) in the form

$$q_j = \sum_{k=1}^{n} x_k v_{jk} e^{i\omega_k t}$$

with x_k arbitrary constants, or writing $x_k = \xi_k e^{i\gamma_k}$ and reverting to real quantities:

$$q_j = \sum_{k=1}^{n} \xi_k \rho_{jk} \sin(\omega_k t + \chi_{jk} + \gamma_k). \tag{10.53}$$

If we now consider the non-linear autonomous system:

$$\frac{d^2 q_s}{dt^2} + \sum_{j=1}^{n} b_{sj}\frac{dq_j}{dt} + \sum_{j=1}^{n} c_{sj} q_j = \mu f_s\left(q, \frac{dq}{dt}\right) \tag{10.54}$$

we can try to represent its solutions by means of the formulae (10.53), where we assume that the parameters ξ_k, γ_k vary slowly in time if μ is small, while ω_k, $\rho_{jk}e^{i\chi_{jk}} = v_{jk}$ are an eigenvalue and its associated eigenvector of $\mathscr{L}$, and satisfy (10.50).

We postulate that:

$$\frac{dq_j}{dt} = \sum_{k=1}^{n} \omega_k \xi_k \rho_{jk} \cos(\omega_k t + \chi_{jk} + \gamma_k) \tag{10.55}$$

so that by (10.53) we have

$$\sum_{k=1}^{n} \frac{d\xi_k}{dt}\rho_{jk}\sin(\theta_k + \chi_{jk}) + \sum_{k=1}^{n} \xi_k \frac{d\gamma_k}{dt}\rho_{jk}\cos(\theta_k + \chi_{jk}) = 0 \tag{10.56}$$

with $\theta_k = \omega_k t + \gamma_k$.

Calculating $\dfrac{d^2 q_j}{dt^2}$ from (10.55), and inserting the result obtained in (10.54), and taking into account that (10.53) is a solution of (10.47) when ξ_k, γ_k are constants, we arrive at:

(10.57)

$$\sum_{k=1}^{n} \omega_k \frac{d\xi_k}{dt} \rho_{jk} \cos(\theta_k + \chi_{jk}) - \sum_{k=1}^{n} \omega_k \xi_k \rho_{jk} \frac{d\gamma_k}{dt} \sin(\theta_k + \chi_{jk}) = \mu \tilde{f}_j(\xi_{k'}, \theta_{k'}).$$

By solving (10.56), (10.57) with respect to $\frac{d\xi_k}{dt}$, $\xi_k \frac{d\gamma_k}{dt}$, we shall obtain the differential equations for the slowly varying parameters; but for this we have to concern ourselves with the determinant:

$$\det\begin{pmatrix} \rho_{jk} \sin(\theta_k + \chi_{jk}) & \rho_{jk} \cos(\theta_k + \chi_{jk}) \\ \omega_k \rho_{jk} \cos(\theta_k + \chi_{jk}) & -\omega_k \rho_{jk} \sin(\theta_k + \chi_{jk}) \end{pmatrix}. \tag{10.58}$$

We see at once that its partial derivatives with respect to the variables θ_k vanish. The determinant is therefore independent of the θ_k and its constant value, obtained by putting $\theta_k = 0$, and denoted by Δ is, as we saw earlier in (10.51) not zero.

Lastly we deduce from (10.50) and (10.57), that:

$$\begin{aligned} \frac{d\xi_k}{dt} &= \mu A_k(\xi_{k'}, \theta_{k'}) \\ \xi_k \frac{d\gamma_k}{dt} &= \mu B_k(\xi_{k'}, \theta_{k'}) \\ \frac{d\theta_k}{dt} &= \omega_k + \mu \xi_k^{-1} B_k(\xi_{k'}, \theta_{k'}), \end{aligned} \tag{10.59}$$

for $1 \leqslant k \leqslant n$, $1 \leqslant k' \leqslant n$, the functions A_k, B_k being multiperiodic with respect to $\theta_{k'}$ of period 2π; and so one can apply the approximation theorem to the reduced form (10.59).

Non-Autonomous Non-Linear System in the General Case; Examination of the Case When Certain Eigenmodes Are Evanescent

We propose to discuss the case where the perturbation terms in (10.54) depend explicitly on time; thus we consider:

$$\frac{d^2 q_s}{dt^2} + \sum_{s'=1}^{n} b_{ss'} \frac{dq_{s'}}{dt} + \sum_{s'=1}^{n} c_{ss'} q_{s'} = \mu f_s\left(q_{s'}, \frac{dq_{s'}}{dt}, \omega_r t\right), \tag{10.60}$$

$1 \leqslant s \leqslant n$, $1 \leqslant s' \leqslant n$, $n+1 \leqslant r \leqslant m$, where we assume that the linear system has n distinct modes of vibration corresponding to the n real eigenvalues $\omega_1, \ldots, \omega_n$ and the excitation terms are multiperiodic of period 2π with respect to each variable $\theta_r = \omega_r t$, $n+1 \leqslant r \leqslant m$, the complete set of frequencies $\omega_1, \omega_2, \ldots, \omega_n, \omega_{n+1}, \ldots, \omega_m$ being supposed independent.

We are led via the transformation (10.53), subject to (10.56), to the system (10.59) which governs the slow variables, with this modification however, that

A_k, B_k depend on $\theta_r = \omega_r t$, as well as on the variables ξ_k, θ_k. Thus we can adjoin to (10.59) the equation $\dfrac{d\theta_r}{dt} = \omega_r$ and we see that we need to calculate the means:

$$A_k^0(\xi_{k'}) = \frac{1}{(2\pi)^m} \int_0^{2\pi} \cdots \int_0^{2\pi} A_k(\xi_{k'}, \theta_{k'}, \theta_r) d\theta_1 d\theta_2 \cdots d\theta_m$$

and then solve the set of equations:

$$A_k^0(\xi_{k'}) = 0, \quad 1 \leqslant \overset{k}{\underset{k'}{}} \leqslant n.$$

If there is a vector $a \in \mathbb{R}^n$, with no zero components such that $A_k^0(a_{k'}) = 0$ and for which the matrix $H = \left.\dfrac{\partial A_k^0}{\partial \xi_{k'}}\right|_a$ is stable, then we can usefully apply the approximation theorem.

But this result is no longer self-evident if some of the components of a are zero, even though the calculations indicated above are still meaningful, at least in a formal sense. Our next task therefore is to prove that the result still retains its full validity. It is clear that without any real loss of generality we may suppose for convenience that the first p components of a are non-zero and the remaining $n-p$ are all zero. We then introduce polar variables for the first p degrees of freedom and Cartesian variables for the last $n-p$. In other words we adopt a representation

$$(10.61) \qquad q_s = \sum_{j=1}^{p} \xi_j \rho_{sj} \sin(\theta_j + \chi_{sj}) + \sum_{k=p+1}^{n} \xi_k \rho_{sk} \sin(\theta_k + \chi_{sk}) + \sum_{k=p+1}^{n} \eta_k \rho_{sk} \cos(\theta_k + \chi_{sk})$$

with

$$(10.62) \qquad \theta_j = \omega_j t + \gamma_j, \quad \theta_k = \omega_k t, \quad 1 \leqslant j \leqslant p, \quad p+1 \leqslant k \leqslant n$$

and remembering that

$$(10.63) \qquad \theta_r = \omega_r t, \quad n+1 \leqslant r \leqslant m.$$

We impose

$$(10.64) \qquad \frac{dq_s}{dt} = \sum_{j=1}^{p} \omega_j \xi_j \rho_{sj} \cos(\theta_j + \chi_{sj}) + \sum_{k=p+1}^{n} \omega_k \xi_k \rho_{sk} \cos(\theta_k + \chi_{sk}) - \sum_{k=p+1}^{n} \omega_k \eta_k \rho_{sk} \sin(\theta_k + \chi_{sk})$$

so that $\xi_j, \theta_j, \xi_k, \eta_k$ have to satisfy:

$$(10.65) \qquad \sum_{j=1}^{p} \frac{d\xi_j}{dt} \rho_{sj} \sin(\theta_j + \chi_{sj}) + \sum_{k=p+1}^{n} \frac{d\xi_k}{dt} \rho_{sk} \sin(\theta_k + \chi_{sk}) + \sum_{j=1}^{p} \xi_j \frac{d\gamma_j}{dt} \rho_{sj} \cos(\theta_j + \chi_{sj}) + \sum_{k=p+1}^{n} \frac{d\eta_k}{dt} \rho_{sk} \cos(\theta_k + \chi_{sk}) = 0.$$

Lastly, we obtain from (10.64):

(10.66)

$$\frac{d^2 q_s}{dt^2} = \sum_{j=1}^{p} \omega_j \frac{d\xi_j}{dt} \rho_{sj} \cos(\theta_j + \chi_{sj}) + \sum_{k=p+1}^{n} \omega_k \frac{d\xi_k}{dt} \rho_{sk} \cos(\theta_k + \chi_{sk})$$
$$- \sum_{j=1}^{p} \omega_j \xi_j \rho_{sj} \frac{d\gamma_j}{dt} \sin(\theta_j + \chi_{sj}) - \sum_{k=p+1}^{n} \omega_k \frac{d\eta_k}{dt} \rho_{sk} \sin(\theta_k + \chi_{sk}) + \cdots$$

the terms omitted being those which would have appeared had $\xi_j, \gamma_j, \xi_k, \eta_k$ been constants and which are destined to disappear in the course of substitution in (10.60). We find in fact in this way:

$$\sum_{j=1}^{p} \omega_j \frac{d\xi_j}{dt} \rho_{sj} \cos(\theta_j + \chi_{sj}) + \sum_{k=p+1}^{n} \omega_k \frac{d\xi_k}{dt} \rho_{sk} \cos(\theta_k + \chi_{sk})$$
$$- \sum_{j=1}^{p} \omega_j \xi_j \rho_{sj} \frac{d\gamma_j}{dt} \sin(\theta_j + \chi_{sj}) - \sum_{k=p+1}^{n} \omega_k \frac{d\eta_k}{dt} \rho_{sk} \sin(\theta_k + \chi_{sk})$$
$$= \mu \tilde{f}_s(\xi_{j'}, \xi_{k'}, \eta_{k'}, \theta_{j'}, \theta_{k'}, \theta_r) \tag{10.67}$$

with

$$\tilde{f}_s = f_s\Bigg(\sum_{j=1}^{p} \xi_j \rho_{s'j} \sin(\theta_j + \chi_{s'j}) + \sum_{k=p+1}^{n} \xi_k \rho_{s'k} \sin(\theta_k + \chi_{s'k})$$
$$+ \sum_{k=p+1}^{n} \eta_k \rho_{s'k} \cos(\theta_k + \chi_{s'k}), \sum_{j=1}^{p} \omega_j \xi_j \rho_{s'j} \cos(\theta_j + \chi_{s'j})$$
$$+ \sum_{k=p+1}^{n} \omega_k \xi_k \rho_{s'k} \cos(\theta_k + \chi_{s'k}) - \sum_{k=p+1}^{n} \omega_k \eta_k \rho_{s'k} \sin(\theta_k + \chi_{s'k}), \theta_r \Bigg). \tag{10.68}$$

To solve the system (10.65), (10.67) with respect to $\frac{d\xi_j}{dt}, \frac{d\xi_k}{dt}, \xi_j \frac{d\gamma_j}{dt}, \frac{d\eta_k}{dt}$ we note that the associated determinant takes the form:

(10.69)

$$\begin{array}{c} \\ 1 \leqslant s \leqslant n \\ 1 \leqslant s \leqslant n \end{array} \begin{pmatrix} \scriptstyle 1 \leqslant j \leqslant p & \scriptstyle p+1 \leqslant k \leqslant n & \scriptstyle 1 \leqslant j \leqslant p & \scriptstyle p+1 \leqslant k \leqslant n \\ \rho_{sj} \sin(\theta_j + \chi_{sj}) & \rho_{sk} \sin(\theta_k + \chi_{sk}) & \rho_{sj} \cos(\theta_j + \chi_{sj}) & \rho_{sk} \cos(\theta_k + \chi_{sk}) \\ \omega_j \rho_{sj} \cos(\theta_j + \chi_{sj}) & \omega_k \rho_{sk} \cos(\theta_k + \chi_{sk}) & -\omega_j \rho_{sj} \sin(\theta_j + \chi_{sj}) & -\omega_k \rho_{sk} \sin(\theta_k + \chi_{sk}) \end{pmatrix}$$

and is the same as the one already discussed in (10.58). It therefore has a value $\Delta \neq 0$, which is constant, i.e. independent of θ_j, θ_k.

Finally, after solving (10.65), (10.67) we arrive at the system:

$$\frac{d\xi_j}{dt} = \mu A_j(\xi_{j'}, \xi_{k'}, \eta_{k'}, \theta_{j'}, \theta_{k'}, \theta_r) \tag{10.70}$$

$$\frac{d\xi_k}{dt} = \mu A_k(\xi_{j'}, \xi_{k'}, \eta_{k'}, \theta_{j'}, \theta_{k'}, \theta_r)$$

$$\xi_j \frac{d\gamma_j}{dt} = \mu B_j(\xi_{j'}, \xi_{k'}, \eta_{k'}, \theta_{j'}, \theta_{k'}, \theta_r) \quad \text{or} \quad \frac{d\theta_j}{dt} = \omega_j + \mu \xi_j^{-1} B_j$$

$$\frac{d\eta_k}{dt} = \mu B_k(\xi_{j'}, \xi_{k'}, \eta_{k'}, \theta_{j'}, \theta_{k'}, \theta_r)$$

$$\frac{d\theta_k}{dt} = \omega_k, \quad \frac{d\theta_r}{dt} = \omega_r.$$

We introduce the means:

$$A_v^0(\xi_{j'}, \xi_{k'}, \eta_{k'}) = \frac{1}{(2\pi)^m} \int_0^{2\pi} \cdots \int_0^{2\pi} A_v(\xi_{j'}, \xi_{k'}, \eta_{k'}, \theta_{j'}, \theta_{k'}, \theta_r)\, d\theta_1 \cdots d\theta_m$$

with $v = j$ or $v = k$:

$$B_k^0(\xi_{j'}, \xi_{k'}, \eta_{k'}) = \frac{1}{(2\pi)^m} \int_0^{2\pi} \cdots \int_0^{2\pi} B_k(\xi_{j'}, \xi_{k'}, \eta_{k'}, \theta_{j'}, \theta_{k'}, \theta_r)\, d\theta_1 \cdots d\theta_m$$

and note that $A_v^0(\xi_{j'}, \xi_{k'}, 0)$, $1 \leqslant v \leqslant n$, is exactly the same mean of corresponding rank as would have been obtained if we had introduced polar variables only.

Now the assumption which led us to introduce p polar variables and $n-p$ Cartesian variables was based on the existence of a vector $a \in \mathbb{R}^p$ whose components are all non-zero and such that $A_v^0(a_{j'}, 0, 0) = 0$, $1 \leqslant v \leqslant n$.

Instead of assuming this let us now make the weaker assumption (whose truth is implied by the former) that $a \in \mathbb{R}^p$ has all its components non-zero and is such that $A_j^0(a_{j'}, 0, 0) = 0$, $1 \leqslant j \leqslant p$. Before going any further let us describe precisely the rule which enables us to calculate A_j, A_k, B_k.

A_k or B_k is clearly the quotient by Δ of the determinant obtained by replacing in (10.69) the column corresponding to $\frac{d\xi_k}{dt}$, or $\frac{d\eta_k}{dt}$ respectively by the column on the right-hand side of the set of equations comprising the system (10.65), (10.67), i.e. the column represented by the column vector $\begin{pmatrix} 0 \\ \tilde{f}_s \end{pmatrix}$, with $\tilde{f}_s$ defined by (10.68), $1 \leqslant s \leqslant n$. Under the preceding assumptions it is clear that $A_k^0(a_{j'}, 0, 0) = B_k^0(a_{j'}, 0, 0) = 0$, because, when $\xi_k = \eta_k = 0$, $\tilde{f}_s$ no longer depends on θ_k, and the substitution operation described earlier leaves only one column containing a θ_k in the determinant to be evaluated, and consequently the mean value will be zero, as we see immediately by first integrating with respect to θ_k.

We now discuss the stability matrix which can be written:

$$S = \begin{pmatrix} \dfrac{\partial A_j^0}{\partial \xi_{j'}} & \dfrac{\partial A_j^0}{\partial \xi_{k'}} & \dfrac{\partial A_j^0}{\partial \eta_{k'}} \\[2ex] \dfrac{\partial A_k^0}{\partial \xi_{j'}} & \dfrac{\partial A_k^0}{\partial \xi_{k'}} & \dfrac{\partial A_k^0}{\partial \eta_{k'}} \\[2ex] \dfrac{\partial B_k^0}{\partial \xi_{j'}} & \dfrac{\partial B_k^0}{\partial \xi_{k'}} & \dfrac{\partial B_k^0}{\partial \eta_{k'}} \end{pmatrix} \tag{10.71}$$

of order $p + 2(n-p)$.

Let us begin by calculating $\frac{\partial A_j^0}{\partial \xi_{k'}}, \frac{\partial A_j^0}{\partial \eta_{k'}}$; ignoring the denominator Δ which is a numerical constant, we first note that A_j is obtained from (10.69) by replacing the column $\begin{pmatrix} \rho_{sj} \sin(\theta_j + \chi_{sj}) \\ \omega_j \rho_{sj} \cos(\theta_j + \chi_{sj}) \end{pmatrix}$ by $\begin{pmatrix} 0 \\ \tilde{f}_s \end{pmatrix}$. When we differentiate the terms of this column with respect to $\xi_{k'}$, $\eta_{k'}$ we find successively:

$$(10.72)\quad \frac{\partial \tilde{f}_s}{\partial \xi_{k'}} = \sum_{s'=1}^{n} \left[\rho_{s'k'} \sin(\theta_{k'} + \chi_{s'k'}) \cdot \frac{\partial f_s}{\partial q_{s'}} + \omega_{k'} \rho_{s'k'} \cos(\theta_{k'} + \chi_{s'k'}) \cdot \frac{\partial f_s}{\partial q'_{s'}} \right]$$

$$\frac{\partial \tilde{f}_s}{\partial \eta_{k'}} = \sum_{s'=1}^{n} \left[\rho_{s'k'} \cos(\theta_{k'} + \chi_{s'k'}) \cdot \frac{\partial f_s}{\partial q_{s'}} - \omega_{k'} \rho_{s'k'} \sin(\theta_{k'} + \chi_{s'k'}) \cdot \frac{\partial f_s}{\partial q'_{s'}} \right].$$

When $\xi_{j'} = a_{j'}$, $\xi_{k'} = 0$, $\eta_{k'} = 0$, the expressions $\frac{\partial f_s}{\partial q_{s'}}, \frac{\partial f_s}{\partial q'_{s'}}$ no longer depend on $\theta_{k'}$; now to work out the determinants which represent $\frac{\partial A_j^0}{\partial \xi_{k'}}, \frac{\partial A_j^0}{\partial \eta_{k'}}$ we can reduce the calculation of the sums (10.72) to their first term, then to their second term and so on, and in this way obtain a sum of $2n$ determinants, those identified by the suffix s' containing as a factor $\sin(\theta_{k'} + \chi_{s'k'})$ or $\cos(\theta_{k'} + \chi_{s'k'})$, but also involving in two other columns, namely the k'th and $(k' + n)$th circular functions of the arc $\theta_{k'} + \chi_{sk'}$. This is tantamount to saying that each term in the expansion of the determinant in terms of $\theta_{k'}$ involves products of the type

$$\sin^3 \theta_{k'}, \sin^2 \theta_{k'} \cos \theta_{k'}, \sin \theta_{k'} \cos^2 \theta_{k'}, \cos^3 \theta_{k'},$$

all of which have a mean value zero with respect to $\theta_{k'}$. It has thus been proved that

$$\frac{\partial A_j^0}{\partial \xi_{k'}} = \frac{\partial A_j^0}{\partial \eta_{k'}} = 0.$$

The same argument enables us to show that

$$\frac{\partial A_k^0}{\partial \xi_{k'}} = \frac{\partial A_k^0}{\partial \eta_{k'}} = 0 \quad \text{if} \quad k \neq k' \quad \text{and} \quad \frac{\partial B_k^0}{\partial \xi_{k'}} = \frac{\partial B_k^0}{\partial \eta_{k'}} = 0, \quad \text{if} \quad k \neq k'.$$

It remains therefore to calculate the terms $\frac{\partial A_k^0}{\partial \eta_k}, \frac{\partial B_k^0}{\partial \xi_k}$ and then $\frac{\partial A_k^0}{\partial \xi_k}, \frac{\partial B_k^0}{\partial \eta_k}$. To compare $\frac{\partial A_k^0}{\partial \eta_k}$ and $\frac{\partial B_k^0}{\partial \xi_k}$ it suffices to note that, before taking means, the determinants representing $\frac{\partial A_k}{\partial \eta_k}, \frac{\partial B_k}{\partial \xi_k}$ are the same apart from their kth and $(k + n)$th columns which are, respectively:

$$\begin{matrix} k & & k+n \\ 0 & & \rho_{sk} \cos(\theta_k + \chi_{sk}) \\ \sum_{s'} \left[\rho_{s'k} \cos(\theta_k + \chi_{s'k}) \cdot \frac{\partial f_s}{\partial q_{s'}} - \omega_k \rho_{s'k} \sin(\theta_k + \chi_{s'k}) \cdot \frac{\partial f_s}{\partial q'_{s'}} \right] & & -\omega_k \rho_{sk} \sin(\theta_k + \chi_{sk}) \end{matrix}$$

for $\dfrac{\partial A_k}{\partial \eta_k}$, and

$$\begin{array}{cc} k & k+n \\ \rho_{sk}\sin(\theta_k+\chi_{sk}) & 0 \\ \omega_k\rho_{sk}\cos(\theta_k+\chi_{sk}) & \sum_{s'}\left[\rho_{s'k}\sin(\theta_k+\chi_{s'k})\cdot\dfrac{\partial f_s}{\partial q_{s'}}+\omega_k\rho_{s'k}\cos(\theta_k+\chi_{s'k})\cdot\dfrac{\partial f_s}{\partial q'_{s'}}\right] \end{array}$$

for $\dfrac{\partial B_k}{\partial \xi_k}$.

It will be seen that by interchanging the kth and $(k+n)$th columns and changing θ_k to $\theta_k+\dfrac{\pi}{2}$ (which cannot affect the terms $\dfrac{\partial f_s}{\partial q_{s'}}, \dfrac{\partial f_s}{\partial q'_{s'}}$ since these calculated for $\xi_j=a_j$, $\xi_k=\eta_k=0$ no longer depend on θ_k), the second determinant changes into the first, and therefore $\dfrac{\partial A_k^0}{\partial \eta_k}=-\dfrac{\partial B_k^0}{\partial \xi_k}$.

The expressions $\dfrac{\partial A_k^0}{\partial \xi_k}$ and $\dfrac{\partial B_k^0}{\partial \eta_k}$ can be dealt with in a similar way. Here again we see that before the averaging operation, the determinants differ only in their kth and $(k+n)$th columns which are respectively:

$$\begin{array}{cc} k & k+n \\ 0 & \rho_{sk}\cos(\theta_k+\chi_{sk}) \\ \sum_{s'}\left[\rho_{s'k}\sin(\theta_k+\chi_{s'k})\cdot\dfrac{\partial f_s}{\partial q_{s'}}+\omega_k\rho_{s'k}\cos(\theta_k+\chi_{s'k})\cdot\dfrac{\partial f_s}{\partial q'_{s'}}\right] & -\omega_k\rho_{sk}\sin(\theta_k+\chi_{sk}) \end{array}$$

for $\dfrac{\partial A_k}{\partial \xi_k}$ and

$$\begin{array}{cc} k & k+n \\ \rho_{sk}\sin(\theta_k+\chi_{sk}) & 0 \\ \omega_k\rho_{sk}\cos(\theta_k+\chi_{sk}) & \sum_{s'}\left[\rho_{s'k}\cos(\theta_k+\chi_{s'k})\dfrac{\partial f_s}{\partial q_{s'}}-\omega_k\rho_{s'k}\sin(\theta_k+\chi_{s'k})\cdot\dfrac{\partial f_s}{\partial q'_{s'}}\right] \end{array}$$

for $\dfrac{\partial B_k}{\partial \eta_k}$, from which we easily deduce that:

$$\frac{\partial A_k^0}{\partial \xi_k}=\frac{\partial B_k^0}{\partial \eta_k}.$$

We thus find ourselves again in a situation completely analogous to that encountered earlier, in dealing with the case where there is no coupling between the linear terms. Here, as elsewhere, we can use polar variables exclusively to describe the motion and to analyse its stability.

Gyroscopic Stabiliser with Non-Linear Servomechanism

We shall consider again, with a slightly modified notation, the set of equations obtained in Chapter 4 describing the behaviour of a monorail car, stabilised by a servo device, but this time assuming a non-linear law.

Since the damping and the correcting couple are both weak, we can write the equations of motion in the form:

$$\begin{aligned} &x_1'' + \sigma x_2' + p x_1 = \mu(-c_1 x_1' + (\lambda - \alpha x_1'^2) x_1') \\ &x_2'' - \sigma x_1' + q x_2 = -\mu c_2 x_2' \end{aligned} \tag{10.73}$$

x_1 and x_2 replacing the variables θ, ψ, and where μ denotes a small positive parameter, $c_1, c_2, \lambda, \alpha$ are constants which are $O(1)$, with $c_1 > 0$, $c_2 > 0$. We suppose $p < 0$, $q < 0$ and σ large enough to ensure stability of the stationary solution when $\mu = 0$, i.e.

$$\sigma^2 + p + q > 2\sqrt{pq};$$

in this case the eigenfrequencies ω_1, ω_2 are the positive roots of:

$$\omega^4 - (\sigma^2 + p + q)\omega^2 + pq = 0 \tag{10.74}$$

and the general solution of the linear system with $\mu = 0$ can be written:

$$\begin{aligned} x_1 &= \xi_1 \sin\theta_1 + \xi_2 \sin\theta_2, \qquad \theta_1 = \omega_1 t + \gamma_1 \\ x_2 &= k\xi_1 \cos\theta_1 + r\xi_2 \cos\theta_2, \quad \theta_2 = \omega_2 t + \gamma_2 \end{aligned} \tag{10.75}$$

where the amplitudes ξ_1, ξ_2 and phases γ_1, γ_2 are arbitrary constants, and k and r are given by:

$$k = \frac{\sigma\omega_1}{q - \omega_1^2} = \frac{p - \omega_1^2}{\sigma\omega_1}, \quad r = \frac{p - \omega_2^2}{\sigma\omega_2} = \frac{\sigma\omega_2}{q - \omega_2^2}. \tag{10.76}$$

To describe the solutions of the non-linear system (10.73), we start from the representation (10.75), supposing $\xi_1, \xi_2, \gamma_1, \gamma_2$ to be slowly varying functions of time such that:

$$\begin{aligned} x_1' &= \xi_1\omega_1 \cos\theta_1 + \xi_2\omega_2 \cos\theta_2 \\ x_2' &= -k\xi_1\omega_1 \sin\theta_1 - r\xi_2\omega_2 \sin\theta_2 \end{aligned} \tag{10.77}$$

which necessitates the equations:

$$\begin{aligned} \xi_1' \sin\theta_1 + \xi_2' \sin\theta_2 + \xi_1\gamma_1' \cos\theta_1 + \xi_2\gamma_2' \cos\theta_2 &= 0 \\ k\xi_1' \cos\theta_1 + r\xi_2' \cos\theta_2 - k\xi_1\gamma_1' \sin\theta_1 - r\xi_2\gamma_2' \sin\theta_2 &= 0. \end{aligned} \tag{10.78}$$

Using (10.75), (10.77) we obtain from (10.73):

$$\begin{aligned} \xi_1'\omega_1 \cos\theta_1 + \xi_2'\omega_2 \cos\theta_2 - \xi_1\omega_1\gamma_1' \sin\theta_1 - \xi_2\omega_2\gamma_2' \sin\theta_2 &= \mu f^* \\ k\xi_1'\omega_1 \sin\theta_1 + r\xi_2'\omega_2 \sin\theta_2 + k\xi_1\omega_1\gamma_1' \cos\theta_1 + r\xi_2\omega_2\gamma_2' \cos\theta_2 &= \mu g^* \end{aligned} \tag{10.79}$$

with $\quad f(x_1') = (\lambda - c_1)x_1' - \alpha x_1'^3, \quad g(x_2') = c_2 x_2'$

and

(10.80)
$$f^* = f(\xi_1\omega_1 \cos\theta_1 + \xi_2\omega_2 \cos\theta_2)$$
$$g^* = g(-k\xi_1\omega_1 \sin\theta_1 - r\xi_2\omega_2 \sin\theta_2).$$

The system (10.78), (10.79) which is linear in $\xi_1', \xi_2', \xi_1\gamma_1', \xi_2\gamma_2'$ can be solved algebraically; we thus find as expression for the determinant

$$\Delta = (k\omega_2 - r\omega_1)(r\omega_2 - k\omega_1),$$

and then the equations:

(10.81)
$$\frac{d\xi_1}{dt} = \mu\left(\frac{r\cos\theta_1}{r\omega_1 - k\omega_2} f^* + \frac{\sin\theta_1}{k\omega_1 - r\omega_2} g^*\right)$$
$$\frac{d\xi_2}{dt} = \mu\left(\frac{k\cos\theta_2}{k\omega_2 - r\omega_1} f^* + \frac{\sin\theta_2}{r\omega_2 - k\omega_1} g^*\right)$$

with

(10.82)
$$k\omega_2 - r\omega_1 = \frac{\sigma\omega_1\omega_2(\omega_1^2 - \omega_2^2)}{(q-\omega_1^2)(q-\omega_2^2)},$$
$$r\omega_2 - k\omega_1 = \sigma q \frac{\omega_2^2 - \omega_1^2}{(q-\omega_1^2)(q-\omega_2^2)}$$

and, in line with the general theory, it is unnecessary to consider the equations which take account of $\xi_1\gamma_1', \xi_2\gamma_2'$.

We must now substitute for (10.81) the system obtained by taking the mean values of the right-hand sides with respect to θ_1, θ_2. Operating in this way and introducing the new variables

(10.83)
$$u = (\xi_1\omega_1)^2, \quad v = (\xi_2\omega_2)^2$$

we obtain

(10.84)
$$\frac{du}{dt} = \mu A u(a - u - 2v)$$
$$\frac{dv}{dt} = \mu B v(b + v + 2u)$$

with

(10.85)
$$A = \frac{3\alpha}{4}\frac{q-\omega_1^2}{(\omega_2^2 - \omega_1^2)\omega_1^2},$$
$$a = \frac{4}{3\alpha}\left(\lambda - c_1 + \frac{\omega_1^2}{q}\cdot\frac{q-\omega_2^2}{q-\omega_1^2}c_2\right)$$
$$B = \frac{3\alpha}{4}\frac{q-\omega_2^2}{(\omega_2^2 - \omega_1^2)\omega_2^2},$$
$$b = \frac{4}{3\alpha}\left(-\lambda + c_1 - \frac{\omega_2^2}{q}\frac{q-\omega_1^2}{q-\omega_2^2}c_2\right).$$

We shall assume $\omega_2 > \omega_1$ and incidentally we may note that interchanging ω_1, ω_2 is equivalent to changing A, B, a, b, into $-B, -A, -b, -a$ respectively. The parameter α will be assumed to be positive and hence:

$$A < 0, \quad B < 0 \tag{10.86}$$

The nature of the possible solutions to the equation of motion which can arise depends on the singularities of (10.84) and on their stability, and the investigation of these points must clearly be confined to the quadrant $u \geqslant 0$, $v \geqslant 0$.

1. $u = 0$, $v = 0$.

with

$$\lambda_1 = c_1 - \frac{\omega_1^2}{q}\frac{q - \omega_2^2}{q - \omega_1^2} c_2, \quad \lambda_2 = c_1 - \frac{\omega_2^2}{q}\frac{q - \omega_1^2}{q - \omega_2^2} c_2 \tag{10.87}$$

$$a = \frac{4}{3\alpha}(\lambda - \lambda_1), \quad b = \frac{4}{3\alpha}(\lambda_2 - \lambda) \tag{10.88}$$

The stability condition is $a > 0$, $b > 0$ or:

$$\lambda_1 < \lambda < \lambda_2 \tag{10.89}$$

This makes sense only if

$$\lambda_2 - \lambda_1 = c_2 \frac{(\omega_2^2 - \omega_1^2)(p - q)}{(q - \omega_1^2)(q - \omega_2^2)} \tag{10.90}$$

is positive, i.e. if $p - q > 0$. We shall assume from now on that this condition is always satisfied. (If the sign were reversed the servomechanism, and in particular, the correcting couple, would have to act on the x_2 co-ordinate.)

2. $u = a$, $v = 0$, subject to the condition

$$a > 0. \tag{10.91}$$

The stability matrix in the general case is:

$$\begin{pmatrix} -2A(u+v) + aA & -2Au \\ 2Bv & 2B(u+v) + Bb \end{pmatrix} \tag{10.92}$$

and for $u = a$, $v = 0$ it has two eigenvalues, namely $-Aa$ and $B(b + 2a)$. For stability a would have to be negative and this would contradict (10.91). It is therefore impossible to get harmonic motion at the frequency ω_1.

3. $u = 0$, $v = -b$, subject to the condition

$$b < 0. \tag{10.93}$$

The eigenvalues of the stability matrix are $A(a + 2b)$ and $-Bb$, from which we derive the conditions:

$$\begin{aligned} a + 2b &> 0 \\ b &< 0 \end{aligned} \tag{10.94}$$

which by (10.88) are equivalent to

(10.95) $$\lambda_2 < \lambda < 2\lambda_2 - \lambda_1$$

and these are compatible, since $\lambda_2 > \lambda_1$.

Accordingly for a λ satisfying (10.95), a stable motion of frequency ω_2 can be observed, of amplitude $\xi_2\omega_2 = \sqrt{-b}$.

4. $u + 2v = a$, $v + 2u = -b$ or $u = -\frac{a+2b}{3}$, $v = \frac{2a+b}{3}$.

For this point to lie in the first quadrant, we must have $a + 2b < 0$, $2a + b > 0$, or $2\lambda_2 - \lambda_1 < \lambda$ and $2\lambda_1 - \lambda_2 < \lambda$, which reduces, in view of $\lambda_2 > \lambda_1$, to:

(10.96) $$2\lambda_2 - \lambda_1 < \lambda.$$

Examination of the stability matrix, which in this case is:

$$\begin{pmatrix} \frac{A}{3}(a+2b) & \frac{2A}{3}(2b+a) \\ \frac{2B}{3}(2a+b) & \frac{B}{3}(2a+b) \end{pmatrix}$$

leads to the supplementary condition:

$$A(a+2b) + B(2a+b) < 0.$$

This yields without difficulty

(10.97) $$A(a+2b) + B(2a+b) = -p^{-1}\cdot(\lambda - \lambda_3),$$

$$\lambda_3 = c_1 + \frac{(p-2q)(\omega_1^2 + \omega_2^2) - 3p^2 + 5pq}{(q-\omega_1^2)(q-\omega_2^2)} c_2.$$

It is easily verified that $2\lambda_2 - \lambda_1 < \lambda_3$ and we conclude that a stable biharmonic motion of frequencies ω_1, ω_2 is possible provided $2\lambda_2 - \lambda_1 < \lambda < \lambda_3$.

In conclusion we see that as λ increases from λ_1 to λ_3 passing in turn through the values λ_2, $2\lambda_2 - \lambda_1$, we first obtain a position of equilibrium, and then there is a bifurcation to a harmonic motion of frequency ω_2 and finally to a biharmonic motion of frequencies ω_1, ω_2.

At last we note that the limiting values of λ_1, λ_2, $2\lambda_2 - \lambda_1$, λ_3 when $\sigma \to \infty$, i.e. for very high speeds of rotation of the gyroscope, can easily be calculated from the observation that $\omega_1 \to 0$, $\omega_2 \to \infty$ and $\omega_1^2\omega_2^2 = pq$, so that we have

$$\lambda_1 \sim c_1 + \frac{p}{q}c_2, \quad \lambda_2 \sim c_1 + c_2,$$

$$2\lambda_2 - \lambda_1 \sim c_1 + \left(2 - \frac{p}{q}\right)c_2, \quad \lambda_3 \sim c_1 + \left(2 - \frac{p}{q}\right)c_2.$$

Thus the length of the last interval, the one corresponding to those values of λ for which a biharmonic motion can exist, tends to zero as $\sigma \to \infty$.

Chapter XI. Rotating Machinery

Vibratory phenomena which can affect the smooth running of rotating machinery can have many causes. For clarity of investigation it is desirable to consider these separately. Certain effects, for example those due to coupling between the energy source and the rotating shaft, have already been discussed in Chapter 8, and so we shall assume in what follows that the speed of rotation can be kept constant in time.

Under the effect of the dynamic forces exerted by discs or rotors driven by it, a shaft made of some elastic material behaves like a beam under flexure, and its own true rotational motion has, in general, to be compounded with a whirling motion (also sometimes known as vortical or swirling motion) whose precessional velocity can be calculated, if need be, after making due allowance for any gyroscopic terms which may have to be taken into account.

Excitation induced by want of balance can give rise to forced vibrations which can become dangerous at speeds close to resonance or to critical velocities. In certain cases however the rated speed of rotation of the machinery may lie above a critical speed and yet the situation still be acceptable because of damping, provided the run-up to full rated speed is sufficiently fast, particularly at the moment of crossing the critical velocity threshold.

The stresses exerted on the shaft by oil-film or gas lubricated bearings can be calculated by analysing the flow phenomena taking place between the shaft and the bush of the bearing, assuming thin-film flow characteristics. The stresses can be fairly accurately represented by forces of visco-elastic type and are more often than not anisotropics. The elastic response of bearings is favourable to stability of operation because it tends to raise the critical velocity of the rotating shaft in regard to flexure. When the shaft possesses anisotropic features, it is subjected to parametric excitation, of a frequency twice that of the angular velocity of rotation and the resonant frequencies and the corresponding zones of instability can be determined either by Fourier analysis and the approximate evaluation of determinants of infinite order, or by the method of variation of amplitude.

This approach of an asymptotic nature is based on assuming, as is true in most cases, that the departure from isotropy of the shaft material is not great, but the particular approach chosen may take one of several directions depending on the order of magnitude of the anisotropy in suspension of the bearings.

These methods of calculation can be used profitably for solving other problems such as finding the resonances and zones of instability of the gyroscopic-type movements of an asymmetric rotor mounted in flexible anisotropic bearings, or the vibrations of the rotor-blade system of a helicopter on the ground.

The effects of flexural deformations on the whirling motion of a shaft in rotation depend to a certain extent on the physical law of behaviour of the material. We can get some idea of what happens in the case of a non-linear rheological law with hysteresis effect, for which the stability conditions have been determined.

The chapter ends with a brief account of some applications to magnetic suspension.

I. The Simplified Model with Frictionless Bearings

Preliminary Study of the Static Bending of a Shaft with Circular Cross-Section

1. The shaft is assumed to be supported by bearings with fixed centres A and B, and the neutral axis in its undeformed state is the straight-line segment AB of length l. We introduce an orthonormal set of axes $Axyz$, with Ax passing through B, and we shall suppose that external forces are exerted on the shaft at a point O on the neutral axis of abscissa $x = a \in]0, l[$, which can be resolved into a force of resultant $P\vec{y}$ and a couple of moment $M\vec{z}$ about O. We use the notation $y = y(x)$ for the Cartesian equation of the curve representing the deformed neutral axis after loading and

(11.1) $$W = \int_0^l EIy''^2 \,\mathrm{d}x$$

for the deformation-energy of the elastically deformed material (see Chapter 6).

The function $y(x)$ satisfies, inside the open intervals $]0, a[$ and $]a, l[$ the equation of equilibrium

(11.2) $$(EIy'')'' = 0$$

(ignoring the effects of gravity), and also the boundary conditions:

(11.3) $$y(0) = y(l) = 0$$

(11.4) $$[EIy'']_{x=a} = M, \quad [(EIy'')']_{x=a} = -P,$$

$[f]_{x=a}$ denoting the jump in the value of f as x passes through a, that is $f(a+0) - f(a-0)$.

It follows from these equations by the classical arguments that by giving $y(x)$ a kinematically admissible virtual increment $\delta y(x)$, in other words one satisfying $\delta y(0) = \delta y(l) = 0$, we have

$$\delta W = P\delta\xi + M\delta\eta \quad \text{with} \quad \xi = y(a), \quad \eta = y'(a),$$

whence
$$P = \frac{\partial W}{\partial \xi}, \quad M = \frac{\partial W}{\partial \eta},$$

W being the positive definite quadratic form in ξ, η obtained by putting $y = \xi y_1 + \eta y_2$ in (11.1), where $y_1(x)$, $y_2(x)$, are solutions of (11.2), (11.3) satisfying $y_1(a) = 1$,

$y_1'(a)=0$, $y_2(a)=0$, $y_2'(a)=1$. Thus

$$2W=\beta_{11}\xi^2+2\beta_{12}\xi\eta+\beta_{22}\eta^2$$

and we can write

$$\begin{pmatrix}P\\M\end{pmatrix}=\beta\begin{pmatrix}\xi\\\eta\end{pmatrix},\quad \beta=\begin{pmatrix}\beta_{11}&\beta_{12}\\\beta_{21}&\beta_{22}\end{pmatrix}$$

where β is a positive definite matrix, or

(11.5)
$$\begin{pmatrix}\xi\\\eta\end{pmatrix}=\alpha\begin{pmatrix}P\\M\end{pmatrix},$$

where $\alpha=\beta^{-1}$ is positive definite.

2. Denoting by $P_1\vec{y}$, $P_2\vec{y}$ the stresses exerted by the bearings on the shaft and writing b for $l-a$, the equations of equilibrium give:

$$P_1+P_2+P=0,\quad M-P_1a+P_2b=0$$

whence

(11.6)
$$\begin{aligned}P_1&=(M-Pb)l^{-1}\\P_2&=-(M+Pa)l^{-1}\end{aligned}$$

and the differential equations for the deformed shaft are:

$$EIy''-P_1x=0,\quad x\in]0,a[$$

$$-EIy''+P_2(l-x)=0,\quad x\in]a,l[.$$

Assuming EI to be independent of x, we obtain after integrating and taking into account the kinematic conditions at the bearings:

$$EIy=P_1\frac{x^3}{6}+\lambda x,\qquad x\in]0,a[$$

$$EIy=P_2\frac{(l-x)^3}{6}+\mu(l-x),\quad x\in]a,l[$$

the coefficients λ,μ being found by using the condition that $y(x)$ and $y'(x)$ must be continuous at $x=a$. Remembering (11.6) we thus find

(11.7)
$$\begin{aligned}y(a)&=\alpha_{11}P+\alpha_{12}M\\y'(a)&=\alpha_{21}P+\alpha_{22}M\end{aligned}$$

with

(11.8)
$$\alpha_{11}=\frac{a^2b^2}{3lHI},\quad \alpha_{12}=\alpha_{21}=\frac{ab(b-a)}{3lEI},\quad \alpha_{22}=\frac{a^2+b^2-ab}{3lEI}.$$

Figure 11.1

If the stresses are exerted at a point O outside the segment AB, i.e. if $l < a$, we obtain by a similar calculation expressions for the α_{ij} of the form

(11.9)
$$\alpha_{11} = \frac{ab^3}{3EI}, \quad \alpha_{12} = \alpha_{21} = \frac{b(3b-2l)}{6EI},$$
$$\alpha_{22} = \frac{l-3b}{3EI} \quad \text{with} \quad b = l - a < 0.$$

Steady Motion of a Disc Rotating on a Flexible Shaft

The disc of centre O, of mass m, and the shaft of circular cross-section to which it is solidly attached constitute, in their undeformed state, a solid of revolution about the axis Ax passing through the centres A, B of the bearings. However when set in rotation, the shaft is liable to bend under the influence of inertial effects. We shall assume that its neutral line is plane, stationary, and contained in a plane ABO, moving with a uniform rotation $\Omega\vec{x}$, relative to the orthonormal frame of reference $Axyz$. The intrinsic rotation of a rectangular cross-section of the shaft, relative to this frame, evaluated at a point immediately to the right of the bearing A, where the neutral line may be taken to coincide with a segment of the axis is $\omega\vec{x}$, while relative to the orthonormal frame of reference $Axvw$, whose two Axy axes are contained in the plane AOB, it is $(\omega - \Omega)\vec{x}$. The rotation vector of the rectangular cross-section at O, the centre of the disc, relative to this same set of moving axes (viz. $Axvw$) has the same magnitude $\omega - \Omega$, i.e. the rotation vector is $(\omega - \Omega)\vec{x}_1$, where $\vec{x}_1$ is the unit vector which is tangent at O to the neutral line, and consequently the rotation of the disc relative to $Axyz$ is represented by the vector:

(11.10)
$$(\omega - \Omega)\vec{x}_1 + \Omega\vec{x} = \omega\vec{x}_1 - \Omega\varphi\vec{y}_1$$

where $\vec{y}$ is the unit vector perpendicular to $\vec{x}_1$ in the plane ABO and φ is $(\vec{x}, \vec{x}_1)$.

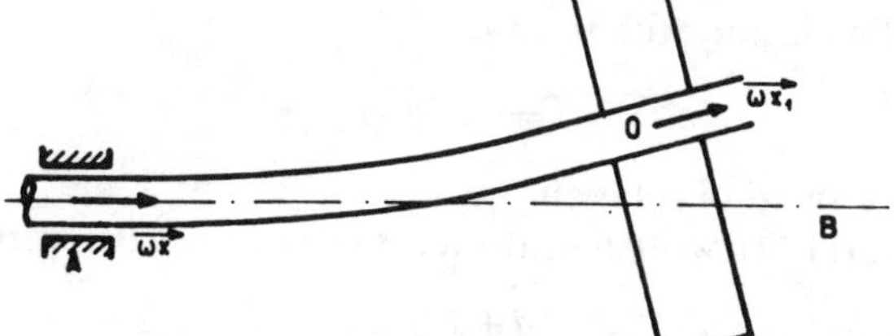

Figure 11.2

The angular momentum vector of the disc at O, with respect to the frame of reference $Oxyz$ is

$$\vec{H} = I\omega\vec{x}_1 - I_1\Omega\varphi\vec{y}_1$$

where I, I_1 are the moments of inertia of the disc with respect to its axis and to a diameter respectively, and the stresses exerted on the shaft due to the motion of the disc, have as resultant $\vec{P}$ and moment $\vec{M}$ at O:

$$-\vec{P} = -m\Omega^2 v \cdot \vec{v}$$

$$-\vec{M} = \frac{d\vec{H}}{dt} = \Omega \vec{x} \wedge \vec{H} = (I\omega\Omega - I_1\Omega^2)\varphi\vec{w}$$

with $(a, v, 0)$ being the co-ordinates of the point O with respect to the axes $Axvw$. Thus, writing $\vec{P} = P\vec{v}$, $\vec{M} = M\vec{w}$, we have:

$$P = m\Omega^2 v$$

$$M = (I_1\Omega - I\omega)\Omega\varphi.$$

Substituting these values in (11.7) and (11.8) and assuming the mass of the shaft to be negligible, we obtain in the case $I = 2I_1$, corresponding to a flat disc:

$$v = \alpha_{11} m\Omega^2 v - \alpha_{12} I_1 \Omega(2\omega - \Omega)\varphi$$

$$\varphi = \alpha_{21} m\Omega^2 v - \alpha_{22} I_1 \Omega(2\omega - \Omega)\varphi$$

so that the equation which defines the admissible whirling or precessional speeds Ω, obtained by eliminating v and φ, can be written:

$$\begin{aligned} &m I_1(\alpha_{11}\alpha_{22} - \alpha_{21}\alpha_{12})\Omega^4 - 2m I_1 \omega(\alpha_{11}\alpha_{22} - \alpha_{21}\alpha_{12})\Omega^3 \\ &- (m\alpha_{11} + I_1\alpha_{22})\Omega^2 + 2\omega\alpha_{22} I_1 \Omega + 1 = 0. \end{aligned} \tag{11.11}$$

It is convenient to introduce the following dimensionless variables:

$$f = \Omega\sqrt{\alpha_{11} m},$$

the dimensionless speed of precession,

$$d = \frac{I_1\alpha_{22}}{m\alpha_{11}},$$

the disc effect,

$$e = \frac{\alpha_{12}^2}{\alpha_{11}\alpha_{22}},$$

the elastic coupling factor, with $0 < e < 1$,

and
$$s = \omega\sqrt{\alpha_{11} m},$$

the dimensionless speed of rotation.

The equation (11.11), written in the form solved for s becomes:

$$s = \frac{f^4 + \dfrac{d+1}{d(e-1)} f^2 - \dfrac{1}{d(e-1)}}{2f^3 + \dfrac{2f}{e-1}} \tag{11.12}$$

and lends itself to a simple discussion if one represents s graphically as a function of f in the s, f plane.

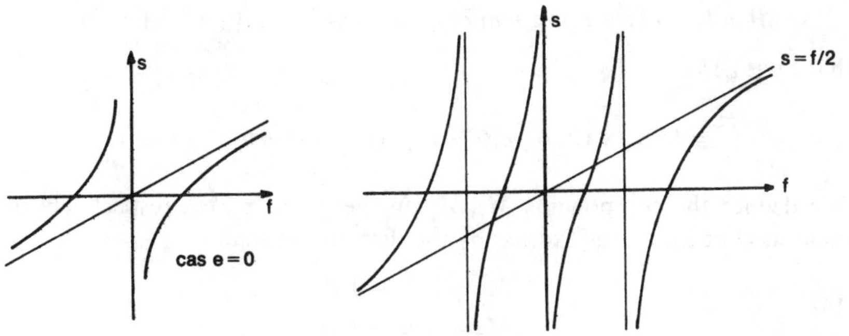

Figure 11.3

It will be seen that for a given s there are four possible values for the whirl velocity Ω. In the symmetric case $a = b$, we have $e = 0$ and simply:

$$s = \frac{f^2 - 1/d}{2f}$$

and in this case there are only two possible precessional speeds for the given s.

Flexural Vibrations When Shaft Is in Rotation

We again consider the preceding model and represent the unit vector $\vec{x}_1$ tangent to the neutral line of the shaft at O, the centre of the disc, by:

$$\vec{x}_1 = \vec{x} + \varphi\vec{y} + \psi\vec{z} \tag{11.13}$$

and the instantaneous rotation of the disc by:

$$\vec{r} = (\omega + \eta)\vec{x}_1 + \lambda\vec{y} + \mu\vec{z} \tag{11.14}$$

where ω is the magnitude of the rotational velocity (assumed constant) of the shaft and η, λ, μ are perturbation terms, of small amplitude, as are the angles φ, ψ resulting from the deformation of the shaft which need not be stationary.

We have on the one hand $\dfrac{d\vec{x}_1}{dt} = \vec{r} \wedge \vec{x}_1 = \mu\vec{y} - \lambda\vec{z}$, taking account of (11.14) and carrying out the calculations to the first order; while on the other hand we have, directly from (11.13):

$$\frac{d\vec{x}_1}{dt} = \varphi'\vec{y} + \psi'\vec{z}, \quad \text{whence} \quad \lambda = -\psi', \mu = \varphi', \tag{11.15}$$

so that finally:

$$\vec{r} = (\omega + \eta)\vec{x}_1 - \psi'\vec{y} + \varphi'\vec{z}.$$

The angular momentum of the disc at O, again to the same order of approximation, is:

$$\vec{H} = I_1\vec{r} + (I - I_1)(\omega + \eta)\vec{x}_1 = I(\omega + \eta)\vec{x}_1 + I_1(-\psi'\vec{y} + \varphi'\vec{z})$$

so that by (11.15):

$$\frac{d\vec{H}}{dt} = I\eta'\vec{x}_1 + (I(\omega + \eta)\varphi' - I_1\psi'')\vec{y} + (I(\omega + \eta)\psi' + I_1\varphi'')\vec{z}.$$

We deduce the components M_y, M_z on the y and z axes respectively of the moment at O of the forces exerted by the disc on the shaft:

$$\begin{aligned} -M_y &= I\omega\varphi' - I_1\psi'' \\ -M_z &= I\omega\psi' + I_1\varphi'' \end{aligned} \tag{11.16}$$

and also the components P_y, P_z of the resultant of these same forces:

$$\begin{aligned} -P_y &= my'' \\ -P_z &= mz''. \end{aligned} \tag{11.17}$$

Associating the system $P_y\vec{y}$, $M_z\vec{z}$ with the projection of the deformation of the shaft on the x, y plane, we can write:

$$\begin{aligned} y &= \alpha_{11}P_y + \alpha_{12}M_z \\ \varphi &= \alpha_{21}P_y + \alpha_{22}M_z \end{aligned} \tag{11.18}$$

and doing the same thing with $P_z\vec{z}$, $M_y\vec{y}$:

$$\begin{aligned} z &= \alpha_{11}P_z - \alpha_{12}M_y \\ \psi &= \alpha_{21}P_z - \alpha_{22}M_y \end{aligned} \tag{11.19}$$

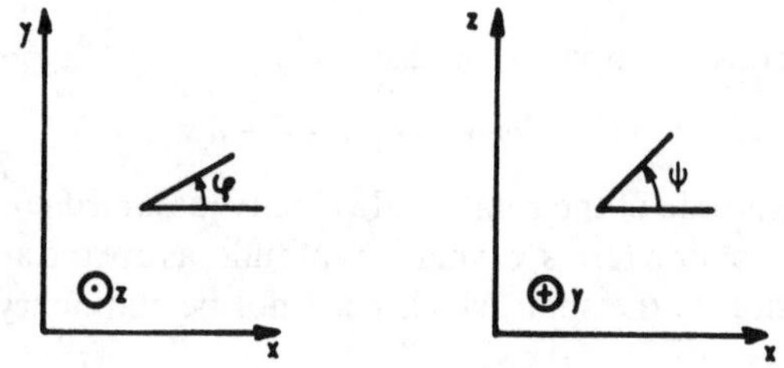

Figure 11.4

the coefficients α_{ij} being defined by (11.8) or (11.9) according to the position of the disc in relation to the bearings.

Solving (11.18), (11.19) we obtain:

$$\begin{aligned} P_y &= (\alpha_{22}y - \alpha_{12}\varphi)q^{-1} \\ M_z &= (-\alpha_{21}y + \alpha_{11}\varphi)q^{-1} \\ P_z &= (\alpha_{22}z - \alpha_{12}\psi)q^{-1}, \quad q = \alpha_{11}\alpha_{22} - \alpha_{12}\alpha_{21} \\ M_y &= (\alpha_{21}z - \alpha_{11}\psi)q^{-1} \end{aligned}$$

and by (11.16), (11.17), the equations of motion:

$$
\begin{aligned}
my'' + q^{-1}(\alpha_{22}y - \alpha_{12}\varphi) &= 0\\
mz'' + q^{-1}(\alpha_{22}z - \alpha_{12}\psi) &= 0\\
I_1\varphi'' + I\omega\psi' + q^{-1}(-\alpha_{21}y + \alpha_{11}\varphi) &= 0\\
I_1\psi'' - I\omega\varphi' + q^{-1}(-\alpha_{21}z + \alpha_{11}\psi) &= 0
\end{aligned}
\tag{11.20}
$$

where we see appear forces of gyroscopic type and elastic forces of conservative type. The matrix associated with the latter is indeed symmetric and positive definite since the associated quadratic form is, disregarding a factor q^{-1}, the positive definite form

$$U = \alpha_{11}(\varphi^2 + \psi^2) + \alpha_{22}(y^2 + z^2) - 2\alpha_{12}\varphi y - 2\alpha_{12}\psi z$$

The system (11.20) therefore has characteristic values which are all pure imaginary numbers and the motion is stable.

Setting

$$y + iz = \xi, \quad \varphi + i\psi = \zeta \tag{11.21}$$

we can rewrite (11.20) in the form:

$$
\begin{aligned}
m\xi'' + q^{-1}(\alpha_{22}\xi - \alpha_{12}\zeta) &= 0\\
I_1\zeta'' - iI\omega\zeta' + q^{-1}(-\alpha_{21}\xi + \alpha_{11}\zeta) &= 0
\end{aligned}
$$

from which it will be seen that the eigenfrequencies Ω are roots of the equation

$$\det\begin{vmatrix} q^{-1}\alpha_{22} - m\Omega^2 & -q^{-1}\alpha_{12} \\ -q^{-1}\alpha_{21} & -I_1\Omega^2 + I\omega\Omega + q^{-1}\alpha_{11} \end{vmatrix} = 0$$

which is identical to (11.11).

Forced Vibrations

Let us suppose the disc to be imperfectly balanced, or what amounts to the same thing, an additional mass μ to be attached to it at a distance e from O.

The additional inertial force due to the motion of this mass has the following components along the axes $Axyz$:

$$0, \quad -\mu(y'' - e\omega^2\cos\omega t), \quad -\mu(z'' - e\omega^2\sin\omega t)$$

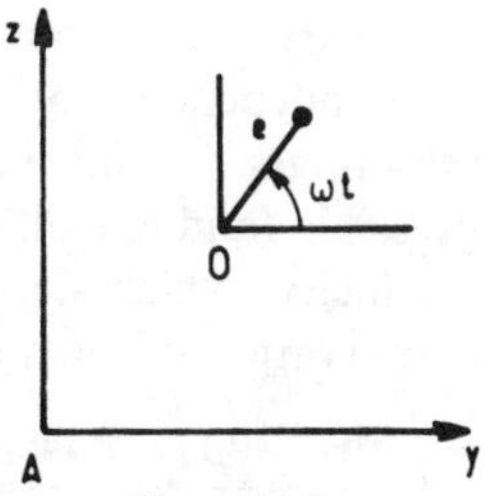

Figure 11.5

and its moment about O can be neglected assuming e is small. Only the first two equations therefore of (11.20) are changed, and using the notations (11.21) we have:

$$(m+\mu)\xi'' + q^{-1}(\alpha_{22}\xi - \alpha_{12}\zeta) = \mu e\omega^2 e^{i\omega t}$$
$$I_1\zeta'' - iI\omega\zeta' + q^{-1}(-\alpha_{21}\xi + \alpha_{11}\zeta) = 0.$$

We represent the forced vibration by $\xi = \tilde{\xi}e^{i\omega t}$, $\zeta = \tilde{\zeta}e^{i\omega t}$, the complex amplitudes $\tilde{\xi}, \tilde{\zeta}$ being the solution of:

$$(q^{-1}\alpha_{22} - m\omega^2)\tilde{\xi} - q^{-1}\alpha_{12}\tilde{\zeta} = \mu e\omega^2$$
$$-q^{-1}\alpha_{21}\tilde{\xi} + (-\omega^2 I_1 + I\omega^2 + q^{-1}\alpha_{11})\tilde{\zeta} = 0$$

after replacing $m+\mu$ by m which differs little from it.

The condition for resonance is obtained by writing down that the determinant of this system vanishes, and this leads to:

$$m(I - I_1)\omega^4 - q^{-1}(\alpha_{22}(I - I_1) - m\alpha_{11})\omega^2 - q^{-1} = 0.$$

Since $q > 0$, we see that in the case $I > I_1$ (flat disc) there is only one critical velocity.

If $I < I_1$, we find two critical velocities ω_1, ω_2 separated by $\omega_0 = \dfrac{1}{\sqrt{\alpha_{11}m}}$, which is the precessional velocity of the shaft in the absence of proper rotation.

II. Effects of Flexibility of the Bearings

Hydrodynamics of Thin Films and Reynold's Equation

The rotating shaft of a machine mounted in bearings is not normally in direct contact with the bearing surface. To reduce friction and the wear of the parts in movement relative to each other, the annular space between the shaft and the bushing is filled with lubricant, e.g. oil or gas, whose effects on the shaft cannot be properly described except by analysing the mechanism of the flow occurring in this space of very small thickness. The lubricant may be regarded as a viscous Newtonian fluid, adhering to the walls, and since the clearance between the walls of the bearing is very small in comparison with their radii of curvature we may, following O. Reynolds, adopt the following model: the flow, assumed to be laminar, fills the region $0 < z < h(x, y, t)$ defined in terms of an orthonormal frame of reference $Oxyz$, and the pressure does not vary within the thickness of the film. In other words the pressure $p = p(x, y, t)$ is independent of z, and we shall make a similar assumption regarding the density (mass per unit volume) $\rho = \rho(x, y, t)$ in the case where the lubricant is a compressible fluid (gas-filled bearing). Writing u, v, w for the components of the fluid velocity, which depend on x, y, z, t, the equation representing the principle of conservation of mass can be written:

$$\frac{\partial\rho}{\partial t} + \frac{\partial\rho u}{\partial x} + \frac{\partial\rho v}{\partial y} + \frac{\partial\rho w}{\partial z} = 0$$

or after integrating with respect to z, from 0 to h:

$$h\frac{\partial\rho}{\partial t}+\frac{\partial}{\partial x}\left(\int_0^h \rho u\,\mathrm{d}z\right)+\frac{\partial}{\partial y}\left(\int_0^h \rho v\,\mathrm{d}z\right)+\rho\left(w-u\frac{\partial h}{\partial x}-v\frac{\partial h}{\partial y}\right)\bigg|_{z=h}-\rho w\bigg|_{z=0}=0,$$

i.e. with

(11.22)
$$q_x=\int_0^h u\,\mathrm{d}z,\quad q_y=\int_0^h v\,\mathrm{d}z$$

$$\frac{\partial\rho h}{\partial t}+\frac{\partial\rho q_x}{\partial x}+\frac{\partial\rho q_y}{\partial y}+\rho\left(w-u\frac{\partial h}{\partial x}-v\frac{\partial h}{\partial y}-\frac{\partial h}{\partial t}\right)\bigg|_{z=h}-\rho w\bigg|_{z=0}=0.$$

If the walls are impermeable we obtain simply:

(11.23)
$$\frac{\partial\rho h}{\partial t}+\frac{\partial\rho q_x}{\partial x}+\frac{\partial\rho q_y}{\partial y}=0$$

or in the case of an incompressible fluid

(11.24)
$$\frac{\partial h}{\partial t}+\frac{\partial q_x}{\partial x}+\frac{\partial q_y}{\partial y}=0.$$

On the other hand the components τ_x, τ_y of the shearing stresses per unit area parallel to the plane Oxy are obtained from the classical viscosity law:

$$\tau_x=\mu\frac{\partial u}{\partial z},\quad \tau_y=\mu\frac{\partial v}{\partial z},$$

where μ is the coefficient of viscosity. Writing down equilibrium conditions for a volume element of sides dx, dy, dz within the oil-film, on the assumption that inertial effects can be ignored, we are led to:

(11.25)
$$-\frac{\partial p}{\partial x}+\frac{\partial}{\partial z}\left(\mu\frac{\partial u}{\partial z}\right)=0$$
$$-\frac{\partial p}{\partial y}+\frac{\partial}{\partial z}\left(\mu\frac{\partial v}{\partial z}\right)=0.$$

These equations are easily integrated, since p is independent of z and we shall assume that the same is true of μ. Taking into account the adherence condition, which means that the velocities of the fluid at the points $(x,y,0)$ and (x,y,h) are the same as those of the points on the walls with which they coincide, and whose x and y components we denote by U_1, V_1 and U_2, V_2, we obtain:

(11.26)
$$\mu u=\frac{z^2}{2}\frac{\partial p}{\partial x}+\left(\frac{\mu}{h}(U_2-U_1)-\frac{h}{2}\frac{\partial p}{\partial x}\right)z+\mu U_1$$
$$\mu v=\frac{z^2}{2}\frac{\partial p}{\partial y}+\left(\frac{\mu}{h}(V_2-V_1)-\frac{h}{2}\frac{\partial p}{\partial y}\right)z+\mu V_1$$

whence we deduce by (11.22)

$$q_x = -\frac{h^3}{12\mu}\frac{\partial p}{\partial x} + \frac{U_1 + U_2}{2}\cdot h$$

$$q_y = -\frac{h^3}{12\mu}\frac{\partial p}{\partial y} + \frac{V_1 + V_2}{2}\cdot h$$

and lastly by (11.23)

$$\frac{\partial}{\partial x}\left(\rho\frac{h^3}{\mu}\frac{\partial p}{\partial x}\right) + \frac{\partial}{\partial y}\left(\rho\frac{h^3}{\mu}\frac{\partial p}{\partial y}\right) = 6\frac{\partial}{\partial x}(\rho(U_1 + U_2)h) + 6\frac{\partial}{\partial y}(\rho(V_1 + V_2)h) + 12\frac{\partial h\rho}{\partial t}.$$

We can simplify this equation by writing $U = U_1 + U_2$, $V = V_1 + V_2$, and noting that by a suitable choice of axes we can, in general make $U = 0$. Accordingly in the case of impermeable walls and an incompressible lubricant, we obtain finally, assuming V and μ to be constants, the Reynolds equation [12]:

(11.27) $$\frac{\partial}{\partial x}\left(h^3\frac{\partial p}{\partial x}\right) + \frac{\partial}{\partial y}\left(h^3\frac{\partial p}{\partial y}\right) = 6\mu V\frac{\partial h}{\partial y} + 12\mu\frac{\partial h}{\partial t}$$

which can be further simplified, in the stationary case, by the omission of the term containing $\frac{\partial h}{\partial t}$.

Application to Circular Bearings [12]

Consider the case where the shaft turns in bearings of circular cross-section. We shall assume that the axis of the shaft always remains parallel to that of the bearing, and in the common direction x. In a plane normal to this axis the internal cross-section of the bearing is a circle of centre O, radius R_1, while that of the shaft is a circle of centre A, radius $R < R_1$. We refer the plane of this cross-section to orthonormal axes $A\xi\eta$, where $A\eta$ is borne by $\overrightarrow{AO}$ and we write $e = OA$ for the eccentricity (departure from centre) which clearly satisfies:

(11.28) $$e < R - R_1.$$

In this plane the position of a point M on the shaft can be defined by $\theta = (A\eta, AM)$. Producing the radius AM to N on the inner wall of the bearing, $h = MN$ measures the thickness of the film of oil between the two cylindrical surfaces (of the shaft and bearing respectively).

By considering the triangle OAN we have:

$$\sin\alpha = \frac{e}{R_1}\sin\theta,$$

and by projection of $\overrightarrow{NO} + \overrightarrow{OA} = \overrightarrow{NA}$ on NA:

$$R_1\cos\alpha + e\cos\theta = R + h$$

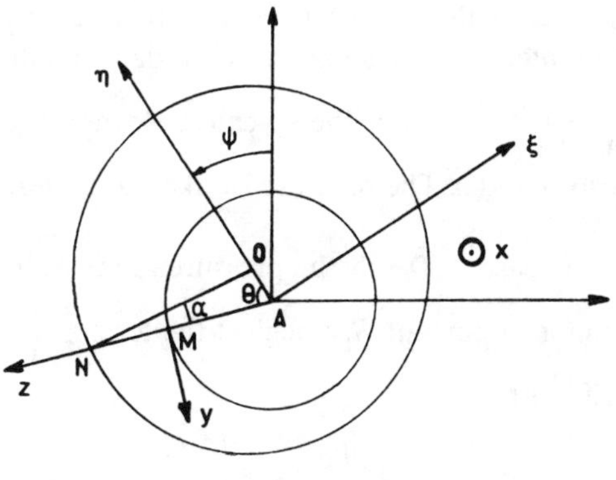

Figure 11.6

whence we deduce:

$$h = e\cos\theta - R + R_1\sqrt{1 - \frac{e^2}{R_1^2}\sin^2\theta}.$$

But $\dfrac{e}{R_1} < \dfrac{R_1 - R}{R_1} \ll 1$ and, to within the second order in $\dfrac{e}{R_1}$ we can write:

(11.29) $$h = c(1 + \varepsilon\cos\theta)$$

(11.30) $$c = R_1 - R, \quad \varepsilon = \frac{e}{c}.$$

To calculate the stresses exerted on the shaft by the film of oil and thus analyse the performance of the system we need to work out the pressure p, the solution of Reynold's equation.

We first seek to define the operating conditions in the stationary mode, i.e. the mode in which the eccentricity ε, and the angle of attitude between $A\eta$ and the upward vertical, are constant. We can then leave out the term $12\mu\dfrac{\partial h}{\partial t}$ in Reynold's equation (11.27), and replace the variable y by θ, $y = R\theta$. It is also however appropriate to make certain assumptions about the geometrical design of the bearings, and in particular about the ratio of the bearing-length L (measured in the direction of the x-axis) to the shaft-diameter $D = 2R$.

In the long bearing case where L/D is large, it is usually assumed that p does not depend on x; the relevant equation [12, 53] is:

$$\frac{\partial}{\partial y}\left(h^3\frac{\partial p}{\partial y}\right) = 6\mu V\frac{\partial h}{\partial y} \quad \text{with} \quad V = R\omega,$$

being the tangential velocity of the shaft, so that by integration:

$$\frac{dp}{dy} = 6\mu V\frac{h - \bar{h}}{h^3}$$

$\bar{h}$ denoting the clearance at the point where the pressure gradient is zero. To be able to complete the integration we take as boundary conditions $p = 0$ at $y = 0$ or $\theta = 0$ and $p = \dfrac{dp}{dy} = 0$ at $\theta = \pi + \alpha$, the so-called Reynold's conditions, designed to give simultaneously α and $\bar{h}$. The pressure is taken to be zero for $\pi + \alpha < \theta < 2\pi$, [12].

In the short bearing case, $L/D \sim 1$, the pressure gradient in a direction parallel to the axis of the shaft is important. We neglect the term $\dfrac{\partial}{\partial y}\left(h^3 \dfrac{\partial p}{\partial y}\right)$, in Reynold's equation, and are left with

$$\frac{\partial}{\partial x}\left(h^3 \frac{\partial p}{\partial x}\right) = 6\mu V \frac{\partial h}{\partial y}$$

or, since h does not depend on x:

$$\frac{\partial^2 p}{\partial x^2} = \frac{6\mu V}{h^3} \frac{dh}{dy}$$

an equation which is easily integrated, having regard to the boundary conditions $p = 0$ at $y = \pm \dfrac{L}{2}$, ($A\xi\eta$ a plane of symmetry of the bearing) to give:

$$p = 3\mu V \frac{dh}{dy} \cdot \left(x^2 - \frac{L^2}{4}\right) h^{-3}. \tag{11.31}$$

This formula represents fairly well what happens in the region where the pressure is positive, i.e. where the bearing clearance h is decreasing in the direction of rotation of the shaft [34]; a satisfactory approximation is to adopt the formula (11.31) in the region $0 < \theta < \pi$, and to take $p = 0$ in $\pi < \theta < 2\pi$ (Figure 11.6). We can easily deduce from this the stresses due to the pressure exerted by the lubricant on the shaft, which can obviously be reduced to a single force passing through A, whose components along the axes $A\xi\eta$ are:

$$P_\xi = \int_0^\pi \int_{-L/2}^{L/2} pR \sin\theta \, d\theta \, dx = \frac{\mu V L^3}{4c^2} \cdot \frac{\pi\varepsilon}{(1-\varepsilon^2)^{3/2}}$$

$$P_\eta = -\int_0^\pi \int_{-L/2}^{L/2} pR \cos\theta \, d\theta \, dx = \frac{\mu V L^3}{c^2} \cdot \frac{\varepsilon^2}{(1-\varepsilon^2)^2}.$$

Let us now examine the effects due to shear, i.e. the tangential stresses caused by the film of oil on the shaft. Because of the symmetry of the flow in relation to the plane $x = 0$, the shearing stresses have no component along the x-axis; to calculate the shear tangential to the circular cross-section of the shaft, we recall (11.26) with $V_1 = V$, $V_2 = 0$,

$$\mu v = \frac{z^2}{2} \frac{\partial p}{\partial y} - \left(\frac{\mu}{h} V + \frac{h}{2} \frac{\partial p}{\partial y}\right) z + \mu V$$

whence we deduce that the shearing stress per unit area at $z = 0$ is:

$$\tau = \mu \frac{\partial v}{\partial z} = -\frac{h}{2}\frac{\partial p}{\partial y} - \frac{\mu}{h} V, \quad 0 < \theta < \pi$$

$$= -\frac{\mu}{h} V, \qquad \pi < \theta < 2\pi$$

and by integrating over the surface of the shaft in contact with the bearing, we obtain the following expressions $\mathscr{C}_\xi$, $\mathscr{C}_\eta$, of the components along the axes $A\xi$, $A\eta$ of the total stress due to friction:

$$\mathscr{C}_\xi = \frac{1}{2}\int_0^\pi \int_{-L/2}^{L/2} h \frac{\partial p}{\partial y} \cos\theta \cdot R\,\mathrm{d}\theta\,\mathrm{d}x + \mu L \int_0^{2\pi} \frac{V}{h} \cos\theta \cdot R\,\mathrm{d}\theta$$

$$\mathscr{C}_\eta = \frac{1}{2}\int_0^\pi \int_{-L/2}^{L/2} h \frac{\partial p}{\partial y} \sin\theta \cdot R\,\mathrm{d}\theta\,\mathrm{d}x + \mu L \int_0^{2\pi} \frac{V}{h} \sin\theta \cdot R\,\mathrm{d}\theta.$$

We see, on integrating by parts, that the double integral in $\mathscr{C}_\xi$ can be replaced by:

$$-\frac{1}{2}\int_0^\pi \int_{-L/2}^{L/2} p \frac{\mathrm{d}}{\mathrm{d}\theta}(h\cos\theta)\,\mathrm{d}\theta\,\mathrm{d}x,$$

and the double integral in $\mathscr{C}_\eta$ by a similar expression. Remembering that $L/R \sim 1$, we can show that these terms are of the order of c/R relatively to P_ξ, P_η; and so, assuming a bearing with horizontal axis, we can say that the "load-bearing property" of the bearing is essentially due to the pressure of the lubricant. Denoting by P the proportion of the weight of the shaft supported by the bearing concerned, in the stationary mode, we see that:

$$P = \sqrt{P_\xi^2 + P_\eta^2} = \frac{\pi\mu V L^3}{4c^2}\frac{\varepsilon}{(1-\varepsilon^2)^2}\left(\left(\frac{16}{\pi^2} - 1\right)\varepsilon^2 + 1\right)^{1/2} \tag{11.32}$$

and

$$\tan\psi = \frac{P_\xi}{P_\eta} = \frac{\pi}{4\varepsilon}(1-\varepsilon^2)^{1/2}. \tag{11.33}$$

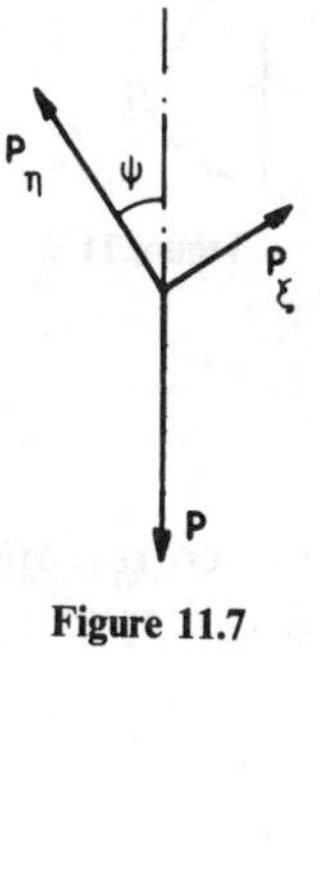

Figure 11.7

Sommerfeld's parameter S, defined by:

$$S = \mu \frac{DLN}{P}\left(\frac{D}{c}\right)^2, \tag{11.34}$$

where N is the number of revolutions per second so that $V = R\omega = 2\pi RN$, fits naturally into the formula (11.32) which can be written:

$$1 = \frac{\pi^2}{4}\cdot\left(\frac{L}{D}\right)^2 \cdot S \cdot \frac{\varepsilon}{(1-\varepsilon^2)^2}\left(\left(\frac{16}{\pi^2} - 1\right)\varepsilon^2 + 1\right)^{1/2} \tag{11.35}$$

showing that in a steady regime (mode of running) the value of ε depends on Sommerfeld's number and on the parameter of the form L/D. The attitude angle ψ, calculated from (11.33), is also a function of these same variables. This is a general result, the eccentricity ε, and the attitude angle ψ depend only on S for bearings of the same geometric form (circular bearings, roller bearings, multilobular bearings...).

The moment about Ax of the frictional forces on the shaft is:

$$-\int_0^{\pi}\int_{-L/2}^{L/2} \frac{h}{2}\frac{\partial p}{\partial y} R^2\, d\theta\, dx - \int_0^{2\pi} \mu \frac{V}{h} R^2\, d\theta\, dx \sim -2\pi\mu \frac{VR^2L}{c}$$

with the approximation $h \sim c$, and the power loss, i.e. the energy dissipated in friction, obtained by multiplying this moment by $\omega = 2\pi N$, is

$$\mathscr{P} = (2\pi)^3 \frac{\mu L R^3 N^2}{c^2} \quad \text{(Petroff loss).}$$

Coming back to the formula (11.33) where $\varepsilon = \frac{e}{c}$, we note that it defines the attitude diagram that is to say the set of admissible positions of A in the steady running mode, when the load P varies.

Using the approximation $\pi/4 \sim 1$, it will be seen that this diagram is very nearly semicircular in shape [53].

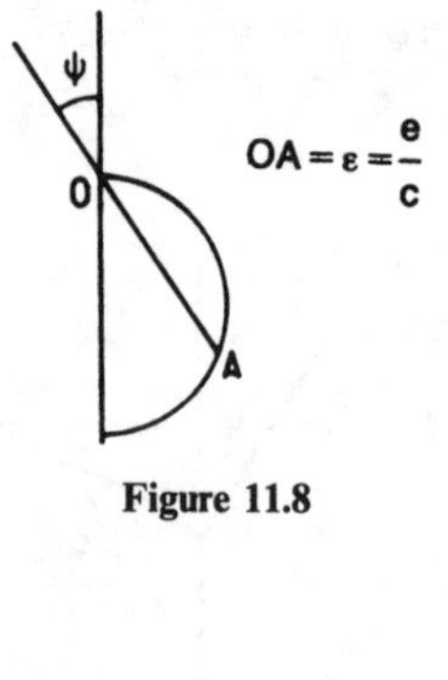

Figure 11.8

Unsteady Regime

We have seen that a steady regime, corresponding to well-defined values of the eccentricity ε and attitude angle ψ can appear for a given load P. If the latter is altered by a perturbation, the centre of the shaft will be displaced from the point A corresponding to the steady state to a new point A_1. To investigate this in more detail let us introduce the orthonormal set of axes $Axyz$, Az being the upward vertical, and suppose the resultant of the forces acting on the shaft, initially represented by $(0, 0, -P)$ to become $(0, F_y^*, -P + F_z^*)$. Introducing the dimensionless variables:

$$F_y = \frac{F_y^*}{P}, \quad F_z = \frac{F_z^*}{P} \quad \text{and} \quad y = \frac{y^*}{c}, \quad z = \frac{z^*}{c}$$

where y^*, z^* are the components of the vector $\overrightarrow{AA_1}$, we can show [53] that the perturbation forces can be expressed in terms of the displacement $\overrightarrow{AA_1}$ and the velocity of A_1:

(11.36)
$$F_y = f\left(y, z, \frac{y'}{\omega}, \frac{z'}{\omega}\right)$$
$$F_z = g\left(y, z, \frac{y'}{\omega}, \frac{z'}{\omega}\right)$$

the structure of the function f, g depending only on Sommerfeld's parameter and a form parameter (depending on the particular form of bearing).

In the context of a linearised theory the formulae (11.36) can be written:

$$F_y = a_{11}y + a_{12}z + b_{11}\frac{y'}{\omega} + b_{12}\frac{z'}{\omega}$$

$$F_z = a_{21}y + a_{22}z + b_{21}\frac{y'}{\omega} + b_{22}\frac{z'}{\omega}$$

where the coefficients a_{ij}, b_{ij} depend on S and the form parameter. However as they show considerable variations in terms of the latter they are better represented as functions of S and ε [53].

Gas Lubricated Bearings

These are advantageous for small machines rotating at high speed. The "take-off" speed is quite low, being of the order of 300 r.p.m. but for very high rates of revolution, of the order of 10,000 r.p.m. experience shows that the shaft may take on a precessional motion of angular velocity $\Omega \sim \omega/2$ (Ω being the angular velocity of the radius OA about O). Actually, one can distinguish two distinct critical

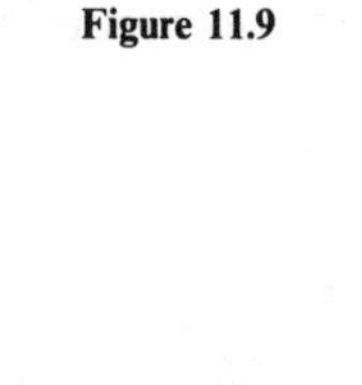

Figure 11.9

velocities ω_1 and ω_2. The angular velocity ω_1 is the speed of revolution of the shaft at which the shaft begins to swirl when its speed of revolution is gradually increased, while the angular velocity ω_2 which is around 10% smaller than ω_1 is the speed at which the shaft stops swirling when the speed of revolution is gradually decreased again after the whirling motion had set in. The limiting velocity for stable operation can be increased by arranging for a groove to be provided in the bearing bush through which gas can be fed to create additional pressure at that point. By means of suitable devices of this kind speeds of the order of 15,000 r.p.m. have been successfully attained [20].

It is customary to make use when dealing with bearings lubricated by gas of Sommerfeld's parameter $G=\dfrac{6\mu VR}{e^2 p_a}$ where p_a is the ambient pressure between shaft and bearing. If compressibility effects are ignored it can be shown, in the same way as with oil-lubricated bearings, that

$$W=Gf\left(\varepsilon,\frac{L}{R}\right),\quad W=\frac{P}{LRp_a}$$

a formula corresponding to (11.35). However as G increases, this representation is no longer valid after a certain threshold has been passed, W increasing more slowly than G with constant ε and L/R.

The representation of W in terms of G, i.e. broadly speaking the load-bearing capacity for a given r.p.m., depends on the geometric parameters ε, L/R, but also on the coefficient of viscosity μ, and the ratio of the specific heats of the gas (at constant pressure and at constant volume) and can be obtained only by experiment [20].

In conclusion it is worth emphasizing that the viscosity of a gas increases with temperature, which is the opposite of what happens with a liquid lubricant.

Effects of Bearing Flexibility on the Stability of Rotation of a Disc

We consider a shaft mounted horizontally on two identical bearings, carrying a disc of mass m at its midpoint and revolving with an angular velocity ω. We have described in the preceding sections the forces exerted by the lubricant on the shaft, on the assumption of fixed bearings; in particular the terms corresponding to perturbations of the relative equilibrium configuration are amenable to computation, for any bearing geometry, by numerical integration of the Reynold's equation, and so are the coefficients of their representation in the context of a linearised theory. However we have underlined that these coefficients depend on the speed of rotation, and this fact tends to complicate somewhat the analysis of the stability of the motion.

The bearings themselves however are in general endowed with a certain flexibility which has to be taken into account. We shall confine ourselves in what follows to analysing the effect on stability of the restoring forces exerted by the bearings when they depart from their equilibrium configuration and of the forces arising out of the bending of the shaft carrying the disc. We shall consequently ignore any perturbations of the eccentricity and the angle of attitude inside the bearings which might be caused by unsteadiness in the motion of the lubricant.

We denote by O, O' the centres of the bearings in their undeformed state, by S the centre of OO', by $Sxyz$ an orthonormal frame of reference, Sx being carried by OO', Sz being the upward vertical. We then write $\overline{OS} = \overline{SO'} = l$, $(\pm l, y_1, z_1)$ for the co-ordinates of the points O_1, O'_1 the centres of the bearings in their deformed state, and $(0, y_2, z_2)$ the components of the vector representing the flexure of the shaft at its midpoint, i.e. $\overrightarrow{S_1C}$, with S_1 the midpoint of $O_1O'_1$ and C the point on the natural axis which is also the centre of the disc, and finally

$$y = y_1 + y_2 \qquad (11.37)$$
$$z = z_1 + z_2$$

which with $x = 0$ are the co-ordinates of C, the centre of the disc.

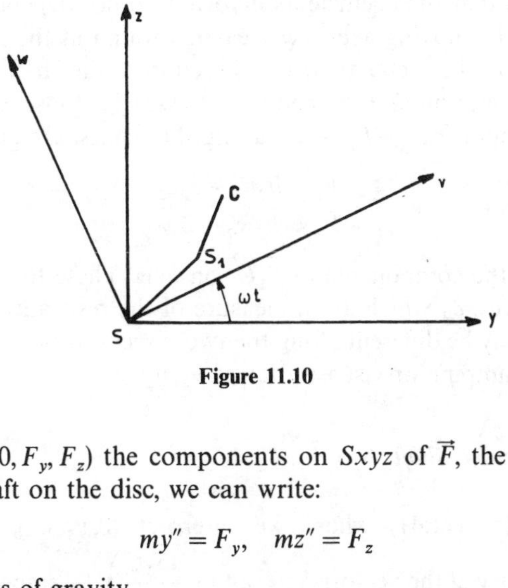

Figure 11.10

Denoting by $(0, F_y, F_z)$ the components on $Sxyz$ of $\vec{F}$, the sum of the forces exerted by the shaft on the disc, we can write:

$$my'' = F_y, \quad mz'' = F_z \qquad (11.38)$$

ignoring the effects of gravity.

The vector $(0, -F_y, -F_z)$ represents the resultant of the forces exerted by the disc on the shaft at its centre and it is clear by symmetry that the shaft, assumed to be of negligible mass, exerts on each bearing forces of resultant $(0, -F_y/2, -F_z/2)$. We shall assume that the response of the bearing to this solicitation is linear, anisotropic, and represented by:

$$-F_y = b_1 y'_1 + c_1 y_1, \quad -F_z = \tilde{b}_1 z'_1 + \tilde{c}_1 z_1 \qquad (11.39)$$

$c_1, \tilde{c}_1$ being coefficients of elasticity (stiffness factors), and $b_1, \tilde{b}_1$ damping factors.

To take account of the effects of the oil-film, we shall find it convenient to express the perturbation of the displacement of the centre of the shaft to the right of the bearing in the form:

$$y_1 = y_{10} + y_{11}, \quad z_1 = z_{10} + z_{11} \qquad (11.40)$$

thus bringing out the distinction between (y_{10}, z_{10}) the perturbation of the shaft relative to the bearing due to the effect of the oil-film, and (y_{11}, z_{11}), that of the centre of the bearing relative to the frame of reference. We can express the stresses by:

$$-F_y = b_{10}y'_{10} + c_{10}y_{10} = b_{11}y'_{11} + c_{11}y_{11}$$
$$-F_z = \tilde{b}_{10}z'_{10} + \tilde{c}_{10}z_{10} = \tilde{b}_{11}z'_{11} + \tilde{c}_{11}z_{11}$$

or, if necessary, more completely, by means of non-diagonal matrices $\{b\}$, $\{c\}$. The elements of these matrices would in any event have to be determined empirically beforehand and it is here that we should perhaps remind the reader that those relating to the oil-film, with suffix zero, depend on the rotational velocity of the shaft. As will readily be appreciated an analysis of the stability of the motion along these lines can be carried out only by a numerical approach.

To simplify the analysis we shall use in what follows the representation (11.39) to describe globally the overall effect of the interaction between shaft and bearing, on the assumption that the coefficients involved do not depend on ω.

In relation to the moving axes $Sxvw$ carried round at the angular velocity ω, so that $(Sy, Sv) = \omega t$, the rectangular cross-section of the shaft in the plane $x = 0$ has a translational motion and the bending stresses, i.e. those exerted on the shaft by the disc, of components $-F_v$, $-F_w$ along these axes, are given by:

$$\text{(11.41)} \qquad \begin{aligned} -F_v &= b_2 v'_2 + c_2 v_2 \\ -F_w &= \tilde{b}_2 w'_2 + \tilde{c}_2 w_2 \end{aligned}$$

where (v_2, w_2) are the components of $\overrightarrow{S_1C}$ on Svw. These formulae show up the role of coefficients c_2, $\tilde{c}_2$ which are a measure of the resistance to bending of the shaft, and which may be different along the two directions tied to the moving axes, and the internal damping or viscosity terms $b_2, \tilde{b}_2$.

1. *Case of an Isotropic Shaft:* $b_2 = \tilde{b}_2$, $c_2 = \tilde{c}_2$

From the formulae (11.41) where we interpret $(0, v_2, w_2)$, $(0, v'_2, w'_2)$, as the components on $Sxvw$ of the vectors $\overrightarrow{S_1C}$ and $\left(\dfrac{d\overrightarrow{S_1C}}{dt}\right)_{Sxvw}$ respectively, we deduce, from the expression for these same vectors with respect to $Sxyz$:

$$\text{(11.42)} \qquad \begin{aligned} -F_y &= b_2(y'_2 + \omega z_2) + c_2 y_2 \\ -F_z &= b_2(z'_2 - \omega y_2) + c_2 z_2. \end{aligned}$$

Eliminating F_y, F_z from the equations (11.38), (11.39), (11.42), we obtain, using (11.37), a set of six linear differential equations in the variables y_1, y_2, z_1, z_2, y, z which we can study by classical methods. However the approach may be very considerably simplified, [52], by making the realistic assumption, that the damping factors $b_1, \tilde{b}_1, b_2, \tilde{b}_2$ are small in comparison with the stiffness c_1/ω, $\tilde{c}_1/\omega\ldots$

Thus, from

$$y_1 + y_2 = y$$
$$c_1 y_1 + b_1 y'_1 = c_2 y_2 + b_2(y'_2 + \omega z_2)$$

and

$$z_1 + z_2 = z$$
$$\tilde{c}_1 z_1 + \tilde{b}_1 z_1' = c_2 z_2 + b_2(z_2' - \omega y_2)$$

which we solve for y_1, y_2 and z_1, z_2 respectively, we deduce

$$(c_1 + c_2)y_1 = c_2 y - b_1 y_1' + b_2(y_2' + \omega z_2)$$
$$(c_1 + c_2)y_2 = c_1 y + b_1 y_1' - b_2(y_2' + \omega z_2) \tag{11.43}$$

and in particular, to order 0:

$$y_1 = \frac{c_2}{c_1 + c_2} y$$
$$y_2 = \frac{c_1}{c_1 + c_2} y \tag{11.44}$$

then

$$(\tilde{c}_1 + c_2)z_1 = c_2 z - \tilde{b}_1 z_1' + b_2(z_2' - \omega y_2)$$
$$(\tilde{c}_1 + c_2)z_2 = \tilde{c}_1 z + \tilde{b}_1 z_1' - b_2(z_2' - \omega y_2) \tag{11.45}$$

and, to order 0:

$$z_1 = \frac{c_2}{\tilde{c}_1 + c_2} z$$
$$z_2 = \frac{\tilde{c}_1}{\tilde{c}_1 + c_2} z. \tag{11.46}$$

Coming back to (11.43), (11.45) we obtain the expressions for y_1, z_1 to order 1:

$$(c_1 + c_2)y_1 = c_2 y - \frac{b_1 c_2}{c_1 + c_2} y' + b_2\left(\frac{c_1}{c_1 + c_2} y' + \frac{\omega \tilde{c}_1}{\tilde{c}_1 + c_2} z\right)$$
$$(\tilde{c}_1 + c_2)z_1 = c_2 z - \frac{\tilde{b}_1 c_2}{\tilde{c}_1 + c_2} z' + b_2\left(\frac{\tilde{c}_1}{\tilde{c}_1 + c_2} z' - \frac{\omega c_1}{c_1 + c_2} y\right)$$

and finally by (11.38), (11.39) the equations:

$$my'' = -\frac{b_1 c_2}{c_1 + c_2} y' - \frac{c_1}{c_1 + c_2}\left(c_2 y - \frac{b_1 c_2}{c_1 + c_2} y' + b_2\left(\frac{c_1}{c_1 + c_2} y' + \frac{\omega \tilde{c}_1}{\tilde{c}_1 + c_2} z\right)\right)$$
$$mz'' = -\frac{\tilde{b}_1 c_2}{\tilde{c}_1 + c_2} z' - \frac{\tilde{c}_1}{\tilde{c}_1 + c_2}\left(c_2 z - \frac{\tilde{b}_1 c_2}{\tilde{c}_1 + c_2} z' + b_2\left(\frac{\tilde{c}_1}{\tilde{c}_1 + c_2} z' - \frac{\omega c_1}{c_1 + c_2} y\right)\right). \tag{11.47}$$

Let us first look at the case of an isotropic bearing, i.e. the case $c_1 = \tilde{c}_1$, $b_1 = \tilde{b}_1$.

Here we can write (11.47) in the form:

$$y'' + ay' + b(y' + \omega z) + ky = 0$$
$$z'' + az' + b(z' - \omega y) + kz = 0 \qquad (11.48)$$

with $a = \dfrac{b_1 c_2^2}{m(c_1 + c_2)^2}$, $b = \dfrac{b_2 c_1^2}{m(c_1 + c_2)^2}$, $k = \dfrac{c_1 c_2}{m(c_1 + c_2)}$,

a being an external friction term connected with damping in the bearings, and b being an internal friction term relating to the shaft.

The characteristic equation associated with (11.48) can be written in the form

$$\sigma^2 + (a+b)\sigma + k \pm ib\omega = 0$$

or with

$$\sigma = \zeta \mp \frac{ib\omega}{a+b}$$

$$\zeta^2 + \left(a + b \mp \frac{2ib\omega}{a+b}\right)\zeta + k - \left(\frac{b\omega}{a+b}\right)^2 = 0.$$

To have rotational stability it is necessary and sufficient that the real parts of the roots ζ be negative. As this is certainly true of their sum we need only write down the condition that their product be positive, and we thus derive the condition for stability:

$$\omega < \left(1 + \frac{a}{b}\right)\sqrt{k}. \qquad (11.49)$$

Accordingly for rigid bearings, where $a = 0$, the limiting velocity is $\sqrt{k}$; it is increased by having a flexible bearing with damping. We may note that, if we took account of a want of balance in the disc, which can be represented by an additional point mass μ at a distance e from the centre C, we should be led, for the purposes of calculating the forced vibration, to the equations:

$$y'' + ay' + b(y' + \omega z) + ky = \frac{\mu}{m} e\omega^2 \cos \omega t$$

$$z'' + az' + b(z' - \omega y) + kz = \frac{\mu}{m} e\omega^2 \sin \omega t$$

where m is the total mass of the disc.

Let us now examine the case of an anisotropic bearing and let us introduce into the equations (11.47) the notations:

$$mk = \frac{c_1 c_2}{c_1 + c_2}, \qquad m\tilde{k} = \frac{\tilde{c}_1 c_2}{\tilde{c}_1 + c_2}$$

$$mf = \frac{b_1 c_2^2}{(c_1 + c_2)^2}, \quad m\tilde{f} = \frac{\tilde{b}_1 c_2^2}{(\tilde{c}_1 + c_2)^2},$$

$$mh = \frac{b_2 c_1^2}{(c_1 + c_2)^2}, \quad m\tilde{h} = \frac{b_2 \tilde{c}_1^2}{(\tilde{c}_1 + c_2)^2}$$

so that

$$(11.50)\qquad \begin{aligned} &y'' + (f+h)y' + \sqrt{h\tilde{h}}\cdot\omega z + ky = 0\\ &z'' + (\tilde{f}+\tilde{h})z' - \sqrt{h\tilde{h}}\cdot\omega y + \tilde{k}z = 0. \end{aligned}$$

We easily obtain the characteristic equation in the form:

$$\sigma^4 + (f+h+\tilde{f}+\tilde{h})\sigma^3 + (k+\tilde{k}+(f+h)(\tilde{f}+\tilde{h}))\sigma^2 + (k(\tilde{f}+\tilde{h}) + \tilde{k}(f+h))\sigma + k\tilde{k} + h\tilde{h}\omega^2 = 0$$

and Routh's stability conditions, bearing in mind that all the coefficients of this equation are positive, reduce to:

$$\delta = (f+h+\tilde{f}+\tilde{h})(k+\tilde{k}+(f+h)(\tilde{f}+\tilde{h})) - (k(\tilde{f}+\tilde{h}) + \tilde{k}(f+h)) > 0$$
$$\delta\cdot(k(\tilde{f}+\tilde{h}) + \tilde{k}(f+h)) - (f+h+\tilde{f}+\tilde{h})^2(k\tilde{k} + h\tilde{h}\omega^2) > 0.$$

The first inequality is always satisfied and the second leads to:

$$(11.51)\qquad h\tilde{h}(f+h+\tilde{f}+\tilde{h})\omega^2 < (f+h)(\tilde{f}+\tilde{h})[(k-\tilde{k})^2 + (k(\tilde{f}+\tilde{h}) + \tilde{k}(f+h))(f+h+\tilde{f}+\tilde{h})]$$

which shows that anisotropy of bearings has the effect of raising the limit of stability.

2. *Case Where Shaft and Bearings Are Both Anisotropic*

To simplify matters we shall ignore damping in the calculations; accordingly we shall write (11.39) and (11.41) as:

$$(11.52)\qquad \begin{aligned} y_1 &= -s_1 F_y\\ z_1 &= -\tilde{s}_1 F_z, \end{aligned}$$

$$(11.53)\qquad \begin{aligned} v_2 &= -s_2 F_v = -s_2[F_y\cos\omega t + F_z \sin\omega t]\\ w_2 &= -\tilde{s}_2 F_w = -\tilde{s}_2[-F_y \sin\omega t + F_z\cos\omega t] \end{aligned}$$

with $s = c^{-1}$, from which we deduce:

$$y_2 = v_2\cos\omega t - w_2\sin\omega t$$

or

$$(11.54)\qquad y_2 = -[s_2\cos^2\omega t + \tilde{s}_2\sin^2\omega t]F_y - (s_2 - \tilde{s}_2)\sin\omega t\cos\omega t\cdot F_z$$

and similarly:

$$(11.55)\qquad z_2 = -(s_2-\tilde{s}_2)\sin\omega t\cos\omega t\cdot F_y - (s_2\sin^2\omega t + \tilde{s}_2\cos^2\omega t)F_z.$$

Eliminating $y_1, y_2, z_1, z_2, F_y, F_z$ from (11.52), (11.54), (11.55), (11.38) and (11.37) we obtain after a few calculations:

$$(11.56)\qquad m\left[s_1 + \frac{s_2+\tilde{s}_2}{2} + \frac{s_2-\tilde{s}_2}{2}\cos 2\omega t\right]y'' + m\frac{s_2-\tilde{s}_2}{2}\sin 2\omega t\cdot z'' + y = 0$$

$$m\left[\tilde{s}_1 + \frac{s_2+\tilde{s}_2}{2} - \frac{s_2-\tilde{s}_2}{2}\cos 2\omega t\right]z'' + m\frac{s_2-\tilde{s}_2}{2}\sin 2\omega t\cdot y'' + z = 0.$$

Let us now put

$$s_1+\tilde{s}_1+s_2+\tilde{s}_2=s,\quad \alpha=\frac{s_1-\tilde{s}_1}{s},\quad \eta=\frac{s_2-\tilde{s}_2}{s} \tag{11.57}$$

so that $|\alpha|<1$, $|\eta|<1$, and then:

$$\tilde{\omega}^2=\frac{2}{ms},\quad \omega_1^2=\frac{\tilde{\omega}^2}{1+\alpha},\quad \omega_2^2=\frac{\tilde{\omega}^2}{1-\alpha}. \tag{11.58}$$

With these notations we can write (11.56) as:

$$\begin{aligned}(1+\alpha+\eta\cos 2\omega t)y''+\eta\sin 2\omega t\cdot z''+\tilde{\omega}^2 y=0\\ (1-\alpha-\eta\cos 2\omega t)z''+\eta\sin 2\omega t\cdot y''+\tilde{\omega}^2 z=0.\end{aligned} \tag{11.59}$$

The parameter η is associated with the anisotropy of the shaft and we shall suppose it to be small, whereas α corresponds to the anisotropy of the bearing and may be of a different order of magnitude.

Finally solving (11.59) for y'', z'' and restricting ourselves to the terms of the first order in η, we obtain the modified system:

$$\begin{aligned}y''+\omega_1^2 y=\eta(1+\alpha)^{-1}\{\omega_1^2\cos 2\omega t\cdot y+\omega_2^2\sin 2\omega t\cdot z\}\\ z''+\omega_2^2 z=\eta(1-\alpha)^{-1}\{-\omega_2^2\cos 2\omega t\cdot z+\omega_1^2\sin 2\omega t\cdot y\}\end{aligned} \tag{11.60}$$

on which we shall base our discussion of the conditions for stability. We could complete the right-hand sides of (11.60) by including the terms of higher order in η but we shall not do so as we shall remain satisfied in the following treatment with results obtained from the first-order approximation. The system (11.60)—and also (11.59)—has an interesting property of reciprocity, which asserts that to every solution $y(t)$, $z(t)$ corresponds another solution $y(-t)$, $-z(-t)$.

Periodic Linear Differential Equation with Reciprocity Property

Consider the differential equation

$$\frac{dx}{dt}=A(t)x \tag{11.61}$$

where $x\in\mathbb{R}^n$ and $A(t)$ is a real, locally integrable, periodic $n\times n$ matrix of period T.

We shall say that (11.61) is reciprocal or reversible if there exists an $n\times n$ invertible constant matrix S, and a real number γ such that for any solution $x(t)$ of (11.61), $Sx(-t+\gamma)$ is also a solution.

We can interpret this condition by the remark that if $X(t)$ is the resolvent matrix of (11.61) defined by:

$$\frac{dX}{dt}=A(t)X,\ X(0)=I,$$

every solution of (11.61) is represented by $x(t)=X(t)b$, $b\in\mathbb{R}^n$ and consequently, there is a $b \in \mathbb{R}^n$ such that for a given $c \in \mathbb{R}^n$

$$SX(-t+\gamma)c = X(t)b$$

i.e.

$$b = SX(\gamma)c$$

and

$$SX(-t+\gamma)c = X(t)SX(\gamma)c, \quad \forall c \in \mathbb{R}^n$$

or

$$SX(-t+\gamma) = X(t)SX(\gamma), \quad \forall t \in \mathbb{R} \tag{11.62}$$

and conversely.

Noting that $\dfrac{\mathrm{d}X(-t+\gamma)}{\mathrm{d}t} = -A(-t+\gamma)X(-t+\gamma)$ we deduce from the identity:

$$\frac{\mathrm{d}}{\mathrm{d}t}(SX(-t+\gamma) - X(t)SX(\gamma)) = -SA(-t+\gamma)X(-t+\gamma) - A(t)X(t)SX(\gamma)$$

that (11.62) is equivalent to

$$A(t)S + SA(-t+\gamma) = 0. \tag{11.63}$$

The eigenvalues ρ of the matrix $C = X(T)$ are called the characteristic multipliers, and associated with these are the characteristic exponents σ, defined $\left(\text{modulo}\,\dfrac{2i\pi}{T}\right)$ by $\rho = e^{\sigma T}$. It is indeed a known fact [42] that C is invertible and from the periodicity of $A(t)$ that,

$$X(t+T) = X(t)C, \quad \forall t \in \mathbb{R}.$$

From this result for $t = \gamma - T$ and from (11.62) for $t = T$, we obtain:

$$SX(\gamma)C^{-1} = CSX(\gamma)$$

or

$$SX(\gamma)C^{-1}(SX(\gamma))^{-1} = C$$

which shows that C and C^{-1} have the same eigenvalues with the same order of multiplicity. Since $X(T)$ is a real-valued matrix, its eigenvalues occur in conjugate pairs and the property of reciprocity shows that to every eigenvalue ρ corresponds an eigenvalue $(\bar{\rho})^{-1}$. Clearly the zero solution of (11.61) can be stable only if the eigenvalues of $C = X(T)$ have modulus unity [25].

We now consider, instead of (11.61), the equation

$$\frac{\mathrm{d}x}{\mathrm{d}t} = A(t,\mu)x \tag{11.64}$$

in which the matrix $A(t,\mu)$ is t-periodic of period T, depends on a real parameter μ, is continuous with respect to (t,μ) and is such that (11.64) is a reciprocal equation for every value of μ.

Let us now suppose that, for $\mu = 0$, the solution $x = 0$ of the equation (11.64) is stable. In that case the characteristic exponents are pure imaginaries $\pm i\sigma_k$, with

σ_k real, and the characteristic multipliers are, for $\mu=0$,

$$\exp\left(\pm\frac{2i\pi\rho_k}{\omega}\right), \quad \text{with} \quad T=\frac{2\pi}{\omega}.$$

Suppose they are all distinct, i.e.

$$2\sigma_j \neq 0, \quad \sigma_j \pm \sigma_k \neq 0, \quad \bmod \omega, \quad \forall j, k. \tag{11.65}$$

Then we can assert: if the characteristic multipliers of (11.64), for $\mu=0$, are all of modulus unity and all distinct, then the same is true for μ in a neighbourhood of zero.

The fact that they remain distinct for μ in a suitable neighbourhood of 0 is obvious from continuity considerations. As for the remaining part of the assertion let ρ_0 be a characteristic multiplier of $A(t,0)$ and let $|\rho_0|=1$. For any disc D in the complex plane of centre ρ_0 and arbitrarily small radius δ, we can always find a $\mu_1>0$ such that whenever $|\mu|<\mu_1$, there is one and only one $\rho(\mu)\in D$, which is a characteristic multiplier of $A(t,\mu)$. But then

$$(\bar{\rho}(\mu))^{-1}=\frac{\rho(\mu)}{|\rho(\mu)|^2}$$

is also a characteristic multiplier and by continuity will also be in D for $|\mu|<\mu_1$, if μ_1 has been chosen small enough. It follows therefore that if $|\rho(\mu)|$ were different from 1 there would be two distinct multipliers contrary to hypothesis. We have therefore proved that

$$|\mu|<\mu_1, \quad |\rho(\mu)|=1.$$

Thus provided the conditions (11.65) are satisfied, we can be certain that $x=0$ is a stable solution of (11.64) for small enough $|\mu|$.

The same type of argument allows us to prove that if, for some value μ_0, the characteristic multipliers are all distinct and of modulus unity, then the same will be true for all μ in an open interval of centre μ_0. In particular if the conditions (11.65) are fulfilled, the characteristic multipliers are distinct and of modulus unity for $\mu=0$ and this property remains valid throughout a maximal interval $]-h, l[$; at $\mu=l$ (or $-h$) all the multipliers will still be of modulus unity but at least two of them will coincide.

On the other hand if, at $\mu=\mu_0$, there is at least one characteristic multiplier of modulus greater than 1, the same will apply, by continuity, in an open neighbourhood of μ_0. We can therefore define a maximal open interval $]\mu_1, \mu_2[$ containing μ_0 and throughout which the following property holds: for $\mu=\mu_1$ (or μ_2) the characteristic multipliers are all of modulus 1, but are not all distinct, because if they were the same would be true in a neighbourhood of μ_1 (or μ_2). At $\mu=\mu_1$ or μ_2 there is a merging of identity between two or more characteristic multipliers of modulus 1. Accordingly we can predict that the boundary points on the μ-axis separating the zones of stability from those of instability will be precisely those where the characteristic multipliers lose their identity.

Since $X(T)$ is real, its multipliers are in conjugate pairs and therefore when they lose their identity (i.e. two become coincident) we must have one of the three following cases:

1. A multiplier $\rho = +1$, which is double.
In this case there exists a non-zero $c \in \mathbb{R}^n$, such that $X(T)c = c$ and $X(t)c$ is a periodic solution of (11.64) of period T;

2. A multiplier $\rho = -1$, which is double.
In this case there exists a non-zero $c \in \mathbb{R}^n$ such that $X(T)c = -c$ and $X(t)c$ is a periodic solution of (11.64) of period $2T$;

3. Two multipliers $\rho, \bar{\rho}$ both double.

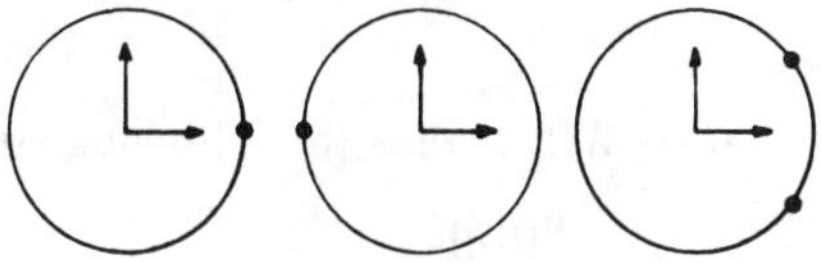

Figure 11.11

Stability of Rotation of Disc Where the System Has Anisotropic Flexibilities

We again consider the equations (11.60),

(11.66)
$$\begin{aligned} y'' + \omega_1^2 y &= \eta(1+\alpha)^{-1}\{\omega_1^2 \cos 2\omega t \cdot y + \omega_2^2 \sin 2\omega t \cdot z\} \\ z'' + \omega_2^2 z &= \eta(1-\alpha)^{-1}\{-\omega_2^2 \cos 2\omega t \cdot z + \omega_1^2 \sin 2\omega t \cdot y\} \end{aligned}$$

which can easily be put into the form of a system of differential equations of the first order by introducing the variables v, w defined by $y' = v$, $z' = w$. The resolvent matrix of this system for $\eta = 0$ is:

$$X(t) = \begin{pmatrix} X_1(t) & 0 \\ 0 & X_2(t) \end{pmatrix},$$

with
$$X_j(t) = \begin{pmatrix} \cos \omega_j t & \omega_j^{-1} \sin \omega_j t \\ -\omega_j \sin \omega_j t & \cos \omega_j t \end{pmatrix}, \quad j = 1, 2$$

and the characteristic multipliers or eigenvalues of $X\left(\frac{\pi}{\omega}\right)$ are

(11.67)
$$\exp\left(\pm i\pi \frac{\omega_1}{\omega}\right), \quad \exp\left(\pm i\pi \frac{\omega_2}{\omega}\right),$$

which are distinct if and only if

(11.68)
$$2\omega_1 \neq 0, \quad 2\omega_2 \neq 0, \quad \omega_1 \pm \omega_2 \neq 0 \bmod 2\omega.$$

If these conditions are satisfied the null solution of the system (11.66) is stable for small enough $|\eta|$.

For $\eta \neq 0$, $\rho = \exp\left(i\frac{\pi\sigma}{\omega}\right)$ is a characteristic multiplier of (11.66) if and only if there is a $c \in \mathbb{R}^4$ such that:

$$X\left(\frac{\pi}{\omega}, \eta\right)c = \rho c,$$

$X(t, \eta)$ being the resolvent matrix associated with (11.66).

Now the latter, because of the periodicity of the system, satisfies

$$X\left(t + \frac{\pi}{\omega}, \eta\right) = X(t, \eta) \cdot X\left(\frac{\pi}{\omega}, \eta\right), \quad \forall t$$

and consequently the solution $X(t, \eta)c$ has the following structure:

$$X(t, \eta)c = e^{i\sigma t}u(t)$$

where $u(t)$ is periodic of period $\frac{\pi}{\omega}$. This solution is obviously infinitely differentiable with respect to t, and $u(t)$ can be expanded in a Fourier series, i.e.

$$y = \sum_{-\infty}^{+\infty} a_k e^{i(\sigma + 2k\omega)t}, \quad z = \sum_{-\infty}^{+\infty} b_k e^{i(\sigma + 2k\omega)t}$$

and these expansions can be differentiated term-by-term with respect to t arbitrarily often without ceasing to be convergent and to represent the corresponding derivatives of y and z. By substituting in (11.66) we see that the Fourier coefficients a_k, b_k are the solutions of the infinite set of homogeneous linear equations:

(11.69)
$$(\omega_1^2 - (\sigma + 2k\omega)^2)a_k = \frac{\eta(1+\alpha)^{-1}}{2}[(\omega_1^2 a_{k-1} - i\omega_2^2 b_{k-1}) + (\omega_1^2 a_{k+1} + i\omega_2^2 b_{k+1})]$$

$$(\omega_2^2 - (\sigma + 2k\omega)^2)b_k = \frac{\eta(1-\alpha)^{-1}}{2}[-(i\omega_1^2 a_{k-1} + \omega_2^2 b_{k-1}) + (i\omega_1^2 a_{k+1} - \omega_2^2 b_{k+1})]$$

whose infinite determinant can be put into normal form (see Chapter 9) and regarded as a function $\Delta(\sigma, \omega, \eta)$.

Let us now examine the conditions for resonance in the neighbourhood of $\eta = 0$, corresponding to the following cases:

1. $\eta = 0$, $\omega = \omega_1$.

By (11.67) we have a double characteristic-multiplier $\rho = -1$ and a pair $\rho = \exp\left(\pm i\pi\frac{\omega_2}{\omega_1}\right)$. We shall suppose in what follows that ω_1, ω_2 are independent.

If we draw the separatrix in the (η, ω) plane emanating from the point $(0, \omega_1)$ we see at once that $\rho = -1$ must remain a characteristic multiplier, because otherwise it would have to "undouble" itself and we should then have four distinct multipliers. We can therefore take on this separatrix $\sigma = \omega$, and obtain a periodic solution of period $2T = \dfrac{2\pi}{\omega}$.

2. $\eta = 0$, $\omega = \omega_2$; we can draw the same conclusion as in §1, with $\sigma = \omega$.

3. $\eta = 0$, $\omega = \dfrac{\omega_1}{2}$; by (11.67) we have for $\eta = 0$ the values $\rho = +1$ $\rho = \exp\left(\pm 2i\pi \dfrac{\omega_2}{\omega_1}\right)$.

Along the separatrix emanating from the point $\left(0, \dfrac{\omega_1}{2}\right)$ we shall have $\rho = +1$ or $\sigma = 0$, and a periodic solution of period $T = \dfrac{\pi}{\omega}$.

4. $\eta = 0$, $\omega = \dfrac{\omega_2}{2}$; this is similar to the case of §3.

5. $\eta = 0$, $\omega = \dfrac{\omega_1 + \omega_2}{2}$.

By (11.67) we have two pairs of equal characteristic multipliers:

$$\rho = \exp\left(i\pi \frac{\omega_1}{\omega}\right) = \exp\left(-i\pi \frac{\omega_2}{\omega}\right), \quad \sigma = -\omega_2 \quad \text{or} \quad \sigma = \omega_1$$

$$\rho = \exp\left(-i\pi \frac{\omega_1}{\omega}\right) = \exp\left(i\pi \frac{\omega_2}{\omega}\right), \quad \sigma = \omega_2 \quad \text{or} \quad \sigma = -\omega_1.$$

We shall seek to solve the equation $\Delta = 0$ by using for σ and ω a representation near $\eta = 0$ defined by:

(11.70)
$$\begin{aligned} \sigma &= -\omega_2 + \eta p(\eta) & & \qquad \sigma = \omega_1 + \eta p(\eta) \\ \omega &= \frac{\omega_1 + \omega_2}{2} + \eta q(\eta) & \text{or} & \qquad \omega = \frac{\omega_1 + \omega_2}{2} + \eta q(\eta). \end{aligned}$$

6. $\eta = 0$, $\omega = \dfrac{\omega_2 - \omega_1}{2}$.

We again have two pairs of equal characteristic multipliers:

$$\rho = \exp\left(-i\pi \frac{\omega_1}{\omega}\right) = \exp\left(-i\pi \frac{\omega_2}{\omega}\right), \quad \sigma = -\omega_2 \quad \text{or} \quad \sigma = -\omega_1$$

$$\rho = \exp\left(i\pi \frac{\omega_1}{\omega}\right) = \exp\left(i\pi \frac{\omega_2}{\omega}\right), \quad \sigma = \omega_2 \quad \text{or} \quad \sigma = \omega_1$$

and in this case also, to solve $\Delta = 0$, we shall look for a representation:

(11.71)
$$\begin{aligned} \sigma &= -\omega_2 + \eta p(\eta) \\ \omega &= \frac{\omega_2 - \omega_1}{2} + \eta q(\eta). \end{aligned}$$

(11.72)

$$
\begin{array}{ccccc}
a_{-2} & b_{-2} & a_{-1} & b_{-1} & a_0 \\
\frac{-\eta(1+\alpha)^{-1}}{2}\omega_1^2 & \frac{i\eta(1+\alpha)^{-1}}{2}\omega_2^2 & \omega_1^2-(\sigma-2\omega)^2 & 0 & \frac{-\eta(1+\alpha)^{-1}}{2}\omega_1^2 \\
\frac{i\eta(1-\alpha)^{-1}}{2}\omega_1^2 & \frac{\eta(1-\alpha)^{-1}}{2}\omega_2^2 & 0 & \omega_2^2-(\sigma-2\omega)^2 & \frac{-i\eta(1-\alpha)^{-1}}{2}\omega_1^2 \\
0 & 0 & \frac{-\eta(1+\alpha)^{-1}}{2}\omega_1^2 & \frac{i\eta(1+\alpha)^{-1}}{2}\omega_2^2 & \omega_1^2-\sigma^2 \\
0 & 0 & \frac{i\eta(1-\alpha)^{-1}}{2}\omega_1^2 & \frac{\eta(1-\alpha)^{-1}}{2}\omega_2^2 & 0 \\
0 & 0 & 0 & 0 & \frac{-\eta(1+\alpha)^{-1}}{2}\omega_1^2 \\
0 & 0 & 0 & 0 & \frac{i\eta(1-\alpha)^{-1}}{2}\omega_1^2 \\
\hdashline
b_0 & a_1 & b_1 & a_2 & b_2 \\
\frac{-i\eta(1+\alpha)^{-1}}{2}\omega_2^2 & 0 & 0 & 0 & 0 \\
\frac{\eta(1-\alpha)^{-1}}{2}\omega_2^2 & 0 & 0 & 0 & 0 \\
0 & \frac{-\eta(1+\alpha)^{-1}}{2}\omega_1^2 & \frac{-i\eta(1+\alpha)^{-1}}{2}\omega_2^2 & 0 & 0 \\
\omega_2^2-\sigma^2 & \frac{-i\eta(1-\alpha)^{-1}}{2}\omega_1^2 & \frac{\eta(1-\alpha)^{-1}}{2}\omega_2^2 & 0 & 0 \\
\frac{i\eta(1+\alpha)^{-1}}{2}\omega_2^2 & \omega_1^2-(\sigma+2\omega)^2 & 0 & \frac{-\eta(1+\alpha)^{-1}}{2}\omega_1^2 & \frac{-i\eta(1+\alpha)^{-1}}{2}\omega_2^2 \\
\frac{\eta(1-\alpha)^{-1}}{2}\omega_2^2 & 0 & \omega_2^2-(\sigma+2\omega)^2 & \frac{-i\eta(1-\alpha)^{-1}}{2}\omega_1^2 & \frac{\eta(1-\alpha)^{-1}}{2}\omega_2^2
\end{array}
$$

Let us now study more closely the infinite determinant associated with (11.69), or rather the part of it corresponding to the equations of order $k = -1, k = 0, k = 1$.

The diagonal containing the elements $\omega_1^2 - (\sigma + 2\omega k)^2$, $\omega_2^2 - (\sigma + 2\omega k)^2$, $k \in \mathbb{Z}$, will be regarded as the leading diagonal; by attaching the index $(0, 0)$ to the element $\omega_1^2 - \sigma^2$, we can easily number the rows and columns, and furthermore it is clear that the elements not on the leading diagonal are $O(\eta)$. We now proceed to analyse the cases of resonance considered above.

1. $\sigma = \omega$ and $\omega \sim \omega_1$ for $\eta \sim 0$.

We retain from (11.72) the determinant whose elements are at the intersection of the rows numbered -2 and 0 and the columns numbered -2 and 0, or in other words, in view of $\sigma = \omega$:

$$\det \begin{pmatrix} \omega_1^2 - \omega^2 & \dfrac{-\eta(1+\alpha)^{-1}}{2}\omega_1^2 \\ \dfrac{-\eta(1+\alpha)^{-1}}{2}\omega_1^2 & \omega_1^2 - \omega^2 \end{pmatrix} = 0$$

or

$$\omega = \omega_1\left(1 \pm \frac{\eta}{4}(1+\alpha)^{-1}\right). \tag{11.73}$$

Taking higher order determinants into account would not affect the result to the first order of approximation, because if we write $\omega = \omega_1(1 \pm \eta f(\eta))$, we can eliminate the factor η from the elements of the rows -2 and 0; the elements in the other rows vanish when $\eta = 0$ except for those on the leading diagonal, and we are left with

$$f(0) = \pm \frac{(1+\alpha)^{-1}}{4}.$$

2. $\sigma = \omega$, $\omega \sim \omega_2$ for $\eta \sim 0$.

We consider the elements at the intersections of rows -1, $+1$ and columns -1, $+1$ obtaining in this way:

$$\det \begin{pmatrix} \omega_2^2 - \omega^2 & \dfrac{\eta(1-\alpha)^{-1}}{2}\omega_2^2 \\ \dfrac{\eta(1-\alpha)^{-1}}{2}\omega_2^2 & \omega_2^2 - \omega^2 \end{pmatrix} = 0$$

and the first-order approximation:

$$\omega = \omega_2\left(1 \pm \frac{\eta}{4}(1-\alpha)^{-1}\right). \tag{11.74}$$

3. $\sigma = 0$, $\omega \sim \dfrac{\omega_1}{2}$ for $\eta \sim 0$, or $\omega = \dfrac{\omega_1}{2}(1 + \eta f(\eta))$.

We can eliminate the factor η from the elements in rows -2, $+2$ of the determinant (11.72). When we put $\eta = 0$ all the elements except for those in rows ± 2 vanish with the exception of those on the leading diagonal. As for the elements $(-2, -2)$

and $(+2, +2)$ they have the same value $-2f(0)\omega_1^2$ and the equation $\Delta = 0$ imposes $f(0) = 0$.

This means that in the case under consideration the branches of the separatrix emanating from $\left(0, \dfrac{\omega_1}{2}\right)$ have a representation of the form

$$\omega = \frac{\omega_1}{2}(1 + O(\eta^2));$$

for a given small η the width of the stability band will be $O(\eta^2)$ and consequently the resonance $\omega \sim \dfrac{\omega_1}{2}$ is of less importance than the preceding ones.

4. We obtain a result analogous to 3 above.

5. Starting from the representation (11.70) we isolate in (11.72) the determinant of order 2 whose diagonal elements vanish when $\eta = 0$, that is

$$\det\begin{pmatrix} \omega_2^2 - \sigma^2 & -\dfrac{i\eta(1-\alpha)^{-1}}{2}\omega_1^2 \\ \dfrac{i\eta(1+\alpha)^{-1}}{2}\omega_2^2 & \omega_1^2 - (\sigma + 2\omega)^2 \end{pmatrix} = 0 \tag{11.75}$$

or

$$p(2\omega_2 - \eta p)(p + 2q)(2\omega_1 + \eta p + 2\eta q) + \frac{(1-\alpha^2)^{-1}}{4}\omega_1^2\omega_2^2 = 0$$

and for $\eta = 0$:

$$p^2 + 2pq + \frac{(1-\alpha^2)^{-1}}{16}\omega_1\omega_2 = 0$$

which expresses $p(0)$ in terms of $q(0)$.

There will be instability if p is imaginary, i.e. if:

$$q^2 - \frac{(1-\alpha^2)^{-1}}{16}\omega_1\omega_2 < 0.$$

The zone of instability may thus be represented, to the first order of approximation, by:

$$\omega = \frac{\omega_1 + \omega_2}{2} + \eta q, \tag{11.76}$$

$$|q| < \frac{1}{4}\sqrt{\frac{\omega_1\omega_2}{1-\alpha^2}}.$$

6. The representation (11.71) leads to considering in (11.72) the determinant of order 2, whose diagonal elements vanish for $\eta = 0$, which is in fact identical to (11.75). By taking the limit as $\eta \to 0$ we obtain the equation

$$p^2 + 2pq - \frac{(1-\alpha^2)^{-1}}{16}\omega_1\omega_2 = 0$$

which shows that $p(0)$ is always real for a given $q(0)$. There is therefore first-order stability, i.e. the width of the instability band in the neighbourhood of this resonance is at least $O(\eta^2)$.

It follows from the foregoing discussion that the main resonances are found in the vicinity of $\omega_1, \omega_2, \dfrac{\omega_1 + \omega_2}{2}$.

An Alternative Approach to the Stability Problem

We consider the differential equation

$$\text{(11.77)} \qquad \begin{aligned} \frac{da}{dt} &= \mu(A(\theta) + \mu R(\theta, \mu))a, \\ \frac{d\theta}{dt} &= \omega, \quad \omega = \{\omega_1, \omega_2, \ldots, \omega_q\}, \end{aligned}$$

for which we make no assumption of reciprocity, and in which $A(\theta)$, $R(\theta, \mu)$ are $n \times n$ matrices depending continuously on

$$\theta = \{\theta_1, \theta_2, \ldots, \theta_q\},$$

and periodic with respect to each θ_j, $1 \leqslant j \leqslant q$, and of period 2π; and where the frequencies ω_j are independent and μ is a small parameter. This equation may be discussed, in regard to the stability of its zero solution, by the amplitude variation method described in Chapter 10. In particular, if the matrix

$$S = \frac{1}{(2\pi)^q} \int_0^{2\pi} \cdots \int_0^{2\pi} A(\theta)\, d\theta_1 \cdots d\theta_q$$

is stable, i.e. if the real parts of its eigenvalues are all negative, then $a = 0$ is an asymptotically stable solution of (11.77) for $\mu > 0$ small enough and $t \to +\infty$.

We can supplement this result, in the linear case which concerns us here, by an instability condition: if at least one of the eigenvalues has a positive real part, then $a = 0$ is an unstable solution of (11.77) for $\mu > 0$ small enough and $t \to +\infty$.

In practice however it frequently happens that $A(\theta)$, $R(\theta, \mu)$ are multiperiodic polynomials (abbreviated to m.p.p.) i.e. polynomials in $e^{i\theta_j}$, $i \leqslant j \leqslant q$. In this case we can develop an iterative theory which enables us to discuss the stability problem when the eigenvalues of the matrix S have negative or zero real parts. To explain briefly the method we shall base our reasoning on the typical equation:

$$\text{(11.78)} \qquad \frac{da}{dt} = \mu[A(\theta) + \mu A_1(\theta) + \mu^2 R(\theta, \mu)]a,$$

$$\theta_j = \omega_j t, \quad 1 \leqslant j \leqslant q$$

where $A(\theta)$, $A_1(\theta)$ are m.p.p. matrices and $R(\theta, \mu)$ is a bounded matrix.

Clearly we can define m.p.p. matrices $U(\theta)$, $U_1(\theta)$ such that:

$$\sum_{j=1}^{q} \omega_j \frac{\partial U}{\partial \theta_j} = A(\theta) - S \tag{11.79}$$

$$\sum_{j=1}^{q} \omega_j \frac{\partial U_1}{\partial \theta_j} = P_1(\theta) - S_1 \tag{11.80}$$

where S and S_1 are the means

$$S = \frac{1}{(2\pi)^q} \int_0^{2\pi} \cdots \int_0^{2\pi} A(\theta)\, d\theta_1 \cdots d\theta_q$$

$$S_1 = \frac{1}{(2\pi)^q} \int_0^{2\pi} \cdots \int_0^{2\pi} P_1(\theta)\, d\theta_1 \cdots d\theta_q$$

where the precise definition of the m.p.p. matrix $P_1(\theta)$, will be given later on.

We make in (11.78) the change of variable:

$$a = (I + \mu U + \mu^2 U_1)x \tag{11.81}$$

which, in view of (11.79), (11.80) gives:

$$\frac{dx}{dt} = \mu[S + \mu(AU + A_1 - P_1 + S_1 - US) + O(\mu^2)]x.$$

Thus, defining P_1 by:

$$P_1 = AU + A_1 - US \tag{11.82}$$

we obtain:

$$\frac{dx}{dt} = \mu(S + \mu S_1 + O(\mu^2))x.$$

Obviously the $O(\mu^2)$ term will, in general, depend on θ, but if we have been unable to decide the stability of $a = 0$ from a knowledge of S, or in other words if S has some eigenvalues with a zero real part, the others having negative real parts, then we can examine the eigenvalues of $S + \mu S_1$ in the neighbourhood of $\mu = 0$, which when $\mu \to 0$ tend to the pure imaginary eigenvalues of S, and this investigation will throw some light on the question of stability. It is also worth noting that it is possible to define an iterative procedure starting from the equation:

$$\frac{da}{dt} = \mu\left[A(\theta) + \mu \sum_{k=1}^{l} \mu^{k-1} A_k(\theta) + \mu^{l+1} R(\theta, \mu)\right] a \tag{11.83}$$

where the matrices $A_k(\theta)$ are m.p.p. and $R(\theta, \mu)$ is bounded. We can, by the change of variable:

$$a = \left[I + \mu \sum_{k=1}^{l} \mu^{k-1} U_k(\theta)\right] x$$

transform (11.83) into:

$$\frac{dx}{dt} = \mu\left[S + \mu \sum_{k=1}^{l} \mu^{k-1} S_k + O(\mu^{l+1})\right] x$$

where the matrices S_k have constant coefficients. To do so we need to calculate the m.p.p. matrix $U_k(\theta)$, such that:

$$\sum_{j=1}^{q} \omega_j \frac{\partial U_k}{\partial \theta_j} = P_k(\theta) - S_k,$$

where S_k is the mean value of the m.p.p. matrix $P_k(\theta)$ obtained by algebraic operations on terms already calculated, i.e. of rank less than k.

Application to the Problem of the Stability of a Rotating Shaft

1. We propose to investigate the conditions for the instability of the zero solution of the system (11.66) in the neighbourhood of the resonance $\omega = \omega_1$.

We put

$$\omega_1 = \omega(1 + \eta\gamma)^{1/2}, \tag{11.84}$$

with a given γ and

$$2\omega t = \tau. \tag{11.85}$$

We can therefore rewrite (11.66), using the dot notation to denote differentiation with respect to τ, as $\frac{\mathrm{d}y}{\mathrm{d}\tau} = \dot{y}$, $\frac{\mathrm{d}^2 y}{\mathrm{d}\tau^2} = \ddot{y}$, etc.:

$$\ddot{y} + \frac{y}{4} = \frac{\eta}{4}\left[-\gamma y + (1+\alpha)^{-1} \cos\tau \cdot y + \left(\frac{\omega_2}{\omega_1}\right)^2 (1+\alpha)^{-1} \sin\tau \cdot z \right] + O(\eta^2) \tag{11.86}$$

$$\ddot{z} + \left(\frac{\omega_2}{\omega_1}\right)^2 \frac{z}{4} = \frac{\eta}{4}\left[-\left(\frac{\omega_2}{\omega_1}\right)^2 \gamma z + (1-\alpha)^{-1} \sin\tau \cdot y - \left(\frac{\omega_2}{\omega_1}\right)^2 (1-\alpha)^{-1} \cos\tau \cdot z \right] + O(\eta^2)$$

where the $O(\eta^2)$ terms represent linear expressions in y, z with coefficients periodic in τ of period 2π, which are of the order of magnitude of η^2.

We then make, in (11.86), the change of variable:

$$\dot{y} + i\frac{y}{2} = u_1 e^{i\tau/2}, \quad \dot{y} - i\frac{y}{2} = u_2 e^{-i\tau/2}$$

whence

$$\ddot{y} + \frac{y}{4} = \dot{u}_1 e^{i\tau/2} = \dot{u}_2 e^{-i\tau/2}$$

and finally, by writing $\dot{z} = \zeta$, bring the system into the standard form:

$$\dot{z} = \zeta$$

$$\dot{\zeta} = -\left(\frac{\omega_2}{\omega_1}\right)^2 \frac{z}{4} + \frac{\eta}{4}\left[i(1-\alpha)^{-1} \sin\tau \cdot (u_2 e^{-i\tau/2} - u_1 e^{i\tau/2}) - \left(\frac{\omega_2}{\omega_1}\right)^2 (\gamma + (1-\alpha)^{-1} \cos\tau) z \right] + O(\eta^2)$$

$$\dot{u}_1 = \frac{\eta}{4}\left[i(\gamma - (1+\alpha)^{-1}\cos\tau)(u_1 - u_2 e^{-i\tau}) + \left(\frac{\omega_2}{\omega_1}\right)^2 (1+\alpha)^{-1} e^{-i\tau/2}\sin\tau\cdot z \right] + O(\eta^2) \tag{11.87}$$

$$\dot{u}_2 = \frac{\eta}{4}\left[i(\gamma - (1+\alpha)^{-1}\cos\tau)(u_1 e^{i\tau} - u_2) + \left(\frac{\omega_2}{\omega_1}\right)^2 (1+\alpha)^{-1} e^{i\tau/2}\sin\tau\cdot z \right] + O(\eta^2)$$

where the coefficients are periodic in τ of period 4π.

We can rewrite (11.87) in matrix form

$$(\dot{z}, \dot{\zeta}, \dot{u}_1, \dot{u}_2)^T = \left[A + \frac{\eta}{4} A_1(\tau) + O(\eta^2) \right]\cdot(z, \zeta, u_1, u_2)^T$$

with

$$A = \left(\begin{array}{cc|c} 0 & 1 & 0 \\ -\left(\dfrac{\omega_2}{2\omega_1}\right)^2 & 0 & \\ \hline 0 & & 0 \end{array}\right)$$

Now we find without difficulty:

$$\frac{1}{4\pi}\int_0^{4\pi} A_1(\tau)\,d\tau = \begin{pmatrix} 0 & 0 & 0 & \\ -\left(\dfrac{\omega_2}{\omega_1}\right)^2 & 0 & & \\ & & i\gamma & i\dfrac{(1+\alpha)^{-1}}{2} \\ 0 & & -i\dfrac{(1+\alpha)^{-1}}{2} & -i\gamma \end{pmatrix}$$

and there will be instability if among the eigenvalues of $A + \frac{\eta}{4}\int_0^{4\pi} A_1(\tau)\,d\tau$, there are one or more which have a positive real part; for small enough $|\eta|$ these can only be the roots of:

$$\det\left| \frac{i\eta}{4}\begin{pmatrix} \gamma & \dfrac{(1+\alpha)^{-1}}{2} \\ -\dfrac{(1+\alpha)^{-1}}{2} & -\gamma \end{pmatrix} - \lambda I \right| = 0$$

i.e. of

$$\lambda^2 + \frac{\eta^2}{16}\left(\gamma^2 - \frac{(1+\alpha)^{-2}}{4}\right) = 0.$$

Thus a sufficient condition for instability, for sufficiently small $|\eta|$, is $|\gamma| < \dfrac{(1+\alpha)^{-1}}{2}$.

Returning to (11.84) we find once again the instability limits obtained earlier:

$$\omega = \omega_1 \left(1 \pm \frac{(1+\alpha)^{-1}}{4} \eta \right).$$

2. We now consider the resonance $\omega \sim \frac{\omega_1 + \omega_2}{2}$. To this end we represent ω by:

$$\frac{\omega_1 + \omega_2}{2} = \omega(1 + \eta\gamma)^{1/2}, \tag{11.88}$$

where γ is given, and rewrite the differential equations (11.66) in terms of the variable $\tau = 2\omega t$, i.e. with:

$$\sigma_1 = \frac{\omega_1}{\omega_1 + \omega_2}, \quad \sigma_2 = \frac{\omega_2}{\omega_1 + \omega_2}.$$

$$\ddot{y} + \sigma_1^2 y = \eta[\sigma_1^2((1+\alpha)^{-1}\cos\tau - \gamma)y + \sigma_2^2(1+\alpha)^{-1}\sin\tau \cdot z] + O(\eta^2)$$
$$\ddot{z} + \sigma_2^2 z = \eta[\sigma_1^2(1-\alpha)^{-1}\sin\tau \cdot y - \sigma_2^2((1-\alpha)^{-1}\cos\tau + \gamma)z] + O(\eta^2)$$

which define two weakly coupled oscillators in resonance since $\sigma_1 + \sigma_2 = 1$.

We introduce the variables u_1, u_2, v_1, v_2 by:

$$\dot{y} + i\sigma_1 y = u_1 e^{i\sigma_1\tau}, \quad \dot{y} - i\sigma_1 y = u_2 e^{-i\sigma_1\tau}$$
$$\dot{z} + i\sigma_2 z = v_1 e^{i\sigma_2\tau}, \quad \dot{z} - i\sigma_2 z = v_2 e^{-i\sigma_2\tau}$$

and with $\sigma_1\tau = \theta_1$, $\sigma_2\tau = \theta_2$, $\tau = \theta_1 + \theta_2$, obtain the set of differential equations:

$$\dot{u}_1 = \eta(1+\alpha)^{-1}\left[\frac{i\sigma_1}{2}(\cos(\theta_1+\theta_2) - \gamma(1+\alpha))(u_2 e^{-2i\theta_1} - u_1)\right.$$
$$\left. + \frac{i\sigma_2}{2}\sin(\theta_1+\theta_2)\cdot(v_2 e^{-i(\theta_1+\theta_2)} - v_1 e^{i(\theta_2-\theta_1)})\right]$$

$$\dot{u}_2 = \eta(1+\alpha)^{-1}\left[\frac{i\sigma_1}{2}(\cos(\theta_1+\theta_2) - \gamma(1+\alpha))(u_2 - u_1 e^{2i\theta_1})\right.$$
$$\left. + \frac{i\sigma_2}{2}\sin(\theta_1+\theta_2)\cdot(v_2 e^{i(\theta_1-\theta_2)} - v_1 e^{i(\theta_1+\theta_2)})\right]$$

$$\dot{v}_1 = \eta(1-\alpha)^{-1}\left[\frac{i\sigma_1}{2}\sin(\theta_1+\theta_2)\cdot(u_2 e^{-i(\theta_1+\theta_2)} - u_1 e^{i(\theta_1-\theta_2)})\right.$$
$$\left. - \frac{i\sigma_2}{2}(\cos(\theta_1+\theta_2) + \gamma(1-\alpha))(v_2 e^{-2i\theta_2} - v_1)\right]$$

$$\dot{v}_2 = \eta(1-\alpha)^{-1}\left[\frac{i\sigma_1}{2}\sin(\theta_1+\theta_2)\cdot(u_2 e^{i(\theta_2-\theta_1)} - u_1 e^{i(\theta_1+\theta_2)})\right.$$
$$\left. - \frac{i\sigma_2}{2}(\cos(\theta_1+\theta_2) + \gamma(1-\alpha))(v_2 - v_1 e^{2i\theta_2})\right]$$

(11.89)
$$\frac{d\theta_1}{dt} = \sigma_1, \quad \frac{d\theta_2}{dt} = \sigma_2$$

in which the $O(\eta^2)$ terms have been omitted.

Taking the means with respect to θ_1, θ_2 of the coefficients of the variables u, v in (11.88) we can replace (11.89) by the system with constant coefficients:

$$\dot{u}_1 = \eta(1+\alpha)^{-1}\left[\frac{i\sigma_1}{2}\gamma(1+\alpha)u_1 + \frac{\sigma_2}{4}v_2\right]$$

$$\dot{u}_2 = \eta(1+\alpha)^{-1}\left[-\frac{i\sigma_1}{2}\gamma(1+\alpha)u_2 + \frac{\sigma_2}{4}v_1\right]$$

$$\dot{v}_1 = \eta(1-\alpha)^{-1}\left[\frac{\sigma_1}{4}u_2 + \frac{i\sigma_2}{2}\gamma(1-\alpha)v_1\right]$$

$$\dot{v}_2 = \eta(1-\alpha)^{-1}\left[\frac{\sigma_1}{4}u_1 - \frac{i\sigma_2}{2}\gamma(1-\alpha)v_2\right]$$

and it now remains to discuss the stability of its zero solution. For this purpose we need to calculate the eigenvalues λ of the matrix:

$$\begin{pmatrix} \frac{i\sigma_1\gamma}{2} & 0 & 0 & \frac{\sigma_2(1+\alpha)^{-1}}{4} \\ 0 & -\frac{i\sigma_1\gamma}{2} & \frac{\sigma_2(1+\alpha)^{-1}}{4} & 0 \\ 0 & \frac{\sigma_1(1-\alpha)^{-1}}{4} & \frac{i\sigma_2}{2}\gamma & 0 \\ \frac{\sigma_1(1-\alpha)^{-1}}{4} & 0 & 0 & -\frac{i\sigma_2}{2}\gamma \end{pmatrix}$$

or in other words the roots of the equation:

$$\lambda^2 \pm i\gamma\frac{\sigma_1-\sigma_2}{2}\lambda + \left(\gamma^2 - \frac{1}{4(1-\alpha^2)}\right)\frac{\sigma_1\sigma_2}{4} = 0.$$

Thus there will be instability, for small enough η, if the equation for $i\lambda$ which has real coefficients has imaginary roots, i.e. if:

$$\gamma^2 < \frac{\sigma_1\sigma_2}{1-\alpha^2}.$$

Referring back to (11.88) $\omega = \frac{\omega_1+\omega_2}{2} - \frac{\omega_1+\omega_2}{4}\eta\gamma$, we see that the limits of the zone of instability correspond to

$$\omega = \frac{\omega_1+\omega_2}{2} \pm \frac{\eta}{4}\sqrt{\frac{\omega_1\omega_2}{1-\alpha^2}}$$

which is precisely the result already obtained in (11.76).

3. Case where the anisotropy of the bearing is weak. Let us go back to the equations (11.59) and assume that η and α are small parameters, so that setting $\alpha = \eta a$ and retaining only the first order terms in η we can write them in the form:

$$y'' + \tilde{\omega}^2 y = \eta\tilde{\omega}^2[(a + \cos 2\omega t)y + \sin 2\omega t \cdot z] + O(\eta^2)$$
$$z'' + \tilde{\omega}^2 z = \eta\tilde{\omega}^2[-(a + \cos 2\omega t)z + \sin 2\omega t \cdot y] + O(\eta^2).$$

Now the characteristic multipliers of this system, when $\eta = 0$, are not distinct and we cannot therefore immediately conclude that the zero solution is stable for small enough η in the absence of resonance. But suppose ω is close to $\tilde{\omega}$, so that $\tilde{\omega} = \omega(1 + \eta\gamma)^{1/2}$ with γ given, and let $2\omega t = \tau$; then we are led to:

$$\ddot{y} + \frac{y}{4} = \frac{\eta}{4}[(a - \gamma + \cos\tau)y + \sin\tau \cdot z] + O(\eta^2)$$

$$\ddot{z} + \frac{z}{4} = \frac{\eta}{4}[-(a + \gamma + \cos\tau)z + \sin\tau \cdot y] + O(\eta^2)$$

and after the change of variables:

$$\dot{y} + i\frac{y}{2} = u_1 e^{i\tau/2}, \quad \dot{y} - i\frac{y}{2} = u_2 e^{-i\tau/2}$$
$$\dot{z} + i\frac{z}{2} = v_1 e^{i\tau/2}, \quad \dot{z} - i\frac{z}{2} = v_2 e^{-i\tau/2}$$

to the equations:

$$\dot{u}_1 = \frac{i\eta}{4}[(a - \gamma + \cos\tau)(u_2 e^{-i\tau} - u_1) + \sin\tau \cdot (v_2 e^{-i\tau} - v_1)] + O(\eta^2)$$

$$\dot{u}_2 = \frac{i\eta}{4}[(a - \gamma + \cos\tau)(u_2 - u_1 e^{i\tau}) + \sin\tau \cdot (v_2 - v_1 e^{i\tau})] + O(\eta^2)$$

$$\dot{v}_1 = \frac{i\eta}{4}[-(a + \gamma + \cos\tau)(v_2 e^{-i\tau} - v_1) + \sin\tau \cdot (u_2 e^{-i\tau} - u_1)] + O(\eta^2)$$

$$\dot{v}_2 = \frac{i\eta}{4}[-(a + \gamma + \cos\tau)(v_2 - v_1 e^{i\tau}) + \sin\tau \cdot (u_2 - u_1 e^{i\tau})] + O(\eta^2)$$

which is a linear system with coefficients periodic in τ of period 2π. The matrix of mean values, to the first order of approximation, is:

$$S = \begin{pmatrix} i(\gamma - a) & \frac{i}{2} & 0 & \frac{1}{2} \\ -\frac{i}{2} & i(a-\gamma) & \frac{1}{2} & 0 \\ 0 & \frac{1}{2} & i(a+\gamma) & -\frac{i}{2} \\ \frac{1}{2} & 0 & \frac{i}{2} & -i(a+\gamma) \end{pmatrix}$$

whose eigenvalues are the roots of

$$\lambda^4 + (2a^2 + 2\gamma^2 - 1)\lambda^2 + (a^2 - \gamma^2)^2 - \gamma^2 = 0.$$

For instability it is sufficient that the transformed equation in $\lambda^2 = \Lambda$ should not have both its roots negative, i.e. that one of the three following conditions hold:

(11.90) $$2a^2 + 2\gamma^2 - 1 < 0$$

(11.91) $$\text{or} \quad (a^2 - \gamma^2)^2 - \gamma^2 < 0$$

(11.92) $$\text{or} \quad (2a^2 + 2\gamma^2 - 1)^2 - 4(a^2 - \gamma^2)^2 + 4\gamma^2 < 0$$

conditions which we can express as:

$$\gamma^2 < \frac{1 - 2a^2}{2}, \qquad \text{cf. curve I}$$

$$\text{or} \quad \frac{2a^2 + 1 - \sqrt{4a^2 + 1}}{2} < \gamma^2 < \frac{2a^2 + 1 + \sqrt{4a^2 + 1}}{2}, \qquad \text{cf. curves II and III}$$

$$\text{or} \quad \gamma^2 < \frac{1}{4} - \frac{1}{16a^2}, \qquad \text{cf. curve IV}$$

A closer study of the graphs in the (γ^2, a^2) plane corresponding to the equations obtained by replacing an inequality sign in these conditions by an equality sign enables us to add a few details.

Noting in particular that the point

$$a^2 = \frac{1 + \sqrt{5}}{8}, \quad \gamma^2 = \frac{3 - \sqrt{5}}{8}$$

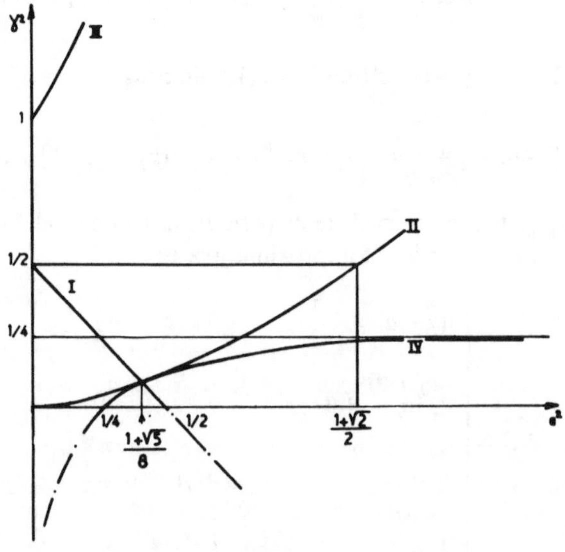

Figure 11.12

is common to the three curves

$$\gamma^2 = \frac{1-2a^2}{2}, \quad \gamma^2 = \frac{1}{4} - \frac{1}{16a^2}, \quad \gamma^2 = \frac{2a^2+1-\sqrt{4a^2+1}}{2},$$

the two latter being tangent, we can express the conditions for instability in the final form:

$$\gamma^2 < \frac{2a^2+1+\sqrt{4a^2+1}}{2}, \quad \text{if} \quad a^2 < \frac{1+\sqrt{5}}{8},$$

and $\gamma^2 < \dfrac{1}{4} - \dfrac{1}{16a^2}$, or else $\dfrac{2a^2+1-\sqrt{4a^2+1}}{2} < \gamma^2 < \dfrac{2a^2+1+\sqrt{4a^2+1}}{2}$,

if $a^2 > \dfrac{1+\sqrt{5}}{8}$.

III. Stability of Motion of a Rigid Rotor on Flexible Bearings. Gyroscopic Effects and Stability

Notation and Equations of Motion

1. We shall study the motion of the rotor relative to an orthonormal frame of reference $Oxyz$, in which Oz is the upward vertical, and the axis Ox passes through the geometrical centres A_1, A_2 of the bearings in their undeformed configuration.

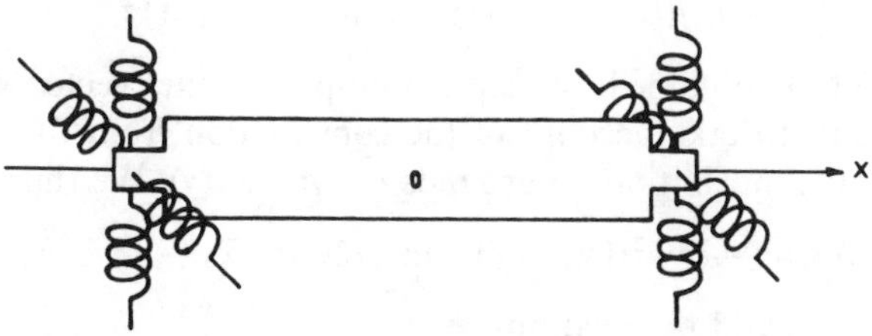

Figure 11.13

We shall suppose the rotor to be rigid but the bearings, which are subject to deformation, will be assumed to exert elastic restoring forces on the shaft which tend to bring the axis of the shaft back into coincidence with Ox. The moduli of rigidity of the bearings need not be the same in the y and z directions. We denote by $Oxvw$ the orthonormal set of axes whose instantaneous rotation with respect to $Oxyz$ in $\omega\vec{x}$ and $O_1x_1v_1w_1$ the orthonormal set of the principal axes of the rotor whose motion is assumed to be such that its centre of inertia O_1 coincides with O while $Ox_1v_1w_1$ departs little from $Oxvw$. Accordingly with respect to the moving axes $Oxvw$ the position of the rotor can be defined by three parameters φ, ψ, χ which represent respectively the three rotations specified by:

(i) $Oxvw \to O\xi\eta w$, which is a rotation of angle φ around Ow
(ii) $O\xi\eta w \to Ox_1\eta\zeta$, which is a rotation of angle $-\psi$ around $O\eta$
(iii) $Ox_1\eta\zeta \to Ox_1v_1w_1$ which is a rotation of angle χ around Ox_1.

On the assumption that these angles are all small, it is easy to see that we can write, to a first order of approximation:

$$\vec{\eta} - \vec{v} = \varphi\vec{w} \wedge \vec{v} = -\varphi\vec{x} = -\varphi\vec{x}_1$$
$$\vec{v}_1 - \vec{\eta} = \chi\vec{x}_1 \wedge \vec{\eta} = \chi\vec{\zeta} = \chi\vec{w}_1$$

so that
$$\vec{v} = \vec{v}_1 + \varphi\vec{x}_1 - \chi\vec{w}_1$$

and also
$$\vec{v}_1 = \vec{v} - \varphi\vec{x} + \chi\vec{w}.$$

Similar calculations allow us to evaluate $\vec{x}_1 - \vec{x}$, $\vec{w}_1 - \vec{w}$ so that finally we can write

$$\begin{aligned} \vec{x}_1 &= \vec{x} + \varphi\vec{v} + \psi\vec{w} \\ \vec{v}_1 &= \vec{v} - \varphi\vec{x} + \chi\vec{w} \\ \vec{w}_1 &= \vec{w} - \psi\vec{x} - \chi\vec{v} \end{aligned} \quad \text{or} \quad \begin{aligned} \vec{x} &= \vec{x}_1 - \varphi\vec{v}_1 - \psi\vec{w}_1 \\ \vec{v} &= \vec{v}_1 + \varphi\vec{x}_1 - \chi\vec{w}_1 \\ \vec{w} &= \vec{w}_1 + \psi\vec{x}_1 + \chi\vec{v}_1 \end{aligned} \tag{11.93}$$

to the first order in φ, ψ, χ: and the instantaneous angular velocity of the rotor with respect to $Oxvw$ is $\chi'\vec{x} - \psi'\vec{v} + \varphi'\vec{w}$.

2. We shall obtain the equations of motion by applying the theorem on angular momentum to the rotor at O. The moment $\vec{\mathcal{M}}_0$ at O of the restoring forces exerted by the bearings can be calculated as follows: with O as centre of A_1A_2, $2l = A_1A_2$, and k_1, k_2 the moduli of rigidity in the directions y and z respectively, the restoring force exerted by the bearing at $(l, 0, 0)$ is:

$$-k_1 l((\vec{x}_1 - \vec{x})\cdot\vec{y})\vec{y} - k_2 l((\vec{x}_1 - \vec{x})\cdot\vec{z})\vec{z}$$

and the moment is obtained by cross-multiplying the vector $l\vec{x}_1 \sim l\vec{x}$ and doubling the result to take account of the contribution from the other bearing (which is an equal contribution on grounds of symmetry). We thus find

$$\vec{\mathcal{M}}_0 = -2k_1l^2((\vec{x}_1 - \vec{x})\cdot\vec{y})\vec{z} + 2k_2l^2((\vec{x}_1 - \vec{x})\cdot\vec{z})\vec{y}$$

i.e. in view of (11.93) and the relations

$$\vec{y} = \vec{v}\cos\omega t - \vec{w}\sin\omega t, \quad \vec{z} = \vec{v}\sin\omega t + \vec{w}\cos\omega t: \tag{11.94}$$

$$\begin{aligned} \vec{\mathcal{M}}_0 = {} & l^2[-(k_1 - k_2)\sin 2\omega t\cdot\varphi + ((k_1 + k_2) + (k_2 - k_1)\cos 2\omega t)\psi]\vec{v} \\ & + l^2[-((k_1 + k_2) + (k_1 - k_2)\cos 2\omega t)\varphi + (k_1 - k_2)\sin 2\omega t\cdot\psi]\vec{w}. \end{aligned}$$

Let $\begin{pmatrix} I_1 & 0 & 0 \\ 0 & I_2 & 0 \\ 0 & 0 & I_3 \end{pmatrix}$ be the inertia tensor of the rotor with respect to the principal axes, with $I_3 > I_2$. It is clear from (11.93) that $\alpha + \beta\varphi + \gamma\psi$, $\beta - \alpha\varphi + \gamma\chi$, $\gamma - \alpha\psi - \beta\chi$ are the x_1, v_1, w_1 components of the vector $\alpha\vec{x} + \beta\vec{v} + \gamma\vec{w}$ and the invariant form (ellipsoid of inertia) associated with the inertia tensor is expressed by:

$$2I(\alpha,\beta,\gamma)=I_1(\alpha+\beta\varphi+\gamma\psi)^2+I_2(\beta-\alpha\varphi+\gamma\chi)^2+I_3(\gamma-\alpha\psi-\beta\chi)^2.$$

The partial derivatives $\frac{\partial I}{\partial\alpha},\frac{\partial I}{\partial\beta},\frac{\partial I}{\partial\gamma}$, calculated for $\alpha=\omega+\chi'$, $\beta=-\psi'$, $\gamma=\varphi'$, the principal values of the components on x, v, w of the angular velocity of the rotor, define the components on these same axes of the angular momentum $\vec{H}$ of the rotor at O, so that:

$$\text{(11.95)}\quad \vec{H}=I_1(\omega+\chi')\vec{x}-(I_2\psi'+(I_2-I_1)\omega\varphi)\vec{v}+(I_3\varphi'-(I_3-I_1)\omega\psi)\vec{w}$$

to a first-order approximation.

The equation of motion $\left(\frac{\mathrm{d}\vec{H}}{\mathrm{d}t}\right)_{xyz}=\vec{\mathscr{M}}_0+C\vec{x}$, where C is the driving torque which keeps the shaft revolving at angular velocity ω can be written:

$$\left(\frac{\mathrm{d}\vec{H}}{\mathrm{d}t}\right)_{xvw}+\omega\vec{x}\wedge\vec{H}=\vec{\mathscr{M}}_0+C\vec{x}$$

whence, with (11.94), (11.95) we obtain by projection on to the v, w axes the linearised equations governing the angular co-ordinates φ,ψ:

$$\text{(11.96)}\quad \begin{aligned}&-I_2\psi''-\omega(I_2-I_1)\varphi'-\omega(I_3\varphi'-(I_3-I_1)\omega\psi)\\&\quad=l^2[-(k_1-k_2)\sin 2\omega t\cdot\varphi+((k_1+k_2)+(k_2-k_1)\cos 2\omega t)\psi]\\&I_3\varphi''-\omega(I_3-I_1)\psi'-\omega(I_2\psi'+(I_2-I_1)\omega\varphi)\\&\quad=l^2[-((k_1+k_2)+(k_1-k_2)\cos 2\omega t)\varphi+(k_1-k_2)\sin 2\omega t\cdot\psi].\end{aligned}$$

We put:

$$\text{(11.97)}\quad \delta=\frac{I_3-I_2}{I_3+I_2},\quad \lambda=\frac{2I_1}{I_2+I_3},\quad \omega_0^2=2l^2\frac{k_1+k_2}{I_2+I_3},\quad \mu=\frac{k_2-k_1}{k_1+k_2},$$

noting that

$$\text{(11.98)}\quad 0<\delta<1,\quad 0<\lambda<2$$

We replace φ,ψ by the new variables:

$$\Phi=(1-\delta)^{-1/2}\varphi,\quad \Psi=(1+\delta)^{-1/2}\psi$$

and thus obtain, from (11.96):

$$\text{(11.99)}\quad \begin{aligned}&\Psi''+\frac{2-\lambda}{\sqrt{1-\delta^2}}\omega\Phi'+(\omega_0^2-(1+\delta-\lambda)\omega^2)\frac{\Psi}{1-\delta}\\&\quad+\mu\omega_0^2\left\{\frac{\Psi}{1-\delta}\cos 2\omega t+\frac{\Phi}{\sqrt{1-\delta^2}}\sin 2\omega t\right\}=0\\&\Phi''-\frac{2-\lambda}{\sqrt{1-\delta^2}}\omega\Psi'+(\omega_0^2-(1-\delta-\lambda)\omega^2)\frac{\Phi}{1+\delta}\\&\quad+\mu\omega_0^2\left\{\frac{\Psi}{\sqrt{1-\delta^2}}\sin 2\omega t-\frac{\Phi}{1+\delta}\cos 2\omega t\right\}=0\end{aligned}$$

where we see that the anisotropy of the suspension causes parametric excitation [24].

Analysis of Stability in the Isotropic Case

Suppose $\mu = 0$. The equations (11.97) reduce to a system with constant coefficients and gyroscopic term, for which it is easy to discuss the stability of its zero solution.

We shall note in advance that $1 + \delta - \lambda = 2\frac{I_3 - I_1}{I_2 + I_3}$, $1 - \delta - \lambda = 2\frac{I_2 - I_1}{I_2 + I_3}$ and we shall assume that $I_2 > I_1$, $I_3 > I_1$, (which is true for a long rotor), i.e. that

$$1 + \delta - \lambda > 1 - \delta - \lambda > 0. \tag{11.100}$$

For example, in the case of a homogeneous cylindrical rotor of radius R and length $2l$, this condition is equivalent to $2l > \sqrt{3}R$.

For $\mu = 0$ it is easily verified that the zero solution of (11.99) is stable on condition that $\frac{\omega^2}{\omega_0^2}$ lies outside the interval

$$[(1 + \delta - \lambda)^{-1}, (1 - \delta - \lambda)^{-1}].$$

Because of dissipatives forces, which are never completely absent, but which we have so far ignored, it is reasonable to suppose that the zero solution will be unstable if $\frac{\omega^2}{\omega_0^2}$ exceeds $(1 - \delta - \lambda)^{-1}$.

Thus for $\mu = 0$, the zone of stability for the rotor is defined by:

$$\frac{\omega^2}{\omega_0^2} < (1 + \delta - \lambda)^{-1} = \frac{I_2 + I_3}{2(I_3 - I_1)}. \tag{11.101}$$

If $\omega > \omega_0 \sqrt{\frac{I_2 + I_3}{2(I_3 - I_1)}}$, the zero solution will be unstable for $\mu \neq 0$, in a suitable neighbourhood of zero. Furthermore instability zones may appear in the frequency band $\omega \in \left]0, \sqrt{\frac{I_2 + I_3}{2(I_3 - I_1)}}\right[$, as we shall explain in more detail below. In this connection it is worth remarking that (11.99) is a reciprocal system because to every solution $\Phi(t)$, $\Psi(t)$ there obviously corresponds another solution $-\Phi(-t)$, $\Psi(-t)$.

Calculating the Critical Speeds of the Rotor

When there is parametric excitation of frequency 2ω, then resonances appear if $2\omega_1 = 0$, $2\omega_2 = 0$, $\omega_1 \pm \omega_2 \equiv 0 \bmod 2\omega$, where ω_1, ω_2 denote the eigenfrequencies of the gyroscopic system (11.99) when $\mu = 0$.

On the other hand it must not be forgotten that ω_1 and ω_2 depend on ω, so that it is first necessary to determine the values of ω which correspond to the above-mentioned conditions, these values having in any case to satisfy (11.101).

With p, q defined by:

$$\left(\frac{\omega_0}{\omega}\right)^2 - (1+\delta-\lambda) = 4(1-\delta)p \tag{11.102}$$
$$\left(\frac{\omega_0}{\omega}\right)^2 - (1-\delta-\lambda) = 4(1+\delta)q$$

and

$$h = \frac{2-\lambda}{2\sqrt{1-\delta^2}} > 0, \tag{11.103}$$

the equation which defines the eigenfrequencies ω_1, ω_2 is:

$$\det\begin{vmatrix} -\Omega^2 + 4p\omega^2 & 2ih\omega\Omega \\ -2ih\omega\Omega & -\Omega^2 + 4q\omega^2 \end{vmatrix} = 0$$

or

$$\Omega^4 - 4\omega^2(p+q+h^2)\Omega^2 + 16pq\omega^4 = 0, \tag{11.104}$$

which always has two positive roots which may conveniently be denoted by $\omega_1 < \omega_2$.

There is resonance $\omega = \frac{\omega_2 - \omega_1}{2}$ or $\omega = \frac{\omega_1 + \omega_2}{2}$ if and only if:

$$(\omega_1^2 + \omega_2^2 - 4\omega^2)^2 - 4\omega_1^2\omega_2^2 = 0. \tag{11.105}$$

But by (11.104) we have:

$$\omega_1^2 + \omega_2^2 = 4\omega^2(p+q+h^2) \tag{11.106}$$
$$\omega_1^2\omega_2^2 = 16pq\omega^4$$

and putting

$$\left(\frac{\omega_0}{\omega}\right)^2 = x, \; y = x + \lambda - 2 \tag{11.107}$$

and using (11.102), this becomes:

$$p+q = \frac{y}{2(1-\delta^2)} + \frac{1}{2}, \quad 16pq = \frac{(y+1)^2 - \delta^2}{1-\delta^2}, \quad p-q = \frac{y\delta}{2(1-\delta^2)}. \tag{11.108}$$

By substituting the expressions (11.106), calculated in terms of y, into the equation (11.105) we obtain:

$$f(y) = \frac{\delta^2}{4(1-\delta^2)^2}y^2 + \frac{h^2-1}{1-\delta^2}y + h^2(h^2-1) = 0. \tag{11.109}$$

To tell which resonance corresponds to which root of this last equation it suffices to notice that, by (11.105) we get the resonance $\omega = \frac{\omega_2 - \omega_1}{2}$ or $\omega = \frac{\omega_1 + \omega_2}{2}$ according to whether $\omega_1^2 + \omega_2^2 - 4\omega^2$ is positive or negative, or in other words, in view of (11.106), (11.107), (11.108) according to whether y is greater than or less than $(1-\delta^2)(1-2h^2)$.

In addition a necessary condition for ω to lie within the band of stability (11.101) is that

$$y > \delta - 1. \tag{11.110}$$

In what follows we shall assume that $\delta - 1 < (1-\delta^2)(1-2h^2)$ i.e. that

$$h^2 < \frac{2+\delta}{2(1+\delta)} \tag{11.111}$$

which is equivalent, in view of (11.103), to $\delta(\delta+1) < 2\lambda - \frac{\lambda^2}{2}$, an inequality which may be taken in conjunction with the assumption $\delta + \lambda < 1$. These two inequalities together imply in particular:

$$\delta < \frac{-2+\sqrt{13}}{3} \sim 0.5.$$

Let us add that by reason of $\delta + \lambda < 1$ we also have:

$$h = \frac{2-\lambda}{2\sqrt{1-\delta^2}} > \frac{1}{2}\sqrt{\frac{1+\delta}{1-\delta}} > 1/2 \quad \text{and by (11.111): } h < 1.$$

It follows that (11.109) has two real roots y_1, y_2 separated by 0. It is easily checked that $\delta - 1$ lies outside, while $(1-\delta^2)(1-2h^2)$ lies inside the interval y_1, y_2:

$$\delta - 1 < y_1 < 0 < y_2$$
$$y_1 < (1-\delta^2)(1-2h^2) < y_2.$$

To y_1 corresponds a rotational velocity ω_+, with resonance

$$\omega_+ = \frac{\omega_1(\omega_+) + \omega_2(\omega_+)}{2}$$

and to y_2 a velocity ω_-, $\omega_- < \omega_+$, with resonance

$$\omega_- = \frac{\omega_2(\omega_-) - \omega_1(\omega_-)}{2}.$$

To obtain the resonances $\omega = \omega_1$ or $\omega = \omega_2$, it suffices to write $\Omega = \omega$ in (11.104) which leads to:

$$1 - 4(p+q+h^2) + 16pq = 0 \tag{11.112}$$

from which in view of (11.108) we have $y = \pm 2h\sqrt{1-\delta^2}$.

The solution which corresponds to the negative sign is inadmissible since by (11.103) and (11.107) it leads to an infinite value of ω, and so we shall take $y = 2h\sqrt{1-\delta^2} = 2 - \lambda$, to which corresponds a value ω such that:

$$\left(\frac{\omega_0}{\omega}\right)^2 = 4h\sqrt{1-\delta^2} \tag{11.113}$$

inside the stability band (11.101).

If we write ω_* for the other eigenfrequency, which is with ω also a root of (11.104) we have

$$\omega_*^2 + \omega^2 = 4\omega^2(p + q + h^2),$$

and then by (11.108)

$$p + q = \frac{h}{\sqrt{1 - \delta^2}} + \frac{1}{2}$$

and thus $\omega_*^2 + \omega^2 > 2\omega^2$ or $\omega_*^2 > \omega^2$, which means that the resonance defined by (11.113) corresponds to $\omega = \omega_1$.

It remains to locate $y = 2h\sqrt{1 - \delta^2}$ in relation to y_1, y_2 the roots of (11.109). Now, using (11.103) we can calculate:

$$f(2h\sqrt{1 - \delta^2}) = \frac{h^2}{4(1 - \delta^2)(2 - \lambda)} \cdot (4 - \lambda)(8\delta^2 - 6\lambda + \lambda^2)$$

whose sign is that of $8\delta^2 - 6\lambda + \lambda^2$.

By examining the regions of the λ, δ plane where the inequalities

$$\delta(\delta + 1) - 2\lambda + \frac{\lambda^2}{2} < 0, \quad \delta + \lambda < 1, \quad \lambda > 0, \quad \delta > 0,$$

are simultaneously satisfied, it is easily verified that these conditions imply that $8\delta^2 - 6\lambda + \lambda^2$ must be negative. Thus we can write

$$y_1 < 2h\sqrt{1 - \delta^2} < y_2$$

and we conclude that as ω increases from zero through the stability band (11.101), we encounter successively the resonances $\omega = \dfrac{\omega_2 - \omega_1}{2}$, then $\omega = \omega_1$, and finally $\omega = \dfrac{\omega_1 + \omega_2}{2}$.

Resonant Instability Near $\boldsymbol{\omega = \dfrac{\omega_1 + \omega_2}{2}}$

We suppose ω to be in the vicinity of the value ω_+ which corresponds, when $\mu = 0$ to the resonance $\omega_+ = \dfrac{\omega_1 + \omega_2}{2}$. We have seen that

$$y_1 = \left(\frac{\omega_0}{\omega_+}\right)^2 + \lambda - 2$$

is the smallest negative root of (11.109). We put

$$\omega = \omega_+(1 + \mu\gamma)^{-1/2}, \tag{11.114}$$

where γ is a given parameter, which in a certain sense, measures the relative discrepancy $\dfrac{\omega-\omega_+}{\omega_+}$, and then $\dfrac{\mu}{4}\left(\dfrac{\omega_0}{\omega_+}\right)^2=\nu$ with p,q defined by (11.102) where $\omega=\omega_+$. With these notations and with the introduction of the variable $\tau=2\omega t$, the equations (11.99) can be written

$$\ddot\psi+h\dot\varphi+p\psi=\nu\left[-\frac{\gamma+\cos\tau}{1-\delta}\psi-\frac{\sin\tau}{\sqrt{1-\delta^2}}\varphi+O(\nu)\right]=\nu f \tag{11.115}$$

$$\ddot\varphi-h\dot\psi+q\varphi=\nu\left[\frac{\cos\tau-\gamma}{1+\delta}\varphi-\frac{\sin\tau}{\sqrt{1-\delta^2}}\psi+O(\nu)\right]=\nu g$$

the derivatives being taken with respect to τ, and reverting to the use of small letters for the variables φ,ψ. Taking into account the change of scaling factor for the time variable, the eigenfrequencies of the system (11.113) when $\nu=0$ are the positive roots σ_1,σ_2 of:

$$\sigma^4-(p+q+h^2)\sigma^2+pq=0 \tag{11.116}$$

and they satisfy the resonance condition:

$$\sigma_1+\sigma_2=1. \tag{11.117}$$

It will be noted that p and q are positive, as ω_+ belongs to the stability band (11.101) and we have

$$p<q \tag{11.118}$$

because of (11.108), since $y=y_1<0$.

Furthermore it is clear from (11.116) that

$$\sigma_1^2<p<q<\sigma_2^2. \tag{11.119}$$

We apply the method of variation of amplitude to (11.115) (on the hypothesis that σ_1,σ_2 are independent, i.e. σ_1/σ_2 is irrational). We take as starting point a representation of the solution of (11.115) in the form

$$\psi=\xi_1\sin\sigma_1\tau+\eta_1\cos\sigma_1\tau+\xi_2\sin\sigma_2\tau+\eta_2\cos\sigma_2\tau \tag{11.120}$$
$$\varphi=k(\xi_1\cos\sigma_1\tau-\eta_1\sin\sigma_1\tau)+r(\xi_2\cos\sigma_2\tau-\eta_2\sin\sigma_2\tau)$$

where

$$k=\frac{h\sigma_1}{q-\sigma_1^2}=\frac{p-\sigma_1^2}{h\sigma_1},\qquad r=\frac{p-\sigma_2^2}{h\sigma_2}=\frac{h\sigma_2}{q-\sigma_2^2}, \tag{11.121}$$

so that it satisfies (11.115) when $\nu=0$, for $\xi_1,\eta_1,\xi_2,\eta_2$ constants.

But in the case $\nu\neq0$ we shall assume, in accordance with the general theory that $\xi_1,\eta_1,\xi_2,\eta_2$ vary slowly with τ, and we impose the condition that the equations

$$\dot\psi=\xi_1\sigma_1\cos\sigma_1\tau-\eta_1\sigma_1\sin\sigma_1\tau+\xi_2\sigma_2\cos\sigma_2\tau-\eta_2\sigma_2\sin\sigma_2\tau \tag{11.122}$$
$$\dot\varphi=-k\sigma_1(\xi_1\sin\sigma_1\tau+\eta_1\cos\sigma_1\tau)-r\sigma_2(\xi_2\sin\sigma_2\tau+\eta_2\cos\sigma_2\tau)$$

should be satisfied, or in other words that:

(11.123)
$$\dot{\xi}_1 \sin\sigma_1\tau + \dot{\eta}_1 \cos\sigma_1\tau + \dot{\xi}_2 \sin\sigma_2\tau + \dot{\eta}_2 \cos\sigma_2\tau = 0$$
$$k(\dot{\xi}_1 \cos\sigma_1\tau - \dot{\eta}_1 \sin\sigma_1\tau) + r(\dot{\xi}_2 \cos\sigma_2\tau - \dot{\eta}_2 \sin\sigma_2\tau) = 0.$$

Calculating the second derivatives $\ddot{\psi}$, $\ddot{\varphi}$ from (11.122) and substituting in (11.115) we obtain:

(11.124)
$$\dot{\xi}_1\sigma_1 \cos\sigma_1\tau - \dot{\eta}_1\sigma_1 \sin\sigma_1\tau + \dot{\xi}_2\sigma_2 \cos\sigma_2\tau - \dot{\eta}_2\sigma_2 \sin\sigma_2\tau = vf$$
$$-k\sigma_1(\dot{\xi}_1 \sin\sigma_1\tau + \dot{\eta}_1 \cos\sigma_1\tau) - r\sigma_2(\dot{\xi}_2 \sin\sigma_2\tau + \dot{\eta}_2 \cos\sigma_2\tau) = vg.$$

We solve that set of equations (11.123), (11.124) with respect to the derivatives $\dot{\xi}_1, \dot{\eta}_1, \dot{\xi}_2, \dot{\eta}_2$ and after introducing the variables

(11.125)
$$\theta_1 = \sigma_1\tau, \quad \theta_2 = \sigma_2\tau,$$

and

(11.126)
$$\Delta = (k\sigma_2 - r\sigma_1)(r\sigma_2 - k\sigma_1),$$

we obtain after some calculation:

(11.127)
$$\dot{\xi}_1 = v\left[\frac{r\cos\theta_1}{r\sigma_1 - k\sigma_2}\cdot f - \frac{\sin\theta_1}{k\sigma_1 - r\sigma_2}\cdot g\right]$$
$$\dot{\xi}_2 = v\left[-\frac{k\cos\theta_2}{r\sigma_1 - k\sigma_2}\cdot f + \frac{\sin\theta_2}{k\sigma_1 - r\sigma_2}\cdot g\right]$$
$$\dot{\eta}_1 = v\left[-\frac{r\sin\theta_1}{r\sigma_1 - k\sigma_2}\cdot f - \frac{\cos\theta_1}{k\sigma_1 - r\sigma_2}\cdot g\right]$$
$$\dot{\eta}_2 = v\left[\frac{k\sin\theta_2}{r\sigma_1 - k\sigma_2}\cdot f + \frac{\cos\theta_2}{k\sigma_1 - r\sigma_2}\cdot g\right]$$
$$\dot{\theta}_1 = \sigma_1, \quad \dot{\theta}_2 = \sigma_2.$$

It remains to find more explicit expressions for f and g defined by (11.115) but it will be more convenient to write these in terms of the new variables introduced in (11.120) and (11.125). Accordingly, omitting the $O(v)$ terms, we obtain

(11.128)
$$f = -\left\{\left(\frac{\gamma + \cos(\theta_1+\theta_2)}{1-\delta}\sin\theta_1 + \frac{k\sin(\theta_1+\theta_2)}{\sqrt{1-\delta^2}}\cos\theta_1\right)\xi_1\right.$$
$$+\left(\frac{\gamma + \cos(\theta_1+\theta_2)}{1-\delta}\cos\theta_1 - \frac{k\sin(\theta_1+\theta_2)\sin\theta_1}{\sqrt{1-\delta^2}}\right)\eta_1$$
$$+\left(\frac{\gamma + \cos(\theta_1+\theta_2)}{1-\delta}\sin\theta_2 + \frac{r\sin(\theta_1+\theta_2)}{\sqrt{1-\delta^2}}\cos\theta_2\right)\xi_2$$
$$\left.+\left(\frac{\gamma + \cos(\theta_1+\theta_2)}{1-\delta}\cos\theta_2 - \frac{r\sin(\theta_1+\theta_2)}{\sqrt{1-\delta^2}}\sin\theta_2\right)\eta_2\right\}$$
$$g = -\left\{\left(k\frac{\gamma - \cos(\theta_1+\theta_2)}{1+\delta}\cos\theta_1 + \frac{\sin(\theta_1+\theta_2)}{\sqrt{1-\delta^2}}\sin\theta_1\right)\xi_1\right.$$

$$+\left(-k\frac{\gamma-\cos(\theta_1+\theta_2)}{1+\delta}\sin\theta_1+\frac{\sin(\theta_1+\theta_2)}{\sqrt{1-\delta^2}}\cos\theta_1\right)\eta_1$$
$$+\left(r\frac{\gamma-\cos(\theta_1+\theta_2)}{1+\delta}\cos\theta_2+\frac{\sin(\theta_1+\theta_2)}{\sqrt{1-\delta^2}}\sin\theta_2\right)\xi_2$$
$$\left.+\left(-r\frac{\gamma-\cos(\theta_1+\theta_2)}{1+\delta}\sin\theta_2+\frac{\sin(\theta_1+\theta_2)}{\sqrt{1-\delta^2}}\cos\theta_2\right)\eta_2\right\}.$$

To calculate the means of the right-hand sides of (11.127) with respect to θ_1, θ_2 we first carry out the averaging operation on the terms $f\cos\theta_1, \ldots$, thus obtaining

$$\overline{f\cos\theta_1} = -\frac{\gamma}{2(1-\delta)}\eta_1 - \frac{1}{4\sqrt{1-\delta^2}}\left(\sqrt{\frac{1+\delta}{1-\delta}}-r\right)\eta_2$$

$$\overline{f\sin\theta_1} = -\frac{\gamma}{2(1-\delta)}\xi_1 + \frac{1}{4\sqrt{1-\delta^2}}\left(\sqrt{\frac{1+\delta}{1-\delta}}-r\right)\xi_2$$

$$\overline{f\cos\theta_2} = -\frac{1}{4\sqrt{1-\delta^2}}\left(\sqrt{\frac{1+\delta}{1-\delta}}-k\right)\eta_1 - \frac{\gamma}{2(1-\delta)}\eta_2$$

$$\overline{f\sin\theta_2} = \frac{1}{4\sqrt{1-\delta^2}}\left(\sqrt{\frac{1+\delta}{1-\delta}}-k\right)\xi_1 - \frac{\gamma}{2(1-\delta)}\xi_2$$

$$\overline{g\cos\theta_1} = -\frac{k\gamma}{2(1+\delta)}\xi_1 - \frac{1}{4(1+\delta)}\left(\sqrt{\frac{1+\delta}{1-\delta}}-r\right)\xi_2$$

$$\overline{g\sin\theta_1} = \frac{k\gamma}{2(1+\delta)}\eta_1 - \frac{1}{4(1+\delta)}\left(\sqrt{\frac{1+\delta}{1-\delta}}-r\right)\eta_2$$

$$\overline{g\cos\theta_2} = -\frac{1}{4(1+\delta)}\left(\sqrt{\frac{1+\delta}{1-\delta}}-k\right)\xi_1 - \frac{r\gamma}{2(1+\delta)}\xi_2$$

$$\overline{g\sin\theta_2} = -\frac{1}{4(1+\delta)}\left(\sqrt{\frac{1+\delta}{1-\delta}}-k\right)\eta_1 + \frac{r\gamma}{2(1+\delta)}\eta_2$$

and we can now write down the equations derived from (11.127) by the averaging process, in the form

(11.129)
$$\begin{aligned}\dot{\xi}_1 &= \nu(-\gamma\alpha\eta_1 + a\eta_2)\\ \dot{\xi}_2 &= \nu(b\eta_1 + \gamma\beta\eta_2)\\ \dot{\eta}_1 &= \nu(\gamma\alpha\xi_1 + a\xi_2)\\ \dot{\eta}_2 &= \nu(b\xi_1 - \gamma\beta\xi_2)\end{aligned}$$

with

(11.130)
$$\begin{aligned}\alpha &= \frac{1}{2}\left[\frac{r}{(1-\delta)(r\sigma_1-k\sigma_2)} + \frac{k}{(1+\delta)(k\sigma_1-r\sigma_2)}\right]\\ \beta &= \frac{1}{2}\left[\frac{k}{(1-\delta)(r\sigma_1-k\sigma_2)} + \frac{r}{(1+\delta)(k\sigma_1-r\sigma_2)}\right]\end{aligned}$$

and

(11.131)
$$a=\frac{1}{4}\left(\sqrt{\frac{1+\delta}{1-\delta}}-r\right)\left(-\frac{r}{(r\sigma_1-k\sigma_2)\sqrt{1-\delta^2}}+\frac{1}{(1+\delta)(k\sigma_1-r\sigma_2)}\right)$$
$$b=\frac{1}{4}\left(\sqrt{\frac{1+\delta}{1-\delta}}-k\right)\left(\frac{k}{(r\sigma_1-k\sigma_2)\sqrt{1-\delta^2}}-\frac{1}{(1+\delta)(k\sigma_1-r\sigma_2)}\right).$$

The eigenvalues of the matrix of the system (11.129) are the roots of:

(11.132)
$$\lambda^4-2\lambda^2\left(ab-\gamma^2\frac{\alpha^2+\beta^2}{2}\right)+(ab+\gamma^2\alpha\beta)^2=0,$$

and the null solution will be unstable, for small enough $|v|$ if the equation (11.132) considered as an equation in λ^2, does not have both its roots real and negative, i.e. if:

$$ab-\gamma^2\frac{\alpha^2+\beta^2}{2}>0,\quad\text{or if}\quad\left(ab-\gamma^2\frac{\alpha^2+\beta^2}{2}\right)^2-(ab+\gamma^2\alpha\beta)^2<0;$$

conditions which lead to

(11.133)
$$\gamma^2<\frac{4ab}{(\alpha-\beta)^2}$$

provided that $ab>0$, which as we shall show later is in fact true.

We can calculate ab from (11.131) but it is clear from (11.121) and (11.119) that $k>0$, $r<0$ and:

(11.134)
$$r\sigma_1-k\sigma_2=p\frac{\sigma_1^2-\sigma_2^2}{h\sigma_1\sigma_2}<0$$
$$k\sigma_1-r\sigma_2=\frac{\sigma_2^2-\sigma_1^2}{h}>0$$

so that the sign of ab is the same as that of:

$$c=\frac{1}{4}\left(k-\sqrt{\frac{1+\delta}{1-\delta}}\right)\cdot\left(\frac{1}{(1+\delta)(k\sigma_1-r\sigma_2)}-\frac{r}{(r\sigma_1-k\sigma_2)\sqrt{1-\delta^2}}\right).$$

Now by (11.121) and (11.116) we have:

(11.135)
$$kr=\frac{(p-\sigma_1^2)(p-\sigma_2^2)}{k^2\sigma_1\sigma_2}=-\sqrt{\frac{p}{q}}$$

and by (11.134):

(11.136)
$$\frac{r\sigma_1-k\sigma_2}{k\sigma_1-r\sigma_2}=-\frac{p}{\sigma_1\sigma_2}=-\sqrt{\frac{p}{q}}$$

and hence

(11.137)
$$c=\frac{1}{4k(k\sigma_1-r\sigma_2)(1+\delta)}\cdot\left(k-\sqrt{\frac{1+\delta}{1-\delta}}\right)^2\geqslant 0.$$

To complete the proof (that $ab > 0$), let d be such that $ab = cd$, so that:

$$d = \frac{1}{4}\left(\sqrt{\frac{1+\delta}{1-\delta}} - r\right)\left(-\frac{k}{(r\sigma_1 - k\sigma_2)\sqrt{1-\delta^2}} + \frac{1}{(1+\delta)(k\sigma_1 - r\sigma_2)}\right)$$

and by (11.135), (11.136):

$$d = -\frac{1}{4r(k\sigma_1 - r\sigma_2)(1+\delta)} \cdot \left(\sqrt{\frac{1+\delta}{1-\delta}} - r\right)^2.$$

Then we have

(11.138)

$$ab = -(16kr)^{-1}(k\sigma_1 - r\sigma_2)^{-2}(1+\delta)^{-2}\left(\sqrt{\frac{1+\delta}{1-\delta}} - r\right)^2\left(k - \sqrt{\frac{1+\delta}{1-\delta}}\right)^2.$$

Note that

$$k^2 = \frac{(p-\sigma_1^2)^2}{h^2\sigma_1^2} = 1 + \frac{(p-q)(p-\sigma_1^2)}{h^2\sigma_1^2}$$

by (11.116), and $k^2 < 1$ since $p - q < 0$, and it follows that ab is strictly positive.

Let us now calculate $\alpha - \beta$. By (11.130), (11.135), (11.136) we have:

(11.139)
$$\alpha = \frac{1}{2(k\sigma_1 - r\sigma_2)}\left(\frac{1}{k(1-\delta)} + \frac{k}{1+\delta}\right)$$

$$\beta = \frac{1}{2(k\sigma_1 - r\sigma_2)}\left(\frac{1}{r(1-\delta)} + \frac{r}{1+\delta}\right)$$

whence

$$\alpha - \beta = \frac{k-r}{2(k\sigma_1 - r\sigma_2)}\left(\frac{1}{1+\delta} - \frac{1}{kr(1-\delta)}\right) > 0$$

and the zone of instability is defined by:

(11.140)

$$\gamma^2 < \frac{4ab}{(\alpha-\beta)^2} = -\frac{1}{kr(k-r)^2(1+\delta)^2} \cdot \frac{\left(\sqrt{\frac{1+\delta}{1-\delta}} - r\right)^2\left(r - \sqrt{\frac{1+\delta}{1-\delta}}\right)^2}{\left(\frac{1}{1+\delta} - \frac{1}{kr(1-\delta)}\right)^2}.$$

We deduce from (11.121) that:

$$k - r = \frac{(p + \sigma_1\sigma_2)(\sigma_2 - \sigma_1)}{h\sigma_1\sigma_2}$$

or with $\qquad \sigma_1\sigma_2 = \sqrt{pq},\ \sigma_1 + \sigma_2 = 1,\ \sigma_2 - \sigma_1 = (1 - 4\sqrt{pq})^{1/2},$

$$k - r = \frac{1}{h}\left(\sqrt{\frac{p}{q}} + 1\right)(1 - 4\sqrt{pq})^{1/2}.$$

Similarly we find:

$$k + r = \frac{p - \sigma_1\sigma_2}{h\sigma_1\sigma_2} = \frac{1}{h}\left(\sqrt{\frac{p}{q}} - 1\right)$$

and we can now make explicit (11.140):

$$\gamma^* = \frac{2\sqrt{ab}}{|\alpha - \beta|} = \frac{\left(h\dfrac{1+\delta}{1-\delta} - \left(\sqrt{\dfrac{p}{q}} - 1\right)\sqrt{\dfrac{1+\delta}{1-\delta}} - h\sqrt{\dfrac{p}{q}}\right)\left(\dfrac{p}{q}\right)^{1/4}}{\left(\sqrt{\dfrac{p}{q}} + 1\right)(1 - 4\sqrt{pq})^{1/2}\left(\dfrac{1+\delta}{1-\delta} + \sqrt{\dfrac{p}{q}}\right)},$$

which gives the zone of instability defined to the 1st order by $|\gamma| < \gamma^*$. We recall that p, q are calculated from (11.102), with $\omega = \omega_+$ associated by (11.107) with the negative root y_1 of (11.109).

Instability Near the Resonance $\omega = \omega_1$

Let us remember that this corresponds to the value $y = 2h\sqrt{1-\delta^2} = \left(\frac{\omega_0}{\omega_1}\right)^2 + \lambda - 2$. We put $\omega = \omega_1(1 + \mu\gamma)^{1/2}$, $\frac{\mu}{4}\left(\frac{\omega_0}{\omega_1}\right)^2 = \nu$, with p, q as before being defined by (11.102) with $\omega = \omega_1$, or in other words:

$$p = \frac{y + 1 - \delta}{4(1-\delta)}, \quad q = \frac{y + 1 + \delta}{4(1+\delta)}, \quad 0 < q < p. \tag{11.141}$$

We introduce $\tau = 2\omega t$ into the equations (11.99) which thus assume a form identical to that of (11.115). The eigenfrequencies of the new system (11.115) for $\nu = 0$ are however, taking into account the scale-change of the time variable, the positive roots of $\sigma^4 - (h^2 + p + q)\sigma^2 + pq = 0$, or by (11.112) $\sigma_1 = 1/2$ and $\sigma_2 > \sigma_1$ Also we have

$$\sigma_1^2 < q < p < \sigma_2^2. \tag{11.142}$$

With these new values of σ_1, σ_2 we make the change of variable defined by (11.120), satisfying (11.121), (11.122), (11.123) where $\theta_1 = \sigma_1\tau = \frac{\tau}{2}$, $\theta_2 = \sigma_2\tau$ replace (11.125), and we arrive at the system (11.127), where the functions f and g, still defined by (11.115), but expressed in terms of the new variables θ_1, θ_2 are

$$\begin{aligned} f = -\Bigg\{&\left(\frac{\gamma + \cos 2\theta_1}{1-\delta}\sin\theta_1 + \frac{k\sin 2\theta_1\cos\theta_1}{\sqrt{1-\delta^2}}\right)\xi_1 \\ &+ \left(\frac{\gamma + \cos 2\theta_1}{1-\delta}\cos\theta_1 - \frac{k\sin 2\theta_1\sin\theta_1}{\sqrt{1-\delta^2}}\right)\eta_1 \\ &+ \left(\frac{\gamma + \cos 2\theta_1}{1-\delta}\sin\theta_2 + \frac{r\sin 2\theta_1}{\sqrt{1-\delta^2}}\cos\theta_2\right)\xi_2 \end{aligned} \tag{11.143}$$

$$+\left(\frac{\gamma+\cos 2\theta_1}{1-\delta}\cos\theta_2-\frac{r\sin 2\theta_1\sin\theta_2}{\sqrt{1-\delta^2}}\right)\eta_2\Bigg\}$$

$$g=-\Bigg\{\left(k\frac{\gamma-\cos 2\theta_1}{1+\delta}\cos\theta_1+\frac{\sin 2\theta_1}{\sqrt{1-\delta^2}}\sin\theta_1\right)\xi_1$$

$$+\left(-k\frac{\gamma-\cos 2\theta_1}{1+\delta}\sin\theta_1+\frac{\sin 2\theta_1}{\sqrt{1-\delta^2}}\cos\theta_1\right)\eta_1$$

$$+\left(r\frac{\gamma-\cos 2\theta_1}{1+\delta}\cos\theta_2+\frac{\sin 2\theta_1}{\sqrt{1-\delta^2}}\sin\theta_2\right)\xi_2$$

$$+\left(-r\frac{\gamma-\cos 2\theta_1}{1+\delta}\sin\theta_2+\frac{\sin 2\theta_1}{\sqrt{1+\delta^2}}\cos\theta_2\right)\eta_2\Bigg\}.$$

We deduce from these expressions the mean values:

$$\overline{f\cos\theta_1}=-\left(\frac{\gamma}{2(1-\delta)}-\frac{1}{4\sqrt{1-\delta^2}}\left(k-\sqrt{\frac{1+\delta}{1-\delta}}\right)\right)\eta_1$$

$$\overline{f\sin\theta_1}=-\left(\frac{\gamma}{2(1-\delta)}+\frac{1}{4\sqrt{1-\delta^2}}\left(k-\sqrt{\frac{1+\delta}{1-\delta}}\right)\right)\xi_1$$

$$\overline{f\cos\theta_2}=-\frac{\gamma}{2(1-\delta)}\eta_2$$

$$\overline{f\sin\theta_2}=-\frac{\gamma}{2(1-\delta)}\xi_2$$

$$\overline{g\cos\theta_1}=-\frac{1}{2(1+\delta)}\left(k\gamma-\frac{1}{2}\left(k-\sqrt{\frac{1+\delta}{1-\delta}}\right)\right)\xi_1$$

$$\overline{g\sin\theta_1}=\frac{1}{2(1+\delta)}\left(k\gamma+\frac{1}{2}\left(k-\sqrt{\frac{1+\delta}{1-\delta}}\right)\right)\eta_1$$

$$\overline{g\cos\theta_2}=-\frac{r\gamma}{2(1+\delta)}\xi_2$$

$$\overline{g\sin\theta_2}=\frac{r\gamma}{2(1+\delta)}\eta_2$$

and from (11.127) the "averaged" equations, in the form:

$$\begin{aligned}\dot\xi_1&=\nu(-c-\alpha\gamma)\eta_1\\ \dot\xi_2&=\nu\beta\gamma\eta_2\\ \dot\eta_1&=\nu(-c+\alpha\gamma)\xi_1\\ \dot\eta_2&=-\nu\beta\gamma\xi_2\end{aligned}\tag{11.144}$$

in which α, β are still defined by (11.130) or (11.139) and c by (11.137), with of course the specific values of k, r, σ_1, σ_2.

The eigenvalues of the matrix associated with the system (11.144) are the roots of

$$(\lambda^2+\beta^2\gamma^2)(\lambda^2+\alpha^2\gamma^2-c^2)=0.$$

However we have, from (11.121) with $\sigma_1 = 1/2$:

$$k=\frac{p-\sigma_1^2}{h\sigma_1}=\frac{1}{2h}(4p-1)=\sqrt{\frac{1+\delta}{1-\delta}} \tag{11.145}$$

taking into account (11.141) where $y=2h\sqrt{1-\delta^2}$.

Furthermore we can see from (11.120), (11.142) that $r<0$ and we can therefore guarantee after (11.137) that $c=0$.

Thus the equation (11.145) has all its roots pure imaginaries and we can conclude from this that the width of the zone of instability corresponding to the resonance $\omega=\omega_1$ is of at least the second order in μ.

The calculation leads to an analogous result for the resonance $\omega=\frac{\omega_2-\omega_1}{2}$; accordingly the resonances $\omega=\frac{\omega_2-\omega_1}{2}$, $\omega=\omega_1$ will be hardly noticeable in contrast to the one corresponding to $\omega=\frac{\omega_1+\omega_2}{2}$ which can be termed the principal resonance.

Ground Resonance of the Helicopter Blade Rotor System

1. We can describe the rotary motion of the blades of a helicopter on the ground, on the following assumptions:

a) The blades move with a zero angle of incidence, the motion being in a horizontal plane O_1xy forming part of an orthonormal set of axes of which O_1z is the upward vertical.

b) The system consists of $q(q=2, 3 \text{ or } 4)$ identical blades of mass m, identified by the suffix j, $1\leqslant j\leqslant q$, and treated as a segment A_jB_j with centre of gravity G_j, hinged at A_j to an arm OA_j which forms an integral part of the rotor whose centre is at O. These rigid arms, assumed to be non-deformable, are uniformly spaced, i.e. the angle between $\overrightarrow{OA_j}$ and $\overrightarrow{OA_{j+1}}$ is $\frac{2\pi}{q}$. We introduce the angular variables $\theta_j=(\overrightarrow{OA_j}, \overrightarrow{A_jB_j})$ to specify the angle between the blade and its arm, and we shall assume that the hinge exerts an elastic restoring couple of moment $-C\theta_j$ which tends to bring the two parts (i.e. blade and arm) back into alignment.

c) Elastic deformations of the structure of which the rotor bearings form an integral part are taken into account. We shall effectively assume that these deformations give rise to stresses which are equivalent to a single restoring force exerted on the rotor at its centre O whose components are $-K_1x$, $-K_2y$ on O_1xy, where x, y are the co-ordinates of O and K_1, K_2 the moduli of rigidity in the x, y directions respectively.

d) The rotor of mass m_1 revolves with an uniform angular velocity ω. This description [25] which leads to a system having $q+2$ degrees of freedom will serve as a basis for establishing linearised equations of motion which will enable us to analyse the instabilities of the system.

2. Let us first apply the theorem on the centre of inertia to the whole system. With $OA_j = a$, $A_jG_j = b$, $\alpha_j = \dfrac{2\pi j}{q}$ and $\xi = x + iy$, the position of the centre of inertia G_j of the jth blade is represented by the complex number

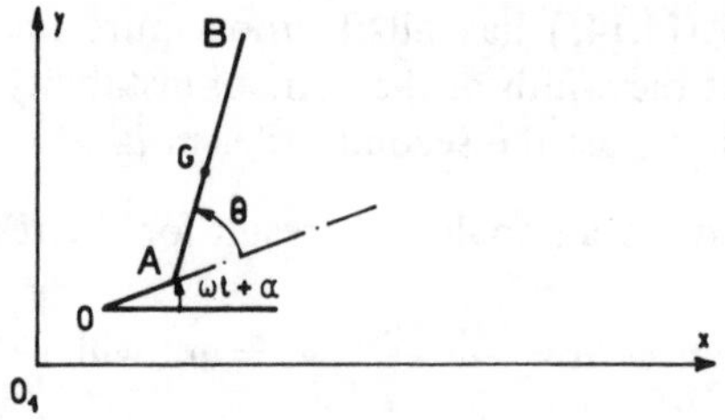

Figure 11.14

$$\xi_j = \xi + a e^{i(\omega t + \alpha_j)} + b e^{i(\omega t + \alpha_j + \theta_j)} \tag{11.146}$$

and the acceleration of this same point by:

$$\xi_j'' = \xi'' - a\omega^2 e^{i(\omega t + \alpha_j)} + b(i\theta_j'' - (\omega + \theta_j')^2) e^{i(\omega t + \alpha_j + \theta_j)}. \tag{11.147}$$

The rate of change of momentum is therefore:

$$m_1 \xi'' + \sum_{j=1}^{q} m\xi_j'' = (qm + m_1)\xi'' + mb \sum_{j=1}^{q} (i\theta_j'' - (\omega + \theta_j')^2) e^{i(\omega t + \alpha_j + \theta_j)}$$

or after linearisation:

$$m_1 \xi'' + \sum_{j=1}^{q} m\xi_j'' = M\xi'' + imb \sum_{j=1}^{q} \left(\frac{d}{dt} + i\omega\right)^2 \theta_j \cdot e^{i(\omega t + \alpha_j)}$$

where $M = qm + m_1$ and having regard to $\sum_{j=1}^{q} e^{i\alpha_j} = 0$.

The elastic restoring force being given by

$$-K_1 x - iK_2 y = -K\xi - \Delta K \cdot \bar{\xi} \quad \text{where} \quad K = \frac{K_1 + K_2}{2}, \Delta K = \frac{K_1 - K_2}{2}$$

we obtain, with the new variable:

$$\zeta = \xi e^{-i\omega t}, \tag{11.148}$$

the equation

$$\left(\frac{d}{dt} + i\omega\right)^2 \zeta + i\frac{m}{M} b \sum_{j=1}^{q} \left(\frac{d}{dt} + i\omega\right)^2 \theta_j e^{i\alpha_j} + \frac{K}{M}\zeta + \frac{\Delta K}{M}\bar{\zeta} e^{-2i\omega t} = 0. \tag{11.149}$$

We now apply the angular momentum theorem to each blade. Giving up the suffix notation for the moment, we can express the angular momentum of the blade AB at A, in its absolute motion by:

$$\overrightarrow{H_A} = \overrightarrow{H_G} + m\overrightarrow{AG} \wedge \overrightarrow{V_G}.$$

Now we know that $\overrightarrow{H_G} = I(\theta' + \omega)\vec{z}$, where I is the moment of inertia of the blade with respect to the axis Gz, and we calculate without difficulty:

$$m\overrightarrow{AG} \wedge \overrightarrow{V_G} = m\,\mathrm{Im}\left\{be^{-i(\omega t+\alpha+\theta)}.\frac{\mathrm{d}\xi}{\mathrm{d}t} + i\omega abe^{-i\theta} + i(\omega+\theta')b^2\right\}\vec{z}.$$

with $AG = b$, or after linearisation:

$$m\overrightarrow{AG} \wedge \overrightarrow{V_G} = m\,\mathrm{Im}\left\{b\frac{\mathrm{d}\xi}{\mathrm{d}t}e^{-i(\omega t+\alpha)} + i\omega ab(1-i\theta) + i(\omega+\theta')b^2\right\}\vec{z}.$$

Thus

$$\frac{\mathrm{d}\overrightarrow{H_A}}{\mathrm{d}t} = [(I+mb^2)\theta'' + mb\,\mathrm{Im}\,(e^{-i(\omega t+\alpha)}.\xi')']\vec{z}$$

and by the angular momentum theorem:

$$\frac{\mathrm{d}\overrightarrow{H_A}}{\mathrm{d}t} = \overrightarrow{M_A} + m\overrightarrow{V_G} \wedge \overrightarrow{V_A} \tag{11.150}$$

where $\overrightarrow{M_A}$ is the moment at A of the external forces acting on the blade, i.e. $\overrightarrow{M_A} = -C\theta\vec{z}$.

We also find without difficulty that:

$$\overrightarrow{V_A} \wedge m\overrightarrow{V_G} = m[ab\omega^2\theta + b\,\mathrm{Im}\,(i\omega e^{-i(\omega t+\alpha)}\xi')]\vec{z}$$

and we are thus in a position to express (11.150) in explicit terms.

We shall introduce, as we did earlier, $\zeta = \xi e^{-i\omega t}$, so that the corresponding equation for the jth blade becomes:

$$(I+mb^2)\theta_j'' + (C+mab\omega^2)\theta_j + mb\,\mathrm{Im}\left[e^{-i\alpha_j}\left(\frac{\mathrm{d}}{\mathrm{d}t}+i\omega\right)^2\zeta\right] = 0. \tag{11.151}$$

The combination of (11.149) and (11.151) gives us the $q+2$ differential equations allowing us to describe the oscillations of the system. They are linear equations with periodic coefficients of period π/ω. It can be shown that they constitute a reciprocal system, and thanks to this property we shall be able to discuss the possible resonances for small ΔK.

To establish the reciprocity property we first construct a real $q \times q$ matrix, say S, such that:

$$S\cdot(e^{i\alpha_1},\ldots,e^{i\alpha_q})^T = (e^{-i\alpha_1},\ldots,e^{-i\alpha_q})^T. \tag{11.152}$$

It is easy to check, because of the definition $\alpha_j = 2\pi \frac{j}{q}$ that this condition can be satisfied with

$$S = -\begin{pmatrix} 1 & 1 & \cdots & 1 & 0 & 1 \\ 1 & 1 & \cdots & 0 & 1 & 1 \\ \cdots & \cdots & \cdots & \cdots & \cdots & \cdots \\ 0 & 1 & \cdots & 1 & 1 & 1 \\ 1 & 1 & \cdots & 1 & 1 & 0 \end{pmatrix}$$

or $s_{rj} = 0$ if $r + j = q$, $s_{qq} = 0$ and $s_{rj} = -1$ for all other elements.

This matrix is invertible as is easily verified directly since $S\theta = 0$ implies $\theta = 0$. Besides which the determinant of S (a circulant) is easily calculated and we find $\det S = (1-q)(-1)^q \neq 0$.

However we can also verify that to every solution $\zeta(t)$, $\theta(t) = \{\theta_1(t), \ldots, \theta_q(t)\}$ of the equations (11.149), (11.151) corresponds $-\bar{\zeta}(-t)$, $S\theta(-t)$ which is also a solution of the same system, and this establishes the property of reciprocity.

To investigate the nature of the instability of the mechanical system we shall begin by calculating the eigenvalues $\omega_j(\omega)$, $0 \leqslant j \leqslant q+2$ of the system with constant coefficients obtained for $\Delta K = 0$; they depend on ω and the stability condition requires that they should all be real, with corresponding limitations for ω.

With this proviso and if ΔK is not too large we can predict that the motion will be stable except for values of ω in the vicinity of the resonant frequencies which we can calculate by solving the equations $\omega_j(\omega) = 0 \bmod \omega$, $\omega_j(\omega) \pm \omega_l(\omega) = 0 \bmod 2\omega$, $1 \leqslant j \leqslant q+2$, $1 \leqslant l \leqslant q+2$.

IV. Whirling Motion of a Shaft in Rotation with Non-Linear Law of Physical Behaviour

The shaft in its undeformed state is a cylinder of revolution of radius r supported by two circular bearings of centres O_1, O_1' at a distance of $2l$ apart. We denote by S_1xyz the orthonormal frame of reference in which the axis S_1x lies along the line O_1O_1'; S_1 is the midpoint of this line and S_1z is the upward vertical.

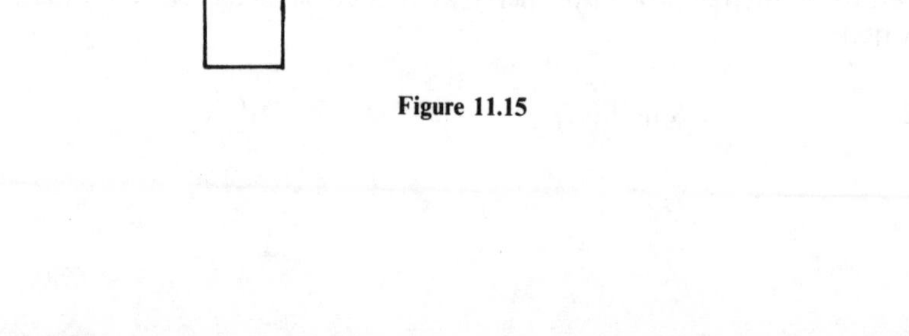

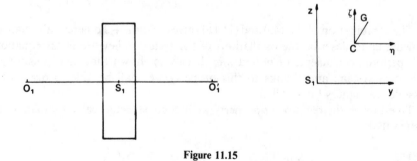

Figure 11.15

The shaft carries at its centre a rigidly attached disc of mass m and the whole revolves with an uniform angular velocity represented by the rotation vector $\omega\vec{x}$. Under the effect of inertial forces the shaft can undergo deformation which we assume to be flexural, discarding by hypothesis any torsion or elongation.

Because of the symmetry of the whole configuration with respect to the vertical plane S_1yz, it is unnecessary to take account of gyroscopic effects of the disc and the motion can be described completely with the help of the co-ordinates y, z of C, the point of intersection with the plane S_1yz of the neutral axis of the shaft, i.e. of the line joining the centres of the circular sections. The latter is a plane curve, which is the image under deformation of the axis $O_1S_1O_1'$. With y, z as co-ordinates of C in the plane S_1yz, we write

(11.153) $$\rho^2 = y^2 + z^2, \quad \varphi = \operatorname{Arctan}\frac{z}{y}$$

so that φ is the angle of precession and describes the whirling or orbital motion.

The bent shaft exerts on the disc a force $-\vec{T}$ which, by symmetry, lies in the plane S_1yz. It may happen that the disc is imperfectly balanced; its centre of inertia G is then distinct from C and $CG = e$ is the eccentricity, which in any case will be small.

Noting that $(\overrightarrow{S_1y}, \overrightarrow{CG}) = \omega t$ and applying the theorem on the motion of the centre of inertia to the disc we obtain:

(11.154) $$\begin{aligned} m(y'' - e\omega^2 \cos\omega t) &= -T_y \\ m(z'' - e\omega^2 \sin\omega t) &= -T_z \end{aligned}$$

where T_y, T_z are the components of $\vec{T}$ on S_1yz.

Neglecting the mass of the shaft, it is clear that $-\dfrac{\vec{T}}{2}$ represents the stress extерted by the bearings, and denoting by $\vec{\mathcal{M}}$ the moment at C of the forces exerted through the cross-section $x = 0$ by the material elements of the shaft contained in the half-space $x > 0$ on these in $x < 0$, we can write

$$\vec{\mathcal{M}} - \overrightarrow{CO_1} \wedge \frac{\vec{T}}{2} = 0$$

or

(11.155) $$\vec{\mathcal{M}} = \tfrac{1}{2}(zT_y - yT_z)\vec{x} + \frac{l}{2}T_z\vec{y} - \frac{l}{2}T_y\vec{z}.$$

If $\mathcal{D}$ denotes the cross-section of the shaft in the plane $x = 0$ we can express $\vec{\mathcal{M}}$ by

(11.156) $$\vec{\mathcal{M}} = \iint_{\mathcal{D}} (\overrightarrow{CP} \wedge \tau\vec{x})\, d\eta\, d\zeta$$

τ being the component of the constraint force per unit area in the direction normal to the cross-section.

The components of $\vec{\mathcal{M}}$ on S_1x is small, by (11.155), because so is the deflection of the shaft. Comparing (11.155) and (11.156) we obtain:

(11.157) $$T_y = \frac{2}{l}\iint_{\mathscr{D}} \eta\tau \, d\eta \, d\zeta, \quad T_z = \frac{2}{l}\iint_{\mathscr{D}} \zeta\tau \, d\eta \, d\zeta.$$

Following [32] we shall assume that the constitutive law is:

(11.158) $$\tau = E\varepsilon + H\dot{\varepsilon} + f_1(\varepsilon) + f_2(\varepsilon, \dot{\varepsilon})$$

where ε denotes the rate of elongation, and $\dot{\varepsilon}$ the particular derivative of ε. We shall calculate ε on the basis of the linear approximation, which is reasonable considering that the terms other than $E\varepsilon$ in (11.158) involve small perturbations only.

For a fibre intersecting the Cvw plane in the point P of co-ordinates v, w with respect to the frame of reference Cvw, in which the axis Cv is in the direction S_1C, the extension or relative elongation is $\varepsilon = -\frac{v}{R}$, where R is the radius of curvature of the neutral line at C. In accordance with the elementary theory on the bending of beams (Section I of this chapter), the deflection $\delta(x)$ along the neutral line is represented in each interval $[-l, 0]$ and $[0, l]$ by polynomials of degree 3 in x whose coefficients can be obtained simply by expressing the kinematic conditions $\delta(\pm l) = 0$, the static conditions (zero moment at the bearings) $\frac{d^2}{dx^2}\delta(\pm l) = 0$, and finally the boundary conditions $\delta(0) = \rho$, $\frac{d}{dx}\delta(0) = 0$. Accordingly we have, for $x \in [-l, 0]$

$$\delta(x) = -\frac{\rho}{2l^3}(x + l)^3 + \frac{3\rho}{2l}(x + l)$$

and in particular, as the radius of curvature at C has the value $R = \left(\frac{d^2}{dx^2}\delta(0)\right)^{-1} = -\frac{l^2}{3\rho}$ it follows that:

(11.159) $$\varepsilon = \frac{3v\rho}{l^2}.$$

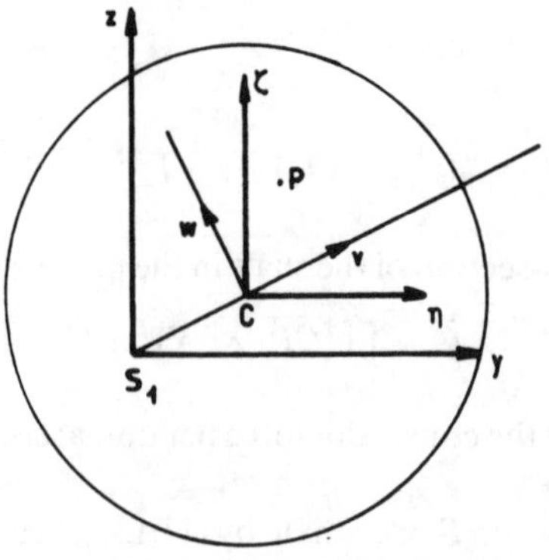

Figure 11.16

Calculation of T_y, T_z

We use (11.157), (11.158) and (11.159).

1. Contribution from $E\varepsilon$
We find

$$T_y = ky \quad T_z = kz \quad \text{with} \quad k = \frac{6EI}{l^3} \quad \text{and} \quad I = \frac{\pi r^4}{4} \tag{11.160}$$

the geometric moment of inertia of the section with respect to a diameter.

2. Contribution from $f_1(\varepsilon)$
We obtain

$$T_y = \Phi(\rho)\cdot y \quad T_z = \Phi(\rho)\cdot z \tag{11.161}$$

with
$$\Phi(\rho) = \frac{4l^5}{27\rho^4}\int_{-h}^{+h} \varepsilon\sqrt{h^2-\varepsilon^2}\, f_1(\varepsilon)\,\mathrm{d}\varepsilon, \quad h = \frac{3r\rho}{l^2}.$$

3. Contribution from $f_2(\varepsilon, \dot{\varepsilon})$
To calculate the particular derivative $\dot{\varepsilon}$, of the relative elongation at the point $P \in \mathscr{D}$, we have, under the usual assumptions of beam theory, to consider the point P as being rigidly bound to the cross-section $\mathscr{D}$. Since the angular velocity of $\mathscr{D}$ relative to Cvw is $\omega - \varphi'$, we can thus calculate

$$\dot{v} = -(\omega - \varphi')w$$

and by (11.159)

$$\dot{\varepsilon} = \frac{3\dot{v}\rho}{l^2} + \frac{3v\dot{\rho}}{l^2} = -\frac{3\rho}{l^2}(\omega - \varphi')w + \frac{3v\rho'}{l^2}.$$

The result, after some calculation, is that we find for the contribution from $f_2(\varepsilon, \dot{\varepsilon})$ to T_y, T_z

$$\begin{aligned} T_y &= yF_1(\rho, \rho', \omega - \varphi') - zF_2(\rho, \rho', \omega - \varphi') \\ T_z &= yF_2(\rho, \rho', \omega - \varphi') + zF_1(\rho, \rho', \omega - \varphi') \end{aligned} \tag{11.162}$$

$$F_j = \frac{2l^5}{27\rho^4}\int_{-h}^{+h} \mathrm{d}\zeta_1 \int_{-(h^2-\zeta_1^2)^{1/2}}^{(h^2-\zeta_1^2)^{1/2}} \zeta_j f_2\left(\zeta_1, \frac{\rho'}{\rho}\zeta_1 - (\omega - \varphi')\zeta_2\right)\mathrm{d}\zeta_2, \tag{11.163}$$

$$h = \frac{3r\rho}{l^2}.$$

In particular for $f_2 = H\dot{\varepsilon}$ we have:

$$F_1 = b\frac{\rho'}{\rho}, \quad b = \frac{6HI}{l^3}$$
$$F_2 = -b(\omega - \varphi')$$

whence

$$yF_1 - zF_2 = b(y' + \omega z)$$
$$yF_2 + zF_1 = b(z' - \omega y) \tag{11.164}$$

and finally after bringing together the contributions (11.160), (11.161), (11.162), (11.164):

$$\begin{aligned} T_y &= ky + \{b(y' + \omega z) + y\Phi(\rho) + yF_1(\rho, \rho', \omega - \varphi') \\ &\quad - zF_2(\rho, \rho', \omega - \varphi')\} \\ T_z &= kz + \{b(z' - \omega y) + z\Phi(\rho) + yF_2(\rho, \rho', \omega - \varphi') \\ &\quad + zF_1(\rho, \rho', \omega - \varphi')\}. \end{aligned} \tag{11.165}$$

The Equations of Motion

However to write down the equations of motion we have to take account of the effect of the bearings. If we denote by y_1, z_1 the components of the displacement of their centres, which are the same for O_1 and O_1' by symmetry, the total displacement of the point C can be assessed by the co-ordinates:

$$Y = y_1 + y \quad Z = z_1 + z \tag{11.166}$$

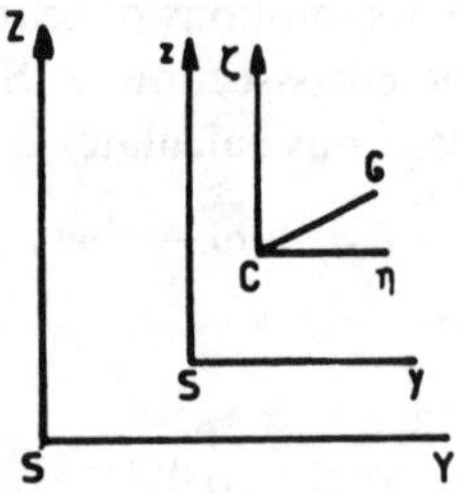

Figure 11.17

and the equations of motion modified as follows:

$$\begin{aligned} m(Y'' - e\omega^2 \cos \omega t) &= -T_y \\ m(Z'' - e\omega^2 \sin \omega t) &= -T_z. \end{aligned} \tag{11.167}$$

Assuming that the bearing response, of visco-elastic type, can be represented by:

$$\begin{aligned} T_y &= c_1 y_1 + b_1 y_1', \\ T_z &= \tilde{c}_1 z_1 + \tilde{b}_1 z_1' \end{aligned} \qquad c > 0, b > 0 \tag{11.168}$$

we shall suppose that the elasticity terms predominate over the viscous damping terms. Adopting a simplified notation we write (11.165) in the form:

$$\begin{aligned} T_y &= ky + p(y, z, y', z') \\ T_z &= kz + q(y, z, y', z') \end{aligned} \tag{11.169}$$

where p and q are likewise perturbation terms small in comparison with ky, kz

and we obtain from (11.166), (11.168), (11.169):

$$ky + p = c_1 y_1 + b_1 y_1'$$
$$y_1 + y = Y$$

which we solve for y, y_1:

$$y = \frac{c_1}{k+c_1} Y + \left(\frac{b_1}{k+c_1} y_1' - \frac{p}{k+c_1}\right), \quad y_1 = \frac{k}{k+c_1} Y + \left(-\frac{b_1}{k+c_1} y_1' + \frac{p}{k+c_1}\right),$$

on the understanding that in the perturbation terms containing $b_1 y_1'$ and p, we shall replace y_1' and y, z, y', z' by their principal values:

$$y_1' = \frac{k}{k+c_1} Y', \quad y = \frac{c_1}{k+c_1} Y, \quad y' = \frac{c_1}{k+c_1} Y', \dots$$

i.e. confining ourselves to y_1, for example, we have

$$y_1 = \frac{k}{k+c_1} Y + \left(-\frac{b_1 k}{(k+c_1)^2} Y' + \frac{1}{k+c_1} p\left(\frac{c_1}{k+c_1} Y, \dots\right)\right), \dots$$

Carrying out the analogous calculations for z_1, we can write (11.168) in the form:

(11.170) $$T_y = \frac{c_1 k}{k+c_1} Y + \frac{b_1 k^2}{(k+c_1)^2} Y' + \frac{c_1}{k+c_1} p\left(\frac{c_1}{k+c_1} Y, \dots\right)$$
$$T_z = \frac{\tilde{c}_1 k}{k+\tilde{c}_1} Z + \frac{\tilde{b}_1 k^2}{(k+\tilde{c}_1)^2} Z' + \frac{\tilde{c}_1}{k+\tilde{c}_1} q\left(\frac{c_1}{k+\tilde{c}_1} Y, \frac{\tilde{c}_1}{k+\tilde{c}_1} Z, \dots\right).$$

With the help of these representations we can express the equations of motion (11.167) in explicit form, or returning to the variables y, z

$$y = \frac{c_1}{k+c_1} Y, \quad z = \frac{\tilde{c}_1}{k+\tilde{c}_1} Z:$$

$$my'' + \frac{b_1 k^2}{(k+c_1)^2} y' + \frac{c_1 k}{k+c_1} y + \left(\frac{c_1}{k+c_1}\right)^2 p(y, z, y', z') = \frac{m c_1 e \omega^2}{k+c_1} \cos \omega t$$
$$mz'' + \frac{\tilde{b}_1 k^2}{(k+\tilde{c}_1)^2} z' + \frac{\tilde{c}_1 k}{k+\tilde{c}_1} z + \left(\frac{\tilde{c}_1}{k+\tilde{c}_1}\right)^2 q(y, z, y', z') = \frac{m \tilde{c}_1 e \omega^2}{k+\tilde{c}_1} \sin \omega t.$$

Thus in the isotropic case $b_1 = \tilde{b}_1$, $c_1 = \tilde{c}_1$, after making p, q explicit we have:

(11.171) $$my'' + \frac{b_1 k^2}{(k+c_1)^2} y' + \frac{b c_1^2}{(k+c_1)^2} (y' + \omega z) + \frac{c_1 k}{k+c_1} y$$
$$+ \left(\frac{c_1}{k+c_1}\right)^2 [y\Phi(\rho) + yF_1(\rho, \rho', \omega - \varphi') - zF_2(\rho, \rho', \omega - \rho')]$$
$$= \frac{m c_1 e \omega^2}{k+c_1} \cos \omega t$$

$$mz'' + \frac{b_1 k^2}{(k+c_1)^2} z' + \frac{b c_1^2}{(k+c_1)^2} (z' - \omega y) + \frac{c_1 k}{k+c_1} z$$

$$+\left(\frac{c_1}{k+c_1}\right)^2 [z\Phi(\rho)+yF_2(\rho,\rho',\omega-\varphi')+zF_1(\rho,\rho',\omega-\varphi')]$$
$$=\frac{mc_1e\omega^2}{k+c_1}\sin\omega t.$$

Effect of Hysteresis on Whirling

Let us take as constitutive law:

$$\tau = E\varepsilon + H\dot{\varepsilon} + \varkappa \operatorname{sgn}\varepsilon, \quad \text{where} \quad \operatorname{sgn} x = +1 \text{ if } x>0 \quad \text{and} \quad -1 \text{ if } x<0.$$

Going back to the formulae (11.163) we obtain in the present case after a few calculations:

$$F_1=\frac{8}{3}\frac{\varkappa r^3}{\rho l}\frac{\rho'/\rho}{\sqrt{(\rho'/\rho)^2+(\omega-\varphi')^2}}, \quad F_2=-\frac{8}{3}\frac{\varkappa r^3}{\rho l}\frac{\omega-\varphi'}{\sqrt{(\rho'/\rho)^2+(\omega-\varphi')^2}}$$

and

$$yF_1-zF_2=\frac{8\varkappa r^3}{3l}\frac{y'+\omega z}{\sqrt{(y'+\omega z)^2+(z'-\omega y)^2}}$$
$$yF_2+zF_1=\frac{8\varkappa r^3}{3l}\frac{z'-\omega y}{\sqrt{(y'+\omega z)^2+(z'-\omega y)^2}}$$

and by (11.171) the equations of motion:

(11.172)

$$y''+\omega_0^2 y=\mu\left(f\omega^2\cos\omega t-\alpha y'-\beta(y'+\omega z)-\gamma\frac{y'+\omega z}{\sqrt{(y'+\omega z)^2+(z'-\omega y)^2}}\right)$$
$$z''+\omega_0^2 z=\mu\left(f\omega^2\sin\omega t-\alpha z'-\beta(z'-\omega y)-\gamma\frac{z'-\omega y}{\sqrt{(y'+\omega z)^2+(z'-\omega y)^2}}\right)$$

with

(11.173)
$$\omega_0^2=\frac{c_1k}{m(k+c_1)}, \quad \mu\alpha=\frac{b_1k^2}{m(k+c_1)^2}, \quad \mu\beta=\frac{bc_1^2}{m(k+c_1)^2},$$
$$\mu\gamma=\frac{8\varkappa r^3}{3ml}\left(\frac{c_1}{k+c_1}\right)^2, \quad \mu f=\frac{c_1e}{k+c_1},$$

where μ is a small positive dimensionless parameter.

Stability of the Regime $\omega<\omega_0$

1. Case of zero eccentricity, $e=0$ or $f=0$: here $y=z=0$ is a solution of (11.172), and after multiplying these equations by $y'+\omega z$, $z'-\omega y$ and adding, we easily

obtain:

(11.174) $$[y'^2 + z'^2 + 2\omega(y'z - z'y) + \omega_0^2(y^2 + z^2)]' + 2\mu\mathcal{R} = 0$$

with

$$\mathcal{R} > [(y' + \omega z)^2 + (z' - \omega y)]^{1/2} \cdot [\gamma + \beta((y' + \omega z)^2 + (z' - \omega y)^2)^{1/2} - \alpha(y'^2 + z'^2)^{1/2}].$$

The quadratic form $U = y'^2 + z'^2 + 2\omega(y'z - z'y) + \omega_0^2(y^2 + z^2)$ is positive definite for $\omega < \omega_0$. There is therefore an $\eta > 0$ such that:

(11.175) $$U < \eta \quad \text{implies} \quad \mathcal{R} > 0.$$

Thus if the data at $t = 0$ are such that $U < \eta$, it follows from (11.174) that U will be non-increasing for $t > 0$, which ensures the permanence of (11.175) and proves that the equilibrium solution $y = z = 0$ is stable.

2. Case $f \neq 0$. By treating the equations (11.172) as in the preceding case it becomes apparent that $U(y, z, y', z') < \eta$, (η depending only on $\alpha, \beta, \gamma, \omega$) implies

$$[U(y, z, y', z') - 2\mu f \omega^2(y \cos \omega t + z \sin \omega t)]' < 0$$

or, in other words that with the variables $v = y \cos \omega t + z \sin \omega t$, $w = y \sin \omega t - z \cos \omega t$,

(11.176) $$U = v'^2 + w'^2 + (\omega_0^2 - \omega^2)(v^2 + w^2) < \eta$$

entails

(11.177) $$[U(v, w, v', w') - 2\mu f \omega^2 v]' < 0$$

This can be interpreted very simply if we change from v to ξ by the translation $v = \dfrac{\mu f \omega^2}{\omega_0^2 - \omega^2} + \xi$ so that

(11.178) $$\frac{d}{dt} U(\xi, w, \xi', w') < 0,$$

which is valid provided (11.176) is true.

Noting that

$$U(v, w, v', w') = U(\xi, w, \xi', w') + 2\mu f \omega^2 \xi + \frac{\mu^2 f^2 \omega^4}{\omega_0^2 - \omega^2}$$

we now stipulate that the initial conditions ($t = 0$) must satisfy:

(11.179) $$U(\xi, w, \xi', w') < \frac{\eta}{2}$$

which implies $|\xi| < \left(\dfrac{\eta}{2(\omega_0^2 - \omega^2)}\right)^{1/2}$, so that restricting μ by:

(11.180) $$\mu\left(2f\omega^2\left(\frac{\eta}{2(\omega_0^2 - \omega^2)}\right)^{1/2} + \frac{\mu f^2 \omega^4}{\omega_0^2 - \omega^2}\right) < \frac{\eta}{2}$$

we see that (11.179), (11.180) have as consequence (11.176) which by (11.178) implies that (11.179) continues to hold at any subsequent point of time, and the same is true of (11.176).

Thus for μ satisfying (11.180) the solution of equations (11.172) remains in the neighbourhood (11.179) of the forced vibration corresponding to $v = \mu \frac{f\omega^2}{\omega_0^2 - \omega^2}$, $w = 0$, if it does so at the initial instant.

Analysis of the Rotatory Regime When $\omega > \omega_0$

In the case where the disc is perfectly balanced, $f = 0$, the equations (11.172) have a simple solution:

$$y = \rho \sin \omega_0 t, \quad z = -\rho \cos \omega_0 t$$

with ρ such that:

$$((\alpha+\beta)\omega_0 - \beta\omega)\rho + \gamma \operatorname{sgn} \rho(\omega_0 - \omega) = 0$$

provided

(11.181) $$\omega_0 < \omega < \left(1 + \frac{\alpha}{\beta}\right)\omega_0$$

we thus find:

(11.182) $$\rho = \frac{\gamma}{(\alpha+\beta)\omega_0 - \beta\omega}$$

which corresponds to a whirling motion whose angular velocity of precession is ω_0.

In the case $f \neq 0$, we cannot predict so easily what form the rotatory regime will take. It should be noted however that the following analysis and its conclusions would not be essentially altered if we were to add a gravity term $-\mu g$ to the right-hand side of the equation for z in (11.172).

We shall apply to (11.172) the method of amplitude variation which, as we know, leads to putting

$$\begin{matrix} y = \rho_1 \sin \theta_1 & z = \rho_2 \sin \theta_2 \\ y' = \rho_1 \omega_0 \cos \theta_1 & z' = \rho_2 \omega_0 \cos \theta_2 \end{matrix}, \quad \theta = \omega t$$

whence

$$\frac{d\rho_1}{dt} = \mu A_1, \quad \frac{d\rho_2}{dt} = \mu A_2$$

$$\frac{d\theta_1}{dt} = \omega_0 + \mu B_1, \quad \frac{d\theta_2}{dt} = \omega_0 + \mu B_2, \quad \frac{d\theta}{dt} = \omega$$

with $\Delta = (y' + \omega z)^2 + (z' - \omega y)^2$ et:

$$A_1 = -\left[(\alpha+\beta)\rho_1 \cos\theta_1 + \beta\frac{\omega}{\omega_0}\rho_2 \sin\theta_2 \right.$$
$$\left. + \gamma\left(\rho_1 \cos\theta_1 + \rho_2 \frac{\omega}{\omega_0}\sin\theta_2\right)\Delta^{-1/2} + f\frac{\omega^2}{\omega_0}\cos\theta\right]\cos\theta_1$$

$$A_2 = -\left[(\alpha+\beta)\rho_2\cos\theta_2 - \beta\frac{\omega}{\omega_0}\rho_1\sin\theta_1 + \gamma\left(\rho_2\cos\theta_2 - \rho_1\frac{\omega}{\omega_0}\sin\theta_1\right)\Delta^{-1/2} + f\frac{\omega^2}{\omega_0}\sin\theta\right]\cos\theta_2$$

$$B_1 = \left[(\alpha+\beta)\cos\theta_1 + \beta\frac{\omega}{\omega_0}\frac{\rho_2}{\rho_1}\sin\theta_2 + \gamma\left(\cos\theta_1 + \frac{\rho_2}{\rho_1}\frac{\omega}{\omega_0}\sin\theta_2\right)\Delta^{-1/2} + f\frac{\omega^2}{\omega_0\rho_1}\cos\theta\right]\sin\theta_1$$

$$B_2 = \left[(\alpha+\beta)\cos\theta_2 - \beta\frac{\omega}{\omega_0}\frac{\rho_1}{\rho_2}\sin\theta_1 + \gamma\left(\cos\theta_2 - \frac{\rho_1}{\rho_2}\frac{\omega}{\omega_0}\sin\theta_1\right)\Delta^{-1/2} + f\frac{\omega^2}{\omega_0\rho_2}\cos\theta\right]\sin\theta_2.$$

However the variables θ_1, θ_2 are in resonance and for this reason we shall introduce, in place of θ_2 the variable $\chi = \theta_2 - \theta_1$ and with the amplitude equations as

$$\text{(11.183)} \qquad \frac{d\rho_1}{dt} = \mu A_1, \quad \frac{d\rho_2}{dt} = \mu A_2, \quad \frac{d\chi}{dt} = \mu(B_2 - B_1) = \mu C$$

with

(11.184)

$$C = \frac{\alpha+\beta}{2}[\sin 2(\theta_1+\chi) - \sin 2\theta_1] - \beta\frac{\omega}{\omega_0}\left(\frac{\rho_1}{\rho_2}+\frac{\rho_2}{\rho_1}\right)\sin\theta_1\cdot\sin(\theta_1+\chi)$$
$$+\gamma\left\{\tfrac{1}{2}(\sin 2(\theta_1+\chi) - \sin 2\theta_1) - \left(\frac{\rho_1}{\rho_2}+\frac{\rho_2}{\rho_1}\right)\frac{\omega}{\omega_0}\sin\theta_1\cdot\sin(\theta_1+\chi)\right\}\Delta^{-1/2}$$
$$+f\frac{\omega^2}{\omega_0}\cos\theta\cdot\left(\frac{\sin(\theta_1+\chi)}{\rho_2} - \frac{\sin\theta_1}{\rho_1}\right)$$

$$\Delta = (\rho_1\omega_0\cos\theta_1 + \rho_2\omega\sin(\theta_1+\chi))^2 + (\rho_2\omega_0\cos(\theta_1+\chi) - \rho_1\omega\sin\theta_1)^2$$

on the understanding that the angular variables are henceforth to be θ_1 and θ. We therefore have to calculate the means A_1, A_2, C with respect to θ_1 and θ. If we use a superscript 0 to indicate the result of the averaging operation, we have:

$$\text{(11.185)} \quad A_1^0 = -\left[(\alpha+\beta)\frac{\rho_1}{2} + \frac{\beta\omega}{2\omega_0}\rho_2\sin\chi + \frac{\gamma}{2\pi}\int_0^{2\pi}\left[\rho_1\cos^2\theta_1 + \rho_2\frac{\omega}{\omega_0}\sin(\theta_1+\chi)\cos\theta_1\right]\Delta^{-1/2}\,d\theta_1\right]$$

$$A_2^0 = -\Bigg[(\alpha+\beta)\frac{\rho_2}{2} + \frac{\beta\omega}{2\omega_0}\rho_1 \sin\chi$$

$$+ \frac{\gamma}{2\pi}\int_0^{2\pi} \Bigg[\rho_2 \cos^2(\theta_1+\chi) - \rho_1 \frac{\omega}{\omega_0} \sin\theta_1 \cos(\theta_1+\chi) \Bigg] \Delta^{-1/2}\, d\theta_1 \Bigg]$$

$$C^0 = -\frac{\beta\omega}{\omega_0}\left(\frac{\rho_1}{\rho_2}+\frac{\rho_2}{\rho_1}\right)\frac{\cos\chi}{2} + \frac{\gamma}{2\pi}\int_0^{2\pi}\Bigg[\tfrac{1}{2}(\sin 2(\theta_1+\chi) - \sin 2\theta_1)$$

$$-\left(\frac{\rho_1}{\rho_2}+\frac{\rho_2}{\rho_1}\right)\frac{\omega}{\omega_0}\sin\theta_1 \sin(\theta_1+\chi) \Bigg]\Delta^{-1/2}\, d\theta_1$$

and we have to solve:

$$A_1^0 = 0, \quad A_2^0 = 0, \quad C^0 = 0 \tag{11.186}$$

for ρ_1, ρ_2, χ.

For each solution obtained we shall need to calculate the associated Jacobian matrix and its eigenvalues, in order to know whether there corresponds a solution of (11.172) which remains bounded for all future time.

We look for a solution of (11.186) taking

$$\rho_1 = \rho_2 = \rho, \quad \rho > 0, \qquad \chi = -\frac{\pi}{2} \tag{11.187}$$

so that
$$\Delta^{1/2} = \rho|\omega_0 - \omega|, \quad C^0 = 0$$

and (11.186) is reduced to:

$$\rho((\alpha+\beta)\omega_0 - \beta\omega) + \gamma \,\mathrm{sgn}\,(\omega_0 - \omega) = 0$$

which has a solution meeting the required conditions if

$$\omega_0 < \omega < \left(1 + \frac{\alpha}{\beta}\right)\omega_0 \quad \text{or in other words} \tag{11.188}$$

$$\rho = \frac{\gamma}{(\alpha+\beta)\omega_0 - \beta\omega}. \tag{11.189}$$

To obtain the Jacobian matrix associated with the solution (11.187), (11.189), we have to calculate, for these values, the partial derivatives of (11.185) with respect to ρ_1, ρ_2, χ. Omitting details we have:

$$\frac{\partial A_1^0}{\partial \rho_1} = \frac{\partial A_2^0}{\partial \rho_2} = -\frac{\alpha+\beta}{8}\frac{5\omega - 3\omega_0}{\omega - \omega_0} + \frac{\beta\omega}{8\omega_0}\frac{\omega+\omega_0}{\omega-\omega_0} = p$$

$$\frac{\partial A_1^0}{\partial \rho_2} = \frac{\partial A_2^0}{\partial \rho_1} = \frac{\alpha+\beta}{8}\frac{\omega+\omega_0}{\omega-\omega_0} - \frac{\beta\omega}{8\omega_0}\frac{5\omega_0 - 3\omega}{\omega-\omega_0} = q$$

$$\frac{\partial A_1^0}{\partial \chi} = \frac{\partial A_2^0}{\partial \chi} = 0$$

$$\frac{\partial C^0}{\partial \chi} = -\alpha\frac{3\omega - \omega_0}{4\omega_0(\omega-\omega_0)} - \frac{\beta}{4\omega_0}(\omega+\omega_0) < 0.$$

The eigenvalues to be found, apart from $\frac{\partial C^0}{\partial \chi}$ which is negative, are the roots of $(\lambda - p)^2 - q^2 = 0$, i.e. $p \pm q$. Now

$$p + q = \beta \frac{\omega}{\omega_0} - (\alpha + \beta) < 0$$

$$p - q = -\alpha \frac{3\omega - \omega_0}{4\omega_0(\omega - \omega_0)} - \frac{\beta}{4\omega_0}(\omega + \omega_0) < 0$$

and in conclusion, by virtue of the amplitude-variation theorem, we can be sure that there exists, for ω satisfying (11.188), a solution of the equations (11.172) defined and bounded for all future time, which is stable if μ is small enough.

V. Suspension of Rotating Machinery in Magnetic Bearings

Principle of Magnetic Suspension

The shaft of the rotor is provided with a ring of magnetic material suspended between two electromagnets connected to the fixed part of the machinery.

Two position sensors continuously monitor the centring error (i.e. the departure of the magnetic ring from its true central position in relation to the stator) and control the variations of magnetisation in the electromagnets required to make the necessary corrections [50].

In actual practice two such suspensions at right angles to each other in the same plane are used to ensure radial rigidity of the system. To prevent movement of the shaft in the direction of its axis a magnetic system is included at each end of the shaft which plays the role of a buffer or end-stop.

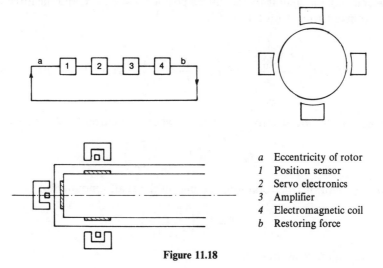

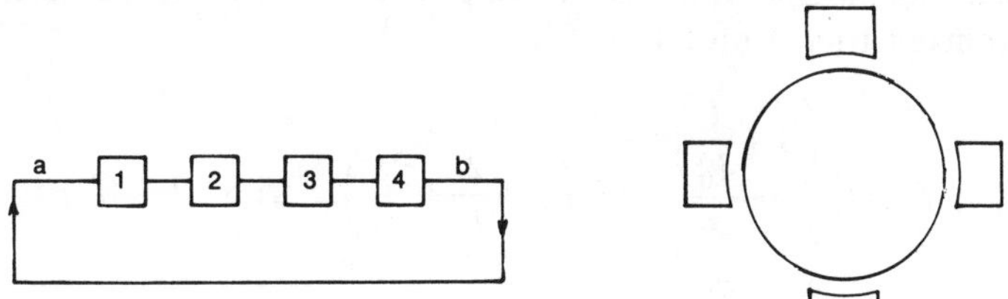

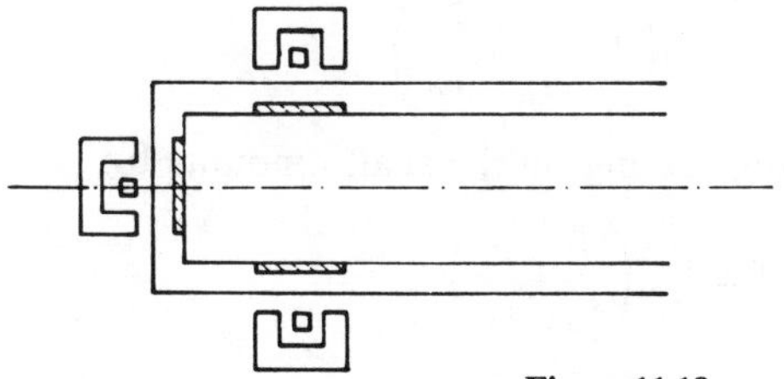

a Eccentricity of rotor
1 Position sensor
2 Servo electronics
3 Amplifier
4 Electromagnetic coil
b Restoring force

Figure 11.18

In the simplest case the system has five degrees of freedom, so that the equations of motion expressed in terms of displacement and velocity involve ten co-ordinates, e.g. those for the centre of inertia of the shaft, the angles of precession and of nutation, and their derivatives with respect to time. Thus displacements and velocities have to be measured in real time by appropriate sensors in order to be able to ensure control of the system.

We may illustrate the operation of the principle by a simple model with one degree of freedom, represented diagrammatically in Figure 11.19.

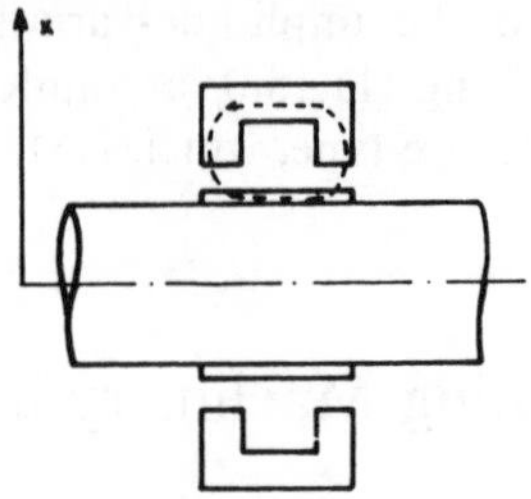

Figure 11.19

The force of attraction is proportional to the square of the magnetic induction in the air gap. If n is the number of turns and I the current intensity in the coil of the electromagnet, the magnetic flux Φ is given by $\mathcal{R}\Phi = nI$, where $\mathcal{R} = \int \frac{dl}{\mu S}$ is the reluctance of the magnetic circuit. In fact the major contribution to $\mathcal{R}$ is from the air-gap so that we can use the approximation $\mathcal{R} = \frac{2e}{\mu_0 S}$, where e is the distance separating the magnetic ring from the poles of the electromagnet, S is the area of each pole, and $\mu_0 = 1$ is the magnetic permeability of air. If x measures the displacement of the shaft from its centred position, $e = e_0 - x$, and the force exerted by electromagnet no. 1 on the shaft is:

$$f_1 = f_0 \frac{\left(1 + \frac{i}{I_0}\right)^2}{\left(1 - \frac{x}{e_0}\right)^2} = f_0\left(1 + \frac{2i}{I_0} + \frac{2x}{e_0}\right), \quad \text{with} \quad I = I_0 + i,$$

while that exerted by electromagnet no. 2, where the current is $I = I_0 - i$, is:

$$f_2 = -f_0\left(1 - \frac{2i}{I_0} - \frac{2x}{e_0}\right)$$

so that, by addition, the total force acting on the shaft amounts to:

$$f = 4f_0\left(\frac{i}{I_0} + \frac{x}{e_0}\right)$$

and the equation of motion is:

$$mx'' = 4f_0\left(\frac{i}{I_0} + \frac{x}{e_0}\right)$$

which we put into the form:

$$(11.190) \qquad \begin{aligned} x' &= y \\ y' &= ax + gi \end{aligned} \quad \text{with} \quad a = \frac{4f_0}{me_0}, \quad g = \frac{4f_0}{mI_0}.$$

The sensors allow the values of x and y on which depends the value of the control input variable i, to be kept monitored at all times. The value of $i(t)$ at time t, for a given value of $x(t)$ and $y(t)$, must be such as to allow the system to return to the centred position $x = y = 0$, and is usually designed so as to minimise a quadratic functional

$$(11.191) \qquad J = \int_0^{+\infty} (x^2(t) + y^2(t))\,dt + r\int_0^{+\infty} i^2(t)\,dt,$$

where r is a given positive parameter representing a compromise between the desired performance of the system (in terms of an optimum in regard to the speed of return to the centred position) and the cost in terms of the energy expenditure required from the servomechanism.

In the general case, and within the context of a linearised theory, we see that the equation of motion can be written in matrix form as

$$(11.192) \qquad x' = Ax + Gu$$

where x is the vector which defines the position and velocity of the rotor, u the control vector and A, G are constant matrices.

Denoting by y the vector defining displacements and velocities at the right of the bearings, we may postulate a control vector of the form:

$$(11.193) \qquad u = K_1 y + K_2 i,$$

where i is the vector representing the current intensity in the coils, which is tied to the voltage v by:

$$(11.194) \qquad i = K_3 v,$$

K_3 being the gain matrix.

Lastly there is a matrix C such that:

$$(11.195) \qquad y = Cx.$$

By substituting in (11.192) we arrive at:

$$x' = (A + GK_1C)x + GK_2K_3v$$

and the optimal control problem which we have to solve is to determine how $v(t)$ has to be defined in order to minimise the integral:

$$\int_0^{+\infty} (x^T Qx + v^T Rv)\,dt$$

where Q, R are given symmetric positive definite matrices.

It is this problem which we shall now consider in a slightly more general context.

Quadratic Functionals and Optimal Control [19]

Let x be the variable defining the state of a system governed by the differential equation:

(11.196)
$$\frac{dx}{dt} = A(t)x$$

where $A(t)$ is an $n \times n$ matrix, which is an integrable function of t.

We recall that the resolvent or solution matrix $X(t)$ defined by $\frac{dX}{dt} = A(t)X$, $X(0) = I$, (I being the identity matrix), satisfies:

(11.197)
$$\frac{dX^{-1}}{dt} = -X^{-1}A(t), \quad \frac{d(X^{-1})^T}{dt} = -A^T(t)(X^{-1})^T.$$

Let us introduce the functional:

(11.198)
$$J = \int_t^{t_1} x^T(s)Q(s)x(s)\,ds + x^T(t_1)Fx(t_1)$$

where $F, Q(s)$ are symmetric matrices, F with constant elements and $x(s)$ is a solution of (11.196). With $x(t) = c$ and having regard to $x(s) = X(s)X^{-1}(t)c$ we can write (11.198) as

$$J = \int_t^{t_1} c^T(X^{-1}(t))^T X^T(s)Q(s)X(s)X^{-1}(t)c\,ds + c^T(X^{-1}(t))^T X^T(t_1)FX(t_1)X^{-1}(t)c$$

i.e.

(11.199)
$$J = c^T P(t)c$$

with

(11.200)
$$P(t) = \int_t^{t_1} X^{-1}(t)^T X^T(s)Q(s)X(s)X^{-1}(t)\,ds + (X^{-1}(t))^T X^T(t_1)FX(t_1)X^{-1}(t).$$

Now it is easily verified by (11.197) that:

(11.201)
$$\frac{dP}{dt} + A^T P + PA + Q = 0$$
$$P(t_1) = F$$

and these equations, in turn, characterise $P(t)$.

We now introduce a control u, that is to say we consider:

(11.202)
$$\frac{dx}{dt} = Ax + Gu$$

where $G(t)$ is a rectangular $n \times p$ matrix, and associate with (11.202) the functional

(11.203) $$J = \int_t^{t_1} (x^T(s), u^T(s)) \begin{pmatrix} Q & S \\ S^T & R \end{pmatrix} \begin{pmatrix} x(s) \\ u(s) \end{pmatrix} \mathrm{d}s + x^T(t_1) F x(t_1)$$

where $Q(s)$, $R(s)$ are respectively $n \times n$ and $p \times p$ matrices both of which are symmetric and positive, while $S(s)$ is an $n \times p$ matrix; and lastly we associate with (11.202) the vector u (the "control" vector):

(11.204) $$u = Kx \quad \text{or} \quad u^T = x^T K^T$$

so that

$$J = \int_t^{t_1} x^T(s)(I, K^T(s)) \begin{pmatrix} Q(s) & S(s) \\ S^T(s) & R(s) \end{pmatrix} \begin{pmatrix} I \\ K(s) \end{pmatrix} x(s)\,\mathrm{d}s + x^T(t_1) F x(t_1)$$

or

(11.205) $$J = \int_t^{t_1} x^T(s)(Q + K^T S^T + SK + K^T R K) x(s)\,\mathrm{d}x + x^T(t_1) F x(t_1).$$

However (11.202) and (11.204) imply:

$$\frac{\mathrm{d}x}{\mathrm{d}t} = (A + GK)x$$

and we can apply (11.199), (11.200) changing A into $A + GK$ and Q into $Q + K^T S^T + SK + K^T R K$, so that we have:

(11.206) $$J = c^T P(t) c, \quad c = x(t),$$

$P(t)$ being the unique solution of

(11.207) $$\frac{\mathrm{d}P}{\mathrm{d}t} + (A^T + K^T G^T)P + P(A + GK) + Q + K^T S^T + SK + K^T R K = 0$$

with $$P(t_1) = F.$$

Assuming R to be positive definite and hence invertible, we can write: (11.207):

(11.208) $$\begin{aligned} &\frac{\mathrm{d}P}{\mathrm{d}t} + A^T P + PA + Q - (PG + S)R^{-1}(G^T P + S^T) \\ &\quad + (K^T R + PG + S)R^{-1}(RK + G^T P + S^T) = 0 \\ &P(t_1) = F. \end{aligned}$$

We can satisfy (11.208), which characterises P for a given control instruction K, as follows:

(11.209) $$\frac{\mathrm{d}P}{\mathrm{d}t} + A^T P + PA + Q - (PG + S)R^{-1}(G^T P + S^T) = 0$$

$$P(t_1) = F$$

(11.210) $$K = -R^{-1}(G^T P + S^T).$$

In fact, by the existence theorem on the solutions of differential equations, the above system defines $P(t)$ uniquely for all t, even if we assume only that the matrices A, Q, G, S, R^{-1} are locally bounded and measurable with respect to t, and (11.210) defines the associated control instruction:

$$u^*(.) = -R^{-1}(G^T P + S^T)\cdot x. \tag{11.211}$$

Now let $x(s)$ be a solution of (11.202) for a given control instruction $u(s)$ and let us again write $x(t) = c$. We define $P(s)$ by (11.209) and $u^*(s)$ associated by (11.211), for this matrix $P(s)$, with the solution $x(s)$ under consideration.

We now proceed to calculate J for the input-output pair $u(s)$, $x(s)$. We first note that:

$$(x^T, u^T)\begin{pmatrix} O & S \\ S^T & R \end{pmatrix}\begin{pmatrix} x \\ u \end{pmatrix} = x^T Q x + u^T S^T x + x^T S u + u^T R u$$

or by (11.209):

$$(x^T, u^T)\begin{pmatrix} Q & S \\ S^T & R \end{pmatrix}\begin{pmatrix} x \\ u \end{pmatrix} = -x^T \frac{dP}{ds} x - x^T A^T P x - x^T P A x$$
$$+ x^T (PG + S) R^{-1} (G^T P + S^T) x + u^T S^T x + x^T S u + u^T R u$$

and then that:

$$x^T (PG + S) R^{-1} (G^T P + S^T) x = u^{*T} R u^*, \quad \text{by (11.211)},$$
$$-x^T \frac{dP}{ds} x - x^T A^T P x - x^T P A x = -\frac{d}{ds}(x^T P x) + u^T G^T P x + x^T P G u$$

i.e.

$$(x^T, u^T)\begin{pmatrix} Q & S \\ S^T & R \end{pmatrix}\begin{pmatrix} x \\ u \end{pmatrix} = -\frac{d}{ds}(x^T P x) + u^T (G^T P + S^T) x \tag{11.212}$$
$$+ x^T (PG + S) u + u^{*T} R u^* + u^T R u = -\frac{d}{ds}(x^T P x) + (u - u^*)^T R (u - u^*),$$

and thus finally arrive at:

$$J = c^T P(t) c + \int_t^{t_1} (u - u^*)^T R (u - u^*)\, ds, \tag{11.213}$$

from which we see that the optimum control function, i.e. the one which minimises J on $[t, t_1]$ corresponds to $u = u^*$.

Suppose now that the matrices A, Q, G, S, R are constant, and that P is a solution of the algebraic equation:

$$A^T P + PA + Q - (PG + S) R^{-1} (G^T P + S^T) = 0. \tag{11.214}$$

By considering the formula (11.212) we can see that:

$$J = c^T P c - x^T(t_1)(P - F) x(t_1) + \int_t^{t_1} (u - u^*) R (u - u^*)\, ds,$$

with u^* still defined by (11.211).

If the condition

$$A - GR^{-1}(G^T P + S^T) \quad \text{asymptotically stable} \tag{11.215}$$

is fulfilled, then clearly the optimum control function on $(t, +\infty)$ corresponds to $u = u^*$ and the corresponding value of J is

$$J = c^T P c. \tag{11.216}$$

We can in fact go rather further and assert that the equation (11.214), also known as Riccati's equation, has at most one solution P satisfying (11.215). For suppose there were two solution $P, \tilde{P}$ of (11.214), (11.215) and that $u^*, \tilde{u}^*$ were the corresponding optimal control functions with which could be associated J and $\tilde{J}$; then by (11.126) we should have $\tilde{J} \geqslant J = c^T P c$ and also $J \geqslant \tilde{J} = c^T \tilde{P} c$ from which it would follow that $c^T P c = c^T \tilde{P} c$, $\forall c$, and hence $P = \tilde{P}$.

Application to the Model with One Degree of Freedom

We consider afresh the system (11.190) written in matrix form:

$$\begin{pmatrix} x' \\ y' \end{pmatrix} = A \begin{pmatrix} x \\ y \end{pmatrix} + Gi \tag{11.217}$$

with $A = \begin{pmatrix} 0 & 1 \\ a & 0 \end{pmatrix}$, $G = \begin{pmatrix} 0 \\ g \end{pmatrix}$; we choose:

$$Q = \begin{pmatrix} 1 & 0 \\ 0 & 1 \end{pmatrix}, \quad S = 0, \quad R = r, \quad r > 0$$

and look for a solution $P = \begin{pmatrix} \alpha & \gamma \\ \gamma & \beta \end{pmatrix}$, for these data, of the equation (11.214). After some calculation we find:

$$\begin{aligned} &2\gamma a + 1 - v(\alpha^2 + \gamma^2) = 0 \\ &2\gamma + 1 - v(\beta^2 + \gamma^2) = 0, \quad \text{with} \quad v = r^{-1} g^2 > 0 \\ &\alpha + \beta a - v(\alpha + \beta)\gamma = 0. \end{aligned} \tag{11.218}$$

We solve the first two equations for α^2, β^2 in terms of γ and eliminate α, β with the help of the third equation written in the form:

$$\alpha(1 - v\gamma) = -\beta(a - v\gamma). \tag{11.219}$$

We thus find:

$$v\gamma^2 + 2\frac{v - a}{1 + a}\gamma - 1 = 0 \tag{11.220}$$

which has two equal roots of opposite sign.

We then obtain α^2 and β^2, by (11.218), (11.220):

$$\nu\alpha^2 = 2\gamma\frac{\nu + a^2}{1+a} \tag{11.221}$$

$$\nu\beta^2 = 2\gamma\frac{1+\nu}{1+a}$$

from which it follows that only the positive root γ_+ is acceptable. To specify precisely α, β from (11.221) we have to assign the signs of α, β so that (11.219) is satisfied and this leaves at most two possible solutions. However we also have to satisfy the condition (11.215), or in other words the matrix

$$A - g^2 r^{-1} P = \begin{pmatrix} -\nu\alpha & 1-\nu\gamma \\ a-\nu\gamma & -\nu\beta \end{pmatrix}$$

must be asymptotically stable. Now its eigenvalues are defined as the roots of

$$\lambda^2 + \nu(\alpha+\beta)\lambda + \nu^2\alpha\beta - (a-\nu\gamma)(1-\nu\gamma) = 0$$

and since $\alpha\beta$ and $-(a-\nu\gamma)(1-\nu\gamma)$ have, by (11.129), the same sign, we see that we have to ensure that

$$(a-\nu\gamma)(1-\nu\gamma) < 0 \tag{11.222}$$

and hence choose α and β positive, which means that there is an unique solution if (11.222) is satisfied for $\gamma = \gamma_+$. By expressing that this must be so, we obtain $\nu < a^2 + a$. Thus the optimal control problem has an unique solution if:

$$r > \left(\frac{f_0}{mI_0}\right)^2 \cdot \left(\frac{f_0}{e_0 m}\right)^{-1} \left(\frac{f_0}{e_0 m} + \frac{1}{4}\right)^{-1}$$

Characteristics and Applications of Magnetic Bearings

The load capacities vary depending on the nature of the magnetic materials utilised. With ordinary materials (silicon iron) the load capacity, for an induction of 10,000 gauss is of the order of 25 N per cm^2 of projected active surface. This capacity can be increased to 56 N per cm^2 for an induction of 15,000 gauss, if magnetic alloys with 50% cobalt content are used.

The advantages of the new technique of magnetic suspension are obvious:

- absence of all mechanical contact with consequent possibility of very high speeds of rotation, limited only by the capacity of the system to resist the stresses due to centrifugal forces;
- no wear, very long life, negligible heating, extremely slight braking torque;
- low power consumption, absence of vibration, silent operation;
- no lubrication required and consequently no pollution (very desirable feature when operating in a vacuum environment), no corrosion, very wide range of temperature tolerance;

- toleration of imperfectly balanced rotor, the latter revolving around its principal axis of inertia.

Among the very numerous possible applications may be mentioned inertial guidance systems for satellites, navigational gyroscopes, ultracentrifuges for separation of isotopes, vacuum pumps; and in the area of medium or highpowered equipment, compressors for cryogenic or corrosive fluids, and compressors or turbomachinery used in the liquefaction of natural gas, and generators in nuclear power stations, etc.

Chapter XII. Non-Linear Waves and Solitons

The role which the Korteweg-de Vries equation can play in describing the effects of non-linear interactions in periodic chains has already been underlined in Chapter 2 in connection with vibrations in lattices. Its importance appears just as clearly in the analysis of disturbances which can arise in a continuous one-dimensional medium, governed by conservation of mass and momentum equations when the "pressure" forces depend not only on the variables specifying the state of the system, but on their derivatives with respect to space and time and thus cause dissipative or dispersive effects.

An ingenious method, known by the name of inverse scattering has recently been developed to solve problems connected with the Korteweg-de Vries equation or even equations of more general type [21, 28]. The unknown function $q(x, t)$ is first considered as the potential of a second-order differential equation of Schrödinger type for which it is possible to conduct a detailed analysis of the behaviour of the wave-solutions for constant t. This constitutes the direct problem which is used as a starting point to bring out clearly the eigenvalues $\varkappa_1, \varkappa_2, \ldots, \varkappa_N$, the corresponding eigenfunctions or bound states, and the amplitude spectrum $\rho(k)$ of the reflected wave. The inverse problem consists in showing that to reconstruct the potential of the Schrödinger equation, it is sufficient to be given the spectral function $\rho(k)$, the eigenvalues $\varkappa_1, \ldots, \varkappa_N$ and the scaling factors $c_1, \ldots, c_N$ of the normalised eigenfunctions.

This operation can of course be carried out for any value of t, and at this point has to be introduced the essential idea of knowing what condition needs to be imposed on $q(x, t)$ so that, on the one hand the spectral value $\varkappa_j$ shall be invariant with respect to t, and on the other hand, the evolution of the spectral function $\rho(k)$ and its scaling factors c_j, in terms of time should be governed by linear differential equations which can be easily integrated. It emerges that this latter condition holds when the potential $q(x, t)$ obeys the Korteweg-de Vries equation, so that starting from the initial given $q(x, 0)$ we can calculate by the direct method the elements $\varkappa_1, \ldots, \varkappa_N$, $c_1, \ldots, c_N$, $\rho(k)$ at $t = 0$, and then their values at a later instant of time t by the rules of evolution, and then finally, by the inverse method get back to $q(x, t)$.

The theory developed here in its various aspects, with a new approach on the question of a finite number of eigenvalues, the solving of the inverse problem by means of the Gelfand-Levitan equation, and the calculation of the invariant integrals, allow us to understand the mechanism of the interaction of solitary waves or solitons and to calculate the asymptotic elements outside the zone of mixing.

1. Waves in Dispersive or Dissipative Media

For a large class of physical systems, the evolution of their state can often be described (in the case of unidirectional motion) by the conservation equations:

(12.1) $$n_t + (nu)_x = 0$$

(12.2) $$n(u_t + uu_x) + P_x = 0$$

where n is the number of particles per unit volume or "density", P is a "pressure", u a velocity, and x, t the space and time variables, all of these variables being scalars.

To these equations has to be added a knowledge of P as a function of the variables defining the state of the system. The latter "state" variables may include, apart from n and u, another variable f, admittedly connected to n and u by an additional relation $F = 0$. The dependence of P and F on these state variables usually involves the partial derivatives of these variables so that:

(12.3) $$P = P(n, u, f, n_x, n_t, u_x, \ldots, n_{xx}, \ldots)$$

(12.4) $$F = F(n, u, f, n_x, n_t, u_x, \ldots, n_{xx}, \ldots).$$

We shall assume that P, F are invariant under any Galilean transformation, which imposes certain structural conditions. Thus with the change of variable:

(12.5) $$x' = x - ct, t' = t$$

and taking account of:

(12.6) $$n' = n, f' = f, u' = u - c$$

the invariance condition can be written as:

$$P(n, u, f, n_x, \ldots) = P(n', u', f', n'_{x'}, \ldots)$$

i.e. with (12.6), $f(x, t) = f(x' + ct', t') = f'(x', t')$ whence $f'_{x'} = f_x$, $f'_{t'} = cf_x + f_t$ and analogous formulae for n, u:

(12.7) $$P(n, u, f, n_x, n_t, \ldots) = P(n, u - c, f, n_x, n_t + cn_x, \ldots).$$

In particular if n, u, f are constants, associated with a stationary state, we must have for all c:

$$P(n, u, f, 0, 0, \ldots) = P(n, u - c, f, 0, 0, \ldots)$$

and consequently:

(12.8) $$\frac{\partial P}{\partial u}(n, u, f, 0, 0, \ldots) = 0$$

and more generally

(12.9) $$\frac{\partial^{\alpha+\beta+\gamma} P}{\partial u^\alpha \partial n^\beta \partial f^\gamma}(n, u, f, 0, 0, \ldots) = 0, \quad \forall \alpha > 0$$

and an analogous result for F.

A stationary state may be defined by the constant values $n=n_0$, $u=0$, $f=f_0$. If we ignore for the time being the dependence of P and F on the derivatives $u_x,\ldots$, linearisation near the stationary solution $(n_0, 0, f_0)$ leads us, in view of (12.8), which is valid for that state, to:

$$\begin{aligned} P &= P^{(0)} + P_{n_0}\cdot(n-n_0) + P_{f_0}\cdot(f-f_0) \\ F &= F_{n_0}\cdot(n-n_0) + F_{f_0}\cdot(f-f_0) = 0, \end{aligned} \qquad \text{since } F^{(0)}=0 \tag{12.10}$$

and then to:

$$\begin{aligned} n_t + n_0 u_x &= 0 \\ n_0 u_t + P_x &= 0 \end{aligned}$$

n, u, f denoting the perturbed state.

We get from (12.10):

$$P = P^{(0)} + a_0^2(n-n_0)$$

under the assumptions that $F_{f_0} \neq 0$ and:

$$a_0^2 = P_{n_0} - \frac{P_{f_0}\cdot F_{n_0}}{F_{f_0}} > 0, \tag{12.11}$$

so that the linearised equations are

$$n_t + n_0 u_x = 0, \quad n_0 u_t + a_0^2 n_x = 0$$

or, after eliminating n:

$$u_{tt} - a_0^2 u_{xx} = 0. \tag{12.12}$$

The behaviour of u and of n is therefore governed by a wave equation of velocity $\pm a_0$.

We propose to discuss the variations of the wave profiles caused by non-linear effects on the one hand and by dispersive or dissipative effects characterised by a dependence of P and F on the derivatives of the "state" variables, on the other. To this end we use, following [54], a frame of reference moving at an uniform velocity of a_0 (or $-a_0$) and we introduce the scaled-up variables

$$\xi = \varepsilon^{\alpha}(x - a_0 t), \quad \tau = \varepsilon^{\alpha+1}\cdot t \tag{12.13}$$

in which α is an as yet unspecified positive constant, and ε a small parameter, which in some sense measures the size of the initial disturbance.

The equations (12.1), (12.2) expressed in terms of those variables become

$$\begin{aligned} \varepsilon n_\tau + (u-a_0)n_\xi + n u_\xi &= 0 \\ \varepsilon u_\tau + (u-a_0)u_\xi + n^{-1}P_\xi &= 0. \end{aligned} \tag{12.14}$$

We shall assume that the state variables n, u, f can be represented asymptotically in the neighbourhood of the stationary state by:

$$\begin{aligned} n &= n_0 + \varepsilon n^{(1)} + \varepsilon^2 n^{(2)} + \cdots \\ u &= \varepsilon u^{(1)} + \cdots \\ f &= f_0 + \varepsilon f^{(1)} + \cdots. \end{aligned} \tag{12.15}$$

On substituting these expansions in the expressions for P and F, after having expressed the arguments of these two functions in terms of the new co-ordinates ξ and τ we are led to:

$$\begin{aligned}&P(n, u, f, n_x, n_t, \ldots, n_{xx}, n_{xt}, n_{tt}, \ldots)\\&\quad = P(n, u, f, \varepsilon^{\alpha} n_{\xi}, \varepsilon^{\alpha+1} n_{\tau} - a_0 \varepsilon^{\alpha} n_{\xi}, \ldots, \varepsilon^{2\alpha} n_{\xi\xi}, \varepsilon^{2\alpha+1} n_{\tau\xi} - a_0 \varepsilon^{2\alpha} n_{\xi\xi},\\&\quad\quad \varepsilon^{2\alpha+2} n_{\tau\tau} - 2a_0 \varepsilon^{2\alpha+1} n_{\xi\tau} + a_0^2 \varepsilon^{2\alpha} n_{\xi\xi}, \ldots)\end{aligned}$$

where the terms not written down involve on the one hand the partial derivatives of the first and second order of u, f with structure analogous to that of the derivatives of n and on the other hand the partial derivatives of order greater than 2, the latter terms not needing to appear in an approximation of the order concerned.

The Non-Linear Perturbation Equations

Let us try to find the expansion of P in powers of ε in the neighbourhood of the stationary state. Taking account of (12.8) in the stationary state, we obtain the contribution up to order ε in the form:

$$\begin{aligned}P &= P^{(0)} + \varepsilon(P_{n_0} \cdot n^{(1)} + P_{f_0} \cdot f^{(1)}) = P^{(0)} + \varepsilon P^{(1)}\\F &= \varepsilon(F_{n_0} \cdot n^{(1)} + F_{f_0} \cdot f^{(1)}) + \cdots = 0\end{aligned}$$

whence

$$P^{(1)} = \left(P_{n_0} - \frac{F_{n_0}}{F_{f_0}} P_{f_0} \right) n^{(1)} = a_0^2 n^{(1)} \tag{12.16}$$

and $\dfrac{\partial P^{(1)}}{\partial \xi} = a_0^2 \dfrac{\partial n^{(1)}}{\partial \xi}$, whence by (12.14):

$$a_0 n_{\xi}^{(1)} = n_0 u_{\xi}^{(1)}$$

$$u_{\xi}^{(1)} = \frac{a_0}{n_0} n_{\xi}^{(1)},$$

equations which, after integration and taking account of $n^{(1)}, u^{(1)}$ vanishing at infinity, lead to

$$a_0 n^{(1)} = n_0 u^{(1)} \tag{12.17}$$

and

$$f^{(1)} = -\frac{F_{n_0}}{F_{f_0}} \cdot n^{(1)} = -\frac{F_{n_0}}{F_{f_0}} \cdot \frac{n_0}{a_0} u^{(1)} \tag{12.18}$$

which allows the calculation of the first-order perturbations to be reduced to the finding of a single unknown, for example $n^{(1)}$. To go further, i.e. to obtain the equation which governs the way in which $n^{(1)}$ evolves, we have to consider the terms of the expansion of P in powers of ε of order higher than 1 and to analyse them.

Excluding, for the moment, the effects on the calculation of the arguments of P and F which are partial derivatives of the state variables, we obtain to the second order in ε, on the one hand:

$$\varepsilon^2(P_{n_0}\cdot n^{(2)} + P_{f_0}\cdot f^{(2)}) \tag{12.19}$$

and on the other:

$$\frac{\varepsilon^2}{2}(P_{n_0 n_0}n^{(1)2} + 2P_{n_0 f_0}n^{(1)}f^{(1)} + P_{f_0 f_0}f^{(1)2}), \tag{12.20}$$

noting that the derivatives P_{nu}, P_{u^2}, P_{fu} calculated in the stationary state are zero. Remembering (12.17), (12.18), we see that the contribution (12.20) can be written $(\varepsilon^2/2)pn^{(1)2}$ with

$$p = P_{n_0 n_0} - 2P_{n_0 f_0}\cdot\frac{F_{n_0}}{F_{f_0}} + P_{f_0 f_0}\left(\frac{F_{n_0}}{F_{f_0}}\right)^2. \tag{12.21}$$

We next consider the contribution from the terms of order $\varepsilon^{\alpha+1}$ or $\varepsilon^{2\alpha+1}$ which appear in connection with the derivatives of the state variables; and we find in this way:

$$\begin{aligned}\varepsilon^{\alpha+1}n_\xi^{(1)}P_{n_x} - a_0\varepsilon^{\alpha+1}n_\xi^{(1)}P_{n_t} + \varepsilon^{2\alpha+1}n_{\xi\xi}^{(1)}P_{n_{xx}} - a_0\varepsilon^{2\alpha+1}n_{\xi\xi}^{(1)}P_{n_{xt}} \\ + a_0^2\varepsilon^{2\alpha+1}n_{\xi\xi}^{(1)}P_{n_{tt}} = \varepsilon^{\alpha+1}(P_{n_x} - a_0P_{n_t})n_\xi^{(1)} + \varepsilon^{2\alpha+1}(P_{n_{xx}} - a_0P_{n_{xt}} + a_0^2P_{n_{tt}})n_{\xi\xi}^{(1)}\end{aligned}$$

with corresponding terms associated with the variables u, f. In actual fact we have kept the terms in $\varepsilon^{\alpha+1}$ and $\varepsilon^{2\alpha+1}$ and left out those in $\varepsilon^{\alpha+2}$ which would be derived from the contribution from $\varepsilon^{\alpha+2}n_\tau^{(1)}$, for the following reason: if at the end of our calculations the term in $\varepsilon^{\alpha+1}$ does not vanish, it has to be associated with the terms in ε^2 already calculated in (12.19) and (12.20) or (12.21); i.e. we shall take $\alpha = 1$ and the terms in $\varepsilon^{2\alpha+1}$ and $\varepsilon^{\alpha+2}$ will not be involved at this stage. If however the terms in $\varepsilon^{\alpha+1}$ disappear (for example if P, F are independent of the first derivatives $n_x, n_t, \ldots$) then the terms in $\varepsilon^{2\alpha+1}$ are relevant and to assimilate them to those of order ε^2 means having to put $\alpha = 1/2$; and in that case as well the terms in $\varepsilon^{\alpha+2}$ are of higher order and it would therefore be natural to omit them. In the final analysis we see that we shall be able to write:

$$P = P^{(0)} + \varepsilon P^{(1)} + \varepsilon^2 P^{(2)}$$

$$\begin{aligned}P^{(2)} = P_{n_0}n^{(2)} + P_{f_0}f^{(2)} + \frac{p}{2}n^{(1)2} + \varepsilon^{\alpha-1}\sum_{n,f,u}(P_{n_x} - a_0P_{n_t})n_\xi^{(1)} \\ + \varepsilon^{2\alpha-1}\sum_{n,f,u}(P_{n_{xx}} - a_0P_{n_{xt}} + a_0^2P_{n_{tt}})n_{\xi\xi}^{(1)}\end{aligned} \tag{12.22}$$

with an analogous expression for $F^{(2)}$ which must be equated to zero:

$$\begin{aligned}F^{(2)} = F_{n_0}n^{(2)} + F_{f_0}f^{(2)} + \frac{q}{2}n^{(1)2} + \varepsilon^{\alpha-1}\sum_{n,f,u}(F_{n_x} - a_0F_{n_t})n_\xi^{(1)} \\ + \varepsilon^{2\alpha-1}\sum_{n,f,u}(F_{n_{xx}} - a_0F_{n_{xt}} + a_0^2F_{n_{tt}})n_{\xi\xi}^{(1)} = 0\end{aligned} \tag{12.23}$$

and this equation allows us to express $f^{(2)}$ in terms of $n^{(2)}$ and $n^{(1)}$. Lastly we can write (12.22) as follows:

$$P^{(2)} = a_0^2 n^{(2)} + \frac{A}{2} n^{(1)2} + \varepsilon^{\alpha-1} B n_\xi^{(1)} + \varepsilon^{2\alpha-1} C n_{\xi\xi}^{(1)}$$

where A, B, C are numerical constants which can easily be calculated from the preceding formulae, and finally:

$$P_\xi^{(2)} = a_0^2 n_\xi^{(2)} + A n^{(1)} n_\xi^{(1)} + \varepsilon^{\alpha-1} B n_{\xi\xi}^{(1)} + \varepsilon^{2\alpha-1} C n_{\xi\xi\xi}^{(1)}.$$

Coming back to (12.13), (12.15), (12.17) the equations to a second order of approximation are:

$$n_\tau^{(1)} + \frac{a_0}{n_0} n^{(1)} n_\xi^{(1)} - a_0 n_\xi^{(2)} + n_0 u_\xi^{(2)} + \frac{a_0}{n_0} n^{(1)} n_\xi^{(1)} = 0$$

$$\frac{a_0}{n_0} n_\tau^{(1)} + \left(\frac{a_0}{n_0}\right)^2 n^{(1)} n_\xi^{(1)} - a_0 u_\xi^{(2)} + \frac{a_0^2}{n_0} n_\xi^{(2)} + \frac{A}{n_0} n^{(1)} n_\xi^{(1)} + \varepsilon^{\alpha-1} \frac{B}{n_0} n_{\xi\xi}^{(1)}$$

$$+ \varepsilon^{2\alpha-1} \frac{C}{n_0} n_{\xi\xi\xi}^{(1)} - \frac{n^{(1)}}{n_0^2} P_\xi^{(1)} = 0$$

with, by (12.16): $P_\xi^{(1)} = a_0^2 n_\xi^{(1)}$.

It will be seen that the term $- a_0 n_\xi^{(2)} + n_0 u_\xi^{(2)}$, can be eliminated from these two equations, giving:

$$n_\tau^{(1)} + \left(\frac{A}{2a_0} + \frac{a_0}{n_0}\right) n^{(1)} n_\xi^{(1)} + \varepsilon^{\alpha-1} \frac{B}{2a_0} n_{\xi\xi}^{(1)} + \varepsilon^{2\alpha-1} \frac{C}{2a_0} n_{\xi\xi\xi}^{(1)} = 0.$$

If $B \neq 0$ we take $\alpha = 1$ and ignore the last term, thus obtaining the Burgers equation.

If $B = 0$ we take $\alpha = 1/2$ and obtain the Korteweg-de Vries equation.

An Example: Gravity Waves in Shallow Water

On the assumption of a level bottom at a depth h below the level of the free surface, a flow-velocity of u, and no dissipation of energy, the stationary state corresponds to $h = h_0$, $u = 0$. It is known [41] that we can take

$$P = \tfrac{1}{2} g h^2 + \tfrac{1}{3} g h_0^2 (h h_{xx} - \tfrac{1}{2}(h_x)^2),$$

and $F \equiv 0$, where g is the acceleration of gravity, so that

$$a_0^2 = \left.\frac{\partial P}{\partial h}\right|_0 = g h_0, \quad h^{(1)} = \frac{h_0}{a_0} u^{(1)}$$

and we find:

$$A = g, \quad B = 0, \quad C = \tfrac{1}{3} g h_0^3$$

whence:

$$h_\tau^{(1)} + \frac{3}{2}\sqrt{\frac{g}{h_0}}\, h^{(1)} h_\xi^{(1)} + \frac{1}{6}\sqrt{\frac{g}{h_0}}\, h_0^3 h_{\xi\xi\xi}^{(1)} = 0$$

and an analogous equation for $u^{(1)}$.

By a simple change of variable, we can rewrite this equation (Korteweg-de Vries) in the form:

$$u_\tau + uu_\xi + \delta^2 u_{\xi\xi\xi} = 0. \tag{12.24}$$

It has a well-known solution, which has already been discussed in Chapter 2, representing the solitary wave:

$$u = u_\infty + \frac{u_0 - u_\infty}{\mathrm{ch}^2[(\xi - c\tau)\Gamma^{-1}]} \tag{12.25}$$

with

$$c = u_\infty + \frac{u_0 - u_\infty}{3}, \quad \Gamma = \delta \sqrt{\frac{12}{u_0 - u_\infty}}$$

where $u_0, u_\infty, u_0 > u_\infty$, are two parameters which can assume any value satisfying $u_0 > u_\infty$. The non-linear character is clearly marked in the sense that the amplitude, velocity and shape of the wave are closely interdependent. Zabusky and Kruskal [21] have investigated numerically the process of interaction of several solitary waves which are solutions of the equation (12.24) and have discovered that they retain their individuality. After colliding the waves emerge with the same shape, amplitude and velocity as before, and only the phase is changed. This phenomenon has since then been the subject of theoretical investigations [26, 21] based on so-called inverse scattering method, the principle of which we shall now describe.

2. The Inverse Scattering Method

It is useful to recall here a few essential facts on the Fourier transformation.

Consider the linear partial differential equation

$$\frac{\partial u}{\partial t}(x,t) = -i\omega\left(-i\frac{\partial}{\partial x}\right)u(x,t), \quad -\infty < x < +\infty \tag{12.26}$$

with the initial condition

$$u(x,0) = u_0(x) \tag{12.27}$$

where $u_0(x)$ is given and $\omega = \omega(k)$, a given polynomial in k, represents the law of dispersion, as can be recognised from the fact that $e^{i(kx-\omega t)}$ with k, ω constants, is a solution of (12.26) if and only if $\omega = \omega(k)$. To solve (12.26), (12.27) we can introduce the Fourier transform defined by

$$\hat{u}(k,t) = \frac{1}{\sqrt{2\pi}} \int_{-\infty}^{+\infty} u(x,t) e^{-ikx}\,dx$$

so that

$$\hat{u}_t(k,t) = -i\omega(k)\hat{u}(k,t)$$

or in other words: $\hat{u}(k,t) = \exp[-i\omega(k)t] \cdot \hat{u}(k,0)$ whence we derive the following scheme to find the solution:

$$u(x,0) = u_0(x) \to \hat{u}(k,0) \to \hat{u}(k,t) \to u(x,t)$$

which calls for two remarks.

1. There is a separability property which ensures that the passage from $\hat{u}(k,0)$ to $\hat{u}(k,t)$ can be effected by an operation which depends only on the value of k, the wave-number, which can therefore be regarded at this stage as a constant parameter.

2. The set of wave-numbers, i.e. the spectrum, needed to represent the solution $u(x,t)$ at any instant is invariant with respect to time, which here is trivially obvious since the set in question is $\mathbb{R}$.

The inverse-scattering method is designed to be a generalisation of this process which can be applied to non-linear problems.

Let us consider the Schrödinger equation:

(12.28)
$$\ddot{u} + (k^2 - q(x)) = 0 \quad \ddot{u} = \frac{\mathrm{d}^2 u}{\mathrm{d}x^2}$$

where the real-valued perturbation term $q(x)$ is assumed to vanish for $x = \pm\infty$ (more precise assumptions will be formulated in due course). As will be established later, we can define, for real k, a solution $u(x)$ of (12.28) characterised by the following asymptotic behaviour:

$$\begin{aligned} &u \sim e^{ikx} + \rho(k)e^{-ikx} \\ &\dot{u} \sim ik(e^{ikx} - \rho(k)e^{-ikx}), \quad && x \to -\infty \\ &u \sim \tau(k)e^{ikx}, \\ &\dot{u} \sim ik\tau(k)e^{ikx}, && x \to +\infty \end{aligned}$$

with $\rho(k)$, $\tau(k)$ reflection and transmission factors, depending only on the differential equation, i.e. on $q(x)$.

In addition there may exist, for certain complex values of k, with $\operatorname{Im} k > 0$, solutions which tend to zero as $x \to \pm\infty$. The corresponding values of k, or eigenvalues, are situated on the positive imaginary axis, $k = i\varkappa$, $\varkappa > 0$, and, under suitable assumptions regarding $q(x)$, are finite in number.

If we write $i\varkappa_p$, $\varkappa_p > 0$, $1 \leqslant p \leqslant N$ for the set of eigenvalues, and $u_p(x)$ for the corresponding eigenmodes, the latter are real and can be normalised so that

$$\int_{-\infty}^{+\infty} u_p^2(x)\,\mathrm{d}x = 1 \quad \text{and} \quad c_p = \lim_{x \to -\infty} (u_p(x)e^{-\varkappa_p x}) > 0.$$

The direct problem consists in determining, from (12.28), the functions $\rho(k)$, $\tau(k)$, the eigenvalues $i\varkappa_p$ and the numbers c_p. It is natural however to ask one's self to what extent appropriate information on all of these would enable us to invert the process and recover the perturbation term $q(x)$. The inversion problem accordingly consists in showing that one can, in a certain fashion, obtain $q(x)$ from known $\rho(k)$, $\varkappa_p$ and c_p.

Suppose now that q depends not only on x but also on another variable t, which here plays the role of a parameter, and let us consider once more, in these conditions, the equation (12.28). Obviously we can, for each value of t, study the direct problem and enquire as to what conditions would have to be satisfied by $q(x, t)$ in order that:

1. The spectrum should be invariant, i.e. the eigenvalues $i\varkappa_p$ should be independent of t.

2. The separability condition should be satisfied, that is to say ρ and c_p qua functions of t are capable of being described for each wave-number, by an equation of the form:

(12.29)
$$\rho_t = -2ik\Omega(k^2)\rho$$
$$(c_p^2)_t = -2ik_p\Omega(k_p^2)(c_p^2), \quad k_p = i\varkappa_p,$$

with $\Omega = \Omega_1/\Omega_2$ being a rational function of k^2.

It can be shown that if $\Omega(k^2)$ is given, then for the above conditions to be fulfilled, it will suffice that $q(x, t)$ satisfy the equation

(12.30)
$$\Omega_2(L^*)q_t = \Omega_1(L^*)q_x$$

where L^* is the operator

(12.31)
$$L^*g = -\tfrac{1}{4}g_{xx} + qg + \tfrac{1}{2}q_x \int_{-\infty}^{x} g\,\mathrm{d}x$$

adjoint to L:

(12.32)
$$Lf = \tfrac{1}{2}qf - \tfrac{1}{4}f_{xx} - \tfrac{1}{2}\int_{x}^{+\infty} qf_x\,\mathrm{d}x.$$

For example, with $\Omega(k^2) = 4k^2$, we get $q_t = 4L^*q_x$ that is

(12.33)
$$q_t = 6qq_x - q_{xxx}$$

which by changing q into $-q/6$ reduces to the Korteweg-de Vries equation.

The Method of Solution

If q satisfy an equation of type (12.30) then the properties of invariance and separability hold good as do the equations (12.29). The process of solving (12.30) once the initial value $q(x, 0)$ has been prescribed, may be summarised by the scheme:

$$q(x, 0) \to \{\rho(k, 0), \varkappa_p, c_p(0)\} \to \{\rho(k, t), \varkappa_p, c_p(t)\} \to q(x, t)$$

In other words, the direct problem has first to be solved at $t = 0$, and then the values of ρ, c_p at time t are found using (12.29), and then the inverse problem has to be solved for this value of t in order to obtain $q(x, t)$.

3. The Direct Problem

1. We consider the one-dimensional wave equation:

(12.34)
$$\ddot{u} + (k^2 - q(x))u = 0$$

where k is the real or complex wave-number, $q(x)$ is a real-valued locally integrable function which vanishes at infinity in the sense that

(12.35) $$\int_{-\infty}^{+\infty} x^2|q(x)|\,\mathrm{d}x < +\infty.$$

By the transformation $u = e^{ikx}v$, (12.34) becomes:

(12.36) $$\ddot{v} + 2ik\dot{v} - q(x)v = 0$$

and we seek the solution of this equation for which

(12.37) $$v(+\infty) = 1, \quad \dot{v}(+\infty) = 0.$$

Using a notation which brings out the fact that v depends on the wave-number, we can replace the system of equations (12.36) and (12.37) by the integral equation:

(12.38) $$v(x,k) = 1 - \int_x^{+\infty} (2ik)^{-1}\cdot(1 - e^{2ik(x'-x)})\cdot q(x')v(x',k)\,\mathrm{d}x',$$

because it is clear that any continuous solution of (12.38) that is bounded near $+\infty$ must satisfy (12.36) and (12.37). The solving of (12.38) is based on the classical method of successive approximations:

(12.39) $$v_{p+1}(x,k) = 1 - \int_x^{+\infty} (2ik)^{-1}\cdot(1 - e^{2ik(x'-x)})\cdot q(x')v_p(x',k)\,\mathrm{d}x',$$

$$v_0(x,k) = 1,$$

and makes use of the observation that:

(12.40) $$(2ik)^{-1}(1 - e^{2ik(x'-x)}) = -\int_0^{x'-x} e^{2iks}\,\mathrm{d}s$$

is holomorphic with respect to k throughout the whole complex plane and satisfies

(12.41) $$\left|\int_0^{x'-x} e^{2iks}\,\mathrm{d}s\right| \leqslant x'-x, \quad \text{for } x'-x \geqslant 0, \quad \operatorname{Im} k \geqslant 0.$$

In view of this, with $\beta > 0$ chosen so that

(12.42) $$\int_\beta^{+\infty} x'|q(x')|\,\mathrm{d}x' < 1/2,$$

which can always be done in view of (12.35), it is easy to show that the approximations $v_p(x,k)$ are continuous in x, k and satisfy:

(12.43) $$\begin{aligned} &|v_p(x,k)| \leqslant 2, \\ &|v_{p+1}(x,k) - v_p(x,k)| \leqslant \tfrac{1}{2} \sup_{x' \geqslant \beta} |v_p(x',k) - v_{p-1}(x',k)| \end{aligned}$$

in $x \geqslant \beta$, $\operatorname{Im} k \geqslant 0$. These estimates ensure that $v_p(x,k)$ converges uniformly to a solution $v(x,k)$ of (12.38) which is continuous in (x,k) in the region $x \geqslant \beta$, $\operatorname{Im} k \geqslant 0$ and is furthermore unique.

2. Suppose $v_p(x,k)$ to be holomorphic with respect to k in $\operatorname{Im} k \geqslant 0$, for all $x \geqslant \beta$, meaning by this, if $\operatorname{Im} k = 0$, that

$$\lim_{\Delta k \to 0, \operatorname{Im} \Delta k \geqslant 0} \frac{v_p(x, k+\Delta k) - v_p(x,k)}{\Delta k}$$

exists; suppose also $\dfrac{\partial v_p}{\partial k}(x,k)$ to be continuous in (x,k) throughout the region $x \geqslant \beta$ and $\operatorname{Im} k \geqslant 0$, and to satisfy within this region

(12.44) $$\left|\frac{\partial v_p}{\partial k}(x,k)\right| < 2.$$

It is then easy to see, on the hypothesis that β satisfy (12.42) and

(12.45) $$\int_{\beta}^{+\infty} x'^2 |q(x')| \mathrm{d}x' < \tfrac{1}{4}$$

made possible by (12.35), that the properties which have been assumed for $v_p(x,k)$ are also true when p is replaced by $p+1$, the partial derivatives $\dfrac{\partial v_{p+1}}{\partial k}$ being defined by:

(12.46) $$\begin{aligned}\frac{\partial v_{p+1}}{\partial k}(x,k) = &- \int_x^{+\infty} \frac{\partial}{\partial k}[(2ik)^{-1}\cdot(1-e^{2ik(x'-x)})]\cdot q(x') v_p(x',k)\,\mathrm{d}x' \\ &- \int_x^{+\infty} (2ik)^{-1}\cdot(1-e^{2ik(x'-x)})\cdot q(x') \frac{\partial v_p}{\partial k}(x',k)\,\mathrm{d}x'.\end{aligned}$$

We shall make use of the identity

$$\frac{\partial}{\partial k}[(2ik)^{-1}\cdot(1-e^{2ik(x'-x)})] = -2i \int_0^{x'-x} s e^{2iks}\,\mathrm{d}s$$

giving

(12.47) $$\left|\frac{\partial}{\partial k}[(2ik)^{-1}\cdot(1-e^{2ik(x'-x)})]\right| \leqslant |x'-x|^2, \; x'-x \geqslant 0, \operatorname{Im} k \geqslant 0$$

and also of the properties of $v_p(x,k)$ already found or assumed in order to establish that the integrals on the right of (12.46) converge uniformly with respect to (x,k) in $x \geqslant \beta$, $\operatorname{Im} k \geqslant 0$. This property of v_p ensures firstly the continuity in (x,k) of the right-hand side and secondly that v_{p+1} is holomorphic in $\operatorname{Im} k \geqslant 0$ and has a partial derivative $\partial v_{p+1}/\partial k$ defined by (12.46) which is continuous in (x,k) within the region $x \geqslant \beta$, $\operatorname{Im} k \geqslant 0$ as already mentioned and which satisfies in this region the inequality $|(\partial v_{p+1}/\partial k)(x,k)| < 2$.

Provided that the conditions (12.42), (12.45) imposed on β are satisfied, it can easily be shown, starting from the recurrence relation (12.46), that the sequence $(\partial v_p/\partial k)(x,k)$ converges uniformly for $p \to \infty$ in $\operatorname{Im} k \geqslant 0$, $x \geqslant \beta$, to a limit, evidently continuous in x, k, which is none other than the derivative $\dfrac{\partial v}{\partial k}$ as is seen by considering the limit as $p \to \infty$ in the relation

$$v_p(x,k) - v_p(x,\tilde{k}) = \int_{\tilde{k}}^{k} (\partial v_p/\partial k)\,\mathrm{d}k, \quad \operatorname{Im} k \geqslant 0, \quad \operatorname{Im} \tilde{k} \geqslant 0$$

Thus for all $x \geqslant \beta$, the solution $v(x,k)$ is holomorphic in k for $\operatorname{Im} k \geqslant 0$, and its derivative $\partial v/\partial k$ is continuous in (x,k) and satisfies the equation:

$$\frac{\partial v}{\partial k} = -\int_x^{+\infty} \frac{\partial}{\partial k}[(2ik)^{-1}\cdot(1-e^{2ik(x'-x)})]\cdot q(x')v(x',k)\,dx'$$
$$-\int_x^{+\infty}(2ik)^{-1}\cdot(1-e^{2ik(x'-x)})\cdot q(x')\frac{\partial v}{\partial k}(x',k)\,dx'$$

and moreover, by (12.44) satisfies:

(12.48) $$\left|\frac{\partial v}{\partial k}\right| \leqslant 2, \quad \text{in} \quad x \geqslant \beta, \operatorname{Im} k \geqslant 0.$$

3. From the equation derived by differentiating (12.38) with respect to x, we have:

(12.49) $$\dot{v}(x,k) = -\int_x^{+\infty} e^{2ik(x'-x)}q(x')v(x',k)\,dx'$$

and it is easily verified that $\dot{v}(x,k)$ is continuous in (x,k) in the region $x \geqslant \beta$, $\operatorname{Im} k \geqslant 0$, holomorphic in k in $\operatorname{Im} k \geqslant 0$ for all $x \geqslant \beta$, and that the derivative $\partial v/\partial k$ is continuous in $x \geqslant \beta$, $\operatorname{Im} k \geqslant 0$.

Lastly it is essential to note that the solution $v(x,k)$ can be extended by analytic continuation for all x, starting from the initial values $v(x_0,k)$, $\dot{v}(x_0,k)$ taken at a point $x_0 > \beta$. The continuity and differentiability with respect to k of $v(x,k)$ and $\dot{v}(x,k)$ continue to hold for the extended solution, i.e. for all real x and $\operatorname{Im} k \geqslant 0$.

4. It should be added that we could just as well have defined a solution of (12.34) by specifying its asymptotic behaviour as $x \to -\infty$.

To make the necessary distinctions, let us agree to denote by $u^+(x,k)$ the solution, viz.:

(12.50) $$u^+(x,k) = v^+(x,k)e^{ikx}, \quad v^+(+\infty,k) = 1, \quad \dot{v}^+(+\infty,k) = 0,$$

and by $u^-(x,k)$ the one defined by the conditions:

(12.51) $$u^-(x,k) = v^-(x,k)e^{-ikx}, \quad v^-(-\infty,k) = 1, \quad \dot{v}^{-1}(-\infty,k) = 0.$$

One can verify that $v^-(x,k)$ is a solution of the integral equation:

(12.52) $$v^-(x,k) = 1 - \int_{-\infty}^{x}(2ik)^{-1}\cdot(1-e^{-2ik(x'-x)})q(x')v^-(x',k)\,dx'$$

and show that $v^-(x,k)$ is holomorphic with respect to k in $\operatorname{Im} k \geqslant 0$, and that $(\partial v^-/\partial k)(x,k)$ is continuous in (x,k) in the region $\operatorname{Im} k \geqslant 0$.

3.1. The Eigenvalue Problem

To construct a solution of (12.34) which vanishes at $\pm\infty$ we can try to join up the solutions u^+ and u^- at $x=0$. The problem thus consists of finding a k with $\operatorname{Im} k > 0$ such that

$$u^+(0,k) = \sigma u^-(0,k)$$
$$\dot{u}^+(0,k) = \sigma \dot{u}^-(0,k)$$

for a finite non-zero real or complex σ, or in other words with the help of the $v^{\pm}$:

$$\begin{gathered} v^+(0,k) = \sigma v^-(0,k) \\ ikv^+(0,k) + \dot{v}^+(0,k) = \sigma(-ikv^-(0,k) + \dot{v}^-(0,k)) \end{gathered} \tag{12.53}$$

i.e. eliminating σ:

$$f(k) = 0 \tag{12.54}$$

with

$$\begin{aligned} f(k) = 2ikv^+(0,k)v^-(0,k) &+ \dot{v}^+(0,k)v^-(0,k) \\ &- \dot{v}^-(0,k)v^+(0,k) \end{aligned} \tag{12.55}$$

holomorphic with respect to k in $\operatorname{Im} k \geqslant 0$.

In point of fact a simple calculation shows that $f(k)$ is simply the Wronskian $\dot{u}^+u^- - \dot{u}^-u^+$ which is known to be independent of x, so that one can write:

$$\begin{aligned} f(k) = 2ikv^+(x,k)v^-(x,k) &+ \dot{v}^+(x,k)v^-(x,k) \\ &- \dot{v}^-(x,k)v^+(x,k). \end{aligned} \tag{12.56}$$

Lastly if u_1, u_2 are solutions vanishing at $\pm\infty$, which correspond to the eigenvalues k_1, k_2 satisfying (12.54) in $\operatorname{Im} k > 0$, we can deduce by integration along the real axis of $u_1\ddot{u}_2 - \ddot{u}_1u_2 = (k_2^2 - k_1^2)u_1u_2$ that:

$$(k_2^2 - k_1^2)\int_{-\infty}^{+\infty} u_1u_2\,dx = 0. \tag{12.57}$$

If (k_1, u_1) is one eigenmode then $(-\overline{k_1}, \overline{u_1})$ is another because $q(x)$ is real valued (a bar being used to denote the complex conjugate) and applying (12.57) we find:

$$(\overline{k_1^2} - k_1^2)\int_{-\infty}^{+\infty} |u_1|^2\,dx = 0$$

whence it follows that k_1 must necessarily be of the form $i\varkappa$ with $\varkappa$ real and positive. The eigenvalues are therefore pure imaginaries, the associated eigenmodes are real and those corresponding to two distinct eigenvalues are orthogonal.

On Some Estimates

1. It follows from (12.38) that:

$$|v^+(x,k)| \leqslant 1 + |k|^{-1}\cdot\int_x^{+\infty} |q(x')||v^+(x',k)|\,dx', \quad \operatorname{Im} k \geqslant, k \neq 0$$

hence by Gronwall's Lemma [42]:

$$|v^+(x,k)| \leqslant \exp\left(|k|^{-1}\cdot\int_x^{+\infty} |q(x')|\,dx'\right), \quad \operatorname{Im} k \geqslant 0, k \neq 0. \tag{12.58}$$

In particular $v^+(x,k)$ is bounded with respect to $x \in \mathbb{R}$, for fixed k, $\operatorname{Im} k \geqslant 0$, $k \neq 0$. Using (12.38) once more, we obtain the result that, for $\operatorname{Im} k > 0$, $\lim\limits_{x \to -\infty} v^+(x,k)$

exists, is finite, and is given by:

$$\lim_{x\to-\infty} v^+(x,k) = 1 - \frac{1}{2ik}\cdot\int_{-\infty}^{+\infty} q(x')v^+(x',k)\,\mathrm{d}x', \quad \operatorname{Im} k > 0 \tag{12.59}$$

and in view of (12.49):

$$\lim_{x\to-\infty} \dot{v}^+(x,k) = 0, \quad \operatorname{Im} k > 0 \tag{12.60}$$

whence the following representation of $f(k)$ in $\operatorname{Im} k > 0$, obtained by letting x tend to $-\infty$ in (12.56):

$$f(k) = 2ik - \int_{-\infty}^{+\infty} q(x')v^+(x',k)\,\mathrm{d}x', \quad \operatorname{Im} k > 0 \tag{12.61}$$

which holds moreover in the region $\operatorname{Im} k \geqslant 0$, $k \neq 0$, in consequence of (12.58) and the continuity properties of $v^+(x,k)$ and $f(k)$.

Analogous considerations could be developed for $v^-(x,k)$ and would lead to

$$f(k) = 2ik - \int_{-\infty}^{+\infty} q(x')v^-(x',k)\,\mathrm{d}x', \quad \operatorname{Im} k \geqslant 0, k \neq 0. \tag{12.62}$$

2. We deduce from (12.38) with the help of (12.40), (12.41) the inequality

$$|v^+(x,k)| \leqslant 1 + \int_x^{+\infty} (x'-x)|q(x')|\cdot|v^+(x',k)|\,\mathrm{d}x', \quad \operatorname{Im} k \geqslant 0$$

for $x \geqslant -l$, where $l > 0$ is given,

$$|v^+(x,k)| \leqslant 1 + \int_x^{+\infty} (|x'|+l)|q(x')||v^+(x',k)|\,\mathrm{d}x'$$

i.e.

$$|v^+(x,k)| \leqslant \exp \int_x^{+\infty} (|x'|+l)|q(x')|\,\mathrm{d}x', \quad \forall x \geqslant -l, \operatorname{Im} k \geqslant 0. \tag{12.63}$$

Similarly we would obtain from (12.52)

$$|v^-(x,k)| \leqslant 1 + \int_{-\infty}^{x} (|x'|+l)|q(x')||v^-(x',k)|\,\mathrm{d}x', \quad x \leqslant l, \operatorname{Im} k \geqslant 0$$

or:

$$|v^-(x,k)| \leqslant \exp \int_{-\infty}^{x} (|x'|+l)|q(x')|\,\mathrm{d}x', \quad \forall x < l, \operatorname{Im} k \geqslant 0. \tag{12.64}$$

3. Starting from (12.38) written in the form:

$$\begin{aligned} v^+(x,k) - 1 = &- \int_x^{+\infty} (2ik)^{-1}\cdot(1 - e^{2ik(x'-x)})q(x')\,\mathrm{d}x' \\ &- \int_x^{+\infty} (2ik)^{-1}\cdot(1 - e^{2ik(x'-x)})q(x')(v^+(x',k) - 1)\,\mathrm{d}x' \end{aligned} \tag{12.65}$$

we obtain for $\operatorname{Im} k \geqslant 0$, $k \neq 0$:

$$|v^+(x,k)-1| \leqslant |k|^{-1} \int_{-\infty}^{+\infty} |q(x')|\,\mathrm{d}x' + |k|^{-1} \int_{x}^{+\infty} |q(x')|\cdot|v^+(x',k)-1|\,\mathrm{d}x'$$

and by Gronwall's Lemma:

$$|v^+(x,k)-1| \leqslant |k|^{-1}\cdot \int_{-\infty}^{+\infty} |q(x')|\,\mathrm{d}x' \cdot \exp\left(|k|^{-1}\cdot \int_{x}^{+\infty} |q(x')|\,\mathrm{d}x' \right)$$

and in particular

$$(12.66) \qquad |v^+(x,k)-1| \leqslant \frac{C}{|k|}, \quad \text{in} \quad \operatorname{Im} k \geqslant 0, |k| \geqslant 1$$

with C a numerical constant,

$$C = \left(\int_{-\infty}^{+\infty} |q(x')|\,\mathrm{d}x' \right) \exp \int_{-\infty}^{+\infty} |q(x')|\,\mathrm{d}x'.$$

Returning to (12.65) we can write:

$$(12.67) \qquad v^+(x,k) - 1 = -(2ik)^{-1}\cdot \int_{x}^{+\infty} (1 - e^{2ik(x'-x)}q(x')\,\mathrm{d}x' + \delta^+(x,k)$$

$$|\delta^+(x,k)| < C_1 |k|^{-2}, \quad \operatorname{Im} k \geqslant 0, |k| \geqslant 1$$

with C_1 a numerical constant; incidentally in view of (12.63), $\delta^+(x,k)$ is bounded in $\operatorname{Im} k \geqslant 0$, $x \geqslant -l$.

Similar results can be worked out for $v^-(x,k)$ from (12.52), viz.:

$$(12.68) \qquad v^-(x,k) - 1 = -(2ik)^{-1} \int_{-\infty}^{x} (1 - e^{-2ik(x'-x)})q(x')\,\mathrm{d}x' + \delta^-(x,k)$$

$$|v^-(x,k)-1| \leqslant C|k|^{-1} \quad \text{and} \quad |\delta^-(x,k)| \leqslant C_1|k|^{-2}, \quad \operatorname{Im} k \geqslant 0, |k| \geqslant 1$$

with C, C_1 numerical constants, and $\delta^-(x,k)$ bounded in $\operatorname{Im} k \geqslant 0$, $x < l$.

4. We obtain from (12.65) again making use of (12.40), (12.41), the result that, for $x \geqslant 0$:

$$|v^+(x,k)-1| \leqslant \int_{x}^{+\infty} x'|q(x')|\,\mathrm{d}x' + \int_{x}^{+\infty} x'|q(x')|\cdot|v^+(x',k)-1|\,\mathrm{d}x'$$

or

$$(12.69) \quad |v^+(x,k)-1| \leqslant \int_{x}^{+\infty} x'|q(x')|\,\mathrm{d}x' \cdot \exp\left(\int_{x}^{+\infty} x'|q(x')|\,\mathrm{d}x' \right), \quad x \geqslant 0, \operatorname{Im} k \geqslant 0.$$

We would similarly obtain:

$$(12.70) \qquad |v^-(x,k)-1| \leqslant \int_{-\infty}^{x} |x'||q(x')|\,\mathrm{d}x' \cdot \exp\left(\int_{-\infty}^{x} |x'|\cdot|q(x')|\,\mathrm{d}x' \right),$$

$$x \leqslant 0, \operatorname{Im} k \geqslant 0.$$

5. It is important to observe, for the sequel, that by replacing (12.35) by the weaker hypothesis $\int_{-\infty}^{+\infty} |q(x')|\,\mathrm{d}x' < +\infty$, we can still solve the equations (12.38) and (12.52) defining $v^+(x, k)$ and $v^-(x, k)$ respectively; it would suffice to use in the successive approximation process, estimates of the type:

$$\left| \int_x^{+\infty} (2ik)^{-1}(1 - e^{2ik(x'-x)})q(x')v(x', k)\,\mathrm{d}x' \right| \leqslant |k|^{-1}$$
$$\cdot \int_x^{+\infty} |q(x')|\,|v(x', k)|\,\mathrm{d}x', \quad \operatorname{Im} k \geqslant 0, k \neq 0.$$

These, as can be seen, depend on k, whereas the condition (12.35) enabled us to free ourselves of this dependence, but this does not jeopardise the construction of the solutions v^+, v^-. The latter are defined for all k in $\operatorname{Im} k \geqslant 0$, $k \neq 0$; the holomorphic properties hold in $\operatorname{Im} k > 0$ (the real axis excluded) and among the estimates obeyed by these solutions, those established under the conditions $\operatorname{Im} k \geqslant 0$, k not belonging to a neighbourhood of zero, remain valid. In particular this will apply to the estimates (12.58), (12.66), (12.69).

In actual fact the hypotheses (12.35) is really only essential to establish that the number of eigenvalues is finite.

The Finiteness of the Set of Eigenvalues

We have seen that the eigenvalues are numbers $i\varkappa$, where i is the complex unity and $\varkappa$ real positive. It is clear from (12.58) and (12.61) that $f(i\varkappa) \sim -2\varkappa$ when $\varkappa \to +\infty$ and consequently the set of eigenvalues is bounded; each of these being a zero of a function holomorphic in $\operatorname{Im} k > 0$, is isolated and if it can be established that 0 is not a limit point of the spectrum, it will then follow that the set of distinct eigenvalues is finite and can therefore be denoted thenceforth by $i\varkappa_1, i\varkappa_2, \ldots, i\varkappa_N$ [29].

We give a new proof of this result based on a reductio ad absurdum argument. Suppose that 0 were a limit point of the spectrum; there would then be an infinity of eigenvalues $i\varkappa_n$, $\varkappa_n > 0$, with $\lim_{n \to \infty} \varkappa_n = 0$.

Since $f(k)$ defined by (12.55) is continuous in the region $\operatorname{Im} k \geqslant 0$ we should have $f(0) = \lim_{n \to \infty} f(i\varkappa_n) = 0$, which would imply that 0 would be an eigenvalue; thus using the notation $\omega^+(x) = u^+(x, 0) = v^+(x, 0)$, $\omega^-(x) = u^-(x, 0) = v^-(x, 0)$, there would be a finite, non-zero number σ such that

$$\omega^+(0) = \sigma\omega^-(0)$$
$$\dot{\omega}^+(0) = \sigma\dot{\omega}^-(0)$$

and for every integer n, a finite non-zero σ_n such that

$$u^+(0, i\varkappa_n) = \sigma_n u^-(0, i\varkappa_n)$$
$$\dot{u}^+(0, i\varkappa_n) = \sigma_n \dot{u}^-(0, i\varkappa_n).$$

By reason of the continuity of $u^\pm(x, k)$, $\dot{u}^\pm(x, k)$ with respect to k in the region $\operatorname{Im} k \geqslant 0$ and also because $\omega^-(0)$, $\dot{\omega}^-(0)$ cannot both be zero simultaneously, it is clear that:

(12.71) $$\lim_{n\to\infty} \sigma_n = \sigma \neq 0.$$

By (12.69) we can find a positive constant b not depending on k such that:

(12.72) $$|v^+(x,k) - 1| < \tfrac{1}{2}, \quad \forall x \geqslant b, \quad \operatorname{Im} k \geqslant 0$$

and similarly by (12.70) a negative constant a independent of k such that:

(12.73) $$|v^-(x,k) - 1| < \tfrac{1}{2}, \quad \forall x \leqslant a, \quad \operatorname{Im} k \geqslant 0.$$

Let us denote by u_n the real eigenmode which corresponds to the eigenvalue $i\varkappa_n$ and which can be written in the form:

$$u_n(x) = v^+(x, i\varkappa_n) e^{-\varkappa_n x}.$$

It follows from (12.72) that

$$\frac{e^{-\varkappa_n x}}{2} < u_n(x) < \tfrac{3}{2} \cdot e^{-\varkappa_n x}, \quad \forall x \geqslant b, \quad \forall n$$

whence:

(12.74) $$\int_b^{+\infty} u_n(x) u_m(x)\,\mathrm{d}x > \frac{1}{4} \int_b^{+\infty} e^{-(\varkappa_n + \varkappa_m)x}\,\mathrm{d}x = \frac{e^{-(\varkappa_n+\varkappa_m)b}}{4(\varkappa_n + \varkappa_m)}.$$

For $a < x < b$, one sees, taking $l = -a$ in (12.63), that $v^+(x, i\varkappa_n)$ and also $u_n(x)$ are uniformly bounded with respect to n, so that

(12.75) $$\operatorname*{Sup}_{n,m} \left| \int_a^b u_n(x) u_m(x)\,\mathrm{d}x \right| < +\infty.$$

For $x \leqslant a$ and using the representation: $u_n(x) = \sigma_n v_n^-(x, i\varkappa_n) e^{\varkappa_n x}$ we obtain by (12.73)

$$\frac{\sigma_n}{2} e^{\varkappa_n x} < u_n(x) < \tfrac{3}{2} \sigma_n e^{\varkappa_n x}, \quad \text{if} \quad \sigma_n > 0$$

the inequalities being reversed if $\sigma_n < 0$. It is also clear by (12.71) that the σ_n are all of the same sign from a certain point onwards, so that one can write, whatever be the sign

$$u_n(x) u_m(x) > \frac{\sigma_n \sigma_m}{4} e^{(\varkappa_n + \varkappa_m)x}, \quad \forall x \leqslant a$$

and hence by integration:

(12.76) $$\int_{-\infty}^{a} u_n(x) u_m(x)\,\mathrm{d}x > \frac{\sigma_n \sigma_m}{4} \frac{e^{(\varkappa_n+\varkappa_m)a}}{\varkappa_n + \varkappa_m}.$$

Bearing in mind that $\sigma_n \sigma_m \to \sigma^2 > 0$, if n and m become infinite, and that $\lim_{n\to\infty} \varkappa_n = 0$, we see by bringing together (12.74), (12.75), and (12.76) that $\int_{-\infty}^{+\infty} u_n(x) u_m(x)\,\mathrm{d}x$ would become infinite along with n and m, and this contradicts for $n \neq m$ the orthogonality property of the eigenmodes.

Thus the point 0 cannot be a limit point of the spectrum, and the set of eigenvalues must be finite.

3.2. Transmission and Reflection Coefficients

In parallel with the solution $u^+(x,k)$ of (12.34), represented by (12.50) and which we shall in future denote by:

(12.77) $u_1^+(x,k)=e^{ikx}v_1^+(x,k),\quad v_1^+(+\infty,k)=1,\quad \dot{v}_1^+(+\infty,k)=0,\quad \operatorname{Im}k\geqslant 0$

we can, by changing k into $-k$, introduce the solution:

(12.78)
$$u_2^+(x,k)=u_1^+(x,-k)$$

that is:

(12.79) $u_2^+(x,k)=e^{-ikx}v_2^+(x,k),\quad v_2^+(+\infty,k)=1,\quad \dot{v}_2^+(+\infty,k)=0,$

defined this time in $\operatorname{Im}k\leqslant 0$, and enjoying in this half-plane the same properties of regularity as $u_1^+(x,k)$ does in $\operatorname{Im}k\geqslant 0$. It can moreover also be seen, by (12.77), (12.79), that:

$$u_2^+(x,k)=\overline{u_1^+(x,k)},\quad \text{for real } k,$$

and by (12.78):

(12.80)
$$\begin{aligned} v_1^+(x,-k)&=\overline{v_1^+(x,k)}\\ v_2^+(x,-k)&=\overline{v_2^+(x,k)},\quad k \text{ real.}\end{aligned}$$

Similar operations can be carried out starting from the solution $u^-(x,k)$ represented by (12.51) and denoted in what follows by:

(12.81) $u_2^-(x,k)=e^{-ikx}v_2^-(x,k),\quad v_2^-(-\infty,k)=1,\quad \dot{v}_2^-(-\infty,k)=0,\quad \operatorname{Im}k\geqslant 0.$

We thus define

$$u_1^-(x,k)=u_2^-(x,-k)$$

or

(12.82) $u_1^-(x,k)=e^{ikx}v_1^-(x,k),\quad v_1^-(-\infty,k)=1,\quad \dot{v}_1^-(-\infty,k)=0,\quad \operatorname{Im}k\leqslant 0$

and obtain for real k

$$u_2^-(x,k)=\overline{u_1^-(x,k)}$$

and hence:

(12.83)
$$\begin{aligned} v_1^-(x,-k)&=\overline{v_1^-(x,k)}\\ v_2^-(x,-k)&=\overline{v_2^-(x,k)},\quad k \text{ real.}\end{aligned}$$

When k is real the solutions (12.77), (12.81), (12.82) are simultaneously defined and are not independent, i.e. there are functions $\rho(k)$, $\tau(k)$ not depending on x, such that one can write:

$$(12.84)\qquad \begin{aligned} u_1^-(x,k)+\rho(k)u_2^-(x,k)&=\tau(k)u_1^+(x,k)\\ \dot u_1^-(x,k)+\rho(k)\dot u_2^-(x,k)&=\tau(k)\dot u_1^+(x,k)\end{aligned}$$

for all x. We can interpret $\rho(k)$, $\tau(k)$ physically, by noting that with any incident wave represented by $u_1^-(x,k)$, and, asymptotically at $x=-\infty$, by e^{ikx}, there is associated a wave $\rho(k)u_2^-(x,k)\sim\rho(k)e^{-ikx}$ at $x=-\infty$, which can be qualified as a reflected wave, with $\rho(k)$ representing the amplitude at $x=-\infty$. This combination is equivalent to $\tau(k)u_1^+(x,k)$ and its asymptotic behaviour at $x=+\infty$ where it behaves like $\tau(k)e^{ikx}$ shows clearly enough that we have a transmitted wave of amplitude $\tau(k)$. For these reasons we shall call $\rho(k)$ the reflection coefficient and $\tau(k)$ the transmission coefficient.

We see also from (12.84) that $\tau u_1^+\sim e^{ikx}+\rho e^{-ikx}$, and $\tau\dot u_1^+\sim ik(e^{ikx}-\rho e^{-ikx})$ at $x\sim-\infty$ so that:

$$(12.85)\qquad \begin{aligned} \tau(iku_1^+ +\dot u_1^+)&\sim 2ike^{ikx}, && x\to-\infty\\ \tau(iku_1^+ -\dot u_1^+)&\sim 2ik\rho e^{-ikx}, && x\to-\infty,\quad \text{for } k \text{ real}\end{aligned}$$

On the other hand it follows from (12.77), (12.38), (12.49) that:

$$iku_1^+ +\dot u_1^+ =(2ikv_1^+ +\dot v_1^+)e^{ikx}=\left(2ik-\int_x^{+\infty} q(x')v_1^+(x',k)\,dx'\right)e^{ikx}$$

$$iku_1^+ -\dot u_1^+ =-\dot v_1^+ e^{ikx}=e^{-ikx}\cdot\int_x^{+\infty} e^{2ikx'}q(x')v_1^+(x',k)\,dx'$$

whence by (12.85), we have for real non-zero k:

$$(12.86)\qquad \tau(k)=2ik\cdot\left(2ik-\int_{-\infty}^{+\infty} q(x')v_1^+(x',k)\,dx'\right)^{-1}$$

$$(12.87)\qquad \rho(k)=\frac{\tau(k)}{2ik}\cdot\int_{-\infty}^{+\infty} e^{2ikx'}q(x')v_1^+(x',k)\,dx'$$

and we note incidentally that, as a consequence of (12.80)

$$(12.88)\qquad \tau(-k)=\overline{\tau(k)},\quad \rho(-k)=\overline{\rho(k)},$$

again with k real.

However we see from (12.61) that we can write:

$$(12.89)\qquad \tau(k)=\frac{2ik}{f(k)}$$

and since we know by (12.55) that $f(k)$ is holomorphic in $\operatorname{Im} k\geqslant 0$, formula (12.89) allows us to define $\tau(k)$ in this half-plane. Thus $\tau(k)$ is a meromorphic function of k in $\operatorname{Im} k\geqslant 0$, except perhaps in $k=0$, its singularities being poles at a finite number of points corresponding to the zeros of $f(k)$. The nature of the point $k=0$ will be clarified later. Note also that, in the general case, the definition of $\rho(k)$ cannot be extended to points not on the real axis.

With $u(x,k)=\tau(k)u_1^+(x,k)$, k real, we see from (12.84) that:

(12.90)
$$u \sim e^{ikx}+\rho(k)e^{-ikx}$$
$$\dot{u} \sim ik(e^{ikx}-\rho(k)e^{-ikx}), \quad \text{when} \quad x \to -\infty$$

and

(12.91)
$$u \sim \tau(k)e^{ikx}$$
$$\dot{u} \sim ik\tau(k)e^{ikx}, \quad \text{when} \quad x \to +\infty.$$

Since $u(x,k)$ and $\overline{u(x,k)}$ are solutions of the same differential equation (12.34) with real coefficients, their Wronskian is independent of x and in particular:

$$\lim_{x\to+\infty}(u\dot{\bar{u}}-\dot{u}\bar{u}) = \lim_{x\to-\infty}(u\dot{\bar{u}}-\dot{u}\bar{u})$$

i.e. in view of the estimates (12.90), (12.91):

(12.92)
$$|\rho(k)|^2+|\tau(k)|^2=1, \quad k \text{ real}.$$

Since $\tau(k)=0$ for a real k can be excluded as this would imply by (12.84) a linear dependence between $u_1^-(x,k)$ and $u_2^-(x,k)$ which is impossible, we can see as a consequence of (12.92) that $|\rho(k)|<1$, for all real k.

Reverting to (12.89) and remembering that $f(k)$ defined by (12.55) is holomorphic in $\operatorname{Im} k \geqslant 0$, and in particular at $k=0$, we see that on the hypothesis of $f(0)=0$ we should have:

$$\lim_{k\to 0}\tau(k) = \lim_{k\to 0, \operatorname{Im} k \geqslant 0}\frac{2ik}{f(k)} = \frac{2i}{f'(0)}$$

and since, by (12.92), $|\tau(k)|<1$, for real k, it is clear that $f'(0)$ is necessarily non-zero. This result shows that we can, in all circumstances, extend the domain of definition of the function $\tau(k)$ by continuity to include the point $k=0$, the value assigned to the function at this point being finite.

Finally it is clear from (12.89) and from (12.61), (12.58) that:

(12.93)
$$\tau(k)=1+O(k^{-1}),$$

in $\operatorname{Im} k \geqslant 0$, for $|k|$ large.

As for $\rho(k)$, we can say, in view of (12.87), that:

$$\lim_{k\in\mathbb{R}, |k|\to\infty} |k\rho(k)|=0$$

or more precisely, by (12.66), that:

$$k^2\left(\rho(k)-(2ik)^{-1}\cdot\int_{-\infty}^{+\infty} e^{2ikx'}q(x')\,dx'\right)$$

is bounded for real k tending to $\pm\infty$.

Eigenvalues (Continued)

Consider the solutions of type (12.77) associated with distinct values k_1, k_2 in the half-plane $\operatorname{Im} k \geqslant 0$, denoted by

$$w_1(x) = u_1^+(x, k_1), \quad w_2(x) = u_1^+(x, k_2).$$

By integrating the equation $w_2 \ddot{w}_1 - w_1 \ddot{w}_2 = (k_2^2 - k_1^2) w_1 w_2$ with respect to x over the interval $(0, b)$, $b > 0$, we obtain:

$$(k_1^2 - k_2^2) \int_0^b w_1(x) w_2(x)\, dx + (w_2 \dot{w}_1 - w_1 \dot{w}_2)\Big|_0^b = 0 \tag{12.94}$$

Let us choose $k_1 = i\varkappa + \varepsilon$, $k_2 = i\varkappa - \varepsilon$, $\varkappa$ with $\varkappa$ any positive real number and note that we can write

$$w_1(x) = u_1^+(x, i\varkappa) + \varepsilon \frac{\partial}{\partial k} u_1^+(x, i\varkappa) + O(\varepsilon^2)$$

$$\dot{w}_1(x) = \dot{u}_1^+(x, i\varkappa) + \varepsilon \frac{\partial}{\partial k} \dot{u}_1^+(x, i\varkappa) + O(\varepsilon^2)$$

with analogous expressions for $w_2(x)$, $\dot{w}_2(x)$ obtained by changing ε into $-\varepsilon$, which are all valid uniformly in x on any compact interval.

After substituting these in (12.94) and taking the limit as $\varepsilon \to 0$, we get

$$2ik \int_0^b (u_1^+(x, i\varkappa))^2\, dx + \left[u_1^+(x, i\varkappa) \frac{\partial}{\partial k} \dot{u}_1^+(x, i\varkappa) + \dot{u}_1^+(x, i\varkappa) \frac{\partial}{\partial k} u_1^+(x, i\varkappa) \right]_0^b = 0$$

or making b tend to $+\infty$, and expressing the integrated part by means of (12.77) in terms of v^+:

$$2i\varkappa \int_0^{+\infty} (u_1^+(x, i\varkappa))^2\, dx = \left(v_1^+ \frac{\partial \dot{v}_1^+}{\partial k} - \dot{v}_1^+ \frac{\partial v_1^+}{\partial k} + i v_1^{+2} \right)\Bigg|_{k = i\varkappa, x = 0}. \tag{12.95}$$

A similar argument can obviously be used for the solution $u_2^- = e^{-ikx} v_2^-(x, k)$, defined by (12.81), for an arbitrary x in the interval $(a, 0)$, $a < 0$, in the neighbourhood of the point $k = i\varkappa$, where $\varkappa$ is a given positive number. Thus we find, similarly to (12.95):

$$2i\varkappa \int_{-\infty}^0 (u_2^-(x, i\varkappa))^2\, dx = \left(\dot{v}_2^- \frac{\partial v_2^-}{\partial k} - v_2^- \frac{\partial \dot{v}_2^-}{\partial k} + i v_2^{-2} \right)\Bigg|_{k = i\varkappa, x = 0}. \tag{12.96}$$

From now on suppose $i\varkappa$ to be an eigenvalue, so that:

$$f(i\varkappa) = 0, \quad \varkappa > 0. \tag{12.97}$$

Then there exists a non-zero constant σ such that:

$$u_1^+(x, i\varkappa) = \sigma u_2^-(x, i\varkappa) \tag{12.98}$$

for all real x, or:

$$e^{-\varkappa x} v_1^+(x, i\varkappa) = \sigma e^{\varkappa x} v_2^-(x, i\varkappa)$$

and we can define the associated eigenmode by:

$$u(x) = u_1^+(x, i\varkappa) = e^{-\varkappa x} v_1^+(x, i\varkappa)$$

or

(12.99) $$u(x)=\sigma u_2^-(x,i\varkappa)=\sigma e^{\varkappa x}v_2^-(x,i\varkappa),\quad \forall x\in\mathbb{R}.$$

We deduce from (12.95), (12.96) that:

$$2i\varkappa\int_{-\infty}^{+\infty}(u(x))^2\,\mathrm{d}x$$
$$=\left[v_1^+\frac{\partial\dot v_1^+}{\partial k}-\dot v_1^+\frac{\partial v_1^+}{\partial k}+iv_1^{+2}+\sigma^2\left(\dot v_2^-\frac{\partial v_2^-}{\partial k}-v_2^-\frac{\partial\dot v_2^-}{\partial k}+iv_2^{-2}\right)\right]_{\substack{k=i\varkappa\\x=0}}$$

and by (12.98), (12.56):

(12.100) $$2i\varkappa\int_{-\infty}^{+\infty}(u(x))^2\,\mathrm{d}x=\sigma\frac{\mathrm{d}f}{\mathrm{d}k}(i\varkappa).$$

In particular we see that $\frac{\mathrm{d}f}{\mathrm{d}k}(i\varkappa)\neq 0$, i.e. that the eigenvalues or in other words the zeros of the function $f(k)$, which is holomorphic in $\operatorname{Im}k>0$, are all simple.

We can agree to normalise the eigenmode, defining it by:

(12.101) $$w(x)=\frac{c}{\sigma}u(x),$$

where the constant c is chosen so that:

(12.102) $$\int_{-\infty}^{+\infty}(w(x))^2\,\mathrm{d}x=1$$

i.e. after (12.100), so that

(12.103) $$2i\varkappa\sigma=c^2\frac{\mathrm{d}f}{\mathrm{d}k}(i\varkappa).$$

The eigenmode $w(x)$ so normalised has the asymptotic behaviour described by:

(12.104) $$\begin{aligned}w(x)&\sim\frac{c}{\sigma}e^{-\varkappa x},\quad x\to+\infty\\&\sim ce^{\varkappa x},\quad x\to-\infty.\end{aligned}$$

Lastly the residue of $\tau(k)$ at $k=i\varkappa$ is, by (12.89):

(12.105) $$\lim_{k\to i\varkappa}\tau(k)(k-i\varkappa)=-\frac{2\varkappa}{f'(i\varkappa)}=i\frac{c^2}{\sigma}.$$

4. The Inverse Problem

Let us recall the fundamental equation (12.84):

(12.106) $$u_1^-(x,k)+\rho(k)u_2^-(x,k)=\tau(k)u_1^+(x,k),\quad k\text{ real}.$$

We multiply both sides by $e^{-iky}\varphi(k)$, where $\varphi(k)$ is a holomorphic function of k in $\operatorname{Im} k \geqslant 0$, such that $k^2\varphi(k)$ is bounded in this half-plane, and which will be defined more precisely later on. The resulting product is integrated with respect to k over the real axis, it being understood that y is to be regarded as a real variable. We thus obtain:

$$\text{(12.107)}\qquad \int_{-\infty}^{+\infty} e^{ik(x-y)}\varphi(k)v_1^-(x,k)\,dk + \int_{-\infty}^{+\infty} e^{-ik(x+y)}\varphi(k)\rho(k)v_2^-(x,k)\,dk = \int_{-\infty}^{+\infty} e^{ik(x-y)}\varphi(k)\tau(k)v_1^+(x,k)\,dk$$

and we shall now proceed to interpret the various terms in the half-plane $x-y>0$. We introduce:

$$\text{(12.108)}\qquad K(x,y) = \frac{1}{2\pi}\int_{-\infty}^{+\infty} e^{iky}\cdot e^{-ikx}(v_2^-(x,k)-1)\,dk$$

the Fourier transform of the function $e^{-ikx}(v_2^-(x,k)-1)$ which is, by (12.68), square-integrable with respect to k, when k runs through $\mathbb{R}$. Accordingly $K(x,y)$ is, for each x, defined almost everywhere with respect to y and is square-integrable with respect to the latter variable running through $\mathbb{R}$.

Moreover since $v_2^-(x,k)-1$ is holomorphic in $\operatorname{Im} k \geqslant 0$ and $O(1/k)$ at infinity in the half-plane, it is clear by Cauchy's theorem that:

$$\text{(12.109)}\qquad K(x,y)=0 \quad \text{for} \quad x-y<0.$$

Let us begin by calculating the terms on the left of (12.107).

1. We write

$$\int_{-\infty}^{+\infty} \varphi(k)\rho(k)e^{-ik(x+y)}v_2^-(x,k)\,dx = \int_{-\infty}^{+\infty} \varphi(k)\rho(k)e^{-ik(x+y)}\,dk + \int_{-\infty}^{+\infty} \varphi(k)\rho(k)e^{-ik(x+y)}(v_2^-(x,k)-1)\,dk$$

and define:

$$\text{(12.110)}\qquad \tilde{R}(\xi) = \frac{1}{2\pi}\int_{-\infty}^{+\infty} \varphi(k)\rho(k)e^{-ik\xi}\,dk$$

$$\text{(12.111)}\qquad \chi(\xi) = \frac{1}{2\pi}\int_{-\infty}^{+\infty} \varphi(k)e^{-ik\xi}\,dk.$$

Since $\varphi(k)$ is holomorphic in $\operatorname{Im} k \geqslant 0$ and is $O(1/k^2)$ at infinity in this half-plane, we have:

$$\text{(12.112)}\qquad \chi(\xi)=0 \quad \text{for} \quad \xi<0.$$

We shall make the further assumption in regard to $\varphi(k)$ that $\varphi(-k)=\overline{\varphi(k)}$, for real k, from which it will follow that $\tilde{R}(\xi)$, $\chi(\xi)$ are real-valued functions of the real variable ξ.

Similarly $K(x, y)$ defined by (12.108) is real, by (12.83), and we can equally well write:

$$K(x, \xi) = \frac{1}{2\pi} \int_{-\infty}^{+\infty} e^{-ik\xi} \cdot \overline{e^{-ikx}(v_2^-(x, k) - 1)}\, dk. \tag{12.113}$$

Applying Parseval's theorem, we see at once that:

$$\begin{aligned} &\int_{-\infty}^{+\infty} \varphi(k)\rho(k)e^{-iky} \cdot e^{-ikx}(v_2^-(x, k) - 1)\, dk \\ &\quad = 2\pi \int_{-\infty}^{+\infty} \tilde{R}(\xi + y)K(x, \xi)\, d\xi = 2\pi \int_{-\infty}^{x} K(x, z)\tilde{R}(z + y)\, dz \end{aligned} \tag{12.114}$$

where we have made use of the obvious result:

$$\int_{-\infty}^{+\infty} \varphi(k)\rho(k)e^{-ik(x+y)}\, dk = 2\pi\tilde{R}(x + y). \tag{12.115}$$

2. We now turn to the calculation of:

$$\begin{aligned} &\int_{-\infty}^{+\infty} e^{ik(x-y)}\varphi(k)v_1^-(x, k)\, dk \\ &\quad = \int_{-\infty}^{+\infty} e^{ik(x-y)}\varphi(k)\overline{v_2^-(x, k)}\, dx \\ &\quad = \int_{-\infty}^{+\infty} e^{ik(x-y)}\varphi(k)\, dk + \int_{-\infty}^{+\infty} \overline{e^{-ik(x-y)}\varphi(k)(v_2^-(x, k) - 1)}\, dk, \end{aligned} \tag{12.116}$$

noting that $v_1^-(x, k) = \overline{v_2^-(x, k)}$, for real k, and that the integral is real-valued.

The first integral in the last line of (12.116) vanishes for $x - y > 0$. Noticing that

$$\chi(y - \xi) = \frac{1}{2\pi} \int_{-\infty}^{+\infty} e^{-ik\xi} \cdot e^{iky}\overline{\varphi(k)}\, dk$$

and that $\chi(y - \xi) = 0$ for $y - \xi < 0$, and using Parseval's theorem to evaluate the last integral in (12.116):

$$\int_{-\infty}^{+\infty} e^{iky}\overline{\varphi(k)} \cdot e^{-ikx}(v_2^-(x, k) - 1)\, dk = 2\pi \int_{-\infty}^{+\infty} \chi(y - \xi)K(x, \xi)\, d\xi$$

we are led to:

$$\int_{-\infty}^{+\infty} e^{ik(x-y)}\varphi(k)v_1^-(x, k)\, dk = 2\pi \int_{-\infty}^{y} K(x, z)\chi(y - z)\, dz \tag{12.117}$$

in $x > y$.

3. It remains to calculate:

$$W(x, y) = \int_{-\infty}^{+\infty} e^{ik(x-y)}\varphi(k)\tau(k)v_1^+(x, k)\, dk. \tag{12.118}$$

We know that $v_1^+(x, k)$ is a holomorphic function of k in $\operatorname{Im} k \geqslant 0$, which is bounded in this half-plane, and that in addition $\tau(k)$ is meromorphic, has a finite number of simple poles $i\varkappa_j$, $1 \leqslant j \leqslant N$, $\varkappa_j > 0$, and tends to 1 as $|k|$ tends to ∞, in the half-plane $\operatorname{Im} k \geqslant 0$. Lastly $\varphi(k)$ is holomorphic in $\operatorname{Im} k \geqslant 0$, and is of order $1/k^2$ at infinity, so that by applying the residue-theorem to (12.118) and using (12.89) we arrive at:

$$W(x, y) = -4\pi i \sum_{j=1}^{N} \frac{\varkappa_j}{f'(i\varkappa_j)} \cdot \varphi(i\varkappa_j) v_1^+(x, i\varkappa_j) e^{-\varkappa_j(x-y)}$$

in $x - y > 0$.

On the other hand it follows from (12.98) that:

$$v_1^+(x, i\varkappa_j) e^{-\varkappa_j x} = \sigma_j v_2^-(x, i\varkappa_j) e^{\varkappa_j x}$$

and from (12.105) that:

$$2i\varkappa_j \sigma_j = c_j^2 f'(i\varkappa_j)$$

and consequently:

$$W(x, y) = -2\pi \sum_{j=1}^{N} \varphi(i\varkappa_j) c_j^2 v_2^-(x, i\varkappa_j) e^{\varkappa_j(x+y)}. \tag{12.119}$$

Introducing:

$$\tilde{B}(\xi) = \tilde{R}(\xi) + \sum_{j=1}^{N} \varphi(i\varkappa_j) c_j^2 e^{\varkappa_j \xi} \tag{12.120}$$

and inserting the results expressed by: (12.114), (12.115), (12.117), (12.119) into the equation (12.107), we get, after dividing by 2π:

$$\begin{aligned} \tilde{B}(x+y) + \int_{-\infty}^{x} K(x, z) \tilde{R}(z+y)\, dz + \int_{-\infty}^{y} K(x, z) \chi(y-z)\, dz \\ = -\sum_{j=1}^{N} \varphi(i\varkappa_j) c_j^2 e^{\varkappa_j(x+y)} (v_2^-(x, i\varkappa_j) - 1), \quad \text{for } x - y > 0. \end{aligned} \tag{12.121}$$

Since $v_2^-(x, k) - 1$ is holomorphic in $\operatorname{Im} k \geqslant 0$ and $O(1/k)$ when $|k| \to \infty$, we can, by the theorem on residues, write:

$$v_2^-(x, i\varkappa_j) - 1 = \frac{1}{2\pi i} \int_{-\infty}^{+\infty} \frac{v_2^-(x, k) - 1}{k - i\varkappa_j}\, dk$$

and interpret:

$$e^{\varkappa_j x}(v_2^-(x, i\varkappa_j) - 1) = \frac{1}{2\pi} \int_{-\infty}^{+\infty} \frac{e^{i(k - i\varkappa_j)x}}{i(k - i\varkappa_j)} \cdot e^{-ikx}(v_2^-(x, k) - 1)\, dk$$

by Parseval's theorem, noting that:

$$\frac{1}{2\pi} \int_{-\infty}^{+\infty} e^{-ik\xi} \cdot \frac{e^{i(k - i\varkappa_j)x}}{i(k - i\varkappa_j)}\, dk = \begin{cases} 0, & x < \xi \\ e^{\varkappa_j \xi}, & x > \xi \end{cases}$$

and recalling (12.113), so that:

$$e^{\varkappa_j x}(v_2^-(x, i\varkappa_j) - 1) = \int_{-\infty}^{x} e^{\varkappa_j \xi} K(x, \xi)\, d\xi.$$

We can thus rewrite the equation (12.121) as:

(12.122) $$\tilde{B}(x+y)+\int_{-\infty}^{x} K(x,z)\tilde{B}(z+y)\,dz+\int_{-\infty}^{y} K(x,z)\chi(y-z)\,dz=0$$

for $x-y>0$.

The Kernel $K(x,y)$ (Continued)

To study the behaviour of the kernel $K(x,y)$ defined by (12.108), we can make use of the representation of $v_2^-(x,k)-1$ given by (12.68), indexing where necessary the terms $v^-(x,k)$ and $\delta^-(x,k)$: with the suffix 2. We are thus led to write:

(12.123) $$K(x,y)=U(x,y)+V(x,y)+H(x,y)$$

where
$$U(x,y)=-\frac{1}{2\pi}\int_{-\infty}^{+\infty} e^{ik(y-x)}\frac{dk}{2ik}\int_{-\infty}^{x} q(x')\,dx'$$

$$V(x,y)=\frac{1}{2\pi}\int_{-\infty}^{+\infty} e^{ik(y-x)}\cdot\frac{1}{2ik}\int_{-\infty}^{x} e^{-2ik(x'-x)}q(x')\,dx'\cdot dk$$

$$H(x,y)=\frac{1}{2\pi}\int_{-\infty}^{+\infty} e^{ik(y-x)}\cdot\delta_2^-(x,k)\,dk.$$

Since the integral defining $H(x,y)$ is, by virtue of (12.68), uniformly convergent in x and y, it is clear that $H(x,y)$ is a continuous function in the plane $(x,y)\in\mathbb{R}^2$.

We now proceed to calculate U and V:

1. $U(x,y)$ is equivalent to:

(12.124) $$U=\pm\frac{1}{4}\int_{-\infty}^{x} q(x')\,dx',$$

the sign being + if $x-y>0$ and − if $x-y<0$.

2. We can write:

$$V=\lim_{A\to+\infty}\frac{1}{2\pi}\int_{-A}^{+A}\int_{-\infty}^{x}\frac{e^{ik(x+y-2x')}}{2ik}\cdot q(x')\,dx'\,dk$$

$$=\lim_{A\to+\infty}\frac{1}{2\pi}\int_{0}^{A}\int_{-\infty}^{x}\frac{\sin k(x+y-2x')}{k}q(x')\,dx'\,dk$$

or after interchanging the order of integration, which is allowable because of the properties of $q(x')$:

(12.125) $$V(x,y)=\lim_{A\to+\infty}\frac{1}{2\pi}\int_{-\infty}^{x}\left(\int_0^A\frac{\sin k(x+y-2x')}{k}\,dk\right)q(x')\,dx'.$$

However as $\lim_{X\to+\infty}\int_0^X\frac{\sin u}{u}\,du=\pi$, it is clear that

$$\int_0^A\frac{\sin k(x+y-2x')}{k}\,dk=\int_0^{A(x+y-2x')}\frac{\sin u}{u}\,du$$

is bounded uniformly with respect to the three variables x, y, x' and tends to $+\pi$ or $-\pi$ as A tends to $+\infty$, depending on whether $x+y-2x'$ is positive or negative respectively. Since $q(x')$ is integrable we can apply the theorem on dominated convergence [37] to (12.125) and obtain:

$$V(x,y)=\frac{1}{2}\int_{-\infty}^{x} h(x+y-2x')q(x')\,\mathrm{d}x', \quad \text{with} \quad \begin{matrix} h(u)=1, & u>0 \\ h(u)=-1, & u<0 \end{matrix}$$

or

$$V(x,y)=\frac{1}{2}\int_{-\infty}^{(x+y)/2} q(x')\,\mathrm{d}x' - \frac{1}{2}\int_{(x+y)/2}^{x} q(x')\,\mathrm{d}x', \quad \text{for } x>y$$

(12.126)
$$V(x,y)=\frac{1}{2}\int_{-\infty}^{x} q(x')\,\mathrm{d}x', \quad \text{for } x<y$$

Thus $H(x,y)$, $V(x,y)$ are continuous throughout the plane $(x,y)\in\mathbb{R}^2$, while $U(x,y)$ is discontinuous on crossing the line $x=y$; and lastly we know that $K(x,y)=0$ for $x<y$. Consequently we can write:

$$\lim_{y\uparrow x} K(x,y)=\frac{1}{4}\int_{-\infty}^{x} q(x')\,\mathrm{d}x' + V(x,x)+H(x,x)$$

$$0=\lim_{y\downarrow x} K(x,y)=-\frac{1}{4}\int_{-\infty}^{x} q(x')\,\mathrm{d}x' + V(x,x)+H(x,x)$$

whence
$$\lim_{y\uparrow x} K(x,y)=\frac{1}{2}\int_{-\infty}^{x} q(x')\,\mathrm{d}x'.$$

To summarise $K(x,y)$ is continuous in the half-plane $x\geqslant y$, bounded in $x\geqslant y$, $x\leqslant l$ for all l, and its value at the boundary is:

$$K(x,x)=\frac{1}{2}\int_{-\infty}^{x} q(x')\mathrm{d}x'.$$

or

(12.127)
$$q(x)=2\frac{\mathrm{d}}{\mathrm{d}x}K(x,x).$$

The Gelfand-Levitan Integral Equation

Returning to the equation (12.122), we introduce an infinite sequence of regularising functions $\varphi_n(k)$ defined by:

$$\varphi_n(k)=\left(\frac{in}{k+in}\right)^2, \quad n=1,2,\ldots$$

with which are associated:

(12.128)
$$R_n(\xi)=\frac{1}{2\pi}\int_{-\infty}^{+\infty} \varphi_n(k)\rho(k)e^{-ik\xi}\,\mathrm{d}k$$

$$B_n(\xi)=R_n(\xi)+\sum_{j=1}^{N} \varphi_n(i\varkappa_j)c_j^2 e^{\varkappa_j\xi}.$$

Now it is clear from (12.87), and (12.92) that the reflection coefficient $\rho(k)$ is square-integrable on the real axis, so that this allows us to introduce its Fourier transform:

$$R(\xi)=\frac{1}{2\pi}\int_{-\infty}^{+\infty}\rho(k)e^{-ik\xi}\,dk \tag{12.129}$$

$$B(\xi)=R(\xi)+\sum_{j=1}^{N}c_j^2e^{\varkappa_j\xi}.$$

We may note incidentally that by starting from (12.87), (12.66) and conducting an analysis similar to the one undertaken when studying the $V(x, y)$ term in (12.123), we could show that $R(\xi)$ is a continuous function of ξ.

By Parseval's formula we can write:

$$2\pi\int_{-\infty}^{+\infty}|R_n(\xi)-R(\xi)|^2\,d\xi=\int_{-\infty}^{+\infty}|\varphi_n(k)-1|^2|\rho(k)|^2\,dk$$

and observing that:

$$|\varphi_n(k)-1|^2|\rho(k)|^2<4|\rho(k)|^2,\quad \lim_{n\to\infty}|\varphi_n(k)-1|=0,$$

conclude, using the theorem on dominated convergence that $R_n(\xi)$ converges to $R(\xi)$, when convergence is defined in terms of the norm in $L^2(-\infty,+\infty)$. We can deduce from (12.128) that:

$$\lim_{n\to\infty}\int_{-\infty}^{x}|B_n(\xi)-B(\xi)|^2\,d\xi=0,\quad \forall x$$

and

$$\lim_{n\to\infty}\int_{-\infty}^{x}K(x,z)B_n(y+z)\,dz=\int_{-\infty}^{x}K(x,z)B(y+z)\,dz.$$

It should be added that we can extract from the sequence of positive integers a subsequence (which may be renumbered so that we can retain the suffix n to indicate the nth term of the subsequence) with the property that

$$\lim_{n\to\infty}R_n(\xi)=R(\xi)\ \text{a.e.}$$

and

$$\lim_{n\to\infty}B_n(\xi)=B(\xi)\ \text{a.e.}\quad\text{as well.}$$

A simple calculation shows that:

$$\chi_n(\xi)=\frac{1}{2\pi}\int_{-\infty}^{+\infty}\varphi_n(k)e^{-ik\xi}\,dk=n^2\xi e^{-n\xi},\quad \xi>0$$

$$=0,\quad \xi<0$$

from which may be deduced:

$$\int_{-\infty}^{y}K(x,z)\chi_n(y-z)\,dz=n^2\int_{-\infty}^{y}(K(x,z)-K(x,y))(y-z)e^{-n(y-z)}\,dz+K(x,y)$$

with $n^2\int_0^{+\infty} te^{-nt}\,\mathrm{d}t = 1$, and in view of the continuity of $K(x,z)$ in $x \geqslant z$:

$$\lim_{n\to\infty} \int_{-\infty}^{y} K(x,z)\chi_n(y-z)\,\mathrm{d}z = K(x,y)$$

and finally, the Gelfand-Levitan integral equation [22]:

$$B(x+y) + \int_{-\infty}^{x} K(x,z)B(z+y)\mathrm{d}z + K(x,y) = 0, \quad x-y \geqslant 0. \tag{12.130}$$

An Alternative Definition of the Kernel $K(x,y)$

From (12.108), with the variables x, y replaced by ξ, η we may deduce, after multiplying by $q(\xi)$ and integrating over the interior of the semi-infinite rectangle $\mathscr{R}_\varepsilon$ (shown in Figure 12.1) whose two long sides are the half-lines parallel to the line $\xi = \eta$ bisecting the principal axes, and passing through the two vertices (x, y) and $\left(\frac{x+y+\varepsilon}{2}, \frac{x+y-\varepsilon}{2}\right)$, with ε satisfying $0 < \varepsilon < x - y$:

$$\iint_{\mathscr{R}_\varepsilon} q(\xi)K(\xi,\eta)\,\mathrm{d}\xi\,\mathrm{d}\eta = \frac{1}{2\pi}\iint_{\mathscr{R}_\varepsilon} q(\xi)\,\mathrm{d}\xi\,\mathrm{d}\eta \int_{-\infty}^{+\infty} e^{ik(\eta-\xi)}\cdot(v_2^-(\xi,k)-1)\,\mathrm{d}k.$$

However we can write:

$$\iint_{\mathscr{R}_\varepsilon} q(\xi)\,\mathrm{d}\xi\,\mathrm{d}\eta \int_{-A}^{+A} e^{ik(\eta-\xi)}\cdot(v_2^-(\xi,k)-1)\,\mathrm{d}k = \int_{-A}^{+A} \mathrm{d}k \iint_{\mathscr{R}_\varepsilon} q(\xi)e^{ik(\eta-\xi)} \cdot(v_2^-(\xi,k)-1)\,\mathrm{d}\xi\,\mathrm{d}\eta,$$

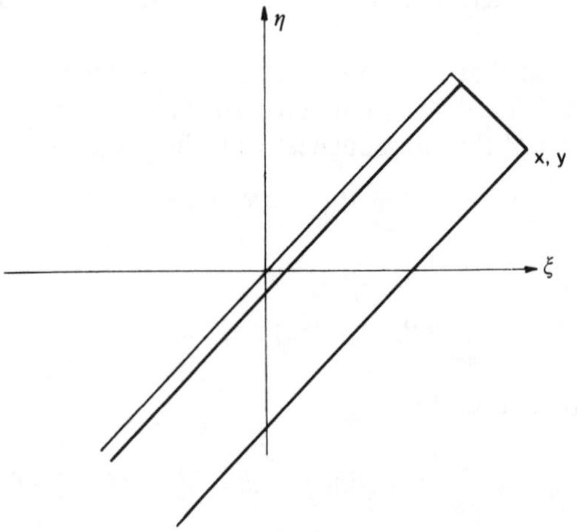

Figure 12.1

the change in the order of integration being justified because $q(\xi)$ is integrable over $\mathbb{R}$ and $v_2^-(\xi,k)$ is bounded over the domain of integration, by virtue of (12.64).

On the other hand, by the analysis which followed the decomposition in (12.123), it is apparent that:

$$\int_{-A}^{+A} e^{ik(\eta-\xi)}\cdot(v_2^-(\xi,k)-1)\,\mathrm{d}k$$

is bounded for $A>0$, $(\xi,\eta)\in\mathscr{R}_\varepsilon$ and since $q(\xi)$ is integrable on $\mathscr{R}_\varepsilon$ we can apply the theorem on dominated convergence, that is to say we can take the limit as $A\to+\infty$ under the integral sign, which gives us:

(12.131)
$$\iint_{\mathscr{R}_\varepsilon} q(\xi)K(\xi,\eta)\,\mathrm{d}\xi\,\mathrm{d}\eta = \frac{1}{2\pi}\int_{-\infty}^{+\infty}\mathrm{d}k\iint_{\mathscr{R}_\varepsilon} e^{ik(\eta-\xi)}q(\xi)(v_2^-(\xi,k)-1)\,\mathrm{d}\xi\,\mathrm{d}\eta.$$

With $w(\xi,k)=v_2^-(\xi,k)-1$ and remembering that $\ddot{w}-2ik\dot{w}-q=qw$, by reason of (12.81) and (12.34), we see that we need to calculate:

(12.132)
$$I=\frac{1}{2\pi}\int_{-\infty}^{+\infty}\mathrm{d}k\iint_{\mathscr{R}_\varepsilon} e^{ik(\eta-\xi)}(\ddot{w}-2ik\dot{w})\,\mathrm{d}\xi\,\mathrm{d}\eta$$
$$J=-\frac{1}{2\pi}\int_{-\infty}^{+\infty}\mathrm{d}k\iint_{\mathscr{R}_\varepsilon} e^{ik(\eta-\xi)}q(\xi)\,\mathrm{d}\xi\,\mathrm{d}\eta.$$

To calculate I we write the inner double integral in the form

$$\iint_{\mathscr{R}_\varepsilon} e^{ik(\xi+\eta)}\cdot(\dot{w}e^{-2ik\xi})\cdot\mathrm{d}\xi\,\mathrm{d}\eta$$

which we evaluate as a repeated integral integrating first with respect to $\sigma=(\xi-\eta)/\sqrt{2}$ with $s=(\xi+\eta)/\sqrt{2}$ constant, and then with respect to s from $-\infty$ to $(x+y)/\sqrt{2}$, noting that for any function h of the single variable ξ we have:

$$\dot{h}(\xi)=\sqrt{2}\frac{\partial}{\partial\sigma}h((\sigma+s)/\sqrt{2})=\sqrt{2}\frac{\partial}{\partial s}h((\sigma+s)/\sqrt{2}).$$

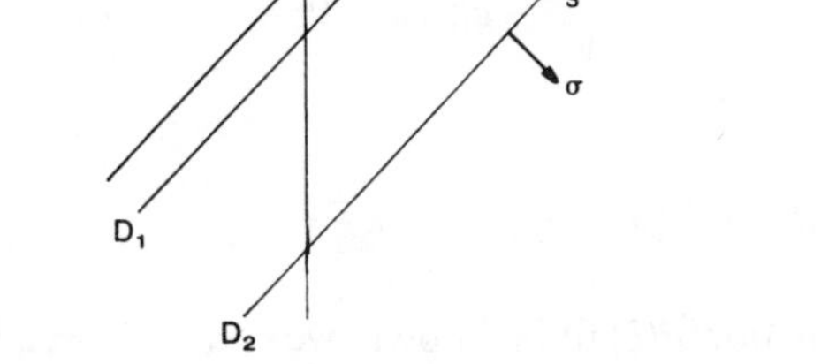

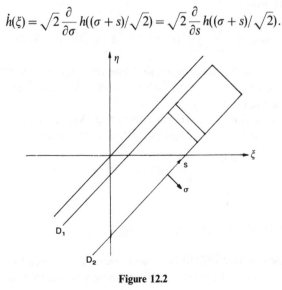

Figure 12.2

Accordingly we write successively:

$$\iint_{\mathcal{R}_\varepsilon} e^{ik(\xi+\eta)}(\dot{w}e^{-2ik\xi})\,d\xi\,d\eta = \sqrt{2}\int_{D_1} e^{ik(\xi+\eta)}\dot{w}e^{-2ik\xi}\,ds$$

$$-\sqrt{2}\int_{D_2} e^{ik}(\xi+\eta)\dot{w}e^{-2ik\xi}\,ds = 2e^{ik(y-x)}w(x,k) - 2e^{-ik\varepsilon}w\left(\frac{x+y+\varepsilon}{2},k\right)$$

and

$$I = \frac{1}{\pi}\int_{-\infty}^{+\infty} e^{ik(y-x)}w(x,k)\,dk - \frac{1}{\pi}\int_{-\infty}^{+\infty} e^{-ik\varepsilon}w\left(\frac{x+y+\varepsilon}{2},k\right)$$

that is:

$$I = 2K(x,y) - 2K\left(\frac{x+y+\varepsilon}{2},\frac{x+y-\varepsilon}{2}\right). \tag{12.133}$$

On the other hand we can represent the inner double integral in J with the help of (12.127), as:

$$\iint_{\mathcal{R}_\varepsilon} e^{ik(\eta-\xi)}q(\xi)\,d\xi\,d\eta = 2\int_{\varepsilon}^{x-y} e^{-iku}K\left(\frac{u+x+y}{2},\frac{u+x+y}{2}\right)du$$

which we can interpret as the Fourier transform of a function $g(u)$ of compact support in $(\varepsilon, x-y)$, whose values in that interval are $K\left(\frac{u+x+y}{2},\frac{u+x+y}{2}\right)$. This function $g(u)$ is of bounded variation by reason of (12.79) and hence by the theorem on the inversion of the Fourier transformation of a function of bounded variation, integrable over $\mathbb{R}$, we see that J, which is none other than the value at $u=0$ of the inverse of this transform, must vanish.

Accordingly $J=0$ and it follows from (12.131), (12.132), (12.133) and (12.127), by making $\varepsilon\to 0$, that

$$K(x,y) = \frac{1}{2}\int_{-\infty}^{(x+y)/2} q(x')\,dx' + \frac{1}{2}\iint_{\mathcal{R}} q(\xi)K(\xi,\eta)\,d\xi\,d\eta, \tag{12.134}$$

for $x-y>0$, where $\mathcal{R}$ denotes the semi-infinite rectangle which extends to infinity in the direction of negative x, one of whose sides lies on the line $\xi=\eta$ and one of whose vertices is the point (x,y).

The integral equation (12.134) characterises the kernel $K(x,y)$ in the half-plane $x>y$; the kernel satisfies in this half-plane the wave equation:

$$\frac{\partial^2 K}{\partial x^2} - \frac{\partial^2 K}{\partial y^2} - q(x)K = 0, \quad x-y>0. \tag{12.135}$$

Solving Gelfand-Levitan's Equation

If we suppose the function $B(\xi)$ to be known, we can interpret, for each given value $x=a$, the equation (12.130) as an integral equation of which $K(a,y)=H(y)$ would be a solution belonging to the function-space of functions square-integrable over $(-\infty, a)$:

(12.136) $$B(a+y)+H(y)+\int_{-\infty}^{a} B(y+z)H(z)\,\mathrm{d}z=0, \quad -\infty<y<a.$$

In this connection it is convenient to introduce the operator

$$(I+\mathscr{C})H = H(y)+\int_{-\infty}^{a} B(y+z)H(z)\,\mathrm{d}z, \quad y<a$$
$$= 0, \quad y>a$$

acting on any real function $H \in L^2(-\infty, a)$.

The contribution from the terms $e^{\varkappa_j \xi}$ of $B(\xi)$ is $e^{\varkappa_j y}\int_{-\infty}^{a} e^{\varkappa_j z}H(z)\,\mathrm{d}z$ which belongs to $L^2(-\infty, a)$. To investigate the contribution from $R(\xi)$ we write $h(k)$ for the Fourier transform of $H(y)$:

$$h(k)=\int_{-\infty}^{a} e^{ik\xi}H(\xi)\,\mathrm{d}\xi \quad \text{or} \quad H(\xi)=\frac{1}{2\pi}\int_{-\infty}^{+\infty} e^{-ik\xi}h(k)\,\mathrm{d}k, \quad \xi<a$$
$$= 0, \quad \xi>a.$$

By Parseval's theorem we have

(12.137) $$\int_{-\infty}^{a} R(y+z)H(z)\,\mathrm{d}z=\frac{1}{2\pi}\int_{-\infty}^{+\infty} e^{-iky}\rho(k)\overline{h(k)}\,\mathrm{d}k$$

and since $\rho(k)h(k)$ belongs to $L^2(-\infty, +\infty)$ since $|\rho(k)|<1$, its Fourier transform has the same property. The linear operator $\mathscr{C}$ is therefore a mapping of $L^2(-\infty, a)$ into itself and writing $\|\cdot\|$ for the usual norm in that space we can assert that:

$$\left\| \int_{-\infty}^{a} R(y+z)H(z)\,\mathrm{d}z \right\| \leqslant \|H\|.$$

It is clear that $\mathscr{C}$ is a symmetric operator, and furthermore $I+\mathscr{C}$ is invertible. For if $H \in L^2(-\infty, a)$ we have:

$$\begin{aligned}
&((I+\mathscr{C})H, H)\\
&= \int_{-\infty}^{a} |H(y)|^2\,\mathrm{d}y+\int_{-\infty}^{a}\int_{-\infty}^{a}\left(R(y+z)+\sum_{1}^{N} c_j^2 e^{\varkappa_j(y+z)}\right)H(z)\,\mathrm{d}z\cdot H(y)\,\mathrm{d}y\\
&= \int_{-\infty}^{a} |H(y)|^2\,\mathrm{d}y+\sum_{1}^{N} c_j^2\left|\int_{-\infty}^{a} e^{\varkappa_j z}H(z)\,\mathrm{d}z\right|^2\\
&\quad+\int_{-\infty}^{a}\left(\int_{-\infty}^{a} R(y+z)H(z)\,\mathrm{d}z\right)H(y)\,\mathrm{d}y
\end{aligned}$$

and by Parseval's theorem

$$\int_{-\infty}^{a}\left(\int_{-\infty}^{a} R(y+z)H(z)\,\mathrm{d}z\right)H(y)\,\mathrm{d}y=\frac{1}{2\pi}\int_{-\infty}^{+\infty} (h(k))^2\overline{\rho(k)}\,\mathrm{d}k.$$

Finally we can write

$$((I+\mathscr{C})H,H)=\frac{1}{2\pi}\int_{-\infty}^{+\infty}(|h(k)|^2+\overline{\rho(k)}(h(k))^2)\,\mathrm{d}k+\sum_1^N c_j^2\left|\int_{-\infty}^{a}e^{\varkappa_j z}H(z)\,\mathrm{d}z\right|^2.$$

The integral $\int_{-\infty}^{+\infty}\overline{\rho(k)(h(k))^2}\,\mathrm{d}k$ is obviously real, because $\overline{\rho(k)}=\rho(-k)$, $\overline{h(k)}=h(-k)$; lastly from $|\rho(k)|<1$, for all real k, we deduce that $((I+\mathscr{C})H,H)>0$ for any non-zero H.

Consequently $(I+\mathscr{C})H=0$ implies $H=0$, which proves that $I+\mathscr{C}$ is invertible (Fredholm's alternative [37]).

These results show that, if we consider $B(\xi)$ to be constructed by (12.129) starting from the elements $\rho(k)$, $c_j, \varkappa_j$ obtained by analysis of the direct problem, for a perturbation potential $q(x)$ satisfying (12.35) then the equation (12.136) or better still (12.130) has, for each value x, an unique solution $K(x,z)$ which is square-integrable with respect to $z\in(-\infty,x)$ and which can be identified with the kernel K introduced earlier by (12.108) and for which (12.127) has been established.

However the inverse problem poses itself in slightly different terms; having been given a priori the positive numbers c_j, $\varkappa_j$, $1\leqslant j\leqslant N$, and the function $\rho(k)$ such that $\overline{\rho(k)}=\rho(-k)$, $|\rho(k)|<1$, whose square is integrable over $\mathbb{R}$ (these are of course minimum assumptions), we define $B(\xi)$ by (12.129) and associate with it the Gelfand-Levitan equation which, for each x, has an unique solution denoted by $K(x,y)$.

We can now inquire as to what additional conditions would have to be imposed on $\rho(k)$ to ensure that $K(x,x)$ would be differentiable, that the potential q defined by $q=2\frac{\mathrm{d}}{\mathrm{d}x}K(x,x)$ would satisfy (12.35), and that the reflection coefficient $\rho(k)$, eigenvalues $\varkappa_j$ and gauge coefficients c_j, originally given are in fact those which would emerge from an analysis of the direct problem defined by the potential q.

5. The Inverse Scattering Method

The principle of the method was outlined at the beginning of this chapter; we shall now fill in the details, and we shall assume k to be real unless otherwise indicated.

1. We consider again the one-dimensional Schrödinger equation under the hypothesis that the perturbation term $q(x,t)$ is a real-valued function of the space-variable x and time-variable t:

$$\frac{\mathrm{d}^2u}{\mathrm{d}x^2}+(k^2-q(x,t))u=0 \tag{12.138}$$

which is an ordinary differential equation in which t appears as a parameter, and for which we can introduce for each value of t the solutions $u_1^+, u_2^+, u_1^-, u_2^-$ defined by (12.77), (12.79), (12.81), (12.82) and then the reflection and transmission coefficients, which are functions of the real variables k and t, such that:

(12.139)
$$\begin{aligned} u_1^-(x,t,k) + \rho(k,t)u_2^-(x,t,k) &= \tau(k,t)u_1^+(x,t,k) \\ u_{1x}^-(x,t,k) + \rho(k,t)u_{2x}^-(x,t,k) &= \tau(k,t)u_{1x}^+(x,t,k) \end{aligned}$$

from which we deduce, noting that $u_{1x}^- u_2^- - u_1^- u_{2x}^-$, the Wronskian of (12.138), is independent of x and has the value $2ik$ at $x = -\infty$:

(12.140)
$$\tau = \frac{2ik}{u_{1x}^+ u_2^- - u_1^+ u_{2x}^-}, \quad \rho = \frac{u_1^+ u_{1x}^- - u_1^- u_{1x}^+}{u_{1x}^+ u_2^- - u_1^+ u_{2x}^-}.$$

From now on we shall assume that $q(x,t)$ is continuous, with a continuous partial derivative $q_t(x,t)$; that q_t, q are integrable with respect to x on the real axis; and that the convergence of the integrals at $x = \pm\infty$ is uniform with respect to t on every finite interval. These assumptions ensure, in particular, that every solution of (12.138) is differentiable with respect to t, and a simple calculation shows that for any two solutions u and w of (12.138) we have:

(12.141)
$$(uw_x - wu_x)_t = (uw_{xt} - w_t u_x) - (wu_{xt} - u_t w_x)$$

and

(12.142)
$$(uw_{xt} - w_t u_x)_x = uw_{xxt} - w_t u_{xx} = q_t uw$$

or interchanging u and w

$$(wu_{xt} - u_t w_x)_x = q_t uw.$$

Taking $w = u_1^+(x,t,k) = e^{ikx}v_1^+(x,t,k)$, $v_1^+(+\infty,t,k) = 1$, $v_{1x}^+(+\infty,t,k) = 0$,

whence
$$\lim_{x\to+\infty} w_t = \lim_{x\to+\infty} w_{xt} = 0,$$

and
$$u = u_1^-(x,t,k) = e^{ikx}v_1^-(x,t,k),$$

or
$$u = u_2^-(x,t,k) = e^{-ikx}v_2^-(x,t,k),$$
$$v_{1,2}^-(-\infty,t,k) = 1, \quad v_{1,2,x}^-(-\infty,t,k) = 0$$

so that
$$\lim_{x\to-\infty} u_t = \lim_{x\to-\infty} u_{xt} = 0,$$

we obtain by integration of (12.142):

$$uw_{xt} - w_t u_x = -\int_x^{+\infty} q_t uw \, dx$$

$$wu_{xt} - u_t w_x = \int_{-\infty}^{x} q_t uw \, dx$$

and by (12.141):

$$(uw_x - wu_x)_t = -\int_{-\infty}^{+\infty} q_t uw \, dx$$

i.e. with (12.140):

(12.143)
$$2ik(\tau^{-1})_t = -\int_{-\infty}^{+\infty} q_t u_1^+ u_2^- \, dx$$
$$2ik\left(\frac{\rho}{\tau}\right)_t = \int_{-\infty}^{+\infty} q_t u_1^+ u_1^- \, dx$$

and finally, by (12.139):

(12.144)
$$2ik\rho_t \tau^{-2} = \int_{-\infty}^{+\infty} q_t (u_1^+)^2 \, dx.$$

2. If u, w are any solutions of (12.138), we can write:

(12.145)
$$(u_x w_x)_x + k^2 (uw)_x = q(uw)_x$$

whence, by integration:

(12.146)
$$(u_x w_x + k^2 uw)\Big|_{-\infty}^{+\infty} = \int_{-\infty}^{+\infty} q(uw)_x \, dx = -\int_{-\infty}^{+\infty} q_x uw \, dx$$

on assuming that q is continuously differentiable in x and that its derivative q_x is integrable over $(-\infty, +\infty)$, which implies since q is integrable over $(-\infty, +\infty)$, that $\lim_{x \to +\infty} q = \lim_{x \to -\infty} q = 0$.

Applying (12.146) for $u = w = u_1^+$, and noting that $\lim_{x \to +\infty} ((u_{1x}^+)^2 + k^2 (u_1^+)^2) = 0$ by reason of (12.77), we are left with:

(12.147)
$$\lim_{x \to -\infty} ((u_{1x}^+)^2 + k^2 (u_1^+)^2) = \int_{-\infty}^{+\infty} q_x (u_1^+)^2 \, dx.$$

On the other hand we have from (12.139):

$$\tau^2 ((u_{1x}^+)^2 + k^2 (u_1^+)^2) = (u_{1x}^- + \rho u_{2x}^-)^2 + k^2 (u_1^- + \rho u_2^-)^2$$

and, using (12.81), (12.82), to estimate the asymptotic behaviour of the right-hand side at $x = -\infty$,

$$\lim_{x \to -\infty} \tau^2 ((u_{1x}^+)^2 + k^2 (u_1^+)^2) = 4\rho k^2$$

that is, by (12.147):

(12.148)
$$\frac{4k^2\rho}{\tau^2} = \int_{-\infty}^{+\infty} q_x (u_1^+)^2 \, dx.$$

3. By integrating (12.145) with respect to x, with $u = w = u_1^{\pm}$ and taking account of the behaviour at infinity implied by (12.77), (12.82) we have:

$$(u_{1x}^-)^2 + k^2 (u_1^-)^2 = \int_{-\infty}^{x} q(u_1^{-2})_x \, dx$$
$$(u_{1x}^+)^2 + k^2 (u_1^+)^2 = -\int_{x}^{+\infty} q(u_1^{+2})_x \, dx$$

and as every solution u of (12.138) satisfies the equation:

$$k^2u^2 = qu^2 - \tfrac{1}{2}(u^2)_{xx} + (u_x)^2$$

we see, by taking $u = u_1^+$ and combining with the preceding result, that:

$$(12.149) \qquad k^2(u_1^+)^2 = L(u_1^+)^2 = \frac{q}{2}\cdot(u_1^+)^2 - \tfrac{1}{4}\cdot((u_1^+)^2)_{xx} - \frac{1}{2}\int_x^{+\infty} q(u_1^{+2})_x \,\mathrm{d}x$$

where L is the linear operator defined [28] by:

$$(12.150) \qquad L = \frac{q}{2}\cdot - \frac{1}{4}\frac{\partial^2}{\partial x^2} - \frac{1}{2}\int_x^{+\infty} q\cdot(\quad)_x \,\mathrm{d}x.$$

4. We can define Lf by (12.150) for any function $f(x)$ which is twice continuously differentiable, and is, along with its derivatives, bounded on the real axis, and the function so obtained is bounded.

We can similarly define, for any function $g(x)$ which is twice continuously-differentiable, integrable over $(-\infty, +\infty)$, and in addition is such that g and g_x both tend to zero when $x \to \pm\infty$:

$$(12.151) \qquad L^*g = -\tfrac{1}{4}\cdot g_{xx} + qg + \tfrac{1}{2}q_x\cdot \int_{-\infty}^{x} g \,\mathrm{d}x$$

and a simple calculation shows that:

$$(12.152) \qquad gLf - fL^*g = -\tfrac{1}{4}\cdot(gf_x - fg_x)_x + \frac{1}{2}\left(\int_{-\infty}^{x} g\,\mathrm{d}x\cdot \int_x^{+\infty} fq_x \,\mathrm{d}x\right)_x$$

whence

$$\int_{-\infty}^{+\infty} g\cdot Lf \,\mathrm{d}x = \int_{-\infty}^{+\infty} L^*g\cdot f \,\mathrm{d}x$$

which makes it clear that L^* is the operator adjoint to L and justifies the notation.

Since we shall later on need to calculate certain powers of the operators L and L^*, it seems appropriate to examine certain conditions for existence.

We shall say that a function $h(x)$ is of class $\mathscr{T}_n$ if $h(x)$ is differentiable to order n, and together with its derivatives up to the nth order, integrable over $(-\infty, +\infty)$, which implies among other things that $h(x)$ and its derivatives up to order $n-1$ tend to zero as $x \to \pm\infty$.

We shall define the class $\mathscr{T}_n^*$ as the class consisting of the functions $h(x)$ which are differentiable up to order n, and, together with its derivatives up to that order, are bounded on the real axis.

We see immediately by (12.150) that $q \in \mathscr{T}_N$, $f \in \mathscr{T}_{N'}^*$, $N' \leqslant N$ together imply $Lf \in \mathscr{T}_{N-2}^*$; in particular $q \in \mathscr{T}_{2n-1}$ implies $L(1) = (q/2) \in \mathscr{T}_{2n-2}^*$ and $L^n(1) \in \mathscr{T}_0^*$, i.e. $L^n(1)$ is defined and bounded.

Similarly, $q \in \mathscr{T}_N$, $g \in \mathscr{T}_{N'}$, $N' \leqslant N$ imply $L^*g \in \mathscr{T}_{N'-2}$; in particular if $q \in \mathscr{T}_{2n+1}$, then $q_x \in \mathscr{T}_{2n}$ and $L^*q_x \in \mathscr{T}_{2n-2}, \ldots, L^{*n}q_x \in \mathscr{T}_0$.

Finally, if $f \in \mathscr{T}_{2n}^*$, $g \in \mathscr{T}_{2n}$, $q \in \mathscr{T}_{2n}$ we can write:

$$\int_{-\infty}^{+\infty} g\cdot L^n f \,\mathrm{d}x = \int_{-\infty}^{+\infty} L^{*n} g\cdot f \,\mathrm{d}x.$$

The classes $\mathscr{T}_\infty$, $\mathscr{T}^*_\infty$ are then defined in the obvious way.

5. By integrating (12.145) with $u = u_1^+$, $w = u_2^-$ we have:

$$u_{1x}^+ u_{2x}^- + k^2 u_1^+ u_2^- = -\int_x^{+\infty} q(u_1^+ u_2^-)_x \, dx + \lim_{x \to +\infty} (u_{1x}^+ u_{2x}^- + k^2 u_1^+ u_2^-) \tag{12.153}$$

the integral converging since q is integrable over $(-\infty, +\infty)$ and $(u_1^+ u_2^-)_x$ is bounded on $\mathbb{R}$.

To calculate the limit on the right-hand side, we first observe that by taking the conjugate values in (12.139) and noting that $\bar{u}_1^+ = u_2^+$, $\bar{u}_1^- = u_2^-$, we can write $u_2^- + \bar{\rho} u_1^- = \bar{\tau} u_2^+$, which taken in conjunction with $u_1^- + \rho u_2^- = \tau u_1^+$ and $1 = |\rho|^2 + |\tau|^2$, allows us to obtain:

$$u_2^- = \tau^{-1} u_2^+ - \frac{\bar{\rho}}{\bar{\tau}} u_1^+$$

and also:

$$u_{2x}^- = \tau^{-1} u_{2x}^+ - \frac{\bar{\rho}}{\bar{\tau}} u_{1x}^+$$

so that:

$$\lim_{x \to +\infty} (u_{1x}^+ u_{2x}^- + k^2 u_1^+ u_2^-) = \lim_{x \to +\infty} \left[-ike^{ikx} \left(\frac{ik}{\tau} e^{-ikx} + \frac{ik\bar{\rho}}{\bar{\tau}} e^{ikx} \right) + k^2 e^{ikx} \left(\tau^{-1} e^{-ikx} - \frac{\bar{\rho}}{\bar{\tau}} e^{ikx} \right) \right] = 2k^2 \tau^{-1}$$

and similarly:

$$\lim_{x \to -\infty} (u_{1x}^+ u_{2x}^- + k^2 u_1^+ u_2^-) = \lim_{x \to -\infty} \left[-ike^{-ikx} \left(\frac{ik}{\tau} e^{ikx} - \frac{ik\rho}{\tau} e^{-ikx} \right) + k^2 e^{-ikx} \left(\tau^{-1} e^{ikx} + \frac{\rho}{\tau} e^{-ikx} \right) \right] = 2k^2 \tau^{-1}.$$

We can thus rewrite (12.153) as:

$$u_{1x}^+ u_{2x}^- + k^2 u_1^+ u_2^- = -\int_x^{+\infty} q(u_1^+ u_2^-)_x \, dx + 2k^2 \tau^{-1} \tag{12.154}$$

and making x tend to $-\infty$:

$$\int_{-\infty}^{+\infty} q(u_1^+ u_2^-)_x \, dx = 0$$

or:

$$\int_{-\infty}^{+\infty} q_x \cdot u_1^+ u_2^- \, dx = 0. \tag{12.155}$$

Remembering that for all solutions, u, w of (12.138) we have:

$$k^2 uw = quw - \tfrac{1}{2} \cdot (uw)_{xx} + u_x w_x$$

we obtain, by taking $u = u_1^+$, $w = u_2^-$ and using (12.154):

$$k^2 u_1^+ u_2^- = \frac{q}{2} \cdot u_1^+ u_2^- - \tfrac{1}{4} \cdot (u_1^+ u_2^-)_{xx} - \frac{1}{2} \int_x^{+\infty} q (u_1^+ u_2^-)_x \, dx + k^2 \tau^{-1}$$

or:

$$L(u_1^+ u_2^-) = k^2 (u_1^+ u_2^-) - k^2 \tau^{-1}. \tag{12.156}$$

6. *An identity*. We shall show that, assuming $q \in \mathscr{T}_{2n-1}$:

$$\int_{-\infty}^{+\infty} q_x L^m(1) \, dx = 0, \quad \text{for all integers } m \leqslant n. \tag{12.157}$$

This identity obviously holds for $m = 0$ and $m = 1$, since $L(1) = q/2$. Suppose (12.157) is true up to $m - 1$; since by (12.156):

$$(L - k^2)(u_1^+ u_2^-) = -k^2 \tau^{-1} \cdot 1$$

it is equally true that:

$$(L^{m+1} - k^{2m+2})(u_1^+ u_2^-) = -k^2 \tau^{-1} [L^m + k^2 L^{m-1} + k^4 L^{m-2} + \cdots + k^{2m}](1).$$

Multiplying by q_x and integrating with respect to $x \in \mathbb{R}$, we arrive, having regard to the inductive hypothesis and to (12.155), at:

$$\int_{-\infty}^{+\infty} q_x L^m(1) \, dx = -\tau \cdot \int_{-\infty}^{+\infty} q_x L^{m+1} \left(\frac{u_1^+ u_2^-}{k^2} \right) dx$$

that is to say:

$$\int_{-\infty}^{+\infty} q_x L^m(1) \, dx = -\tau \cdot \int_{-\infty}^{+\infty} L^{*m-1} q_x \cdot \left(k^2 u_1^+ u_2^- - k^2 \tau^{-1} - \frac{q \tau^{-1}}{2} \right) dx \tag{12.158}$$

where we have used (12.156) to calculate $L^2(u_1^+ u_2^-)$.

The left-hand side of (12.158) is independent of k; since $L^{*m-1} q_x$ is integrable over $\mathbb{R}$ and $\lim_{k \to \infty} \tau = 1$, if it can be proved that $k^2 u_1^+ u_2^- - k^2 \tau^{-1} - q \tau^{-1}/2$ tends to 0 as $k \to \infty$ through real values, uniformly with respect to x, then we shall be able to conclude that the right-hand side of (12.158) also tends to 0 under these same conditions, that is to say is zero since it is known to be independent of k.

To assess the asymptotic behaviour of

$$k^2 u_1^+ u_2^- - k^2 \tau^{-1} - \frac{q \tau^{-1}}{2} = k^2 (v_1^+ v_2^- - 1) - \frac{q}{2} - \left(k^2 + \frac{q}{2} \right) (\tau^{-1} - 1)$$

we need to calculate the expressions for v_1^+, v_2^-, τ^{-1} to at least the second order in k^{-2}. These expressions can be easily derived from the formulae (12.65), (12.52), (12.68) and (12.86) by suitable iterations and by taking advantage of the fact that the Fourier integrals involving $q(x)$ can be integrated by parts since q is differentiable.

We thus obtain:

$$v_1^+(x, k) - 1 = -\frac{1}{2ik} \cdot \int_x^{+\infty} q(x') \, dx' + \frac{1}{4k^2} \left[q(x) - \frac{1}{2} \left(\int_x^{+\infty} q(x') \, dx' \right)^2 \right] + O(k^{-3})$$

$$v_2^-(x,k)-1=-\frac{1}{2ik}\cdot\int_{-\infty}^{x} q(x')\,\mathrm{d}x'+\frac{1}{4k^2}\left[q(x)-\frac{1}{2}\left(\int_{-\infty}^{x} q(x')\,\mathrm{d}x'\right)^2\right]+O(k^{-3})$$

$$\tau^{-1}(k)-1=-\frac{1}{2ik}\cdot\int_{-\infty}^{+\infty} q(x')\,\mathrm{d}x'-\frac{1}{8k^2}\left(\int_{-\infty}^{+\infty} q(x')\,\mathrm{d}x'\right)^2+O(k^{-3})$$

for real k, the terms $O(1/k^3)$ being of order k^{-3} when $k\to\infty$, uniformly with respect to $x\in\mathbb{R}$.

Thus the assertion that $k^2u_1^+u_2^- - k^2\tau^{-1} - \dfrac{q\tau^{-1}}{2}$ tends to 0 as $k\to\infty$, uniformly with respect to x is justified and so is the formula (12.157).

Lastly we note that with $L(1)=q/2$, we can write:

(12.159)
$$\int_{-\infty}^{+\infty} q_x L^{n-1} q\,\mathrm{d}x=0,\quad \text{if } q\in\mathcal{T}_{2n-1}.$$

The Evolution Equation

1. Going back to the formulae (12.144) and (12.148) we see that for a given function $\Omega(k^2)$, we could write:

(12.160)
$$\rho_t=-2ik\Omega(k^2)\rho$$

if it were true that:

$$\int_{-\infty}^{+\infty} q_t(u_1^+)^2\,\mathrm{d}x=\Omega(k^2)\int_{-\infty}^{+\infty} q_x(u_1^+)^2\,\mathrm{d}x$$

i.e. by a formal calculation allowing us, starting from (12.149), to write:

$$\Omega(k^2)(u_1^+)^2=\Omega(L)(u_1^+)^2,$$

$$\int_{-\infty}^{+\infty}(q_t-\Omega(L^*)q_x)(u_1^+)^2\,\mathrm{d}x=0.$$

It would therefore suffice that q should obey the equation

$$q_t-\Omega(L^*)q_x=0$$

to ensure that the "evolution" of the reflection coefficient ρ (i.e. its value considered as a function of time) will be governed by (12.160).

Note that with $\Omega(k^2)=4k^2$ we would be led to $q_t=4L^*q_x$, that is to say in accordance with the definition of the operator L^*, $q_t=6qq_x-q_{xxx}$, which is equivalent after changing q into $-q/6$ to $q_t+qq_x+q_{xxx}=0$, which is none other than the normalised Korteweg-de Vries equation.

We shall now specify in more detail the preceding formal calculations, and to this end, we shall suppose that with $\Omega_1(k^2)$, $\Omega_2(k^2)$ being given polynomials in k^2, the function $q(x,t)$ satisfies the equation:

(12.161)
$$\Omega_2(L^*)q_t-\Omega_1(L^*)q_x=0,$$

and we shall assume that q and q_t are of class $\mathcal{T}_{2n+1}$, $\mathcal{T}_{2p}$ respectively where n and p are the degrees of the polynomials $\Omega_1(s)$ $\Omega_2(s)$, and $n\geqslant p$.

The following sequence of operations can then be shown to be perfectly legitimate:

$$\int_{-\infty}^{+\infty} (\Omega_2(L^*)q_t - \Omega_1(L^*)q_x)(u_1^+)^2\,dx = 0$$

or:

$$\int_{-\infty}^{+\infty} q_t\Omega_2(L)(u_1^+)^2\,dx = \int_{-\infty}^{+\infty} q_x\Omega_1(L)(u_1^+)^2\,dx$$

and by (12.149):

$$\Omega_2(k^2)\int_{-\infty}^{+\infty} q_t(u_1^+)^2\,dx = \Omega_1(k^2)\int_{-\infty}^{+\infty} q_x(u_1^+)^2\,dx$$

i.e. with (12.144), (12.148):

(12.162) $$\rho_t = -2ik\Omega(k^2)\rho, \quad \Omega = \frac{\Omega_1}{\Omega_2},$$

assuming $\Omega_2(k^2) \neq 0$, for all real k.

2. Always assuming that q satisfies the evolution equation (12.161), and starting from:

$$\int_{-\infty}^{+\infty} \Omega_2(L)u_1^+u_2^- \cdot q_t\,dx = \int_{-\infty}^{+\infty} u_1^+u_2^- \cdot \Omega_2(L^*)q_t\,dx$$

$$= \int_{-\infty}^{+\infty} u_1^+u_2^-\Omega_1(L^*)q_x\,dx = \int_{-\infty}^{+\infty} \Omega_1(L)u_1^+u_2^- \cdot q_x\,dx$$

and observing, in view of (12.155), (12.156), (12.157) that

$$\int_{-\infty}^{+\infty} q_xL^m(u_1^+u_2^-)\,dx = 0 \quad \text{if} \quad m \leqslant n,$$

we obtain:

(12.163) $$\int_{-\infty}^{+\infty} \Omega_2(L)u_1^+u_2^- \cdot q_t\,dx = 0.$$

On the other hand, it is easy to justify, starting from (12.156), that:

$$L^j(u_1^+u_2^-) = k^{2j}u_1^+u_2^- - \tau^{-1}\sum_{s=1}^{j} k^{2s}L^{j-s}(1), \quad \forall j \leqslant n$$

so that (12.163) can be written in the form:

(12.164) $$\Omega_2(k^2)\int_{-\infty}^{+\infty} u_1^+u_2^- \cdot q_t\,dx - \tau^{-1}\cdot k^2G(k^2, t) = 0$$

where $G(k^2, t)$ is a polynomial in k^2 whose degree is at most equal to that of $\Omega_2(k^2)$ diminished by 1.

We note that the relation (12.164) is still valid throughout the half-plane $\operatorname{Im} k \geqslant 0$, all the operations described above being meaningful within this region. Let us suppose that $\Omega_2(0) \neq 0$ and that at time $t = t_0$, τ^{-1} and $\Omega_2(k^2)$ have no zero in common in $\operatorname{Im} k > 0$, a property which will still be valid, by continuity, in some neighbourhood of t_0.

It is easy to deduce from this that $G(k^2, t)$ is identically zero; for if not, as the degree of the polynomial $G(k^2, t)$ is strictly less than that of $\Omega_2(k^2)$, there would exist at least one zero (say k_0) of $\Omega_2(k^2)$, with $k_0 \neq 0$ since $\Omega_2(0) \neq 0$, $\operatorname{Im} k_0 \geqslant 0$ with the property that G/Ω_2 would become infinite as k approached k_0 in the half-plane $\operatorname{Im} k \geqslant 0$. But this cannot happen with:

$$\frac{\tau}{k^2} \int_{-\infty}^{+\infty} u_1^+ u_2^- q_t \, dx$$

which remains finite.

We have therefore proved that $G(k^2, t)$ is identically zero and consequently:

$$\int_{-\infty}^{+\infty} u_1^+ u_2^- q_t \, dx = 0 \tag{12.165}$$

from which we deduce by (12.143):

$$\tau_t = 0 \tag{12.166}$$

in the neighbourhood of t_0. This conclusion means in particular that the poles of τ are independent of t, subject to the reservation already made that none of them coincide with a zero of $\Omega_2(k^2)$.

3. By (12.165), and with q satisfying (12.161) and (12.156) we can write:

$$\int_{-\infty}^{+\infty} q_t \cdot L(u_1^+ u_2^-) \, dx = -k^2 \tau^{-1} \cdot \int_{-\infty}^{+\infty} q_t \, dx$$

or

$$-\tau \int_{-\infty}^{+\infty} L^* q_t \cdot \frac{u_1^+ u_2^-}{k^2} \, dx = \int_{-\infty}^{+\infty} q_t \, dx$$

an equation whose right-hand side is independent of k. If we make k tend to infinity, and note that $(u_1^+ u_2^- / k^2)$ tends to zero uniformly in x, and that $\tau \to 1$, we conclude that:

$$\int_{-\infty}^{+\infty} q_t \, dx = 0 \tag{12.167}$$

and it is easy to establish from this, that:

$$\int_{-\infty}^{+\infty} q_t \cdot L^m(1) \, dx = 0, \quad \forall m \leqslant p + 1 \leqslant n + 1. \tag{12.168}$$

For suppose this formula were true up to $m - 1$; then by a process similar to that which led to (12.157), we obtain by (12.156) and (12.165):

$$\begin{aligned} \int_{-+}^{+\infty} q_t L^m(1) \, dx &= -\tau \int_{-\infty}^{+\infty} q_t L^{m+1}\left(\frac{u_1^+ u_2^-}{k^2}\right) dx \\ &= -\tau \int_{-\infty}^{+\infty} L^{*m-1} q_t \cdot L^2\left(\frac{u_1^+ u_2^-}{k^2}\right) dx \end{aligned}$$

$$= -\tau \int_{-\infty}^{+\infty} L^{*m-1} q_t \cdot \left[k^2(u_1^+ u_2^- - \tau^{-1}) - \frac{\tau^{-1} q}{2} \right] dx$$

and, making k tend to infinity, we are left with (12.168).

It should be emphasized that, in contrast to (12.157), the validity of (12.168) is conditional on q being a solution of the evolution equation (12.161).

If however such is not the case, that is to say for a function $q(x, t)$ which is simply subject to the regularity conditions $q \in \mathscr{T}_{2j-1}$, $q_t \in \mathscr{T}_{2j-2}$ we obtain by integration of (12.143) after multiplication by $\tau(k, t)$:

(12.169) $$\int_{-\infty}^{+\infty} \tau(k, t) \left(\int_{-\infty}^{+\infty} u_1^+ u_2^- q_t \, dx \right) dt = 0$$

for large enough real k, under the hypothesis that:

(12.170) $$\lim_{t \to \pm\infty} \int_{-\infty}^{+\infty} |q(x, t)| \, dx = 0,$$

which ensures, by (12.66), (12.86) that $\lim_{k \to \infty} |\tau(k, t) - 1| = 0$, uniformly with respect to $t \in \mathbb{R}$, and $\lim_{t \to \pm\infty} \tau(k, t) = 1$, so that $\int_{-\infty}^{+\infty} \frac{\tau_t}{\tau} dt = 0$.

Using an argument similar to the one which allowed us to establish (12.157), we can easily deduce from (12.169), assuming $q(x, t)$ has compact support in $\mathbb{R}^2$:

(12.171) $$\int_{-\infty}^{+\infty} \int_{-\infty}^{+\infty} q_t L^m(1) \, dx \, dt = 0 \quad m \leqslant j.$$

Integral Invariants

1. Consider the operator L defined by (12.151). We have $L(1) = (\frac{1}{2})q$, and

(12.172) $$L^2(1) = \tfrac{1}{2} Lq = \tfrac{1}{2} \left[\tfrac{1}{2} q^2 - \tfrac{1}{4} q_{xx} - \frac{1}{2} \int_x^{+\infty} q q_x \, dx \right] = \tfrac{1}{8}(3q^2 - q_{xx}).$$

Thus $L(1)$, $L^2(1)$ are polynomials in q and its successive partial derivatives with respect to x. This is a property which can be extended to the iterates of L of any order. We can in effect show, assuming $q \in \mathscr{T}_{2m-1}$, that $L^m(1)$ is a polynomial in q and its repeated derivatives $q_{x^\alpha} = q^{(\alpha)}$ of order up to $2(m-1)$. We shall prove this by induction, assuming the assertion true for all $m \leqslant j$; in particular for $\forall q \in \mathscr{T}_{2j-1}$ we have the representation:

(12.173) $$q_x L^j(1) = Q_j(q, q^{(1)}, \ldots, q^{(2j-2)})$$

where Q_j is a polynomial, and remembering (12.157), we can write:

(12.174) $$\int_{-\infty}^{+\infty} Q_j(q, q^{(1)}, \ldots, q^{(2j-2)}) \, dx = 0.$$

With any element $q(\xi) \in \mathscr{T}_{2j-1}$ given in this way we can associate a $\tilde{q}(\xi)$ of the same class defined by:

(12.175)

$$\tilde{q}(\xi) = q(\xi), \quad \xi \leqslant x$$
$$\tilde{q}(\xi) = \left[q(x) + (\xi - x)q_x(x) + \cdots + \frac{(\xi - x)^{2j-1}}{(2j-1)!} q_{x^{2j-1}}(x) \right] \varphi(\xi - x), \quad \xi \geqslant x$$

where $\varphi(s)$, of compact support on $\mathbb{R}$, and differentiable up to order $2j$ at least, is a given function such that $\varphi(s) = 1$, for $0 \leqslant s \leqslant 1$ and fixed $x \in \mathbb{R}$.

Applying (12.174) to $\tilde{q}$ leads to:

$$\int_{-\infty}^{x} Q_j(\tilde{q}, \ldots, \tilde{q}^{(2j-2)}) \mathrm{d}x = - \int_{x}^{+\infty} Q_j(\tilde{q}, \ldots, \tilde{q}^{(2j-2)}) \mathrm{d}x.$$

But it follows from the structure of $\tilde{q}(\xi)$ for $\xi \geqslant x$ that the integral on the right is a polynomial function of the values of q and its derivatives in x, say $P_j(q(x), \ldots, q^{(2j-1)}(x))$, whereas the one on the left is a primitive of $Q_j(q(\xi), \ldots)$, i.e. of $q_x L^j(1)$. Thus we obtain:

$$q_x L^j(1) = Q_j(q, q^{(1)}, \ldots, q^{(2j-2)}) = \frac{\mathrm{d}P_j}{\mathrm{d}x} \tag{12.176}$$

and in particular it emerges that P_j is a polynomial in q and its derivatives of order $2j-3$ at most.

Assuming now that $q \in \mathscr{T}_{2j+1}$ we calculate:

$$L^{j+1}(1) = LL^j(1) = \tfrac{1}{2} q L^j(1) - \tfrac{1}{4}(L^j(1))_{xx} - \frac{1}{2} \int_{x}^{+\infty} q (L^j(1))_x \mathrm{d}x$$

or, integrating by parts:

$$L^{j+1}(1) = q L^j(1) - \tfrac{1}{4}(L^j(1))_{xx} + \frac{1}{2} \int_{x}^{+\infty} q_x L^j(1) \mathrm{d}x$$

and by (12.176)

$$L^{j+1}(1) = q L^j(1) - \tfrac{1}{4}(L^j(1))_{xx} - \tfrac{1}{2} P_j(q, q^{(1)}, \ldots, q^{(2j-3)})$$

which proves that $L^{j+1}(1)$ is a polynomial in $q, \ldots, q^{(2j)}$. Furthermore it is clear that $L^j(1) = (1/2)(-1/4)^{j-1} q^{(2j-2)} + \cdots$, the unwritten part of the polynomial containing q and its derivatives only up to order $2j-4$ at most.

2. We can likewise deal with the representation of $L^{*j} q_x$. We shall show in fact that if $q \in \mathscr{T}_{2j+1}$, then there exists a polynomial $M_j(q, \ldots, q^{(2j)})$ such that $L^{*j} q_x = (\mathrm{d}M_j/\mathrm{d}x)$. This is obvious for $j=0$ and $j=1$ with the polynomials $M_0 = q$, $M_1 = (1/4)(3q^2 - q_{xx})$ respectively. We use an inductive argument, assuming the assertion to be true for all non-negative integer m less than j and write:

$$1 \cdot L^{*j} q_x - L^j 1 \cdot q_x = \sum_{\alpha=0}^{j-1} (L^{\alpha} 1 \cdot L^{*j-\alpha} q_x - L^{\alpha+1} 1 \cdot L^{*j-\alpha-1} q_x).$$

Applying (12.152) with $f = L^{\alpha}1$, $g = L^{*j-\alpha-1}q_x$, we see by virtue of the inductive hypothesis and the already established fact that $q_x L^{\alpha}1$ is the derivative with respect to x of a polynomial in $q, q^{(1)}, \ldots$, that:

$$L^{\alpha}1 \cdot L^{*j-\alpha}q_x - L^{\alpha+1}1 \cdot L^{*j-\alpha-1}q_x,$$

is, in turn, the derivative with respect to x of a polynomial in q and its successive derivatives, from which follows an analogous result for $L^{*j}q_x$ and the proof of the proposition.

We shall also show by induction, having regard to the definition of L^*, that

$$L^{*j}q_x = (-\tfrac{1}{4})^j q^{(2j+1)} + \frac{\mathrm{d}}{\mathrm{d}x} N_j(q, \ldots, q^{(2j-2)})$$

where N_j is a polynomial.

In particular we can ensure that $\lim\limits_{x \to \pm\infty} L^{*j}q_x = 0$ for $q \in \mathscr{T}_{2j+1}$ if one assumes in addition that $\lim\limits_{x \to \pm\infty} q^{(2j+1)} = 0$.

3. From now on we recall the dependence of q on t and the formula (12.171) which is valid for every sufficiently regular function $q(x, t)$. In particular, for every function $q(x, t)$ with compact support in $\mathbb{R}^2$, such that $q \in \mathscr{T}_{2j-1}$, $q_t \in \mathscr{T}_{2j-2}$, we can write:

$$\int_{-\infty}^{+\infty} \int_{-\infty}^{+\infty} q_t L^j(1) \, \mathrm{d}x \, \mathrm{d}t = 0. \tag{12.177}$$

Let us associate with a function q of this type, the function $\tilde{q}$ of the same type defined by:

$$\begin{aligned} &\tilde{q}(\xi, t) = q(\xi, t), \quad \xi \leqslant x \\ &\tilde{q}(\xi, t) = \Big[q(x, t) + (\xi - x) q_x(x, t) + \cdots \\ &\qquad + \frac{(\xi - x)^{2j-1}}{(2j-1)!} q_{x^{2j-1}}(x, t) \Big] \varphi(\xi - x), \quad \xi \geqslant x, \end{aligned} \tag{12.178}$$

$\varphi(s)$ being of compact support in $\mathbb{R}$, differentiable up to order $2j$ at least, and such that $\varphi(s) = 1$, $\forall s \in [0, 1]$.

We know that $L^j(1)$ is a polynomial in $q, q^{(1)}, \ldots, q^{(2j-2)}$, so that by applying (12.177) to $\tilde{q}$, we can write in view of the structure (12.178):

$$\begin{aligned} &\int_{-\infty}^{+\infty} \left(\int_{\xi \leqslant x} q_t L^j(1) \, \mathrm{d}\xi \right) \mathrm{d}t \\ &\quad = - \int_{-\infty}^{+\infty} \int_{\xi \geqslant x} H_j(q(x, t), q_x(x, t), \ldots, q_{tx^{2j-1}}(x, t), \xi) \, \mathrm{d}\xi \, \mathrm{d}t \\ &\quad = \int_{-\infty}^{+\infty} R_j(q(x, t), \ldots, q_{tx^{2j-1}}(x, t)) \, \mathrm{d}t \end{aligned} \tag{12.179}$$

where H_j, R_j are polynomials in $q, q_x, \ldots, q_{tx^{2j-1}}$, the first having coefficients which depend on ξ, and the second having constant coefficients.

Reversing the order of integration in the repeated integral on the left of (12.179) and differentiating with respect to x, assuming from now on that q, with compact support, is continuously differentiable in (t, x) to order 1 in t and to order $2j$ in x, we can write:

$$\int_{-\infty}^{+\infty}\left[q_t L^j(1) - \frac{\partial R_j}{\partial x}(q, \dots, q_{tx^{2j-1}})\right] \mathrm{d}t = 0.$$

Putting

$$q_t L^j(1) - \frac{\partial R_j}{\partial x}(q, \dots, q_{tx^{2j-1}}) = S_j \tag{12.180}$$

S_j appears as a polynomial in $q, q_t, \dots, q_{tx^{2j}}$ at most, satisfying:

$$\int_{-\infty}^{+\infty} S_j(q, \dots)\, \mathrm{d}t = 0, \quad \forall x. \tag{12.181}$$

This relation is in fact an identity which is true for every function q which satisfies the above-mentioned regularity conditions. To each such element $q(x, s)$ corresponds an associated $\tilde{q}(x, s)$

$$\begin{aligned} \tilde{q}(x, s) &= q(x, s), \quad s \leqslant t \\ \tilde{q}(x, s) &= [q(x, t) + (s - t) q_t(x, t)] \varphi(s - t), \quad s \geqslant t \end{aligned} \tag{12.182}$$

with $\varphi(\sigma)$ having compact support, continuously differentiable and with $\varphi(\sigma) = 1$, $\sigma \in [0, 1]$, so that $\tilde{q}(x, s)$ thus satisfies the regularity conditions which are required to enable the identity (12.181) to be applied to it. In other words, taking into account the structure (12.182):

$$\int_{-\infty}^{t} S_j(q, \dots)\, \mathrm{d}s = -\int_{t}^{+\infty} S_j(\tilde{q}, \dots)\, \mathrm{d}s = T_j(q, \dots, q_{tx^{2j}})$$

and $S_j(q, \dots) = (\partial T_j / \partial t)$, a result which implies in particular that T_j cannot contain any derivative of q with respect to t.

Coming back to (12.180) we obtain:

$$q_t L^j(1) = \frac{\partial R_j}{\partial x} + \frac{\partial T_j}{\partial t} \tag{12.183}$$

where R_j and T_j are polynomials. We can obviously assume this representation to be irreducible, which is tantamount to saying that in reality R_j is a polynomial in $q, \dots, q_{x^{2j-3}}, q_t, \dots, q_{tx^{2j-3}}$ at most and T_j a polynomial in $q, q_x, \dots, q_{x^{2j-2}}$ at most.

If q is not of compact support but fulfils the necessary conditions to ensure that $L^j(1)$ is a polynomial in q and its derivative q_x then the formula (12.183) remains valid, since its two components are essentially locally-valued.

Finally if q obeys the evolution equation (12.161), we know that $\int_{-\infty}^{+\infty} q_t L^j(1)\, \mathrm{d}x = 0$ and we can deduce from (12.183) that

$$\int_{-\infty}^{+\infty} \frac{\partial T_j}{\partial t}\, \mathrm{d}x = 0 \quad \text{or} \quad \int_{-\infty}^{+\infty} T_j(q, \dots)\, \mathrm{d}x = C^{te}.$$

In particular if q, q_t belong to $\mathscr{T}_\infty$ we obtain an infinity of integral invariants. Let us illustrate (12.183) for $j = 0, 1, 2$:

$$j=0, \quad q_t L^0(1) = q_t, \quad R_0 = 0, \quad T_0 = q, \quad \int_{-\infty}^{+\infty} q\,\mathrm{d}x = C^{te}$$

$$j=1, \quad q_t L(1) = \frac{q q_t}{2}, \quad R_1 = 0, \quad T_1 = \frac{q^2}{4}, \quad \int_{-\infty}^{+\infty} q^2\,\mathrm{d}x = C^{te}$$

$$j=2, \quad q_t L^2(1) = (1/8)(3q^2 q_t - q_{xx} q_t),$$

$$R_1 = -\frac{q_x q_t}{8}, \quad T_1 = \tfrac{1}{16}\cdot(2q^3 + q_x^2), \quad \int_{-\infty}^{+\infty} (2q^3 + q_x^2)\,\mathrm{d}x = C^{te}.$$

Another Approach to the Evolution Equation

1. Supposing $u(x,t)$ to be a solution of the differential equation (12.138), let us seek to define functions A, B, C, D of x, t and k such that

(12.184) $$w = u_t - Au - Bu_x$$

will satisfy

(12.185) $$w_x = u_{tx} - Cu - Du_x$$

and will also be a solution of (12.138).

Substituting the value of w given by (12.184) in (12.185) and using the fact that u satisfies (12.138), we obtain:

(12.186) $$\begin{aligned} A + B_x &= D \\ A_x + B(q - k^2) &= C. \end{aligned}$$

The fact that w is a solution of (12.138) now gives us:

(12.187) $$\begin{aligned} q_t - C_x + (A - D)(q - k^2) &= 0 \\ D_x + C &= (q - k^2)B. \end{aligned}$$

It follows from (12.186), (12.187) that $A_x + D_x = 0$, which suggests that A, D should satisfy:

(12.188) $$A + D = 0$$

and we are left with:

(12.189) $$\begin{aligned} 2A + B_x &= 0 \\ A_x + B(q - k^2) &= C \\ q_t &= C_x - 2A(q - k^2). \end{aligned}$$

We can consider (12.189) as a system of first order differential equations and try to define its solution A, B, C as a function of x on the real axis, with t appearing as a parameter. However to accord with certain earlier results, we shall look for solutions of a particular structure described by:

$$A=\frac{\sum_{j=0}^{n} A_j k^{2j}}{\Omega_2(k^2)},\quad B=\frac{\sum_{j=0}^{n} B_j k^{2j}}{\Omega_2(k^2)},\quad C=\frac{C_{-1}+k^2\sum_{j=0}^{n} C_j k^{2j}}{\Omega_2(k^2)} \tag{12.190}$$

with $\Omega_2(k^2)=\sum_{j=0}^{p} a_j k^{2j}$ being a polynomial with given numerical coefficients and A_j, B_j, C_{-1}, C_j being the unknowns of the problem, i.e. functions of the variables x and t only, which satisfy the following limit conditions:

$$\lim_{x\to-\infty} A_j=0,\quad \lim_{x\to-\infty}(B_j+C_j)=0,\quad \lim_{x\to-\infty} B_j=b_j,\quad \lim_{x\to-\infty} C_{-1}=0 \tag{12.191}$$

where the b_j denote given numerical constants for $0\leqslant j\leqslant n$.

By substituting from (12.190) in (12.189) we obtain:

$$\begin{aligned} 2A_0+B_{0,x}&=0\\ A_{0,x}+qB_0&=C_{-1}\\ a_0q_t+2A_0q&=C_{-1,x} \end{aligned} \tag{12.192}$$

to order 0 in k^2, and then

$$\begin{aligned} 2A_j+B_{j,x}&=0\\ A_{j,x}+qB_j&=B_{j-1}+C_{j-1},\quad 1\leqslant j\leqslant n\\ a_jq_t+2A_jq&=C_{j-1,x}+2A_{j-1} \end{aligned} \tag{12.193}$$

it being understood that $a_j=0$ if $j>p$, and lastly

$$\begin{aligned} B_n+C_n&=0\\ a_{n+1}q_t&=C_{n,x}+2A_n \end{aligned} \tag{12.194}$$

to order $(n+1)$ in k^2 and

$$a_jq_t=0,\quad j>n+1\quad \text{or}\quad a_j=0,\quad j>n+1. \tag{12.195}$$

We can solve this system by descending induction from j to $j-1$. We first satisfy (12.194) by taking:

$$a_{n+1}=0,\ A_n=0,\ B_n=b_n,\ C_n=-b_n \tag{12.196}$$

from which it follows, by (12.195), that $n\geqslant p$.

We next deduce from (12.193) and (12.192):

$$\begin{aligned} &A_{j-1}=-\tfrac{1}{2}\cdot B_{j-1,x}\\ &B_{j-1}+C_{j-1}=A_{j,x}+qB_j=-\tfrac{1}{2}\cdot B_{j,xx}+qB_j,\ 1\leqslant j\leqslant n\\ &(C_{j-1}-B_{j-1})_x=a_jq_t+2A_jq=a_jq_t-qB_{j,x} \end{aligned}$$

i.e. with (12.191):

$$2B_{j-1}=-\tfrac{1}{2}\cdot B_{j,xx}+qB_j+\int_{-\infty}^{x} qB_{j,x}\,\mathrm{d}x-a_j\int_{-\infty}^{x} q_t\,\mathrm{d}x+2b_{j-1}$$

$$2C_{j-1}=-\tfrac{1}{2}\cdot B_{j,xx}+qB_j-\int_{-\infty}^{x} qB_{j,x}\,\mathrm{d}x+a_j\int_{-\infty}^{x} q_t\,\mathrm{d}x-2b_{j-1}.$$

However it is more convenient to work with $B_{j-1,x}$:

$$B_{j-1,x} = -\tfrac{1}{4}(B_{j,x})_{xx} + qB_{j,x} + \tfrac{1}{2}q_x \int_{-\infty}^{x} B_{j,x}\mathrm{d}x - \frac{a_j}{2}q_t + \frac{b_j}{2}q_x$$

or

$$B_{j-1,x} = L^* B_{j,x} - \frac{a_j}{2}q_t + \frac{b_j}{2}q_x$$

equivalent to:

(12.197) $$A_{j-1} = L^* A_j + \frac{a_j}{4}q_t - \frac{b_j}{4}q_x.$$

Starting from $A_n = 0$, we obtain successively $A_{n-1}, A_{n-2}, \ldots, A_0$ and the closure condition:

(12.198) $$0 = A_{-1} = L^* A_0 + \frac{a_0}{4}q_t - \frac{b_0}{4}q_x.$$

It will thus be seen that generally:

(12.199) $$A_{n-j-1} = G_j(L^*)q_t - K_j(L^*)q_x$$

where G_j, K_j are polynomials in the operator L^*, of degree j if $j \leqslant p$, and of degree p and n respectively if $p < j \leqslant n$. It is also clear by the closure condition, that:

(12.200) $$\Omega_2(L^*)q_t - \Omega_1(L^*)q_x = 0$$

with $$\Omega_1(k^2) = \sum_{j=0}^{n} b_j k^{2j},$$

an equation which has to be satisfied by $q(x,t)$ if the process is to terminate.

We shall suppose $q \in \mathscr{T}_{2n+1}$, $q_t \in \mathscr{T}_{2p}$, and that q is a solution of (12.200), which will give point to all the preceding calculations.

Now on these assumptions it is known that:

$$\int_{-\infty}^{+\infty} q_x L^j(1)\mathrm{d}x = 0,\ 0 \leqslant j \leqslant n+1, \quad \text{or} \quad \int_{-\infty}^{+\infty} q_x L^j q\,\mathrm{d}x = 0,\ 0 \leqslant j \leqslant n$$

whence

(12.201) $$\begin{aligned} &\int_{-\infty}^{+\infty} L^{*j} q_x \mathrm{d}x = 0 \\ &\int_{-\infty}^{+\infty} L^{*j} q_x \cdot q\,\mathrm{d}x = 0, \quad 0 \leqslant j \leqslant n \end{aligned}$$

and, with q satisfying (12.200), we also have:

$$\int_{-\infty}^{+\infty} q_t L^j(1)\mathrm{d}x = 0,\ 0 \leqslant j \leqslant p+1 \quad \text{or} \quad \int_{-\infty}^{+\infty} q_t L^j q\,\mathrm{d}x = 0,\ 0 \leqslant j \leqslant p$$

whence

$$(12.202)\qquad \begin{aligned} &\int_{-\infty}^{+\infty} L^{*j} q_t \mathrm{d}x = 0 \\ &\int_{-\infty}^{+\infty} L^{*j} q_t \cdot q \mathrm{d}x = 0, \quad 0 \leqslant j \leqslant p \end{aligned}$$

These results, applied to (12.199) lead to:

$$(12.203)\qquad \int_{-\infty}^{+\infty} A_j \mathrm{d}x = 0, \quad \int_{-\infty}^{+\infty} q A_j \mathrm{d}x = 0, \quad 0 \leqslant j \leqslant n-1$$

and under the additional assumption that $\lim_{x\to\pm\infty} q^{(2n+1)} = 0$, implying. $\lim_{x\to\pm\infty} L^{*n} q_x = 0$ and, by (12.200) $\lim_{x\to\pm\infty} L^{*p} q_t = 0$, in view of the fact that we already know that $L^{*m} q_x$, $L^{*j} q_t$ tend to zero as x tends to infinity, if $m < n$ and $j < p$, we finally obtain:

$$(12.204)\qquad \lim_{x\to+\infty} A_j = \lim_{x\to-\infty} A_j = 0, \quad 0 \leqslant j \leqslant n-1.$$

On the other hand it follows from (12.193), (12.191) that:

$$B_j = b_j - \frac{1}{2} \int_{-\infty}^{x} A_j \mathrm{d}x$$

and consequently:

$$\lim_{x\to+\infty} B_j = \lim_{x\to-\infty} B_j = b_j, \quad 0 \leqslant j \leqslant n-1.$$

By integrating the equations (12.192), (12.193), (12.194) containing $C_{-1,x}$, $C_{j-1,x}$ and $C_{n,x}$, we see, from (12.203) and $\int_{-\infty}^{+\infty} q_t \mathrm{d}x = 0$:

$$\lim_{x\to+\infty} C_{-1} = 0, \quad \lim_{x\to+\infty} C_j = \lim_{x\to-\infty} C_j = -b_j, \quad 1 \leqslant j \leqslant n.$$

It is therefore apparent that:

$$(12.205)\qquad \lim_{x\to+\infty} B = B_+ = \lim_{x\to-\infty} B = B_-$$

with

$$(12.206)\qquad B = \frac{\Omega_1(k^2)}{\Omega_2(k^2)} = \Omega(k^2)$$

and a similar result holds for C, namely:

$$(12.207)\qquad \begin{gathered} \lim_{x\to+\infty} C = C_+ = \lim_{x\to-\infty} C = C_- \\ C_+ = C_- = -k^2 B_+. \end{gathered}$$

2. Having calculated the coefficients A, B, C, D by the process just described, for a suitable solution $q(x, t)$ of (12.200), we can represent the solution (12.184) of

(12.138) calculated for $u=u_1^+$, by:

$$\begin{aligned} u_{1t}^+ - Au_1^+ - Bu_{1x}^+ &= \lambda u_1^+ + \mu u_2^+ \\ u_{1xt}^+ - Cu_1^+ - Du_{1x}^+ &= \lambda u_{1x}^+ + \mu u_{2x}^+ \end{aligned} \tag{12.208}$$

λ, μ being independent of x, since u_1^+, u_2^+ is a fundamental system of solutions of (12.138).

However an examination of the asymptotic behaviour of (12.208) at $x=+\infty$, on the basis of the results (12.188), (12.205), (12.206), (12.207) reveals that $\lambda=0$, $\mu=-ikB_+$ and hence:

$$\begin{aligned} u_{1t}^+ - Au_1^+ - Bu_{1x}^+ &= -ikB_+ u_1^+ \\ u_{1xt}^+ - Cu_1^+ - Du_{1x}^+ &= -ikB_+ u_{1,x}^+ \end{aligned} \tag{12.209}$$

and the formulae which have been established for real k remain valid in $\operatorname{Im} k \geqslant 0$, by analytic continuation.

Similarly we can write:

$$\begin{aligned} u_{2t}^- - Au_2^- - Bu_{2x}^- &= \lambda u_1^- + \mu u_2^- \\ u_{2tx}^- - Cu_2^- - Du_{2x}^- &= \lambda u_{1x}^- + \mu u_{2x}^-, \end{aligned}$$

where the new coefficients λ, μ which are still independent of x can be found by analysing the asymptotic behaviour at $x=-\infty$, which leads to:

$$\begin{aligned} u_{2t}^- - Au_2^- - Bu_{2x}^- &= ikB_- u_2^- \\ u_{2tx}^- - Cu_2^- - Du_{2x}^- &= ikB_- u_{2x}^- \end{aligned} \tag{12.210}$$

formulae which remain valid throughout the half-plane $\operatorname{Im} k \geqslant 0$ by analytic continuation.

Now let $k=i\varkappa, \varkappa>0$ be an eigenvalue and σ_k the number associated with k by the extension condition (12.98) which we write in the form:

$$\begin{aligned} u_1^+(x,t,k) &= \sigma_k u_2^-(x,t,k) \\ u_{1x}^+(x,t,k) &= \sigma_k u_{2x}^-(x,t,k) \end{aligned} \tag{12.211}$$

where $\sigma_k(t)$ is a function of t alone, while as we saw earlier, the eigenvalue $k=i\varkappa$ is independent of t. Differentiating the first equation of (12.211) with respect to t gives:

$$u_{1t}^+ = \sigma_k u_{2t}^- + (\sigma_k)_t u_2^-$$

and by (12.209), (12.210), assuming that $k=i\varkappa$ is not a zero of $\Omega_2(k^2)$, i.e. a pole of $B_+ = B_-$:

$$Au_1^+ + Bu_{1x}^+ - ikB_+u_1^+ = \sigma_k(Au_2^- + Bu_{2x}^- + ikB_-u_2^-) + (\sigma_k)_t u_2^-$$

or:

$$(\sigma_k)_t = -2ik\Omega(k^2)\sigma_k$$

and by (12.105), remembering that $\tau(k)$ does not depend on t:

$$(c_k^2)_t = -2ik\Omega(k^2)c_k^2 \tag{12.212}$$

a linear differential equation which, for $k=i\varkappa$, determines the evolution of the gauge factor as a function of time.

6. Solution of the Inverse Problem in the Case Where the Reflection Coefficient is Zero

1. With an eye to some further applications of the Korteweg-de Vries equation we propose, following [21], [26], to calculate the perturbation term $q(x)$ of equation (12.34) such that the function $B(\xi)$ associated with it by the direct problem shall be of the form:

(12.213)
$$B(\xi)=\sum_{j=1}^{N} c_j^2 e^{\varkappa_j \xi}$$

with $c_1, c_2, \ldots, c_N, \varkappa_1, \ldots, \varkappa_N$ given positive real numbers, and the $\varkappa_j$ all distinct. It will be noted that the structure postulated for $B(\xi)$ implies $R(\xi)=0$, in other words no reflection.

By the Gelfand-Levitan equation (12.130), the kernel $K(x, y)$ is defined in the half-plane $x \geqslant y$ by:

(12.214)
$$K(x, y)+\sum_{1}^{N} c_i^2 e^{\varkappa_i(x+y)}+\sum_{1}^{N} c_i^2 e^{\varkappa_i y} \cdot \int_{-\infty}^{x} e^{\varkappa_i z} K(x, z) \mathrm{d} z=0,$$

so that $K(x, y)$ can be written:

(12.215)
$$K(x, y)=-\sum_{j=1}^{N} c_j \psi_j(x) e^{\varkappa_j y}, \quad x \geqslant y$$

the $\psi_j(x)$ being functions of x alone, which remain to be determined.

In fact, by substituting (12.215) in (12.214) we end up with a set of N linear equations:

(12.216)
$$\psi_i(x)+\sum_{j=1}^{N} c_i c_j \psi_j(x) \frac{e^{(\varkappa_i+\varkappa_j) x}}{\varkappa_i+\varkappa_j}=c_i e^{\varkappa_i x}, \quad 1 \leqslant i \leqslant N$$

which can be put in matrix form:

(12.217)
$$(I+C) \psi=e$$

with

$$I=(\delta_{ij}), C=\left(\frac{c_i c_j e^{(\varkappa_i+\varkappa_j) x}}{\varkappa_i+\varkappa_j}\right), \quad \psi=\begin{pmatrix}\psi_1 \\ \psi_2 \\ \vdots \\ \psi_N\end{pmatrix}, \quad e=\begin{pmatrix}c_1 e^{\varkappa_1 x} \\ c_2 e^{\varkappa_2 x} \\ \vdots \\ c_N e^{\varkappa_N x}\end{pmatrix}.$$

Now the matrix C is symmetric, real and positive because for any non-zero vector $b=\begin{pmatrix}b_1 \\ \vdots \\ b_N\end{pmatrix}$ we have:

$$\sum_i \sum_j \frac{c_i c_j e^{(\varkappa_i+\varkappa_j) x}}{\varkappa_i+\varkappa_j} b_i b_j=\int_{-\infty}^{x}\left[\sum_{1}^{N} b_i c_i e^{\varkappa_i z}\right]^2 \mathrm{d} z>0,$$

so that the matrix $I+C$ is invertible and (12.127) has an unique solution.

We note in passing that $\det C = \det(1/(\varkappa_i + \varkappa_j)).\ (\prod_{i=1}^N c_i^2) e^{2(\sum_{i=1}^N \varkappa_i)x}$ which implies that $\det(1/(\varkappa_i + \varkappa_j))$ and $\det C$ both have the same (positive) sign.

If we denote by A_{ij} the co-factor of the (i, j)th element in the expansion of $\det(I + C)$, we can write, on the one hand

$$\Delta = \det(I + C) = \sum_{i=1}^N \left(\delta_{ij} + \frac{c_i c_j e^{(\varkappa_i + \varkappa_j)x}}{\varkappa_i + \varkappa_j}\right) A_{ij}$$

and on the other hand by Cramer's rule:

$$\psi_j(x) = \Delta^{-1} \sum_{i=1}^N c_i e^{\varkappa_i x} A_{ij} \tag{12.218}$$

from which we deduce that:

$$K(x, x) = -\sum_{j=1}^N c_j \psi_j(x) e^{\varkappa_j x} = -\Delta^{-1} \sum_i \sum_j c_i c_j e^{(\varkappa_i + \varkappa_j)x} A_{ij}$$

or, using the rule for differentiating a determinant:

$$K(x, x) = -\Delta^{-1} \frac{d\Delta}{dx}.$$

Lastly we see from (12.127) that the function $q(x)$ is necessarily expressible by:

$$q = 2 \frac{d}{dx} K(x, x) = -2 \frac{d^2}{dx^2} \log \det(I + C) \tag{12.219}$$

and, in particular, that it is the quotient of two polynomials in the exponential terms $e^{\varkappa_j x}$.

2. Let us try to determine more precisely the behaviour of $\psi_i(x)$ at infinity. For x near $-\infty$, we can put $\psi_i(x) = e^{\varkappa_i x} \theta_i(x)$, the θ_i after (12.216) being solutions of:

$$\theta_i(x) + \sum_{j=1}^N \frac{c_i c_j e^{2\varkappa_j x}}{\varkappa_i + \varkappa_j} \theta_j(x) = c_i.$$

When x approaches $-\infty$, the matrix of this set of equations tends towards the identity matrix and consequently $\lim_{x \to -\infty} \theta_i(x) = c_i$, whence:

$$\lim_{x \to -\infty} \psi_i(x) e^{-\varkappa_i x} = c_i. \tag{12.220}$$

Suppose now that $x \to +\infty$; we shall put $\psi_i(x) = e^{-\varkappa_i x} \theta_i(x)$, the system of linear equations defining the θ_i being:

$$\theta_i(x) e^{-2\varkappa_i x} + \sum_{j=1}^N \frac{c_i c_j}{\varkappa_i + \varkappa_j} \theta_j(x) = c_i.$$

The matrix of this system has the limit $C(0) = c_i c_j / (\varkappa_i + \varkappa_j)$, when $x \to +\infty$, and we know that this matrix in invertible. Consequently we can be certain that:

$$\lim_{x \to +\infty} \psi_i(x) e^{\varkappa_i x} = d_i \tag{12.221}$$

where the numbers d_i are defined by:

$$\sum_{j=1}^{N} \frac{c_j d_j}{\varkappa_i + \varkappa_j} = 1, \quad 1 \leqslant i \leqslant N.$$

The determinant of this system has, as its (i, j)th element the number $c_j/(\varkappa_i + \varkappa_j)$ and its value is:

$$\delta = \prod_{s=1}^{N} c_s \cdot \det \frac{1}{\varkappa_i + \varkappa_j} > 0.$$

By Cramer's rule we obtain:

$$d_j = \delta^{-1} \cdot \delta_j$$

with

$$\delta_j = \det \begin{pmatrix} \dfrac{1}{\varkappa_1 + \varkappa_1} \cdots 1 \cdots \dfrac{1}{\varkappa_1 + \varkappa_N} \\ \vdots \quad\quad \vdots \quad\quad \vdots \\ \dfrac{1}{\varkappa_N + \varkappa_1} \cdots 1 \cdots \dfrac{1}{\varkappa_N + \varkappa_N} \\ j \end{pmatrix} \prod_{s=1}^{N} c_s \cdot c_j^{-1}$$

By substracting the jth row from each of the remaining rows we can obtain a determinant of order $N-1$ whose (m, n)th element is:

$$\frac{\varkappa_j - \varkappa_m}{(\varkappa_m + \varkappa_n)(\varkappa_j + \varkappa_n)},$$

m, n taking the values $1, 2, \ldots, j-1, j+1, \ldots, N$.

We thus obtain:

$$\delta_j = c_j^{-1} \cdot \prod_{s=1}^{N} c_s \cdot \frac{\prod_{m \neq j} (\varkappa_j - \varkappa_m)}{\prod_{n \neq j} (\varkappa_j + \varkappa_n)} \det_{\substack{m \neq j \\ n \neq j}} \left(\frac{1}{\varkappa_m + \varkappa_n} \right)$$

and

$$d_j = c_j^{-1} \cdot \prod_{\substack{m=1 \\ \neq j}}^{N} \frac{\varkappa_j - \varkappa_m}{\varkappa_j + \varkappa_m} \cdot \frac{\det_{\substack{m \neq j \\ n \neq j}} \left(\dfrac{1}{\varkappa_m + \varkappa_n} \right)}{\det \left(\dfrac{1}{\varkappa_m + \varkappa_n} \right)}. \tag{12.222}$$

It will be seen that $d_j \neq 0$, and that its sign is that of the product $\prod_{m=1, \neq j}^{N} (\varkappa_j - \varkappa_m)$, i.e. $(-1)^{N-j}$, assuming the eigenvalues are numbered in order of increasing magnitude: $\varkappa_1 < \varkappa_2 < \cdots < \varkappa_N$.

Returning to the calculation of $q(x)$ we see from (12.215) that:

$$q(x) = 2 \frac{\mathrm{d}}{\mathrm{d}x} K(x, x) = -2 \sum_{1}^{N} c_j \dot{\psi}_j e^{\varkappa_j x} - 2 \sum_{1}^{N} c_j \varkappa_j \psi_j e^{\varkappa_j x}. \tag{12.223}$$

Multiplying (12.216) by $4\varkappa_i\psi_i + 2\dot{\psi}_i$ and summing over i we get:

$$\sum_i (4\varkappa_i\psi_i^2 + 2\dot{\psi}_i\psi_i) + \sum_i\sum_j c_ic_j(4\varkappa_i\psi_i + 2\dot{\psi}_i)\psi_j \cdot \frac{e^{(\varkappa_i+\varkappa_j)x}}{\varkappa_i + \varkappa_j} = \sum_i c_i(4\varkappa_i\psi_i + 2\dot{\psi}_i)e^{\varkappa_i x}$$

and differentiating (12.216) with respect to x and multiplying the equation so obtained by $-2\psi_i$ and then summing over i, we have:

$$-2\sum_i \dot{\psi}_i\psi_i - 2\sum_i\sum_j c_ic_j(\dot{\psi}_j + (\varkappa_i + \varkappa_j)\psi_j)\psi_i \frac{e^{(\varkappa_i+\varkappa_j)x}}{\varkappa_i + \varkappa_j} = -2\sum_i c_i\varkappa_i\psi_i e^{\varkappa_i x}.$$

Finally by adding these two results and using (12.223) we arrive at:

$$q(x) = -4\sum_{i=1}^{N} \varkappa_i\psi_i^2 \tag{12.224}$$

an expression which shows, on the one hand, that $q(x)$ is always negative, and on the other hand, that it tends exponentially to zero, as do its derivatives of all orders, when $x \to \pm\infty$, by virtue of the behaviour of $\psi_i(x)$ described by (12.220) and (12.221).

3. The function $\psi_i(x)$ and the number $\varkappa_i$ are associated eigenfunctions and eigenvalues, or in other words we have:

$$\ddot{\psi}_i - (q(x) + \varkappa_i^2)\psi_i = 0. \tag{12.225}$$

To see this we have only to apply the operator $L_i = (\mathrm{d}^2/\mathrm{d}x^2) - (q(x) + \varkappa_i^2)$, to (12.216), taking account of (12.223) and of $L_i = L_j + (\varkappa_j^2 - \varkappa_i^2)$, thus obtaining:

$$L_i\psi_i + \sum_{j=1}^{N} c_ic_j\left(\frac{L_j\psi_j}{\varkappa_i + \varkappa_j} + (\varkappa_j - \varkappa_i)\psi_j\right)e^{(\varkappa_i+\varkappa_j)x}$$
$$+ \sum_{j=1}^{N} (2c_ic_j\dot{\psi}_j + c_ic_j\psi_j(\varkappa_i + \varkappa_j))e^{(\varkappa_i+\varkappa_j)x} = 2c_ie^{\varkappa_i x}\sum_{j=1}^{N} c_j(\dot{\psi}_j + \varkappa_j\psi_j)e^{\varkappa_j x}$$

or:

$$L_i\psi_i + \sum_{j=1}^{N} c_ic_j\frac{e^{(\varkappa_i+\varkappa_j)x}}{\varkappa_i + \varkappa_j}L_j\psi_j = 0$$

and finally $L_i\psi_i = 0$, $\forall i \in [1, 2, \ldots, N]$, since the matrix $I + C$ is invertible. The asymptotic behaviour described by (12.220), (12.221) now ensures that $\psi_i(x)$ is an eigenfunction.

We may add that it follows from (12.225) that $\psi_i(x)e^{\varkappa_i y} = w_i(x, y)$ is a solution of the partial differential equation:

$$\frac{\partial^2 w}{\partial x^2} - \frac{\partial^2 w}{\partial y^2} - qw = 0$$

and hence we see that the kernel $K(x, y)$ defined by (12.215) certainly satisfies, throughout the half-plane $x \geqslant y$, the wave equation:

$$\frac{\partial^2 K}{\partial x^2} - \frac{\partial^2 K}{\partial y^2} - q(x)K = 0. \tag{12.226}$$

4. If we apply the direct method to (12.34), with $q(x)$ defined by (12.219) or (12.224), which is legitimate because of the good behaviour of $q(x)$ at $x = \pm\infty$, we shall get, starting from the eigenvalues, eigenmodes and corresponding reflection coefficient, a function defined by (12.129) which we shall denote provisionally by $\mathscr{B}(\xi)$, since there is as yet nothing to allow us to identify this function with $B(\xi)$ represented by (12.213).

We are going to show however that $B = \mathscr{B}$ and this will entail several consequences:

1° – the reflection coefficient associated with the equation (12.34), when $q(x)$ is defined by (12.219) or (12.224) is zero.

2° – the only eigenvalues are the prescribed numbers $\varkappa_j$, $1 \leqslant j \leqslant N$, and as the coefficients c_i are the limits as x tends to infinity of the products by $e^{\varkappa_i x}$ of the eigenfunction associated with $\varkappa_i$, the latter, calculated from the equation (12.218), are normalised or in other words:

$$\int_{-\infty}^{+\infty} \psi_i^2(x)\mathrm{d}x = 1, \quad \forall i.$$

To prove the result just announced, let us introduce the kernel $\mathscr{K}(x, y)$ associated, by the direct method, with the potential of the perturbation $q(x)$. This kernel has the following properties which define it uniquely:

- it is a solution of (12.226) in $x > y$.
- it satisfies the condition $\mathscr{K}(x, x) = \frac{1}{2}\int_{-\infty}^{x} q(x')\mathrm{d}x'$. On the boundary $x = y$.
- it is bounded in every sector $x \geqslant y$, $x \leqslant l$ for all real l.

Since these conditions are equally satisfied by the kernel $\mathscr{K}(x, y)$ defined by (12.215) the uniqueness result allows us to deduce that the kernels K and $\mathscr{K}$, are identical.

The Gelfand-Levitan equations associated with the functions $B(\xi)$, $\mathscr{B}(\xi)$ have a common solution $K(x, y)$ so that with $D(\xi) = B(\xi) - \mathscr{B}(\xi)$, being square-integrable near $-\infty$, we can write:

$$D(x + y) + \int_{-\infty}^{x} K(x, z)D(z + y)\mathrm{d}z = 0, \quad x \geqslant y \tag{12.227}$$

with $K(x, y)$ still being defined by (12.215).

Accordingly we can write:

$$\begin{aligned} D(x + y) &= \sum_{i=1}^{N} c_i\psi_i(x) \int_{-\infty}^{x} e^{\varkappa_i z}D(z + y)\mathrm{d}z, \quad x \geqslant y \\ &= \sum_{i=1}^{N} c_i\psi_i(x)e^{-\varkappa_i y} \cdot \int_{-\infty}^{x+y} e^{\varkappa_i s}D(s)\mathrm{d}s, \quad x \geqslant y \end{aligned}$$

or with $x = y$ and changing x into $x/2$:

$$D(x) = \sum_{i=1}^{N} c_i\psi_i\left(\frac{x}{2}\right)e^{-\varkappa_i x/2} \int_{-\infty}^{x} e^{\varkappa_i s}D(s)\mathrm{d}s, \quad \forall x \in \mathbb{R}, \tag{12.228}$$

whence we see by (12.220) that $\lim_{x\to-\infty} D(x)=0$ and consequently $D(x)$ is bounded in any neighbourhood of $-\infty$. Let a be a real number such that:

$$\operatorname*{Sup}_{x\leqslant a}\sum_{i=1}^{N}\frac{c_i}{\varkappa_i}\left|\psi_i\left(\frac{x}{2}\right)\right|e^{\varkappa_i x/2}<\tfrac{1}{2}$$

and let $M=\operatorname{Sup}_{x\leqslant a}|D(x)|$.

We deduce from the equation (12.228) that, for $x\leqslant a$

$$|D(x)|\leqslant M\sum_{i=1}^{N}\frac{c_i}{\varkappa_i}\left|\psi_i\left(\frac{x}{2}\right)\right|e^{\varkappa_i x/2}<\frac{M}{2}$$

and iterating in the obvious way, we see that $D(x)=0$, $\forall x\leqslant a$.

We can therefore rewrite (12.228) for $x\geqslant a$ as:

$$D(x)=\sum_{i=1}^{N}c_i\psi_i\left(\frac{x}{2}\right)e^{\varkappa_i x/2}\int_a^x e^{\varkappa_i(s-x)}D(s)\,\mathrm{d}s,\quad x\geqslant a$$

and noting that

$$\operatorname*{Sup}_{x\geqslant a}\left(\sum_{i=1}^{N}c_i\left|\psi_i\left(\frac{x}{2}\right)\right|e^{\varkappa_i x/2}\right)=\alpha$$

is finite, by (12.221), we can write with $\varkappa=\varkappa_1<\varkappa_2<\cdots<\varkappa_N$:

$$|D(x)|\leqslant\alpha\int_a^x e^{\varkappa(s-x)}D(s)\mathrm{d}s,\quad x\geqslant a$$

from which it follows that $D(x)=0$, $\forall x\geqslant a$.

7. The Korteweg-de Vries Equation. Interaction of Solitary Waves

The method of inverse scattering, applied to the Korteweg-de Vries equation, provides an interesting illustration of the mechanism of interaction between solitary waves [21, 26].

We shall suppose $q(x,t)$ to be a solution of (12.161), with:

$$\Omega_1(k^2)=4k^2,\quad \Omega_2(k^2)=1$$

which leads to:

(12.229) $$q_t=6qq_x-q_{xxx}$$

which can be transformed into the Korteweg-de Vries equation by changing q in $-q/6$. For convenience however we shall retain the form (12.229) in the discussion.

We shall suppose that $q(x,0)$ is expressed by (12.219) or (12.224), with c_1, $c_2,\dots,c_N$, $\varkappa_1,\dots,\varkappa_N$ given positive numbers, which means that at time $t=0$, the result of the calculations involved in applying the direct method to $q(x,0)$ is a function $B(\xi)$ expressed by (12.213).

When t varies the $\varkappa_1 \cdots \varkappa_N$ are invariant, the reflection coefficient remains zero because it was zero at $t=0$, while the $c_i(t)$ evolve according to the law (12.212) or $(c_i^2)_t = -8\varkappa_i^3(c_i^2)$, that is to say:

(12.230) $$c_i^2(t) = \exp(-8\varkappa_i^3 t) \cdot c_i^2(0).$$

Thus the solution $q(x,t)$ of (12.229) is obtained from (12.223), or:

(12.231) $$q(x,t) = 2\frac{\partial}{\partial x}K(x,x,t) = -2\sum_{j=1}^{N} \dot{\theta}_j(x,t)$$

with

$$\theta_j(x,t) = c_j(t)\psi_j(x,t)e^{\varkappa_j x}, \quad \dot{\theta}_j(x,t) = \frac{\partial \theta_j}{\partial x}(x,t)$$

the $\psi_j(x,t)$ being calculated from the system $c_i(t)$, $\varkappa_i$, $1 \leqslant i \leqslant N$.

It is convenient to rewrite (12.216) in the form:

(12.232) $$c_i^{-2}(t)\exp(-2\varkappa_i x) \cdot \theta_i(x,t) + \sum_{j=1}^{N} \frac{\theta_j(x,t)}{\varkappa_i + \varkappa_j} = 1$$

with $c_i(t)$ defined by (12.230), or after differentiation with respect to x:

(12.233)
$$c_i^{-2}(t)\exp(-2\varkappa_i x) \cdot \dot{\theta}_i(x,t) + \sum_{j=1}^{N} \frac{\dot{\theta}_j(x,t)}{\varkappa_i + \varkappa_j} = 2\varkappa_i c_i^{-2}(t)\exp(-2\varkappa_i x) \cdot \theta_i(x,t).$$

Thus we can, from (12.232) and (12.233) calculate the functions $\dot{\theta}_j(x,t)$ and obtain the solution $q(x,t)$ by means of (12.231) in a form which enables us to discuss completely its asymptotic behaviour at $x = \pm\infty$.

In actual fact it is better to carry out the analysis using a moving frame of reference in which the axes move at the speed of one of the solitary waves.

Thus if we use instead of x, the space co-ordinate:

(12.234) $$\xi = x - 4\varkappa_p^2 t$$

the integer p taking one of the values $1, 2, \ldots, N$, and write:

(12.235) $$\begin{aligned} c_i^{-2}(t)\exp(-2\varkappa_i x) &= c_i^{-2}(0)\exp\{8\varkappa_i(\varkappa_i^2 - \varkappa_p^2)t - 2\varkappa_i\xi\} \\ &= \gamma_i(\xi)\exp\{8\varkappa_i(\varkappa_i^2 - \varkappa_p^2)t\} \end{aligned}$$

with

(12.236) $$\gamma_i(\xi) = c_i^{-2}(0)\exp(-2\varkappa_i\xi)$$

we see that the system (12.232), (12.233) becomes:

(12.237) $$\gamma_i \exp\{8\varkappa_i(\varkappa_i^2 - \varkappa_p^2)t\} \cdot \theta_i + \sum_{j=1}^{N} \frac{\theta_j}{\varkappa_i + \varkappa_j} = 1$$

(12.238) $$\gamma_i \exp\{8\varkappa_i(\varkappa_i^2 - \varkappa_p^2)t\} \cdot \dot{\theta}_i + \sum_{j=1}^{N} \frac{\dot{\theta}_j}{\varkappa_i + \varkappa_j} = 2\varkappa_i\gamma_i \exp\{8\varkappa_i(\varkappa_i^2 - \varkappa_p^2)t\} \cdot \theta_i.$$

Investigation of Asymptotic Behaviour for $t \to +\infty$

We assume the $\varkappa_i$ arranged in order of increasing magnitude, $\varkappa_1 < \varkappa_2 < \cdots < \varkappa_N$, so that for $t \to +\infty$ the system (12.237) reduces to:

$$\sum_{j=1}^{N} \frac{\theta_j}{\varkappa_i + \varkappa_j} = 1, \quad i = 1, 2, \quad p-1$$

$$\gamma_p \theta_p + \sum_{j=1}^{N} \frac{\theta_j}{\varkappa_p + \varkappa_j} = 1, \quad i = p$$

$$\theta_i = 0, \quad i = p+1, \ldots, N$$

equations which can be combined in the formula:

$$(12.239) \qquad \begin{aligned} &\sum_{j=1}^{p} \frac{\theta_j}{\varkappa_i + \varkappa_p} = 1 - \gamma_p \delta_{ip} \theta_p, \quad && i = 1, 2, \ldots, p \\ &\theta_i = 0, && i = p+1, \ldots N \end{aligned}$$

and similarly (12.238) leads to:

$$(12.240) \qquad \begin{aligned} &\sum_{j=1}^{p} \frac{\dot{\theta}_j}{\varkappa_i + \varkappa_j} = \gamma_p \delta_{ip} (2\varkappa_p \theta_p - \dot{\theta}_p), \quad && i = 1, 2, \ldots, p \\ &\dot{\theta}_i = 0, && i = p+1, \ldots N. \end{aligned}$$

We introduce the matrix $M_p = (1/(\varkappa_i + \varkappa_j))$, i, j taking their values from the set $1, 2, \ldots, p$. This matrix is positive by virtue of an earlier remark, and we denote by $M_{p;ij}$ the cofactor of the (i, j)th element of the determinant of M_p, and by L_p the matrix derived from M_p by replacing the elements of its last column by units. We deduce from (12.239), (12.240) by applying Cramer's rule, that:

$$(12.241) \qquad \begin{aligned} &\theta_j \det M_p = \sum_{i=1}^{p} M_{p;ij} - \gamma_p M_{p;pj} \theta_p \\ &\dot{\theta}_j \det M_p = \gamma_p M_{p;pj} (2\varkappa_p \theta_p - \dot{\theta}_p), \quad j = 1, 2, \ldots, p. \end{aligned}$$

With $i = p$ we obtain:

$$(12.242) \qquad \theta_p = \frac{\det L_p}{\det M_p + \gamma_p \det M_{p-1}}$$

$$(12.243) \qquad \dot{\theta}_p = \frac{2\varkappa_p \gamma_p \theta_p \det M_{p-1}}{\det M_p + \gamma_p \det M_{p-1}}$$

(Note that γ_p, $\det M_p$, $\det M_{p-1}$, are positive.)

By summing (12.241) over j from 1 to p and observing that $M_{p;pj} = M_{p;jp}$, we obtain, by using (12.242) and (12.243) and remembering that our results are asymptotic in nature:

$$(12.244) \qquad \lim_{\substack{t \to +\infty \\ \text{fixed } \xi}} \sum_{j=1}^{p} \dot{\theta}_j = \frac{2\gamma_p \varkappa_p}{\left(\dfrac{\det M_p}{\det L_p} + \gamma_p \dfrac{\det M_{p-1}}{\det L_p} \right)^2}.$$

By subtracting the last row from the other rows of L_p, it is easy to see that:

$$\det L_p = \prod_{j=1}^{p-1} \frac{\varkappa_p - \varkappa_j}{\varkappa_p + \varkappa_j} \det M_{p-1} \tag{12.245}$$

and similarly by subtracting the last column of M_p, from its other columns, we arrive at:

$$\det M_p = \frac{\prod_{i=1}^{p-1} (\varkappa_p - \varkappa_i)}{\prod_{i=1}^{p} (\varkappa_i + \varkappa_p)} \det L_p. \tag{12.246}$$

Lastly we see from (12.231), (12.244), (12.245), (12.246) that:

$$\lim_{\substack{t \to +\infty \\ \text{fixed } \xi}} q = -16\varkappa_p^3 \gamma_p \cdot \prod_{i=1}^{p-1} \left(\frac{\varkappa_p + \varkappa_i}{\varkappa_p - \varkappa_i} \right)^2 \cdot \left[1 + 2\varkappa_p \gamma_p \prod_{i=1}^{p-1} \left(\frac{\varkappa_p + \varkappa_i}{\varkappa_p - \varkappa_i} \right)^2 \right]^{-2}.$$

Introducing ξ_p defined by:

$$\exp(2\varkappa_p \xi_p) = 2\varkappa_p c_p^{-2}(0) \prod_{i=1}^{p-1} \left(\frac{\varkappa_p + \varkappa_i}{\varkappa_p - \varkappa_i} \right)^2$$

we have, by (12.236):

$$2\varkappa_p \gamma_p \prod_{i=1}^{p-1} \left(\frac{\varkappa_p + \varkappa_i}{\varkappa_p - \varkappa_i} \right)^2 = \exp[2\varkappa_p(\xi_p - \xi)]$$

and finally:

$$\lim_{\substack{t \to +\infty \\ \text{fixed } \xi}} q = -8\varkappa_p^2 \exp[2\varkappa_p(\xi_p - \xi)] \cdot (1 + \exp[2\varkappa_p(\xi_p - \xi)])^{-2}$$

or

$$\lim_{\substack{t \to +\infty \\ \text{fixed } \xi}} q = -\frac{2\varkappa_p^2}{\operatorname{ch}^2[\varkappa_p(x - 4\varkappa_p^2 t - \xi_p)]}$$

representing a solitary wave of amplitude $2\varkappa_p^2$ moving in the direction of increasing x at a velocity $4\varkappa_p^2$.

Asymptotic Behaviour for $t \to -\infty$

The equations (12.237), (12.238), after taking the limit as $t \to -\infty$ lead to:

$$\sum_{j=p}^{N} \frac{\theta_j}{\varkappa_i + \varkappa_j} = 1 - \gamma_p \delta_{ip} \theta_p, \qquad i = p, \dots, N$$

$$\sum_{j=p}^{N} \frac{\dot{\theta}_j}{\varkappa_i + \varkappa_j} = \gamma_p \delta_{ip} (2\varkappa_p \theta_p - \dot{\theta}_p), \quad i = p, \dots, N$$

$$\theta_i = 0, \dot{\theta}_i = 0, \qquad i = 1, 2, \dots, p-1.$$

These relations have the same structure as (12.239), (12.240) the only difference being that the suffixes i, j run from p to N whereas in the first case they ran from 1 to p. Consequently with $\tilde{\xi}_p$ defined by:

$$\exp(2\varkappa_p\tilde{\xi}_p) = 2\varkappa_p c_p^{-2}(0) \prod_{i=p+1}^{N} \left(\frac{\varkappa_p + \varkappa_i}{\varkappa_p - \varkappa_i}\right)^2$$

we obtain:

$$\lim_{\substack{t \to -\infty \\ \text{fixed } \xi}} q(x, t) = -\frac{2\varkappa_p^2}{\text{ch}^2[\varkappa_p(x - 4\varkappa_p^2 t - \tilde{\xi}_p)]}.$$

The interaction therefore has the effect of producing a displacement of the profile of the solitary wave. The displacement is $\xi_p - \tilde{\xi}_p$ relative to axes moving with the wave at a velocity $4\varkappa_p^2$.

References

1. Antman, S. (1983) Oscillations de grande amplitude dans un solide élastique non linéaire. (Communication orale)
2. Arnold, V. (1974) Méthodes mathématiques de la mécanique classique. M.I.R., Moscow
3. Bessis, D., Villani, M. (1975) Perturbative variationnal approximation to the spectral properties of semi bounded Hilbert space operator based on the moment problem with finite or divergent moments. Application to quantum mechanical systems. J. Math. Phys. *16*, no. 3, 462–474
4. Bisplinghoff, R. L., Ashley, H., Halfman, R. L. (1957) Aéroelasticity. Addison Wesley
5. Bogoliouboff, N. N., Mitropolski, J. A., Samoilenko, A. M. (1976) Methods of accelerated convergence in non linear mechanics. Springer, Berlin Heidelberg New York
6. Bourgine, A. (1973) Sur une approche statistique de la dynamique vibratoire des structures. O.N.E.R.A. publ. no. 149
7. Brillouin, L., Parodi, M. (1956) Propagation des ondes dans les milieux périodiques. Masson, Paris
8. Burdess, J. S., Fox, C. H. J. (1978) The dynamics of a multigimbal Hooke's joint gyroscope. J. Mech. Engng. Sci. *20*, no. 5, 255–262
9. Burdess, J. S., Fox, C. H. J. (1978) The dynamics of an imperfect multigimbal Hooke's joint gyroscope. J. Mech. Engng. Sci. *20*, no. 5, 263–269
10. Butenin, N. V. (1965) Elements of the theory of non linear oscillations. Blaisdell, New York
11. Cabannes, H. (1976) Cours de mécanique générale. Dunod, Paris
12. Cameron, A. (1976) Basic lubrication theory. Ellis Horwood, Chichester
13. Cerneau, S., Sanchez-Palencia, E. (1976) Sur les vibrations libres des corps élastiques plongés dans des fluides. J. de Mécanique *15*, 399–425
14. Choquet-Bruhat, Y. (1973) Distributions. Théorie et problèmes. Masson, Paris
15. Ciarlet, P. G. (1982) Introduction à l'analyse numérique matricielle et à l'optimisation. Masson, Paris
16. Colombo, G. (1958) Theoria del regolatore di Bouasse e Sarda. R. C. Semin. Mat. Univ., Padova. 28–2, pp. 338–347
17. Dieudonné, J. (1960) Foundations of modern analysis. Academic Press, New York
18. Duvaut, G., Lions, J. L. (1972) Les inéquations en mécanique et en physique. Dunod, Paris
19. Faurre, P., Clerget, M., Germain, F. (1979) Opérateurs rationnels positifs. Dunod, Paris
20. Ford, G. W. K., Harris, D. M., Pantall, D. (1957) Principlés and applications of hydrodynamics type gas bearing. Proceedings of the Institution of Mechanical Engineers, vol. 171, no. 2, 93–128
21. Gardner, C. S., Greene, J. M., Kruskal, M. D., Miura, R. M. (1974) Korteweg-de Vries equation and generalisation; methods for exact solution. Comm. Pure and Appl. Math., vol. XXVII, pp. 97–133
22. Gelfand, I. M., Levitan, B. M. (1955) On the determination of a differential equation from its spectral function. Amer. Math. Soc. Transl. Ser. 2, no. 1, pp. 253–304
23. Germain, P. (1973) Mécanique des milieux continus, t. 1. Masson, Paris
24. Gladwell, G. M. L., Stammers, C. W. (1966) On the stability of an unsymmetrical rigid rotor supported on unsymmetrical bearings. J. Sound and Vibration *3*, no. 3, 221–232
25. Gladwell, G. M. L., Stammers, C. W. (1968) Prediction of the unstable regions of a reciprocal system governed by a set of linear equations with periodic coefficients. J. Sound and Vibration *8*, no. 3, 457–468
26. Kay, I., Moses, H. E. (1956) Reflectionless transmission through dielectrics and scattering potentials. J. Appl. Phys. *27*, no. 12, 1503–1508

27. Korteweg, D. J., de Vries G. (1895) On the change of form of long waves advancing in a rectangular canal and a new type of long stationnary waves. Phil. Mag. *39*, 422–443
28. Ablowitz, M. J., Kaup, D. J., Newell, A. C., Segur, H. (1974) The inverse scattering transform Fourier for non linear problem. Studies in applied mathematics, vol. LIII, no.4, pp. 249–315
29. Levinson, N. (1949) On the uniqueness of the potential in a Schrodinger equation for a given asymptotic phase. Kgl. Danske Videnskabernes Selskal. Mat. Fys. Medd. *25*, no. 9, 3–29
30. Magnus, W., Winkler, S. (1966) Hill's equation. Interscience tracts in pure and applied mathematics no. 20
31. Mingori, D. L. (1970) A stability theorem for mechanical systems with constraint damping. Trans. of the American Soc. of Mech. Eng., no. 92, 253–258
32. Muszinska, A. (1973) Vibrations non linéaires des arbres tournants. Acad. polonaise des sciences. Conférence fasc. 101. Centre scientifique de Paris. (1983) Rub, an important malfunction in rotating machinery. Bently Nevada Corp. Report.
33. Nečas, J. (1967) Les méthodes directes en théorie des équations elliptiques, Masson, Paris
34. Ocvirk, F. W. (1952) Short bearing approximation for full journal bearings. N. A. C. A. Tech. note 2808
35. Panovko, Y. G., Gubanova, I. I. (1965) Stability and oscillation of elastic systems. Consultant Bureau, New York
36. Prager, W. (1961) Introduction to mechanics of continua. Ginn and Company
37. Riesz, F., Nagy, B. (1955) Leçons d'analyse fonctionnelle. Gauthier Villars, Paris
38. Rocard, Y. (1971) Dynamique générale des vibrations. Masson, Paris
39. Rocard, Y. (1954) L'instabilité en mécanique. Masson, Paris
40. Roseau, M. (1966) Vibrations non linéaires et théorie de la stabilité. Springer, Berlin Heidelberg New York
41. Roseau, M. (1975) Asymptotic wave theory. North-Holland, Amsterdam
42. Roseau, M. (1976) Équations différentielles. Masson, Paris
43. Roseau, M. (1981) La mèthode de modulation d'amplitude et son application à l'étude des oscillateurs couplés. Journal de Mécanique *20*, no. 2, 199–217
44. Roseau, M. (1981) Some case of instability in rotating machinery; an approach based on the theory of singular perturbation. XI International Conference on Non-linear Oscillations. Kiev
45. Roseau, M. (1982) On the coupling between a vibrating mechanical system and the external forces acting upon it. Intern. J. of Non-linear Mech. *17*, no. 3, 211–216
46. Roseau, M. (1982) Stabilité de régime des machines tournantes et problèmes associés. Lecture Notes in Math. 1107, Nonlinear Analysis and Optimization, Bologna, pp. 193–214.
47. Rosenberg, R. M. (1958) On the periodic solution of the forced oscillation equation. Quart. Appl. Math. *15*, 341
48. Sanchez-Palencia, E. (1980) Non homogeneous media and vibration theory. Springer, Berlin Heidelberg New York
49. Sanchez-Palencia, E. (1982) Fréquence de diffusion dans le probléme de vibration d'un corps élastique plongé dans un fluide compressible de petite densité. Comptes Rendus Acad. Sc., Série I, t. 295, p. 197
50. Schweitzer, G., Ulbrich, H. (1980) Magnetic bearings: a novel type of suspension. Second Intern. Conference Vibration in Rotating Machinery. The Institution of Mechanical Engineers, London, p. 151
51. Siegel, S. C., Moser, J. K. (1971) Lectures in celestial mechanics. Springer, Berlin Heidelberg New York
52. Smith, D. M. (1933) The motion of a rotor carried by a flexible shaft in flexible bearings. Proc. Roy. Soc. A, *142*, 92
53. Smith, D. M. (1969) Journal bearings in turbo machinery. Chapman and Hall, London
54. Su, C. H., Gardner, C. S. (1969) Korteweg-de Vries equation and generalisation. Derivation of the Korteweg-de Vries equation and Burger's equation. J. Math. Phys. *10*, 536–539
55. Toda, M. (1967) Wave propagation in anharmonic lattices. J. Phys. Soc. Japan *23*, 501
56. Villat, H. (1930) Leçons sur la théorie des tourbillons. Gauthier Villars, Paris
57. Vorobyev, Y. V. (1965) Method of moments in applied mathematics. Gordon and Breach, New York

58. Wannier, G. H. (1960) Elements of solid state theory. Cambridge University Press
59. Ziegler, H. (1968) Principles of structural stability. Blaisdell, Waltham, Mass.
60. Bourcier de Carbon, C. (May 1951) Sur la stabilité de route des remorques routières, J.S.I.A. pp. 109–112
61. Croix-Marie, F. Coussin d'Air. Techniques de l'Ingénieur B1 190, 1–11; B1 191, 1–7
62. Ehrich, F. F., O'Connor, J. J. (1966) Stator whirl with rotors in bearing clearance. A. S. M. E. paper 66. WA/MD-8

Subject Index